Series in Analysis Vol. 7

Asymptotic Analysis

SERIES IN ANALYSIS **ISSN: 1793-1126**

Series Editor: Professor Roderick Wong
City University of Hong Kong, Hong Kong, China

Published

Vol. 7 *Asymptotic Analysis*
by Xiang-Sheng Wang and Roderick Wong

Vol. 6 *Asymptotic Methods for Integrals*
by Nico M. Temme

Vol. 5 *Lecture Notes on Applied Analysis*
by Roderick Wong

Vol. 4 *Mathematical Foundation and Applications of the P and H-P Finite Elemen Methods*
by Benqi Guo

Vol. 3 *Symmetrization and Applications*
by S Kesavan

Vol. 2 *Differential Equations and Asymptotic Theory in Mathematical Physics*
edited by Hua Chen and Roderick S.C. Wong

Vol. 1 *Wavelet Analysis: Twenty Years' Developments*
edited by Ding-Xuan Zhou

Series in Analysis Vol. 7

Asymptotic Analysis

Xiang-Sheng Wang
University of Louisiana at Lafayette, USA

Roderick Wong
City University of Hong Kong, Hong Kong
Zhengzhou University, China

World Scientific

NEW JERSEY • LONDON • SINGAPORE • BEIJING • SHANGHAI • HONG KONG • TAIPEI • CHENNAI • TOKYO

Published by

World Scientific Publishing Co. Pte. Ltd.
5 Toh Tuck Link, Singapore 596224
USA office: 27 Warren Street, Suite 401-402, Hackensack, NJ 07601
UK office: 57 Shelton Street, Covent Garden, London WC2H 9HE

Library of Congress Cataloging-in-Publication Data
Names: Wang, Xiang-Sheng author | Wong, Roderick, 1944– author
Title: Asymptotic analysis / Xiang-Sheng Wang, University of Louisiana at Lafayette, USA, Roderick Wong, City University of Hong Kong, Hong Kong & Zhengzhou University, China.
Description: Singapore ; Hackensack, NJ : World Scientific Publishing, Co, [2026] | Series: Series in analysis, 1793-1126 ; volume 7 | Includes bibliographical references and index.
Identifiers: LCCN 2026001702 | ISBN 9789819824687 hardcover | ISBN 9789819825486 paperback | ISBN 9789819824694 ebook for institutions | ISBN 9789819824700 ebook for individuals
Subjects: LCSH: Asymptotic expansions | Series, Infinite
Classification: LCC QA297 .W346 2026
LC record available at https://lccn.loc.gov/2026001702

British Library Cataloguing-in-Publication Data
A catalogue record for this book is available from the British Library.

For any available supplementary material, please visit
https://www.worldscientific.com/worldscibooks/10.1142/14629#t=suppl

Desk Editors: Nambirajan Karuppiah/Lai Fun Kwong

Typeset by Stallion Press
Email: enquiries@stallionpress.com

In memory of our mentor

Frank W. J. Olver

Preface

Asymptotic analysis is a branch of classical analysis; see Birkhoff (1973, p. 93). To illustrate our point, we recall the following three results:

$$\ln n! \sim \left(n+\frac{1}{2}\right)\ln n - n + \frac{1}{2}\ln 2\pi + \sum_{s=0}^{\infty}\frac{B_{2s+2}}{(2s+1)(2s+2)\,n^{2s+1}},$$

$$H_n := \sum_{k=1}^{n}\frac{1}{k} \sim \ln n + \gamma + \frac{1}{2n} - \sum_{s=1}^{\infty}\frac{B_{2s}}{(2n)^{2s}},$$

$$L_n \sim \frac{4}{\pi^2}\left\{\ln(2n+1) + A_0 - \sum_{s=1}^{\infty}\frac{A_s}{(2n+1)^{2s}}\right\},$$

where B_{2s} denotes the Bernoulli numbers defined by

$$\frac{z}{e^z-1} = \sum_{n=0}^{\infty} B_n \frac{z^n}{n!},$$

and

$$A_0 = 2\sum_{m=1}^{\infty}\frac{\ln m}{4m^2-1} + 2\ln 2 + \gamma,$$

$$A_s = \frac{(1-2^{2s-1})B_{2s}}{s}\left[1 - \sum_{m=1}^{s}\frac{(-1)^{m-1}}{(2m)!}B_{2m}\pi^{2m}\right], \quad s \geq 1,$$

with γ being the Euler constant. The first two results were discovered nearly 300 years ago, while the third was given by Watson (1930).

In the past, researchers in asymptotics worked mainly in two areas: one concerns ordinary differential equations, and the other deals with integrals. For differential equations, we refer to Olver (1974a); for integrals, see Wong (1989) and Temme (2015). These works formed the foundation of the modern theory and provided many of the essential tools that remain indispensable today.

In recent years, the area concerning difference equations suddenly became active again. Ever since 1930, when Birkhoff (1930) and Birkhoff and Trjitzinsky (1933) published their two papers, research in this area had largely stalled. In general, these two papers were considered too complicated, and there were doubts about the validity of the results in their entirety; see Wimp and Zeilberger (1985), Wimp (1991), and Olver (2014). A simpler approach to handle second-order linear difference equations was given by Wong and Li (1992a,b), and a breakthrough occurred when Wang and Wong (2002) found a way, using a difference equation method, to derive the asymptotic expansion of $J_\nu(\nu a)$ as $\nu \to \infty$, which holds uniformly for all bounded $a > 0$. This work initiated a sequence of papers on difference equations involving a parameter, showing that in difference equations there are results parallel to those already existing for second-order differential equations.

Before the end of the twentieth century, another new method was discovered: the nonlinear steepest descent method, introduced by Deift and Zhou (1993). This method, based on a Riemann-Hilbert formulation, has proven to be remarkably powerful. Its strength lies not only in its applicability to orthogonal polynomials but also in its ability to handle nonlinear differential equations and even partial differential equations.

Classical results in singular perturbation theory were, for a long time, largely formal. The problem with formal results is that it is not even clear whether the error is smaller than the first neglected term. At the beginning of this century, a substantial body of work has appeared; it rigorously justifies many of these earlier formal approximations.

The content of this book is divided into five parts. The first part is devoted to linear difference equations and provides a comprehensive summary of the main results that now constitute a coherent theory. The second part is about how to use the Riemann-Hilbert approach to derive asymptotic expansions of orthogonal polynomials. As illustrations, we have used the Stieltjes-Wigert polynomials, Meixner polynomials, and orthogonal polynomials with Freud weights. The third part is on an older area, namely, differential equations. Here, we include topics such as irregular singular points at infinity, equations with turning points or simple poles. An exceptional case is that we also briefly mention the problem of coalescing of two turning points. Part four is on another older area, namely, asymptotic approximations of integrals. All classical methods, such as Watson's lemma, Laplace's method, the principle of stationary phase, and the method of steepest descent, are briefly mentioned. But there are also some newer results, such as the distribution method, coalescing of stationary points in a double integral, and Berry's idea of smoothing of Stokes discontinuities. Also, we will introduce the idea of Berry and Howls in the construction of error bounds for the steepest descent method. In part five, we show how some of the known results in singular perturbations, once formal, can now be justified rigorously.

The writing of this book took us nearly ten years. Here we must take this opportunity to thank our wives, Edwina and Hongying, for their patience and understanding. Without their continuous encouragement, this project would not have been completed. Finally, we would like to thank Ms. Kwong Lai Fun of World Scientific for her continuous support over the years.

About the Authors

Xiang-Sheng Wang received his BA in Mathematics from the University of Science and Technology of China in 2004 and his PhD in 2009 through a joint program between the University of Science and Technology of China and City University of Hong Kong. From 2009 to 2016, he held academic and research positions at City University of Hong Kong, Memorial University of Newfoundland, Southeast Missouri State University, and York University. He joined the Department of Mathematics at the University of Louisiana at Lafayette in August 2016, where he was promoted to Associate Professor with tenure in August 2022. His research lies primarily in interdisciplinary areas of applied mathematics, with interests including asymptotic analysis, differential equations, mathematical biology, and orthogonal polynomials.

Roderick Wong earned his BA in Mathematics from San Diego State College (1965) and his PhD from the University of Alberta (1969). He began his academic career at the University of Manitoba, where he advanced from Assistant to Full Professor and served as Head of Applied Mathematics (1986–1994). In 1994, he joined City University of Hong Kong as Professor and Head of Mathematics, later founding the Liu Bie Ju Centre for Mathematical Sciences, which he directed until 2019. He also held senior leadership roles, including Dean of Science and Engineering, Dean of Research and Graduate Studies, Vice-President (Research & Technology), and Vice-President for Development & External Relations. Upon retiring in 2019, he was named Emeritus Professor in 2020. Roderick Wong served as President of both the Canadian Applied Mathematics Society and the Hong Kong Mathematical Society, and Vice-President of the Canadian Mathematical Society. He was also active in editorial leadership across numerous international journals. His many honors include the Killam Research Fellowship, Rh Award, Fellowship of the Royal Society of Canada, foreign membership in the Academy of Sciences of Turin, Chevalier dans l'Ordre National de la Légion d'Honneur by the French government, membership in the European Academy of Sciences, and Doctorate Honoris Causa from Ovidius University.

Contents

Part 3: Differential Equations 233

Part 4: Integrals 277

Part 1
Difference Equations

Chapter 1

Asymptotics of Difference Equations with Bounded Coefficients

1.1 Formal asymptotic expansions

As a discrete analogue of the second-order linear differential equation

$$y''(x) + f(x)y'(x) + g(x)y(x) = 0, \tag{1.1}$$

we consider the second-order linear difference equation

$$Y(n+2) + A(n)Y(n+1) + B(n)Y(n) = 0, \tag{1.2}$$

where the coefficients are bounded and have the following asymptotic expansions

$$A(n) \sim \sum_{s=0}^{\infty} \frac{a_s}{n^s} \quad \text{and} \quad B(n) \sim \sum_{s=0}^{\infty} \frac{b_s}{n^s} \tag{1.3}$$

as $n \to \infty$.

The material in this chapter is taken from Wong and Li (1992a). In this chapter, we always assume $b_0 \neq 0$. (The case of $b_0 = 0$ will be studied in the next chapter.) We look for an ansatz solution with the asymptotic expansion:

$$Y(n) \sim \rho^n n^\lambda \sum_{s=0}^{\infty} \frac{y_s}{n^s}, \qquad n \to \infty, \tag{1.4}$$

where, without loss of generality, we take $y_0 = 1$. By substituting (1.3) and (1.4) into (1.2), and then dividing the resulting equation by $\rho^n n^\lambda$, we obtain

$$\rho^2 \left(1 + \frac{2}{n}\right)^\lambda \sum_{s=0}^{\infty} \frac{y_s}{(n+2)^s} + \rho \left(1 + \frac{1}{n}\right)^\lambda \sum_{s=0}^{\infty} \frac{y_s}{(n+1)^s} \cdot \sum_{s=0}^{\infty} \frac{a_s}{n^s} + \sum_{s=0}^{\infty} \frac{y_s}{n^s} \cdot \sum_{s=0}^{\infty} \frac{b_s}{n^s} \sim 0. \tag{1.5}$$

To balance the coefficients of n^{-s} in (1.5), we find a sequence of equations (exercise):

$$\sum_{j=0}^{s}\left\{\rho^2 2^{s-j}\binom{\lambda-j}{s-j}+\rho\sum_{k=0}^{s-j}\binom{\lambda-j}{s-j-k}a_k+b_{s-j}\right\}y_j=0, \tag{1.6}$$

for $s = 0, 1, 2, \cdots$. Recall that $y_0 = 1$. We intend to solve $\rho, \lambda, y_1, y_2, \cdots$ from the above equations in a successive manner. When $s = 0$, equation (1.6) reads

$$\rho^2 + a_0\rho + b_0 = 0. \tag{1.7}$$

This equation is known as the *characteristic equation*, and its solutions, ρ_i, $i = 1, 2$, are called the *characteristic roots.* When $s = 1$, equation (1.6) can be written as

$$(2\rho^2 + a_0\rho)\lambda + a_1\rho + b_1 = 0. \tag{1.8}$$

To solve λ from the above equation, we require $\rho \neq 0$ and $\rho \neq -a_0/2$, which, on account of (1.7), are equivalent to $b_0 \neq 0$ and $a_0^2 \neq 4b_0$. In this case, we find two distinct characteristic roots from the characteristic equation (1.7). For each characteristic root, we solve λ from (1.8) and y_{s-1} from (1.6) for $s = 2, 3, \cdots$. Here, we should note that in (1.6) with $s \geq 2$, the coefficient of y_s vanishes while the coefficient of y_{s-1} is nonzero. Therefore, (1.4) gives two linearly independent asymptotic solutions corresponding to the two distinct characteristic roots. This case (i.e., $b_0 \neq 0$ and $a_0^2 \neq 4b_0$) is called the *normal case* and the asymptotic solutions are called the *normal solutions* or *normal series.*

Next, we consider the case when $b_0 \neq 0$ and $a_0^2 = 4b_0$. From (1.7), we have a double characteristic root $\rho = -a_0/2$. To validate equation (1.8), we require $a_1\rho + b_1 = 0$, which is the same as $a_0a_1 = 2b_1$. We refer to this case ($b_0 \neq 0$, $a_0^2 = 4b_0$ and $a_0a_1 = 2b_1$) as the *exceptional case.* We shall solve λ from equation (1.6) with $s = 2$. On account of $y_0 = 1$, $\rho = -a_0/2$ and $a_0a_1 = 2b_1$, this equation becomes

$$\rho^2\lambda(\lambda - 1) - b_1\lambda + a_2\rho + b_2 = 0, \tag{1.9}$$

which is called the *indicial equation.* There are two subcases to be considered.

Exceptional case (i): the two roots of the indicial equation (1.9) are distinct and the difference of these two roots is not an integer. In this subcase, the two roots give two linearly independent solutions with asymptotic expansions taking the form of (1.4), where the coefficients y_{s-2} can be determined recursively from equation (1.6) with $s = 3, 4, \cdots$. Note that in

(1.6) with $s \geq 3$, the coefficients of y_s and y_{s-1} vanish, while the coefficient of y_{s-2} is nonzero.

Exceptional case (ii): the two roots of the indicial equation (1.9) coincide or the difference of these two roots is an integer. Denote the two roots by λ_1 and λ_2, where $\lambda_2 - \lambda_1$ is a nonnegative integer. If we set $\lambda = \lambda_1$ and solve y_{s-2} recursively from (1.6) with $s \geq 3$, the formula (1.4) gives an asymptotic solution $Y_1(n)$. We need to find a second linearly independent solution. Note that if $\lambda_1 = \lambda_2$, then equation (1.6) gives the same asymptotic solution as $Y_1(n)$; if $\lambda_2 - \lambda_1$ is a positive integer, equation (1.6) may break down at $s = \lambda_2 - \lambda_1$ in the sense that the coefficient of y_{s-2} vanishes. Hence, we look for a second linearly independent solution with following asymptotic expansion:

$$Y_2(n) \sim \rho^n n^{\lambda_2} \sum_{s=0}^{\infty} \frac{\bar{y}_s}{n^s} + c(\log n)\rho^n n^{\lambda_1} \sum_{s=0}^{\infty} \frac{y_s}{n^s}, \tag{1.10}$$

where y_s are the coefficients in the asymptotic expansion of first solution $Y_1(n)$. The coefficients $\bar{y}_s$ and c will satisfy a sequence of equations derived by substituting (1.3) and (1.10) into the difference equation (1.2), and then balancing the coefficients of n^{-s} with $s \geq 0$. The resulting equations are (exercise)

$$\begin{aligned}
&\sum_{j=0}^{s} \left\{ \rho^2 2^{s-j} \binom{\lambda_2 - j}{s-j} + \rho \sum_{k=0}^{s-j} \binom{\lambda_2 - j}{s-j-k} a_k + b_{s-j} \right\} \bar{y}_j \\
&= -c \sum_{j=0}^{\bar{s}} \left\{ \rho^2 \sum_{l=1}^{\bar{s}-j} \frac{(-1)^{l-1} 2^{\bar{s}-j}}{l} \binom{\lambda_1 - j}{\bar{s}-j-l} \right. \\
&\quad \left. + \rho \sum_{l=1}^{\bar{s}-j} \sum_{k=0}^{\bar{s}-j-l} \frac{(-1)^{l-1} a_k}{l} \binom{\lambda_1 - j}{\bar{s}-j-k-l} \right\} y_j,
\end{aligned} \tag{1.11}$$

where, for simplicity, we denote $\bar{s} = s - \lambda_2 + \lambda_1$. Since any linear combination $\alpha Y_1(n) + \beta Y_2(n)$ with $\beta \neq 0$ is a solution independent of $Y_1(n)$, there are two free parameters in finding the coefficients of $Y_2(n)$. Without loss of generality we may choose $\bar{y}_{\lambda_2 - \lambda_1} = 0$. Furthermore, if $\lambda_1 = \lambda_2$, we choose $c = 1$; if $\lambda_2 - \lambda_1$ is a positive integer, we choose $\bar{y}_0 = 1$. Note that on the left-hand side of (1.11), the coefficients of $\bar{y}_s$ and $\bar{y}_{s-1}$ vanish, while the coefficient of $\bar{y}_{s-2}$ is $\rho^2(s-2)(s-2-\lambda_2+\lambda_1)$. On the right-hand side of (1.11), the coefficients of $y_{\bar{s}}$ and $y_{\bar{s}-1}$ vanish, while the coefficient of $y_{\bar{s}-2}$ is $-c\rho^2(4 - s - \bar{s})$. If $\lambda_1 = \lambda_2$, then $\bar{y}_0 = 0$, $c = 1$, and $\bar{y}_{s-2}$ with $s \geq 3$ can be determined from (1.11). If $\lambda_2 - \lambda_1$ is a positive integer, then

$\bar{y}_0 = 1$, $\bar{y}_{\lambda_2-\lambda_1} = 0$, and equation (1.11) with $3 \le s \le \lambda_2 - \lambda_1 + 1$ gives the coefficients $\bar{y}_1, \cdots, \bar{y}_{\lambda_2-\lambda_1-1}$. The constant c can be determined from equation (1.11) with $s = \lambda_2 - \lambda_1 + 2$, where the right-hand side has only one nonzero term $c\rho^2(\lambda_2 - \lambda_1)$; recall that $y_0 = 1$. From equation (1.11), we find the coefficient $\bar{y}_{s-2}$ for all $s \ge \lambda_2 - \lambda_1 + 3$.

Finally, we study the *subnormal case* when $b_0 \ne 0$, $a_0^2 = 4b_0$ and $a_0 a_1 \ne 2b_1$. Note that here, equation (1.8) is not satisfied and the asymptotic expansion (1.4) is no longer valid. We shall look for the *subnormal solutions* of the form

$$Y(n) \sim \rho^n e^{\gamma\sqrt{n}} n^\lambda \sum_{s=0}^{\infty} \frac{y_s}{n^{s/2}}, \quad n \to \infty, \tag{1.12}$$

where, without loss of generality, we choose $y_0 = 1$. We substitute (1.3) and (1.12) into (1.2), and then divide the resulting equation by $\rho^n e^{\gamma\sqrt{n}} n^\lambda$. This yields

$$\rho^2 e^{\gamma(\sqrt{n+2}-\sqrt{n})} \sum_{s=0}^{\infty} \frac{y_s(1+2/n)^\lambda}{(n+2)^{s/2}} + \rho e^{\gamma(\sqrt{n+1}-\sqrt{n})} \sum_{s=0}^{\infty} \frac{y_s(1+1/n)^\lambda}{(n+1)^{s/2}} \sum_{s=0}^{\infty} \frac{a_s}{n^s}$$
$$+ \sum_{s=0}^{\infty} \frac{y_s}{n^{s/2}} \sum_{s=0}^{\infty} \frac{b_s}{n^s} \sim 0. \tag{1.13}$$

For convenience, we denote $t = 1/\sqrt{n}$ and define

$$f(t) = \exp\{\gamma(\sqrt{n+2} - \sqrt{n})\} = \exp\left(\frac{2\gamma t}{1+\sqrt{1+2t^2}}\right) = \sum_{s=0}^{\infty} f_s t^s, \tag{1.14}$$

$$g(t) = \exp\{\gamma(\sqrt{n+1} - \sqrt{n})\} = \exp\left(\frac{\gamma t}{1+\sqrt{1+t^2}}\right) = \sum_{s=0}^{\infty} g_s t^s. \tag{1.15}$$

The first few coefficients in (1.14) and (1.15) are calculated as

$$f_0 = 1,\ f_1 = \gamma,\ f_2 = \gamma^2/2,\ f_3 = (\gamma^3 - 3\gamma)/6, \tag{1.16}$$
$$g_0 = 1,\ g_1 = \gamma/2,\ g_2 = \gamma^2/8,\ g_3 = (\gamma^3 - 6\gamma)/48. \tag{1.17}$$

Substituting the expansions (1.14) and (1.15) into (1.13) gives

$$\rho^2 \sum_{j,k,l=0}^{\infty} (f_k t^k)(y_j t^j) \binom{\lambda - j/2}{l} (2t^2)^l$$
$$+\rho \sum_{i,j,k,l=0}^{\infty} (g_k t^k)(y_j t^j) \binom{\lambda - j/2}{l} t^{2l} (a_i t^{2i})$$
$$+ \sum_{i,j=0}^{\infty} (y_j t^j)(b_i t^{2i}) \sim 0. \tag{1.18}$$

By balancing the coefficients of t^s, we obtain (exercise)

$$\sum_{j=0}^{s}\left\{\rho^2\sum_{k+2l=s-j}2^l f_k\binom{\lambda-j/2}{l}\right.$$
$$\left.+\rho\sum_{k+2l+2i=s-j}a_i g_k\binom{\lambda-j/2}{l}+\sum_{2i=s-j}b_i\right\}y_j=0 \tag{1.19}$$

for $s=0,1,\cdots$. When $s=0$, equation (1.19) gives the characteristic equation (1.7), which together with the condition $a_0^2=4b_0$, implies that there exists a double characteristic root $\rho=-a_0/2$. When $s=2$, equation (1.19) can be written as (exercise)

$$\rho^2\gamma^2/4+a_1\rho+b_1=0. \tag{1.20}$$

Since $a_1\rho+b_1\neq 0$, the above equation gives two distinct solutions for γ, which will correspond to two linearly independent solutions of the form (1.12). Furthermore, from equation (1.19) with $s=3$, we find (exercise)

$$\lambda=\frac{1}{4}+\frac{b_1}{2b_0}. \tag{1.21}$$

In equation (1.19) with $s\geq 4$, the coefficients of y_s, y_{s-1} and y_{s-2} vanish, while the coefficient of y_{s-3} is nonzero. Hence, we can determine the coefficients y_{s-3} for $s\geq 4$ in a successive manner.

In the following sections, we shall make use of the Banach fixed point theorem for a contraction mapping on a certain Banach space to prove the asymptotic natures of the expansions for (1.4) in the normal case, for (1.10) in the second exceptional case, and for (1.12) in the subnormal case, respectively. To conclude this section, we introduce a type of Banach space which will be frequently used in this chapter and the next chapter. Given $p\geq 0$, we introduce a Banach space B_p of sequences consisting of infinite sequences $\{s(n)\}_{n\geq 1}$ such that the equipped norm $\|s\|_p:=\sup_{n\geq 1}n^p|s(n)|$ is bounded. Note that $B_q\subset B_p$ for any $0\leq p<q$. Occasionally, we will also make use of the shifted Banach space $B_p^{(m)}$ for real $p\geq 0$ and positive integer m, which is the collection of all infinite sequences $\{s(n)\}_{n\geq m}$ such that the shifted norm $\|s\|_p^{(m)}:=\sup_{n\geq m}n^p|s(n)|$ is bounded. For simplicity, we will always drop the superscript (m) in the notation of shifted norm. For any $\{s(n)\}_{n\geq m}\in B_p^{(m)}$, we may extend the sequence to the Banach space B_p by arbitrarily choosing any values for $s(n)$ with $1\leq n<m$.

1.2 Normal case

Consider the normal case when $b_0 \neq 0$ and $a_0^2 \neq 4b_0$. In this case, the characteristic equation (1.7) has two distinct roots ρ_1 and ρ_2. We order the characteristic roots so that $|\rho_1| \leq |\rho_2|$. We will prove the following result.

Theorem 1.1. *For any $p \geq 1$, the difference equation (1.2) has a solution*

$$Y_{1,p}(n) = \rho_1^n n^{\lambda_1} \left[\sum_{s=0}^{p-1} \frac{y_s}{n^s} + E_p(n) \right], \tag{1.22}$$

where $y_0 = 1$, $\lambda_1 = -(a_1\rho_1 + b_1)/(2\rho_1^2 + a_0\rho_1)$, and

$$\begin{aligned} y_{s-1} =& \frac{1}{(s-1)(2\rho_1^2 + a_0\rho_1)} \sum_{j=0}^{s-2} \left\{ \rho_1^2 2^{s-j} \binom{\lambda_1 - j}{s-j} \right. \\ & \left. + \rho_1 \sum_{k=0}^{s-j} \binom{\lambda_1 - j}{s-j-k} a_k + b_{s-j} \right\} y_j \end{aligned} \tag{1.23}$$

for $s \geq 2$. The error term $E_p(n) = O(1/n^p)$ as $n \to \infty$.

Proof. Substituting (1.22) into (1.2) yields

$$\begin{aligned} &\rho_1^2 \left(1 + \frac{2}{n}\right)^{\lambda_1} E_p(n+2) + \rho_1 \left(1 + \frac{1}{n}\right)^{\lambda_1} A(n) E_p(n+1) + B(n) E_p(n) \\ &= -\, \varepsilon_{p+1}(n), \end{aligned}$$

where

$$\varepsilon_{p+1}(n) = \rho_1^2 \sum_{s=0}^{p-1} \frac{y_s (1+2/n)^{\lambda_1}}{(n+2)^s} + \rho_1 A(n) \sum_{s=0}^{p-1} \frac{y_s(1+1/n)^{\lambda_1}}{(n+1)^s} + B(n) \sum_{s=0}^{p-1} \frac{y_s}{n^s}.$$

Observe that by (1.3) and the choices of ρ_1, λ_1 and y_s with $s \geq 0$, the coefficient of n^{-s} in the expansion of $\varepsilon_{p+1}(n)$ vanishes for $s = 0, 1, \cdots, p$. Thus, $\varepsilon_{p+1} \in B_{p+1}$. Note from (1.3) that

$$\begin{aligned} \rho_1^2 \left(1 + \frac{2}{n}\right)^{\lambda_1} &= \rho_1^2 + \frac{C_2(n)}{n}, \\ \rho_1 \left(1 + \frac{1}{n}\right)^{\lambda_1} A(n) &= \rho_1 a_0 + \frac{C_1(n)}{n}, \\ B(n) &= b_0 + \frac{C_0(n)}{n}, \end{aligned}$$

where $C_2(n), C_1(n), C_0(n)$ are some bounded infinite sequences; namely, $C_2, C_1, C_0 \in B_0$. We then have

$$\begin{aligned}&\rho_1^2 E_p(n+2) + \rho_1 a_0 E_p(n+1) + b_0 E_p(n)\\ =&- \varepsilon_{p+1}(n) - \frac{C_0(n)E_p(n) + C_1(n)E_p(n+1) + C_2(n)E_p(n+2)}{n}. \quad (1.24)\end{aligned}$$

To show that the above equation has a solution $E_p \in B_p$, we first consider the difference equation

$$\rho_1^2 u(n+2) + \rho_1 a_0 u(n+1) + b_0 u(n) = h(n), \quad (1.25)$$

for a given $h \in B_{p+1}$. A particular solution of the above equation is given by $u = L(h)$, where L takes the following convolution form (exercise)

$$L(h)(n) = \sum_{k=n}^{\infty} G(k-n+1)h(k), \quad (1.26)$$

and the kernel function is given by

$$G(k) = \frac{1 - (\rho_1/\rho_2)^k}{\rho_1(\rho_2 - \rho_1)};$$

see Bender and Orszag (1978, p. 43). Since $|\rho_1| \leq |\rho_2|$, we have $G \in B_0$. It is obvious that $L(B_{p+1}) \subset B_p$. Next, we define $F : B_p \to B_{p+1}$ by

$$F(u)(n) = -\varepsilon_{p+1}(n) - \frac{C_0(n)u(n) + C_1(n)u(n+1) + C_2(n)u(n+2)}{n}.$$

A fixed point of $L \circ F$ on B_p gives a solution of (1.24). We intend to show that for any large p, the composite map $L \circ F$ is a contraction on the shifted Banach space $B_p^{(p)}$. Let u_1 and u_2 be two infinite sequences in $B_p^{(p)}$. For any $n \geq p$, we have

$$|L(F(u_1))(n) - L(F(u_2))(n)| \leq \sum_{k=n}^{\infty} \frac{C\|u_1 - u_2\|_p}{k^{p+1}} \leq \frac{2C}{pn^p}\|u_1 - u_2\|_p,$$

where $C = \|G\|_0(\|C_0\|_0 + \|C_1\|_0 + \|C_2\|_0)$. Hence, $L \circ F$ is a contraction map in $B_p^{(p)}$ for any $p > 2C$, and thus possesses a fixed point $E_p \in B_p^{(p)}$. The original difference equation (1.2) has a solution $Y_{1,p}(n)$ given as in (1.22) for $n \geq p$. The values of $Y_{1,p}(n)$ for $0 \leq n < p$ can be obtained by solving (1.2) backwards. Thus, we have found an asymptotic solution $Y_{1,p}(n)$ for any $p \geq 2C$. If $1 \leq p < 2C$, we simply truncate the solution $Y_{1,2C}(n)$ to obtain the desired asymptotic formula and error estimation. This completes the proof. □

Now, we look for a second linearly independent asymptotic solution of (1.2).

Theorem 1.2. *For any $p \geq 1$, the difference equation* (1.2) *has a solution*

$$Y_{2,p}(n) = \rho_2^n n^{\lambda_2} \left[\sum_{s=0}^{p-1} \frac{\bar{y}_s}{n^s} + \bar{E}_p(n) \right], \tag{1.27}$$

where $\bar{y}_0 = 1$, $\lambda_2 = -(a_1\rho_2 + b_1)/(2\rho_2^2 + a_0\rho_2)$, and

$$\bar{y}_{s-1} = \frac{1}{(s-1)(2\rho_2^2 + a_0\rho_2)} \sum_{j=0}^{s-2} \left\{ \rho_2^2 2^{s-j} \binom{\lambda_2 - j}{s-j} \right.$$

$$\left. + \rho_2 \sum_{k=0}^{s-j} \binom{\lambda_2 - j}{s-j-k} a_k + b_{s-j} \right\} \bar{y}_j, \quad s \geq 2, \tag{1.28}$$

and the error term $\bar{E}_p \in B_p$.

Proof. If $|\rho_1| = |\rho_2|$, we can proceed as in the proof of Theorem 1.1, except that the kernel function is defined by switching the roles of ρ_1 and ρ_2:

$$\bar{G}(k) = \frac{1 - (\rho_2/\rho_1)^k}{\rho_2(\rho_1 - \rho_2)}.$$

Since $|\rho_2/\rho_1| = 1$, we still have $\bar{G} \in B_0$ and use a similar argument to find the second linearly independent solution. If $|\rho_1| < |\rho_2|$, then the above kernel function becomes exponentially large as $k \to \infty$. We need to find a solution of the form

$$\begin{aligned} Y(n) &= Y_{1,p+1}(n) \left(\frac{\rho_2}{\rho_1}\right)^n n^{\lambda_2 - \lambda_1} \left[\sum_{s=0}^{p-1} \frac{\tilde{y}_s}{n^s} + \tilde{E}_p(n) \right] \\ &= \rho_2^n n^{\lambda_2} \left[\sum_{s=0}^{p} \frac{y_s}{n^s} + E_{p+1}(n) \right] \left[\sum_{s=0}^{p-1} \frac{\tilde{y}_s}{n^s} + \tilde{E}_p(n) \right], \end{aligned}$$

where $\tilde{y}_s$ and $\bar{y}_s$ are related by the recurrence relation

$$\sum_{j=0}^{s} y_j \tilde{y}_{s-j} = \bar{y}_s. \tag{1.29}$$

Substituting the expression of $Y(n)$ into the difference equation (1.2) yields

$$\begin{aligned} & Y_{1,p+1}(n+2) \left(\frac{\rho_2}{\rho_1}\right)^2 \left(1 + \frac{2}{n}\right)^{\lambda_2 - \lambda_1} \tilde{E}_p(n+2) \\ & + Y_{1,p+1}(n+1) \frac{\rho_2}{\rho_1} \left(1 + \frac{1}{n}\right)^{\lambda_2 - \lambda_1} \tilde{E}_p(n+1) A(n) \\ & + Y_{1,p+1}(n) \tilde{E}_p(n) B(n) = -\rho_1^n n^{\lambda_1} \tilde{\varepsilon}_{p+1}(n), \end{aligned}$$

where

$$\begin{aligned}\tilde{\varepsilon}_{p+1}(n) =& \left[\sum_{s=0}^{p} \frac{y_s}{(n+2)^s} + E_{p+1}(n+2)\right] \rho_2^2 \left(1+\frac{2}{n}\right)^{\lambda_2} \sum_{s=0}^{p-1} \frac{\tilde{y}_s}{(n+2)^s} \\ &+ \left[\sum_{s=0}^{p} \frac{y_s}{(n+1)^s} + E_{p+1}(n+1)\right] \rho_2 \left(1+\frac{1}{n}\right)^{\lambda_2} \sum_{s=0}^{p-1} \frac{\tilde{y}_s}{(n+1)^s} A(n) \\ &+ \left[\sum_{s=0}^{p} \frac{y_s}{n^s} + E_{p+1}(n)\right] \sum_{s=0}^{p-1} \frac{\tilde{y}_s}{n^s} B(n). \end{aligned} \tag{1.30}$$

It can be shown that $\tilde{\varepsilon}_{p+1} \in B_{p+1}$ (exercise). On account of

$$Y_{1,p+1}(n+1)A(n) = -Y_{1,p+1}(n+2) - Y_{1,p+1}(n)B(n),$$

we may rewrite the difference equation as

$$\begin{aligned} Y_{1,p+1}(n+2)\Bigg[&\left(\frac{\rho_2}{\rho_1}\right)^2 \left(1+\frac{2}{n}\right)^{\lambda_2-\lambda_1} \tilde{E}_p(n+2) \\ &- \frac{\rho_2}{\rho_1}\left(1+\frac{1}{n}\right)^{\lambda_2-\lambda_1} \tilde{E}_p(n+1)\Bigg] \\ - Y_{1,p+1}(n)&\left[\frac{\rho_2}{\rho_1}\left(1+\frac{1}{n}\right)^{\lambda_2-\lambda_1} \tilde{E}_p(n+1) - \tilde{E}_p(n)\right] B(n) \\ = -\,&\rho_1^n n^{\lambda_1} \tilde{\varepsilon}_{p+1}(n). \end{aligned}$$

Define

$$\Delta_p(n) := \frac{\rho_2}{\rho_1}(n+1)^{\lambda_2-\lambda_1}\tilde{E}_p(n+1) - n^{\lambda_2-\lambda_1}\tilde{E}_p(n). \tag{1.31}$$

It is readily seen that

$$\frac{\rho_2}{\rho_1} Y_{1,p+1}(n+2)\Delta_p(n+1) - Y_{1,p+1}(n)\Delta_p(n)B(n) = -\rho_1^n n^{\lambda_2}\tilde{\varepsilon}_{p+1}(n).$$

In view of (1.3), (1.22) and the fact that $\rho_1\rho_2 = b_0$, we obtain

$$\frac{\rho_2}{\rho_1} Y_{1,p+1}(n+2) = b_0 \rho_1^n n^{\lambda_1}\left[1 + \frac{\tilde{C}_1(n)}{n}\right],$$

$$B(n)Y_{1,p+1}(n) = b_0 \rho_1^n n^{\lambda_1}\left[1 + \frac{\tilde{C}_0(n)}{n}\right],$$

where $\tilde{C}_0, \tilde{C}_1 \in B_0$. Consequently, we have

$$\begin{aligned}\Delta_p(n+1) - \Delta_p(n) =& -\frac{1}{b_0} n^{\lambda_2-\lambda_1}\tilde{\varepsilon}_{p+1}(n) \\ &+ \frac{\tilde{C}_0(n)\Delta_p(n) - \tilde{C}_1(n)\Delta_p(n+1)}{n}. \end{aligned} \tag{1.32}$$

For convenience, we denote $\tilde{p} := p - \operatorname{Re}(\lambda_2 - \lambda_1)$. Now, we define a map $\tilde{L}$ on $B_{\tilde{p}}$ as

$$\tilde{L}(u)(n) := -\sum_{k=n}^{\infty}\left[\frac{1}{b_0}k^{\lambda_2-\lambda_1}\tilde{\varepsilon}_{p+1}(k) - \frac{\tilde{C}_0(k)u(k) - \tilde{C}_1(k)u(k+1)}{k}\right].$$

It is easily seen that $\tilde{L}$ maps $B_{\tilde{p}}$ onto itself. Moreover, a fixed point of $\tilde{L}$ in $B_{\tilde{p}}$ is the same as a solution to the difference equation (1.32). We shall show that $\tilde{L}$ is a contraction for large $\tilde{p}$. Actually, for any $u_1, u_2 \in B_{\tilde{p}}$ and any $n \geq \tilde{p}$, we have

$$|\tilde{L}(u_1)(n) - \tilde{L}(u_2)(n)| \leq \sum_{k=n}^{\infty}\frac{\tilde{C}\|u_1 - u_2\|_{\tilde{p}}}{k^{\tilde{p}+1}} \leq \frac{2\tilde{C}}{\tilde{p}n^{\tilde{p}}}\|u_1 - u_2\|_{\tilde{p}},$$

where $\tilde{C} := \|\tilde{C}_0\|_0 + \|\tilde{C}_1\|_1$. Thus, $\tilde{L}$ possesses a fixed point Δ_p in $B_{\tilde{p}}^{(\tilde{p})}$ for any $\tilde{p} > 2\tilde{C}$. Next, we use (1.31) to obtain

$$\tilde{E}_p(n) = \sum_{k=m}^{n-1}\left(\frac{\rho_2}{\rho_1}\right)^{k-n} n^{\lambda_1-\lambda_2}\Delta_p(k) + \left(\frac{\rho_2}{\rho_1}\right)^{m-n}(\frac{m}{n})^{\lambda_2-\lambda_1}\tilde{E}_p(m), \quad n \geq m,$$

where $m > \tilde{p}$ and $\tilde{E}_p(m)$ is independent of n. Note that

$$|n^p\tilde{E}_p(n)| \leq \sum_{k=m}^{n-1}|\frac{\rho_2}{\rho_1}|^{k-n}(\frac{n}{k})^{\tilde{p}}\|\Delta_p\|_{\tilde{p}} + |\frac{\rho_2}{\rho_1}|^{m-n}(\frac{m}{n})^{\lambda_2-\lambda_1}n^p|\tilde{E}_p(m)|. \tag{1.33}$$

The first term on the right-hand side is uniformly bounded (exercise), and the second term tends to zero as $n \to \infty$. Thus, $\tilde{E}_p \in B_p^{(m)}$, and we have found a second linearly independent solution:

$$\begin{aligned} Y_{2,p}(n) =& \rho_2^n n^{\lambda_2}\left[\sum_{s=0}^{p-1}\frac{y_s}{n^s} + E_p(n)\right]\left[\sum_{s=0}^{p-1}\frac{\tilde{y}_s}{n^s} + \tilde{E}_p(n)\right] \\ =& \rho_2^n n^{\lambda_2}\left[\sum_{s=0}^{p-1}\frac{\bar{y}_s}{n^s} + \bar{E}_p(n)\right] \end{aligned}$$

for $n \geq m$. Since $E_p, \tilde{E}_p \in B_p$, we obtain from the recurrence relation (1.29) that $\bar{E}_p \in B_p^{(m)}$. The values of $Y_{2,p}(n)$ for $0 \leq n < m$ can be determined backwards from (1.2). This completes the proof. □

Now, we are ready to state the main theorem in this section.

Theorem 1.3. *The difference equation* (1.2) *has two linearly independent solutions* $Y_1(n)$ *and* $Y_2(n)$ *with following asymptotic expansions:*

$$Y_1(n) \sim \rho_1^n n^{\lambda_1} \sum_{s=0}^{\infty} \frac{y_s}{n^s}, \tag{1.34}$$

$$Y_2(n) \sim \rho_2^n n^{\lambda_2} \sum_{s=0}^{\infty} \frac{\bar{y}_s}{n^s}, \tag{1.35}$$

where ρ_i, λ_i *with* $i = 1, 2$*, and* $y_s, \bar{y}_s$ *with* $s \geq 0$ *are determined as in Theorems 1.1 and 1.2.*

Proof. For a sufficiently large integer p, by Theorems 1.1 and 1.2 we have two solutions $Y_{1,p}(n)$ and $Y_{2,p}(n)$, which are denoted as $Y_1(n)$ and $Y_2(n)$, respectively. We claim $Y_1(n)$ and $Y_2(n)$ have the asymptotic expansions in (1.34)-(1.35). Actually, for any $q > p$, there exist two solutions $Y_{1,q}(n)$ and $Y_{2,q}(n)$ satisfying the asymptotic expansions given in Theorems 1.1 and 1.2, with p replaced by q. It is obvious that $Y_1(n)$ and $Y_2(n)$ are linearly independent since they have different asymptotic expansions. We then have the linear representation:

$$Y_{1,q}(n) = c_{11} Y_1(n) + c_{12} Y_2(n).$$

From the asymptotic expansions of $Y_{1,q}(n), Y_1(n), Y_2(n)$ and $|\lambda_1| \leq |\lambda_2|$, we obtain $c_{11} = 1$ and $c_{12} = 0$. Since q is arbitrarily chosen, we have $Y_1(n) = Y_{1,q}(n)$ for any $q > p$ and thus $Y_1(n)$ has the asymptotic expansion given in (1.34). Next, we investigate the linear representation:

$$Y_{2,q}(n) = c_{21} Y_1(n) + c_{22} Y_2(n).$$

If $|\lambda_1| = |\lambda_2|$, we observe from the asymptotic expansions of $Y_{2,q}(n), Y_1(n), Y_2(n)$ that $c_{21} = 0$ and $c_{22} = 1$. If $|\lambda_1| < |\lambda_2|$, we still have $c_{22} = 1$. But c_{21} cannot be determined from the asymptotic expansions. However, since $Y_1(n)$ is exponentially small compared to $Y_2(n)$, we conclude that $Y_2(n)$ and $Y_{2,q}(n)$ have the same asymptotic expansion for any $q > p$. Thus, $Y_2(n)$ has the asymptotic expansion (1.35). □

1.3 Exceptional cases

In this section, we consider the exceptional cases: $a_0^2 = 4b_0 \neq 0$ and $a_0 a_1 = 2b_1$. The characteristic equation (1.7) has a double root $\rho = -a_0/2$. Denote the roots of indicial equation by λ_1 and λ_2, where $\operatorname{Re}\lambda_1 \le \operatorname{Re}\lambda_2$. We first study the subcase (i): $\lambda_1 - \lambda_2 \notin \mathbf{Z}$.

Theorem 1.4. *If $\lambda_1 - \lambda_2$ is not an integer, then the difference equation* (1.2) *has two linearly independent solutions:*

$$Y_1(n) \sim \rho^n n^{\lambda_1} \sum_{s=0}^{\infty} \frac{y_s}{n^s}, \tag{1.36}$$

$$Y_2(n) \sim \rho^n n^{\lambda_2} \sum_{s=0}^{\infty} \frac{\bar{y}_s}{n^s}, \tag{1.37}$$

where y_s and $\bar{y}_s$ are determined by (1.6) *with $\lambda = \lambda_1$ and $\lambda = \lambda_2$, respectively.*

Proof. Without loss of generality, we may assume

$$\rho = 1, \ a_0 = -2, \ b_0 = 1, \ a_1 + b_1 = 0. \tag{1.38}$$

Otherwise, we consider the difference equation satisfied by $\tilde{P}_n(x) = P_n(x)\rho^{-n}$, where the coefficients are scaled as $\tilde{A}_n = A_n/\rho$ and $\tilde{B}_n = B_n/\rho^2$. The indicial equation (1.9) becomes

$$\lambda(\lambda - 1) - b_1\lambda + a_2 + b_2 = 0. \tag{1.39}$$

First, we claim that for any $p \ge 1$, there exist two solutions

$$Y_{1,p}(n) = n^{\lambda_1} \left[\sum_{s=0}^{p-1} \frac{y_s}{n^s} + E_p(n) \right], \tag{1.40}$$

$$Y_{2,p}(n) = n^{\lambda_2} \left[\sum_{s=0}^{p-1} \frac{\bar{y}_s}{n^s} + \bar{E}_p(n) \right], \tag{1.41}$$

where $E_p, \bar{E}_p \in B_p$. If the claim is true, then $Y_{1,p}(n)$ and $Y_{2,p}(n)$ are linearly independent because $\lambda_1 - \lambda_2 \notin \mathbf{Z}$. Similarly, as in the previous case, we can use linearity to prove that for any $p \ne q$, $Y_{i,p}(n) = Y_{i,q}(n)$ with $i = 1, 2$. We only need to verify that (1.40) gives a solution; a similar argument works for (1.41). Substituting (1.40) into (1.2) yields

$$\left(1 + \frac{2}{n}\right)^{\lambda_1} E_p(n+2) + \left(1 + \frac{1}{n}\right)^{\lambda_1} A(n)E_p(n+1) + B(n)E_p(n)$$
$$= -\,\varepsilon_{p+2}(n),$$

where

$$\varepsilon_{p+2}(n) = \sum_{s=0}^{p-1} \frac{y_s(1+2/n)^{\lambda_1}}{(n+2)^s} + A(n)\sum_{s=0}^{p-1} \frac{y_s(1+1/n)^{\lambda_1}}{(n+1)^s} + B(n)\sum_{s=0}^{p-1} \frac{y_s}{n^s}.$$

On account of (1.3), we have from the above equation

$$\varepsilon_{p+2}(n) = \sum_{s=0}^{\infty} \frac{1}{n^s} \sum_{j=0}^{\min\{s,p-1\}} \left\{ 2^{s-j} \binom{\lambda - j}{s-j} + \sum_{k=0}^{s-j} \binom{\lambda - j}{s-j-k} a_k + b_{s-j} \right\} y_j.$$

With the special values given in (1.38), one can show from (1.6) that the coefficient of n^{-s} in the above infinite series is zero for all $s = 0, 1, \cdots, p-1$. Thus,

$$\varepsilon_{p+2}(n) = \sum_{s=p}^{\infty} \frac{1}{n^s} \sum_{j=0}^{p-1} \left\{ 2^{s-j} \binom{\lambda - j}{s-j} + \sum_{k=0}^{s-j} \binom{\lambda - j}{s-j-k} a_k + b_{s-j} \right\} y_j.$$

Since the coefficients of y_s and y_{s-1} in equation (1.6) both vanish, the coefficients of n^{-p} and n^{-p-1} in the above infinite series are also zero. Hence, $\varepsilon_{p+2}(n) = O(n^{-p-2})$; i.e., $\varepsilon_{p+2}(n) \in B_{p+2}$. Next, we note from (1.3) and (1.38) that

$$\left(1 + \frac{2}{n}\right)^{\lambda_1} = 1 + \frac{2\lambda_1}{n} + \frac{C_2(n)}{n^2},$$

$$\left(1 + \frac{1}{n}\right)^{\lambda_1} A(n) = -2 - \frac{b_1 + 2\lambda_1}{n} + \frac{C_1(n)}{n^2},$$

$$B(n) = 1 + \frac{b_1}{n} + \frac{C_0(n)}{n^2},$$

where $C_0, C_1, C_2 \in B_0$. Therefore, we have

$$\left(1 + \frac{2\lambda_1}{n}\right) E_p(n+2) - \left(2 + \frac{b_1 + 2\lambda_1}{n}\right) E_p(n+1) + \left(1 + \frac{b_1}{n}\right) E_p(n)$$
$$= F(E_p)(n),$$

where

$$F(u)(n) := -\varepsilon_{p+2}(n) - \frac{C_2(n)u(n+2) + C_1(n)u(n+1) + C_0(n)u(n)}{n^2}.$$

Now, we consider the following difference equation

$$\left(1 + \frac{2\lambda_1}{n}\right) u(n+2) - \left(2 + \frac{b_1 + 2\lambda_1}{n}\right) u(n+1) + \left(1 + \frac{b_1}{n}\right) u(n) = h(n), \tag{1.42}$$

where $h \in B_{p+2}$. A particular solution to the above difference equation (exercise) is given by $u = L(h)$ with

$$L(h)(n) = \sum_{l=n}^{\infty} \sum_{k=l}^{\infty} \frac{(1 + \frac{2\lambda_1}{k-1}) \cdots (1 + \frac{2\lambda_1}{l})}{(1 + \frac{b_1}{k}) \cdots (1 + \frac{b_1}{l})} h(k). \tag{1.43}$$

If n is sufficiently large, we have (exercise) for any $k \geq l \geq n$,

$$\left| \frac{(1 + \frac{2\lambda_1}{k-1}) \cdots (1 + \frac{2\lambda_1}{l})}{(1 + \frac{b_1}{k}) \cdots (1 + \frac{b_1}{l})} \right| \leq \left(\frac{2k}{l} \right)^{2|\lambda_1| + |b_1|}. \tag{1.44}$$

It can be shown that $L(B_{p+2}) \subset B_p$ if $p > 2|\lambda_1| + |b_1|$. We intend to verify that the composite map $L \circ F$ is a contraction on $B_p^{(p)}$ for some large p. Let $C = 2^{2|\lambda_1|+|b_1|}(\|C_2\|_0 + \|C_1\|_0 + \|C_0\|_0)$. For any $u_1, u_2 \in B_p^{(p)}$, we have

$$\begin{aligned} |L(F(u_1))(n) - L(F(u_2))(n)| &\leq C\|u_1 - u_2\|_p \sum_{l=n}^{\infty} \sum_{k=l}^{\infty} \frac{k^{2|\lambda_1|+|b_1|-p-2}}{l^{2|\lambda_1|+|b_1|}} \\ &\leq C\|u_1 - u_2\|_p \sum_{l=n}^{\infty} \frac{2l^{-p-1}}{p+1-2|\lambda_1| - |b_1|} \\ &\leq C\|u_1 - u_2\|_p \frac{4n^{-p}}{p(p+1-2|\lambda_1| - |b_1|)} \\ &\leq \|u_1 - u_2\|_p \frac{4C}{pn^p}, \quad n \geq p > 2|\lambda_1| + |b_1|, \end{aligned}$$

where we have made use of the following inequality (exercise) twice:

$$\sum_{l=n}^{\infty} \frac{1}{l^{q+1}} \leq \frac{2}{qn^q}, \quad n \geq q > 0. \tag{1.45}$$

For any $p > 2|\lambda_1| + |b_1| + 4C$, the map $L \circ F$ is a contraction on $B_p^{(p)}$, and hence, it possesses a fixed point, denoted by E_p, in $B_p^{(p)}$. This implies that the difference equation (1.2) has a solution $Y_{1,p}(n)$ satisfying (1.40) for $n \geq p$. The values of $Y_{1,p}(n)$ for $0 \leq n < p$ can be derived from the difference equation (1.2). By a similar argument, we obtain the second solution $Y_{2,p}(n)$ given in (1.41). Since $\lambda_1 - \lambda_2 \notin \mathbf{Z}$, these two solutions are linearly independent. Furthermore, for any $q > p$, we can write $Y_{1,q}(n)$ as a linear combination of $Y_{1,p}(n)$ and $Y_{2,p}(n)$:

$$Y_{1,q}(n) = c_1 Y_{1,p}(n) + c_2 Y_{2,p}(n).$$

By asymptotic matching, we conclude that $c_1 = 1$ and $c_2 = 0$. Thus, $Y_{1,p}(n)$ is independent of p. Similarly, $Y_{2,p}(n)$ is independent of p. We have now found two linearly independent solutions with asymptotic expansions given in (1.36) and (1.37). This completes the proof. □

Now, we consider the subcase (ii): $\lambda_1 - \lambda_2 \in \mathbf{Z}$. Without loss of generality, we assume $\lambda_2 - \lambda_1 \geq 0$.

Theorem 1.5. *Let $\lambda_2 - \lambda_1$ be a nonnegative integer. There exist two linearly independent solutions to the difference equation* (1.2)*:*

$$Y_1(n) \sim \rho^n n^{\lambda_1} \sum_{s=0}^{\infty} \frac{y_s}{n^s}, \tag{1.46}$$

$$Y_2(n) \sim \rho^n n^{\lambda_2} \sum_{s=0}^{\infty} \frac{\bar{y}_s}{n^s} + c(\log n)\rho^n n^{\lambda_1} \sum_{s=0}^{\infty} \frac{y_s}{n^s}, \tag{1.47}$$

where y_s are determined by (1.6) *with $\lambda = \lambda_1$, and $\bar{y}_s, c$ are determined by* (1.11) *together with $\bar{y}_{\lambda_2-\lambda_1} = 0$. Furthermore, if $\lambda_1 = \lambda_2$, we set $c = 1$; if $\lambda_2 - \lambda_1$ is a positive integer, we choose $\bar{y}_0 = 1$.*

Proof. Without loss of generality, we assume $\rho = 1$. By a similar argument as given in the proof of Theorem 1.4, there exists a solution as given in (1.40). One may also find a second solution as given in (1.41). However, since λ_1 and λ_2 differ by an integer, these two solutions may be linearly dependent. So, we look for a solution of the form:

$$Y_{2,p}(n) = n^{\lambda_2} \left[\sum_{s=0}^{p-1} \frac{\bar{y}_s}{n^s} + \bar{E}_p(n) \right] + c(\log n) n^{\lambda_1} \sum_{s=0}^{p-1} \frac{y_s}{n^s}.$$

Substituting this into (1.2) yields

$$\left(1 + \frac{2}{n}\right)^{\lambda_2} \bar{E}_p(n+2) + \left(1 + \frac{1}{n}\right)^{\lambda_2} A(n)\bar{E}_p(n+1) + B(n)\bar{E}_p(n) \\ = -\,\bar{\varepsilon}_{p+2}(n), \tag{1.48}$$

where

$$\begin{aligned} \bar{\varepsilon}_{p+2}(n) = {} & \sum_{s=0}^{p-1} \frac{\bar{y}_s}{n^s} \left(1 + \frac{2}{n}\right)^{\lambda_2 - s} + A(n) \sum_{s=0}^{p-1} \frac{\bar{y}_s}{n^s} \left(1 + \frac{1}{n}\right)^{\lambda_2 - s} + B(n) \sum_{s=0}^{p-1} \frac{\bar{y}_s}{n^s} \\ & + c(\log n) \left[\sum_{s=0}^{p-1} \frac{y_s}{n^{s+\lambda_2-\lambda_1}} \left(1 + \frac{2}{n}\right)^{\lambda_1 - s} \right. \\ & \left. + A(n) \sum_{s=0}^{p-1} \frac{y_s}{n^{s+\lambda_2-\lambda_1}} \left(1 + \frac{1}{n}\right)^{\lambda_1 - s} + B(n) \sum_{s=0}^{p-1} \frac{y_s}{n^{s+\lambda_2-\lambda_1}} \right] \\ & + c \left[\log\left(1 + \frac{2}{n}\right) \sum_{s=0}^{p-1} \frac{y_s}{n^{s+\lambda_2-\lambda_1}} \left(1 + \frac{2}{n}\right)^{\lambda_1 - s} \right. \\ & \left. + A(n) \log\left(1 + \frac{1}{n}\right) \sum_{s=0}^{p-1} \frac{y_s}{n^{s+\lambda_2-\lambda_1}} \left(1 + \frac{1}{n}\right)^{\lambda_1 - s} \right]. \end{aligned} \tag{1.49}$$

On account of (1.6) and (1.11), we have $\bar{\varepsilon}_{p+2}(n) \in B_{p+2}$ (exercise). Similarly, as in the proof of Theorem 1.4, there exists a solution $\bar{E}_p \in B_p$ to equation (1.48). Furthermore, since $\bar{y}_0 = 1$ for $\lambda_2 - \lambda_1 > 0$ and $c = 1$ for $\lambda_2 = \lambda_1$, the two solutions $Y_{1,p}(n)$ and $Y_{2,p}(n)$ are linearly independent. For any $q > p$, we have

$$Y_{1,q}(n) = c_1 Y_{1,p}(n) + c_2 Y_{2,p}(n).$$

By asymptotic matching, $c_1 = 1$ and $c_2 = 0$. Thus, $Y_{1,p}(n) = Y_{1,q}(n)$ for any $q > p$. Similarly, $Y_{2,p}(n) = Y_{2,q}(n)$ for any $q > p$. This completes the proof of the theorem. □

1.4 Subnormal case

In this section, we consider the subnormal case: $a_0^2 = 4b_0 \neq 0$ and $a_0 a_1 \neq 2b_1$. In this case, we have only one characteristic root $\rho = -a_0/2$.

Theorem 1.6. *If $a_0^2 = 4b_0 \neq 0$ and $a_0 a_1 \neq 2b_1$, the difference equation (1.2) has two linearly independent solutions*

$$Y_1(n) \sim \rho^n e^{\gamma_1 \sqrt{n}} n^\lambda \sum_{s=0}^{\infty} \frac{y_s}{n^{s/2}}, \tag{1.50}$$

$$Y_2(n) \sim \rho^n e^{\gamma_2 \sqrt{n}} n^\lambda \sum_{s=0}^{\infty} \frac{\bar{y}_s}{n^{s/2}}, \tag{1.51}$$

where γ_1 and γ_2 are two solutions to equation (1.20), λ is given by (1.21), and y_s and $\bar{y}_s$ are determined from (1.19) with $\gamma = \gamma_1$ and $\gamma = \gamma_2$, respectively.

Proof. Without loss of generality, we may assume $\rho = 1$. Consequently, we have

$$a_0 = -2, \ b_0 = 1, \ a_1 + b_1 \neq 0.$$

For any large p, we look for a solution

$$Y_{1,p}(n) = e^{\gamma_1 \sqrt{n}} n^\lambda \left[\sum_{s=0}^{p-1} \frac{y_s}{n^{s/2}} + E_p(n) \right]. \tag{1.52}$$

Substituting this into (1.2) yields

$$e^{\gamma_1(\sqrt{n+2}-\sqrt{n})} \left(1 + \frac{2}{n}\right)^\lambda E_p(n+2) + A(n) e^{\gamma_1(\sqrt{n+1}-\sqrt{n})} \left(1 + \frac{1}{n}\right)^\lambda E_p(n+1)$$
$$+ B(n) E_p(n) = -\varepsilon_{p+3}(n),$$

where

$$\begin{aligned}\varepsilon_{p+3}(n) =& e^{\gamma_1(\sqrt{n+2}-\sqrt{n})}\sum_{s=0}^{p-1}\frac{y_s}{n^{s/2}}\left(1+\frac{2}{n}\right)^{\lambda-s/2}\\ &+A(n)e^{\gamma_1(\sqrt{n+1}-\sqrt{n})}\sum_{s=0}^{p-1}\frac{y_s}{n^{s/2}}\left(1+\frac{1}{n}\right)^{\lambda-s/2}+B(n)\sum_{s=0}^{p-1}\frac{y_s}{n^{s/2}}.\end{aligned}$$

In view of (1.19), we have $\varepsilon_{p+3}(n) = c_3/n^{(p+3)/2} + \varepsilon_{p+4}(n)$, where c_3 is a constant independent of n and ε_{p+4} is an infinite sequence in $B_{(p+4)/2}$. Note that

$$e^{\gamma_1(\sqrt{n+2}-\sqrt{n})}\left(1+\frac{2}{n}\right)^{\lambda} = \alpha(n) + \frac{C_2(n)}{n^2}, \tag{1.53}$$

$$A(n)e^{\gamma_1(\sqrt{n+1}-\sqrt{n})}\left(1+\frac{1}{n}\right)^{\lambda} = -\alpha(n) - \beta(n) + \frac{C_1(n)}{n^2}, \tag{1.54}$$

$$B(n) = \beta(n) + \frac{C_0(n)}{n^2}, \tag{1.55}$$

where

$$\alpha(n) := 1 + \frac{\gamma_1}{\sqrt{n}} + \frac{\gamma_1^2/2 + b_1 + 1/2}{n} + \frac{\gamma_1^3/6 + b_1\gamma_1}{n^{3/2}}, \quad \beta(n) := 1 + \frac{b_1}{n},$$

and $C_2, C_1, C_0 \in B_0$. Thus, we have

$$\alpha(n)E_p(n+2) - [\alpha(n)+\beta(n)]E_p(n+1) + \beta(n)E_p(n) = F(E_p)(n), \tag{1.56}$$

where

$$\begin{aligned}F(u)(n) :=& -\frac{c_3}{n^{(p+3)/2}} - \varepsilon_{p+4}(n)\\ &-\frac{C_2(n)u(n+2) + C_1(n)u(n+1) + C_0(n)u(n)}{n^2}.\end{aligned} \tag{1.57}$$

For convenience, we denote $f(n) := -c_3 n^{-(p+3)/2}$. It is evident that $F(u) - f \in B_{(p+4)/2}$ if $u \in B_{p/2}$. Consider the difference equation

$$\alpha(n)u(n+2) - [\alpha(n)+\beta(n)]u(n+1) + \beta(n)u(n) = h(n), \tag{1.58}$$

where $h - f \in B_{(p+4)/2}$. A particular solution to the above equation (exercise) is given by $u(n) = L(h)(n)$ with

$$L(h)(n) := \begin{cases}\displaystyle\sum_{l=n}^{\infty}\sum_{k=l}^{\infty} G_-(k,l)h(k), & \operatorname{Re}\gamma_1 \le 0,\\ \displaystyle-\sum_{l=n}^{\infty}\sum_{k=m}^{l-1} G_+(k,l)h(k), & \operatorname{Re}\gamma_1 > 0,\end{cases} \tag{1.59}$$

where

$$G_-(k,l) := \frac{\alpha(k-1)\cdots\alpha(l)}{\beta(k)\cdots\beta(l)}, \quad k \geq l,$$

$$G_+(k,l) := \frac{\beta(l-1)\cdots\beta(k+1)}{\alpha(l-1)\cdots\alpha(k)}, \quad k < l,$$

and m is a fixed integer to be determined later. We shall show that $L \circ F$ is a contraction on $B_{p/2}^{(m)}$. Note that $\gamma_1 \neq 0$, and we consider two cases: $\operatorname{Re}\gamma_1 \leq 0$ with $\gamma_1 \neq 0$, and $\operatorname{Re}\gamma_1 > 0$, respectively.

If $\operatorname{Re}\gamma_1 \leq 0$ and $\gamma_1 \neq 0$, we have (exercise) as $l \to \infty$,

$$\begin{aligned} G_-(k,l) &= e^{2\gamma_1(\sqrt{l}-\sqrt{k})}\left(\frac{k}{l}\right)^{1/2} \\ &\quad \times \left[1 + \frac{\gamma_1}{2}\left(\frac{1}{\sqrt{l}} - \frac{1}{\sqrt{k}}\right) + O\left(\frac{1}{l}\right)\right], \quad k \geq l, \end{aligned} \tag{1.60}$$

$$\sum_{k=l}^{\infty} \frac{e^{2\gamma_1(\sqrt{k}-\sqrt{l})}}{k^q} = O\left(\frac{1}{l^{q-1/2}}\right), \quad q > 0. \tag{1.61}$$

Consequently, as $n \to \infty$,

$$\begin{aligned} L(f)(n) &= \sum_{l=n}^{\infty}\sum_{k=l}^{\infty} G_-(k,l)\frac{-c_3}{k^{(p+3)/2}} \\ &= \sum_{l=n}^{\infty}\sum_{k=l}^{\infty} e^{2\gamma_1(\sqrt{k}-\sqrt{l})}\left(\frac{k}{l}\right)^{1/2}\frac{-c_3}{k^{(p+3)/2}} \\ &\quad \times \left[1 + \frac{\gamma_1}{2}\left(\frac{1}{\sqrt{l}} - \frac{1}{\sqrt{k}}\right) + O\left(\frac{1}{l}\right)\right] \\ &= \sum_{l=n}^{\infty} O\left(\frac{1}{l^{(p+2)/2}}\right) \\ &= O\left(\frac{1}{n^{p/2}}\right). \end{aligned}$$

There exists $m_0 > 0$ such that for any $k \geq l \geq m_0$, we have (exercise)

$$|G_-(k,l)| \leq 2\left(\frac{k}{l}\right)^{2|\gamma_1|^2+3|b_1|+2}. \tag{1.62}$$

If $p > 2(2|\gamma_1|^2 + 3|b_1| + 2)$ and $n > m_0 + p$, then for any $g \in B_{(p+4)/2}$,

$$|L(g)(n)| \leq 2\sum_{l=n}^{\infty}\sum_{k=l}^{\infty}\left(\frac{k}{l}\right)^{2|\gamma_1|^2+3|b_1|+2}\frac{\|g\|_{(p+4)/2}}{k^{(p+4)/2}} \leq \frac{16\|g\|_{(p+4)/2}}{pn^{p/2}}.$$

Especially, by choosing $g = h - f$, we have $L(h) = L(f) + L(h-f) \in B_{p/2}$. On the other hand, given $u_1, u_2 \in B_{p/2}$, we have $F(u_1) - F(u_2) \in B_{(p+4)/2}$ and

$$\|F(u_1) - F(u_2)\|_{(p+4)/2} \leq C\|u_1 - u_2\|_{p/2},$$

where $C = \|C_2\|_0 + \|C_1\|_0 + \|C_0\|_0$. Hence,

$$|L(F(u_1))(n) - L(F(u_2))(n)| \leq \frac{16C\|u_1 - u_2\|_{p/2}}{pn^{p/2}}.$$

Now we choose $p > 2(2|\gamma_1|^2 + 3|b_1| + 2) + 16C$ and $m = m_0 + p$ such that $L \circ F$ is a contraction on $B^{(m)}_{p/2}$.

If $\operatorname{Re}\gamma_1 > 0$, then (exercise) for any $s > 0$, there exist constants $M(s)$ and $m_1(s, \operatorname{Re}\gamma_1)$ such that for any $m \geq m_1(s, \operatorname{Re}\gamma_1)$,

$$|G_+(k,l)| \leq 2e^{-2\operatorname{Re}\gamma_1(\sqrt{l}-\sqrt{k})}, \quad l > k \geq m, \tag{1.63}$$

and we have by Euler-Maclaurin's formula (exercise)

$$\sum_{k=m}^{l-1} \frac{e^{-2\operatorname{Re}\gamma_1(\sqrt{l}-\sqrt{k})}}{k^s} \leq \frac{M(s)}{l^{s-1/2}}, \quad l > m. \tag{1.64}$$

Therefore, by choosing $s_1 = (p+3)/2$ and $m > p/2 + m_1(s_1, \operatorname{Re}\gamma_1)$, it follows from (1.59) that for any $n \geq m$,

$$\begin{aligned} |L(h)(n)| &\leq 2\sum_{l=n}^{\infty}\sum_{k=m}^{l-1} e^{-2\operatorname{Re}\gamma_1(\sqrt{l}-\sqrt{k})}\frac{\|h\|_{(p+3)/2}}{k^{(p+3)/2}} \\ &\leq 2\sum_{l=n}^{\infty} \frac{M(s_1)\|h\|_{(p+3)/2}}{l^{(p+2)/2}} \\ &\leq \frac{8M(s_1)\|h\|_{(p+3)/2}}{pn^{p/2}}. \end{aligned}$$

This implies that $L(h) \in B_{p/2}$. Next, we want to prove that $L \circ F$ is a contraction on $B_{p/2}$ for any large p. Given $u_1, u_2 \in B_{p/2}$, we have $F(u_1) - F(u_2) \in B_{(p+4)/2}$ and

$$\|F(u_1) - F(u_2)\|_{(p+4)/2} \leq C\|u_1 - u_2\|_{p/2},$$

where $C = \|C_2\|_0 + \|C_1\|_0 + \|C_0\|_0$. By choosing $s_2 = (p+4)/2$ and $m > p/2 + m_1(s_1, \operatorname{Re}\gamma_1) + m_1(s_2, \operatorname{Re}\gamma_1) + [16CM(s_2)]^2/p^2$, we have for any $n \geq m$,

$$
\begin{aligned}
&|L(F(u_1))(n) - L(F(u_2))(n)| \\
\leq &2\sum_{l=n}^{\infty}\sum_{k=m}^{l-1} e^{-2\operatorname{Re}\gamma_1(\sqrt{l}-\sqrt{k})} \frac{C\|u_1-u_2\|_{p/2}}{k^{(p+4)/2}} \\
\leq &2\sum_{l=n}^{\infty} \frac{CM(s_2)\|u_1-u_2\|_{p/2}}{l^{(p+3)/2}} \\
\leq &\frac{8CM(s_2)\|u_1-u_2\|_{p/2}}{pn^{(p+1)/2}} \\
\leq &\frac{\|u_1-u_2\|_{p/2}}{2n^{p/2}},
\end{aligned}
$$

which implies that $L \circ F$ is a contraction map on $B_{p/2}^{(m)}$. Thus, for $n \geq m$, the difference equation (1.2) has a solution $Y_{1,p}(n)$ as given in (1.52) with $E_p \in B_{p/2}^{(m)}$. The values $Y_{1,p}(n)$ for $1 \leq n < m$ can be obtained recursively from (1.2). Similarly, there exists another solution

$$
Y_{2,p}(n) = e^{\gamma_2\sqrt{n}} n^{\lambda}[\sum_{s=0}^{p-1} \frac{\bar{y}_s}{n^{s/2}} + \bar{E}_p(n)], \tag{1.65}
$$

where $\bar{y}_s$ are determined by (1.19) with $\gamma = \gamma_2$, and $\bar{E}_p \in B_{p/2}$. Since $\gamma_1 \neq \gamma_2$, the two solutions $Y_{1,p}(n)$ and $Y_{2,p}(n)$ are linearly independent. Moreover, for any $q > p$, if $\operatorname{Re}\gamma_1 = \operatorname{Re}\gamma_2$, we can show that $Y_{1,q}(n) = Y_{1,p}(n)$ and $Y_{2,q}(n) = Y_{2,p}(n)$; if $\operatorname{Re}\gamma_1 < \operatorname{Re}\gamma_2$, we have $Y_{1,q}(n) = Y_{1,p}(n)$, while $Y_{2,q}(n) - Y_{2,p}(n)$ is exponentially small compared to $Y_{2,p}(n)$; if $\operatorname{Re}\gamma_1 > \operatorname{Re}\gamma_2$, we have $Y_{2,q}(n) = Y_{2,p}(n)$, while $Y_{1,q}(n) - Y_{1,p}(n)$ is exponentially small compared to $Y_{1,p}(n)$. In either case, the asymptotic expansions in (1.50) and (1.51) are valid. This completes the proof of the theorem. □

1.5 Exercises

1. Verify (1.6), (1.11) and (1.19).
2. Consider equation (1.6) with $b_0 \neq 0$ and $a_0^2 \neq 4b_0$ (normal case). For any $s \geq 2$, show that the coefficient of y_s vanishes, while the coefficient of y_{s-1} is nonzero.
3. Consider equation (1.6) with $b_0 \neq 0$, $a_0^2 = 4b_0$ and the two roots of the indicial equation (1.9) differing by a non-integer (first exceptional case). For any $s \geq 3$, show that the coefficients of y_s and y_{s-1} vanish, while the coefficient of y_{s-2} is nonzero.

4. Consider the second exceptional case when $b_0 \neq 0$, $a_0^2 = 4b_0$ and the two roots of the indicial equation (1.9) differing by an integer. Show that
   (a) on the left-hand side of (1.11), the coefficients of $\bar{y}_s$ and $\bar{y}_{s-1}$ vanish, while the coefficient of $\bar{y}_{s-2}$ is $\rho^2(s-2)(s-2-\lambda_2+\lambda_1)$;
   (b) on the right-hand side of (1.11), the coefficients of $y_{\bar{s}}$ and $y_{\bar{s}-1}$ vanish, while the coefficient of $y_{\bar{s}-2}$ is $-c\rho^2(4-s-\bar{s})$.
5. Consider the subnormal case when $a_0^2 = 4b_0 \neq 0$ and $a_0 a_1 \neq 2b_1$.
   (a) Derive (1.20) from equation (1.19) with $s = 2$.
   (b) Derive (1.21) from equation (1.19) with $s = 3$.
   (c) For equation (1.19) with $s \geq 4$, show that the coefficients of y_s, y_{s-1}, y_{s-2} vanishes, while the coefficient of y_{s-3} is nonzero.
6. Verify that B_p (as well as $B_p^{(m)}$) defined in the last paragraph of Section 1.1 is a Banach space.
7. Prove that $L(h)(n)$ defined in (1.26) is a particular solution to the difference equation (1.25).
8. Prove that $L(h)(n)$ defined in (1.43) is a particular solution to the difference equation (1.42).
9. Prove that $L(h)(n)$ defined in (1.59) is a particular solution to the difference equation (1.58).
10. Let $p > 0$. Show that

$$\sum_{k=n}^{\infty} \frac{1}{k^{p+1}} \leq \left(\frac{1}{p} + \frac{1}{n}\right) \frac{1}{n^p} \leq \frac{2}{pn^p},$$

    for any $n \geq p$.
11. Verify that $\tilde{\varepsilon}_{p+1}$ defined in (1.30) belongs to B_{p+1}.
12. Verify that $|n^p \tilde{E}_p(n)|$ in (1.33) is uniformly bounded as $n \to \infty$.
13. Verify that $\bar{\varepsilon}_{p+2}$ defined in (1.49) belongs to B_{p+2}.
14. Verify the estimate (1.44).
15. Verify the estimates (1.60), (1.61), (1.62), (1.63) and (1.64).

Chapter 2

Asymptotics of Difference Equations with General Coefficients

2.1 Classification

We consider the linear difference equation

$$Y(n+2) + A(n)Y(n+1) + B(n)Y(n) = 0, \tag{2.1}$$

where the coefficients have the following asymptotic expansions:

$$A(n) \sim n^{\mu} \sum_{s=0}^{\infty} \frac{a_s}{n^s}, \quad B(n) \sim n^{\nu} \sum_{s=0}^{\infty} \frac{b_s}{n^s}, \tag{2.2}$$

as $n \to \infty$. Here μ and ν are two integers.

The material in this chapter is taken from Wong and Li (1992b). There are three different cases depending on the sign of $\mu - \nu/2$ and whether ν is even.

(1) If $\mu - \nu/2 \leq 0$ and ν is even, then we set $\bar{Y}(n) = [(n-2)!]^{-\nu/2} Y(n)$ and obtain an equivalent difference equation:

$$\bar{Y}(n+2) + \bar{A}(n)\bar{Y}(n+1) + \bar{B}(n)\bar{Y}(n) = 0;$$

the new coefficients become

$$\bar{A}(n) = n^{-\nu/2} A(n) \sim \sum_{s=0}^{\infty} \frac{\bar{a}_s}{n^s},$$

$$\bar{B}(n) = [n(n-1)]^{-\nu/2} B(n) \sim \sum_{s=0}^{\infty} \frac{\bar{b}_s}{n^s},$$

where

$$\bar{a}_s = \begin{cases} 0, & s < \nu/2 - \mu, \\ a_{s+\mu-\nu/2}, & s \geq \nu/2 - \mu, \end{cases}$$

and

$$\bar{b}_s = \sum_{k=0}^{s} \binom{-\nu/2}{k} (-1)^k b_{s-k}.$$

Especially, $\bar{b}_0 = b_0 \neq 0$. This case has been considered in the previous chapter.

(2) If $\mu - \nu/2 \leq 0$ and ν is odd, then we still set $\bar{Y}(n) = [(n-2)!]^{-\nu/2} Y(n)$ to obtain an equivalent difference equation:

$$\bar{Y}(n+2) + \bar{A}(n)\bar{Y}(n+1) + \bar{B}(n)\bar{Y}(n) = 0,$$

where

$$\bar{A}(n) = n^{-\nu/2} A(n) \sim n^{-1/2} \sum_{s=0}^{\infty} \frac{\bar{a}_s}{n^s},$$

$$\bar{B}(n) = [n(n-1)]^{-\nu/2} B(n) \sim \sum_{s=0}^{\infty} \frac{\bar{b}_s}{n^s},$$

and

$$\bar{a}_s = \begin{cases} 0, & s < (\nu+1)/2 - \mu, \\ a_{s+\mu-(\nu+1)/2}, & s \geq (\nu+1)/2 - \mu, \end{cases}$$

$$\bar{b}_s = \sum_{k=0}^{s} \binom{-\nu/2}{k} (-1)^k b_{s-k}.$$

Again, $\bar{b}_0 = b_0 \neq 0$. We classify this as *case A* and will study it in the next section.

(3) If $\mu - \nu/2 > 0$, we set $\bar{Y}(n) = [(n-2)!]^{-\mu} Y(n)$ and obtain an equivalent difference equation:

$$\bar{Y}(n+2) + \bar{A}(n)\bar{Y}(n+1) + \bar{B}(n)\bar{Y}(n) = 0;$$

the new coefficients are

$$\bar{A}(n) = n^{-\mu} A(n) \sim \sum_{s=0}^{\infty} \frac{a_s}{n^s},$$

$$\bar{B}(n) = [n(n-1)]^{-\mu} B(n) \sim n^{\nu-2\mu} \sum_{s=0}^{\infty} \frac{\bar{b}_s}{n^s},$$

where

$$\bar{b}_s = \sum_{k=0}^{s} \binom{-\mu}{k} (-1)^k b_{s-k}.$$

Note that $a_0 \neq 0$, $\bar{b}_0 = b_0 \neq 0$, and $\nu - 2\mu < 0$. We classify this as *case B* and will study it later in this chapter.

2.2 Case A

In this section, we study the difference equation (2.1) and assume that the coefficients have the following asymptotic expansions:

$$A(n) \sim n^{-1/2} \sum_{s=0}^{\infty} \frac{a_s}{n^s}, \quad B(n) \sim \sum_{s=0}^{\infty} \frac{b_s}{n^s}, \tag{2.3}$$

as $n \to \infty$, where $b_0 \neq 0$. We look for a formal asymptotic solution

$$Y(n) \sim \rho^n e^{\gamma\sqrt{n}} n^{\lambda} \sum_{s=0}^{\infty} \frac{y_s}{n^{s/2}}, \quad n \to \infty. \tag{2.4}$$

Substituting the above expression into (2.1) and dividing both sides by $\rho^n e^{\gamma\sqrt{n}} n^{\lambda}$ gives

$$\begin{aligned} &\rho^2 e^{\gamma(\sqrt{n+2}-\sqrt{n})} \sum_{s=0}^{\infty} \frac{y_s(1+2/n)^{\lambda-s/2}}{n^{s/2}} \\ &+ A(n)\rho e^{\gamma(\sqrt{n+1}-\sqrt{n})} \sum_{s=0}^{\infty} \frac{y_s(1+1/n)^{\lambda-s/2}}{n^{s/2}} + B(n) \sum_{s=0}^{\infty} \frac{y_s}{n^{s/2}} \sim 0. \end{aligned} \tag{2.5}$$

For convenience, we denote $t = 1/\sqrt{n}$ and define $f(t)$ and $g(t)$ as in (1.14)-(1.15). Substituting the expansions (1.14), (1.15) and (2.3) into (2.5) gives

$$\begin{aligned} &\rho^2 \sum_{j,k,l=0}^{\infty} (f_k t^k)(y_j t^j) \binom{\lambda - j/2}{l} (2t^2)^l \\ &+\rho \sum_{i,j,k,l=0}^{\infty} (g_k t^k)(y_j t^j) \binom{\lambda - j/2}{l} t^{2l} (a_i t^{2i+1}) \\ &+ \sum_{i,j=0}^{\infty} (y_j t^j)(b_i t^{2i}) \sim 0. \end{aligned} \tag{2.6}$$

By balancing the coefficients of t^s, we obtain (exercise)

$$\begin{aligned} &\sum_{j=0}^{s} \Bigg\{ \rho^2 \sum_{k+2l=s-j} 2^l f_k \binom{\lambda - j/2}{l} \\ &+ \rho \sum_{k+2l+2i=s-j-1} a_i g_k \binom{\lambda - j/2}{l} + \sum_{2i=s-j} b_i \Bigg\} y_j = 0 \end{aligned} \tag{2.7}$$

for $s = 0, 1, \cdots$. When $s = 0$, equation (2.7) becomes $\rho^2 + b_0 = 0$, which implies $\rho = \pm\sqrt{-b_0}$. When $s = 1$, equation (2.7) becomes $\rho^2\gamma + \rho a_0 = 0$, which implies $\gamma = -a_0/\rho$. When $s = 2$, equation (2.7) becomes $2\lambda\rho^2 + b_1 = 0$, which implies $\lambda = -b_1/(2\rho^2)$. When $s > 2$, the coefficients of y_s and

y_{s-1} in (2.7) vanish, while the coefficient of y_{s-2} is $-\rho^2(s-2) \neq 0$. By setting $y_0 = 1$, we can compute $y_1, y_2, \cdots$ successively from equation (2.7) with $s = 3, 4, \cdots$.

For $\rho = \pm\sqrt{-b_0}$, we need to prove that there exists a solution to (2.1) with asymptotic expansion as given in (2.4).

Theorem 2.1. *Let $\rho = \sqrt{-b_0}$ or $\rho = -\sqrt{-b_0}$. For any $p \geq 1$, the difference equation (2.1) has a solution*

$$Y_p(n) = \rho^n e^{\gamma\sqrt{n}} n^\lambda \left[\sum_{s=0}^{p-1} \frac{y_s}{n^{s/2}} + E_p(n) \right], \tag{2.8}$$

where $\gamma = -a_0/\rho$, $\lambda = -b_1/(2\rho^2)$, $y_0 = 1$, and

$$\begin{aligned} y_{s-2} = & \frac{1}{\rho^2(s-2)} \sum_{j=0}^{s-3} \left\{ \rho^2 \sum_{k+2l=s-j} 2^l f_k \binom{\lambda - j/2}{l} \right. \\ & \left. + \rho \sum_{k+2l+2i=s-j-1} a_i g_k \binom{\lambda - j/2}{l} + \sum_{2i=s-j} b_i \right\} y_j \end{aligned} \tag{2.9}$$

for $s \geq 3$. The error term $E_p \in B_{p/2}$.

Proof. Without loss of generality, we may assume $\rho = 1$; otherwise we multiply equation (2.1) by ρ^{-n} and consider the difference equation satisfied by $\tilde{Y}(n) := \rho^{-n} Y(n)$. Thus, $b_0 = -1$, $\gamma = -a_0$ and $\lambda = -b_1/2$. By substituting (2.8) into (2.1) and dividing both sides by $e^{\gamma\sqrt{n}} n^\lambda$, we obtain

$$e^{\gamma(\sqrt{n+2}-\sqrt{n})}(1+2/n)^\lambda E_p(n+2) + e^{\gamma(\sqrt{n+1}-\sqrt{n})}(1+1/n)^\lambda A(n) E_p(n+1) + B(n) E_p(n) = -\varepsilon_{p+2}(n),$$

where

$$\begin{aligned} \varepsilon_{p+2}(n) = & e^{\gamma(\sqrt{n+2}-\sqrt{n})} \sum_{s=0}^{p-1} \frac{y_s (1+2/n)^{\lambda - s/2}}{n^{s/2}} \\ & + e^{\gamma(\sqrt{n+1}-\sqrt{n})} A(n) \sum_{s=0}^{p-1} \frac{y_s (1+1/n)^{\lambda - s/2}}{n^{s/2}} + B(n) \sum_{s=0}^{p-1} \frac{y_s}{n^{s/2}}. \end{aligned} \tag{2.10}$$

From the choice of y_s, we obtain $\varepsilon_{p+2} \in B_{(p+2)/2}$ (exercise). Actually, there exist two constants c_2 and c_3 such that $\varepsilon_{p+2}(n) = c_2 n^{-(p+2)/2} + c_3 n^{-(p+3)/2} + \varepsilon_{p+4}(n)$ for some $\varepsilon_{p+4} \in B_{(p+4)/2}$ (exercise). Note that

$$e^{\gamma(\sqrt{n+2}-\sqrt{n})}(1+2/n)^\lambda = \alpha(n) + \frac{C_2(n)}{n^{3/2}}, \tag{2.11}$$

$$e^{\gamma(\sqrt{n+1}-\sqrt{n})}(1+1/n)^\lambda A(n) = \beta(n) - \alpha(n) + \frac{C_1(n)}{n^{3/2}}, \tag{2.12}$$

$$B(n) = -\beta(n) + \frac{C_0(n)}{n^{3/2}}, \tag{2.13}$$

where

$$\alpha(n) = 1 + \frac{\gamma}{\sqrt{n}} + \frac{\gamma^2/2 - b_1}{n}, \;\; \beta(n) = 1 - \frac{b_1}{n},$$

and $C_2(n), C_1(n), C_0(n)$ are uniformly bounded as $n \to \infty$. These formulas are essentially those given in (1.53)-(1.55); the apparent difference comes from the fact that $\lambda = -b_1/2$ in the present case, whereas $\lambda = 1/4 + b_1/2$ in the previous case; see (1.21). We then have the difference equation:

$$\alpha(n)E_p(n+2) + [\beta(n) - \alpha(n)]E_p(n+1) - \beta(n)E_p(n) = F(E_p)(n), \tag{2.14}$$

which is the same as (1.56). Here,

$$F(E_p)(n) := -\varepsilon_{p+2}(n) - \frac{C_2(n)E_p(n+2) + C_1(n)E_p(n+1) + C_0(n)E_p(n)}{n^{3/2}}. \tag{2.15}$$

We need to show that equation (2.14) has a solution in $B_{p/2}$. Set $E_p = S(D_{p+2})$, where

$$S(D_{p+2})(n) := \sum_{k=n}^{\infty} D_{p+2}(k). \tag{2.16}$$

It is obvious that $S(B_{(p+2)/2}) \subset B_{p/2}$. Equation (2.14) becomes

$$\alpha(n)D_{p+2}(n+1) + \beta(n)D_{p+2}(n) = -F(S(D_{p+2}))(n). \tag{2.17}$$

We consider the difference equation

$$\alpha(n)u(n+1) + \beta(n)u(n) = -h(n). \tag{2.18}$$

A particular solution of the above equation is given by $u = L(h)$, where (exercise)

$$L(h)(n) := \begin{cases} \displaystyle\sum_{k=n}^{\infty} (-1)^{n+k} G_-(k,n)h(k), & \operatorname{Re}\gamma \le 0, \\ \displaystyle\sum_{k=m}^{n-1} (-1)^{n+k+1} G_+(k,n)h(k), & \operatorname{Re}\gamma > 0, \end{cases} \tag{2.19}$$

and

$$G_-(k,n) := \frac{\alpha(k-1)\cdots\alpha(n)}{\beta(k)\cdots\beta(n)}, \quad k \ge n,$$

$$G_+(k,n) := \frac{\beta(n-1)\cdots\beta(k+1)}{\alpha(n-1)\cdots\alpha(k)}, \quad k < n,$$

m being a fixed large positive integer to be determined later (exercise). It suffices to prove that the composite operator $L \circ F \circ S$ possesses a fixed point in $B^{(m)}_{(p+2)/2}$. We shall consider two cases, based on whether $\operatorname{Re}\gamma$ is positive or non-positive.

If $\operatorname{Re}\gamma \le 0$, then we have (exercise) as $n \to \infty$,

$$G_-(k,n) = e^{2\gamma(\sqrt{k}-\sqrt{n})}$$
$$\times \left[1 + (\frac{\gamma}{2} - \gamma^3 + 2\gamma b_1)(\frac{1}{\sqrt{n}} - \frac{1}{\sqrt{k}}) + O(n^{-1})\right], \quad k \ge n, \tag{2.20}$$

and

$$\sum_{k=n}^{\infty} \frac{(-1)^k e^{2\gamma\sqrt{k}}}{k^q} = \frac{(-1)^n e^{2\gamma\sqrt{n}}}{2n^q}[1 + O(n^{-1/2})], \quad q > 0. \tag{2.21}$$

Especially, $G_-(k,n)$ is uniformly bounded for $k \ge n$. Based on the above estimates, we have $L(h) \in B_q$ if $h(n) = n^{-q}$ or $h \in B_{q+1}$. For any $u \in B_{(p+2)/2}$, we can rewrite

$$F(S(u))(n) = -\,\varepsilon_{p+2}(n) - \sum_{k=n}^{\infty} \frac{[C_2(n) + C_1(n) + C_0(n)]u(k)}{n^{3/2}}$$
$$+ \frac{C_2(n)[u(n) + u(n+1)] + C_1(n)u(n)}{n^{3/2}}.$$

Recall that $\varepsilon_{p+2}(n) = c_2 n^{-(p+2)/2} + c_3 n^{-(p+3)/2} + \varepsilon_{p+4}(n)$ for some $\varepsilon_{p+4} \in B_{(p+4)/2}$. Thus, we have

$$F(S(u))(n) = \frac{c_2}{n^{(p+2)/2}} + \frac{c_3}{n^{(p+3)/2}} - \sum_{k=n}^{\infty} \frac{c_1 u(k)}{n^{3/2}} + h(n),$$

where $h \in B_{(p+4)/2}$. It follows from (2.20)-(2.21) that $L(g) \in B_q$ if $g(n) = n^{-q-1}$. Moreover, since $G_-(k,n)$ is uniformly bounded for $k \ge n$ by (2.20), we have $L(h) \in B_{(p+2)/2}$. Next, we denote

$$f(n) := \sum_{k=n}^{\infty} \frac{c_1 u(k)}{n^{3/2}},$$

and consider

$$L(f)(n) = \sum_{k=n}^{\infty} \sum_{l=k}^{\infty} (-1)^{n+k} G_-(k,n) \frac{c_1 u(l)}{k^{3/2}}.$$

Again, we obtain from (2.20) that

$$
\begin{aligned}
L(f)(n) &= \sum_{l=n}^{\infty}\sum_{k=n}^{l}(-1)^{n+k}G_{-}(k,n)\frac{c_1}{k^{3/2}}u(l) \\
&= \sum_{l=n}^{\infty}\sum_{k=n}^{l}\frac{(-1)^{n+k}e^{2\gamma(\sqrt{k}-\sqrt{n})}c_1}{k^{3/2}} \\
&\quad\times\left[1+(\frac{\gamma}{2}-\gamma^3+2\gamma b_1)(\frac{1}{\sqrt{n}}-\frac{1}{\sqrt{k}})+O(n^{-1})\right]u(l).
\end{aligned}
$$

For any integer $m>0$, real $q>0$ and complex γ with $\operatorname{Re}\gamma\le 0$, we have (exercise) as $n\to\infty$,

$$
\begin{aligned}
&\sum_{k=n}^{n+2m-1}\frac{(-1)^{n+k}e^{2\gamma\sqrt{k}}}{k^q} \\
&=\sum_{k=0}^{m-1}\frac{e^{2\gamma\sqrt{n+2k}}}{(n+2k)^q}\left[1-e^{2\gamma(\sqrt{n+2k+1}-\sqrt{n+2k})}\left(1+\frac{1}{n+2k}\right)^{-q}\right] \\
&=\frac{e^{2\gamma\sqrt{n}}}{2n^q}[1+O(n^{-1/2})]-\frac{e^{2\gamma\sqrt{n+2m}}}{2(n+2m)^q}[1+O((n+2m)^{-1/2})], \qquad (2.22)
\end{aligned}
$$

and

$$
\begin{aligned}
&\sum_{k=n}^{n+2m}\frac{(-1)^{n+k}e^{2\gamma\sqrt{k}}}{k^q} \\
&=\frac{e^{2\gamma\sqrt{n}}}{2n^q}[1+O(n^{-1/2})]+\frac{e^{2\gamma\sqrt{n+2m}}}{2(n+2m)^q}[1+O((n+2m)^{-1/2})]. \qquad (2.23)
\end{aligned}
$$

Consequently,

$$
|L(f)(n)|\le\frac{O(1)}{n^{3/2}}\cdot\sum_{l=n}^{\infty}\frac{\|u\|_{(p+2)/2}}{l^{(p+2)/2}}=O(n^{-(p+3)/2}).
$$

Combining the above arguments, we have $L(F(S(u)))\in B_{(p+2)/2}$. Next, we want to show that $L\circ F\circ S$ is a contraction on $B^{(m)}_{(p+2)/2}$ for some fixed large integer $m>0$. Given two $u_1,u_2\in B_{(p+2)/2}$, we have

$$
\begin{aligned}
&F(S(u_1))(n)-F(S(u_2))(n) \\
&=-\sum_{k=n}^{\infty}\frac{[C_2(n)+C_1(n)+C_0(n)][u_1(k)-u_2(k)]}{n^{3/2}} \\
&\quad+\frac{[C_2(n)+C_1(n)][u_1(n)-u_2(n)]+C_2(n)[u_1(n+1)-u_2(n+1)]}{n^{3/2}}.
\end{aligned}
$$

For simplicity, we rewrite the above equation as

$$F(S(u_1))(n) - F(S(u_2))(n) = -F_1(n) + F_2(n),$$

where F_1 is the sum and F_2 is the remaining quantity on the right-hand side. From (2.11)-(2.13), we have

$$C_2(n) + C_1(n) + C_0(n) = c_1 + \bar{C}(n)/\sqrt{n}$$

for some $\bar{C} \in B_0$. Thus, we obtain

$$\begin{aligned} L(F_1)(n) = &\sum_{l=n}^{\infty}\sum_{k=n}^{l}(-1)^{n+k}G_-(k,n)\frac{c_1}{k^{3/2}}[u_1(l) - u_2(l)] \\ &+ \sum_{k=n}^{\infty}\sum_{l=k}^{\infty}(-1)^{n+k}G_-(k,n)\frac{\bar{C}(k)[u_1(l) - u_2(l)]}{k^2}. \end{aligned}$$

There exist $m_1 \in \mathbf{N}$ and $M > 0$ such that

$$|\sum_{k=n}^{l}(-1)^{n+k}G_-(k,n)\frac{c_1}{k^{3/2}}| \le \frac{M}{n^{3/2}},$$

and

$$|G_-(k,n)| \le M$$

for any $n \ge m_1$. Consequently,

$$\begin{aligned} |L(F_1)(n)| &\le \sum_{l=n}^{\infty}\frac{M\|u_1 - u_2\|_{(p+2)/2}}{n^{3/2}l^{(p+2)/2}} + \sum_{k=n}^{\infty}\sum_{l=k}^{\infty}\frac{M\|\bar{C}\|_0\|u_1 - u_2\|_{(p+2)/2}}{l^{(p+2)/2}k^2} \\ &\le [\frac{4M}{pn^{(p+3)/2}} + \frac{16M\|\bar{C}\|_0}{p(p+2)n^{(p+2)/2}}]\|u_1 - u_2\|_{(p+2)/2}. \end{aligned}$$

Moreover, we can estimate

$$\begin{aligned} |L(F_2)(n)| &\le \sum_{k=n}^{\infty}\frac{(2\|C_2\|_0 + \|C_1\|_0)\|u_1 - u_2\|_{(p+2)/2}}{k^{(p+5)/2}} \\ &\le \frac{4(2\|C_2\|_0 + \|C_1\|_0)}{(p+3)n^{(p+3)/2}}\|u_1 - u_2\|_{(p+2)/2}. \end{aligned}$$

Therefore,

$$\begin{aligned} &|L(F(S(u_1)))(n) - L(F(S(u_2)))(n)| \le |L(F_1)(n)| + |L(F_2)(n)| \\ &\le \left[\frac{4M}{pn^{(p+3)/2}} + \frac{16M\|\bar{C}\|_0\|}{p(p+2)n^{(p+2)/2}} + \frac{4(2\|C_2\|_0 + \|C_1\|_0)}{(p+3)n^{(p+3)/2}}\right]\|u_1 - u_2\|_{(p+2)/2}. \end{aligned}$$

By choosing sufficiently large p and m, we can make the coefficient on the right-hand side of the above equality strictly less than one, and thus the composite map $L \circ F \circ S$ is a contraction on $B^{(m)}_{(p+2)/2}$.

Next, we consider the case when $\operatorname{Re}\gamma > 0$. Note that (exercise) as $k \to \infty$,

$$G_+(k,n) = e^{-2\gamma(\sqrt{n}-\sqrt{k})}[1 + O(k^{-1/2})], \quad n-1 \geq k. \tag{2.24}$$

Furthermore, for any $q > 0$, there exist $m_1 \in \mathbf{N}$ and $M_1 > 0$ such that (exercise)

$$\sum_{k=m}^{n-1} \frac{e^{-2\operatorname{Re}\gamma(\sqrt{n}-\sqrt{k})}}{k^q} \leq \frac{M_1}{n^{q-1/2}}, \tag{2.25}$$

$$|\sum_{k=m}^{n-1} \frac{(-1)^{n+k+1} e^{-2\gamma(\sqrt{n}-\sqrt{k})}}{k^q}| \leq \frac{M_1}{n^q}, \tag{2.26}$$

for all $n-1 \geq m \geq m_1$. Similarly, as in the previous case, for any $u \in B_{(p+2)/2}$, we rewrite

$$F(S(u))(n) = \frac{c_2}{n^{(p+2)/2}} + h(n)$$

for some $h \in B_{(p+3)/2}$. It then follows from (2.24)-(2.26) that $L(F(S(u))) \in B^{(m)}_{(p+2)/2}$ for sufficiently large m. For any $u_1, u_2 \in B_{(p+2)/2}$, we have

$$F(S(u_1))(n) - F(S(u_2))(n) = -F_1(n) + F_2(n),$$

where

$$F_1(n) := \sum_{k=n}^{\infty} \frac{[C_2(n) + C_1(n) + C_0(n)][u_1(k) - u_2(k)]}{n^{3/2}},$$

$$F_2(n) := \frac{[C_2(n) + C_1(n)][u_1(n) - u_2(n)] + C_2(n)[u_1(n+1) - u_2(n+1)]}{n^{3/2}}.$$

Note that $C_2(n) + C_1(n) + C_0(n) = c_1 + \bar{C}(n)/\sqrt{n}$, where $\bar{C} \in B_0$. In view of the identity

$$\sum_{k=m}^{n-1} \sum_{l=k}^{\infty} \varphi(l,k) = \sum_{l=m}^{\infty} \sum_{k=m}^{\min\{n-1,l\}} \varphi(l,k),$$

we consider

$$L(F_1)(n) = \sum_{l=m}^{\infty} \sum_{k=m}^{\min\{n-1,l\}} (-1)^{n+k+1} G_+(k,n) \times \frac{[c_1 + \bar{C}(k)/\sqrt{k}]}{k^{3/2}}[u_1(l) - u_2(l)].$$

On account of (2.24)-(2.26), we may find $m_2 \in \mathbf{N}$, $M_2 > 0$ and $M_3 > 0$ such that for all $n \geq m \geq m_2$,

$$|L(F_1)(n)| \leq \left[\sum_{l=m}^{n-1} \frac{M_2 e^{-2\,\mathrm{Re}\,\gamma(\sqrt{n}-\sqrt{l})}}{n^{3/2} l^{(p+2)/2}} + \sum_{l=n}^{\infty} \frac{M_2}{n^{3/2} l^{(p+2)/2}}\right] \|u_1 - u_2\|_{(p+2)/2}$$
$$\leq \frac{M_3}{n^{(p+3)/2}} \|u_1 - u_2\|_{(p+2)/2}.$$

On the other hand, we have

$$|L(F_2)(n)| \leq \sum_{k=m}^{n-1} \frac{M_4 e^{-2\,\mathrm{Re}\,\gamma(\sqrt{n}-\sqrt{k})} \|u_1 - u_2\|_{(p+2)/2}}{k^{(p+5)/2}}$$
$$\leq \frac{M_5 \|u_1 - u_2\|_{(p+2)/2}}{n^{(p+4)/2}},$$

for some $M_4 > 0$ and $M_5 > 0$. We choose m sufficiently large such that

$$|L(F(S(u_1)))(n) - L((F(S(u_2)))(n)| \leq |L(F_1)(n)| + |L(F_2)(n)|$$
$$\leq \frac{\|u_1 - u_2\|^{(m)}_{(p+2)/2}}{2n^{(p+2)/2}},$$

for any $n \geq m$. This implies that $L \circ F \circ S$ is a contraction on $B^{(m)}_{(p+2)/2}$.

In either case, we have found a fixed point of $L \circ F \circ S$, denoted by D_{p+2}, in $B^{(m)}_{(p+2)/2}$. We then define

$$E_p(n) := \sum_{k=n}^{\infty} D_{p+2}(k)$$

for $n \geq m$. It is readily seen that $E_p \in B^{(m)}_{p/2}$. To complete the values of $E_p(n)$ with $1 \leq n < m$, we only need to solve the recurrence relation for $E_p(n)$ in a backward manner. This completes the proof.

□

2.3 Case B

Now, we consider the case B:

$$A(n) \sim \sum_{s=0}^{\infty} \frac{a_s}{n^s}, \quad B(n) \sim n^{-\delta} \sum_{s=0}^{\infty} \frac{b_s}{n^s}, \quad n \to \infty, \tag{2.27}$$

where $a_0 \neq 0$, $b_0 \neq 0$ and δ is a positive integer. We look for a formal solution

$$Y(n) \sim \rho^n n^{\lambda} \sum_{s=0}^{\infty} \frac{y_s}{n^s}, \quad n \to \infty. \tag{2.28}$$

Substituting the above formula into (2.1) and dividing both sides by $\rho^n n^\lambda$ give

$$\rho^2 \sum_{s=0}^{\infty} \frac{y_s(1+2/n)^{\lambda-s}}{n^s} + A(n)\rho \sum_{s=0}^{\infty} \frac{y_s(1+1/n)^{\lambda-s}}{n^s} + B(n) \sum_{s=0}^{\infty} \frac{y_s}{n^s} \sim 0. \tag{2.29}$$

Balancing the coefficients of n^{-s} on the left-hand side yields (exercise)

$$\sum_{j=0}^{s} \left\{ \rho^2 2^{s-j} \binom{\lambda - j}{s-j} + \rho \sum_{k=0}^{s-j} \binom{\lambda - j}{s-j-k} a_k + b_{s-j-\delta} \right\} y_j = 0, \tag{2.30}$$

where, for convenience, we set $b_k = 0$ for any negative integer k. When $s = 0$, equation (2.30) gives (exercise)

$$\rho^2 + a_0 \rho = 0. \tag{2.31}$$

The only nonzero root of the above characteristic equation is $\rho = -a_0$. When $s = 1$, we obtain from (2.30) that (exercise)

$$2\rho^2\lambda + \rho\lambda a_0 + \rho a_1 + b_{1-\delta} = 0, \tag{2.32}$$

which together with $\rho = -a_0$, implies that $\lambda = -(\rho a_1 + b_{1-\delta})/\rho^2$. Note that $b_{1-\delta} = 0$ if $\delta > 1$. When $s \geq 2$, we consider the left-hand side of (2.30) and observe that the coefficient of y_s vanishes, while the coefficient of y_{s-1} is a nonzero constant $-\rho^2(s-1)$. By normalizing $y_0 = 1$, we can determine $y_1, \cdots$ successively from (2.30) with $s = 2, \cdots$.

Theorem 2.2. *For any $p \geq 1$, the difference equation* (2.1) *with coefficients* (2.27) *has a solution*

$$Y_p(n) = \rho^n n^\lambda [\sum_{s=0}^{p-1} \frac{y_s}{n^s} + E_p(n)], \tag{2.33}$$

where $\rho = -a_0$, $\lambda = -(\rho a_1 + b_{1-\delta})/\rho^2$, $y_0 = 1$, and

$$y_{s-1} = \frac{1}{\rho^2(s-1)} \sum_{j=0}^{s-2} \left\{ \rho^2 2^{s-j} \binom{\lambda - j}{s-j} + \rho \sum_{k=0}^{s-j} \binom{\lambda - j}{s-j-k} a_k + b_{s-j-\delta} \right\} y_j, \quad s \geq 2. \tag{2.34}$$

Moreover, the error term $E_p \in B_p$.

Proof. Without loss of generality, we may assume that $\rho = 1$; namely, $a_0 = -1$. Substituting (2.33) into (2.1) and dividing both sides by n^λ yield

$$(1+\frac{2}{n})^\lambda E_p(n+2) + (1+\frac{1}{n})^\lambda A(n)E_p(n+1) + B(n)E_p(n) = -\varepsilon_{p+1}(n), \tag{2.35}$$

where

$$\varepsilon_{p+1}(n) = \sum_{s=0}^{p-1} \frac{y_s(1+2/n)^{\lambda-s}}{n^s} + A(n)\sum_{s=0}^{p-1} \frac{y_s(1+1/n)^{\lambda-s}}{n^s} + B(n)\sum_{s=0}^{p-1} \frac{y_s}{n^s}.$$

It is readily seen from the choices of λ and y_s that $\varepsilon_{p+1} \in B_{p+1}$. Note that

$$(1+\frac{2}{n})^\lambda = 1 + \frac{C_2(n)}{n}, \quad (1+\frac{1}{n})^\lambda A(n) = -1 + \frac{C_1(n)}{n}, \quad B(n) = \frac{C_0(n)}{n},$$

where $C_2, C_1, C_0 \in B_0$. Hence, we arrive at the difference equation:

$$E_p(n+2) - E_p(n+1) = -\varepsilon_{p+1}(n) - \frac{C_2(n)E_p(n+2) + C_1(n)E_p(n+1) + C_0(n)E_p(n)}{n}.$$

Let m be a positive integer to be determined later. For any $u \in B_p$, we define a map

$$L(u)(n) := \sum_{k=n-1}^{\infty} \varepsilon_{p+1}(k) + \frac{C_2(k)u(k+2) + C_1(k)u(k+1) + C_0(k)u(k)}{k}$$

for $n > m$, and $L(u)(m) = 0$. When $n > m$, we have

$$\begin{aligned} |L(u)(n)| &\le \sum_{k=n-1}^{\infty} \frac{\|\varepsilon_{p+1}\|_{p+1} + C\|u\|_p}{k^{p+1}} \\ &\le \frac{\|\varepsilon_{p+1}\|_{p+1} + C\|u\|_p}{n^p}\left(\frac{2^{p+1}+1}{n} + \frac{1}{p}\right), \end{aligned}$$

where $C := \|C_2\|_0 + \|C_1\|_0 + \|C_0\|_0$. If $m > p(2^{p+1}+1)$, then

$$|L(u)(n)| \le (\|\varepsilon_{p+1}\|_{p+1} + C\|u\|_p)\frac{2}{pn^p},$$

which implies that $L(B_p^{(m)}) \subset B_p^{(m)}$. Furthermore, for any $u_1, u_2 \in B_p^{(m)}$, we obtain from a similar argument that

$$|L(u_1)(n) - L(u_2)(n)| \le \frac{2C\|u_1 - u_2\|_p}{pn^p}.$$

Now, we choose $p > 2C$ and $m > p(2^{p+1}+1)$ such that L is a contraction map on $B_p^{(m)}$. Let E_p be a fixed point of L. It is readily seen that $E_p(m) = 0$ and the difference equation (2.35) is satisfied for all $n \ge m$. Using (2.35), we can define $E_p(n)$ with $1 \le n < m$ recursively backwards. Thus, we conclude that (2.1) has a solution of the form (2.33). This completes the proof. □

To find a second asymptotic solution, we define $\bar{Y}(n) := ((n-2)!)^{\delta}Y(n)$ and consider the difference equation for $\bar{Y}(n)$:

$$n^{-\delta}\bar{Y}(n+2) + A(n)\bar{Y}(n+1) + \bar{B}(n)\bar{Y}(n) = 0, \tag{2.36}$$

where

$$\bar{B}(n) = n^{\delta}(1-1/n)^{\delta}B(n) \sim \sum_{s=0}^{\infty}\frac{\bar{b}_s}{n^s} \tag{2.37}$$

with

$$\bar{b}_s = \sum_{k=0}^{\delta}(-1)^k\binom{\delta}{k}b_{s-k}. \tag{2.38}$$

Again, we set $b_k = 0$ if $k < 0$. We look for a formal solution:

$$\bar{Y}(n) \sim \bar{\rho}^n n^{\bar{\lambda}}\sum_{s=0}^{\infty}\frac{\bar{y}_s}{n^s}, \quad n \to \infty. \tag{2.39}$$

Substituting the above formula into (2.36) and dividing both sides by $\bar{\rho}^n n^{\bar{\lambda}}$ give

$$\bar{\rho}^2\sum_{s=0}^{\infty}\frac{\bar{y}_s(1+2/n)^{\bar{\lambda}-s}}{n^{s+\delta}} + A(n)\bar{\rho}\sum_{s=0}^{\infty}\frac{\bar{y}_s(1+1/n)^{\bar{\lambda}-s}}{n^s} + \bar{B}(n)\sum_{s=0}^{\infty}\frac{\bar{y}_s}{n^s} \sim 0. \tag{2.40}$$

Balancing the coefficients of n^{-s} on the left-hand side yields (exercise)

$$\sum_{j=0}^{s}\left\{\bar{\rho}^2 2^{s-j-\delta}\binom{\bar{\lambda}-j}{s-j-\delta} + \bar{\rho}\sum_{k=0}^{s-j}\binom{\bar{\lambda}-j}{s-j-k}a_k \right.$$
$$\left. + \bar{b}_{s-j}\right\}\bar{y}_j = 0, \tag{2.41}$$

where, for simplicity, we let $\binom{m}{k} = 0$ for any negative integer k. When $s = 0$, equation (2.41) reads (exercise)

$$\bar{\rho}a_0 + \bar{b}_0 = 0, \tag{2.42}$$

which gives the characteristic root $\bar{\rho} = -\bar{b}_0/a_0$. To solve for $\bar{\lambda}$, we need to look at equation (2.41) with $s = 1$:

$$\bar{\rho}^2 2^{1-\delta}\binom{\bar{\lambda}}{1-\delta} + \bar{\rho}\bar{\lambda}a_0 + \bar{\rho}a_1 + \bar{b}_1 = 0, \tag{2.43}$$

from which we obtain (exercise)

$$\bar{\lambda} = \begin{cases}(\bar{\rho}^2 + \bar{\rho}a_1 + \bar{b}_1)/\bar{b}_0, & \delta = 1,\\ (\bar{\rho}a_1 + \bar{b}_1)/\bar{b}_0, & \delta > 1.\end{cases} \tag{2.44}$$

When $s \geq 2$, on the left-hand side of (2.41), the coefficient of $\bar{y}_s$ vanishes, while the coefficient of $\bar{y}_{s-1}$ is

$$\bar{\rho}^2 2^{1-\delta}\binom{\bar{\lambda}-s+1}{1-\delta} + \bar{\rho}(\bar{\lambda}-s+1)a_0 + \bar{\rho}a_1 + \bar{b}_1 = \bar{b}_0(s-1) \neq 0.$$

Thus, by normalizing $\bar{y}_0 = 1$, we can determine $\bar{y}_{s-1}$ successively from (2.41) with $s = 2, \cdots$.

Theorem 2.3. *For any $p \geq 1$, the difference equation* (2.36) *has a solution*

$$\bar{Y}_p(n) = \bar{\rho}^n n^{\bar{\lambda}}[\sum_{s=0}^{p-1} \frac{\bar{y}_s}{n^s} + \bar{E}_p(n)], \tag{2.45}$$

where $\bar{\rho} = -\bar{b}_0/a_0$, $\bar{\lambda}$ is given in (2.44), *$\bar{y}_0 = 1$, and*

$$\begin{aligned}\bar{y}_{s-1} = &\frac{1}{\bar{b}_0(1-s)} \sum_{j=0}^{s-2}\left\{\bar{\rho}^2 2^{s-j-\delta}\binom{\bar{\lambda}-j}{s-j-\delta}\right.\\ &\left.+\bar{\rho}\sum_{k=0}^{s-j}\binom{\bar{\lambda}-j}{s-j-k}a_k + \bar{b}_{s-j}\right\}\bar{y}_j, \quad s \geq 2.\end{aligned} \tag{2.46}$$

Moreover, the error term $\bar{E}_p \in B_p$.

Proof. Without loss of generality, we may assume that $\bar{\rho} = 1$; namely, $\bar{b}_0 = -a_0$. Substituting (2.45) into (2.36) and dividing both sides by $n^{\bar{\lambda}}$ yield

$$\begin{aligned}&n^{-\delta}(1+\frac{2}{n})^{\bar{\lambda}}\bar{E}_p(n+2) + (1+\frac{1}{n})^{\bar{\lambda}}A(n)\bar{E}_p(n+1) + \bar{B}(n)\bar{E}_p(n)\\ &= -\bar{\varepsilon}_{p+1}(n),\end{aligned} \tag{2.47}$$

where

$$\bar{\varepsilon}_{p+1}(n) = \sum_{s=0}^{p-1}\frac{\bar{y}_s(1+2/n)^{\bar{\lambda}-s}}{n^{s+\delta}} + A(n)\sum_{s=0}^{p-1}\frac{\bar{y}_s(1+1/n)^{\bar{\lambda}-s}}{n^s} + \bar{B}(n)\sum_{s=0}^{p-1}\frac{\bar{y}_s}{n^s}.$$

It is readily seen from the choices of $\bar{\lambda}$ and $\bar{y}_s$ that $\bar{\varepsilon}_{p+1} \in B_{p+1}$. Note that

$$\begin{aligned}n^{-\delta}(1+2/n)^{\bar{\lambda}} &= \frac{\bar{C}_2(n)}{n},\\ (1+1/n)^{\bar{\lambda}}A(n) &= a_0 + \frac{\bar{C}_1(n)}{n},\\ \bar{B}(n) &= -a_0 + \frac{\bar{C}_0(n)}{n},\end{aligned}$$

where $\bar{C}_2, \bar{C}_1, \bar{C}_0 \in B_0$; thus arriving at the difference equation:

$$\bar{E}_p(n+1) - \bar{E}_p(n) = -\frac{\bar{\varepsilon}_{p+1}(n)}{a_0} - \frac{\bar{C}_2(n)\bar{E}_p(n+2) + \bar{C}_1(n)\bar{E}_p(n+1) + \bar{C}_0(n)E_p(n)}{a_0 n}.$$

For any $u \in B_p$, we define a map

$$\bar{L}(u)(n) := \sum_{k=n}^{\infty} \frac{\varepsilon_{p+1}(k)}{a_0} + \frac{\bar{C}_2(k)u(k+2) + \bar{C}_1(k)u(k+1) + \bar{C}_0(k)u(k)}{a_0 k}.$$

For any $n \geq p$, it can be obtained that

$$|\bar{L}(u)(n)| \leq \sum_{k=n}^{\infty} \frac{\|\varepsilon_{p+1}\|_{p+1} + \bar{C}\|u\|_p}{|a_0|k^{p+1}} \leq \frac{\|\varepsilon_{p+1}\|_{p+1} + \bar{C}\|u\|_p}{|a_0|(p/2)n^p},$$

where $\bar{C} := \|\bar{C}_2\|_0 + \|\bar{C}_1\|_0 + \|\bar{C}_0\|_0$. This implies that $\bar{L}(B_p^{(p)}) \subset B_p^{(p)}$. Furthermore, for any $u_1, u_2 \in B_p^{(p)}$, we obtain from a similar argument that

$$|\bar{L}(u_1)(n) - \bar{L}(u_2)(n)| \leq \frac{2\bar{C}\|u_1 - u_2\|_p}{p|a_0|n^p}.$$

Now, we choose $p > 2\bar{C}/|a_0|$ so that $\bar{L}$ is a contraction map on $B_p^{(p)}$. By the contraction mapping principle, we conclude that (2.36) has a solution of the form (2.45) for $n \geq p$. The values of $\bar{Y}_p(n)$ for $1 \leq n < p$ can be obtained reversely from (2.36). This completes the proof. □

Finally, we have the following result.

Theorem 2.4. *The difference equation* (2.1) *with* (2.27) *has two linearly independent solutions which have the asymptotic expansions:*

$$Y_1(n) \sim \rho^n n^\lambda \sum_{s=0}^{\infty} \frac{y_s}{n^s}, \tag{2.48}$$

$$Y_2(n) \sim ((n-2)!)^{-\delta} \bar{\rho}^n n^{\bar{\lambda}} \sum_{s=0}^{\infty} \frac{\bar{y}_s}{n^s}, \tag{2.49}$$

where $\rho, \lambda, y_s, \bar{\rho}, \bar{\lambda}, \bar{y}_s$ are given as in Theorems 2.2 and 2.3.

Proof. For any large integer $p > 0$, there exist two solutions $Y_{1,p}(n) = Y_p(n)$ and $Y_{2,p}(n) = ((n-2)!)^{-\delta}\bar{Y}_p(n)$ by Theorems 2.2 and 2.3. Obviously, these two solutions are linearly independent because, by Stirling's formula, $Y_{2,p}(n)/Y_{1,p}(n)$ tends to zero rapidly as $n \to \infty$. For any other integer $q > p$, we find another pair of solutions $Y_{1,q}(n)$ and $Y_{2,q}(n)$, and conclude from linear independency that $Y_{2,q}(n) = Y_{2,p}(n)$ and $Y_{1,q}(n) - Y_{1,p}(n)$ is a constant multiplication of $Y_{2,p}(n)$. Thus, by defining $Y_1(n) = Y_{1,p}(n)$ and $Y_2(n) = Y_{2,p}(n)$, we obtain two linearly independent solutions which satisfy the desired asymptotic expansions (2.48) and (2.49), respectively. This completes the proof. □

2.4 Exercises

1. Verify (2.7). Prove that
   (a) When $s = 0$, equation (2.7) becomes $\rho^2 + b_0 = 0$, which implies that $\rho = \pm\sqrt{-b_0}$.
   (b) When $s = 1$, equation (2.7) becomes $\rho^2\gamma + \rho a_0 = 0$, which implies that $\gamma = -a_0/\rho$.
   (c) When $s = 2$, equation (2.7) becomes $2\lambda\rho^2 + b_1 = 0$, which implies $\lambda = -b_1/(2\rho^2)$.
   (d) When $s > 2$, the coefficients of y_s and y_{s-1} in (2.7) vanish, while the coefficient of y_{s-2} is $-\rho^2(s-2) \neq 0$.
2. (a) Prove that ε_{p+2} defined in (2.10) belongs to $B_{(p+2)/2}$.
   (b) Show that there are constants c_2 and c_3 such that $\varepsilon_{p+2}(n) = c_2 n^{-(p+2)/2} + c_3 n^{-(p+3)/2} + \varepsilon_{n+4}(n)$ for some $\varepsilon_{n+4}(n) \in B_{(p+4)/2}$.
3. Verify that the difference equation (2.18) has a particular solution as given in (2.19).
4. Verify (2.20) and (2.21).
5. Verify (2.22) and (2.23).
6. Verify (2.24), (2.25) and (2.26).
7. Verify (2.30). Derive (2.31) and (2.32) from (2.30). Consider the left-hand side of (2.30) with $s \geq 2$. Show that the coefficient of y_s vanishes, while the coefficient of y_{s-1} is a nonzero constant $-\rho^2(s-1)$.
8. Verify (2.41). Derive (2.42) and (2.44) from (2.41). Consider the left-hand side of (2.41) with $s \geq 2$. Show that the coefficient of $\bar{y}_s$ vanishes, while the coefficient of $\bar{y}_{s-1}$ is a nonzero constant $\bar{b}_0(s-1)$.

Chapter 3

Asymptotics of Difference Equations with a Parameter

3.1 Introduction

We consider the following difference equation with a parameter x,

$$P_{n+1}(x) - [C_1(n)x + C_0(n)]P_n(x) + P_{n-1}(x) = 0, \quad n \geq 1, \tag{3.1}$$

with initial conditions $P_0(x) = 1$ and $P_1(x) = C_1(0)x + C_0(0)$. Here, the coefficients $C_1(n)$ and $C_0(n)$ have asymptotic expansions of the form

$$C_1(n) \sim n^{-\theta} \sum_{s=0}^{\infty} \frac{a_s}{n^s}, \quad C_0(n) \sim \sum_{s=0}^{\infty} \frac{b_s}{n^s}, \tag{3.2}$$

with $\theta > 0$ and $a_0 \neq 0$. We further assume that $C_1(n) \neq 0$ for all $n \geq 0$. It is obvious that $P_n(x)$ is a polynomial of x with degree n. Actually, Favard's theorem shows that $\{P_n(x)\}_{n=0}^{\infty}$ is a sequence of orthogonal polynomials with respect to some measure on the real line; see Beals and Wong (2016, p. 86).

The material in this chapter is taken from Wang and Wong (2012). We introduce the canonical scale $x = n^\theta t$ and investigate the asymptotic behavior of $P_n(n^\theta t)$ for t in various complex regions. A simple calculation shows that

$$\gamma_n = \prod_{k=0}^{n-1} C_1(k) \tag{3.3}$$

is the leading coefficient of $P_n(x)$. Let $\pi_n(x) := P_n(x)/\gamma_n$ be the monic polynomials. It follows from (3.1) that

$$\pi_{n+1}(x) = (x - A_n)\pi_n(x) - B_n\pi_{n-1}(x), \quad n \geq 1, \tag{3.4}$$

with initial conditions $\pi_0(x) = 1$ and $\pi_1(x) = x - A_0$, where

$$A_n = -\frac{C_0(n)}{C_1(n)} \sim n^\theta \sum_{s=0}^{\infty} \frac{\alpha_s}{n^s}, \quad B_n = \frac{1}{C_1(n)C_1(n-1)} \sim n^{2\theta} \sum_{s=0}^{\infty} \frac{\beta_s}{n^s}. \tag{3.5}$$

Comparing (3.2) with (3.5) yields (exercise)

$$\alpha_0 = -\frac{b_0}{a_0}, \quad \alpha_1 = -\frac{b_0}{a_0}\left(\frac{b_1}{b_0} - \frac{a_1}{a_0}\right), \quad \beta_0 = \frac{1}{a_0^2}, \quad \beta_1 = -\frac{1}{a_0^2}\left(\theta + \frac{2a_1}{a_0}\right). \quad (3.6)$$

Note from (3.2) and (3.3) that the limit of $\gamma_n / \prod_{k=1}^{n-1} k^{-\theta}(a_0 + a_1/k)$ as $n \to \infty$ exists. By Stirling's formula, there exists $\kappa_0 \neq 0$ such that (exercise)

$$\gamma_n \sim \kappa_0 (e/n)^{n\theta} n^{\theta/2 + a_1/a_0} a_0^n, \quad (3.7)$$

as $n \to \infty$. In real applications when the expressions of $C_1(k)$ are explicitly given, we are most likely able to find an asymptotic expansion for the leading coefficient γ_n. Thus, to find the asymptotic formula for $P_n(x)$, it is equivalent to investigate the asymptotic behavior of the monic polynomials $\pi_n(x)$. Define

$$w_k(x) = \pi_k(x)/\pi_{k-1}(x), \quad k \geq 1. \quad (3.8)$$

We have the factorization

$$\pi_n(x) = \prod_{k=1}^{n} w_k(x). \quad (3.9)$$

Moreover, the second-order linear difference equation (3.4) becomes a first-order nonlinear difference equation

$$w_{k+1}(x) = x - A_k - \frac{B_k}{w_k(x)}, \quad k \geq 1, \quad (3.10)$$

with initial condition $w_1(x) = x - A_0$. We shall first study the first-order nonlinear difference equation (3.10) and find an asymptotic formula for $w_k(x)$. Next, we derive an asymptotic formula for $\pi_n(x)$. This formula is valid only when x stays away from the oscillatory interval on which the zeros of $\pi_n(x)$ lie. In a complex neighborhood of the oscillatory interval, we find two asymptotic solutions to the difference equation (3.1) with no initial restrictions. Hence, $P_n(x)$ can be expressed as a linear combination of these two asymptotic solutions. The coefficients in this linear combination will be determined via asymptotic matching.

3.2 Asymptotic formula of $w_k(x)$

We introduce the auxiliary functions

$$u_k(x) = \frac{x - A_{k-1} + \sqrt{(x - A_{k-1})^2 - 4B_k}}{2}, \quad (3.11)$$

$$v_k(x) = \frac{x - A_{k-1} - \sqrt{(x - A_{k-1})^2 - 4B_k}}{2}, \quad (3.12)$$

for $k \geq 1$. For convenience, we set $B_0 = 0$ and $v_0(x) = 0$. Define

$$\delta_k(x) = \frac{v_k(x) - v_{k-1}(x)}{u_k(x) - v_{k-1}(x)}, \quad k \geq 1. \tag{3.13}$$

We intend to prove that, under certain conditions, $w_k(x) = u_k(x)[1+\delta_k(x)+O(1/n^2)]$, uniformly for all $k = 1, \cdots, n$. To see this, we denote

$$w_k(x) = u_k(x)[1 + \delta_k(x) + \varepsilon_k(x)], \quad k \geq 1. \tag{3.14}$$

In view of $x - A_k = u_{k+1}(x) + v_{k+1}(x)$ and $B_k = u_k(x)v_k(x)$, we obtain from (3.10), (3.13) and (3.14)

$$\varepsilon_{k+1}(x) = \frac{v_k(x)}{u_{k+1}(x)}[1 - \delta_{k+1}(x) - (1 + \delta_k(x) + \varepsilon_k(x))^{-1}]. \tag{3.15}$$

The asymptotic nature of $w_k(x) \sim u_k(x)[1 + \delta_k(x)]$ is justified by the following theorem.

Theorem 3.1. *Given the functions $u_k(x)$, $v_k(x)$, $\delta_k(x)$, $\varepsilon_k(x)$ defined in (3.11), (3.12), (3.13) and (3.14), respectively, let $x = n^\theta t$. Assume that there exist $n_0 > 0$, $0 < r < 1$ and $M > 0$ such that*

$$\sup_{1\leq k<n} \left|\frac{v_k(x)}{u_{k+1}(x)}\right| < r, \quad \sup_{1\leq k\leq n} |\delta_k(x)| < \frac{M}{n}, \quad \sup_{1\leq k<n} |\delta_k(x) - \delta_{k+1}(x)| < \frac{M}{n^2}, \tag{3.16}$$

for all $n \geq n_0$. Then there exists $L > 0$ and $n_1 \geq n_0$ such that

$$\sup_{1\leq k\leq n} |\varepsilon_k(x)| \leq \frac{L}{n^2} \tag{3.17}$$

for all $n \geq n_1$.

Proof. Without loss of generality, we may assume $n_0 > 2M + 2$. For any $n \geq n_0$ and $|s| \leq M + 1$, we have from Taylor expansion

$$(1 + s/n)^{-1} - (1 - s/n) = (s/n)^2/(1 + \xi)^3$$

for some ξ between 0 and s/n. Especially, $|\xi| \leq |s|/n \leq (M + 1)/n_0 \leq 1/2$. Hence,

$$n^2|(1 + s/n)^{-1} - (1 - s/n)| \leq 8|s|^2 \leq 8(M + 1)^2,$$

for all $n \geq n_0$ and $|s| \leq M + 1$. Denote $K = 8(M + 1)^2$, $L = r(M + K)/(1 - r)$, and $n_1 = n_0 + L$. For any $n \geq n_1$, we will prove by induction that $|\varepsilon_k(x)| \leq L/n^2$ for all $k = 1, \cdots, n$. This is obvious for $k = 1$ because $\varepsilon_1(x) = 0$. Assume that $|\varepsilon_k(x)| \leq L/n^2$. Then $|\delta_k(x) + \varepsilon_k(x)| \leq M/n + L/n^2 \leq (M + 1)/n$. From the choice of K, we have

$$|1 - \delta_k(x) - \varepsilon_k(x) - (1 + \delta_k(x) + \varepsilon_k(x))^{-1}| \leq K/n^2.$$

Substituting this into (3.15) gives

$$|\varepsilon_{k+1}(x)| \le |\frac{v_k(x)}{u_{k+1}(x)}| \left[|\varepsilon_k(x)| + |\delta_k(x) - \delta_{k+1}(x)| + K/n^2\right].$$

On account of (3.16) and $|\varepsilon_k(x)| \le L/n^2$, we obtain

$$|\varepsilon_{k+1}(x)| \le r(L/n^2 + M/n^2 + K/n^2) = L/n^2.$$

This completes the proof. □

Remark 3.1. The conditions (3.16) should be verified when the coefficients A_k and B_k are explicitly given. As we shall see later in an illustrative example, these conditions are usually satisfied when t stays away from the oscillatory interval.

3.3 Asymptotic formula of $\pi_n(x)$

Throughout this section, we assume that the conditions in Theorem 3.1 are satisfied. It then follows from the factorization formula (3.9) and the asymptotic formula of $w_k(x)$ in (3.14) that

$$\ln \pi_n(x) = \sum_{k=1}^{n} \ln u_k(x) + \sum_{k=1}^{n} \delta_k(x) + O(1/n). \tag{3.18}$$

Lemma 3.1. *Let $x = n^\theta t$ and $u_k(x)$ be given as in (3.11). As $n \to \infty$, we have*

$$\sum_{k=1}^{n} \ln \frac{2u_k(x)}{n^\theta} \sim n \ln[t - \alpha_0 + \sqrt{(t-\alpha_0)^2 - 4\beta_0}] + n\phi(t) - n\theta$$
$$+ (\frac{1}{2} + \frac{\beta_1}{2\beta_0\theta}) \ln \frac{t - \alpha_0 + \sqrt{(t-\alpha_0)^2 - 4\beta_0}}{2t} - (\frac{\alpha_1}{\theta} - \alpha_0 - \frac{\alpha_0\beta_1}{2\theta\beta_0})\eta(t),$$

where

$$\phi(t) := \int_0^1 \frac{ts^{1/\theta - 1}}{\sqrt{(t-\alpha_0 s)^2 - 4\beta_0 s^2}} ds, \tag{3.19}$$

$$\eta(t) := \frac{1}{\sqrt{\alpha_0^2 - 4\beta_0}} \ln \frac{(\alpha_0 + \sqrt{\alpha_0^2 - 4\beta_0})t}{\alpha_0(t-\alpha_0) + 4\beta_0 + \sqrt{\alpha_0^2 - 4\beta_0}\sqrt{(t-\alpha_0)^2 - 4\beta_0}}, \tag{3.20}$$

and the error is of order $O(1/n)$ if $\theta > 1$, $O(\ln n/n)$ if $\theta = 1$, and $O(1/n^\theta)$ if $0 < \theta < 1$.

Proof. In view of (3.5), we obtain

$$A_{k-1} = \alpha_0 k^\theta + (\alpha_1 - \alpha_0\theta)k^{\theta-1} + O(k^{\theta-2}),\ \ B_k = \beta_0 k^{2\theta} + \beta_1 k^{2\theta-1} + O(k^{2\theta-2})$$

as $k \to \infty$. Consequently, we have from (3.11)

$$\begin{aligned}
\frac{2u_k(x)}{n^\theta} &= t - \frac{A_{k-1}}{n^\theta} + \sqrt{(t - \frac{A_{k-1}}{n^\theta})^2 - \frac{4B_k}{n^{2\theta}}} \\
&= t - \alpha_0 r^\theta - \frac{(\alpha_1 - \alpha_0\theta)r^{\theta-1}}{n} + O\left(\frac{k^{\theta-2}}{n^\theta}\right) \\
&\quad + \sqrt{(t - \alpha_0 r^\theta - \frac{(\alpha_1 - \alpha_0\theta)r^{\theta-1}}{n})^2 - (4\beta_0 r^{2\theta} + \frac{4\beta_1 r^{2\theta-1}}{n}) + O\left(\frac{k^{\theta-2}}{n^\theta}\right)} \\
&= t - \alpha_0 r^\theta + \sqrt{(t - \alpha_0 r^\theta)^2 - 4\beta_0 r^{2\theta}} \\
&\quad - \frac{(\alpha_1 - \alpha_0\theta)r^{\theta-1}}{n} - \frac{(t - \alpha_0 r^\theta)(\alpha_1 - \alpha_0\theta)r^{\theta-1} + 2\beta_1 r^{2\theta-1}}{n\sqrt{(t - \alpha_0 r^\theta)^2 - 4\beta_0 r^{2\theta}}} \\
&\quad + O\left(\frac{1}{n^2} + \frac{k^{\theta-2}}{n^\theta}\right),
\end{aligned}$$

where $r = k/n$. This implies that

$$\begin{aligned}
\ln\frac{2u_k(x)}{n^\theta} &= \ln[t - \alpha_0 r^\theta + \sqrt{(t - \alpha_0 r^\theta)^2 - 4\beta_0 r^{2\theta}}] \\
&\quad - \frac{(\alpha_1 - \alpha_0\theta)r^{\theta-1}}{n\sqrt{(t - \alpha_0 r^\theta)^2 - 4\beta_0 r^{2\theta}}} - \frac{\beta_1[t - \alpha_0 r^\theta - \sqrt{(t - \alpha_0 r^\theta)^2 - 4\beta_0 r^{2\theta}}]}{2\beta_0 r n\sqrt{(t - \alpha_0 r^\theta)^2 - 4\beta_0 r^{2\theta}}} \\
&\quad + O\left(\frac{1}{n^2} + \frac{k^{\theta-2}}{n^\theta}\right).
\end{aligned}$$

Note that (exercise)

$$\sum_{k=1}^{n} \frac{k^{\theta-2}}{n^\theta} = \begin{cases} O(\frac{1}{n}), & \theta > 1, \\ O(\frac{\ln n}{n}), & \theta = 1, \\ O(\frac{1}{n^\theta}), & 0 < \theta < 1. \end{cases} \tag{3.21}$$

It then follows from the trapezoidal rule that

$$\begin{aligned}
\sum_{k=1}^{n} \ln\frac{2u_k(x)}{n^\theta} &\sim n\int_0^1 \ln[t - \alpha_0 r^\theta + \sqrt{(t - \alpha_0 r^\theta)^2 - 4\beta_0 r^{2\theta}}]dr \\
&\quad + \frac{1}{2}\ln\frac{t - \alpha_0 + \sqrt{(t - \alpha_0)^2 - 4\beta_0}}{2t} - \int_0^1 \frac{(\alpha_1 - \alpha_0\theta)r^{\theta-1}}{\sqrt{(t - \alpha_0 r^\theta)^2 - 4\beta_0 r^{2\theta}}}dr \\
&\quad - \int_0^1 \frac{\beta_1[t - \alpha_0 r^\theta - \sqrt{(t - \alpha_0 r^\theta)^2 - 4\beta_0 r^{2\theta}}]}{2\beta_0 r\sqrt{(t - \alpha_0 r^\theta)^2 - 4\beta_0 r^{2\theta}}}dr,
\end{aligned}$$

where the error is of order $O(1/n)$ if $\theta > 1$, $O(\ln n/n)$ if $\theta = 1$, and $O(1/n^\theta)$ if $0 < \theta < 1$. In view of the following identities (exercise)

$$\int_0^1 \ln[t - \alpha_0 r^\theta + \sqrt{(t-\alpha_0 r^\theta)^2 - 4\beta_0 r^{2\theta}}]dr$$

$$= \ln[t - \alpha_0 + \sqrt{(t-\alpha_0)^2 - 4\beta_0}] + \int_0^1 \frac{ts^{1/\theta - 1}}{\sqrt{(t-\alpha_0 s)^2 - 4\beta_0 s^2}} ds - \theta, \quad (3.22)$$

$$\eta(t) := \int_0^1 \frac{\theta r^{\theta-1}}{\sqrt{(t-\alpha_0 r^\theta)^2 - 4\beta_0 r^{2\theta}}} dr$$

$$= \frac{1}{\sqrt{\alpha_0^2 - 4\beta_0}} \ln \frac{(\alpha_0 + \sqrt{\alpha_0^2 - 4\beta_0})t}{\alpha_0(t-\alpha_0) + 4\beta_0 + \sqrt{\alpha_0^2 - 4\beta_0}\sqrt{(t-\alpha_0)^2 - 4\beta_0}}, \quad (3.23)$$

$$\int_0^1 \frac{\theta[t - \sqrt{(t-\alpha_0 r^\theta)^2 - 4\beta_0 r^{2\theta}}]}{r\sqrt{(t-\alpha_0 r^\theta)^2 - 4\beta_0 r^{2\theta}}} dr = \ln \frac{2t}{t - \alpha_0 + \sqrt{(t-\alpha_0)^2 - 4\beta_0}}, \quad (3.24)$$

we obtain the desired asymptotic formula in the lemma. □

Lemma 3.2. *Let $x = n^\theta t$ and $\delta_k(x)$ be given as in (3.13). As $n \to \infty$, we have*

$$\sum_{k=1}^n \delta_k(x) \sim -\frac{\alpha_0}{2}\eta(t) + \frac{1}{4}\ln \frac{t^2}{(t-\alpha_0)^2 - 4\beta_0},$$

where $\eta(t)$ is given in (3.20), and the error is of order $O(1/n)$ if $\theta > 1$, $O(\ln n/n)$ if $\theta = 1$, and $O(1/n^\theta)$ if $0 < \theta < 1$.

Proof. In view of (3.11), (3.12) and (3.13), we have

$$\delta_{k+1}(x) = \frac{A_{k-1} - A_k + \sqrt{(x-A_{k-1})^2 - 4B_k} - \sqrt{(x-A_k)^2 - 4B_{k+1}}}{A_{k-1} - A_k + \sqrt{(x-A_{k-1})^2 - 4B_k} + \sqrt{(x-A_k)^2 - 4B_{k+1}}}.$$

Recall that $x = n^\theta t$. Denote $r = k/n$. It then follows from (3.5) that

$$\frac{A_{k-1} - A_k}{n^\theta} \sim \frac{-\alpha_0 \theta r^{\theta-1}}{n},$$

and

$$\frac{\sqrt{(x-A_{k-1})^2 - 4B_k}}{n^\theta}$$

$$\sim \sqrt{\left(t - \frac{\alpha_0 k^\theta + (\alpha_1 - \alpha_0\theta)k^{\theta-1}}{n^\theta}\right)^2 - \frac{4\beta_0 k^{2\theta} + 4\beta_1 k^{2\theta-1}}{n^{2\theta}}}$$

$$\sim \sqrt{(t-\alpha_0 r^\theta)^2 - \frac{2(\alpha_1 - \alpha_0\theta)r^{\theta-1}(t-\alpha_0 r^\theta)}{n} - 4\beta_0 r^{2\theta} - \frac{4\beta_1 r^{2\theta-1}}{n}}$$

$$\sim \sqrt{(t-\alpha_0 r^\theta)^2 - 4\beta_0 r^{2\theta}} - \frac{(\alpha_1 - \alpha_0\theta)r^{\theta-1}(t-\alpha_0 r^\theta) + 2\beta_1 r^{2\theta-1}}{n\sqrt{(t-\alpha_0 r^\theta)^2 - 4\beta_0 r^{2\theta}}},$$

and

$$
\begin{aligned}
&\frac{\sqrt{(x-A_k)^2-4B_{k+1}}}{n^\theta}\\
&\sim\sqrt{\left(t-\frac{\alpha_0 k^\theta+\alpha_1 k^{\theta-1}}{n^\theta}\right)^2-\frac{4\beta_0 k^{2\theta}+4(\beta_1+2\theta\beta_0)k^{2\theta-1}}{n^{2\theta}}}\\
&\sim\sqrt{(t-\alpha_0 r^\theta)^2-\frac{2\alpha_1 r^{\theta-1}(t-\alpha_0 r^\theta)}{n}-4\beta_0 r^{2\theta}-\frac{4(\beta_1+2\theta\beta_0)r^{2\theta-1}}{n}}\\
&\sim\sqrt{(t-\alpha_0 r^\theta)^2-4\beta_0 r^{2\theta}}-\frac{\alpha_1 r^{\theta-1}(t-\alpha_0 r^\theta)+2(\beta_1+2\theta\beta_0)r^{2\theta-1}}{n\sqrt{(t-\alpha_0 r^\theta)^2-4\beta_0 r^{2\theta}}},
\end{aligned}
$$

where the errors are of order $O(1/n^2+k^{\theta-2}/n^\theta)$. Consequently,

$$
\delta_{k+1}(x)=\frac{-\alpha_0\theta r^{\theta-1}+\frac{\alpha_0\theta r^{\theta-1}(t-\alpha_0 r^\theta)+4\theta\beta_0 r^{2\theta-1}}{\sqrt{(t-\alpha_0 r^\theta)^2-4\beta_0 r^{2\theta}}}}{2n\sqrt{(t-\alpha_0 r^\theta)^2-4\beta_0 r^{2\theta}}}+O\left(\frac{1}{n^2}+\frac{k^{\theta-2}}{n^\theta}\right).
$$

A simple application of the trapezoidal rule yields

$$
\begin{aligned}
\sum_{k=1}^{n}\delta_k(x)&\sim\sum_{k=0}^{n}\frac{-\alpha_0\theta r^{\theta-1}+\frac{\alpha_0\theta r^{\theta-1}(t-\alpha_0 r^\theta)+4\theta\beta_0 r^{2\theta-1}}{\sqrt{(t-\alpha_0 r^\theta)^2-4\beta_0 r^{2\theta}}}}{2n\sqrt{(t-\alpha_0 r^\theta)^2-4\beta_0 r^{2\theta}}}\\
&\sim\int_0^1\frac{-\alpha_0\theta r^{\theta-1}+\frac{\alpha_0\theta r^{\theta-1}(t-\alpha_0 r^\theta)+4\theta\beta_0 r^{2\theta-1}}{\sqrt{(t-\alpha_0 r^\theta)^2-4\beta_0 r^{2\theta}}}}{2\sqrt{(t-\alpha_0 r^\theta)^2-4\beta_0 r^{2\theta}}}dr,
\end{aligned}
$$

where the error is of order $O(1/n)$ if $\theta>1$, $O(\ln n/n)$ if $\theta=1$, and $O(1/n^\theta)$ if $0<\theta<1$. Finally, on account of (3.23) and

$$
\begin{aligned}
&\int_0^1\frac{\alpha_0\theta r^{\theta-1}(t-\alpha_0 r^\theta)+4\theta\beta_0 r^{2\theta-1}}{2[(t-\alpha_0 r^\theta)^2-4\beta_0 r^{2\theta}]}dr=\int_0^1\frac{\alpha_0(t-\alpha_0 s)+4\beta_0 s}{2[(t-\alpha_0 s)^2-4\beta_0 s^2]}ds\\
&=-\frac{1}{4}\ln[(t-\alpha_0 s)^2-4\beta_0 s^2]\Big|_0^1=\frac{1}{4}\ln\frac{t^2}{(t-\alpha_0)^2-4\beta_0},
\end{aligned}
$$

we obtain the desired asymptotic formula in the lemma. □

Coupling Lemmas 3.1 and 3.2 gives the following theorem.

Theorem 3.2. *Let $x=n^\theta t$ and assume that all the conditions in Theorem 3.1 are satisfied. As $n\to\infty$, we have*

$$
\begin{aligned}
\pi_n(x)&=(n^\theta/2)^n e^{n\phi(t)-n\theta-(\frac{\alpha_1}{\theta}-\frac{\alpha_0}{2}-\frac{\alpha_0\beta_1}{2\theta\beta_0})\eta(t)}[t-\alpha_0+\sqrt{(t-\alpha_0)^2-4\beta_0}]^n\\
&\times(\frac{t-\alpha_0+\sqrt{(t-\alpha_0)^2-4\beta_0}}{2t})^{\frac{1}{2}+\frac{\beta_1}{2\beta_0\theta}}(\frac{t^2}{(t-\alpha_0)^2-4\beta_0})^{1/4}[1+O(E_\theta(n))],
\end{aligned}
$$

where

$$E_\theta(n) := \begin{cases} 1/n, & \theta > 1, \\ \ln n/n, & \theta = 1, \\ 1/n^\theta, & 0 < \theta < 1. \end{cases} \tag{3.25}$$

3.4 A general asymptotic solution

Now, we come back to equation (3.1) and look for an asymptotic solution

$$P_n(n^\theta t) \sim f(t)\cos[n\varphi(t)] + g(t)\sin[n\varphi(t)]. \tag{3.26}$$

For simplicity, we assume that $|b_0| \geq 2$ (or equivalently, $\alpha_0^2 > 4\beta_0$) throughout this section. The case $|b_0| \leq 2$ can be investigated in a similar manner. By symmetry, we may assume without loss of generality that $a_0 > 0$; otherwise, we consider the difference equation satisfied by the polynomials $(-1)^n P_n(-x)$.

Note that the formula in Theorem 3.2 involves some fractional powers of $t-\alpha_0+\sqrt{(t-\alpha_0)^2-4\beta_0}$. Since the function $\ln[t-\alpha_0+\sqrt{(t-\alpha_0)^2-4\beta_0}]$ is analytic in $\mathbf{C}\setminus[t_*, t^*]$, where $t_* = \alpha_0 - 2\sqrt{\beta_0} = (-2-b_0)/a_0$ and $t^* = \alpha_0 + 2\sqrt{\beta_0} = (2-b_0)/a_0$ are two roots of $(t-\alpha_0)^2 - 4\beta_0 = 0$, the formula in Theorem 3.2 is invalid on the interval $[t_*, t^*]$, which we refer to as the oscillatory interval, because, as we shall see later, the asymptotic solution oscillates on this interval. In this section, we will find a general asymptotic solution in the oscillatory region; namely, when t lies in an open complex neighborhood of (t_*, t^*). The case when t lies in a neighborhood of the critical value t_* or t^* is more challenging and it will be investigated in the next four chapters.

Denote $n_\pm := n\pm 1$ and $t_\pm := (n/n_\pm)^\theta t \sim t \mp \theta t/n$. A simple calculation gives $r(t_\pm) \sim r(t) \mp \theta t r'(t)/n$, and

$$\cos[n_\pm\varphi(t_\pm)] \sim \cos(n\varphi)(\cos\lambda \mp \frac{\mu}{n}\sin\lambda) \mp \sin(n\varphi)(\sin\lambda \pm \frac{\mu}{n}\cos\lambda), \tag{3.27}$$

$$\sin[n_\pm\varphi(t_\pm)] \sim \sin(n\varphi)(\cos\lambda \mp \frac{\mu}{n}\sin\lambda) \pm \cos(n\varphi)(\sin\lambda \pm \frac{\mu}{n}\cos\lambda), \tag{3.28}$$

where

$$\lambda(t) := \varphi(t) - \theta t\varphi'(t), \quad \mu(t) := [\theta(\theta-1)t\varphi'(t) + \theta^2 t^2\varphi''(t)]/2. \tag{3.29}$$

Consequently,

$$\begin{aligned}&P_{n\pm1}(n_{\pm}^{\theta}t_{\pm})\\ \sim&[f\mp\frac{\theta tf'}{n}][\cos(n\varphi)(\cos\lambda\mp\frac{\mu}{n}\sin\lambda)\mp\sin(n\varphi)(\sin\lambda\pm\frac{\mu}{n}\cos\lambda)]\\ &+[g\mp\frac{\theta tg'}{n}][\sin(n\varphi)(\cos\lambda\mp\frac{\mu}{n}\sin\lambda)\pm\cos(n\varphi)(\sin\lambda\pm\frac{\mu}{n}\cos\lambda)].\end{aligned}$$

Substituting this and (3.26) into (3.1), making use of $x = n^{\theta}t$ and (3.2), and ignoring the $O(1/n^2)$ terms, we obtain

$$\begin{aligned}&[\cos(n\varphi)\cos\lambda-\frac{\mu}{n}\sin(n\varphi)\cos\lambda]f+\frac{\theta tf'}{n}\sin(n\varphi)\sin\lambda\\ &+[\sin(n\varphi)\cos\lambda+\frac{\mu}{n}\cos(n\varphi)\cos\lambda]g-\frac{\theta tg'}{n}\cos(n\varphi)\sin\lambda\\ =&(\psi_0+\frac{\psi_1}{n})[f\cos(n\varphi)+g\sin(n\varphi)],\end{aligned}$$

where $\psi_i = (a_it + b_i)/2$ for $i \geq 0$. By comparing the coefficients of $\cos(n\varphi)$ and $\sin(n\varphi)$, respectively, we have

$$f\cos\lambda+\frac{\mu g\cos\lambda-\theta tg'\sin\lambda}{n}=\psi_0 f+\frac{\psi_1 f}{n}, \tag{3.30}$$

$$g\cos\lambda+\frac{-\mu g\cos\lambda+\theta tf'\sin\lambda}{n}=\psi_0 g+\frac{\psi_1 g}{n}. \tag{3.31}$$

The above system of leading term reads

$$\lambda=\arccos\psi_0. \tag{3.32}$$

Recall that $\lambda = \varphi - \theta t\varphi'$. We have

$$(t^{-1/\theta}\varphi)'=t^{-1/\theta}\varphi'-\frac{1}{\theta}t^{-1/\theta-1}\varphi=-\frac{1}{\theta}t^{-1/\theta-1}\arccos\psi_0.$$

Integrating both sides and making use of integration by parts gives (exercise)

$$\varphi(t)=\arccos\psi_0(t)+\int_{t^*}^{t}\frac{a_0(t/s)^{1/\theta}}{2\sqrt{1-\psi_0^2(s)}}ds+t^{1/\theta}\varphi^*, \tag{3.33}$$

where $\varphi^* = \varphi(t^*)$ and $t^* = (2 - b_0)/a_0$ is a turning (transition) point such that $\psi_0(t^*) = 1$. As we shall see later, we may choose $\varphi^* = 0$ without loss of generality. To determine f and g, we need to look into the $O(1/n)$ terms in (3.30)-(3.31) and obtain

$$\psi_1 f=\mu g\cos\lambda-\theta tg'\sin\lambda, \tag{3.34}$$

$$\psi_1 g=-\mu f\cos\lambda+\theta tf'\sin\lambda. \tag{3.35}$$

Note from (3.29) that $\mu = -t\theta\lambda'/2$, and we have

$$[(\sin\lambda)^{1/2}g]' = -(\sin\lambda)^{-1/2}\psi_1 f/(t\theta),$$
$$[(\sin\lambda)^{1/2}f]' = (\sin\lambda)^{-1/2}\psi_1 g/(t\theta).$$

Let $u := (\sin\lambda)^{1/2}f$ and $v := (\sin\lambda)^{1/2}g$. The above system becomes

$$u' = \frac{\psi_1 v}{\theta t\sin\lambda}, \qquad v' = \frac{-\psi_1 u}{\theta t\sin\lambda}. \tag{3.36}$$

Denote $z := u + iv$; we have

$$z' = \frac{-i\psi_1 z}{\theta t\sin\lambda}. \tag{3.37}$$

Recall that $\cos\lambda = \psi_0$. We have $\sin\lambda = \sqrt{1-\psi_0^2}$ and

$$z = \exp[\int \frac{-i\psi_1(t)}{\theta t\sqrt{1-\psi_0^2(t)}}dt]. \tag{3.38}$$

Since $\psi_0 = (a_0 t + b_0)/2$ and $\psi_1 = (a_1 t + b_1)/2$, we obtain

$$\begin{aligned}\int \frac{\psi_1(t)}{t\sqrt{1-\psi_0^2(t)}}dt &= \int (\frac{a_1}{a_0} + \frac{b_1}{2\psi_0 - b_0})\frac{d\psi_0}{\sqrt{1-\psi_0^2}} \\ &= -\frac{a_1}{a_0}\arccos\psi_0 - \frac{b_1}{\sqrt{b_0^2-4}}\arccos\frac{2-b_0\psi_0}{b_0-2\psi_0} + C,\end{aligned}$$

where we have made use of the fact that

$$\frac{d}{d\psi_0}\arccos\frac{2-b_0\psi_0}{b_0-2\psi_0} = \frac{\sqrt{b_0^2-4}}{(b_0-2\psi_0)\sqrt{1-\psi_0^2}}.$$

For convenience, we denote

$$\xi(t) := \frac{1}{\sqrt{b_0^2-4}}\arccos\frac{2-b_0\psi_0(t)}{b_0-2\psi_0(t)}. \tag{3.39}$$

It then follows that

$$z = (C_1 + iC_2)\exp[i(\frac{a_1\lambda}{\theta a_0} + \frac{b_1\xi}{\theta})]. \tag{3.40}$$

Separating the real and imaginary parts yields

$$u = C_1\cos(\frac{a_1\lambda}{\theta a_0} + \frac{b_1\xi}{\theta}) - C_2\sin(\frac{a_1\lambda}{\theta a_0} + \frac{b_1\xi}{\theta}),$$
$$v = C_1\sin(\frac{a_1\lambda}{\theta a_0} + \frac{b_1\xi}{\theta}) + C_2\cos(\frac{a_1\lambda}{\theta a_0} + \frac{b_1\xi}{\theta}).$$

Recall that $u = (\sin\lambda)^{1/2}f$ and $v = (\sin\lambda)^{1/2}g$. The asymptotic solution in (3.26) reads

$$P_n(n^\theta t) \sim (\sin\lambda)^{-1/2}[C_1\cos(n\varphi - \frac{a_1\lambda}{\theta a_0} - \frac{b_1\xi}{\theta}) + C_2\sin(n\varphi - \frac{a_1\lambda}{\theta a_0} - \frac{b_1\xi}{\theta})].$$

Especially, there are two linearly independent asymptotic solutions $(\sin\lambda)^{-1/2}\cos(n\varphi-\frac{a_1\lambda}{\theta a_0}-\frac{b_1\xi}{\theta})$ and $(\sin\lambda)^{-1/2}\sin(n\varphi-\frac{a_1\lambda}{\theta a_0}-\frac{b_1\xi}{\theta})$. If we multiply a solution to the difference equation (3.1) by a function of $x=n^\theta t$, it remains a solution. So, $(\sin\lambda)^{-1/2}\cos(n\varphi-nt^{1/\theta}\varphi^*-\frac{a_1\lambda}{\theta a_0}-\frac{b_1\xi}{\theta})$ and $(\sin\lambda)^{-1/2}\sin(n\varphi-nt^{1/\theta}\varphi^*-\frac{a_1\lambda}{\theta a_0}-\frac{b_1\xi}{\theta})$ are also two linearly independent asymptotic solutions, where $\varphi^*=\varphi(t^*)$ and $t^*=(2-b_0)/a_0$. Without loss of generality, we may assume $\varphi^*=0$ (otherwise we replace the function φ with $\varphi-\varphi^*$). Consequently, any solution of (3.26) with t in an open complex neighborhood of (t_*,t^*) has the asymptotic formula

$$P_n(n^\theta t)\sim\frac{C_1(x)\cos(n\varphi-\frac{a_1\lambda}{\theta a_0}-\frac{b_1\xi}{\theta})+C_2(x)\sin(n\varphi-\frac{a_1\lambda}{\theta a_0}-\frac{b_1\xi}{\theta})}{\sqrt{\sin\lambda}} \tag{3.41}$$

for some functions $C_1(x)$ and $C_2(x)$, where φ is given in (3.33) with $\varphi^*=0$, and λ and ξ are defined in (3.32) and (3.39), respectively.

3.5 Asymptotic matching

Recall that $P_n(x)=\gamma_n\pi_n(x)$, where γ_n is defined in (3.3). Following the previous section, we assume that $a_0>0$ and $|b_0|\geq 2$. Let $x=n^\theta t$, where t lies in an open complex neighborhood of (t_*,t^*) and the conditions in Theorem 3.1 are satisfied. Theorem 3.2 implies that

$$P_n(n^\theta t)\sim\gamma_n(n^\theta/2)^n e^{n\phi(t)-n\theta-(\frac{\alpha_1}{\theta}-\frac{\alpha_0}{2}-\frac{\alpha_0\beta_1}{2\theta\beta_0})\eta(t)}[t-\alpha_0+\sqrt{(t-\alpha_0)^2-4\beta_0}]^n$$
$$\times\Big(\frac{t-\alpha_0+\sqrt{(t-\alpha_0)^2-4\beta_0}}{2t}\Big)^{\frac{1}{2}+\frac{\beta_1}{2\beta_0\theta}}\Big(\frac{t^2}{(t-\alpha_0)^2-4\beta_0}\Big)^{1/4}.$$

We intend to match this formula with (3.41) to determine the coefficient functions $C_1(x)$ and $C_2(x)$. Note from (3.6) that

$$\alpha_0=-\frac{b_0}{a_0},\ \beta_0=\frac{1}{a_0^2},\ \frac{1}{2}+\frac{\beta_1}{2\beta_0\theta}=-\frac{a_1}{\theta a_0}, \tag{3.42}$$

and (exercise)

$$\frac{\alpha_1-\alpha_0\beta_1/(2\beta_0)}{\theta\sqrt{\alpha_0^2-4\beta_0}}-\frac{\alpha_0}{2\sqrt{\alpha_0^2-4\beta_0}}=-\frac{b_1}{\theta\sqrt{b_0^2-4}}. \tag{3.43}$$

We also note that

$$\psi_0=\frac{a_0t+b_0}{2}=\frac{t-\alpha_0}{2\sqrt{\beta_0}}. \tag{3.44}$$

If $\operatorname{Im}t>0$, then the functions ϕ in (3.19) and φ in (3.33) can be related as

$$\begin{aligned}\phi(t)=&\int_0^{t/t^*}\frac{ts^{1/\theta-1}}{\sqrt{(t-\alpha_0s)^2-4\beta_0s^2}}ds+\int_{t/t^*}^1\frac{ts^{1/\theta-1}}{\sqrt{(t-\alpha_0s)^2-4\beta_0s^2}}ds\\=&t^{1/\theta}\phi^*+i[\varphi(t)-\lambda(t)],\end{aligned} \tag{3.45}$$

where

$$\phi^* = \int_0^{1/t^*} \frac{r^{1/\theta-1}}{\sqrt{(1-\alpha_0 r)^2 - 4\beta_0 r^2}} dr. \tag{3.46}$$

The functions η in (3.20) and ξ in (3.39) are connected by the following relation

$$\begin{aligned}\eta(t)\sqrt{\alpha_0^2 - 4\beta_0} &= \ln \frac{\alpha_0 + \sqrt{\alpha_0^2 - 4\beta_0}}{-2\sqrt{\beta_0}} - i \arccos \frac{\alpha_0(t-\alpha_0) + 4\beta_0}{-2\sqrt{\beta_0}t} \\ &= \ln \frac{b_0 - \sqrt{b_0^2 - 4}}{2} - i\xi(t)\sqrt{b_0^2 - 4}.\end{aligned}$$

This together with (3.43) yields

$$(\frac{\alpha_1}{\theta} - \frac{\alpha_0}{2} - \frac{\alpha_0\beta_1}{2\theta\beta_0})\eta(t) = -\frac{b_1}{\theta\sqrt{b_0^2-4}} \ln \frac{b_0 - \sqrt{b_0^2-4}}{2} + \frac{ib_1}{\theta}\xi(t). \tag{3.47}$$

Furthermore, we have

$$\ln[t - \alpha_0 + \sqrt{(t-\alpha_0)^2 - 4\beta_0}] = \ln(2\sqrt{\beta_0}) + i \arccos \frac{t-\alpha_0}{2\sqrt{\beta_0}} = \ln \frac{2}{a_0} + i\lambda(t).$$

It then follows from the above identities that

$$\begin{aligned}P_n(n^\theta t) \sim &\gamma_n (\frac{n^\theta}{a_0})^n e^{in\varphi(t) - n\theta - \frac{ib_1}{\theta}\xi(t) - \frac{ia_1}{\theta a_0}\lambda(t) + nt^{1/\theta}\phi^*} (\frac{b_0 - \sqrt{b_0^2-4}}{2})^{\frac{b_1}{\theta\sqrt{b_0^2-4}}} \\ &\times (a_0 t)^{\frac{a_1}{\theta a_0}} (\frac{a_0 t}{2i\sqrt{1-\psi_0^2}})^{1/2}.\end{aligned}$$

In view of $x = n^\theta t$, $\sin\lambda = \sqrt{1-\psi_0^2}$, and the asymptotic formula of γ_n in (3.7), we obtain

$$P_n(n^\theta t) \sim \kappa_1 \frac{x^{\frac{1}{2} + \frac{a_1}{\theta a_0}}}{\sqrt{\sin\lambda(t)}} e^{in\varphi(t) - \frac{ib_1}{\theta}\xi(t) - \frac{ia_1}{\theta a_0}\lambda(t) - \frac{i\pi}{4} + x^{1/\theta}\phi^*}, \tag{3.48}$$

where

$$\kappa_1 = 2^{-1/2}\kappa_0 a_0^{\frac{1}{2} + \frac{a_1}{\theta a_0}} (\frac{b_0 - \sqrt{b_0^2-4}}{2})^{\frac{b_1}{\theta\sqrt{b_0^2-4}}}. \tag{3.49}$$

Since $\operatorname{Im}\varphi(t) < 0$ for $\operatorname{Im} t > 0$, the asymptotic formula (3.41) is equivalent to

$$P_n(n^\theta t) \sim \frac{C_1(x) - iC_2(x)}{2\sqrt{\sin\lambda(t)}} e^{in\varphi(t) - \frac{ib_1}{\theta}\xi(t) - \frac{ia_1}{\theta a_0}\lambda(t)}. \tag{3.50}$$

Matching this with (3.48) gives

$$\frac{C_1(x) - iC_2(x)}{2} = \kappa_1 e^{-i\pi/4} x^{\frac{1}{2} + \frac{a_1}{\theta a_0}} e^{x^{1/\theta}\phi^*}. \tag{3.51}$$

Next, we consider the case when $\operatorname{Im} t < 0$. In a similar argument, we obtain

$$\phi(t) = t^{1/\theta}\phi^* - i[\varphi(t) - \lambda(t)], \tag{3.52}$$

$$(\frac{\alpha_1}{\theta} - \frac{\alpha_0}{2} - \frac{\alpha_0\beta_1}{2\theta\beta_0})\eta(t) = -\frac{b_1}{\theta\sqrt{b_0^2-4}}\ln\frac{b_0-\sqrt{b_0^2-4}}{2} - \frac{ib_1}{\theta}\xi(t), \tag{3.53}$$

$$\ln[t-\alpha_0+\sqrt{(t-\alpha_0)^2-4\beta_0}] = \ln\frac{2}{a_0} - i\lambda(t). \tag{3.54}$$

Consequently,

$$P_n(n^\theta t) \sim \kappa_1 \frac{x^{\frac{1}{2}+\frac{a_1}{\theta a_0}}}{\sqrt{\sin\lambda(t)}} e^{-in\varphi(t)+\frac{ib_1}{\theta}\xi(t)+\frac{ia_1}{\theta a_0}\lambda(t)+\frac{i\pi}{4}+x^{1/\theta}\phi^*}. \tag{3.55}$$

Now, we have $\operatorname{Im}\varphi > 0$ and the asymptotic formula (3.41) is equivalent to

$$P_n(n^\theta t) \sim \frac{C_1(x)+iC_2(x)}{2\sqrt{\lambda(t)}} e^{-in\varphi(t)+\frac{ib_1}{\theta}\xi(t)+\frac{ia_1}{\theta a_0}\sin\lambda(t)}. \tag{3.56}$$

Matching this with (3.55) gives

$$\frac{C_1(x)+iC_2(x)}{2} = \kappa_1 e^{i\pi/4} x^{\frac{1}{2}+\frac{a_1}{\theta a_0}} e^{x^{1/\theta}\phi^*}. \tag{3.57}$$

Solving (3.51) and (3.57) yields

$$\begin{aligned} C_1(x) = C_2(x) &= \sqrt{2}\kappa_1 x^{\frac{1}{2}+\frac{a_1}{\theta a_0}} e^{x^{1/\theta}\phi^*} \\ &= \kappa_0 (a_0 x)^{\frac{1}{2}+\frac{a_1}{\theta a_0}} (\frac{b_0-\sqrt{b_0^2-4}}{2})^{\frac{b_1}{\theta\sqrt{b_0^2-4}}} e^{x^{1/\theta}\phi^*}, \end{aligned}$$

where we have made use of (3.49). Hence, the asymptotic formula (3.56) reads

$$\begin{aligned} P_n(n^\theta t) \sim & \kappa_0 e^{x^{1/\theta}\phi^*} (a_0 x)^{\frac{1}{2}+\frac{a_1}{\theta a_0}} (\frac{b_0-\sqrt{b_0^2-4}}{2})^{\frac{b_1}{\theta\sqrt{b_0^2-4}}} \sqrt{\frac{2}{\sin\lambda}} \\ & \times \cos(n\varphi - \frac{a_1\lambda}{\theta a_0} - \frac{b_1\xi}{\theta} - \frac{\pi}{4}). \end{aligned} \tag{3.58}$$

Recall that $P_n(x) = \gamma_n \pi_n(x)$. It then follows from (3.7) that

$$\begin{aligned} \pi_n(n^\theta t) \sim & (\frac{n}{e})^{n\theta} e^{nt^{1/\theta}\phi^*} (a_0 t)^{\frac{1}{2}+\frac{a_1}{\theta a_0}} a_0^{-n} (\frac{b_0-\sqrt{b_0^2-4}}{2})^{\frac{b_1}{\theta\sqrt{b_0^2-4}}} \sqrt{\frac{2}{\sin\lambda}} \\ & \times \cos(n\varphi - \frac{a_1\lambda}{\theta a_0} - \frac{b_1\xi}{\theta} - \frac{\pi}{4}). \end{aligned} \tag{3.59}$$

The above formula is valid for all t in an open complex neighborhood of (t_*, t^*), even when the conditions in Theorem 3.1 are not satisfied.

3.6 An example

Every birth and death process leads to a family of orthogonal polynomials. Given positive birth rates $\{\lambda_n\}$ and death rates $\{\mu_n\}$ with $n \geq 1$, the orthogonal polynomials arising from the birth and death process satisfy the difference equation

$$-xQ_n(x) = \lambda_n Q_{n+1}(x) + \mu_n Q_{n-1}(x) - (\lambda_n + \mu_n)Q_n(x), \quad n \geq 1, \tag{3.60}$$

with the initial conditions $Q_0(x) = 1$ and $Q_1(x) = (\lambda_0 + \mu_0 - x)/\lambda_0$, where $\lambda_0 > 0$ and $\mu_0 \geq 0$. Conrad and Flajolet (2006) introduced the following family of cubic rates:

$$\lambda_n = (3n+c+1)(3n+c+2)^2, \quad \mu_n = (3n+c)^2(3n+c+1), \quad c > 0; \tag{3.61}$$

and Dai *et al.* (2014) proved that the moment problem associated with the above rates are indeterminate; that is, the orthogonality measure of the associated polynomials is not unique. In fact, in this case, the orthogonality measure is not even known. To apply the results in the previous sections, we normalize the polynomials as

$$\pi_n(x) = (-1)^n \left(\prod_{k=0}^{n-1} \lambda_k\right) Q_n(x). \tag{3.62}$$

It is readily seen that the monic polynomials $\pi_n(x)$ satisfies the difference equation (3.4) with initial conditions $\pi_0(x) = 1$ and $\pi_1(x) = x - A_0$, where

$$A_n = \lambda_n + \mu_n = (3n+c+1)[(3n+c+2)^2 + (3n+c)^2], \quad n \geq 0, \tag{3.63}$$

$$B_n = \lambda_{n-1}\mu_n = (3n+c-2)(3n+c-1)^2(3n+c)^2(3n+c+1), \quad n \geq 1. \tag{3.64}$$

Hence, $\theta = 3$ and the first few terms in the asymptotic expansions (3.5) are (exercise)

$$\alpha_0 = 54, \quad \alpha_1 = 54(c+1), \quad \beta_0 = 729, \quad \beta_1 = 729(2c-1). \tag{3.65}$$

The turning points are calculated as

$$t_* = \alpha_0 - 2\sqrt{\beta_0} = 0, \quad t^* = \alpha_0 + 2\sqrt{\beta_0} = 108. \tag{3.66}$$

In the following lemmas, we shall verify the conditions (3.16) in Theorem 3.1.

Lemma 3.3. *Let $x = n^3 t$ with $t \in \mathbf{C} \setminus [0, 108]$. We have*

$$\sup_{1 \leq k < n} \frac{|x - A_{k-1} - \sqrt{(x - A_{k-1})^2 - 4B_k}|}{|x - A_k + \sqrt{(x - A_k)^2 - 4B_{k+1}}|} = O(1), \tag{3.67}$$

as $n \to \infty$. The above estimate is uniform for t in any complex region bounded away from the interval $[0, 108]$.

Proof. We prove by contradiction. Assume to the contrary that there exist sequences $\{x_j\}$, $\{n_j\}$, $\{k_j\}$ and $\{t_j\}$ such that $x_j = n_j^3 t_j$, $1 \le k_j < n_j$, and

$$\lim_{j\to\infty} \frac{|x_j - A_{k_j-1} - \sqrt{(x_j - A_{k_j-1})^2 - 4B_{k_j}}|}{|x_j - A_{k_j} + \sqrt{(x_j - A_{k_j})^2 - 4B_{k_j+1}}|} = \infty.$$

First, we claim that t_j are uniformly bounded; otherwise, we may extract a divergent subsequence, which, without loss of generality, is still denoted by t_j, such that $t_j \to \infty$. Since A_{k_j}/n_j^3 and B_{k_j}/n_j^6 are uniformly bounded, we have $A_{k_j-1}/x_j \to 0$, $B_{k_j}/x_j^2 \to 0$, $A_{k_j}/x_j \to 0$ and $B_{k_j+1}/x_j^2 \to 0$ as $j \to \infty$, which imply that

$$\lim_{j\to\infty} \frac{|x_j - A_{k_j-1} - \sqrt{(x_j - A_{k_j-1})^2 - 4B_{k_j}}|}{|x_j - A_{k_j} + \sqrt{(x_j - A_{k_j})^2 - 4B_{k_j+1}}|} = 0;$$

a contradiction to our assumption. Hence, we can find a subsequence, which is also denoted by t_j, that converges to $t_\infty < \infty$. Next, we show that $n_j \to \infty$. If not, then there exists a convergent subsequence, which is still denoted by n_j such that $n_j \to n_\infty < \infty$. Since $k_j < n_j$, by extracting a subsequence, we may assume $k_j \to k_\infty < \infty$. Consequently, we have the following contradiction

$$\begin{aligned}\infty &= \lim_{j\to\infty} \frac{|x_j - A_{k_j-1} - \sqrt{(x_j - A_{k_j-1})^2 - 4B_{k_j}}|}{|x_j - A_{k_j} + \sqrt{(x_j - A_{k_j})^2 - 4B_{k_j+1}}|} \\ &= \frac{|x_\infty - A_{k_\infty-1} - \sqrt{(x_\infty - A_{k_\infty-1})^2 - 4B_{k_\infty}}|}{|x_\infty - A_{k_\infty} + \sqrt{(x_\infty - A_{k_\infty})^2 - 4B_{k_\infty+1}}|},\end{aligned}$$

where $x_\infty = n_\infty^3 t_\infty < \infty$. The denominator on the right-hand side is non-zero because $B_{k_\infty+1} > 0$. Thus, there exists a divergent subsequence, denoted by n_j, such that $n_j \to \infty$. Also, we can show that k_j is unbounded. If k_j is bounded, then $A_{k_j}/n_j^3 \to 0$ and $B_{k_j}/n_j^6 \to 0$, which again lead to the following contradiction

$$\infty = \lim_{j\to\infty} \frac{|x_j - A_{k_j-1} - \sqrt{(x_j - A_{k_j-1})^2 - 4B_{k_j}}|}{|x_j - A_{k_j} + \sqrt{(x_j - A_{k_j})^2 - 4B_{k_j+1}}|} = 0.$$

Finally, we extract a divergent subsequence, still denoted by k_j, such that $k_j \to \infty$. Without loss of generality, we may assume $k_j/n_j \to r \in [0, 1]$. It then follows that $A_{k_j}/n_j^3 \to \alpha_0 r^3$ and $B_{k_j}/n_j^6 \to \beta_0 r^6$. Therefore,

$$\begin{aligned}\infty &= \lim_{j\to\infty} \frac{|x_j - A_{k_j-1} - \sqrt{(x_j - A_{k_j-1})^2 - 4B_{k_j}}|}{|x_j - A_{k_j} + \sqrt{(x_j - A_{k_j})^2 - 4B_{k_j+1}}|} \\ &= \frac{|t_\infty - \alpha_0 r^3 - \sqrt{(t_\infty - \alpha_0 r^3)^2 - 4\beta_0 r^6}|}{|t_\infty - \alpha_0 r^3 + \sqrt{(t_\infty - \alpha_0 r^3)^2 - 4\beta_0 r^6}|},\end{aligned}$$

a contradiction. This completes the proof. □

Lemma 3.4. *Let $x = n^3 t$ with $t \in \mathbf{C} \setminus [0, 108]$. We have*

$$\sup_{1 \le k < n} \frac{|A_{k-1} - A_k + \sqrt{(x - A_{k-1})^2 - 4B_k} - \sqrt{(x - A_k)^2 - 4B_{k+1}}|}{|A_{k-1} - A_k + \sqrt{(x - A_{k-1})^2 - 4B_k} + \sqrt{(x - A_k)^2 - 4B_{k+1}}|}$$
$$= O(\frac{1}{n}), \tag{3.68}$$

as $n \to \infty$. The above estimate is uniform for t in any complex region bounded away from the interval $[0, 108]$.

Proof. First, it is easily seen that $(A_{k-1} - A_k)/n^3 = O(1/n)$. Next, we can show that

$$\frac{\sqrt{(x - A_{k-1})^2 - 4B_k} - \sqrt{(x - A_k)^2 - 4B_{k+1}}}{n^3}$$
$$= \frac{(x - A_{k-1})^2 - 4B_k - (x - A_k)^2 + 4B_{k+1}}{n^3[\sqrt{(x - A_{k-1})^2 - 4B_k} + \sqrt{(x - A_k)^2 - 4B_{k+1}}]} = O(1/n)$$

because (exercise)

$$\frac{(x - A_{k-1})^2 - 4B_k - (x - A_k)^2 + 4B_{k+1}}{n^6} = O(1/n) \tag{3.69}$$

and (exercise)

$$\frac{n^3}{\sqrt{(x - A_{k-1})^2 - 4B_k} + \sqrt{(x - A_k)^2 - 4B_{k+1}}} = O(1). \tag{3.70}$$

Finally, since (exercise)

$$\frac{n^3}{A_{k-1} - A_k + \sqrt{(x - A_{k-1})^2 - 4B_k} + \sqrt{(x - A_k)^2 - 4B_{k+1}}} = O(1), \tag{3.71}$$

we obtain our desired estimate. □

Similarly, we obtain the following lemma whose proof is left as an exercise.

Lemma 3.5. *Let $x = n^3 t$ with $t \in \mathbf{C} \setminus [0, 108]$. We have*

$$\sup_{1 \le k < n} |\delta_k(x) - \delta_{k+1}(x)| = O(\frac{1}{n^2}), \tag{3.72}$$

as $n \to \infty$, where $\delta_k(x)$ is defined as in (3.13). The above estimate is uniform for t in any complex region bounded away from the interval $[0, 108]$.

Now, we are ready to state the asymptotic formula of $\pi_n(x)$ in the outer region. Let $x = n^3 t$ with $t \in \mathbf{C} \setminus [0, 108]$. In view of Lemmas 3.3-3.5, the conditions (3.16) in Theorem 3.1 are satisfied. On account of (3.65), the functions ϕ in (3.19) and η in (3.23) become

$$\phi(t) = \int_0^1 \frac{ts^{-2/3}}{\sqrt{t^2 - 108ts}} ds = \int_0^1 \frac{3}{\sqrt{1 - 108r^3/t}} dr, \tag{3.73}$$

$$\eta(t) = \int_0^1 \frac{3r^2}{\sqrt{t^2 - 108tr^3}} dr = \frac{t - \sqrt{t^2 - 108t}}{54t}. \tag{3.74}$$

The formula for $\eta(t)$ can be also obtained from (3.20) together with L'Hôpital's rule (exercise).

Theorem 3.3. *Let $x = n^3 t$ with $t \in \mathbf{C} \setminus [0, 108]$ and $\phi(t)$ be given as in (3.73). We have*

$$\pi_n(n^3 t) = \left(\frac{n^3 t}{e^3}\right)^n \frac{\exp[n\phi(t) + 2(n + \frac{c+1}{3}) \ln \frac{1+\sqrt{1-108/t}}{2}]}{(1 - 108/t)^{1/4}} [1 + O(\frac{1}{n})]. \tag{3.75}$$

as $n \to \infty$. The above estimate is uniform for t in any complex region bounded away from the interval $[0, 108]$.

Proof. It follows from (3.65) and a simple calculation that

$$\frac{\beta_1}{2\theta\beta_0} = \frac{2c-1}{6}, \quad \frac{\alpha_1}{\theta} - \frac{\alpha_0}{2} - \frac{\alpha_0\beta_1}{2\theta\beta_0} = 0.$$

A combination of (3.73), (3.74) and Theorem 3.2 implies that

$$\pi_n(x) = (\frac{n}{e})^{3n} e^{n\phi(t)} (\frac{t - 54 + \sqrt{t^2 - 108t}}{2})^n$$
$$\times (\frac{t - 54 + \sqrt{t^2 - 108t}}{2t})^{\frac{1}{2} + \frac{2c-1}{6}} (\frac{t^2}{t^2 - 108t})^{1/4} [1 + O(\frac{1}{n})].$$

This together with the identity

$$\frac{t - 54 + \sqrt{t^2 - 108t}}{2t} = (\frac{1 + \sqrt{1 - 108/t}}{2})^2$$

gives (3.75). □

In view of (3.6) and (3.65), we have

$$a_0 = \frac{1}{27}, \quad b_0 = -2, \quad a_1 = -\frac{c+1}{27}, \quad b_1 = 0. \tag{3.76}$$

Now, we let t be in an open complex neighborhood of $(0, 108)$ and apply (3.59). We first calculate ϕ^*, φ and λ from (3.46), (3.33) and (3.32), respectively. A simple calculation yields

$$\phi^* = \int_0^{1/108} \frac{r^{-2/3}}{\sqrt{1-108r}} dr = \frac{\Gamma(1/2)\Gamma(1/3)}{\Gamma(5/6)108^{1/3}}, \tag{3.77}$$

$$\lambda(t) = \arccos \frac{t-54}{54}, \tag{3.78}$$

$$\varphi(t) = \lambda(t) + \int_{108}^{t} \frac{(t/s)^{1/3}}{\sqrt{108s - s^2}} ds. \tag{3.79}$$

There is no need to calculate $\xi(t)$ because $b_1 = 0$. Now, we rewrite (3.59) as

$$\pi(n^3 t) \sim (\frac{n}{e})^{3n} e^{nt^{1/3}\phi^*} (\frac{t}{27})^{\frac{1}{2}-\frac{c+1}{3}} 27^n \sqrt{\frac{2}{\sin\lambda(t)}} \cos[n\varphi(t) + \frac{c+1}{3}\lambda(t) - \frac{\pi}{4}].$$

Note that

$$\sin\lambda(t) = \sqrt{1 - (\frac{t-54}{54})^2} = \frac{\sqrt{108t - t^2}}{54}.$$

We have the following result.

Theorem 3.4. *For any t in an open complex neighborhood of $(0, 108)$, we have as $n \to \infty$,*

$$\pi(n^3 t) \sim \frac{2(n/e)^{3n} 27^{n+\frac{c+1}{3}} e^{nt^{1/3}\phi^*}}{(108t - t^2)^{1/4} t^{\frac{c+1}{3}-\frac{1}{2}}} \cos[n\varphi(t) + \frac{c+1}{3}\lambda(t) - \frac{\pi}{4}], \tag{3.80}$$

where ϕ^*, λ and φ are given in (3.77), (3.78) and (3.79), respectively.

3.7 Exercises

1. Verify (3.6).
2. Verify (3.7).
3. Verify (3.21).
4. Verify (3.22), (3.23) and (3.24).
5. Verify (3.33).
6. Verify (3.43).
7. Verify (3.65).
8. Verify (3.69), (3.70) and (3.71).
9. Prove Lemma 3.5.
10. Prove (3.74) from (3.20) together with L'Hôpital's rule.

Chapter 4

Turning Point Theory: Airy Expansion

4.1 Introduction

Second-order linear difference equations such as (1.2) or (2.1) often depend on an auxiliary parameter. For instance, the Bessel function $J_\nu(x)$ satisfies the difference equation

$$J_{\nu+1}(x) + J_{\nu-1}(x) = \frac{2\nu}{x} J_\nu(x), \tag{4.1}$$

and any three consecutive polynomials of an orthogonal system $\{p_n(x)\}_{n\geq 0}$ satisfy a three-term recurrence relation

$$p_{n+1}(x) = [c_1(n)x + c_0(n)]p_n(x) - c_2(n)p_{n-1}(x), \quad n = 1, 2, \cdots, \tag{4.2}$$

where $c_0(n)$, $c_1(n)$ and $c_2(n)$ are independent of x; see Szegö (1975, p. 42). In (4.1) and (4.2), the variable x plays the role of an auxiliary parameter.

When the parameter varies, the locations of the characteristic roots in (1.7) change. As the parameter approaches a certain critical value, the validity of the asymptotic solution may fail. The objective of this chapter is to derive an asymptotic expansion for the solution which holds in an interval containing the critical value. The material in this chapter is taken from Wang and Wong (2002, 2003).

To proceed further, we return to (4.2) and define a sequence $\{K_n\}$ recursively by $K_{n+1}/K_{n-1} = c_2(n)$, with K_0 and K_1 depending on the particular sequence of functions or polynomials $\{p_n(x)\}$. Put $C_1(n) = c_1(n)K_n/K_{n+1}$, $C_0(n) = c_0(n)K_n/K_{n+1}$ and $P_n(x) = p_n(x)/K_n$. Equation (4.2) now becomes

$$P_{n+1}(x) - [C_1(n)x + C_0(n)]P_n(x) + P_{n-1}(x) = 0. \tag{4.3}$$

Motivated by the assumption in (1.3), we impose the condition that the coefficients $C_1(n)$ and $C_0(n)$ have asymptotic expansions of the form

$$C_1(n) \sim n^{-\theta} \sum_{s=0}^{\infty} \frac{a_s}{n^s}, \quad C_0(n) \sim \sum_{s=0}^{\infty} \frac{b_s}{n^s}, \tag{4.4}$$

where θ is a real number and $a_0 \neq 0$. Let τ_0 be a constant, whose value will be chosen later. Put $N := n + \tau_0$. Clearly, the expansions in (4.4) can be recast in the form

$$C_1(n) \sim N^{-\theta} \sum_{s=0}^{\infty} \frac{a'_s}{N^s}, \quad C_0(n) \sim \sum_{s=0}^{\infty} \frac{b'_s}{N^s}, \tag{4.5}$$

with $a'_0 = a_0$, $a'_1 = a_1 + a_0\theta\tau_0$, $b'_0 = b_0$ and $b'_1 = b_1$. It is natural to introduce the canonical scale $x = N^\theta t$. Similar to (1.7), we obtain the characteristic equation

$$\rho^2 - (a'_0 t + b'_0)\rho + 1 = 0, \tag{4.6}$$

whose roots coincide when $a'_0 t + b'_0 = \pm 2$. We denote $t^* = (2 - b'_0)/a'_0$ and $t_* = (-2 - b'_0)/a'_0$. The values t^* and t_* play an important role in the asymptotic theory of the three-term recurrence relation (4.3), and they correspond to the transition points (i.e., turning points and poles) occurring in differential equations [Olver (1974a), p. 362]. For this reason, we shall also call them *transition points* or *turning points*. We may restrict ourselves to just the case $t = t^*$; the case $t = t_*$ can be treated in a similar manner. As we shall see later in this chapter and the forthcoming chapters, the values of t^* and θ will affect the uniform asymptotic expansions near the transition points. More specifically, three cases will be considered separately: (i) $\theta \neq 0$ and $t^* \neq 0$, (ii) $\theta \neq 0$ and $t^* = 0$, and (iii) $\theta = 0$. Here, θ is the exponent in the expansion of $C_1(n)$ in (4.4) or (4.5).

In this chapter, we shall be concerned with only the first case; namely, case (i). For simplicity, we assume $\theta > 0$. The analysis for the case $\theta < 0$ is very similar. By symmetry, we may assume without loss of generality that $a'_0 = a_0 > 0$; otherwise, we consider the difference equation satisfied by $\{(-1)^n P_n(-x)\}_{n\geq 0}$. To make the presentation simpler, we also assume that $|b_0| < 2$ and $b_1 = 0$. It is readily seen from $|b_0| < 2$ that $t_* < 0 < t^*$. The assumption $b_1 = 0$ implies $b'_1 = 0$ and it is satisfied in most of the classical cases. Actually, Dingle and Morgan (1967a,b) have assumed that $a_{2s+1} = b_{2s+1} = 0$ for $s = 0, 1, 2, \cdots$. Now, we choose $\tau_0 = -a_1/(a_0\theta)$ such that $a'_1 = 0$. Clearly, $\tau_0 = 0$ if $a_1 = 0$.

To see how the Airy function appears in the asymptotic expansion, we formally assume $P_n(N^\theta t) \sim \chi(Z)$ where $Z = N^\lambda \zeta(t)$ with $\zeta(t^*) = 0$. Since $x = N^\theta t$ is fixed, t actually depends on N; i.e., $x = N^\theta t_N$. It is readily shown that $t_{N\pm1} = t(1 \pm \frac{1}{N})^{-\theta}$. If we let $Z_\pm := (N \pm 1)^\lambda \zeta(t(1 \pm \frac{1}{N})^{-\theta})$, then it follows that $Z_\pm = Z(t_{N\pm1})$ and $P_{n\pm1}(N^\theta t) \sim \chi(Z_\pm)$. Assume $t - t^* = O(1/N)$ and let $N \to \infty$. We have $Z_\pm \sim N^\lambda \zeta(t) \pm N^{\lambda-1} t^* \theta \zeta'(t^*)$,

and

$$P_{n\pm1}(N^\theta t) \sim \chi(Z) \pm N^{\lambda-1}t^*\theta\zeta'(t^*)\chi'(Z) + \frac{1}{2}N^{2\lambda-2}[t^*\theta\zeta'(t^*)]^2\chi''(Z).$$

On the other hand, since $\zeta(t) \sim \zeta(t^*) + (t-t^*)\zeta'(t^*)$, we obtain

$$C_1(n)x + C_0(n) \sim a_0t + b_0 = 2 + a_0(t-t^*) \sim 2 + \frac{a_0Z}{N^\lambda\zeta'(t^*)}.$$

Substituting the above formulas into the difference equation (4.3) gives

$$2\chi(Z) + N^{2\lambda-2}[t^*\theta\zeta'(t^*)]^2\chi''(Z) \sim \left[2 + \frac{a_0Z}{N^\lambda\zeta'(t^*)}\right]\chi(Z).$$

The leading terms on both sides are the same. To balance the next order terms on both sides, we require $2\lambda - 2 = -\lambda$; namely, $\lambda = 2/3$. Next, we set $[t^*\theta\zeta'(t^*)]^2 = a_0/\zeta'(t^*)$; that is, $\zeta'(t^*) = a_0^{1/3}(t^*\theta)^{-2/3}$. It then follows that $\chi''(Z) = Z\chi(Z)$, which is exactly the Airy equation. This suggests that we might as well try the formal series solution

$$P_n(N^\theta t) = \chi(N^{2/3}\zeta(t))\sum_{s=0}^{\infty}\frac{A_s(\zeta(t))}{N^s} + \chi'(N^{2/3}\zeta(t))\sum_{s=0}^{\infty}\frac{B_s(\zeta(t))}{N^{s+1/3}}, \quad (4.7)$$

which we have encountered in the differential equation theory [Olver (1974a), p. 409] or integral methods [Wong (1989), p. 370].

4.2 A perturbation lemma for Airy functions

Recall that we have introduced the canonical scale $x = N^\theta t$, where $N = n + \tau_0$ and θ is the exponent in the expansion of $C_1(n)$ in (4.4). Since the variable x in (4.3) is a fixed quantity, if we change n to $n \pm 1$ (or, equivalently, N to $N \pm 1$), we must also change t to $t_\pm = (1 \pm 1/N)^{-\theta}t$; that is, $x = N^\theta t = (N \pm 1)^\theta t_\pm$. Furthermore, the Airy function and its derivative in (4.7) become $\chi(N_\pm^{2/3}\zeta_\pm(t))$ and $\chi'(N_\pm^{2/3}\zeta_\pm(t))$, where $N_\pm = N \pm 1$ and $\zeta_\pm(t) = \zeta(t_\pm)$. To express these two perturbed Airy functions in terms of unperturbed Airy functions, we shall make use of the following perturbation lemma for Airy functions, which was first given by Wang and Wong (2002, 2003).

Lemma 4.1. *Let $\chi(z)$ be a solution of the Airy equation $\chi''(z) = z\chi(z)$; namely, it is a linear combination of the Airy functions* Ai(z) *and* Bi(z). *For any complex z and a, we have the following expressions for the perturbed*

Airy function

$$\chi(z+a) = \chi(z) \sum_{k=0}^{\infty} \frac{X_k(a\sqrt{z})}{z^{3k/2}} + \chi'(z) \sum_{k=0}^{\infty} \frac{Y_k(a\sqrt{z})}{z^{(3k+1)/2}}, \tag{4.8}$$

$$\chi'(z+a) = \chi(z) \sum_{k=0}^{\infty} \frac{X_k'(a\sqrt{z})}{z^{(3k-1)/2}} + \chi'(z) \sum_{k=0}^{\infty} \frac{Y_k'(a\sqrt{z})}{z^{3k/2}}, \tag{4.9}$$

where $X_0(u) = \cosh u$, $Y_0(u) = \sinh u$, *and*

$$X_k(u) = \int_0^u \sinh(u-v) v X_{k-1}(v) dv, \tag{4.10}$$

$$Y_k(u) = \int_0^u \sinh(u-v) v Y_{k-1}(v) dv, \tag{4.11}$$

for $k \geq 1$. *Let* $\zeta(t)$ *be a real analytic and increasing function in a finite or infinite interval* I *containing* t^* *and with* $\zeta(t^*) = 0$. *Denote* $N_\pm = N \pm 1$, $t_\pm = (N/N_\pm)^\theta t$ *and* $\zeta_\pm(t) = \zeta(t_\pm)$. *We have*

$$\chi(N_\pm^{2/3}\zeta_\pm) = \chi(N^{2/3}\zeta) G^\pm(\zeta, N) \pm N^{-1/3} \chi'(N^{2/3}\zeta) H^\pm(\zeta, N), \tag{4.12}$$

$$\chi'(N_\pm^{2/3}\zeta_\pm) = \pm N^{1/3} \chi(N^{2/3}\zeta) L^\pm(\zeta, N) + \chi'(N^{2/3}\zeta) K^\pm(\zeta, N), \tag{4.13}$$

where

$$G^\pm(\zeta, N) = \sum_{k=0}^{\infty} \frac{X_k(\zeta^{1/2} u^\pm)}{N^k \zeta^{3k/2}}, \quad H^\pm(\zeta, N) = \pm \sum_{k=0}^{\infty} \frac{Y_k(\zeta^{1/2} u^\pm)}{N^k \zeta^{(3k+1)/2}}, \tag{4.14}$$

$$L^\pm(\zeta, N) = \pm \sum_{k=0}^{\infty} \frac{X_k'(\zeta^{1/2} u^\pm)}{N^k \zeta^{(3k-1)/2}}, \quad K^\pm(\zeta, N) = \sum_{k=0}^{\infty} \frac{Y_k'(\zeta^{1/2} u^\pm)}{N^k \zeta^{3k/2}}, \tag{4.15}$$

and

$$u^\pm = N^{1/3}(N_\pm^{2/3}\zeta_\pm - N^{2/3}\zeta). \tag{4.16}$$

Proof. It is readily seen from (4.10) and (4.11) that

$$X_k'(u) = \int_0^u \cosh(u-v) v X_{k-1}(v) dv, \quad Y_k'(u) = \int_0^u \cosh(u-v) v Y_{k-1}(v) dv,$$

and

$$X_k''(u) = X_k(u) + u X_{k-1}(u), \quad Y_k''(u) = Y_k(u) + u Y_{k-1}(u),$$

for $k \geq 1$. Actually, $X_k(u)$ and $Y_k(u)$ can be solved recursively from the above inhomogeneous linear second-order differential equation together with initial conditions $X_k(0) = X_k'(0) = 0$ and $Y_k(0) = Y_k'(0) = 0$. Since $|\cosh u| \leq e^{|u|}$ and $|\sinh u| \leq e^{|u|}$, one can further prove by induction that

$$|X_k(u)| \leq \frac{|u|^{2k} e^{|u|}}{2^k k!}, \quad |Y_k(u)| \leq \frac{|u|^{2k} e^{|u|}}{2^k k!} \tag{4.17}$$

for $k \geq 0$. Thus, given $y \in \mathbf{C}$, the function

$$F_y(u) := \chi(y^2) \sum_{k=0}^{\infty} \frac{X_k(u)}{y^{3k}} + \chi'(y^2) \sum_{k=0}^{\infty} \frac{Y_k(u)}{y^{3k+1}}$$

is well-defined as long as $|u/y|$ is bounded. It is easily seen that the above function satisfies the differential equation $F_y''(u) = (1 + u/y^3)F_y(u)$ with initial conditions $F_y(0) = \chi(y^2)$ and $F_y'(0) = \chi'(y^2)/y$. Obviously, $\chi(y^2 + u/y)$ is also a solution to the same initial value problem. By uniqueness, we have $F_y(u) = \chi(y^2 + u/y)$ and $F_y'(u) = \chi'(y^2 + u/y)/y$. Setting $y = \sqrt{z}$ and $u = ay = a\sqrt{z}$ yields (4.8) and (4.9). By choosing $z = N^{2/3}\zeta$ and $a = N_\pm^{2/3}\zeta_\pm - N^{2/3}\zeta = N^{-1/3}u^\pm$, we further obtain (4.12) and (4.13). □

The following proposition will be used in the remaining part of this chapter and its proof is left as an exercise.

Proposition 4.1. *Let X_k, Y_k, ζ, $u^\pm$, $G^\pm$, $H^\pm$, $L^\pm$ and $K^\pm$ be given in Lemma 4.1. We have $X_k(-u) = (-1)^k X_k(u)$, $Y_k(-u) = (-1)^{k+1} Y_k(u)$. Define $h_j = h^{(j)}(0)/j!$ as the coefficient of ε^j in the series expansion of $h(\varepsilon) = (1+\varepsilon)^{2/3}\zeta(t(1+\varepsilon)^{-\theta})$. We have*

$$u^\pm = \sum_{j=0}^{\infty} \frac{(\pm 1)^{j+1} h_{j+1}}{N^j} = \sum_{j=0}^{\infty} \frac{(\pm 1)^{j+1} u_j}{N^j}, \tag{4.18}$$

for any large N such that $t_\pm = (1 \pm 1/N)^{-\theta} t \in I$, where $u_j = h_{j+1}$. For $m \geq 1$ and $l \geq 0$, we denote

$$u_{m,l} = \sum_{\substack{j_1+\cdots+j_l=m \\ j_1,\cdots,j_l \geq 1}} u_{j_1} \cdots u_{j_l} \tag{4.19}$$

be the coefficient of ε^m in the product $(\sum_{j=1}^{\infty} u_j \varepsilon^j)^l$. For consistency, we also set $u_{0,0} = 1$ and $u_{0,l} = 0$ for all $l \geq 1$. We then have

$$(u^\pm \mp u_0)^l = \sum_{m=l}^{\infty} \frac{(\pm 1)^{m+l} u_{m,l}}{N^m}, \tag{4.20}$$

and

$$X_k(\zeta^{1/2}u^{\pm}) = \sum_{m=0}^{\infty}\sum_{l=0}^{m} \frac{(\pm 1)^{m+k} X_k^{(l)}(\zeta^{1/2}u_0)\zeta^{l/2}u_{m,l}}{l!N^m}, \tag{4.21}$$

$$Y_k(\zeta^{1/2}u^{\pm}) = \sum_{m=0}^{\infty}\sum_{l=0}^{m} \frac{(\pm 1)^{m+k+1} Y_k^{(l)}(\zeta^{1/2}u_0)\zeta^{l/2}u_{m,l}}{l!N^m}, \tag{4.22}$$

$$X_k'(\zeta^{1/2}u^{\pm}) = \sum_{m=0}^{\infty}\sum_{l=0}^{m} \frac{(\pm 1)^{m+k+1} X_k^{(l+1)}(\zeta^{1/2}u_0)\zeta^{l/2}u_{m,l}}{l!N^m}, \tag{4.23}$$

$$Y_k'(\zeta^{1/2}u^{\pm}) = \sum_{m=0}^{\infty}\sum_{l=0}^{m} \frac{(\pm 1)^{m+k} Y_k^{(l+1)}(\zeta^{1/2}u_0)\zeta^{l/2}u_{m,l}}{l!N^m}, \tag{4.24}$$

for sufficiently large N, where the superscript (l) denotes the l-th derivative. Furthermore, we have

$$G^{\pm}(\zeta, N) = \sum_{s=0}^{\infty} \frac{(\pm 1)^s G_s(\zeta)}{N^s}, \quad H^{\pm}(\zeta, N) = \sum_{s=0}^{\infty} \frac{(\pm 1)^s H_s(\zeta)}{N^s}, \tag{4.25}$$

$$L^{\pm}(\zeta, N) = \sum_{s=0}^{\infty} \frac{(\pm 1)^s L_s(\zeta)}{N^s}, \quad K^{\pm}(\zeta, N) = \sum_{s=0}^{\infty} \frac{(\pm 1)^s K_s(\zeta)}{N^s}, \tag{4.26}$$

where

$$G_s(\zeta) = \sum_{k=0}^{s}\sum_{l=0}^{s-k} \frac{X_k^{(l)}(\zeta^{1/2}u_0)u_{s-k,l}}{l!\zeta^{(3k-l)/2}}, \tag{4.27}$$

$$H_s(\zeta) = \sum_{k=0}^{s}\sum_{l=0}^{s-k} \frac{Y_k^{(l)}(\zeta^{1/2}u_0)u_{s-k,l}}{l!\zeta^{(3k-l+1)/2}}, \tag{4.28}$$

$$L_s(\zeta) = \sum_{k=0}^{s}\sum_{l=0}^{s-k} \frac{X_k^{(l+1)}(\zeta^{1/2}u_0)u_{s-k,l}}{l!\zeta^{(3k-l-1)/2}}, \tag{4.29}$$

$$K_s(\zeta) = \sum_{k=0}^{s}\sum_{l=0}^{s-k} \frac{Y_k^{(l+1)}(\zeta^{1/2}u_0)u_{s-k,l}}{l!\zeta^{(3k-l)/2}}. \tag{4.30}$$

It follows from a simple calculation (exercise) that

$$u_0 = \frac{2}{3}\zeta - t\theta\zeta', \quad u_1 = -\frac{1}{9}\zeta - \frac{t\theta(1-3\theta)}{6}\zeta' + \frac{t^2\theta^2}{2}\zeta'', \tag{4.31}$$

$$G_0 = K_0 = \cosh(\zeta^{1/2}u_0), \quad \zeta H_0 = L_0 = \zeta^{1/2}\sinh(\zeta^{1/2}u_0), \tag{4.32}$$

$$H_1 = -\frac{t\theta}{2}\frac{d}{dt}H_0 - \frac{1}{6}H_0, \quad L_1 = -\frac{t\theta}{2}\frac{d}{dt}L_0 + \frac{1}{6}L_0. \tag{4.33}$$

4.3 Formal asymptotic solutions

Let $\chi(z)$ be a solution of the Airy equation $\chi''(z) = z\chi(z)$, and $\zeta(t)$ be a real analytic and increasing function in (t_*, ∞) and with $\zeta(t^*) = 0$. We introduce the canonical scale $x = N^\theta t$ and look for a solution to the difference equation (4.3) with the formal series expansion in (4.7); namely,

$$P_n(N^\theta t) = \chi(N^{2/3}\zeta)\mathbb{A}(\zeta, N) + N^{-1/3}\chi'(N^{2/3}\zeta)\mathbb{B}(\zeta, N), \tag{4.34}$$

where

$$\mathbb{A}(\zeta, N) := \sum_{r=0}^{\infty} \frac{A_r(\zeta)}{N^r}, \quad \mathbb{B}(\zeta, N) := \sum_{r=0}^{\infty} \frac{B_r(\zeta)}{N^r}. \tag{4.35}$$

For convenience, we introduce the notation

$$\Psi(t, N) := C_1(n)x + C_0(n) \sim \sum_{r=0}^{\infty} \frac{2\psi_r(t)}{N^r}, \quad \psi_r(t) := (a_r' t + b_r')/2. \tag{4.36}$$

Denote $N_\pm = N \pm 1$, $t_\pm = (N/N_\pm)^\theta t$ and $\zeta_\pm(t) = \zeta(t_\pm)$. It follows from Lemma 4.1 and (4.34) that

$$\begin{aligned} P_{n\pm1}(N^\theta t) &= [\chi(N^{2/3}\zeta)G^\pm(\zeta, N) \pm N^{-1/3}\chi'(N^{2/3}\zeta)H^\pm(\zeta, N)]\mathbb{A}(\zeta_\pm, N_\pm) \\ &\quad + N_\pm^{-1/3} \cdot [\pm N^{1/3}\chi(N^{2/3}\zeta)L^\pm(\zeta, N) + \chi'(N^{2/3}\zeta)K^\pm(\zeta, N)]\mathbb{B}(\zeta_\pm, N_\pm). \end{aligned} \tag{4.37}$$

Substituting (4.34) and (4.37) into (4.3) and equating the coefficients of $\chi(N^{2/3}\zeta)$ and $\chi'(N^{2/3}\zeta)$ to zero, we obtain two equations

$$\begin{aligned} \Psi(t, N)\mathbb{A}(\zeta, N) &= G^+(\zeta, N)\mathbb{A}(\zeta_+, N_+) + G^-(\zeta, N)\mathbb{A}(\zeta_-, N_-) \\ &\quad + (N/N_+)^{1/3}L^+(\zeta, N)\mathbb{B}(\zeta_+, N_+) - (N/N_-)^{1/3}L^-(\zeta, N)\mathbb{B}(\zeta_-, N_-), \end{aligned} \tag{4.38}$$

$$\begin{aligned} \Psi(t, N)\mathbb{B}(\zeta, N) &= H^+(\zeta, N)\mathbb{A}(\zeta_+, N_+) - H^-(\zeta, N)\mathbb{A}(\zeta_-, N_-) \\ &\quad + (N/N_+)^{1/3}K^+(\zeta, N)\mathbb{B}(\zeta_+, N_+) + (N/N_-)^{1/3}K^-(\zeta, N)\mathbb{B}(\zeta_-, N_-). \end{aligned} \tag{4.39}$$

To expand the right-hand sides of the above two equations as power series of $1/N$, we shall first expand $\mathbb{A}(\zeta_\pm, N_\pm)$ and $\mathbb{B}(\zeta_\pm, N_\pm)$. Let D^j denote the j-th derivative with respect to t; namely, $D^j A_r(\zeta) = d^j A_r(\zeta(t))/dt^j$ for $j, r = 0, 1, 2, \cdots$. Define $\gamma_{j,l}$ as the coefficient of ε^l in the power series of

$[(1+\varepsilon)^{-\theta}-1]^j$. We have

$$(t_\pm - t)^j = \sum_{l=0}^{\infty} \frac{\gamma_{j,l} t^j}{(\pm 1)^l N^l}, \quad A_r(\zeta_\pm) = \sum_{j=0}^{\infty}\sum_{l=0}^{\infty} \frac{D^j A_r(\zeta)\gamma_{j,l} t^j}{(\pm 1)^l j! N^l}, \tag{4.40}$$

$$\mathbb{A}(\zeta_\pm, N_\pm) = \sum_{r=0}^{\infty}\sum_{j=0}^{\infty}\sum_{l=0}^{\infty}\sum_{k=0}^{\infty} \binom{-r}{k} \frac{D^j A_r(\zeta)\gamma_{j,l} t^j}{(\pm 1)^{l+k} j! N^{l+r+k}}, \tag{4.41}$$

$$(N/N_\pm)^{1/3}\mathbb{B}(\zeta_\pm, N_\pm) = \sum_{r=0}^{\infty}\sum_{j=0}^{\infty}\sum_{l=0}^{\infty}\sum_{k=0}^{\infty} \binom{-r-1/3}{k} \frac{D^j B_r(\zeta)\gamma_{j,l} t^j}{(\pm 1)^{l+k} j! N^{l+r+k}}, \tag{4.42}$$

Recall the series expansion of $G^\pm$, $H^\pm$, $L^\pm$ and $K^\pm$ in Proposition 4.1. By equating the coefficients of N^{-p} in the series expansions of (4.38) and (4.39), we obtain from (4.35), (4.36), (4.41) and (4.42)

$$\begin{aligned}&\sum_{s\le p} 2\psi_s(t) A_{p-s}(\zeta) \qquad (4.43)\\ &=\sum_{s\le p}\sum_{i+l\le s} \binom{-p+s}{s-i-l}[1+(-1)^s]G_i(\zeta)\sum_{j=0}^{l} \frac{\gamma_{j,l} t^j D^j A_{p-s}(\zeta)}{j!}\\ &\quad+\sum_{s\le p}\sum_{i+l\le s} \binom{-p+s-1/3}{s-i-l}[1-(-1)^s]L_i(\zeta)\sum_{j=0}^{l} \frac{\gamma_{j,l} t^j D^j B_{p-s}(\zeta)}{j!},\end{aligned}$$

and

$$\begin{aligned}&\sum_{s\le p} 2\psi_s(t) B_{p-s}(\zeta) \qquad (4.44)\\ &=\sum_{s\le p}\sum_{i+l\le s} \binom{-p+s}{s-i-l}[1-(-1)^s]H_i(\zeta)\sum_{j=0}^{l} \frac{\gamma_{j,l} t^j D^j A_{p-s}(\zeta)}{j!}\\ &\quad+\sum_{s\le p}\sum_{i+l\le s} \binom{-p+s-1/3}{s-i-l}[1+(-1)^s]K_i(\zeta)\sum_{j=0}^{l} \frac{\gamma_{j,l} t^j D^j B_{p-s}(\zeta)}{j!},\end{aligned}$$

for $p = 0, 1, \cdots$, where we have made use of the fact that $\gamma_{j,l} = 0$ for $j > l$. Setting $p = 0$ gives $[\psi_0(t) - G_0(\zeta)]A_0(\zeta) = [\psi_0(t) - K_0(\zeta)]B_0(\zeta) = 0$. We may assume without loss of generality that either A_0 or B_0 is nonzero (exercise). It then follows from (4.32) that $\cosh(\zeta^{1/2} u_0) = \psi_0$, where $u_0 = (2/3)\zeta - t\theta\zeta'$ and $\psi_0 = (a_0' t + b_0')/2$; see (4.31) and (4.36). Together with the conditions $\zeta(t^*) = 0$ and $\zeta'(t^*) > 0$, this differential equation has a unique solution (exercise)

$$\zeta(t) = \left(\int_{t^*}^{t} \frac{3a_0'(t/s)^{1/\theta}}{4\sqrt{\psi_0^2(s)-1}} ds - \frac{3}{2}\log[\psi_0(t) + \sqrt{\psi_0^2(t)-1}] \right)^{2/3}, \tag{4.45}$$

which is real analytic and increasing for $t \in (0, \infty)$. The above expression for $\zeta(t)$ is analytic for $t \in \mathbf{C} \setminus (-\infty, t^*]$ and can be analytically continued to the interval $(0, t^*]$. The following formulas will be used frequently in the remaining parts of this chapter.

$$u_0 = -\zeta^{-1/2} \operatorname{arccosh} \psi_0, \quad H_0 = -\zeta^{-1/2}(\psi_0^2 - 1)^{1/2}. \tag{4.46}$$

Especially, $u_0 < 0$ and $H_0 < 0$ for all $t > 0$ (exercise).

Now, we come back to equations (4.43) and (4.44) with $p \geq 1$ and solve for A_{p-1} and B_{p-1} recursively. For convenience, we denote

$$\begin{aligned} f_{p-1}(t) := &\sum_{2\leq s\leq p} \psi_s(t) A_{p-s}(\zeta) \\ &- \sum_{2\leq s\leq p} \sum_{i+l\leq s} \binom{-p+s}{s-i-l} [\frac{1+(-1)^s}{2}] G_i(\zeta) \sum_{j=0}^{l} \frac{\gamma_{j,l} t^j D^j A_{p-s}(\zeta)}{j!} \\ &- \sum_{2\leq s\leq p} \sum_{i+l\leq s} \binom{-p+s-1/3}{s-i-l} [\frac{1-(-1)^s}{2}] L_i(\zeta) \sum_{j=0}^{l} \frac{\gamma_{j,l} t^j D^j B_{p-s}(\zeta)}{j!}, \end{aligned} \tag{4.47}$$

and

$$\begin{aligned} g_{p-1}(t) := &\sum_{2\leq s\leq p} \psi_s(t) B_{p-s}(\zeta) \\ &- \sum_{2\leq s\leq p} \sum_{i+l\leq s} \binom{-p+s}{s-i-l} [\frac{1-(-1)^s}{2}] H_i(\zeta) \sum_{j=0}^{l} \frac{\gamma_{j,l} t^j D^j A_{p-s}(\zeta)}{j!} \\ &- \sum_{2\leq s\leq p} \sum_{i+l\leq s} \binom{-p+s-1/3}{s-i-l} [\frac{1+(-1)^s}{2}] K_i(\zeta) \sum_{j=0}^{l} \frac{\gamma_{j,l} t^j D^j B_{p-s}(\zeta)}{j!}, \end{aligned} \tag{4.48}$$

where the empty sum is zero; namely, $f_0(t) = g_0(t) = 0$. We can simplify (4.43) and (4.44) as (exercise)

$$[L_1(\zeta) + (-p + 2/3)L_0(\zeta)]B_{p-1}(\zeta) - \theta t L_0(\zeta) \frac{d}{dt} B_{p-1}(\zeta) = f_{p-1}(t), \tag{4.49}$$

$$[H_1(\zeta) + (-p + 1)H_0(\zeta)]A_{p-1}(\zeta) - \theta t H_0(\zeta) \frac{d}{dt} A_{p-1}(\zeta) = g_{p-1}(t), \tag{4.50}$$

where we have made use of the facts that $\psi_0 = G_0 = K_0$ and $\psi_1 = 0$. Denote $\Lambda(t) := t^{1/(6\theta)}[-H_0(\zeta)]^{1/2}$. In view of (4.32) and (4.33), we can further rewrite the above two differential equations as

$$\frac{d}{dt}[\zeta^{1/2} t^{(p-1)/\theta} \Lambda B_{p-1}] = -\frac{t^{(p-1)/\theta - 1} \Lambda f_{p-1}}{\theta \zeta^{1/2} H_0}, \tag{4.51}$$

$$\frac{d}{dt}[t^{(p-1)/\theta} \Lambda A_{p-1}] = -\frac{t^{(p-1)/\theta - 1} \Lambda g_{p-1}}{\theta H_0}. \tag{4.52}$$

When $p = 1$, it follows from $f_0 = g_0 = 0$ that $\zeta^{1/2}\Lambda B_0$ and ΛA_0 are two constants independent of t. Since $\zeta(t^*) = 0$, we have $\zeta^{1/2}\Lambda B_0 = 0$. Without loss of generality, we may choose $\Lambda A_0 = 1$. When $p > 1$, we integrate the above two equations to obtain

$$\zeta^{1/2} t^{(p-1)/\theta} \Lambda B_{p-1} = -\int_{t^*}^{t} \frac{s^{(p-1)/\theta - 1} \Lambda f_{p-1}}{\theta \zeta^{1/2} H_0} ds, \tag{4.53}$$

$$t^{(p-1)/\theta} \Lambda A_{p-1} = -\int_{0}^{t} \frac{s^{(p-1)/\theta - 1} \Lambda g_{p-1}}{\theta H_0} ds. \tag{4.54}$$

For each $p > 1$, we can successively determine $A_p(\zeta)$ and $B_p(\zeta)$ from their predecessors $A_0(\zeta)$, $B_0(\zeta), \cdots$, $A_{p-1}(\zeta)$ and $B_{p-1}(\zeta)$ via the above two formulas together with (4.47) and (4.48).

4.4 Bounds for the coefficients

To establish global uniformity of our anticipated expansion, we need to estimate the coefficients $A_s(\zeta)$ and $B_s(\zeta)$ in (4.7). First, we need the following lemma.

Lemma 4.2. *Let $\zeta(t)$ be given as in* (4.45), *and let $G_s(\zeta)$, $H_s(\zeta)$, $L_s(\zeta)$ and $K_s(\zeta)$ be given as in Proposition 4.1. Then there exists a positive constant N_s independent of t such that*

$$|G_s(\zeta)| \le N_s(|t| + 1), \quad |H_s(\zeta)| \le N_s |H_0(\zeta)|, \tag{4.55}$$

$$|K_s(\zeta)| \le N_s(|t| + 1), \quad |L_s(\zeta)| \le N_s |H_0(\zeta)|(1 + |\zeta|), \tag{4.56}$$

for all $t > 0$.

Proof. We only prove the estimate for $G_s(\zeta)$; the other three estimates can be obtained in a similar manner. Note that $\zeta(t)$ is right continuous at $t = 0$. It is easily seen that $G_s(\zeta)$ is real analytic for $t \in (0, \infty)$ and right continuous at $t = 0$. It suffices to show that $G_s(\zeta)/t$ is bounded as $t \to \infty$. To achieve this, we have to find another series expansion of $G_s(\zeta)$. By choosing χ in (4.12) to be the Airy functions $\mathrm{Ai}(z)$ and $\mathrm{Bi}(z)$, we arrive at a linear system

$$\begin{pmatrix} \mathrm{Ai}(N^{2/3}\zeta) & \pm N^{-1/3}\,\mathrm{Ai}'(N^{2/3}\zeta) \\ \mathrm{Bi}(N^{2/3}\zeta) & \pm N^{-1/3}\,\mathrm{Bi}'(N^{2/3}\zeta) \end{pmatrix} \begin{pmatrix} G^{\pm}(\zeta, N) \\ H^{\pm}(\zeta, N) \end{pmatrix} = \begin{pmatrix} \mathrm{Ai}(N_{\pm}^{2/3}\zeta_{\pm}) \\ \mathrm{Bi}(N_{\pm}^{2/3}\zeta_{\pm}) \end{pmatrix}.$$

In view of the identity $\mathrm{Ai}(z)\,\mathrm{Bi}'(z) - \mathrm{Ai}'(z)\,\mathrm{Bi}(z) = 1/\pi$ [Olver (1974a), p. 393], we have

$$G^{\pm}(\zeta, N) = \pi[\mathrm{Ai}(N_{\pm}^{2/3}\zeta_{\pm})\,\mathrm{Bi}'(N^{2/3}\zeta) - \mathrm{Ai}'(N_{\pm}^{2/3}\zeta_{\pm})\,\mathrm{Bi}(N^{2/3}\zeta)]. \tag{4.57}$$

Recall the asymptotic expansion for Airy functions [Olver (1974a), p. 392-393]

$$\mathrm{Ai}(z) \sim \frac{z^{-1/4}}{2\sqrt{\pi}} e^{-(2/3)z^{3/2}} \sum_{k=0}^{\infty} \frac{(-1)^k r_k}{[(2/3)z^{3/2}]^k}, \tag{4.58}$$

$$\mathrm{Bi}(z) \sim \frac{z^{-1/4}}{\sqrt{\pi}} e^{(2/3)z^{3/2}} \sum_{k=0}^{\infty} \frac{r_k}{[(2/3)z^{3/2}]^k}, \tag{4.59}$$

$$\mathrm{Ai}'(z) \sim \frac{-z^{1/4}}{2\sqrt{\pi}} e^{-(2/3)z^{3/2}} \sum_{k=0}^{\infty} \frac{(-1)^k s_k}{[(2/3)z^{3/2}]^k}, \tag{4.60}$$

$$\mathrm{Bi}'(z) \sim \frac{z^{1/4}}{\sqrt{\pi}} e^{(2/3)z^{3/2}} \sum_{k=0}^{\infty} \frac{s_k}{[(2/3)z^{3/2}]^k}, \tag{4.61}$$

as $z \to \infty$, where

$$r_k = \frac{(k+1/2)_{2k}}{54^k k!}, \quad s_k = \frac{1+6k}{1-6k} r_k. \tag{4.62}$$

We have the asymptotic expansion

$$G^{\pm}(\zeta, N) \sim \frac{(\xi/\xi_{\pm})^{1/6}}{2} \sum_{k=0}^{\infty}\sum_{j=0}^{\infty} \frac{[(-1)^k e^{\xi_{\pm}-\xi} + (-1)^j e^{\xi-\xi_{\pm}}] r_j s_k}{\xi_{\pm}^j \xi^k}, \tag{4.63}$$

as $N \to \infty$, where $\xi := (2/3)N\zeta^{3/2}$ and $\xi_{\pm} := (2/3)N_{\pm}\zeta_{\pm}^{3/2}$. On account of Proposition 4.1, the coefficient of N^{-s} in the above asymptotic expansion is the same as $(\pm 1)^s G_s(\zeta)$. We now expand $\xi_+ - \xi$ as follows (exercise)

$$\begin{aligned}
\xi_+ - \xi &= -\int_0^1 \operatorname{arccosh}(\psi_0(t(1+\lambda/N)^{-\theta}))d\lambda \\
&= \operatorname{arccosh}(\psi_0(t)) + \frac{a_0' t\theta}{2N} \int_0^1 \frac{(1-\lambda)(1+\lambda/N)^{-\theta-1}}{\sqrt{\psi_0^2(t(1+\lambda/N)^{-\theta})-1}} d\lambda \\
&= \sum_{j=0}^{\infty} \frac{v_j(t)}{N^j},
\end{aligned} \tag{4.64}$$

where $v_0(t) = \operatorname{arccosh}(\psi_0(t)) = O(\log t)$, and

$$v_j(t) = \frac{(-1)^{j-1}\theta}{j(j+1)} + O(1/t), \quad j \geq 1, \tag{4.65}$$

as $t \to \infty$. It can be further shown (exercise) that $\zeta(t) = O(t^{2/(3\theta)})$ as $t \to \infty$, and, consequently,

$$e^{\xi_+ - \xi} = e^{v_0(t)} \sum_{l=0}^{\infty} \frac{w_{1,l}(t)}{N^l}, e^{\xi - \xi_+} = e^{-v_0(t)} \sum_{l=0}^{\infty} \frac{w_{2,l}(t)}{N^l}, \tag{4.66}$$

$$(\xi_+/\xi)^{-j-1/6} = \sum_{m=0}^{\infty} \frac{w_{3,j,m}(t)}{N^m}, \tag{4.67}$$

where $w_{1,l}(t)$, $w_{2,l}(t)$ and $w_{3,j,m}(t)$ are bounded as $t \to \infty$. Substituting the above expansions into the right-hand side of (4.63) and collecting the coefficient of N^{-s} yields

$$G_s(\zeta) = \sum_{j+k+l+m=s} \frac{[(-1)^k e^{v_0(t)} w_{1,l}(t) + (-1)^j e^{-v_0(t)} w_{2,l}(t)] r_j s_k w_{3,j,m}(t)}{2[(2/3)\zeta^{3/2}(t)]^{j+k}}.$$

Since $e^{v_0(t)}$ and $e^{-v_0(t)}$ are of order $O(t)$ as $t \to \infty$, so is $G_s(\zeta)$. This gives the first estimate in (4.55). The other three estimates on $H_s(\zeta)$, $K_s(\zeta)$ and $L_s(\zeta)$ can be proved similarly. □

Lemma 4.3. *Denote* $\Lambda(t) = t^{1/(6\theta)}(-H_0)^{1/2}$. *For* $s, j = 0, 1, 2, \cdots$, *there exists a constant* $M_{s,j}$ *independent of* t *such that*

$$|t^j \Lambda(t) D^j A_s(\zeta)| + |t^j \Lambda(t)[1 + |\zeta(t)|^{1/2}] D^j B_s(\zeta)| \leq M_{s,j}(1 + t^{-s/\theta}), \tag{4.68}$$

for $t > 0$, *where* D^j *denotes the* j*-th derivative with respect to* t.

Proof. We will prove by induction. First, we note that the estimates in Lemma 4.2 hold for all complex t in the right half-plane, because the corresponding functions are all analytic for $t \in \mathbf{C} \setminus (-\infty, 0]$. By using Cauchy integral formula, it can be proved that

$$|t^j D^j \zeta| \leq \tilde{N}_j(1 + |\zeta|), \quad |t^j D^j G_s| + |t^j D^j K_s| \leq N_{s,j}(1 + t), \tag{4.69}$$

$$|t^j D^j H_s| \leq N_{s,j}|H_0|, \quad |t^j D^j L_s| \leq N_{s,j}|H_0|(1 + |\zeta|), \tag{4.70}$$

for some constants $\tilde{N}_j$ and $N_{s,j}$. When $s = 0$, since $B_0 = 0$ and $\Lambda A_0 = 1$, the estimate in (4.68) holds for $j = 0$. By Cauchy integral formula, it also holds for $j \geq 1$. Now, we assume that the estimate in (4.68) holds for $s = 0, 1, \cdots, p-2$ with $p \geq 2$ and for $j = 0, 1, \cdots$. It follows from (4.47), (4.48) and the above estimates that

$$|\Lambda f_{p-1}| + |(1 + |\zeta|^{1/2})\Lambda g_{p-1}| \leq C_p(t + t^{-(p-2)/\theta}),$$

for some constant C_p. Substituting this into (4.53) and (4.54) gives (exercise)

$$|\Lambda A_{p-1}| + |(1 + |\zeta|^{1/2})\Lambda B_{p-1}| \leq M_{p-1,0}(1 + t^{-(p-1)/\theta}),$$

for some constant $M_{p-1,0}$. Finally, (4.68) with $s = p - 1$ and $j \geq 1$ follows from a direct application of Cauchy integral formula. □

4.5 Asymptotic nature of the expansion

Note that the variable x in the difference equation (4.3) is fixed. A solution to (4.3) multiplied by any function of x remains a solution. For each $p \geq 0$, we truncate the expansion (4.7), multiply it by $x^{1/(6\theta)} = N^{1/6}t^{1/(6\theta)}$, and replace χ by Ai and Bi, respectively. This gives us two asymptotic solutions

$$r_n^p(N^\theta t) := \mathrm{Ai}(N^{3/2}\zeta)\mathbb{A}_p(\zeta, N) + N^{-1/3}\,\mathrm{Ai}'(N^{2/3}\zeta)\mathbb{B}_p(\zeta, N), \tag{4.71}$$

$$s_n^p(N^\theta t) := \mathrm{Bi}(N^{3/2}\zeta)\mathbb{A}_p(\zeta, N) + N^{-1/3}\,\mathrm{Bi}'(N^{2/3}\zeta)\mathbb{B}_p(\zeta, N), \tag{4.72}$$

where

$$\mathbb{A}_p(\zeta, N) = x^{1/(6\theta)} \sum_{s=0}^{p} \frac{A_s(\zeta)}{N^s} = (-H_0)^{-1/2} \sum_{s=0}^{p} \frac{\tilde{A}_s(\zeta)}{N^{s-1/6}}, \tag{4.73}$$

$$\mathbb{B}_p(\zeta, N) = x^{1/(6\theta)} \sum_{s=0}^{p} \frac{B_s(\zeta)}{N^s} = (-H_0)^{-1/2} \sum_{s=0}^{p} \frac{\tilde{B}_s(\zeta)}{N^{s-1/6}}, \tag{4.74}$$

with $\tilde{A}_s(\zeta) := \Lambda(t)A_s(\zeta)$, $\tilde{B}_s(\zeta) := \Lambda(t)B_s(\zeta)$ and $\Lambda(t) := t^{1/(6\theta)}(-H_0)^{1/2}$. Especially, $\tilde{A}_0(\zeta) = 1$ and $\tilde{B}_0(\zeta) = 0$. It is worth recalling that $H_0 = \zeta^{-1/2}\sinh(\zeta^{1/2}u_0) = -[(\psi_0^2 - 1)/\zeta]^{1/2}$ with $2\psi_0 = a_0' t + b_0'$. We also recall that $x = N^\theta t$ and $\Psi(t, N) := C_1(n)x + C_0(n)$. It follows from Lemma 4.1 that (exercise)

$$\begin{aligned}\frac{R_n^p(x)}{N^{p+11/6}} &:= r_{n+1}^p(x) - [C_1(n)x + C_0(n)]r_n^p(x) + r_{n-1}^p(x)\\ &= \mathrm{Ai}(N^{3/2}\zeta)F_{1,n}^p(x) + N^{-1/3}\,\mathrm{Ai}'(N^{2/3}\zeta)F_{2,n}^p(x),\end{aligned} \tag{4.75}$$

$$\begin{aligned}\frac{S_n^p(x)}{N^{p+11/6}} &:= s_{n+1}^p(x) - [C_1(n)x + C_0(n)]s_n^p(x) + s_{n-1}^p(x)\\ &= \mathrm{Bi}(N^{3/2}\zeta)F_{1,n}^p(x) + N^{-1/3}\,\mathrm{Bi}'(N^{2/3}\zeta)F_{2,n}^p(x),\end{aligned} \tag{4.76}$$

where

$$\begin{aligned}F_{1,n}^p(x) = {}& G^+(\zeta, N)\mathbb{A}_p(\zeta_+, N_+) + G^-(\zeta, N)\mathbb{A}_p(\zeta_-, N_-) - \Psi(t, N)\mathbb{A}_p(\zeta, N)\\ &+ (N/N_+)^{1/3}L^+(\zeta, N)\mathbb{B}_p(\zeta_+, N_+) - (N/N_-)^{1/3}L^-(\zeta, N)\mathbb{B}_p(\zeta_-, N_-),\end{aligned} \tag{4.77}$$

$$\begin{aligned}F_{2,n}^p(x) = {}& H^+(\zeta, N)\mathbb{A}_p(\zeta_+, N_+) - H^-(\zeta, N)\mathbb{A}_p(\zeta_-, N_-) - \Psi(t, N)\mathbb{B}_p(\zeta, N)\\ &+ (N/N_+)^{1/3}K^+(\zeta, N)\mathbb{B}_p(\zeta_+, N_+) + (N/N_-)^{1/3}K^-(\zeta, N)\mathbb{B}_p(\zeta_-, N_-).\end{aligned} \tag{4.78}$$

On account of Lemma 4.3, we have

$$|t^j D^j \tilde{A}_s(\zeta)| + |t^j[1 + |\zeta(t)|^{1/2}]D^j \tilde{B}_s(\zeta)| \leq \tilde{M}_{s,j}(1 + t^{-s/\theta}), \tag{4.79}$$

for some constant $\tilde{M}_{s,j}$. The key step is to improve the above estimate and prove by induction

$$|\tilde{A}_s(\zeta)| + |[1+|\zeta(t)|^{1/2}]\tilde{B}_s(\zeta)| \leq \tilde{M}_s, \tag{4.80}$$

for some constant $\tilde{M}_s$. Obviously, the above estimate holds for $s = 0$. Assume it holds for $s = 0, \cdots, p$. It can be proved that (exercise)

$$|\frac{\Lambda(t)F^p_{1,n}(x)}{x^{1/(6\theta)}}| + |\frac{\Lambda(t)[1+|\zeta(t)|^{1/2}]F^p_{2,n}(x)}{x^{1/(6\theta)}}| \leq \frac{C_p(1+t)}{N^{p+2}}, \tag{4.81}$$

for some constant C_p. Substituting this into (4.75) and (4.76) yields

$$|R^p_n(N^\theta t)| \leq C_p(1+t)\alpha_n(N^\theta t), \quad |S^p_n(N^\theta t)| \leq C_p(1+t)\beta_n(N^\theta t), \tag{4.82}$$

for $t > 0$, where

$$\alpha_n(N^\theta t) := (-H_0)^{-1/2}\widetilde{\mathrm{Ai}}(N^\theta t), \quad \beta_n(N^\theta t) := (-H_0)^{-1/2}\widetilde{\mathrm{Bi}}(N^\theta t), \tag{4.83}$$

with $\widetilde{\mathrm{Ai}}$ and $\widetilde{\mathrm{Bi}}$ being the modulus functions defined by

$$\widetilde{\mathrm{Ai}}(z) := \begin{cases} \mathrm{Ai}(z), & z \geq 0, \\ [\mathrm{Ai}^2(z) + \mathrm{Bi}^2(z)]^{1/2}, & z < 0, \end{cases} \tag{4.84}$$

$$\widetilde{\mathrm{Bi}}(z) := \begin{cases} \mathrm{Bi}(z), & z \geq 0, \\ [\mathrm{Ai}^2(z) + \mathrm{Bi}^2(z)]^{1/2}, & z < 0. \end{cases} \tag{4.85}$$

Lemma 4.4. *Fix $p \geq 0$. Assume $\tilde{A}_s(\zeta)$ and $[1+|\zeta(t)|^{1/2}]\tilde{B}_s(\zeta)$ are bounded for $s = 0, \cdots, p$. There exist $\{\varepsilon^p_n(x)\}_{n=1}^\infty$ and $\{\delta^p_n(x)\}_{n=1}^\infty$ such that*

$$\varepsilon^p_{n+1}(x) - [C_1(n)x + C_0(n)]\varepsilon^p_n(x) + \varepsilon^p_{n-1}(x) = -\frac{R^p_n(x)}{N^{p+11/6}}, \tag{4.86}$$

$$\delta^p_{n+1}(x) - [C_1(n)x + C_0(n)]\delta^p_n(x) + \delta^p_{n-1}(x) = -\frac{S^p_n(x)}{N^{p+11/6}}, \tag{4.87}$$

with

$$|\delta^p_n(x)| \leq \frac{M'_p\beta_n(x)}{N^{p+5/6}}, \quad |\varepsilon^p_n(x)| \leq \frac{M''_p\alpha_n(x)}{N^{p+5/6}}, \tag{4.88}$$

for some constants M'_p and M''_p.

Proof. It can be shown that the solution of the difference equation (4.86) with asymptotic estimate in (4.88) is the same as the solution to the following summation equation with the same asymptotic estimate:

$$\varepsilon^p_n(x) = \sum_{j=n+1}^{\infty} \frac{[r^p_n(x)s^p_j(x) - s^p_n(x)r^p_j(x)]R^p_j(x)}{[r^p_{n+1}(x)s^p_n(x) - r^p_n(x)s^p_{n+1}(x)](j+\tau_0)^{p+11/6}}$$
$$+ \sum_{j=n+1}^{\infty} \frac{[r^p_n(x)S^p_j(x) - s^p_n(x)R^p_j(x)]\varepsilon^p_j(x)}{[r^p_{n+1}(x)s^p_n(x) - r^p_n(x)s^p_{n+1}(x)](j+\tau_0)^{p+11/6}}. \tag{4.89}$$

We will construct such a solution by successive approximation. Starting with $\varepsilon_{n,0}^p = 0$, we define $\varepsilon_{n,k}^p(x)$ recursively by

$$\varepsilon_{n,k}^p(x) = \sum_{j=n+1}^{\infty} \frac{[r_n^p(x)s_j^p(x) - s_n^p(x)r_j^p(x)]R_j^p(x)}{[r_{n+1}^p(x)s_n^p(x) - r_n^p(x)s_{n+1}^p(x)](j+\tau_0)^{p+11/6}} + \sum_{j=n+1}^{\infty} \frac{[r_n^p(x)S_j^p(x) - s_n^p(x)R_j^p(x)]\varepsilon_{j,k-1}^p(x)}{[r_{n+1}^p(x)s_n^p(x) - r_n^p(x)s_{n+1}^p(x)](j+\tau_0)^{p+11/6}}. \tag{4.90}$$

Since $\tilde{A}_s(\zeta)$ and $[1+|\zeta(t)|^{1/2}]\tilde{B}_s(\zeta)$ are bounded for $s = 0, \cdots, p$, it follows from (4.71) and (4.72) that

$$|r_n^p(x)| \le CN^{1/6}\alpha_n(x), \quad |s_n^p(x)| \le CN^{1/6}\beta_n(x), \tag{4.91}$$

for some constant C. Moreover, since $(1+|z|^{1/2})\widetilde{\mathrm{Ai}}(z)\widetilde{\mathrm{Bi}}(z)$ is uniformly bounded for $z \in \mathbf{R}$, we have from (4.82)

$$|R_n^p(N^\theta t)\beta_n(N^\theta t)| + |S_n^p(N^\theta t)\alpha_n(N^\theta t)| \le \frac{M'(1+|\zeta(t)|^{1/2})}{1+N^{1/3}|\zeta(t)|^{1/2}}, \tag{4.92}$$

for some constant M'. By (4.83), we have $\beta_n(x)/\alpha_n(x) = \mathrm{Bi}(N^{3/2}\zeta(N^{-\theta}x))/\mathrm{Ai}(N^{3/2}\zeta(N^{-\theta}x))$, whose partial derivatives with respect to n (while treating x as a constant) is (exercise)

$$\frac{\partial}{\partial n}[\frac{\beta_n(x)}{\alpha_n(x)}] = \frac{N^{-1/3}u_0}{\pi\,\mathrm{Ai}^2(N^{3/2}\zeta(N^{-\theta}x))} < 0. \tag{4.93}$$

This implies that

$$\alpha_n(x)\beta_j(x) < \alpha_j(x)\beta_n(x) \tag{4.94}$$

for $j > n$. It can be verified (exercise) that

$$|r_{n+1}^p(x)s_n^p(x) - r_n^p(x)s_{n+1}^p(x) - \frac{1}{\pi}| \le \frac{M''}{N}, \tag{4.95}$$

for some constant M''. Especially,

$$|r_{n+1}^p(x)s_n^p(x) - r_n^p(x)s_{n+1}^p(x)| > \frac{1}{2\pi}, \tag{4.96}$$

for all sufficiently large n. Substituting this into (4.90), we obtain from (4.91) and (4.94) that

$$|\varepsilon_{n,1}^p(x)| \le 4\pi \sum_{j=n+1}^{\infty} \frac{C^2N^{1/6}\beta_n(x)M'[1+|\zeta(x/(j+\tau_0)^\theta)|^{1/2}]}{[1+(j+\tau_0)^{1/3}|\zeta(x/(j+\tau_0)^\theta)|^{1/2}](j+\tau_0)^{p+5/3}}. \tag{4.97}$$

A further estimate of the sum on the right-hand side of the above inequality gives (exercise)

$$|\varepsilon^p_{n,1}(x)| \leq \frac{C''}{N^{p+5/6}}\beta_n(x), \tag{4.98}$$

for some constant C''. Similarly, we obtain

$$|\varepsilon^p_{n,k}(x) - \varepsilon^p_{n,k-1}(x)| \leq \frac{C''}{N^{p+5/6}}\beta_n(x) \sup_{j\geq n} \frac{|\varepsilon^p_{j,k-1}(x) - \varepsilon^p_{j,k-2}(x)|}{(j+\tau_0)^{1/6}\beta_j(x)}. \tag{4.99}$$

By induction, we have from the above two estimates

$$\frac{|\varepsilon^p_{n,k}(x) - \varepsilon^p_{n,k-1}(x)|}{N^{1/6}\beta_n(x)} \leq \left(\frac{C''}{N^{p+1}}\right)^k, \tag{4.100}$$

from which it follows that, for all $n \geq 2C'' - \tau_0$,

$$\varepsilon^p_{n,k}(x) = \sum_{j=1}^{k}[\varepsilon^p_{n,j}(x) - \varepsilon^p_{n,j-1}(x)] \tag{4.101}$$

converges as $k \to \infty$. Clearly, the limit function $\varepsilon^p_n(x)$ satisfies (4.89) as well as (4.86). Moreover, $|\varepsilon^p_n(x)| \leq 2N^{1/6}\beta_n(x)$. Using a similar argument as in the proof of (4.98) and (4.99), we obtain

$$|\varepsilon^p_n(x)| \leq \frac{M'_p\beta_n(x)}{N^{p+5/6}}, \tag{4.102}$$

for some constant M'_p. Similarly, we can construct $\delta^p_n(x)$ which satisfies (4.87) and the first estimate in (4.88). It remains to prove the second estimate in (4.88). To this end, we first define $P_n(x) := r^p_n(x) + \varepsilon^p_n(x)$ and $Q_n(x) := s^p_n(x) + \delta^p_n(x)$. It is obvious that $P_n(x)$ and $Q_n(x)$ are two solutions to the difference equation (4.3). Making use of the fact that $\varepsilon^0_n(x) \to 0$ and $\delta^0_n(x) \to 0$ as $n \to \infty$, one can prove that (exercise)

$$P_{n+1}(x)Q_n(x) - P_n(x)Q_{n+1}(x) = \pi. \tag{4.103}$$

Especially, $P_n(x)$ and $Q_n(x)$ are two linearly independent solutions to (4.3). Now we choose $m = \lfloor n/2 \rfloor$ and derive another identity from the difference equations for ε^p_n, $P_n(x)$ and $Q_n(x)$ (exercise):

$$\begin{aligned}\frac{\varepsilon^p_n(x)}{\pi} =& [r^p_{m+1}(x)\varepsilon^p_m(x) - r^p_m(x)\varepsilon^p_{m+1}(x)]Q_n(x) \\ &+ \sum_{j=n+1}^{\infty} \frac{[r^p_n(x) + \varepsilon^p_n(x)]R^p_j(x)Q_j(x)}{(j+\tau_0)^{p+11/6}} \\ &+ \sum_{j=m+1}^{n} \frac{Q_n(x)[R^p_j(x)r^p_j(x) + R^p_j(x)\varepsilon^p_j(x)]}{(j+\tau_0)^{p+11/6}}.\end{aligned} \tag{4.104}$$

The second estimate in (4.88) can again be proved by using the successive approximation method. □

To complete the induction argument, we need to prove that $\tilde{A}_{p+1}$ and $(1+\zeta^{1/2})\tilde{B}_{p+1}$ are bounded.

Lemma 4.5. *Fix $p \geq 0$. If (4.80) holds for $s = 0, 1, \cdots, p$, then it also holds for $s = p+1$.*

Proof. First, we obtain from (4.47)-(4.48) and (4.80) with $s = 0, 1, \cdots, p$ that

$$|\Lambda(t) f_{p+1}(t)| + |\Lambda(t)[1+\zeta(t)^{1/2}] g_{p+1}(t)| \leq C_{p+1,0}(1+t), \tag{4.105}$$

for some constant $C_{p+1,0}$. This together with (4.54) implies that $\tilde{A}_{p+1}$ is bounded. Recall the notations in (4.71)-(4.78), and change p to $p+1$. For all $s \geq 0$, we denote

$$\varepsilon_s(t, N) := (-H_0)^{1/2} \varepsilon_n^s(N^\theta t), \quad \delta_s(t, N) := (-H_0)^{1/2} \delta_n^s(N^\theta t). \tag{4.106}$$

Similar to (4.92), we have (exercise)

$$\begin{aligned} &|R_n^{p+1}(N^\theta t)\beta_n(N^\theta t)| + |S_n^{p+1}(N^\theta t)\alpha_n(N^\theta t)| \\ \leq &\frac{M^*(1+t^{-(p+1)/\theta})(1+|\zeta(t)|^{1/2})}{1+N^{1/3}|\zeta(t)|^{1/2}}, \end{aligned} \tag{4.107}$$

for all $t > 0$, where M^* is a positive constant. Using the similar argument in the proof of Lemma 4.4, we construct a sequence δ_n^{p+1} satisfying the difference equation (4.87) with p replaced by $p+1$ and

$$|\delta_n^{p+1}(N^\theta t)| \leq \frac{M_{p+1}^*(1+t^{-(p+1)/\theta})\beta_n(N^\theta t)}{N^{p+11/6}} \tag{4.108}$$

for some positive constant M_{p+1}^*. Especially, $Q_n^*(x) = s_n^{p+1}(x) + \delta_n^{p+1}(x)$ is a solution to (4.3). One can prove that $Q_n^*(x) = Q_n(x)$ (exercise); namely,

$$\delta_n^{p+1}(N^\theta t) = Q_n(N^\theta t) - s_n^{p+1}(N^\theta t) = (-H_0)^{-1/2}\delta_{p+1}(t, N). \tag{4.109}$$

By considering two separate cases (i) $0 < t < t^*/2$ and (ii) $t \geq t^*/2$, we obtain from the above three formulas that

$$|\delta_p(t,N) - \frac{\tilde{A}_{p+1}(\zeta)}{N^{p+5/6}}\operatorname{Bi}(N^{2/3}\zeta)| \leq \frac{\tilde{M}_p \widetilde{\operatorname{Bi}}'(N^{2/3}\zeta)}{[1+|\zeta(t)|^{1/2}]N^{p+7/6}}, \tag{4.110}$$

for all $t > 0$, where the function $\widetilde{\operatorname{Bi}}'(z)$ is defined by

$$\widetilde{\operatorname{Bi}}'(z) = \begin{cases} \operatorname{Bi}'(z), & z \geq 0, \\ [\operatorname{Ai}'(z)^2 + \operatorname{Bi}'(z)^2]^{1/2}, & z < 0. \end{cases} \tag{4.111}$$

Coupling (4.108) with (4.110) yields

$$|\,\mathrm{Bi}'(N^{2/3}\zeta)\tilde{B}_{p+1}(\zeta)| \leq \frac{\tilde{M}_p \widetilde{\mathrm{Bi}}'(N^{2/3}\zeta)}{1+|\zeta(t)|^{1/2}} + \frac{M_{p+1}^*(1+t^{-(p+1)/\theta})\widetilde{\mathrm{Bi}}(N^{2/3}\zeta)}{N^{2/3}}, \tag{4.112}$$

for $t > 0$. Note that for any fixed $t \in (0, t^*)$,

$$\mathrm{Bi}'(N^{2/3}\zeta(t)) \sim \frac{N^{1/6}[-\zeta(t)]^{1/4}}{\sqrt{\pi}} \cos\left\{\frac{2}{3}N[-\zeta(t)]^{3/2} - \frac{\pi}{4}\right\}.$$

Since $0 < (2/3)[-\zeta(t)]^{3/2} < (2/3)[-\zeta(0)]^{3/2} < \pi$ for $0 < t < t_+$, we can choose a subsequence $\{N_k\}_{k=1}^{\infty}$ such that

$$\cos\left\{\frac{2}{3}N_k[-\zeta(t)]^{3/2} - \frac{\pi}{4}\right\} > \frac{1}{2},$$

from which it follows that

$$|\,\mathrm{Bi}'(N_k^{2/3}\zeta(t))| > \frac{1}{2}\widetilde{\mathrm{Bi}}'(N_k^{2/3}\zeta(t)).$$

Divide both sides of (4.112) by $|\,\mathrm{Bi}'\,|$ and let $N = N_k \to \infty$. The result is

$$|\tilde{B}_{p+1}(\zeta)| \leq \frac{2\tilde{M}_p}{1+|\zeta(t)|^{1/2}} \tag{4.113}$$

for $0 < t < t^*$. This together with (4.79) implies that $[1+|\zeta(t)|^{1/2}]\tilde{B}_{p+1}(\zeta)$ is bounded for all $t > 0$. The proof is complete. □

Combining the above lemmas, we obtain the following theorem.

Theorem 4.1. *Assume that the coefficients $C_1(n)$ and $C_0(n)$ in the difference equation (4.3) have the asymptotic expansions of the form given in (4.4) with $\theta \neq 0$, $|b_0| < 2$ and $b_1 = 0$. Set $N = n + \tau_0$ with $\tau_0 = -a_1/(a_0\theta)$. Let $\zeta(t)$ be given as in (4.45). Then equation (4.3) has, for each integer $p \geq 0$, a pair of linearly independent solutions $P_n(x)$ and $Q_n(x)$ such that*

$$P_n(N^\theta t) = \left(\frac{4\zeta}{(a_0't+b_0')^2-4}\right)^{1/4}\left[\mathrm{Ai}(N^{2/3}\zeta)\sum_{s=0}^{p}\frac{\tilde{A}_s(\zeta)}{N^{s-1/6}} + \mathrm{Ai}'(N^{2/3}\zeta)\sum_{s=0}^{p}\frac{\tilde{B}_s(\zeta)}{N^{s+1/6}} + \varepsilon_p(t,N)\right], \tag{4.114}$$

$$Q_n(N^\theta t) = \left(\frac{4\zeta}{(a_0't+b_0')^2-4}\right)^{1/4}\left[\mathrm{Bi}(N^{2/3}\zeta)\sum_{s=0}^{p}\frac{\tilde{A}_s(\zeta)}{N^{s-1/6}} + \mathrm{Bi}'(N^{2/3}\zeta)\sum_{s=0}^{p}\frac{\tilde{B}_s(\zeta)}{N^{s+1/6}} + \delta_p(t,N)\right], \tag{4.115}$$

where $x = N^\theta t$, *and for* $t > 0$,

$$|\varepsilon_p(t, N)| \le \frac{M_p}{N^{p+5/6}} \widetilde{\mathrm{Ai}}(N^{2/3}\zeta), \quad |\delta_p(t, N)| \le \frac{M_p}{N^{p+5/6}} \widetilde{\mathrm{Bi}}(N^{2/3}\zeta), \quad (4.116)$$

for some constant M_p. *The modulus functions* $\widetilde{\mathrm{Ai}}$ *and* $\widetilde{\mathrm{Bi}}$ *are defined in* (4.84) *and* (4.85), *respectively.*

4.6 Exercises

1. Prove Proposition 4.1.
2. Verify the formulas in (4.31), (4.32) and (4.33).
3. Let $P_n(N^\theta t)$ in (4.34) be a solution to (4.3) with series expansions in (4.35). Assume $A_s = B_s = 0$ for all $s < p$ and either A_p or B_p is nonzero. Show that $x^p P_n(N^\theta t)$ is also a solution to (4.3) and it has a similar expansion as in (4.34) and (4.35). Moreover, show that the leading coefficients in the series expansion of $x^p P_n(N^\theta t)$ are not identically zero.
4. Let $u_0(t) = (2/3)\zeta(t) - t\theta\zeta'(t)$ and $\psi_0(t) = (a_0' t + b_0')/2$, where $\theta > 0$, $a_0' > 0$ and $|b_0'| < 2$. Denote $t^* = (2 - b_0')/a_0'$. Show that the differential equation $\cosh(\zeta^{1/2} u_0) = \psi_0$ with initial conditions $\zeta(t^*) = 0$ and $\zeta'(t^*) > 0$ has a unique solution $\zeta(t)$ given in (4.45). Moreover, prove that ζ is real analytic and increasing for $t > 0$.
5. Verify (4.46) and show that both u_0 and H_0 are real analytic and negative for all $t > 0$.
6. Verify (4.49) and (4.50).
7. Verify (4.64) and (4.65).
8. Let $\zeta(t)$ be given in (4.45). Show that $\zeta(t) = O(t^{2/(3\theta)})$ as $t \to \infty$.
9. Verify (4.75) and (4.76).
10. Prove (4.81).
11. Verify (4.93).
12. Verify (4.95).
13. Prove (4.98).
14. Let $r_n^p(x)$ and $s_n^p(x)$ with $p \ge 0$ be defined as in (4.71) and (4.72). If $P_n(x)$ and $Q_n(x)$ are two solutions to (4.3) such that $P_n(x) - r_n^0(x) \to 0$ and $Q_n(x) - s_n^0(x) \to 0$ as $n \to \infty$, prove that $P_{n+1}(x)Q_n(x) - P_n(x)Q_{n+1}(x) = \pi$. Moreover, show that such pair of solutions are unique; namely, if $P_n^*(x)$ and $Q_n^*(x)$ are two solutions to (4.3) such that $P_n^*(x) - r_n^0(x) \to 0$ and $Q_n^*(x) - s_n^0(x) \to 0$ as $n \to \infty$, then $P_n^*(x) = P_n(x)$ and $Q_n^*(x) = Q_n(x)$.
15. Verify (4.104).
16. Verify (4.107).

Chapter 5

Turning Point Theory: Bessel Expansion

5.1 Introduction

This is a continuation of a topic considered in Chapter 4. Since in that chapter we have discussed case (i), naturally in the present chapter we should discuss case (ii). But, historically, case (iii) was resolved before case (ii). Hence, in this chapter, we will instead consider case (iii); namely, the case when the exponent θ in (4.4) is equal to zero.

The material in this chapter is taken from Wang and Wong (2005). For convenience, we recall the second-order linear difference equation (4.3)

$$P_{n+1}(x) - [C_1(n)x + C_0(n)]P_n(x) + P_{n-1}(x) = 0, \tag{5.1}$$

where the coefficients $C_1(n)$ and $C_0(n)$ have asymptotic expansions

$$C_1(n) \sim \sum_{s=0}^{\infty} \frac{a_s}{n^s}, \quad C_0(n) \sim \sum_{s=0}^{\infty} \frac{b_s}{n^s}, \tag{5.2}$$

with $a_0 \neq 0$. As in (4.6), we obtain the characteristic equation

$$\rho^2 - (a_0 x + b_0)\rho + 1 = 0, \tag{5.3}$$

whose roots coincide when $a_0 x + b_0 = \pm 2$. We denote $x^* = (2 - b_0)/a_0$ and $x_* = (-2 - b_0)/a_0$. The points x^* and x_* are called *transition points* or *turning points*. We may restrict ourselves to just the case $x = x^*$; the case $x = x_*$ can be treated in a similar manner. By symmetry, we may assume without loss of generality that $a_0 > 0$; otherwise, we consider the difference equation satisfied by $\{(-1)^n P_n(-x)\}_{n \geq 0}$. For simplicity, we also assume that $a_1 = b_1 = 0$. This assumption naturally implies the condition $a_1 x^* + b_1 = 0$, which was used in the previous chapter on turning point theory. It is interesting to note that in a study on WKB methods for difference equations, Dingle and Morgan (1967a,b) also assumed this condition. In fact, they assumed the stronger condition that all coefficients a_s and b_s in (5.2) with odd indices vanish. Define $N := n + \tau_0$ with τ_0 to be determined later. We recast the expansions in (5.2) in terms of N,

$$C_1(n) \sim \sum_{s=0}^{\infty} \frac{a'_s}{N^s}, \quad C_0(n) \sim \sum_{s=0}^{\infty} \frac{b'_s}{N^s}, \tag{5.4}$$

with $a'_0 = a_0$, $b'_0 = b_0$, $a'_1 = b'_1 = 0$, $a'_2 = a_2$, $b'_2 = b_2$, $a'_3 = a_3 + 2\tau_0 a_2$ and $b'_3 = b_3 + 2\tau_0 b_2$.

To see how the Bessel function appears in the asymptotic expansion, we formally assume $P_n(x) \sim \chi(Z)$ where $Z = N\zeta^\lambda(x)$ and $\zeta(x^*) = 0$. It is easily seen that

$$P_{n\pm 1}(x) \sim \chi(Z \pm Z/N) \sim \chi(Z) \pm \frac{Z}{N}\chi'(Z) + \frac{Z^2}{2N^2}\chi''(Z).$$

Substituting this into (5.1) gives

$$\frac{Z^2}{N^2}\chi''(Z) \sim \left[a_0(x - x^*) + \frac{a_2 x + b_2}{N^2}\right]\chi(Z),$$

where we have made use of $a_0 x^* + b_0 = 2$ and ignored the terms of order $O(N^{-3})$. To balance the orders of two terms in the coefficient of $\chi(Z)$ on the right-hand side of the above equation, we require $x - x^* = O(N^{-2})$. By Taylor expansion, we obtain $(Z/N)^{1/\lambda} = \zeta(x) \sim \zeta'(x^*)(x - x^*)$, which implies that $\lambda = 1/2$ and $x - x^* \sim Z^2/[N^2\zeta'(x^*)]$. Substituting this into the above equation gives a Bessel equation

$$\chi''(Z) = \left[\frac{a_0}{\zeta'(x^*)} + \frac{a_2 x^* + b_2}{Z^2}\right]\chi(Z).$$

We further let $\zeta'(x^*) = a_0$ and $\nu^2 = a_2 x^* + b_2 + 1/4$. The above equation has two linearly independent solutions $Z^{1/2}I_\nu(Z)$ and $Z^{1/2}e^{i\pi\nu}K_\nu(Z)$. This suggests that we try a formal series solution of the form

$$P_n(x) = \chi_\nu(N\zeta^{1/2})\sum_{s=0}^{\infty}\frac{A_s(\zeta(x))}{N^{s-1/2}} + \zeta^{1/2}\chi_{\nu-1}(N\zeta^{1/2})\sum_{s=0}^{\infty}\frac{B_s(\zeta(x))}{N^{s-1/2}}, \tag{5.5}$$

motivated from the differential equation theory. Here, χ_ν is a linear combination of I_ν and K_ν.

5.2 A perturbation lemma for modified Bessel functions

When we replace n by $n \pm 1$, the two modified Bessel functions $\chi_\nu(N\zeta^{1/2})$ and $\chi_{\nu-1}(N\zeta^{1/2})$ in (5.5) become $\chi_\nu(N_\pm\zeta^{1/2})$ and $\chi_{\nu-1}(N_\pm\zeta^{1/2})$, where $N_\pm := N \pm 1$. To express these two perturbed modified Bessel functions in terms of unperturbed modified Bessel functions, we shall make use of the following perturbation lemma which was first given by Wang and Wong (2005).

Lemma 5.1. *Let* $\chi_\nu(z)$ *be a solution of the modified Bessel equation*

$$\chi_\nu''(z) + (1/z)\chi_\nu'(z) - (1 + \nu^2/z^2)\chi_\nu(z) = 0; \tag{5.6}$$

namely, it is a linear combination of the modified Bessel functions $I_\nu(z)$ *and* $e^{i\pi\nu}K_\nu(z)$. *Denote* $\psi_\nu(z) := z^{1/2}\chi_\nu(z)$. *We have*

$$\psi_\nu(z+u) = \psi_\nu(z)\sum_{k=0}^{\infty}\frac{X_{\nu,k}(u)}{z^k} + \psi_{\nu-1}(z)\sum_{k=0}^{\infty}\frac{Y_{\nu,k}(u)}{z^k}, \tag{5.7}$$

$$\psi_{\nu-1}(z+u) = \psi_\nu(z)\sum_{k=0}^{\infty}\frac{Y_{\nu-1,k}(u)}{z^k} + \psi_{\nu-1}(z)\sum_{k=0}^{\infty}\frac{\tilde{X}_{\nu-1,k}(u)}{z^k}, \tag{5.8}$$

for all $|z| > 2(\nu^2+1)|u|$, *where* $X_{\nu,0}(u) = \cosh u$, $Y_{\nu,0}(u) = \sinh u$, $X_{\nu,1}(u) = (1/2-\nu)\sinh u$, $Y_{\nu,1}(u) = 0$, *and*

$$X_{\nu,k}(u) = (\nu^2 - \frac{1}{4})\sum_{j=0}^{k-2}(j+1)\int_0^u \sinh(u-v)(-v)^j X_{\nu,k-2-j}(v)dv, \tag{5.9}$$

$$Y_{\nu,k}(u) = (\nu^2 - \frac{1}{4})\sum_{j=0}^{k-2}(j+1)\int_0^u \sinh(u-v)(-v)^j Y_{\nu,k-2-j}(v)dv, \tag{5.10}$$

for $k \geq 2$. *Here, we denote* $\tilde{X}_{\nu-1,k}(u) = X_{\nu-1,k}(u) + 2(\nu-1)Y_{\nu-1,k-1}(u)$ *for* $k \geq 1$ *and* $\tilde{X}_{\nu-1,0}(u) = X_{\nu-1,0}(u) = \cosh u$. *For any* $\zeta \in \mathbf{C}$ *and* $N > 2(\nu^2+1)$, *we have*

$$\begin{aligned}&(N_\pm/N)^{1/2}\chi_\nu(N_\pm\zeta^{1/2})\\ =&\chi_\nu(N\zeta^{1/2})G^\pm(\zeta,N) \pm \zeta^{1/2}\chi_{\nu-1}(N\zeta^{1/2})H^\pm(\zeta,N),\end{aligned} \tag{5.11}$$

$$\begin{aligned}&(N_\pm/N)^{1/2}\chi_{\nu-1}(N_\pm\zeta^{1/2})\\ =&\pm\zeta^{1/2}\chi_\nu(N\zeta^{1/2})L^\pm(\zeta,N) + \chi_{\nu-1}(N\zeta^{1/2})K^\pm(\zeta,N),\end{aligned} \tag{5.12}$$

where

$$G^\pm(\zeta,N) = \sum_{s=0}^{\infty}\frac{(\pm1)^s G_s(\zeta)}{N^s}, \quad H^\pm(\zeta,N) = \sum_{s=0}^{\infty}\frac{(\pm1)^s H_s(\zeta)}{N^s}, \tag{5.13}$$

$$L^\pm(\zeta,N) = \sum_{s=0}^{\infty}\frac{(\pm1)^s L_s(\zeta)}{N^s}, \quad K^\pm(\zeta,N) = \sum_{s=0}^{\infty}\frac{(\pm1)^s K_s(\zeta)}{N^s}, \tag{5.14}$$

with

$$G_s(\zeta) = \frac{X_{\nu,k}(\zeta^{1/2})}{\zeta^{k/2}}, \quad H_s(\zeta) = \frac{Y_{\nu,k}(\zeta^{1/2})}{\zeta^{(k+1)/2}}, \tag{5.15}$$

$$L_s(\zeta) = \frac{Y_{\nu-1,k}(\zeta^{1/2})}{\zeta^{(k+1)/2}}, \quad K_s(\zeta) = \frac{\tilde{X}_{\nu-1,k}(\zeta^{1/2})}{\zeta^{k/2}}. \tag{5.16}$$

Moreover, G_s, H_s, L_s, K_s are entire functions in ζ.

Proof. Note that the modified Bessel function satisfies $\chi'_\nu(z) = \chi_{\nu-1}(z) - (\nu/z)\chi_\nu(z)$ and $\chi_{\nu-1}(z) - \chi_{\nu+1}(z) = (2\nu/z)\chi_\nu(z)$; see Olver *et al.* (2010, (10.29.1) and (10.29.2)). It is readily seen that $\psi_\nu(z) := z^{1/2}\chi_\nu(z)$ satisfies

$$\psi''_\nu(z) = (1 + \frac{\nu^2 - 1/4}{z^2})\psi_\nu(z), \tag{5.17}$$

$$\psi'_\nu(z) = (1/2 - \nu)\psi_\nu(z)/z + \psi_{\nu-1}(z), \tag{5.18}$$

$$\psi_{\nu-2}(z) = \psi_\nu(z) + [2(\nu-1)/z]\psi_{\nu-1}(z). \tag{5.19}$$

Substituting (5.7) into (5.17) gives

$$\begin{aligned}&\psi_\nu(z)\sum_{k=0}^{\infty}\frac{X''_{\nu,k}(u)}{z^k} + \psi_{\nu-1}(z)\sum_{k=0}^{\infty}\frac{Y''_{\nu,k}(u)}{z^k}\\ =&[1 + \frac{\nu^2 - 1/4}{(z+u)^2}][\psi_\nu(z)\sum_{k=0}^{\infty}\frac{X_{\nu,k}(u)}{z^k} + \psi_{\nu-1}(z)\sum_{k=0}^{\infty}\frac{Y_{\nu,k}(u)}{z^k}].\end{aligned}$$

We now make use of the expansion $(z+u)^{-2} = \sum_{j=0}^{\infty}(j+1)(-u)^j z^{-j-2}$. For each $k \geq 2$, comparing the coefficients of $\psi_\nu(z)/z^k$ on both sides of the above equation yields

$$X''_{\nu,k}(u) = X_{\nu,k}(u) + (\nu^2 - 1/4)\sum_{j=0}^{k-2}(j+1)(-u)^j X_{\nu,k-2-j}(u).$$

From (5.7) and (5.18) we observe $X_{\nu,k}(0) = 0$ and $X'_{\nu,k}(0) = 0$ for all $k \geq 2$. Solving the above differential equation gives (5.9). In a similar manner, we obtain (5.10) by comparing the coefficients of $\psi_{\nu-1}(z)/z^k$ on both sides of the equation following (5.19) and making use of $Y_{\nu,k}(0) = 0$ and $Y'_{\nu,k}(0) = 0$ for all $k \geq 2$. Replacing ν with $\nu - 1$ in (5.7) and making use of (5.19) give (5.8).

The above formal argument illustrates how we find the integral equations (5.9) and (5.10). To give a rigorous proof, we need to show that the series involved in the formal argument are convergent. Fix z and denote the right-hand side of (5.7) as $F_z(u)$. On account of $|\sinh v| \leq |v|e^{|v|}$, we can prove by induction that (exercise) $X_{\nu,k}(u)$ and $Y_{\nu,k}(u)/u$ are entire functions in u and

$$|X_{\nu,k}(u)| + |Y_{\nu,k}(u)/u| \leq 2[(\nu^2+1)|u|]^k e^{|u|}, \tag{5.20}$$

for $k \geq 0$. This implies that the series expansion in $F_z(u)$ is convergent if $|z| > 2(\nu^2+1)|u|$. It is easy to verify that $F_z(0) = \psi_\nu(z)$, $F_z'(0) = (1/2-\nu)\psi_\nu(z)/z + \psi_{\nu-1}(z) = \psi_\nu'(z)$, and

$$F_z''(u) = \left[1 + \frac{\nu^2 - 1/4}{(z+u)^2}\right] F_z(u).$$

The above differential equation is also satisfied by $\psi_\nu(z+u)$. By uniqueness, we have $\psi_\nu(z+u) = F_z(u)$, thus proving (5.7). Replacing ν by $\nu-1$ in (5.7), we obtain (5.8) from (5.19). It is obvious that $X_{\nu,k}(-u) = (-1)^k X_{\nu,k}(u)$, $Y_{\nu,k}(-u) = (-1)^{k+1} Y_{\nu,k}(u)$, and $\tilde{X}_{\nu,k}(-u) = (-1)^k \tilde{X}_{\nu,k}(u)$. By setting $z = N\zeta^{1/2}$ and $u = \pm\zeta^{1/2}$, we have (5.11) and (5.12) from (5.7) and (5.8), respectively. Clearly, $G_s(\zeta)$ is analytic for $\zeta \in \mathbf{C}\setminus(-\infty, 0]$ because $X_{\nu,k}(u)$ is an entire function in u. Since $X_{\nu,k}(-u) = (-1)^k X_{\nu,k}(u)$, $G_s(\zeta)$ can be analytically continued to the negative real line $(-\infty, 0)$. In view of (5.20), $\zeta = 0$ is a removable singularity for $G_s(\zeta)$. Similarly, we can show that $H_s(\zeta), L_s(\zeta), K_s(\zeta)$ are also entire functions. □

By choosing $\chi_\nu(z) = I_\nu(z)$ and $\chi_\nu(z) = e^{i\pi\nu}K_\nu(z)$, respectively, we obtain from (5.11) a linear system for $G^\pm$ and $H^\pm$, whose solution is

$$G^\pm = \frac{N_\pm^{1/2}[I_\nu(N_\pm\zeta^{1/2})K_{\nu-1}(N\zeta^{1/2}) + I_{\nu-1}(N\zeta^{1/2})K_\nu(N_\pm\zeta^{1/2})]}{N^{1/2}[I_\nu(N\zeta^{1/2})K_{\nu-1}(N\zeta^{1/2}) + I_{\nu-1}(N\zeta^{1/2})K_\nu(N\zeta^{1/2})]},$$

$$H^\pm = \frac{N_\pm^{1/2}[I_\nu(N_\pm\zeta^{1/2})K_\nu(N\zeta^{1/2}) - I_\nu(N\zeta^{1/2})K_\nu(N_\pm\zeta^{1/2})]}{\pm(N\zeta)^{1/2}[I_\nu(N\zeta^{1/2})K_{\nu-1}(N\zeta^{1/2}) + I_{\nu-1}(N\zeta^{1/2})K_\nu(N\zeta^{1/2})]}.$$

Now, we let $\zeta \to 0$. Since, $I_\nu(z) \sim z^\nu/[2^\nu\Gamma(\nu+1)]$ and $K_\nu(z) \sim z^{-\nu}2^{\nu-1}\Gamma(\nu)$ as $z \to 0$, we have

$$G^\pm(0,N) = (\frac{N_\pm}{N})^{1/2-\nu}, \tag{5.21}$$

$$H^\pm(0,N) = \frac{\pm N}{2\nu}[(\frac{N_\pm}{N})^{1/2+\nu} - (\frac{N_\pm}{N})^{1/2-\nu}]. \tag{5.22}$$

Similarly, we obtain from (5.12) that

$$L^\pm = \frac{N_\pm^{1/2}[I_{\nu-1}(N_\pm\zeta^{1/2})K_{\nu-1}(N\zeta^{1/2}) - I_{\nu-1}(N\zeta^{1/2})K_{\nu-1}(N_\pm\zeta^{1/2})]}{\pm(N\zeta)^{1/2}[I_\nu(N\zeta^{1/2})K_{\nu-1}(N\zeta^{1/2}) + I_{\nu-1}(N\zeta^{1/2})K_\nu(N\zeta^{1/2})]},$$

$$K^\pm = \frac{N_\pm^{1/2}[I_\nu(N\zeta^{1/2})K_{\nu-1}(N_\pm\zeta^{1/2}) + I_{\nu-1}(N_\pm\zeta^{1/2})K_\nu(N\zeta^{1/2})]}{N^{1/2}[I_\nu(N\zeta^{1/2})K_{\nu-1}(N\zeta^{1/2}) + I_{\nu-1}(N\zeta^{1/2})K_\nu(N\zeta^{1/2})]},$$

which, by setting $\zeta \to 0$, give

$$L^{\pm}(0,N) = \frac{\pm N}{2(\nu-1)}[(\frac{N_{\pm}}{N})^{\nu-1/2} - (\frac{N_{\pm}}{N})^{3/2-\nu}], \tag{5.23}$$

$$K^{\pm}(0,N) = (\frac{N_{\pm}}{N})^{\nu-1/2}. \tag{5.24}$$

The following proposition will be used in the remaining part of this chapter, and its proof is left as an exercise.

Proposition 5.1. *Let G_s, H_s, L_s and K_s be as given in Lemma 5.1. We have*

$$G_0(\zeta) = K_0(\zeta) = \cosh(\zeta^{1/2}), \quad H_0(\zeta) = L_0(\zeta) = \frac{\sinh(\zeta^{1/2})}{\zeta^{1/2}}, \tag{5.25}$$

and for each $s \geq 0$,

$$\frac{|G_s(\zeta)| + |K_s(\zeta)|}{1+|\zeta|^{1/2}} + |H_s(\zeta)| + |L_s(\zeta)| \leq C_s|H_0(\zeta)|, \tag{5.26}$$

for some constant C_s.

5.3 Formal asymptotic solutions

Let $\chi_\nu(z)$ be a solution of the modified Bessel equation (5.6), and $\zeta(x)$ be a real analytic and increasing function in (x_*, ∞) and with $\zeta(x^*) = 0$. We look for a solution to the difference equation (5.1) with the formal series expansion in (5.5); namely,

$$P_n(x) = N^{1/2}\chi_\nu(N\zeta^{1/2})\mathbb{A}(\zeta,N) + (N\zeta)^{1/2}\chi_{\nu-1}(N\zeta^{1/2})\mathbb{B}(\zeta,N), \tag{5.27}$$

where

$$\mathbb{A}(\zeta,N) := \sum_{r=0}^{\infty}\frac{A_r(\zeta)}{N^r}, \quad \mathbb{B}(\zeta,N) := \sum_{r=0}^{\infty}\frac{B_r(\zeta)}{N^r}. \tag{5.28}$$

For convenience, we introduce the notation

$$\Psi(x,N) := C_1(n)x + C_0(n) \sim \sum_{s=0}^{\infty}\frac{2\psi_s(x)}{N^s}, \quad \psi_s(x) := (a_s'x + b_s')/2. \tag{5.29}$$

Denote $N_{\pm} = N \pm 1$. It follows from Lemma 5.1 and (5.27) that

$$\begin{aligned} P_{n\pm1}(x) = {} & N^{1/2}[\chi_\nu(N\zeta^{1/2})G^{\pm} \pm \zeta^{1/2}\chi_{\nu-1}(N\zeta^{1/2})H^{\pm}]\mathbb{A}(\zeta,N_{\pm}) \\ & + (N\zeta)^{1/2}[\pm\zeta^{1/2}\chi_\nu(N\zeta^{1/2})L^{\pm} + \chi_{\nu-1}(N\zeta^{1/2})K^{\pm}]\mathbb{B}(\zeta,N_{\pm}). \end{aligned} \tag{5.30}$$

Substituting (5.27) and (5.30) into (5.1) and equating the coefficients of $\chi_\nu(N\zeta^{1/2})$ and $\chi_{\nu-1}(N\zeta^{1/2})$ to zero, we obtain two equations

$$\begin{aligned}\Psi(x,N)\mathbb{A}(\zeta,N) =&G^+(\zeta,N)\mathbb{A}(\zeta,N_+)+G^-(\zeta,N)\mathbb{A}(\zeta,N_-)\\ &+\zeta L^+(\zeta,N)\mathbb{B}(\zeta,N_+)-\zeta L^-(\zeta,N)\mathbb{B}(\zeta,N_-),\end{aligned}\tag{5.31}$$

$$\begin{aligned}\Psi(x,N)\mathbb{B}(\zeta,N) =&H^+(\zeta,N)\mathbb{A}(\zeta,N_+)-H^-(\zeta,N)\mathbb{A}(\zeta,N_-)\\ &+K^+(\zeta,N)\mathbb{B}(\zeta,N_+)+K^-(\zeta,N)\mathbb{B}(\zeta,N_-).\end{aligned}\tag{5.32}$$

For $p=0,1,\cdots$, by equating the coefficients of N^{-p} in the series expansions of (5.31) and (5.32), we obtain

$$\begin{aligned}\sum_{s\le p}\psi_s A_{p-s} =& \sum_{2s\le p} A_{p-2s}\sum_{i=0}^{2s}\binom{2s-p}{2s-i}G_i\\ &+\sum_{2s+1\le p} B_{p-2s-1}\sum_{i=0}^{2s+1}\binom{2s+1-p}{2s+1-i}\zeta L_i,\end{aligned}\tag{5.33}$$

$$\begin{aligned}\sum_{s\le p}\psi_s B_{p-s} =& \sum_{2s+1\le p} A_{p-2s-1}\sum_{i=0}^{2s+1}\binom{2s+1-p}{2s+1-i}H_i\\ &+\sum_{2s\le p} B_{p-2s}\sum_{i=0}^{2s}\binom{2s-p}{2s-i}K_i.\end{aligned}\tag{5.34}$$

Setting $p=0$ gives $(\psi_0-G_0)A_0=(\psi_0-K_0)B_0=0$. We may assume without loss of generality that either A_0 or B_0 is nonzero (exercise). It then follows from Proposition 5.1 that $\cosh(\zeta^{1/2})=\psi_0$; namely,

$$\zeta(x)=\left(\log[\psi_0(x)+\sqrt{\psi_0^2(x)-1}]\right)^2,\tag{5.35}$$

which is analytic for $x\in\mathbf{C}\setminus(-\infty,x^*]$ and can be analytically continued to the interval $(x_*,x^*]$.

From equations (5.33) and (5.34) with $p\ge 1$, we obtain

$$\begin{aligned}(p-1)\zeta L_0 B_{p-1} =& \sum_{2\le 2s\le p} A_{p-2s}\sum_{i=0}^{2s}\binom{2s-p}{2s-i}G_i\\ &+\sum_{3\le 2s+1\le p} B_{p-2s-1}\sum_{i=0}^{2s+1}\binom{2s+1-p}{2s+1-i}\zeta L_i-\sum_{2\le s\le p}\psi_s A_{p-s},\end{aligned}\tag{5.36}$$

$$\begin{aligned}(p-1)H_0 A_{p-1} =& \sum_{3\le 2s+1\le p} A_{p-2s-1}\sum_{i=0}^{2s+1}\binom{2s+1-p}{2s+1-i}H_i\\ &+\sum_{2\le 2s\le p} B_{p-2s}\sum_{i=0}^{2s}\binom{2s-p}{2s-i}K_i-\sum_{2\le s\le p}\psi_s B_{p-s},\end{aligned}\tag{5.37}$$

where we have used the facts that $\psi_0=G_0=K_0$ and $\psi_1=L_1=H_1=0$.

5.4 Bounds for the coefficients

To establish global uniformity of our anticipated expansion, we need to estimate the coefficients $A_s(\zeta)$ and $B_s(\zeta)$ in (5.5).

Lemma 5.2. *Let $\zeta(x)$ be as in (5.35), and suppose that $|A_0(\zeta)|$ and $(1+|\zeta|^{1/2})|B_0(\zeta)|$ are bounded for $x \geq x_* + \delta$. Furthermore, let $A_s(\zeta)$ and $B_s(\zeta)$ be successively defined as in (5.36)-(5.37). Then there exists a positive constant C_s independent of x such that*

$$|A_s(\zeta)| + (1+|\zeta|^{1/2})|B_s(\zeta)| \leq C_s, \quad s = 1, 2, \cdots, \tag{5.38}$$

for $x \geq x_* + \delta$, $\delta > 0$.

Proof. On account of the estimates of G_s, H_s, L_s and K_s in Proposition 5.1, we can use (5.36)-(5.37) and an inductive argument to verify (5.38) provided that the functions $A_s(\zeta)$ and $B_s(\zeta)$ are well defined (i.e., $B_p(\zeta)$ is bounded at $\zeta = 0$). To show that $B_p(0)$ is bounded, we divide our discussion into three cases: (i) $2\nu \neq 0, 1, 2, \cdots$, (ii) $2\nu = 1, 2, 3, \cdots$, and (iii) $2\nu = 0$.

In case (i), we first consider the second-order difference equation

$$y_{N+1} + y_{N-1} - \Psi(x^*, N) y_N = 0, \tag{5.39}$$

where Ψ is given in (5.29). Especially, the first few terms in the expansion of $\Psi(x^*, N)$ are $2\psi_0(x^*) = 2$, $2\psi_1(x^*) = 0$, $2\psi_2(x^*) = a_2'x^* + b_2' = \nu^2 - 1/4$, and $2\psi_3(x^*) = a_3'x^* + b_3' = a_3x^* + b_3 + 2\tau_0(a_2x^* + b_2)$. This difference equation corresponds to the exceptional case studied in Chapter 1, and it has two linearly independent solutions with asymptotic expansions (exercise):

$$y_N^{(1)} \sim N^{1/2-\nu} \sum_{s=0}^{\infty} \frac{c_s}{N^s}, \qquad y_N^{(2)} \sim N^{1/2+\nu} \sum_{s=0}^{\infty} \frac{d_s}{N^s}, \tag{5.40}$$

where $c_0 = d_0 = 1$, and

$$c_1 = \frac{a_3'x^* + b_3'}{1+2\nu}, \qquad d_1 = \frac{a_3'x^* + b_3'}{1-2\nu}. \tag{5.41}$$

By choosing

$$\tau_0 = -\frac{a_3'x^* + b_3'}{2(a_2x^* + b_2)} = -\frac{a_3'x^* + b_3'}{2(\nu^2 - 1/4)}, \tag{5.42}$$

we have $c_1 = d_1 = 0$. We shall prove that for all $p \geq 0$,

$$A_p(0) = c_p, \qquad B_p(0) = \frac{d_{p+1} - c_{p+1}}{2\nu}. \tag{5.43}$$

Especially, $B_0(0) = 0$. To this end, we set $\zeta = 0$ in (5.31)-(5.32) and make use of (5.21)-(5.24) to find

$$x_{N+1} + x_{N-1} = \Psi(x^*, N)x_N, \qquad \hat{x}_{N+1} + \hat{x}_{N-1} = \Psi(x^*, N)\hat{x}_N, \tag{5.44}$$

where

$$x_N := N^{1/2-\nu}\mathbb{A}(0, N), \quad \hat{x}_N := N^{\nu-1/2}[N\mathbb{A}(0, N) + 2\nu\mathbb{B}(0, N)]. \tag{5.45}$$

It then follows from (5.40) that $x_N = y_N^{(1)}$ and $\hat{x}_N = y_N^{(2)}$. Consequently, we obtain (5.43) by comparing the coefficients in the asymptotic expansions of $\mathbb{A}(0, N)$, $\mathbb{B}(0, N)$, $y_N^{(1)}$ and $y_N^{(2)}$.

In case (ii), i.e., $2\nu = 1, 2, \cdots$, the two linearly independent asymptotic solutions of (5.39) are (exercise)

$$y_N^{(1)} \sim N^{1/2-\nu} \sum_{s=0}^{\infty} \frac{c_s}{N^s}, \quad y_N^{(2)} \sim N^{1/2+\nu} \sum_{s=0}^{\infty} \frac{d_s}{N^s} + C y_N^{(1)} \log N, \tag{5.46}$$

where C is a constant. In a similar manner, we have $x_N = y_N^{(1)}$ and $\hat{x}_N = 0$. Comparing the coefficients in asymptotic expansions yields

$$A_s(0) = c_s, \quad 2\nu B_s(0) = -A_{s+1}(0), \quad s = 0, 1, \cdots. \tag{5.47}$$

When $\nu \neq 1/2$, we can still choose τ_0 as in (5.42) so that $B_0(0) = 0$. When $\nu = 1/2$, the modified Bessel functions reduce to the hyperbolic functions, and we can choose τ_0 to be any real number.

In case (iii), i.e., $\nu = 0$, the two linearly independent asymptotic solutions of (5.39) are

$$y_N^{(1)} \sim N^{1/2} \sum_{s=0}^{\infty} \frac{c_s}{N^s}, \quad y_N^{(2)} \sim N^{1/2} \sum_{s=0}^{\infty} \frac{d_s}{N^s} + y_N^{(1)} \log N, \tag{5.48}$$

and one can show that

$$N^{1/2}\mathbb{A}(0, N) = y_N^{(1)}, \quad N^{-1/2}[(N \log N)\mathbb{A}(0, N) + \mathbb{B}(0, N)] = y_N^{(2)} - y_N^{(1)},$$

from which it follows

$$A_s(0) = c_s, \quad B_s(0) = d_{s+1} - c_{s+1}, \quad s = 0, 1, \cdots. \tag{5.49}$$

Here, we can also choose τ_0 as in (5.42) so that $B_0(0) = 0$. The proof of the lemma is now complete. □

5.5 Asymptotic nature of the expansion

Note that the variable x in the difference equation (5.1) is fixed. A solution to (5.1) multiplied by any function of x remains a solution. For each $p \geq 0$, we truncate the expansion (5.5), multiply it by

$$\Lambda(x) := H_0^{-1/2}(\zeta) = [\zeta/(\psi_0^2 - 1)]^{1/4}, \tag{5.50}$$

and replace χ_ν with I_ν and $e^{i\pi\nu}K_\nu$, respectively. This gives us two asymptotic solutions

$$r_n^p(x) := \Lambda N^{1/2}[I_\nu(N\zeta^{1/2})\mathbb{A}_p(\zeta, N) + \zeta^{1/2}I_{\nu-1}(N\zeta^{1/2})\mathbb{B}_p(\zeta, N)], \tag{5.51}$$

$$s_n^p(x) := \Lambda N^{1/2}[K_\nu(N\zeta^{1/2})\mathbb{A}_p(\zeta, N) - \zeta^{1/2}K_{\nu-1}(N\zeta^{1/2})\mathbb{B}_p(\zeta, N)], \tag{5.52}$$

where

$$\mathbb{A}_p(\zeta, N) = \sum_{s=0}^{p} \frac{A_s(\zeta)}{N^s}, \quad \mathbb{B}_p(\zeta, N) = \sum_{s=0}^{p} \frac{B_s(\zeta)}{N^s}. \tag{5.53}$$

It is worth recalling that $2\psi_0 = a_0'x + b_0'$ and $\Psi(x, N) := C_1(n)x + C_0(n)$. From Lemma 5.1, it follows that

$$\begin{aligned} \frac{R_n^p(x)}{N^{p+3/2}} :=& r_{n+1}^p(x) - [C_1(n)x + C_0(n)]r_n^p(x) + r_{n-1}^p(x) \\ =& \Lambda N^{1/2}[I_\nu(N\zeta^{1/2})F_{1,n}^p(x) + \zeta^{1/2}I_{\nu-1}(N\zeta^{1/2})F_{2,n}^p(x)], \end{aligned} \tag{5.54}$$

$$\begin{aligned} \frac{S_n^p(x)}{N^{p+3/2}} :=& s_{n+1}^p(x) - [C_1(n)x + C_0(n)]s_n^p(x) + s_{n-1}^p(x) \\ =& \Lambda N^{1/2}[K_\nu(N\zeta^{1/2})F_{1,n}^p(x) - \zeta^{1/2}K_{\nu-1}(N\zeta^{1/2})F_{2,n}^p(x)], \end{aligned} \tag{5.55}$$

where

$$\begin{aligned} F_{1,n}^p(x) =& G^+(\zeta, N)\mathbb{A}_p(\zeta, N_+) + G^-(\zeta, N)\mathbb{A}_p(\zeta, N_-) - \Psi(x, N)\mathbb{A}_p(\zeta, N) \\ &+ \zeta L^+(\zeta, N)\mathbb{B}_p(\zeta, N_+) - \zeta L^-(\zeta, N)\mathbb{B}_p(\zeta, N_-), \end{aligned} \tag{5.56}$$

$$\begin{aligned} F_{2,n}^p(x) =& H^+(\zeta, N)\mathbb{A}_p(\zeta, N_+) - H^-(\zeta, N)\mathbb{A}_p(\zeta, N_-) - \Psi(x, N)\mathbb{B}_p(\zeta, N) \\ &+ K^+(\zeta, N)\mathbb{B}_p(\zeta, N_+) + K^-(\zeta, N)\mathbb{B}_p(\zeta, N_-). \end{aligned} \tag{5.57}$$

For $x \geq x_* + \delta$, on account of Lemma 5.2, we have (exercise)

$$|F_{1,n}^p(x)| + [1 + |\zeta(x)|^{1/2}]|F_{2,n}^p(x)| \leq \frac{C_p(1 + |x|)}{N^{p+2}}, \tag{5.58}$$

for some constant C_p. Substituting this into (5.54) and (5.55) yields

$$|R_n^p(x)| \leq C_p(1 + |x|)\alpha_n(x), \quad |S_n^p(x)| \leq C_p(1 + |x|)\beta_n(x), \tag{5.59}$$

for $x \geq x_* + \delta$, where

$$\alpha_n(x) := \Lambda(x)\left[|I_\nu(N\zeta^{1/2})| + \frac{|\zeta|^{1/2}}{1+|\zeta|^{1/2}}|I_{\nu-1}(N\zeta^{1/2})|\right], \tag{5.60}$$

$$\beta_n(x) := \Lambda(x)\left[|K_\nu(N\zeta^{1/2})| + \frac{|\zeta|^{1/2}}{1+|\zeta|^{1/2}}|K_{\nu-1}(N\zeta^{1/2})|\right]. \tag{5.61}$$

Lemma 5.3. *For each $p \geq 0$, there exist $\{\varepsilon_n^p(x)\}_{n=1}^\infty$ and $\{\delta_n^p(x)\}_{n=1}^\infty$ such that*

$$\varepsilon_{n+1}^p(x) - [C_1(n)x + C_0(n)]\varepsilon_n^p(x) + \varepsilon_{n-1}^p(x) = -\frac{R_n^p(x)}{N^{p+3/2}}, \tag{5.62}$$

$$\delta_{n+1}^p(x) - [C_1(n)x + C_0(n)]\delta_n^p(x) + \delta_{n-1}^p(x) = -\frac{S_n^p(x)}{N^{p+3/2}}, \tag{5.63}$$

with

$$|\delta_n^p(x)| \leq \frac{M_p\tilde{\beta}_n(x)}{N^{p+1/2}}, \quad |\varepsilon_n^p(x)| \leq \frac{M_p\tilde{\alpha}_n(x)}{N^{p+1/2}}, \tag{5.64}$$

for some constant M_p, where

$$\tilde{\alpha}_n(x) := \Lambda(x)[|I_\nu(N\zeta^{1/2})| + |I_{\nu-1}(N\zeta^{1/2})|], \tag{5.65}$$

$$\tilde{\beta}_n(x) := \Lambda(x)[|K_\nu(N\zeta^{1/2})| + |K_{\nu-1}(N\zeta^{1/2})|]. \tag{5.66}$$

Proof. It can be shown that the solution of the difference equation (5.63) with asymptotic bound in (5.64) is the same as the solution to the following summation equation with the same asymptotic bound:

$$\begin{aligned}\delta_n^p(x) = &\sum_{j=n+1}^\infty \frac{[r_n^p(x)s_j^p(x) - s_n^p(x)r_j^p(x)]S_j^p(x)}{[r_{n+1}^p(x)s_n^p(x) - r_n^p(x)s_{n+1}^p(x)](j+\tau_0)^{p+3/2}}\\ &+ \sum_{j=n+1}^\infty \frac{[r_n^p(x)S_j^p(x) - s_n^p(x)R_j^p(x)]\delta_j^p(x)}{[r_{n+1}^p(x)s_n^p(x) - r_n^p(x)s_{n+1}^p(x)](j+\tau_0)^{p+3/2}}.\end{aligned} \tag{5.67}$$

We will construct such a solution by successive approximation. Starting with $\delta_{n,0}^p = 0$, we define $\delta_{n,k}^p(x)$ recursively by

$$\begin{aligned}\delta_{n,k}^p(x) = &\sum_{j=n+1}^\infty \frac{[r_n^p(x)s_j^p(x) - s_n^p(x)r_j^p(x)]S_j^p(x)}{[r_{n+1}^p(x)s_n^p(x) - r_n^p(x)s_{n+1}^p(x)](j+\tau_0)^{p+3/2}}\\ &+ \sum_{j=n+1}^\infty \frac{[r_n^p(x)S_j^p(x) - s_n^p(x)R_j^p(x)]\delta_{j,k-1}^p(x)}{[r_{n+1}^p(x)s_n^p(x) - r_n^p(x)s_{n+1}^p(x)](j+\tau_0)^{p+3/2}}.\end{aligned} \tag{5.68}$$

Since $A_s(\zeta)$ and $(1+|\zeta|^{1/2})B_s(\zeta)$ are bounded for $x \geq x_* + \delta$, it follows from (5.51) and (5.52) that

$$|r_n^p(x)| \leq CN^{1/2}\alpha_n(x), \quad |s_n^p(x)| \leq CN^{1/2}\beta_n(x), \tag{5.69}$$

for some constant C. Moreover, since $(1+|x|)\alpha_n(x)\beta_n(x)$ is uniformly bounded for $x \geq x_* + \delta$ (exercise), we have from (5.59) and (5.69)

$$|R_n^p(x)s_n^p(x)| + |S_n^p(x)r_n^p(x)| \leq M'N^{1/2}, \tag{5.70}$$

for some constant M'. It can be verified (exercise) that

$$|r_{n+1}^p(x)s_n^p(x) - r_n^p(x)s_{n+1}^p(x) - 1| \leq \frac{2M'}{pN^p}, \tag{5.71}$$

which in turn gives

$$|r_{n+1}^p(x)s_n^p(x) - r_n^p(x)s_{n+1}^p(x)| > \frac{1}{2}, \tag{5.72}$$

for all sufficiently large n. Substituting this into (5.68), we obtain from (5.69) and monotonicity properties of modified Bessel functions I_ν and K_ν that (exercise)

$$|\delta_{n,1}^p(x)| \leq M''\tilde{\beta}_n(x) \sum_{j=n+1}^{\infty} \frac{N^{1/2}}{(j+\tau_0)^{p+1}} \leq \frac{M''\tilde{\beta}_n(x)}{pN^{p-1/2}}, \tag{5.73}$$

for some constant M''. Similarly, we obtain

$$|\delta_{n,k}^p(x) - \delta_{n,k-1}^p(x)| \leq \frac{M''\tilde{\beta}_n(x)}{pN^{p-1/2}} \sup_{j\geq n} \frac{|\delta_{j,k-1}^p(x) - \delta_{j,k-2}^p(x)|}{(j+\tau_0)^{1/2}\beta_j(x)}. \tag{5.74}$$

By induction, we have from the above two estimates

$$\frac{|\delta_{n,k}^p(x) - \delta_{n,k-1}^p(x)|}{N^{1/2}\tilde{\beta}_n(x)} \leq \left(\frac{M''}{pN^p}\right)^k, \tag{5.75}$$

from which it follows that, for all $n \geq 2M'' - \tau_0$,

$$\delta_{n,k}^p(x) = \sum_{j=1}^{k} [\delta_{n,j}^p(x) - \delta_{n,j-1}^p(x)] \tag{5.76}$$

converges as $k \to \infty$. Clearly, the limit function $\delta_n^p(x)$ satisfies (5.67) as well as (5.63). Moreover, $|\delta_n^p(x)| \leq 2N^{1/2}\tilde{\beta}_n(x)$. Using a similar argument as in the proof of (5.73) and (5.74), we obtain

$$|\delta_n^p(x)| \leq \frac{M_p'\tilde{\beta}_n(x)}{N^{p-1/2}}, \tag{5.77}$$

for some constant M_p'. Define $Q_n^p(x) := s_n^p(x) + \delta_n^p(x)$. One could prove that $Q_n^p(x)$ is independent of p (exercise). Especially, $\delta_n^p(x) = \delta_n^{p+1}(x) + s_n^{p+1}(x) - s_n^p(x)$, which, together with (5.52) and (5.77), implies the first estimate in (5.64). To prove the second estimate in (5.64), we define $P_n(x) := r_n^p(x) + \varepsilon_n^p(x)$ and $Q_n(x) := s_n^p(x) + \delta_n^p(x)$. It is obvious that $P_n(x)$ and $Q_n(x)$ are two solutions to the difference equation (5.1). Making use of the fact that $\varepsilon_n^0(x) \to 0$ and $\delta_n^0(x) \to 0$ as $n \to \infty$, one can prove that (exercise)

$$P_{n+1}(x)Q_n(x) - P_n(x)Q_{n+1}(x) = 1. \tag{5.78}$$

Especially, $P_n(x)$ and $Q_n(x)$ are two linearly independent solutions to (5.1). Now we choose $m = \lfloor n/2 \rfloor$ and derive another identity from the difference equations for ε_n^p, $P_n(x)$ and $Q_n(x)$ (exercise):

$$\begin{aligned}\varepsilon_n^p(x) =& [r_{m+1}^p(x)\varepsilon_m^p(x) - r_m^p(x)\varepsilon_{m+1}^p(x)]Q_n(x) \\ &+ \sum_{j=n+1}^{\infty} \frac{[r_n^p(x) + \varepsilon_n^p(x)]R_j^p(x)Q_j(x)}{(j+\tau_0)^{p+3/2}} \\ &+ \sum_{j=m+1}^{n} \frac{Q_n(x)[R_j^p(x)r_j^p(x) + R_j^p(x)\varepsilon_j^p(x)]}{(j+\tau_0)^{p+3/2}}.\end{aligned} \tag{5.79}$$

The second estimate in (5.64) can again be proved by using the successive approximation method. □

Summarizing the above results gives the following theorem.

Theorem 5.1. *Assume that the coefficients $C_1(n)$ and $C_0(n)$ in the linear difference equation* (5.1) *have the asymptotic expansions of the form given in* (5.2) *with $a_0 > 0$ and $a_1 = b_1 = 0$. Let $x^* = (2+b_0)/a_0$ and $x_* = (-2+b_0)/a_0$ be the transition points. Set $N = n + \tau_0$ with $\tau_0 = -(a_3x^* + b_3)/(2a_2x^* + 2b_2)$. Let $\nu^2 - 1/4 = a_2x^* + b_2$ and $\zeta(x)$ be given as in* (5.35)*. Then equation* (5.1) *has, for each integer $p \geq 0$, a pair of linearly independent solutions $P_n(x)$ and $Q_n(x)$ such that*

$$\begin{aligned}P_n(x) =& \left(\frac{4\zeta}{(a_0t+b_0)^2-4}\right)^{1/4}\Bigg[N^{1/2}I_\nu(N\zeta^{1/2})\sum_{s=0}^{p}\frac{A_s(\zeta)}{N^s} \\ &+ (N\zeta)^{1/2}I_{\nu-1}(N\zeta^{1/2})\sum_{s=0}^{p}\frac{B_s(\zeta)}{N^s} + \varepsilon_p(x,N)\Bigg],\end{aligned} \tag{5.80}$$

$$\begin{aligned}Q_n(x) =& \left(\frac{4\zeta}{(a_0t+b_0)^2-4}\right)^{1/4}\Bigg[N^{1/2}K_\nu(N\zeta^{1/2})\sum_{s=0}^{p}\frac{A_s(\zeta)}{N^s} \\ &+ (N\zeta)^{1/2}K_{\nu-1}(N\zeta^{1/2})\sum_{s=0}^{p}\frac{B_s(\zeta)}{N^s} + \delta_p(x,N)\Bigg],\end{aligned} \tag{5.81}$$

where the error terms satisfy

$$|\varepsilon_p(t,N)| \leq \frac{M_p}{N^{p+1/2}}[|I_\nu(N\zeta^{1/2})| + |I_{\nu-1}(N\zeta^{1/2})|], \tag{5.82}$$

$$|\delta_p(t,N)| \leq \frac{M_p}{N^{p+1/2}}[|K_\nu(N\zeta^{1/2})| + |K_{\nu-1}(N\zeta^{1/2})|], \tag{5.83}$$

for $x \geq x_* + \delta$. Here, M_p is a positive constant and the coefficients $A_s(\zeta)$ and $B_s(\zeta)$ can be determined successively for any given $A_0(\zeta)$ and $B_0(\zeta)$; see (5.36)-(5.37).

5.6 Exercises

1. Verify (5.20).
2. Prove Proposition 5.1.
3. Let $P_n(x)$ in (5.27) be a solution to (5.1) with series expansions in (5.28). Assume $A_s = B_s = 0$ for all $s < p$ and either A_p or B_p is nonzero. Show that (5.33)-(5.34) become $(\psi_0 - G_0)A_p = (\psi_0 - K_0)B_p = 0$, from which we have $\cosh(\zeta^{1/2}) = \psi_0$.
4. Prove that $\zeta(x)$ defined in (5.35) is real analytic and increasing for $x \geq x_* + \delta$.
5. Show that the difference equation (5.39) has two linearly independent asymptotic solutions (5.40) for case (i): $2\nu \neq 0, 1, 2, \cdots$.
6. Show that the difference equation (5.39) has two linearly independent asymptotic solutions (5.46) for case (ii): $2\nu = 1, 2, \cdots$.
7. Prove (5.58).
8. Let $\alpha_n(x)$ and $\beta_n(x)$ be given as in (5.60)-(5.61). Prove that $(1 + |x|)\alpha_n(x)\beta_n(x)$ is uniformly bounded for $x \geq x_* + \delta$.
9. Verify (5.71).
10. Prove (5.73).
11. Let $r_n^p(x)$ and $s_n^p(x)$ with $p \geq 0$ be defined as in (5.51) and (5.52). If $P_n(x)$ and $Q_n(x)$ are two solutions to (5.1) such that $P_n(x) - r_n^0(x) \to 0$ and $Q_n(x) - s_n^0(x) \to 0$ as $n \to \infty$, prove that $P_{n+1}(x)Q_n(x) - P_n(x)Q_{n+1}(x) = 1$. Moreover, show that such pair of solutions are unique; namely, if $P_n^*(x)$ and $Q_n^*(x)$ are two solutions to (5.1) such that $P_n^*(x) - r_n^0(x) \to 0$ and $Q_n^*(x) - s_n^0(x) \to 0$ as $n \to \infty$, then $P_n^*(x) = P_n(x)$ and $Q_n^*(x) = Q_n(x)$.
12. Verify (5.79).

Chapter 6

Transition Point at the Origin

6.1 Introduction

After detailed discussions of cases (i) and (iii) in Chapters 4 and 5, we finally come to case (ii); namely, the case in which the exponent θ in (4.4) is not equal to zero and one of the transition points (see the line following (4.6)) occurs at the origin, i.e., $\theta \neq 0$ and $t^* = 0$. Again we recall the second-order linear difference equation (4.3):

$$P_{n+1}(x) - [C_1(n)x + C_0(n)]P_n(x) + P_{n-1}(x) = 0, \tag{6.1}$$

where

$$C_1(n) \sim n^{-\theta} \sum_{s=0}^{\infty} \frac{a_s}{n^s}, \quad C_0(n) \sim \sum_{s=0}^{\infty} \frac{b_s}{n^s}, \tag{6.2}$$

with $a_0 \neq 0$. As before, we let τ_0 be a constant, the value of which will be determined later, and put $N := n + \tau_0$. A simple computation shows that the expansions in (6.2) can be rewritten in the form

$$C_1(n) \sim N^{-\theta} \sum_{s=0}^{\infty} \frac{a'_s}{N^s}, \quad C_0(n) \sim \sum_{s=0}^{\infty} \frac{b'_s}{N^s}, \tag{6.3}$$

with $a'_0 = a_0$, $a'_1 = a_1 + a_0\theta\tau_0$, $b'_0 = b_0$ and $b'_1 = b_1$. The characteristic equation associated with (6.1) is the quadratic equation

$$\rho^2 - (a'_0 t + b'_0)\rho + 1 = 0; \tag{6.4}$$

see (4.6) or (5.3). The two roots of equation (6.4) coincide when $t = t_*$ or $t = t^*$, where

$$a'_0 t^* + b'_0 = 2, \quad a'_0 t_* + b'_0 = -2. \tag{6.5}$$

As in Chapters 4 and 5, the values of these two roots are called the transition points (turning points).

In case (i), we found that the dominant approximants in the asymptotic solutions to (6.1) involve Airy functions $\mathrm{Ai}(\zeta)$ and $\mathrm{Bi}(\zeta)$; in case (iii), the dominant approximants are modified Bessel functions $I_\nu(\zeta)$ and $K_\nu(\zeta)$. To have some idea what type of special functions we can anticipate to appear in case (ii), we first consider the three-term recurrence relation

$$(n+1)L_{n+1}^{(\alpha)}(x) + [x-(2n+\alpha+1)]L_n^{(\alpha)}(x) + (n+\alpha)L_{n-1}^{(\alpha)}(x) = 0 \quad (6.6)$$

of the Laguerre polynomials $L_n^{(\alpha)}(x)$. In terms of the notations used in Section 1 of Chapter 4, we have

$$c_1(n) = -\frac{1}{n+1}, \quad c_0(n) = \frac{2n+\alpha+1}{n+1}, \quad c_2(n) = \frac{n+\alpha}{n+1}.$$

To convert equation (6.6) into our standard form (6.1), we introduce the sequence suggested in Chapter 4:

$$\frac{K_{n+1}}{K_{n-1}} = c_2(n) = \frac{n+\alpha}{n+1}.$$

From this equation, one can easily obtain

$$K_{2n} = \frac{2n-1+\alpha}{2n}\cdot\frac{2n-3+\alpha}{2n-2}\cdots\frac{3+\alpha}{4}\cdot\frac{1+\alpha}{2}K_0,$$
$$K_{2n+1} = \frac{2n+\alpha}{2n+1}\cdot\frac{2n-2+\alpha}{2n-1}\cdots\frac{4+\alpha}{5}\cdot\frac{2+\alpha}{3}K_1,$$

where K_0 and K_1 are constants. The coefficients in (6.1) are given by

$$C_1(n) = c_1(n)\frac{K_n}{K_{n+1}} = n^{-1}[-1+O(\frac{1}{n})],$$
$$C_0(n) = c_0(n)\frac{K_n}{K_{n+1}} = 2+O(\frac{1}{n}).$$

Comparing these two equations with the expansions in (6.2) and (6.3) immediately gives the values

$$\theta = 1, \quad a_0' = a_0 = -1, \quad b_0' = b_0 = 2.$$

Plugging these values into the first equation in (6.5), we readily see that the value of the transition point t^* is zero; i.e., $t^* = 0$. This tells us that Laguerre polynomial is just an example of case (ii). Using differential equation theory, Erdélyi (1960) had shown long ago that for $\alpha \geq 0$ and $0 < a \leq t < \infty$, we have

$$L_n^{(\alpha)}(x) \sim \frac{(-1)^n \pi^{1/2} 2^{5/6} N^{N+1/6} e^{x/2}}{n!(\zeta')^{1/2} x^{\alpha/2+1/2} e^N}\,\mathrm{Ai}(\nu^{2/3}\zeta), \quad (6.7)$$

where $\nu = 4N = 4n + 2\alpha + 2$, $t = x/\nu$ and

$$\begin{cases} \frac{2}{3}\zeta^{3/2} = \frac{1}{2}\sqrt{x(x-1)} - \frac{1}{2}\ln(\sqrt{x} + \sqrt{x-1}), & x \geq 1, \\ \frac{2}{3}(-\zeta)^{3/2} = \frac{1}{2}[\arccos\sqrt{x} - \sqrt{x(1-x)}], & 0 < x \leq 1. \end{cases} \tag{6.8}$$

This result has also been derived by using an integral method; see Frenzen and Wong (1988). The presentation of the material in the remaining parts of this chapter is based on the work of Cao and Li (2014).

In general, case (ii) corresponds to $a_0 \neq 0$ and $b_0^2 = 4$. By symmetry, we may assume without loss of generality that $a_0 < 0$ and $b_0 = 2$. If $a_0 > 0$ and $b_0 = 2$, we consider the difference equation satisfied by $P_n(-x)$. If $a_0 > 0$ and $b_0 = -2$, we consider the difference equation satisfied by $(-1)^n P_n(x)$. If $a_0 < 0$ and $b_0 = -2$, we consider the difference equation satisfied by $(-1)^n P_n(-x)$. By choosing $\tau_0 = -a_1/(a_0\theta)$, we obtain $a_1' = a_1 + a_0\theta\tau_0 = 0$, where a_1' is the coefficient in the expansion (6.3). As in Chapters 4 and 5, we further assume $b_1 = 0$ so that $b_1' = b_1 = 0$. Let $x = N^\theta t$ with $N = n + \tau_0$. It follows from (6.3) that

$$C_1(n)x + C_0(n) = \sum_{s=0}^{\infty} \frac{a_s' t + b_s'}{N^s} =: \sum_{s=0}^{\infty} \frac{2\psi_s(t)}{N^s}, \tag{6.9}$$

where $\psi_s(t) := (a_s' t + b_s')/2$ for $s = 0, 1, \cdots$. Especially, $\psi_0(t) = (a_0 t + b_0)/2$ and $\psi_1(t) = 0$.

To determine which special function appears in the uniform asymptotic expansions of the solutions to (6.1) in the present case, we try a solution of the form $P_n(N^\theta t) \sim \chi(Z)$, where $Z = N\eta^\lambda(t)$ with $\eta(0) = 0$. Let $N_\pm = N \pm 1$ and

$$t_\pm = N^\theta t / N_\pm^\theta \sim t \mp \frac{\theta t}{N} + \frac{\theta(\theta+1)t}{2N^2}.$$

We have

$$\eta(t_\pm) \sim \eta(t) \mp \frac{\theta t \eta'(t)}{N} + \frac{\theta(\theta+1)t\eta'(t)}{2N^2} + \frac{\theta^2 t^2 \eta''(t)}{2N^2},$$

and

$$\frac{\eta^\lambda(t_\pm)}{\eta^\lambda(t)} \sim 1 \mp \frac{\lambda\theta t\eta'(t)}{N\eta(t)} + \frac{\lambda\theta(\theta+1)t\eta'(t)}{2N^2\eta(t)} + \frac{\lambda\theta^2 t^2\eta''(t)}{2N^2\eta(t)} + \frac{\lambda(\lambda-1)\theta^2 t^2[\eta'(t)]^2}{2N^2\eta^2(t)}.$$

Now, we assume that $Z = N\eta^\lambda(t) = O(1)$ and $\eta'(0) \neq 0$. Hence, $\eta(t) = O(N^{-1/\lambda})$ and $t \sim \eta(t)/\eta'(0) = O(N^{-1/\lambda})$. Moreover, we have

$t\eta'(t)/\eta(t) = 1 + O(t) = 1 + O(N^{-1/\lambda})$ and $t^2\eta''(t)/\eta(t) = O(N^{-1/\lambda})$. By ignoring the terms of order $o(N^{-2}) + o(N^{-1/\lambda})$, we obtain

$$\frac{\eta^\lambda(t_\pm)}{\eta^\lambda(t)} \sim 1 \mp \frac{\lambda\theta}{N} + \frac{\lambda\theta(\theta+1)}{2N^2} + \frac{\lambda(\lambda-1)\theta^2}{2N^2} = 1 \mp \frac{\lambda\theta}{N} + \frac{\lambda\theta(\lambda\theta+1)}{2N^2},$$

and

$$\frac{N_\pm\eta^\lambda(t_\pm)}{N\eta^\lambda(t)} \sim 1 \pm \frac{1-\lambda\theta}{N} + \frac{\lambda\theta(\lambda\theta-1)}{2N^2}.$$

Consequently,

$$P_{n\pm1}(N^\theta t) \sim \chi \pm \frac{(1-\lambda\theta)Z\chi'}{N} + \frac{\lambda\theta(\lambda\theta-1)Z\chi' + (1-\lambda\theta)^2Z^2\chi''}{2N^2}.$$

Substituting this and $P_n(N^\theta t) \sim \chi(Z)$ into (6.1) gives

$$2\chi + \frac{\lambda\theta(\lambda\theta-1)Z\chi' + (1-\lambda\theta)^2Z^2\chi''}{N^2} \sim \left[\frac{a_0Z^{1/\lambda}}{N^{1/\lambda}\eta'(0)} + 2 + \frac{b_2'}{N^2}\right]\chi,$$

where we have made use of $a_0' = a_0$, $b_0' = b_0 = 2$, $a_1' = b_1' = 0$, $t \sim \eta(t)/\eta'(0) = (Z/N)^{1/\lambda}/\eta'(0)$, and ignored the terms of order $o(N^{-2}) + o(N^{-1/\lambda})$. Clearly, the term 2χ on both sides cancel with each other. To balance the orders of remaining two terms in the coefficient of $\chi(Z)$ on the right-hand side of the above equation, it is natural to require $\lambda = 1/2$. We shall make an additional assumption $\theta \neq 2$ and convert the above equation to a Bessel equation

$$\chi''(Z) + \frac{\theta}{(\theta-2)Z}\chi'(Z) + \left[\frac{-4a_0}{(\theta-2)^2\eta'(0)} - \frac{4b_2'}{(\theta-2)^2Z^2}\right]\chi(Z) = 0.$$

The above equation has two linearly independent solutions $Z^{1/(2-\theta)}J_\nu(Z)$ and $Z^{1/(2-\theta)}Y_\nu(Z)$ by setting

$$\eta'(0) = -4a_0/(\theta-2)^2, \quad \nu^2 = (1+4b_2')/(\theta-2)^2. \tag{6.10}$$

This suggests that we try a formal series solution of the form

$$P_n(x) = \chi_\nu(N\zeta)\sum_{s=0}^\infty \frac{A_s(\zeta(t))}{N^s} + \chi_{\nu+1}(N\zeta)\sum_{s=0}^\infty \frac{B_s(\zeta(t))}{N^s}, \tag{6.11}$$

motivated from the differential equation theory [Olver (1974a), p. 444] or integral methods [Wong (1989), p. 402], where χ_ν is a linear combination of J_ν and Y_ν, and we have set $\zeta(t) = [\eta(t)]^{1/2}$ to simplify the notation. As we shall see later, $\eta(t)$ is analytic for $t \in \mathbf{C} \setminus [t_*, \infty)$ with $\eta(0) = 0$.

Throughout this chapter, we assume $\theta \neq 0, 2$. The case $\theta = 0$ was studied in Chapter 5, and the main result there (Theorem 5.1) holds whether or not the transition point is at the origin. The case $\theta = 2$ is exceptional in our present investigation, because $\theta - 2$ is in the denominator of the index ν of the Bessel functions; see (6.10). However, $\theta = 2$ and $t^* = 0$ is an important case, since it is exactly the case of Wilson polynomials. We shall leave this case to our subsequent chapter.

6.2 A perturbation lemma for Bessel functions

When we replace n by $n \pm 1$, the two Bessel functions $\chi_\nu(N\zeta)$ and $\chi_{\nu+1}(N\zeta)$ in (6.11) become $\chi_\nu(N_\pm\zeta_\pm)$ and $\chi_{\nu+1}(N_\pm\zeta_\pm)$, where $N_\pm := N \pm 1$ and $\zeta_\pm = \zeta(t_\pm)$ with $t_\pm = (N/N_\pm)^\theta t$. To express these two perturbed Bessel functions in terms of unperturbed Bessel functions, we shall make use of the following perturbation lemma which was first given by Cao and Li (2014).

Lemma 6.1. *Let $\chi_\nu(z)$ be a solution of the Bessel equation*

$$\chi_\nu''(z) + (1/z)\chi_\nu'(z) + (1 - \nu^2/z^2)\chi_\nu(z) = 0; \tag{6.12}$$

namely, it is a linear combination of the Bessel functions $J_\nu(z)$ and $Y_\nu(z)$. Denote $\Psi_\nu(z) := z^{1/2}\chi_\nu(z)$. We have

$$\Psi_\nu(z+u) = \Psi_\nu(z)\sum_{k=0}^{\infty}\frac{X_{\nu,k}(u)}{z^k} + \Psi_{\nu+1}(z)\sum_{k=0}^{\infty}\frac{Y_{\nu,k}(u)}{z^k}, \tag{6.13}$$

$$\Psi_{\nu+1}(z+u) = \Psi_\nu(z)\sum_{k=0}^{\infty}\frac{-Y_{\nu+1,k}(u)}{z^k} + \Psi_{\nu+1}(z)\sum_{k=0}^{\infty}\frac{\tilde{X}_{\nu+1,k}(u)}{z^k}, \tag{6.14}$$

for all $|z| > 2(\nu^2+1)|u|$, where $X_{\nu,0}(u) = \cos u$, $Y_{\nu,0}(u) = -\sin u$, $X_{\nu,1}(u) = (1/2+\nu)\sin u$, $Y_{\nu,1}(u) = 0$, and

$$X_{\nu,k}(u) = (\nu^2 - \frac{1}{4})\sum_{j=0}^{k-2}(j+1)\int_0^u \sin(u-v)(-v)^j X_{\nu,k-2-j}(v)dv, \tag{6.15}$$

$$Y_{\nu,k}(u) = (\nu^2 - \frac{1}{4})\sum_{j=0}^{k-2}(j+1)\int_0^u \sin(u-v)(-v)^j Y_{\nu,k-2-j}(v)dv, \tag{6.16}$$

for $k \geq 2$. Here, we denote $\tilde{X}_{\nu+1,k}(u) = X_{\nu+1,k}(u) + 2(\nu+1)Y_{\nu+1,k-1}(u)$ for $k \geq 1$ and $\tilde{X}_{\nu+1,0}(u) = X_{\nu+1,0}(u) = \cos u$. Let $\theta \neq 0$ and $\zeta(t) = [\eta(t)]^{1/2}$, where $\eta(t)$ is a real analytic function in an interval containing 0 with $\eta(0) = 0$. Denote $N_\pm = N \pm 1$, $t_\pm = (N/N_\pm)^\theta t$, $\zeta_\pm(t) = \zeta(t_\pm)$, and

$$u^\pm = N_\pm\zeta_\pm/\zeta - N. \tag{6.17}$$

For any $N > 2(\nu^2+1)|u^\pm|$, we have

$$\chi_\nu(N_\pm\zeta_\pm) = \chi_\nu(N\zeta)G^\pm(\zeta, N) \pm \chi_{\nu+1}(N\zeta)H^\pm(\zeta, N), \tag{6.18}$$

$$\chi_{\nu+1}(N_\pm\zeta_\pm) = \pm\,\chi_\nu(N\zeta)L^\pm(\zeta, N) + \chi_{\nu+1}(N\zeta)K^\pm(\zeta, N), \tag{6.19}$$

where

$$G^{\pm}(\zeta, N) =(1+\frac{u^{\pm}}{N})^{-1/2}\sum_{k=0}^{\infty}\frac{X_{\nu,k}(\zeta u^{\pm})}{N^k\zeta^k}, \tag{6.20}$$

$$H^{\pm}(\zeta, N) =(1+\frac{u^{\pm}}{N})^{-1/2}\sum_{k=0}^{\infty}\frac{\pm Y_{\nu,k}(\zeta u^{\pm})}{N^k\zeta^k}, \tag{6.21}$$

$$L^{\pm}(\zeta, N) =(1+\frac{u^{\pm}}{N})^{-1/2}\sum_{k=0}^{\infty}\frac{\mp Y_{\nu+1,k}(\zeta u^{\pm})}{N^k\zeta^k}, \tag{6.22}$$

$$K^{\pm}(\zeta, N) =(1+\frac{u^{\pm}}{N})^{-1/2}\sum_{k=0}^{\infty}\frac{\tilde{X}_{\nu+1,k}(\zeta u^{\pm})}{N^k\zeta^k}. \tag{6.23}$$

Proof. Note that the Bessel function satisfies $\chi_\nu'(z) = -\chi_{\nu+1}(z) + (\nu/z)\chi_\nu(z)$ and $\chi_{\nu-1}(z) + \chi_{\nu+1}(z) = (2\nu/z)\chi_\nu(z)$; see Olver *et al.* (2010, (10.6.1) and (10.6.2)). It is readily seen that $\Psi_\nu(z) := z^{1/2}\chi_\nu(z)$ satisfies

$$\Psi_\nu''(z) + (1 - \frac{\nu^2 - 1/4}{z^2})\Psi_\nu(z) = 0, \tag{6.24}$$

$$\Psi_\nu'(z) = (\nu + 1/2)\Psi_\nu(z)/z - \Psi_{\nu+1}(z), \tag{6.25}$$

$$\Psi_{\nu+2}(z) = -\Psi_\nu(z) + [2(\nu+1)/z]\Psi_{\nu+1}(z). \tag{6.26}$$

Substituting (6.13) into (6.24) gives

$$\Psi_\nu(z)\sum_{k=0}^{\infty}\frac{X_{\nu,k}''(u)}{z^k} + \Psi_{\nu+1}(z)\sum_{k=0}^{\infty}\frac{Y_{\nu,k}''(u)}{z^k}$$
$$= \left[-1 + \frac{\nu^2 - 1/4}{(z+u)^2}\right]\cdot\left[\Psi_\nu(z)\sum_{k=0}^{\infty}\frac{X_{\nu,k}(u)}{z^k} + \Psi_{\nu+1}(z)\sum_{k=0}^{\infty}\frac{Y_{\nu,k}(u)}{z^k}\right]. \tag{6.27}$$

We now make use of the expansion $(z+u)^{-2} = \sum_{j=0}^{\infty}(j+1)(-u)^j z^{-j-2}$. For each $k \geq 2$, comparing the coefficients of $\Psi_\nu(z)/z^k$ on both sides of the above equation yields

$$X_{\nu,k}''(u) = -X_{\nu,k}(u) + (\nu^2 - 1/4)\sum_{j=0}^{k-2}(j+1)(-u)^j X_{\nu,k-2-j}(u).$$

From (6.13) and (6.25) we observe $X_{\nu,k}(0) = 0$ and $X_{\nu,k}'(0) = 0$ for all $k \geq 2$. Solving the above differential equation gives (6.15). In a similar manner, we obtain (6.16) by comparing the coefficients of $\Psi_{\nu+1}(z)/z^k$ on both sides of (6.27) and making use of $Y_{\nu,k}(0) = 0$ and $Y_{\nu,k}'(0) = 0$ for all $k \geq 2$.

The above formal argument illustrates how we find the integral equations (6.15) and (6.16). To give a rigorous proof, we need to show that the series involved in the formal argument are convergent. Fix z and denote the right-hand side of (6.13) as $F_z(u)$. On account of $|\sin v| \le |v|e^{|v|}$, we can prove by induction that (exercise) $X_{\nu,k}(u)$ and $Y_{\nu,k}(u)/u$ are entire functions in u and

$$|X_{\nu,k}(u)| + |Y_{\nu,k}(u)/u| \le 2[(\nu^2+1)|u|]^k e^{|u|}, \tag{6.28}$$

for $k \ge 0$. This implies that the series expansion in $F_z(u)$ is convergent if $|z| > 2(\nu^2+1)|u|$. It is easy to verify that $F_z(0) = \Psi_\nu(z)$, $F_z'(0) = (1/2+\nu)\Psi_\nu(z)/z - \Psi_{\nu+1}(z) = \Psi_\nu'(z)$, and

$$F_z''(u) = \left[-1 + \frac{\nu^2 - 1/4}{(z+u)^2}\right] F_z(u).$$

The above differential equation is also satisfied by $\Psi_\nu(z+u)$. By uniqueness, we have $\Psi_\nu(z+u) = F_z(u)$, thus proving (6.13). Replacing ν by $\nu+1$ in (6.13), we obtain (6.14) from (6.26). Finally, by choosing $z = N\zeta$ and $u = \zeta u^\pm$ in (6.13) and (6.14), we obtain (6.18) and (6.19). □

The following proposition will be used in the remaining parts of this chapter, and its proof is left as an exercise.

Proposition 6.1. *Let $X_{\nu,k}$, $Y_{\nu,k}$, $\tilde{X}_{\nu,k}$ ζ, $u^\pm$, $G^\pm$, $H^\pm$, $L^\pm$ and $K^\pm$ be as given in Lemma 6.1. We have $X_{\nu,k}(-u) = (-1)^k X_{\nu,k}(u)$, $Y_{\nu,k}(-u) = (-1)^{k+1} Y_{\nu,k}(u)$, and $\tilde{X}_{\nu,k}(-u) = (-1)^k \tilde{X}_{\nu,k}(u)$. Define $h_j = h^{(j)}(0)/j!$ and $u_j = h_{j+1}$, where $h(\varepsilon) = (1+\varepsilon)\zeta(t(1+\varepsilon)^{-\theta})/\zeta(t)$. For any sufficiently large N, we have $h_0 = 1$ and*

$$u^\pm = Nh(\pm\frac{1}{N}) - N = \sum_{j=0}^{\infty} \frac{(\pm 1)^{j+1} h_{j+1}}{N^j} = \sum_{j=0}^{\infty} \frac{(\pm 1)^{j+1} u_j}{N^j}. \tag{6.29}$$

For $m \ge 1$ and $l \ge 0$, let

$$u_{m,l} := \sum_{\substack{j_1+\cdots+j_l=m \\ j_1,\cdots,j_l \ge 1}} u_{j_1} \cdots u_{j_l} \tag{6.30}$$

be the coefficient of ε^m in the product $(\sum_{j=1}^\infty u_j \varepsilon^j)^l$. For consistency, we also set $u_{0,0} = 1$ and $u_{0,l} = 0$ for all $l \ge 1$. We then have

$$(u^\pm \mp u_0)^l = \sum_{m=l}^{\infty} \frac{(\pm 1)^{m+l} u_{m,l}}{N^m}, \tag{6.31}$$

and

$$X_{\nu,k}(\zeta u^{\pm}) = \sum_{m=0}^{\infty}\sum_{l=0}^{m} \frac{(\pm 1)^{m+k} X_{\nu,k}^{(l)}(\zeta u_0)\zeta^l u_{m,l}}{l! N^m}, \tag{6.32}$$

$$Y_{\nu,k}(\zeta u^{\pm}) = \sum_{m=0}^{\infty}\sum_{l=0}^{m} \frac{(\pm 1)^{m+k+1} Y_{\nu,k}^{(l)}(\zeta u_0)\zeta^l u_{m,l}}{l! N^m}, \tag{6.33}$$

$$\tilde{X}_{\nu,k}(\zeta u^{\pm}) = \sum_{m=0}^{\infty}\sum_{l=0}^{m} \frac{(\pm 1)^{m+k} \tilde{X}_{\nu,k}^{(l)}(\zeta u_0)\zeta^l u_{m,l}}{l! N^m}. \tag{6.34}$$

For $r \geq 0$ and $j \geq 0$, let

$$\bar{u}_{r,j} := \sum_{\substack{k_1+\cdots+k_j=r \\ k_1,\cdots,k_j\geq 0}} u_{k_1}\cdots u_{k_j} \tag{6.35}$$

be the coefficient of ε^r in the product $(\sum_{k=0}^{\infty} u_k\varepsilon^k)^j$. Especially, $\bar{u}_{0,0} = 1$, $\bar{u}_{r,0} = 0$ for $r \geq 1$, and $\bar{u}_{r,1} = u_r$ for $r \geq 0$. We obtain

$$(u^{\pm})^j = \sum_{r=0}^{\infty} \frac{(\pm 1)^{r+j}\bar{u}_{r,j}}{N^r}, \tag{6.36}$$

and

$$(1+\frac{u^{\pm}}{N})^{-1/2} = \sum_{j=0}^{\infty}\sum_{r=0}^{\infty} \binom{-1/2}{j} \frac{(\pm 1)^{j+r}\bar{u}_{r,j}}{N^{j+r}}. \tag{6.37}$$

Furthermore, we have

$$G^{\pm}(\zeta,N) = \sum_{s=0}^{\infty} \frac{(\pm 1)^s G_s(\zeta)}{N^s}, \quad H^{\pm}(\zeta,N) = \sum_{s=0}^{\infty} \frac{(\pm 1)^s H_s(\zeta)}{N^s}, \tag{6.38}$$

$$L^{\pm}(\zeta,N) = \sum_{s=0}^{\infty} \frac{(\pm 1)^s L_s(\zeta)}{N^s}, \quad K^{\pm}(\zeta,N) = \sum_{s=0}^{\infty} \frac{(\pm 1)^s K_s(\zeta)}{N^s}, \tag{6.39}$$

where

$$G_s(\zeta) = \sum_{j+r+k+m=s} \binom{-1/2}{j} \frac{\bar{u}_{r,j}}{\zeta^k} \sum_{l=0}^{m} \frac{X_{\nu,k}^{(l)}(\zeta u_0)u_{m,l}\zeta^l}{l!}, \tag{6.40}$$

$$H_s(\zeta) = \sum_{j+r+k+m=s} \binom{-1/2}{j} \frac{\bar{u}_{r,j}}{\zeta^k} \sum_{l=0}^{m} \frac{Y_{\nu,k}^{(l)}(\zeta u_0)u_{m,l}\zeta^l}{l!}, \tag{6.41}$$

$$L_s(\zeta) = \sum_{j+r+k+m=s} \binom{-1/2}{j} \frac{\bar{u}_{r,j}}{\zeta^k} \sum_{l=0}^{m} \frac{-Y_{\nu+1,k}^{(l)}(\zeta u_0)u_{m,l}\zeta^l}{l!}, \tag{6.42}$$

$$K_s(\zeta) = \sum_{j+r+k+m=s} \binom{-1/2}{j} \frac{\bar{u}_{r,j}}{\zeta^k} \sum_{l=0}^{m} \frac{\tilde{X}_{\nu+1,k}^{(l)}(\zeta u_0)u_{m,l}\zeta^l}{l!}. \tag{6.43}$$

It follows from a simple calculation (exercise) that

$$u_0 = \frac{\zeta - t\theta\zeta'}{\zeta}, \quad u_1 = \frac{\theta(\theta-1)t\zeta' + \theta^2 t^2 \zeta''}{2\zeta} = \frac{-t\theta(\zeta u_0)'}{2\zeta}, \tag{6.44}$$

$$G_0 = \cos(\zeta u_0), \quad G_1 = -\frac{u_0\cos(\zeta u_0)}{2} + \frac{(1/2+\nu-u_1\zeta^2)\sin(\zeta u_0)}{\zeta}, \tag{6.45}$$

$$H_0 = -\sin(\zeta u_0), \quad H_1 = \frac{u_0\sin(\zeta u_0)}{2} - u_1\zeta\cos(\zeta u_0), \tag{6.46}$$

$$L_0 = \sin(\zeta u_0), \quad L_1 = -\frac{u_0\sin(\zeta u_0)}{2} + u_1\zeta\cos(\zeta u_0), \tag{6.47}$$

$$K_0 = \cos(\zeta u_0), \quad K_1 = -\frac{u_0\cos(\zeta u_0)}{2} - \frac{(1/2+\nu+u_1\zeta^2)\sin(\zeta u_0)}{\zeta}. \tag{6.48}$$

6.3 Formal asymptotic solutions

Let $\chi_\nu(z)$ be a solution of the Bessel equation (6.12), and $\eta(t)$ be a real analytic and increasing function in $(-\infty, t_*)$ such that $\eta(0) = 0$. For simplicity, we denote $\zeta(t) = [\eta(t)]^{1/2}$. Let $x = N^\theta t$. We look for a solution to the difference equation (6.1) with the formal series expansion in (6.11); namely,

$$P_n(N^\theta t) = \chi_\nu(N\zeta)\mathbb{A}(\zeta, N) + \chi_{\nu+1}(N\zeta)\mathbb{B}(\zeta, N), \tag{6.49}$$

where

$$\mathbb{A}(\zeta, N) := \sum_{s=0}^{\infty} \frac{A_s(\zeta)}{N^s}, \quad \mathbb{B}(\zeta, N) := \sum_{s=0}^{\infty} \frac{B_s(\zeta)}{N^s}. \tag{6.50}$$

For convenience, we introduce the notation

$$\Psi(t, N) := C_1(n)x + C_0(n) \sim \sum_{s=0}^{\infty} \frac{2\psi_s(t)}{N^s}, \quad \psi_s(t) := (a_s' t + b_s')/2. \tag{6.51}$$

Denote $N_\pm = N \pm 1$ and $\zeta_\pm = \zeta(t_\pm)$ with $t_\pm = (N/N_\pm)^\theta t$. It follows from Lemma 6.1 and (6.49) that

$$\begin{aligned} P_{n\pm1}(N^\theta t) = {} & [\chi_\nu(N\zeta)G^\pm(\zeta, N) \pm \chi_{\nu+1}(N\zeta)H^\pm(\zeta, N)]\mathbb{A}(\zeta_\pm, N_\pm) \\ & + [\pm\chi_\nu(N\zeta)L^\pm(\zeta, N) + \chi_{\nu+1}(N\zeta)K^\pm(\zeta, N)]\mathbb{B}(\zeta_\pm, N_\pm). \end{aligned} \tag{6.52}$$

Substituting (6.49) and (6.52) into (6.1) and equating the coefficients of $\chi_\nu(N\zeta)$ and $\chi_{\nu+1}(N\zeta)$ to zero, we obtain two equations

$$\begin{aligned} \Psi(t, N)\mathbb{A}(\zeta, N) = {} & G^+(\zeta, N)\mathbb{A}(\zeta_+, N_+) + G^-(\zeta, N)\mathbb{A}(\zeta_-, N_-) \\ & + L^+(\zeta, N)\mathbb{B}(\zeta_+, N_+) - L^-(\zeta, N)\mathbb{B}(\zeta_-, N_-), \end{aligned} \tag{6.53}$$

$$\begin{aligned} \Psi(t, N)\mathbb{B}(\zeta, N) = {} & H^+(\zeta, N)\mathbb{A}(\zeta_+, N_+) - H^-(\zeta, N)\mathbb{A}(\zeta_-, N_-) \\ & + K^+(\zeta, N)\mathbb{B}(\zeta_+, N_+) + K^-(\zeta, N)\mathbb{B}(\zeta_-, N_-). \end{aligned} \tag{6.54}$$

To expand the right-hand sides of the above two equations into power series in $1/N$, we shall first expand $\mathbb{A}(\zeta_\pm, N_\pm)$ and $\mathbb{B}(\zeta_\pm, N_\pm)$. Denote D^j to be the j-th derivative with respect to t; namely, $D^j A_r(\zeta) = d^j A_r(\zeta(t))/dt^j$ for $j, r = 0, 1, 2, \cdots$. Define $\gamma_{j,l}$ as the coefficient of ε^l in the power series of $[(1+\varepsilon)^{-\theta} - 1]^j$. We have

$$(t_\pm - t)^j = \sum_{l=0}^{\infty} \frac{\gamma_{j,l} t^j}{(\pm 1)^l N^l}, \quad A_r(\zeta_\pm) = \sum_{j=0}^{\infty}\sum_{l=0}^{\infty} \frac{D^j A_r(\zeta)\gamma_{j,l} t^j}{(\pm 1)^l j! N^l}, \tag{6.55}$$

$$\mathbb{A}(\zeta_\pm, N_\pm) = \sum_{r=0}^{\infty}\sum_{j=0}^{\infty}\sum_{l=0}^{\infty}\sum_{k=0}^{\infty} \binom{-r}{k} \frac{D^j A_r(\zeta)\gamma_{j,l} t^j}{(\pm 1)^{l+k} j! N^{l+r+k}}, \tag{6.56}$$

$$\mathbb{B}(\zeta_\pm, N_\pm) = \sum_{r=0}^{\infty}\sum_{j=0}^{\infty}\sum_{l=0}^{\infty}\sum_{k=0}^{\infty} \binom{-r}{k} \frac{D^j B_r(\zeta)\gamma_{j,l} t^j}{(\pm 1)^{l+k} j! N^{l+r+k}}. \tag{6.57}$$

Recall the series expansions of $G^\pm$, $H^\pm$, $L^\pm$ and $K^\pm$ in Proposition 6.1. By equating the coefficients of N^{-p} in the series expansions of (6.53) and (6.54), we obtain from (6.50), (6.51), (6.56) and (6.57)

$$\begin{aligned} &\sum_{s\le p} \psi_s(t) A_{p-s}(\zeta) \qquad (6.58)\\ &= \sum_{s\le p, s \text{ even}} \sum_{i+l\le s} \binom{-p+s}{s-i-l} G_i(\zeta) \sum_{j=0}^{l} \frac{\gamma_{j,l} t^j D^j A_{p-s}(\zeta)}{j!} \\ &\quad + \sum_{s\le p, s \text{ odd}} \sum_{i+l\le s} \binom{-p+s}{s-i-l} L_i(\zeta) \sum_{j=0}^{l} \frac{\gamma_{j,l} t^j D^j B_{p-s}(\zeta)}{j!}, \end{aligned}$$

and

$$\begin{aligned} &\sum_{s\le p} \psi_s(t) B_{p-s}(\zeta) \qquad (6.59)\\ &= \sum_{s\le p, s \text{ odd}} \sum_{i+l\le s} \binom{-p+s}{s-i-l} H_i(\zeta) \sum_{j=0}^{l} \frac{\gamma_{j,l} t^j D^j A_{p-s}(\zeta)}{j!} \\ &\quad + \sum_{s\le p, s \text{ even}} \sum_{i+l\le s} \binom{-p+s}{s-i-l} K_i(\zeta) \sum_{j=0}^{l} \frac{\gamma_{j,l} t^j D^j B_{p-s}(\zeta)}{j!}, \end{aligned}$$

for $p = 0, 1, \cdots$, where we have made use of the fact that $\gamma_{j,l} = 0$ for $j > l$. Setting $p = 0$ gives $[\psi_0(t) - G_0(\zeta)]A_0(\zeta) = [\psi_0(t) - K_0(\zeta)]B_0(\zeta) = 0$. We may assume without loss of generality that either A_0 or B_0 is nonzero (exercise). It then follows from (6.45) or (6.48) that $\cos(\zeta u_0) = \psi_0$, where $\zeta u_0 = \zeta - t\theta\zeta'$ and $\psi_0 = (a_0' t + b_0')/2 = (1 + a_0 t/2)$; see (6.44) and (6.51).

Consequently, $\zeta(t) - t\theta\zeta'(t) = \pm \arccos \psi_0(t)$. Since $\zeta = \eta^{1/2}$ is defined to simplify the notation and the sign of ζ does not affect the value of $\eta = \zeta^2$, we may choose without loss of generality that

$$\zeta(t) - t\theta\zeta'(t) = -\arccos \psi_0(t) \tag{6.60}$$

To solve this differential equation, we first assume $t \in (0, t_*)$ and then use analytic continuation to extend the solution to the complex plane. Note that

$$(t^{-1/\theta}\zeta)' = -(1/\theta)t^{-1/\theta-1}(\zeta - t\theta\zeta') = -(t^{-1/\theta})' \arccos \psi_0(t). \tag{6.61}$$

There is one free parameter, namely, the constant of integration, when we express $t^{-1/\theta}\zeta$ as an integral of the right-hand side. We shall choose this constant of integration carefully so that $\eta = \zeta^2$ can be analytically continued to the complex region $\mathbf{C}\backslash[t_*, \infty)$. For illustration, we first assume $\theta > 2$ or $\theta < 0$ so that the power function $t^{-1/\theta-1/2}$ is integrable at $t = 0$. A simple application of integration by parts gives the following solution of the above differential equation

$$t^{-1/\theta}\zeta(t) = -t^{-1/\theta} \arccos \psi_0(t) + \int_0^t \frac{(-a_0/2)s^{-1/\theta}}{\sqrt{1 - \psi_0(s)^2}} ds. \tag{6.62}$$

Here, we choose the constant of integration such that the lower end point of the definite integral on the right-hand side is 0. To demonstrate that this choice implies the desired analytic property of $\eta = \zeta^2$, we multiply both sides by $t^{1/\theta}$, express the inverse cosine function as a composition of a logarithmic function and a square root function, introduce a change of variable $r = s/t$ for the definite integral, and make use of the identities $\psi_0(t) = (a_0't + b_0')/2 = 1 + a_0t/2$ and $1 - \psi_0(t)^2 = 1 - (1 + a_0t/2)^2 = -a_0t(1 + a_0t/4)$. This gives

$$\zeta(t) = i \log \left[1 + \frac{a_0t}{2} + i\sqrt{-a_0t\left(1 + \frac{a_0t}{4}\right)}\right] + \int_0^1 \frac{(-a_0/2)t^{1/2}r^{-1/\theta}}{\sqrt{-a_0r(1 + a_0tr/4)}} dr.$$

The above function is defined for $t \in (0, t_*)$ and can be analytically continued to $\mathbf{C} \setminus (-\infty, 0] \cup [t_*, \infty)$. For each $t < 0$, we can calculate the one-sided limits of $\zeta(t \pm i\varepsilon)$ as $\varepsilon \to 0^+$ as

$$\begin{aligned}
&\lim_{\varepsilon \to 0^+} \zeta(t \pm i\varepsilon) \\
&= i \log \left[1 + \frac{a_0t}{2} \mp \sqrt{a_0t\left(1 + \frac{a_0t}{4}\right)}\right] \pm i \int_0^1 \frac{(-a_0/2)(-t)^{1/2}r^{-1/\theta}}{\sqrt{-a_0r(1 + a_0tr/4)}} dr \\
&= \pm i \left\{-\log \left[1 + \frac{a_0t}{2} + \sqrt{a_0t\left(1 + \frac{a_0t}{4}\right)}\right] + \int_0^1 \frac{(-a_0/2)(-t)^{1/2}r^{-1/\theta}}{\sqrt{-a_0r(1 + a_0tr/4)}} dr\right\}.
\end{aligned}$$

This implies that $\eta = \zeta^2$ can be analytically continued to $(-\infty, 0]$ and $\eta(t) < 0$ for $t < 0$. For a general $\theta \neq 0, 2$, one may find that $t^{-1/\theta-1/2}$ is not integrable near 0. Let $k_\theta > 0$ be the smallest integer such that $k_\theta - 1/\theta - 1/2 > 0$. Especially, $k_\theta = 1$ when $\theta > 2$ or $\theta < 0$. We then apply integration by parts k_θ times to obtain

$$\int -(t^{-1/\theta})' \arccos \psi_0(t) dt = -t^{-1/\theta} \arccos \psi_0(t) + \int \frac{t^{-1/\theta-1/2}}{(\frac{4}{-a_0} - t)^{1/2}} dt$$

$$= -t^{-1/\theta} \arccos \psi_0(t) + \frac{t^{1/2-1/\theta}}{(1/2 - 1/\theta)(\frac{4}{-a_0} - t)^{1/2}}$$

$$+ \int \frac{(-1/2)t^{1/2-1/\theta}}{(1/2 - 1/\theta)(\frac{4}{-a_0} - t)^{3/2}} dt$$

$$= -t^{-1/\theta} \arccos \psi_0(t) + \sum_{j=1}^{k_\theta - 1} \frac{(-1/2)_{j-1} t^{j-1/\theta-1/2}}{(1/2 - 1/\theta)_j (\frac{4}{-a_0} - t)^{j-1/2}}$$

$$+ \int \frac{(-1/2)_{k_\theta - 1} t^{k_\theta - 1/\theta - 3/2}}{(1/2 - 1/\theta)_{k_\theta - 1} (\frac{4}{-a_0} - t)^{k_\theta - 1/2}} dt,$$

where in the last step we have made use of Pochhammer notation for shifted factorials $(a)_j = \prod_{k=0}^{j-1}(a+k)$. Now, we can choose the constant of integration such that

$$t^{-1/\theta}\zeta(t) = -t^{-1/\theta} \arccos \psi_0(t) + \sum_{j=1}^{k_\theta - 1} \frac{(-1/2)_{j-1} t^{j-1/\theta-1/2}}{(1/2 - 1/\theta)_j (\frac{4}{-a_0} - t)^{j-1/2}}$$

$$+ \int_0^t \frac{(-1/2)_{k_\theta - 1} s^{k_\theta - 1/\theta - 3/2}}{(1/2 - 1/\theta)_{k_\theta - 1} (\frac{4}{-a_0} - s)^{k_\theta - 1/2}} ds.$$

The integral on the right-hand side is well-defined because $k_\theta - 1/\theta - 1/2 > 0$. Similar as in the simple case ($k_\theta = 1$), we obtain

$$\zeta(t) = i \log \left[1 + \frac{a_0 t}{2} + i\sqrt{-a_0 t \left(1 + \frac{a_0 t}{4}\right)} \right]$$

$$+ \sum_{j=1}^{k_\theta - 1} \frac{(-1/2)_{j-1} t^{j-1/2}}{(1/2 - 1/\theta)_j (\frac{4}{-a_0} - t)^{j-1/2}}$$

$$+ \int_0^1 \frac{(-1/2)_{k_\theta - 1} t^{k_\theta - 1/2} r^{k_\theta - 1/\theta - 3/2}}{(1/2 - 1/\theta)_{k_\theta - 1} (\frac{4}{-a_0} - tr)^{k_\theta - 1/2}} dr. \tag{6.63}$$

The right-hand side of the above expression is defined for $t \in (0, t_*)$ and can be analytically continued to the complex region $\mathbf{C} \setminus (-\infty, 0] \cup [t_*, \infty)$.

Moreover, for each $t < 0$, the one-sided limits of $\zeta(t \pm i\varepsilon)$ as $\varepsilon \to 0^+$ differ by a sign. So, $\eta(t) = \zeta(t)^2$ is analytic for $t \in \mathbf{C} \setminus [t_*, \infty)$ with $\eta(t) > 0$ for $t \in (0, t_*)$ and $\eta(t) < 0$ for $t \in (-\infty, 0)$. For convenience, when $t < 0$, we choose the value of $\zeta(t)$ on the negative imaginary line such that $i\zeta(t) > 0$. By symmetry, if we choose the value of $\zeta(t)$ on the positive imaginary line, the values of $\eta(t)$ and other functions involving $\eta(t)$ will not change.

Now, we come back to equations (6.58) and (6.59) with $p \geq 1$ and solve for A_{p-1} and B_{p-1} recursively. For convenience, we denote

$$\begin{aligned} f_{p-1}(t) := &\sum_{2\leq s\leq p} \psi_s(t)A_{p-s}(\zeta) \\ &- \sum_{2\leq s\leq p, s \text{ even}} \sum_{i+l\leq s} \binom{-p+s}{s-i-l} G_i(\zeta) \sum_{j=0}^{l} \frac{\gamma_{j,l}t^j D^j A_{p-s}(\zeta)}{j!} \\ &- \sum_{2\leq s\leq p, s \text{ odd}} \sum_{i+l\leq s} \binom{-p+s}{s-i-l} L_i(\zeta) \sum_{j=0}^{l} \frac{\gamma_{j,l}t^j D^j B_{p-s}(\zeta)}{j!}, \end{aligned} \tag{6.64}$$

and

$$\begin{aligned} g_{p-1}(t) := &\sum_{2\leq s\leq p} \psi_s(t)B_{p-s}(\zeta) \\ &- \sum_{2\leq s\leq p, s \text{ odd}} \sum_{i+l\leq s} \binom{-p+s}{s-i-l} H_i(\zeta) \sum_{j=0}^{l} \frac{\gamma_{j,l}t^j D^j A_{p-s}(\zeta)}{j!} \\ &- \sum_{2\leq s\leq p, s \text{ even}} \sum_{i+l\leq s} \binom{-p+s}{s-i-l} K_i(\zeta) \sum_{j=0}^{l} \frac{\gamma_{j,l}t^j D^j B_{p-s}(\zeta)}{j!}, \end{aligned} \tag{6.65}$$

where the empty sum is zero; namely, $f_0(t) = g_0(t) = 0$. We can simplify (6.58) and (6.59) as (exercise)

$$[L_1(\zeta) + (-p+1)L_0(\zeta)]B_{p-1}(\zeta) - \theta t L_0(\zeta)\frac{d}{dt}B_{p-1}(\zeta) = f_{p-1}(t), \tag{6.66}$$

$$[H_1(\zeta) + (-p+1)H_0(\zeta)]A_{p-1}(\zeta) - \theta t H_0(\zeta)\frac{d}{dt}A_{p-1}(\zeta) = g_{p-1}(t), \tag{6.67}$$

where we have made use of the facts that $\psi_0 = G_0 = K_0$ and $\psi_1 = 0$. Note from (6.44), (6.46) and (6.47) that $L_0 = -H_0$ and

$$H_1 = -\frac{u_0}{2}H_0 - \frac{t\theta}{2}\frac{d}{dt}H_0 = -\frac{1 - t\theta\zeta'/\zeta}{2}H_0 - \frac{t\theta}{2}\frac{d}{dt}H_0, \tag{6.68}$$

$$L_1 = -\frac{u_0}{2}L_0 - \frac{t\theta}{2}\frac{d}{dt}L_0 = -\frac{1 - t\theta\zeta'/\zeta}{2}L_0 - \frac{t\theta}{2}\frac{d}{dt}L_0. \tag{6.69}$$

Substituting (6.69) into (6.66) and then dividing both sides of (6.66) by tL_0B_{p-1} give

$$\begin{aligned}\frac{f_{p-1}}{tL_0B_{p-1}} =& \frac{1/2-p}{t} + \frac{\theta\zeta'}{2\zeta} - \frac{\theta}{2L_0}\frac{d}{dt}L_0 - \frac{\theta}{B_{p-1}}\frac{d}{dt}B_{p-1} \\ =& -\theta\frac{d}{dt}\ln[t^{(p-1/2)/\theta}(L_0/\zeta)^{1/2}B_{p-1}] \\ =& \frac{-\theta}{t^{(p-1/2)/\theta}(L_0/\zeta)^{1/2}B_{p-1}}\frac{d}{dt}[t^{(p-1/2)/\theta}(L_0/\zeta)^{1/2}B_{p-1}].\end{aligned}$$

Denote $\Lambda(t) := t^{1/(2\theta)}[L_0(\zeta)/\zeta]^{1/2} = t^{1/(2\theta)}[-H_0(\zeta)/\zeta]^{1/2}$. We rewrite the above equation as

$$\frac{d}{dt}[t^{(p-1)/\theta}\Lambda(t)B_{p-1}(t)] = \frac{t^{(p-1)/\theta-1}\Lambda(t)}{\theta H_0(\zeta)}f_{p-1}(t), \tag{6.70}$$

where we have made use of the identity $L_0 = -H_0$. In a similar manner, we substitute (6.68) into (6.67) and obtain

$$\frac{d}{dt}[t^{(p-1)/\theta}\Lambda(t)A_{p-1}(t)] = -\frac{t^{(p-1)/\theta-1}\Lambda(t)}{\theta H_0(\zeta)}g_{p-1}(t). \tag{6.71}$$

When $p = 1$, it follows from $f_0 = g_0 = 0$ that ΛB_0 and ΛA_0 are two constants independent of t. We set $\Lambda A_0 = 1$ and $\Lambda B_0 = 0$. When $p > 1$ and $0 < \theta < 2$, we integrate the above two equations to obtain

$$t^{(p-1)/\theta}\Lambda B_{p-1} = \int_0^t \frac{s^{(p-1)/\theta-1}\Lambda(s)f_{p-1}(s)}{\theta H_0(\zeta(s))}ds, \tag{6.72}$$

$$t^{(p-1)/\theta}\Lambda A_{p-1} = -\int_0^t \frac{s^{(p-1)/\theta-1}\Lambda(s)g_{p-1}(s)}{\theta H_0(\zeta(s))}ds. \tag{6.73}$$

For each $p > 1$, we can successively determine $A_p(\zeta)$ and $B_p(\zeta)$ from their predecessors $A_0(\zeta), B_0(\zeta), \cdots, A_{p-1}(\zeta)$ and $B_{p-1}(\zeta)$ via the above two formulas together with (6.64) and (6.65). If, in addition, we have the assumption $a'_{2s+1} = b'_{2s+1} = 0$ for $s = 0, 1, \cdots$ (see Dingle and Morgan (1967a,b)), we can choose the solutions such that $A_{2s+1} = B_{2s} = 0$. For the cases $\theta > 2$ or $\theta < 0$, the lower limits of the integrals in (6.72) and (6.73) are changed to $-\infty$ or t_* so that the integrals are convergent.

6.4 Bounds for the coefficients

To establish global uniformity of our anticipated expansion, we need to estimate the coefficients $A_s(\zeta)$ and $B_s(\zeta)$ in (6.11). First, we need the following lemma.

Lemma 6.2. *Let $\zeta(t)$ be given as in (6.63), and let $G_s(\zeta)$, $H_s(\zeta)$, $L_s(\zeta)$ and $K_s(\zeta)$ be given as in Proposition 6.1. Then there exists a positive constant N_s independent of t such that*

$$|G_s(\zeta)| + |H_s(\zeta)| + |K_s(\zeta)| + |L_s(\zeta)| \leq N_s(|t| + 1), \tag{6.74}$$

for all $t \leq t_* - \sigma$, $\sigma > 0$.

Proof. We only prove the estimate for $G_s(\zeta)$; the other three estimates can be obtained in a similar manner. It follows from (6.15) and an induction argument that $X_{\nu,k}(u) = O(u^k)$ as $u \to 0$. Consequently, $X_{\nu,k}^{(l)}(\zeta u_0)\zeta^{l-k}$ is analytic at $t = 0$. This implies that $G_s(\zeta)$ in (6.40) is bounded for t in any compact subset of $(-\infty, t_*)$. It suffices to show that $G_s(\zeta)/t$ is bounded as $t \to -\infty$. To achieve this, we have to find another series expansion of $G_s(\zeta)$. By choosing χ_ν in (6.18) to be the Bessel functions $J_\nu(z)$ and $Y_\nu(z)$, respectively, we arrive at a linear system

$$\begin{pmatrix} J_\nu(N\zeta) & \pm J_{\nu+1}(N\zeta) \\ Y_\nu(N\zeta) & \pm Y_{\nu+1}(N\zeta) \end{pmatrix} \begin{pmatrix} G^\pm(\zeta, N) \\ H^\pm(\zeta, N) \end{pmatrix} = \begin{pmatrix} J_\nu(N_\pm\zeta_\pm) \\ Y_\nu(N_\pm\zeta_\pm) \end{pmatrix}.$$

In view of the identity $J_{\nu+1}(z)Y_\nu(z) - J_\nu(z)Y_{\nu+1}(z) = 2/(\pi z)$ (see Olver (1974a, p. 244) or Olver *et al.* (2010, (10.5.2))), we have

$$G^\pm(\zeta, N) = \frac{\pi N\zeta}{2}[J_{\nu+1}(N\zeta)Y_\nu(N_\pm\zeta_\pm) - Y_{\nu+1}(N\zeta)J_\nu(N_\pm\zeta_\pm)]. \tag{6.75}$$

Recall the relations between the Bessel functions and the modified Bessel functions [Olver (1974a), p. 60 & p. 251]:

$$J_\nu(ix) = e^{\nu\pi i/2}I_\nu(x), \quad Y_\nu(ix) = e^{(\nu+1)\pi i/2}I_\nu(x) - \frac{2}{\pi}e^{-\nu\pi i/2}K_\nu(x). \tag{6.76}$$

Let $\xi = -iN\zeta$ and $\xi_\pm = -iN_\pm\zeta_\pm$. We can express the right-hand side of (6.75) in terms of modified Bessel functions

$$G^\pm(\zeta, N) = \xi[I_\nu(\xi_\pm)K_{\nu+1}(\xi) + K_\nu(\xi_\pm)I_{\nu+1}(\xi)]. \tag{6.77}$$

Recall the asymptotic expansion for modified Bessel functions of large argument [Olver (1974a), p. 250-251 & p. 238]

$$I_\nu(z) \sim \frac{e^z}{(2\pi z)^{1/2}}\sum_{j=0}^\infty (-1)^j \frac{c_{\nu,j}}{z^j}, \quad K_\nu(z) \sim \frac{\pi^{1/2}e^{-z}}{(2z)^{1/2}}\sum_{j=0}^\infty \frac{c_{\nu,j}}{z^j}, \tag{6.78}$$

as $z \to \infty$, where $c_{\nu,0} = 1$. For simplicity, we denote $r_j := c_{\nu,j}$ and $s_j := c_{\nu+1,j}$, where ν is fixed. Substituting (6.78) into (6.77) gives

$$G^\pm(\zeta, N) \sim \frac{(\xi/\xi_\pm)^{1/2}}{2}\sum_{k=0}^\infty\sum_{j=0}^\infty \frac{[(-1)^j e^{\xi_\pm - \xi} + (-1)^k e^{\xi - \xi_\pm}]r_j s_k}{\xi_\pm^j \xi^k}, \tag{6.79}$$

as $N \to \infty$. On account of Proposition 6.1, the coefficient of N^{-s} in the above asymptotic expansion is the same as $(\pm 1)^s G_s(\zeta)$. To expand $\xi_+ - \xi = -i(N_+\zeta_+ - N\zeta) = -i[(N+1)\zeta(t(1+1/N)^{-\theta}) - N\zeta(t)$, we regard $F(\lambda) := -i(N+\lambda)\zeta(t(1+\lambda/N)^{-\theta})$ as a function of $\lambda \in [0,1]$. Taking derivative with respect to λ gives

$$\begin{aligned} iF'(\lambda) =& \zeta(t(1+\lambda/N)^{-\theta}) - \theta t(1+\lambda/N)^{-\theta}\zeta'(t(1+\lambda/N)^{-\theta}) \\ =& -\arccos[\psi_0(t(1+\lambda/N)^{-\theta})] = i\,\mathrm{arccosh}[\psi_0(t(1+\lambda/N)^{-\theta})], \end{aligned}$$

where, in the last second equation, we have made use of (6.60) with variable $t(1+\lambda/N)^{-\theta}$. On the other hand, note that $F(1) = -i(N+1)\zeta(t(1+1/N)^{-\theta})$ and $F(0) = -iN\zeta(t)$. We obtain (exercise)

$$\begin{aligned} \xi_+ - \xi =& F(1) - F(0) = \int_0^1 F'(\lambda)d\lambda = \int_0^1 \mathrm{arccosh}[\psi_0(t(1+\lambda/N)^{-\theta})]d\lambda \\ =& \mathrm{arccosh}(\psi_0(t)) + \frac{a_0' t\theta}{2N}\int_0^1 \frac{(\lambda-1)(1+\lambda/N)^{-\theta-1}}{\sqrt{\psi_0^2(t(1+\lambda/N)^{-\theta})-1}}d\lambda \\ =& \sum_{j=0}^{\infty}\frac{v_j(t)}{N^j}, \end{aligned} \tag{6.80}$$

where $v_0(t) = \mathrm{arccosh}(\psi_0(t)) = O(\log|t|)$, and

$$v_j(t) = \frac{(-1)^j\theta}{j(j+1)} + O(1/|t|), \quad j \ge 1, \tag{6.81}$$

as $t \to -\infty$. It can be further shown from (6.63) (exercise) that $\zeta(t) = O(\log|t|)$ and $1/\zeta(t) = O(1/\log|t|)$ as $t \to -\infty$, and, consequently,

$$e^{\xi_+ - \xi} = e^{v_0(t)}\sum_{l=0}^{\infty}\frac{w_{1,l}(t)}{N^l}, \qquad e^{\xi - \xi_+} = e^{-v_0(t)}\sum_{l=0}^{\infty}\frac{w_{2,l}(t)}{N^l}, \tag{6.82}$$

$$(\xi_+/\xi)^{-j-1/2} = \sum_{m=0}^{\infty}\frac{w_{3,j,m}(t)}{N^m}, \tag{6.83}$$

where $w_{1,l}(t)$, $w_{2,l}(t)$ and $w_{3,j,m}(t)$ are bounded as $t \to -\infty$, while the bound may depend on the indices j, m, or l. Substituting the above expansions into the right-hand side of (6.79) and collecting the coefficient of N^{-s} yields

$$G_s(\zeta) = \sum_{j+k+l+m=s}\frac{[(-1)^j e^{v_0(t)}w_{1,l}(t) + (-1)^k e^{-v_0(t)}w_{2,l}(t)]r_j s_k w_{3,j,m}(t)}{2[\zeta(t)]^{j+k}}.$$

Since $e^{v_0(t)}$ and $e^{-v_0(t)}$ are of order $O(|t|)$ as $t \to -\infty$, so is $G_s(\zeta)$. This gives the estimate $|G_s(\zeta)| \le (N_s/4)(|t|+1)$ for some positive constant N_s. In a similar manner, we can prove that $|H_s(\zeta)|$, $|K_s(\zeta)|$ and $|L_s(\zeta)|$ are also bounded by $(N_s/4)(|t|+1)$, thus proving (6.74). □

Now, we are ready to provide some bounds for the coefficients $A_s(\zeta)$ and $B_s(\zeta)$ in (6.11).

Lemma 6.3. *Denote* $\Lambda(t) = t^{1/(2\theta)}[-H_0(\zeta)/\zeta]^{1/2}$. *For* $s, j = 0, 1, 2, \cdots$, *there exists a constant* $M_{s,j}$ *independent of* t *such that*

$$|t^j \Lambda(t) D^j A_s(\zeta)| + |t^j \Lambda(t) D^j B_s(\zeta)| \leq M_{s,j}(1 + |t|^{-s/\theta}), \tag{6.84}$$

for $t < 0$, *where* D^j *denotes the* j*-th derivative with respect to* t.

Proof. We will prove by induction. First, we note that the estimates in Lemma 6.2 hold for all complex t in the left half-plane, because the corresponding functions $G_s(\zeta), H_s(\zeta), L_s(\zeta)$ and $K_s(\zeta)$ are all analytic for $t \in \mathbf{C} \setminus [t_*, \infty)$. In view of (6.60) and (6.63), we obtain

$$|t^j D^j \zeta| \leq \tilde{N}_j(1 + |\zeta|), \tag{6.85}$$

for some positive constant $\tilde{N}_j$. By using Cauchy integral formula, one can further prove that

$$|t^j D^j G_s| + |t^j D^j H_s| + |t^j D^j L_s| + |t^j D^j K_s| \leq N_{s,j}(1 + |t|), \tag{6.86}$$

for some positive constant $N_{s,j}$. When $s = 0$, since $B_0 = 0$ and $\Lambda A_0 = 1$, the estimate in (6.84) holds for $j = 0$. By Cauchy integral formula, it also holds for $j \geq 1$. Now, we assume that the estimate in (6.84) holds for $s = 0, 1, \cdots, p-2$ with $p \geq 2$ and for $j = 0, 1, \cdots$. It follows from (6.64), (6.65) and the above estimates that

$$|\Lambda f_{p-1}| + |\Lambda g_{p-1}| \leq C_p(1 + |t|)(1 + |t|^{-(p-2)/\theta}), \tag{6.87}$$

for some constant C_p. Substituting this into (6.72) and (6.73) gives (exercise)

$$|\Lambda A_{p-1}| + |\Lambda B_{p-1}| \leq M_{p-1,0}(1 + |t|^{-(p-1)/\theta}), \tag{6.88}$$

for some constant $M_{p-1,0}$. Finally, (6.84) with $s = p - 1$ and $j \geq 1$ follows from a direct application of Cauchy integral formula. □

6.5 Asymptotic nature of the expansion

Note that the variable x in the difference equation (6.1) is fixed. A solution to (6.1) multiplied by any function of x remains a solution. For each $p \geq 0$, we truncate the expansion (6.11), multiply it by $x^{1/(2\theta)} = N^{1/2} t^{1/(2\theta)}$, and replace χ_ν with J_ν and $W_\nu := Y_\nu - iJ_\nu$, respectively. This gives us two asymptotic solutions

$$r_n^p(N^\theta t) := J_\nu(N\zeta)\mathbb{A}_p(\zeta, N) + J_{\nu+1}(N\zeta)\mathbb{B}_p(\zeta, N), \tag{6.89}$$

$$s_n^p(N^\theta t) := W_\nu(N\zeta)\mathbb{A}_p(\zeta, N) + W_{\nu+1}(N\zeta)\mathbb{B}_p(\zeta, N), \tag{6.90}$$

where

$$\mathbb{A}_p(\zeta, N) = x^{1/(2\theta)} \sum_{s=0}^{p} \frac{A_s(\zeta)}{N^s} = (-H_0/\zeta)^{-1/2} \sum_{s=0}^{p} \frac{\tilde{A}_s(\zeta)}{N^{s-1/2}}, \tag{6.91}$$

$$\mathbb{B}_p(\zeta, N) = x^{1/(2\theta)} \sum_{s=0}^{p} \frac{B_s(\zeta)}{N^s} = (-H_0/\zeta)^{-1/2} \sum_{s=0}^{p} \frac{\tilde{B}_s(\zeta)}{N^{s-1/2}}, \tag{6.92}$$

with $\Lambda(t) := t^{1/(2\theta)}(-H_0/\zeta)^{1/2}$, and

$$\tilde{A}_s(\zeta) := \Lambda(t)A_s(\zeta), \quad \tilde{B}_s(\zeta) := \Lambda(t)B_s(\zeta). \tag{6.93}$$

Especially, $\tilde{A}_0(\zeta) = 1$ and $\tilde{B}_0(\zeta) = 0$. It is worth recalling that $-H_0 = \sin(\zeta u_0) = (1 - \psi_0^2)^{1/2}$ with $2\psi_0 = a_0' t + b_0' = 2 + a_0 t$. We also recall that $x = N^\theta t$ and $\Psi(t, N) := C_1(n)x + C_0(n)$. From Lemma 6.1, it follows that (exercise)

$$\begin{aligned}\frac{R_n^p(x)}{N^{p+3/2}} :=& r_{n+1}^p(x) - [C_1(n)x + C_0(n)]r_n^p(x) + r_{n-1}^p(x)\\ =& J_\nu(N\zeta)F_{1,n}^p(x) + J_{\nu+1}(N\zeta)F_{2,n}^p(x), \end{aligned}\tag{6.94}$$

$$\begin{aligned}\frac{S_n^p(x)}{N^{p+3/2}} :=& s_{n+1}^p(x) - [C_1(n)x + C_0(n)]s_n^p(x) + s_{n-1}^p(x)\\ =& W_\nu(N\zeta)F_{1,n}^p(x) + W_{\nu+1}(N\zeta)F_{2,n}^p(x), \end{aligned}\tag{6.95}$$

where

$$\begin{aligned}F_{1,n}^p(x) =& G^+(\zeta, N)\mathbb{A}_p(\zeta_+, N_+) + G^-(\zeta, N)\mathbb{A}_p(\zeta_-, N_-) - \Psi(t, N)\mathbb{A}_p(\zeta, N)\\ &+ L^+(\zeta, N)\mathbb{B}_p(\zeta_+, N_+) - L^-(\zeta, N)\mathbb{B}_p(\zeta_-, N_-), \end{aligned}\tag{6.96}$$

$$\begin{aligned}F_{2,n}^p(x) =& H^+(\zeta, N)\mathbb{A}_p(\zeta_+, N_+) - H^-(\zeta, N)\mathbb{A}_p(\zeta_-, N_-) - \Psi(t, N)\mathbb{B}_p(\zeta, N)\\ &+ K^+(\zeta, N)\mathbb{B}_p(\zeta_+, N_+) + K^-(\zeta, N)\mathbb{B}_p(\zeta_-, N_-). \end{aligned}\tag{6.97}$$

On account of Lemma 6.3, we have

$$|t^j D^j \tilde{A}_s(\zeta)| + |t^j D^j \tilde{B}_s(\zeta)| \le \tilde{M}_{s,j}(1 + |t|^{-s/\theta}), \tag{6.98}$$

for some constant $\tilde{M}_{s,j}$. Now, we improve the above estimate and prove by induction

$$|\tilde{A}_s(\zeta)| + |\tilde{B}_s(\zeta)| \le \tilde{M}_s, \tag{6.99}$$

for some constant $\tilde{M}_s$ and all $t < t_* - \sigma$. Obviously, the above estimate holds for $s = 0$. Assume it holds for $s = 0, \cdots, p$. Using an argument similar to that given for (6.87) and (6.88), we have from (6.64) and (6.65) that

$$|\Lambda f_{p+1}| + |\Lambda g_{p+1}| \le C_{p+1,0}(1 + |t|).$$

As a consequence, it can be shown that (6.84) can be improved to read

$$|t^j\Lambda(t)D^jA_s(\zeta)|+|t^j\Lambda(t)D^jB_s(\zeta)|\le M_{s,j}$$

for $t<0$ and $s,j\ge 0$. In particular, these two estimates hold for $j=0$ and $s=p+1$; that is, $\tilde{A}_{p+1}(\zeta)$ and $\tilde{B}_{p+1}(\zeta)$ are bounded for $t<0$. By the continuity of $\zeta(t), G_s(\zeta)$ and $H_s(\zeta)$ we conclude that $\tilde{A}_{p+1}(\zeta)$ and $\tilde{B}_{p+1}(\zeta)$ are also bounded for $t<t_*-\sigma$, thus completing the induction argument.

It can be proved that (exercise)

$$|\frac{\Lambda(t)F^p_{1,n}(x)}{x^{1/(2\theta)}}|+|\frac{\Lambda(t)F^p_{2,n}(x)}{x^{1/(2\theta)}}|\le\frac{C_p(1+|t|)}{N^{p+2}},\tag{6.100}$$

for some constant C_p. Substituting this into (6.94) and (6.95) yields

$$|R^p_n(N^\theta t)|\le C_p(1+|t|)\alpha_n(N^\theta t),|S^p_n(N^\theta t)|\le C_p(1+|t|)\beta_n(N^\theta t),\tag{6.101}$$

for $t<t_*-\sigma$, where

$$\alpha_n(N^\theta t):=|H_0/\zeta|^{-1/2}[|J_\nu(N\zeta)|+|J_{\nu+1}(N\zeta)|],\tag{6.102}$$

$$\beta_n(N^\theta t):=|H_0/\zeta|^{-1/2}[|W_\nu(N\zeta)|+|W_{\nu+1}(N\zeta)|].\tag{6.103}$$

We then have the following lemma.

Lemma 6.4. *Fix $p\ge 0$. There exist $\{\varepsilon^p_n(x)\}_{n=1}^\infty$ and $\{\delta^p_n(x)\}_{n=1}^\infty$ such that*

$$\varepsilon^p_{n+1}(x)-[C_1(n)x+C_0(n)]\varepsilon^p_n(x)+\varepsilon^p_{n-1}(x)=-\frac{R^p_n(x)}{N^{p+3/2}},\tag{6.104}$$

$$\delta^p_{n+1}(x)-[C_1(n)x+C_0(n)]\delta^p_n(x)+\delta^p_{n-1}(x)=-\frac{S^p_n(x)}{N^{p+3/2}},\tag{6.105}$$

with

$$|\delta^p_n(x)|\le\frac{M_p\beta_n(x)}{N^{p+1/2}},\quad |\varepsilon^p_n(x)|\le\frac{M_p\alpha_n(x)}{N^{p+1/2}},\tag{6.106}$$

for some positive constant M_p.

Proof. It can be shown that the solution of the difference equation (6.105) with asymptotic estimate in (6.106) is the same as the solution to the following summation equation with the same asymptotic estimate:

$$\begin{aligned}\delta^p_n(x)&=\sum_{j=n+1}^\infty\frac{[s^p_n(x)r^p_j(x)-r^p_n(x)s^p_j(x)]S^p_j(x)}{[s^p_{n+1}(x)r^p_n(x)-s^p_n(x)r^p_{n+1}(x)](j+\tau_0)^{p+3/2}}\\&\quad+\sum_{j=n+1}^\infty\frac{[s^p_n(x)R^p_j(x)-r^p_n(x)S^p_j(x)]\delta^p_j(x)}{[s^p_{n+1}(x)r^p_n(x)-s^p_n(x)r^p_{n+1}(x)](j+\tau_0)^{p+3/2}}.\end{aligned}\tag{6.107}$$

We will construct such a solution by successive approximation. Starting with $\delta^p_{n,0} = 0$, we define $\delta^p_{n,k}(x)$ recursively by

$$\delta^p_{n,k}(x) = \sum_{j=n+1}^{\infty} \frac{[s^p_n(x)r^p_j(x) - r^p_n(x)s^p_j(x)]S^p_j(x)}{[s^p_{n+1}(x)r^p_n(x) - s^p_n(x)r^p_{n+1}(x)](j+\tau_0)^{p+3/2}} + \sum_{j=n+1}^{\infty} \frac{[s^p_n(x)R^p_j(x) - r^p_n(x)S^p_j(x)]\delta^p_{j,k-1}(x)}{[s^p_{n+1}(x)r^p_n(x) - s^p_n(x)r^p_{n+1}(x)](j+\tau_0)^{p+3/2}}. \quad (6.108)$$

Since $\tilde{A}_s(\zeta)$ and $\tilde{B}_s(\zeta)$ are bounded, it follows from (6.89) and (6.90) that

$$|r^p_n(x)| \le CN^{1/2}\alpha_n(x), \quad |s^p_n(x)| \le CN^{1/2}\beta_n(x), \quad (6.109)$$

for some constant C. Moreover, in view of the behavior of J_ν and W_ν, we have from (6.101) and (6.109) that

$$|R^p_n(N^\theta t)s^p_n(N^\theta t)| + |S^p_n(N^\theta t)r^p_n(N^\theta t)| \le M'N^{1/2}, \quad (6.110)$$

for some constant M'. Note from (6.89), (6.90) and the asymptotic behavior of Bessel functions (exercise) that

$$\lim_{n\to\infty}[r^p_{n+1}(x)s^p_n(x) - r^p_n(x)s^p_{n+1}(x)] = \frac{2}{\pi}. \quad (6.111)$$

Especially,

$$|r^p_{n+1}(x)s^p_n(x) - r^p_n(x)s^p_{n+1}(x)| > \frac{1}{\pi}, \quad (6.112)$$

for all sufficiently large n. A combination of (6.108)-(6.110) and (6.112) yields

$$\begin{aligned} |\delta^p_{n,1}(x)| \le& \pi \sum_{j=n+1}^{\infty} \frac{[|S^p_n(x)r^p_j(x)| + |r^p_n(x)S^p_j(x)|]}{(j+\tau_0)^{p+3/2}} \\ \le& M'' \sum_{j=n+1}^{\infty} \frac{N^{1/2}}{(j+\tau_0)^{p+1}}\beta_n(x), \end{aligned}$$

for some positive constant M''. Here, we have made use of the monotonicity properties and the asymptotic behaviors of the modified Bessel functions I_ν and K_ν; see (6.76) and (6.78). A further estimate of the sum on the right-hand side of the above inequality gives (exercise)

$$|\delta^p_{n,1}(x)| \le \frac{M''}{pN^{p-1/2}}\beta_n(x). \quad (6.113)$$

Similarly, we can prove by induction that

$$|\delta^p_{n,k}(x) - \delta^p_{n,k-1}(x)| \le \left(\frac{M''}{pN^{p+1}}\right)^k N^{3/2}\beta_n(x), \quad (6.114)$$

from which it follows that, for all $n \geq 2M'' - \tau_0$,

$$\delta^p_{n,k}(x) = \sum_{j=1}^{k} [\delta^p_{n,j}(x) - \delta^p_{n,j-1}(x)] \tag{6.115}$$

converges as $k \to \infty$. Clearly, the limit function $\delta^p_n(x)$ satisfies (6.107) as well as (6.105). Moreover, we obtain from (6.113)-(6.115)

$$|\delta^p_n(x)| \leq \frac{M'_p}{N^{p-1/2}} \beta_n(x), \tag{6.116}$$

for some positive constant M'_p. In a similar manner, we can prove that

$$|\varepsilon^p_n(x)| \leq \frac{M''_p}{N^{p-1/2}} \alpha_n(x), \tag{6.117}$$

for some positive constant M''_p. To improve the above two estimates, we first define $P^p_n(x) := r^p_n(x) + \varepsilon^p_n(x)$ and $Q^p_n(x) := s^p_n(x) + \delta^p_n(x)$. It is obvious that $P^p_n(x)$ and $Q^p_n(x)$ are two solutions to the difference equation (6.1). Making use of the fact that $\varepsilon^p_n(x) \to 0$ and $\delta^p_n(x) \to 0$ as $n \to \infty$, one can prove that (exercise)

$$P^p_{n+1}(x)Q^p_n(x) - P^p_n(x)Q^p_{n+1}(x) = \frac{2}{\pi}. \tag{6.118}$$

Especially, $P^p_n(x)$ and $Q^p_n(x)$ are two linearly independent solutions to (6.1). For any $p \neq q$, it can be shown that (exercise) $P^p_n(x) = P^q_n(x)$ and $Q^p_n(x) = Q^q_n(x)$. Hence, $P_n(x) = P^p_n(x)$ and $Q_n(x) = Q^p_n(x)$ are independent of p. Especially, we have

$$\begin{aligned} &\delta^p_n(x) - \delta^{p+1}_n(x) = s^{p+1}_n(x) - s^p_n(x) \\ =&W_\nu(N\zeta)(-H_0/\zeta)^{-1/2} \frac{\tilde{A}_{p+1}(\zeta)}{N^{p+1/2}} + W_{\nu+1}(N\zeta)(-H_0/\zeta)^{-1/2} \frac{\tilde{B}_{p+1}(\zeta)}{N^{p+1/2}}. \end{aligned}$$

It then follows from (6.99), (6.103) and (6.116) that

$$|\delta^p_n(x)| \leq |\delta^{p+1}_n(x)| + |\delta^p_n(x) - \delta^{p+1}_n(x)| \leq \frac{M'_{p+1}}{N^{p+1/2}} \beta_n(x) + \frac{\tilde{M}_{p+1}}{N^{p+1/2}} \beta_n(x).$$

Choosing $M_p = M'_{p+1} + \tilde{M}_{p+1}$ gives the first estimate in (6.106). The second estimate in (6.106) can be obtained similarly from (6.99), (6.102), (6.117), and the identity

$$\begin{aligned} &\varepsilon^p_n(x) - \varepsilon^{p+1}_n(x) = r^{p+1}_n(x) - r^p_n(x) \\ =&J_\nu(N\zeta)(-H_0/\zeta)^{-1/2} \frac{\tilde{A}_{p+1}(\zeta)}{N^{p+1/2}} + J_{\nu+1}(N\zeta)(-H_0/\zeta)^{-1/2} \frac{\tilde{B}_{p+1}(\zeta)}{N^{p+1/2}}. \end{aligned}$$

This completes the proof. □

Now, we obtain the following theorem.

Theorem 6.1. *Assume that the coefficients $C_1(n)$ and $C_0(n)$ in the linear difference equation (6.1) have asymptotic expansions of the form given in (6.2) with $\theta \neq 0, 2$, $a_0 < 0$, $b_0 = 2$ and $b_1 = 0$. Set $N = n + \tau_0$ with $\tau_0 = -a_1/(a_0\theta)$. Let $\zeta(t)$ be given as in (6.63). Let $\tilde{A}_s(\zeta)$ and $\tilde{B}_s(\zeta)$ be defined as in (6.93) with $\tilde{A}_0(\zeta) = 1$ and $\tilde{B}_0(\zeta) = 0$. Then equation (6.1) has, for each integer $p \geq 0$, a pair of linearly independent solutions $P_n(x)$ and $Q_n(x)$ such that*

$$P_n(N^\theta t) = \left(\frac{4\zeta^2}{4 - (a_0' t + b_0')^2}\right)^{1/4} \left[J_\nu(N\zeta) \sum_{s=0}^{p} \frac{\tilde{A}_s(\zeta)}{N^{s-1/2}} + J_{\nu+1}(N\zeta) \sum_{s=0}^{p} \frac{\tilde{B}_s(\zeta)}{N^{s-1/2}} + \varepsilon_p(t, N)\right], \tag{6.119}$$

$$Q_n(N^\theta t) = \left(\frac{4\zeta^2}{4 - (a_0' t + b_0')^2}\right)^{1/4} \left[W_\nu(N\zeta) \sum_{s=0}^{p} \frac{\tilde{A}_s(\zeta)}{N^{s-1/2}} + W_{\nu+1}(N\zeta) \sum_{s=0}^{p} \frac{\tilde{B}_s(\zeta)}{N^{s-1/2}} + \delta_p(t, N)\right], \tag{6.120}$$

where $x = N^\theta t$ and $W_\nu(N\zeta) = Y_\nu(N\zeta) - iJ_\nu(N\zeta)$. Let $\sigma > 0$ be an arbitrarily small positive constant. For $t < t_* - \sigma$,

$$|\varepsilon_p(t, N)| \leq \frac{M_p}{N^{p+1/2}} [|J_\nu(N\zeta)| + |J_{\nu+1}(N\zeta)|], \tag{6.121}$$

$$|\delta_p(t, N)| \leq \frac{M_p}{N^{p+1/2}} [|W_\nu(N\zeta)| + |W_{\nu+1}(N\zeta)|], \tag{6.122}$$

for some positive constant M_p.

Proof. Note that $\varepsilon_n^p(x) = (-H_0/\zeta)^{-1/2} \varepsilon_p(t, N)$ and $\delta_n^p(x) = (-H_0/\zeta)^{-1/2} \delta_p(t, N)$ with $(-H_0/\zeta)^{-1/2} = [4\zeta^2/(4 - (a_0' t + b_0')^2)]^{1/4}$. The theorem is a direct consequence of Lemma 6.4. □

6.6 Exercises

1. Verify (6.28).
2. Prove Proposition 6.1.
3. Prove (6.44)-(6.48).
4. Let $P_n(x)$ in (6.49) be a solution to (6.1) with series expansions in (6.50). Assume $A_s = B_s = 0$ for all $s < p$ and either A_p or B_p is nonzero. Show that (6.58)-(6.59) become $(\psi_0 - G_0)A_p = (\psi_0 - K_0)B_p = 0$, from which we have $\cos(\zeta u_0) = \psi_0$.

5. Prove that $\zeta(t)$ defined in (6.63) is analytic for $t \in \mathbf{C} \setminus (-\infty, 0] \cup [t_*, \infty)$. If $t < 0$, show that $\lim_{\varepsilon \to 0^+} \zeta(t + i\varepsilon) = -\lim_{\varepsilon \to 0^+} \zeta(t - i\varepsilon)$.
6. Verify (6.66) and (6.67).
7. Verify (6.80) and (6.81).
8. Let $\zeta(t)$ be defined in (6.63). When $t < 0$, we define $\zeta(t)$ as the one-sided limit $\lim_{\varepsilon \to 0^+} \zeta(t + i\varepsilon)$. Show that $\zeta(t) = O(\log |t|)$ and $1/\zeta(t) = O(1/\log |t|)$ as $t \to -\infty$.
9. Verify (6.94) and (6.95).
10. Prove (6.100).
11. Let $r_n^p(x)$ and $s_n^p(x)$ be given as in (6.89)-(6.90). Show that $\lim_{n \to \infty} [r_{n+1}^p(x) s_n^p(x) - s_{n+1}^p(x) r_n^p(x) = \frac{2}{\pi}$. Verify (6.111).
12. Prove (6.113).
13. Let $r_n^p(x)$ and $s_n^p(x)$ with $p \geq 0$ be defined as in (6.89) and (6.90). If $P_n(x)$ and $Q_n(x)$ are two solutions to (6.1) such that $P_n(x) - r_n^0(x) \to 0$ and $Q_n(x) - s_n^0(x) \to 0$ as $n \to \infty$, prove that $P_{n+1}(x) Q_n(x) - P_n(x) Q_{n+1}(x) = 2/\pi$. Moreover, show that such pair of solutions are unique; namely, if $P_n^*(x)$ and $Q_n^*(x)$ are two solutions to (6.1) such that $P_n^*(x) - r_n^0(x) \to 0$ and $Q_n^*(x) - s_n^0(x) \to 0$ as $n \to \infty$, then $P_n^*(x) = P_n(x)$ and $Q_n^*(x) = Q_n(x)$.

Chapter 7

Transition Point at the Origin Revisited: A Critical Case

7.1 Introduction

In Chapter 6, we considered case (ii) when $\theta \neq 0$ and one of the transition points occurs at the origin ($t^* = 0$), and the uniform asymptotic solutions were given in terms of Bessel functions. Since a factor $\theta - 2$ appeared in the denominator of the index ν for the Bessel functions, we made an additional assumption that $\theta \neq 2$. In this chapter, we will address this critical case: $\theta = 2$ and $t^* = 0$. We recall the second-order linear difference equation (4.3):

$$P_{n+1}(x) - [C_1(n)x + C_0(n)]P_n(x) + P_{n-1}(x) = 0, \tag{7.1}$$

where

$$C_1(n) \sim n^{-\theta} \sum_{s=0}^{\infty} \frac{a_s}{n^s}, \quad C_0(n) \sim \sum_{s=0}^{\infty} \frac{b_s}{n^s}, \tag{7.2}$$

with $\theta = 2$ and $a_0 \neq 0$. As before, we let τ_0 be a constant, the value of which will be determined later, and put $N := n + \tau_0$. A simple computation shows that the expansions in (7.2) can be rewritten in the form

$$C_1(n) \sim N^{-2} \sum_{s=0}^{\infty} \frac{a'_s}{N^s}, \quad C_0(n) \sim \sum_{s=0}^{\infty} \frac{b'_s}{N^s}, \tag{7.3}$$

with $a'_0 = a_0$, $a'_1 = a_1 + 2a_0\tau_0$, $b'_0 = b_0$ and $b'_1 = b_1$. The characteristic equation associated with (7.1) is the quadratic equation

$$\rho^2 - (a'_0 t + b'_0)\rho + 1 = 0; \tag{7.4}$$

see (4.6), (5.3), or (6.4). The two roots of equation (7.4) coincide when $t = t_*$ or $t = t^*$, where

$$a'_0 t^* + b'_0 = 2, \quad a'_0 t_* + b'_0 = -2. \tag{7.5}$$

As in Chapters 4, 5 and 6, the values of these two roots are called the transition points (turning points).

To motivate our study of this critical case $\theta = 2$ and $t^* = 0$, we first consider the three-term recurrence relation [Koekoek *et al.* (2010), (9.1.4)]

$$-(z+a^2)\tilde{W}_n(z) = A_n\tilde{W}_{n+1}(z) - (A_n+C_n)\tilde{W}_n(z) + C_n\tilde{W}_{n-1}(z) \quad (7.6)$$

of the Wilson polynomials [Koekoek *et al.* (2010), (9.1.1)]

$$\begin{aligned}\tilde{W}_n(z) =& \frac{W_n(z)}{(a+b)_n(a+c)_n(a+d)_n} \\ =& {}_4F_3\left(\begin{matrix}-n, n+a+b+c+d-1, a+i\sqrt{z}, a-i\sqrt{z} \\ a+b, a+c, a+d\end{matrix}; 1\right), \end{aligned} \quad (7.7)$$

where a, b, c, d are parameters and

$$A_n = \frac{(n+a+b+c+d-1)(n+a+b)(n+a+c)(n+a+d)}{(2n+a+b+c+d-1)(2n+a+b+c+d)}, \quad (7.8)$$

$$C_n = \frac{n(n+b+c-1)(n+b+d-1)(n+c+d-1)}{(2n+a+b+c+d-2)(2n+a+b+c+d-1)}. \quad (7.9)$$

In terms of the notations used in Section 1 of Chapter 4, we have

$$c_1(n) = -\frac{1}{A_n}, \quad c_0(n) = \frac{A_n+C_n-a^2}{A_n}, \quad c_2(n) = \frac{C_n}{A_n}.$$

To convert equation (7.6) into our standard form (7.1), we introduce the sequence suggested in Chapter 4:

$$\begin{aligned}\frac{K_{n+1}}{K_{n-1}} = c_2(n) = \frac{C_n}{A_n} =& \frac{(2n+a+b+c+d)}{(2n+a+b+c+d-2)} \\ &\times \frac{n(n+b+c-1)(n+b+d-1)(n+c+d-1)}{(n+a+b+c+d-1)(n+a+b)(n+a+c)(n+a+d)}.\end{aligned}$$

From this equation, one can easily obtain

$$\frac{K_{2n}}{K_0} = \frac{(\frac{a+b+c+d+2}{4})_n(\frac{1}{2})_n(\frac{b+c}{2})_n(\frac{b+d}{2})_n(\frac{c+d}{2})_n}{(\frac{a+b+c+d}{4})_n(\frac{a+b+c+d}{2})_n(\frac{a+b+1}{2})_n(\frac{a+c+1}{2})_n(\frac{a+d+1}{2})_n},$$

$$\frac{K_{2n+1}}{K_1} = \frac{(\frac{a+b+c+d+4}{4})_n(1)_n(\frac{b+c+1}{2})_n(\frac{b+d+1}{2})_n(\frac{c+d+1}{2})_n}{(\frac{a+b+c+d+2}{4})_n(\frac{a+b+c+d+1}{2})_n(\frac{a+b+2}{2})_n(\frac{a+c+2}{2})_n(\frac{a+d+2}{2})_n},$$

where K_0 and K_1 are constants. By choosing

$$K_0 = \frac{\Gamma(\frac{a+b+c+d+2}{4})\Gamma(\frac{1}{2})\Gamma(\frac{b+c}{2})\Gamma(\frac{b+d}{2})\Gamma(\frac{c+d}{2})}{\Gamma(\frac{a+b+c+d}{4})\Gamma(\frac{a+b+c+d}{2})\Gamma(\frac{a+b+1}{2})\Gamma(\frac{a+c+1}{2})\Gamma(\frac{a+d+1}{2})},$$

$$K_1 = \frac{\Gamma(\frac{a+b+c+d+4}{4})\Gamma(1)\Gamma(\frac{b+c+1}{2})\Gamma(\frac{b+d+1}{2})\Gamma(\frac{c+d+1}{2})}{\Gamma(\frac{a+b+c+d+2}{4})\Gamma(\frac{a+b+c+d+1}{2})\Gamma(\frac{a+b+2}{2})\Gamma(\frac{a+c+2}{2})\Gamma(\frac{a+d+2}{2})},$$

we have

$$\begin{aligned}K_{2n} =& \frac{\Gamma(n+\frac{a+b+c+d+2}{4})\Gamma(n+\frac{1}{2})}{\Gamma(n+\frac{a+b+c+d}{4})\Gamma(n+\frac{a+b+c+d}{2})} \\ &\times \frac{\Gamma(n+\frac{b+c}{2})\Gamma(n+\frac{b+d}{2})\Gamma(n+\frac{c+d}{2})}{\Gamma(n+\frac{a+b+1}{2})\Gamma(n+\frac{a+c+1}{2})\Gamma(n+\frac{a+d+1}{2})}, \\ K_{2n+1} =& \frac{\Gamma(n+\frac{a+b+c+d+4}{4})\Gamma(n+1)}{\Gamma(n+\frac{a+b+c+d+2}{4})\Gamma(n+\frac{a+b+c+d+1}{2})} \\ &\times \frac{\Gamma(n+\frac{b+c+1}{2})\Gamma(n+\frac{b+d+1}{2})\Gamma(n+\frac{c+d+1}{2})}{\Gamma(n+\frac{a+b+2}{2})\Gamma(n+\frac{a+c+2}{2})\Gamma(n+\frac{a+d+2}{2})}.\end{aligned}$$

As $n \to \infty$, it follows from Stirling's formula and the definitions of $c_1(n)$ and $c_0(n)$ after (7.9) that

$$K_{2n} \sim n^{-2a-1/2}, \quad K_{2n+1} \sim n^{-2a-1/2}, \quad c_1(n) \sim -4n^{-2}, \quad c_0(n) \sim 2.$$

The coefficients in (7.1) are given by

$$\begin{aligned}C_1(n) &= c_1(n)\frac{K_n}{K_{n+1}} = n^{-2}\left[-4 + O\left(\frac{1}{n}\right)\right], \\ C_0(n) &= c_0(n)\frac{K_n}{K_{n+1}} = 2 + O\left(\frac{1}{n}\right).\end{aligned}$$

Comparing these two equations with the expansions in (7.2) and (7.3) immediately gives the values

$$\theta = 2, \quad a_0' = a_0 = -4, \quad b_0' = b_0 = 2.$$

Plugging these values into the first equation in (7.5), we readily see that the value of the transition point t^* is zero; i.e., $t^* = 0$. This tells us that Wilson polynomial is just an example of this critical case. The presentation of the material in the remaining parts of this chapter is based on the work in Li *et al.* (2020).

In general, we shall consider the case when $a_0 \neq 0$ and $b_0^2 = 4$. By symmetry, we may assume without loss of generality that $a_0 < 0$ and $b_0 = 2$. If $a_0 > 0$ and $b_0 = 2$, we consider the difference equation satisfied by $P_n(-x)$. If $a_0 > 0$ and $b_0 = -2$, we consider the difference equation satisfied by $(-1)^n P_n(x)$. If $a_0 < 0$ and $b_0 = -2$, we consider the difference equation satisfied by $(-1)^n P_n(-x)$. By choosing $\tau_0 = -a_1/(2a_0)$, we obtain $a_1' = a_1 + 2a_0\tau_0 = 0$, where a_1' is the coefficient in the expansion (7.3). As in Chapters 4, 5 and 6, we further assume $b_1 = 0$ so that $b_1' = b_1 = 0$. Let $x = N^2 t$ with $N = n + \tau_0$. It follows from (7.3) that

$$C_1(n)x + C_0(n) = \sum_{s=0}^{\infty} \frac{a_s' t + b_s'}{N^s} =: \sum_{s=0}^{\infty} \frac{2\psi_s(t)}{N^s}, \tag{7.10}$$

where $\psi_s(t) := (a'_s t + b'_s)/2$ for $s = 0, 1, \cdots$. Especially, $\psi_0(t) = (a_0 t + b_0)/2$ and $\psi_1(t) = 0$.

To determine which function appears in the uniform asymptotic expansions of the solutions to (7.1) in the present case, we try a solution of the form $P_n(N^2 t) \sim \chi(Z)$, where $Z = N\eta(t)$ with $\eta(0) = 0$. Assume that $t = o(1)$. Let $N_\pm = N \pm 1$ and

$$t_\pm = \frac{N^2 t}{N_\pm^2} = t \mp \frac{2t}{N} + o\left(\frac{1}{N^2}\right).$$

We have

$$\eta(t_\pm) = \eta(t) \mp \frac{2t\eta'(t)}{N} + o\left(\frac{1}{N^2}\right),$$

and

$$N_\pm \eta(t_\pm) = N\eta(t) \pm [\eta(t) - 2t\eta'(t)] + o\left(\frac{1}{N}\right)$$

Denote $u_0 = \eta(t) - 2t\eta'(t)$. We have $u_0 \to 0$ as $t \to 0$. By ignoring terms of order $o(u_0^2) + o(1/N)$, we obtain

$$P_{n\pm 1}(N^2 t) \sim \chi(N_\pm \eta(t_\pm)) \sim \chi(Z \pm u_0) \sim \chi(Z) \pm u_0 \chi'(Z) + \frac{u_0^2 \chi''(Z)}{2}.$$

Substituting this and $P_n(N^2 t) \sim \chi(Z)$ into (7.1) gives

$$2\chi(Z) + u_0^2 \chi''(Z) \sim (2 + a_0 t)\chi(Z).$$

To balance the asymptotic orders on both sides of the above formula, we assume that $u_0^2/t \to -a_0 > 0$ as $t \to 0$. By ignoring the higher-order terms, the above equation becomes $\chi''(Z) = -\chi(Z)$, which has two linearly independent solutions $\cos Z$ and $\sin Z$. This suggests us to try a formal series solution of the form

$$P_n(x) = \chi(N\eta) \sum_{s=0}^{\infty} \frac{A_s(\eta(t))}{N^s} + \chi'(N\eta) \sum_{s=0}^{\infty} \frac{B_s(\eta(t))}{N^s}, \tag{7.11}$$

where $\chi(Z)$ is a solution of $\chi''(Z) = -\chi(Z)$.

7.2 A perturbation lemma for trigonometric functions

As in Chapters 4, 5 and 6, we first establish the following perturbation lemma (see Li *et al.* (2020)) for trigonometric functions when n is replaced by $n \pm 1$.

Lemma 7.1. *Denote $N_\pm = N \pm 1$ and $t_\pm = (N/N_\pm)^2 t$. Let $\eta(t)$ be a real analytic function on an open interval. Set $\eta_\pm = \eta(t_\pm)$. We have*

$$\cos(N_\pm\eta_\pm) = \cos(N\eta)G^\pm(\eta, N) - \sin(N\eta)H^\pm(\eta, N), \tag{7.12}$$

$$\sin(N_\pm\eta_\pm) = \cos(N\eta)H^\pm(\eta, N) + \sin(N\eta)G^\pm(\eta, N), \tag{7.13}$$

where

$$G^\pm(\eta, N) = \cos(u^\pm), \quad H^\pm(\eta, N) = \sin(u^\pm), \tag{7.14}$$

with

$$u^\pm = N_\pm\eta_\pm - N\eta. \tag{7.15}$$

If $\chi(z)$ is a solution to the differential equation $\chi''(z) = -\chi(z)$; namely, $\chi(z)$ is a linear combination of $\cos z$ and $\sin z$, then

$$\chi(N_\pm\eta_\pm) = \chi(N\eta)G^\pm(\eta, N) + \chi'(N\eta)H^\pm(\eta, N), \tag{7.16}$$

$$\chi'(N_\pm\eta_\pm) = -\chi(N\eta)H^\pm(\eta, N) + \chi'(N\eta)G^\pm(\eta, N). \tag{7.17}$$

Proof. A direct application of trigonometric identities gives (7.12) and (7.13). If $\chi(z) = \cos z$, then (7.16) is the same as (7.12), and (7.17) is obtained by multiplying both sides of (7.13) by -1. If $\chi(z) = \sin z$, then (7.16) is the same as (7.13), and (7.17) is the same as (7.12). By linearity, (7.16) and (7.17) hold for any solution of $\chi''(z) = -\chi(z)$. □

The following proposition will be used in the remaining parts of this chapter, and its proof is left as an exercise.

Proposition 7.1. *Let η, $u^\pm$, $G^\pm$ and $H^\pm$ be as given in Lemma 7.1. Define $h_j = h^{(j)}(0)/j!$ and $u_j = h_{j+1}$, where $h(\varepsilon) = (1+\varepsilon)\eta(t(1+\varepsilon)^{-2})$. For any sufficiently large N, we have*

$$u^\pm = Nh(\pm\frac{1}{N}) - Nh(0) = \sum_{j=0}^{\infty} \frac{(\pm 1)^{j+1} h_{j+1}}{N^j} = \sum_{j=0}^{\infty} \frac{(\pm 1)^{j+1} u_j}{N^j}. \tag{7.18}$$

For $m \ge 1$ and $l \ge 0$, let

$$u_{m,l} := \sum_{\substack{j_1+\cdots+j_l=m \\ j_1,\cdots,j_l \ge 1}} u_{j_1}\cdots u_{j_l} \tag{7.19}$$

be the coefficient of ε^m in the product $(\sum_{j=1}^{\infty} u_j\varepsilon^j)^l$. For consistency, we also set $u_{0,0} = 1$ and $u_{0,l} = 0$ for all $l \ge 1$. We then have

$$(u^\pm \mp u_0)^l = \sum_{m=l}^{\infty} \frac{(\pm 1)^{m+l} u_{m,l}}{N^m}. \tag{7.20}$$

Denote

$$\Phi_l(u) := \frac{d^l}{du^l} \cos u. \tag{7.21}$$

We have $\Phi_l(\pm u) = (\pm 1)^l \Phi_l(u)$ *and*

$$\begin{aligned} G^{\pm}(\eta, N) =& \Phi_0(u^{\pm}) = \sum_{l=0}^{\infty} \frac{\Phi_l(\pm u_0)}{l!} \sum_{m=l}^{\infty} \frac{(\pm 1)^{m+l} u_{m,l}}{N^m} \\ =& \sum_{m=0}^{\infty} \frac{(\pm 1)^m G_m(\eta)}{N^m}, \end{aligned} \tag{7.22}$$

$$\begin{aligned} H^{\pm}(\eta, N) =& -\Phi_1(u^{\pm}) = -\sum_{l=0}^{\infty} \frac{\Phi_{l+1}(\pm u_0)}{l!} \sum_{m=l}^{\infty} \frac{(\pm 1)^{m+l} u_{m,l}}{N^m} \\ =& \sum_{m=0}^{\infty} \frac{(\pm 1)^{m+1} H_m(\eta)}{N^m}, \end{aligned} \tag{7.23}$$

where

$$G_m(\eta) = \sum_{l=0}^{m} \frac{\Phi_l(u_0) u_{m,l}}{l!}, \quad H_m(\eta) = -\sum_{l=0}^{m} \frac{\Phi_{l+1}(u_0) u_{m,l}}{l!}. \tag{7.24}$$

It follows from a simple calculation (exercise) that

$$u_0 = \eta - 2t\eta', \quad u_1 = t\eta' + 2t^2\eta'' = -t\frac{d}{dt}u_0, \tag{7.25}$$

$$G_0 = \cos u_0, \quad G_1 = -u_1 \sin u_0 = -t\frac{d}{dt}G_0, \tag{7.26}$$

$$H_0 = \sin u_0, \quad H_1 = u_1 \cos u_0 = -t\frac{d}{dt}H_0. \tag{7.27}$$

7.3 Formal asymptotic solutions

Let $\chi(u)$ be a linear combination of $\cos u$ and $\sin u$. Let $x = N^2 t$ and $\eta(t)$ be a function of t. We look for a solution to the difference equation (7.1) with the formal series expansion in (7.11); namely,

$$P_n(N^2 t) = \chi(N\eta)\mathbb{A}(\eta, N) + \chi'(N\eta)\mathbb{B}(\eta, N), \tag{7.28}$$

where

$$\mathbb{A}(\eta, N) := \sum_{s=0}^{\infty} \frac{A_s(\eta)}{N^s}, \quad \mathbb{B}(\eta, N) := \sum_{s=0}^{\infty} \frac{B_s(\eta)}{N^s}. \tag{7.29}$$

For convenience, we introduce the notation

$$\Psi(x, N) := C_1(n)x + C_0(n) \sim \sum_{s=0}^{\infty} \frac{2\psi_s(t)}{N^s}, \quad \psi_s(t) := (a_s' t + b_s')/2. \tag{7.30}$$

Denote $N_\pm = N \pm 1$ and $\eta_\pm = \eta(t_\pm)$ with $t_\pm = (N/N_\pm)^2 t$. It follows from Lemma 7.1 and (7.28) that

$$\begin{aligned}P_{n\pm 1}(N^2 t) = & [\chi(N\eta)G^\pm(\eta, N) + \chi'(N\eta)H^\pm(\eta, N)]\mathbb{A}(\eta_\pm, N_\pm) \\ & +[-\chi(N\eta)H^\pm(\eta, N) + \chi'(N\eta)G^\pm(\eta, N)]\mathbb{B}(\eta_\pm, N_\pm). \end{aligned} \tag{7.31}$$

Substituting (7.28) and (7.31) into (7.1) and equating the coefficients of $\chi(N\eta)$ and $\chi'(N\eta)$ to zero, we obtain two equations

$$\begin{aligned}\Psi(x, N)\mathbb{A}(\eta, N) = & G^+(\eta, N)\mathbb{A}(\eta_+, N_+) + G^-(\eta, N)\mathbb{A}(\eta_-, N_-) \\ & - H^+(\eta, N)\mathbb{B}(\eta_+, N_+) - H^-(\eta, N)\mathbb{B}(\eta_-, N_-),\end{aligned} \tag{7.32}$$

$$\begin{aligned}\Psi(x, N)\mathbb{B}(\eta, N) = & H^+(\eta, N)\mathbb{A}(\eta_+, N_+) + H^-(\eta, N)\mathbb{A}(\eta_-, N_-) \\ & + G^+(\eta, N)\mathbb{B}(\eta_+, N_+) + G^-(\eta, N)\mathbb{B}(\eta_-, N_-).\end{aligned} \tag{7.33}$$

To expand the right-hand sides of the above two equations as power series of $1/N$, we shall first expand $\mathbb{A}(\eta_\pm, N_\pm)$ and $\mathbb{B}(\eta_\pm, N_\pm)$. Denote D^j to be the j-th derivative with respect to t; namely, $D^j A_r(\eta) = d^j A_r(\eta(t))/dt^j$ for $j, r = 0, 1, 2, \cdots$. Define $\gamma_{j,l}$ as the coefficient of ε^l in the power series of $[(1+\varepsilon)^{-2} - 1]^j$. We have

$$(t_\pm - t)^j = \sum_{l=0}^{\infty} \frac{\gamma_{j,l} t^j}{(\pm 1)^l N^l}, \quad A_r(\eta_\pm) = \sum_{j=0}^{\infty}\sum_{l=0}^{\infty} \frac{D^j A_r(\eta)\gamma_{j,l} t^j}{(\pm 1)^l j! N^l}, \tag{7.34}$$

$$\mathbb{A}(\eta_\pm, N_\pm) = \sum_{r=0}^{\infty}\sum_{j=0}^{\infty}\sum_{l=0}^{\infty}\sum_{k=0}^{\infty} \binom{-r}{k} \frac{D^j A_r(\eta)\gamma_{j,l} t^j}{(\pm 1)^{l+k} j! N^{l+r+k}}, \tag{7.35}$$

$$\mathbb{B}(\eta_\pm, N_\pm) = \sum_{r=0}^{\infty}\sum_{j=0}^{\infty}\sum_{l=0}^{\infty}\sum_{k=0}^{\infty} \binom{-r}{k} \frac{D^j B_r(\eta)\gamma_{j,l} t^j}{(\pm 1)^{l+k} j! N^{l+r+k}}. \tag{7.36}$$

Recall the series expansions of $G^\pm$ and $H^\pm$ in Proposition 7.1. By equating the coefficients of N^{-p} in the series expansions of (7.32) and (7.33), we obtain from (7.29), (7.30), (7.35) and (7.36)

$$\begin{aligned}&\sum_{s\le p} \psi_s(t) A_{p-s}(\eta) \qquad\qquad\qquad (7.37)\\ &= \sum_{s\le p, s \text{ even}}\ \sum_{i+l\le s} \binom{-p+s}{s-i-l} G_i(\eta) \sum_{j=0}^{l} \frac{\gamma_{j,l} t^j D^j A_{p-s}(\eta)}{j!} \\ &\quad - \sum_{s\le p, s \text{ odd}}\ \sum_{i+l\le s} \binom{-p+s}{s-i-l} H_i(\eta) \sum_{j=0}^{l} \frac{\gamma_{j,l} t^j D^j B_{p-s}(\eta)}{j!},\end{aligned}$$

and

$$\sum_{s\le p}\psi_s(t)B_{p-s}(\eta) \tag{7.38}$$

$$= \sum_{s\le p, s \text{ odd}}\sum_{i+l\le s}\binom{-p+s}{s-i-l}H_i(\eta)\sum_{j=0}^{l}\frac{\gamma_{j,l}t^jD^jA_{p-s}(\eta)}{j!}$$

$$+ \sum_{s\le p, s \text{ even}}\sum_{i+l\le s}\binom{-p+s}{s-i-l}G_i(\eta)\sum_{j=0}^{l}\frac{\gamma_{j,l}t^jD^jB_{p-s}(\eta)}{j!},$$

for $p = 0, 1, \cdots$, where we have made use of the fact that $\gamma_{j,l} = 0$ for $j > l$. Setting $p = 0$ gives $[\psi_0(t) - G_0(\eta)]A_0(\eta) = [\psi_0(t) - G_0(\eta)]B_0(\eta) = 0$. We may assume without loss of generality that either A_0 or B_0 is nonzero (exercise). It then follows from (7.26) that $\cos u_0 = \psi_0$, where $u_0 = \eta - 2t\eta'$ and $\psi_0 = (a_0't + b_0')/2 = (1 + a_0t/2)$; see (7.25) and (7.30). Consequently,

$$\eta(t) - 2t\eta'(t) = \arccos\psi_0(t). \tag{7.39}$$

Solving this differential equation gives (exercise)

$$\eta(t) = \arccos\left(1 + \frac{a_0't}{2}\right) + \frac{\sqrt{-a_0't}}{2}\log\frac{2 + a_0't/4 + 2\sqrt{1 + a_0't/4}}{-a_0't/4}. \tag{7.40}$$

For any constant c_0, the function $\tilde{\eta}(t) = \eta(t) + c_0\sqrt{t}$ also solves the differential equation $\tilde{\eta}(t) - 2t\tilde{\eta}'(t) = \arccos\psi_0(t)$. However, since $\cos(N\tilde{\eta})$ and $\sin(N\tilde{\eta})$ can be expressed as a linear combination of $\cos(N\eta)$ and $\sin(N\eta)$ with coefficients depending only on $x = N^2t$, we may assume without loss of generality that the constant of integration $c_0 = 0$. It is obvious that $u_0 = \eta(t) - 2t\eta'(t) = \arccos\psi_0(t)$ is analytic for $t \in \mathbf{C} \setminus (-\infty, 0] \cup [t_*, \infty)$. For $t < 0$, the one-sided limits of u_0 obtained from upper and lower half-planes differ by a sign. Without loss of generality, we choose

$$u_0 = -i\,\text{arccosh}\,\psi_0(t) = -i\,\text{arccosh}\left(1 + \frac{a_0't}{2}\right), \quad t < 0. \tag{7.41}$$

Now, we come back to equations (7.37) and (7.38) with $p \ge 1$ and solve for A_{p-1} and B_{p-1} recursively. For convenience, we denote

$$f_{p-1}(t) := \sum_{2\le s\le p}\psi_s(t)A_{p-s}(\eta) \tag{7.42}$$

$$- \sum_{2\le s\le p, s \text{ even}}\sum_{i+l\le s}\binom{-p+s}{s-i-l}G_i(\eta)\sum_{j=0}^{l}\frac{\gamma_{j,l}t^jD^jA_{p-s}(\eta)}{j!}$$

$$+ \sum_{2\le s\le p, s \text{ odd}}\sum_{i+l\le s}\binom{-p+s}{s-i-l}H_i(\eta)\sum_{j=0}^{l}\frac{\gamma_{j,l}t^jD^jB_{p-s}(\eta)}{j!},$$

and

$$g_{p-1}(t) := \sum_{2\le s\le p} \psi_s(t)B_{p-s}(\eta) \tag{7.43}$$

$$- \sum_{2\le s\le p, s \text{ odd}} \sum_{i+l\le s} \binom{-p+s}{s-i-l} H_i(\eta) \sum_{j=0}^{l} \frac{\gamma_{j,l}t^j D^j A_{p-s}(\eta)}{j!}$$

$$- \sum_{2\le s\le p, s \text{ even}} \sum_{i+l\le s} \binom{-p+s}{s-i-l} G_i(\eta) \sum_{j=0}^{l} \frac{\gamma_{j,l}t^j D^j B_{p-s}(\eta)}{j!},$$

where the empty sum is zero; namely, $f_0(t) = g_0(t) = 0$. We can simplify (7.37) and (7.38) as (exercise)

$$-[H_1(\eta) + (1-p)H_0(\eta)]B_{p-1}(\eta) + 2tH_0(\eta)\frac{d}{dt}B_{p-1}(\eta) = f_{p-1}(t), \tag{7.44}$$

$$[H_1(\eta) + (1-p)H_0(\eta)]A_{p-1}(\eta) - 2tH_0(\eta)\frac{d}{dt}A_{p-1}(\eta) = g_{p-1}(t), \tag{7.45}$$

where we have made use of the facts that $\psi_0 = G_0$ and $\psi_1 = 0$. There was a typo in Li *et al.* (2020, (4.30)); namely, either the left-hand side or the right-hand side of the equation should be multiplied by -1. Recall from (7.27) that $H_1 = -t\frac{d}{dt}H_0$. Substituting this into (7.44) and then dividing both sides of (7.44) by $-tH_0B_{p-1}$ give

$$\frac{f_{p-1}}{tH_0B_{p-1}} = \frac{1}{H_0}\frac{dH_0}{dt} + \frac{p-1}{t} + \frac{2}{B_{p-1}}\frac{dB_{p-1}}{dt} = 2\frac{d}{dt}\ln[t^{(p-1)/2}(H_0)^{1/2}B_{p-1}]$$

$$= \frac{2}{t^{(p-1)/2}(H_0)^{1/2}B_{p-1}}\frac{d}{dt}[t^{(p-1)/2}(H_0)^{1/2}B_{p-1}].$$

Denote $\Lambda(t) := [H_0(\eta)]^{1/2}$. The above equation can be rewritten as

$$\frac{d}{dt}[t^{(p-1)/2}\Lambda(t)B_{p-1}(t)] = \frac{t^{(p-1)/2-1}}{2\Lambda(t)}f_{p-1}(t). \tag{7.46}$$

There was a typo in Li *et al.* (2020, (4.35)); namely, the negative sign on the right-hand side of the first equation should be replaced by a positive sign. In a similar manner, we substitute $H_1 = -t\frac{d}{dt}H_0$ into (7.45) and obtain

$$\frac{d}{dt}[t^{(p-1)/2}\Lambda(t)A_{p-1}(t)] = -\frac{t^{(p-1)/2-1}}{2\Lambda(t)}g_{p-1}(t). \tag{7.47}$$

When $p = 1$, it follows from $f_0 = g_0 = 0$ that ΛB_0 and ΛA_0 are two constants independent of t. We set $\Lambda A_0 = 1$ and $\Lambda B_0 = 0$. When $p > 1$, we integrate the above two equations to obtain

$$t^{(p-1)/2}\Lambda B_{p-1} = \int_0^t \frac{s^{(p-1)/2-1}f_{p-1}(s)}{2\Lambda(s)}ds, \tag{7.48}$$

$$t^{(p-1)/2}\Lambda A_{p-1} = -\int_0^t \frac{s^{(p-1)/2-1}g_{p-1}(s)}{2\Lambda(s)}ds. \tag{7.49}$$

There was a typo in Li *et al.* (2020, (4.38)); namely, the negative sign on the right-hand side of the equation should be replaced with a positive sign. For each $p > 1$, we can successively determine $A_p(\eta)$ and $B_p(\eta)$ from their predecessors $A_0(\eta)$, $B_0(\eta), \cdots$, $A_{p-1}(\eta)$ and $B_{p-1}(\eta)$ via the above two formulas together with (7.42) and (7.43).

7.4 Bounds for the coefficients

To establish global uniformity of our anticipated expansion, we need to estimate the coefficients $A_s(\eta)$ and $B_s(\eta)$ in (7.11). First, we need the following lemma.

Lemma 7.2. *Let $\eta(t)$ be given as in (7.40), and let $G_s(\eta)$ and $H_s(\eta)$ be given as in Proposition 7.1. Then there exists a positive constant N_s independent of t such that*

$$|G_s(\eta)| + |H_s(\eta)| \leq N_s(|t| + 1), \tag{7.50}$$

for all $t \leq t_* - \sigma$, $\sigma > 0$.

Proof. We only prove the estimate for $G_s(\eta)$; the estimate for $H_s(\eta)$ can be obtained in a similar manner. Recall from Proposition 7.1 that $u_j = h_{j+1}$ and $h_j = h^{(j)}(0)/j!$, where

$$h(\varepsilon) = (1+\varepsilon)\eta(t(1+\varepsilon)^{-2}).$$

It then follows from (7.39) that

$$\begin{aligned} h'(\varepsilon) =& \eta(t(1+\varepsilon)^{-2}) - 2t(1+\varepsilon)^{-2}\eta'(t(1+\varepsilon)^{-2}) \\ =& \arccos[\psi_0(t(1+\varepsilon)^{-2})], \end{aligned} \tag{7.51}$$

where $\psi_0(t) = 1 + a_0' t/2$. Hence, $u_j = u_j(t)$ are analytic functions for $t \in \mathbf{C} \setminus (-\infty, 0] \cup [t_*, \infty)$. Let $u_{m,l}$ and Φ_l be given as in (7.19) and (7.21), respectively. One can further prove that $\Phi_l(u_0)u_{m,l}$ and $\Phi_{l+1}(u_0)u_{m,l}$ can be analytically continued to $(-\infty, 0]$ (exercise). Hence, both $G_s(\eta)$ and $H_s(\eta)$ are analytic for $t \in \mathbf{C}\setminus[t_*, \infty)$. Especially, $G_s(\eta)$ in (7.24) is bounded for t in any compact subset of $(-\infty, t_*)$. It suffices to show that $G_s(\eta)/t$ is bounded as $t \to -\infty$. To achieve this, we have to find another series expansion of $G_s(\eta)$. Denote $\xi = -iN\eta$ and $\xi_\pm = -iN_\pm\eta_\pm$. We obtain from Lemma 7.1

$$G^\pm(\eta, N) = \cos[i(\xi_\pm - \xi)] = \frac{e^{\xi_\pm - \xi} + e^{\xi - \xi_\pm}}{2}. \tag{7.52}$$

Note that (exercise)

$$\begin{aligned}\xi_+ - \xi &= -iN[h(1/N) - h(0)] = -iN\int_0^{1/N} \arccos[\psi_0(t(1+\varepsilon)^{-2})]d\varepsilon \\ &= \int_0^1 \operatorname{arccosh}[\psi_0(t(1+\lambda/N)^{-2})]d\lambda \\ &= \operatorname{arccosh}(\psi_0(t)) + \frac{a_0' t}{N}\int_0^1 \frac{(\lambda-1)(1+\lambda/N)^{-3}}{\sqrt{\psi_0^2(t(1+\lambda/N)^{-2})-1}}d\lambda \\ &= \sum_{j=0}^{\infty} \frac{v_j(t)}{N^j}, \end{aligned} \tag{7.53}$$

where $v_0(t) = \operatorname{arccosh}(\psi_0(t)) = O(\log|t|)$, and

$$v_j(t) = \frac{2(-1)^j}{j(j+1)} + O(1/|t|), \quad j \geq 1, \tag{7.54}$$

as $t \to -\infty$. Consequently,

$$e^{\xi_+ - \xi} = e^{v_0(t)} \sum_{l=0}^{\infty} \frac{w_{1,l}(t)}{N^l}, \qquad e^{\xi - \xi_+} = e^{-v_0(t)} \sum_{l=0}^{\infty} \frac{w_{2,l}(t)}{N^l}, \tag{7.55}$$

where $w_{1,l}(t)$ and $w_{2,l}(t)$ are bounded as $t \to -\infty$, while the bound may depend on the index l. Substituting the above expansions into the right-hand side of (7.52) and collecting the coefficient of N^{-s} yields

$$G_s(\eta) = \frac{e^{v_0(t)} w_{1,s}(t) + e^{-v_0(t)} w_{2,s}(t)}{2}.$$

Since $e^{v_0(t)}$ and $e^{-v_0(t)}$ are of order $O(|t|)$ as $t \to -\infty$, so is $G_s(\eta)$. This gives the estimate $|G_s(\eta)| \leq (N_s/2)(|t|+1)$ for some positive constant N_s. In a similar manner, we can prove that $|H_s(\eta)| \leq (N_s/2)(|t|+1)$, thus proving (7.50). □

Now, we are ready to provide some bounds for the coefficients $A_s(\eta)$ and $B_s(\eta)$ in (7.11).

Lemma 7.3. *Denote $\Lambda(t) = [H_0(\eta)]^{1/2}$. For $s, j = 0, 1, 2, \cdots$, there exists a constant $M_{s,j}$ independent of t such that*

$$|t^j \Lambda(t) D^j A_s(\eta)| + |t^j \Lambda(t) D^j B_s(\eta)| \leq M_{s,j}(1 + |t|^{-s/2}), \tag{7.56}$$

for $t < 0$, where D^j denotes the j-th derivative with respect to t.

Proof. We will prove by induction. First, we note that the estimates in Lemma 7.2 hold for all complex t in the left half-plane, because the corresponding functions $G_s(\eta)$ and $H_s(\eta)$ are analytic for $t \in \mathbf{C} \setminus [t_*, \infty)$. In view of (7.39) and (7.40), we obtain

$$|t^j D^j \eta| \le \tilde{N}_j(1 + |\eta|), \tag{7.57}$$

for some positive constant $\tilde{N}_j$. By using Cauchy integral formula, one can further prove that

$$|t^j D^j G_s| + |t^j D^j H_s| \le N_{s,j}(1 + |t|), \tag{7.58}$$

for some positive constant $N_{s,j}$. When $s = 0$, since $\Lambda A_0 = 1$ and $\Lambda B_0 = 0$, the estimate in (7.56) holds for $j = 0$. By Cauchy integral formula, it also holds for $j \ge 1$. Now, we assume that the estimate in (7.56) holds for $s = 0, 1, \cdots, p-2$ with $p \ge 2$ and for $j = 0, 1, \cdots$. It follows from (7.42), (7.43) and the above estimates that

$$|\Lambda f_{p-1}| + |\Lambda g_{p-1}| \le M'_p(1 + |t|)(1 + |t|^{-(p-2)/2}), \tag{7.59}$$

for some constant M'_p. Substituting this into (7.48) and (7.49) gives (exercise)

$$|\Lambda A_{p-1}| + |\Lambda B_{p-1}| \le M_{p-1,0}(1 + |t|^{-(p-1)/2}), \tag{7.60}$$

for some constant $M_{p-1,0}$. Finally, (7.56) with $s = p-1$ and $j \ge 1$ follows from a direct application of the Cauchy integral formula. □

7.5 Asymptotic nature of the expansion

Note that e^{iz} and e^{-iz} are two linearly independent solutions of $\chi''(z) = -\chi(z)$. We choose these two solutions, instead of the cosine and sine functions, because $|e^{iz}| \cdot |e^{-iz}| = e^{-\operatorname{Im} z} \cdot e^{\operatorname{Im} z} = 1$ for all $z \in \mathbf{C}$. For each $p \ge 0$, we truncate the expansion (7.11), and replace $\chi(N\eta)$ with $e^{iN\eta}$ and $e^{-iN\eta}$, respectively. This gives us two asymptotic solutions

$$r_n^p(N^2 t) := e^{iN\eta}\mathbb{A}_p(\eta, N) + ie^{iN\eta}\mathbb{B}_p(\eta, N), \tag{7.61}$$
$$s_n^p(N^2 t) := e^{-iN\eta}\mathbb{A}_p(\eta, N) - ie^{-iN\eta}\mathbb{B}_p(\eta, N), \tag{7.62}$$

where

$$\mathbb{A}_p(\eta, N) = \sum_{s=0}^{p} \frac{A_s(\eta)}{N^s} = (H_0)^{-1/2} \sum_{s=0}^{p} \frac{\tilde{A}_s(\eta)}{N^s}, \tag{7.63}$$

$$\mathbb{B}_p(\eta, N) = \sum_{s=0}^{p} \frac{B_s(\eta)}{N^s} = (H_0)^{-1/2} \sum_{s=0}^{p} \frac{\tilde{B}_s(\eta)}{N^s}, \tag{7.64}$$

where $\tilde{A}_s(\eta) := \Lambda(t)A_s(\eta)$ and $\tilde{B}_s(\eta) := \Lambda(t)B_s(\eta)$ with $\Lambda(t) := (H_0)^{1/2}$. Especially, $\tilde{A}_0(\eta) = 1$ and $\tilde{B}_0(\eta) = 0$. It is worth recalling that $H_0 = \sin u_0 = (1-\psi_0^2)^{1/2}$ with $2\psi_0 = a_0't + b_0' = 2 + a_0't$. We also recall that $x = N^2 t$ and $\Psi(t, N) := C_1(n)x + C_0(n)$. From Lemma 7.1, it follows that

$$\begin{aligned}\frac{R_n^p(x)}{N^{p+2}} :=& r_{n+1}^p(x) - [C_1(n)x + C_0(n)]r_n^p(x) + r_{n-1}^p(x)\\ =& e^{iN\eta}F_{1,n}^p(x) + ie^{iN\eta}F_{2,n}^p(x), \qquad (7.65)\\ \frac{S_n^p(x)}{N^{p+2}} :=& s_{n+1}^p(x) - [C_1(n)x + C_0(n)]s_n^p(x) + s_{n-1}^p(x)\\ =& e^{-iN\eta}F_{1,n}^p(x) - ie^{-iN\eta}F_{2,n}^p(x), \qquad (7.66)\end{aligned}$$

where

$$\begin{aligned}F_{1,n}^p(x) =& G^+(\eta, N)\mathbb{A}_p(\eta_+, N_+) + G^-(\eta, N)\mathbb{A}_p(\eta_-, N_-) - \Psi(t, N)\mathbb{A}_p(\eta, N)\\ &- H^+(\eta, N)\mathbb{B}_p(\eta_+, N_+) - H^-(\eta, N)\mathbb{B}_p(\eta_-, N_-), \qquad (7.67)\\ F_{2,n}^p(x) =& H^+(\eta, N)\mathbb{A}_p(\eta_+, N_+) + H^-(\eta, N)\mathbb{A}_p(\eta_-, N_-) - \Psi(t, N)\mathbb{B}_p(\eta, N)\\ &+ G^+(\eta, N)\mathbb{B}_p(\eta_+, N_+) + G^-(\eta, N)\mathbb{B}_p(\eta_-, N_-). \qquad (7.68)\end{aligned}$$

On account of Lemma 7.3, we have

$$|t^j D^j \tilde{A}_s(\eta)| + |t^j D^j \tilde{B}_s(\eta)| \le \tilde{M}_{s,j}(1 + |t|^{-s/2}), \qquad (7.69)$$

for some constant $\tilde{M}_{s,j}$. Now, we improve the above estimate and prove by induction

$$|\tilde{A}_s(\eta)| + |\tilde{B}_s(\eta)| \le \tilde{M}_s, \qquad (7.70)$$

for some constant $\tilde{M}_s$ and all $t < t_* - \sigma$. Obviously, the above estimate holds for $s = 0$. Assume it holds for $s = 0, \cdots, p$. Using an argument similar to that given for (7.59) and (7.60), we have from (7.42) and (7.43) that

$$|\Lambda f_{p+1}| + |\Lambda g_{p+1}| \le C_{p+1,0}(1 + |t|).$$

As a consequence, it can be shown that (7.56) can be improved to read

$$|t^j \Lambda(t) D^j A_s(\eta)| + |t^j \Lambda(t) D^j B_s(\eta)| \le M_{s,j}$$

for $t < 0$ and $s, j \ge 0$. In particular, these two estimates hold for $j = 0$ and $s = p+1$; that is, $\tilde{A}_{p+1}(\eta)$ and $\tilde{B}_{p+1}(\eta)$ are bounded for $t < 0$. By the continuity of $\eta(t), G_s(\eta)$ and $H_s(\eta)$, we conclude that $\tilde{A}_{p+1}(\eta)$ and $\tilde{B}_{p+1}(\eta)$ are also bounded for $t < t_* - \sigma$, thus completing the induction argument.

It can be proved that (exercise)

$$|\Lambda(t)F_{1,n}^p(x)| + |\Lambda(t)F_{2,n}^p(x)| \le \frac{C_p'(1+|t|)}{N^{p+2}}, \tag{7.71}$$

for some constant C_p'. Substituting this into (7.65) and (7.66) yields

$$|R_n^p(x)| \le C_p'(1+|t|)\alpha_n(x), \qquad |S_n^p(x)| \le C_p'(1+|t|)\beta_n(x), \tag{7.72}$$

for $t < t_* - \sigma$, where

$$\alpha_n(x) := |H_0|^{-1/2}|e^{iN\eta}|, \qquad \beta_n(x) := |H_0|^{-1/2}|e^{-iN\eta}|. \tag{7.73}$$

Note that $\beta_n(x)/\alpha_n(x) = e^{2N\,\mathrm{Im}\,\eta(N^{-2}x)}$ can be regarded as a function for all positive real n. We fix x and take partial derivative with respect to n. It follows from (7.41) that (exercise)

$$\frac{\partial}{\partial n}\left[\frac{\beta_n(x)}{\alpha_n(x)}\right] = e^{2N\,\mathrm{Im}\,\eta(N^{-2}x)} \cdot 2\,\mathrm{Im}\,u_0 \le 0 \tag{7.74}$$

for $t = N^{-2}x < 0$. If $t \in [0, t_*)$, then $\eta(t)$ is real and $\beta_n(x)/\alpha_n(x) = 1$. Coupling these two cases yields

$$\frac{\beta_n(x)}{\alpha_n(x)} \ge \frac{\beta_j(x)}{\alpha_j(x)} \tag{7.75}$$

for all $j \ge n$ and $t < t_*$. We then have the following lemma.

Lemma 7.4. *Fix $p \ge 0$. There exist $\{\varepsilon_n^p(x)\}_{n=1}^\infty$ and $\{\delta_n^p(x)\}_{n=1}^\infty$ such that*

$$\varepsilon_{n+1}^p(x) - [C_1(n)x + C_0(n)]\varepsilon_n^p(x) + \varepsilon_{n-1}^p(x) = -\frac{R_n^p(x)}{N^{p+2}}, \tag{7.76}$$

$$\delta_{n+1}^p(x) - [C_1(n)x + C_0(n)]\delta_n^p(x) + \delta_{n-1}^p(x) = -\frac{S_n^p(x)}{N^{p+2}}, \tag{7.77}$$

with

$$|\delta_n^p(x)| \le \frac{M_p\beta_n(x)}{N^{p+1}}, \qquad |\varepsilon_n^p(x)| \le \frac{M_p\alpha_n(x)}{N^{p+1}}, \tag{7.78}$$

for some positive constant M_p.

Proof. It can be shown that the solution of the difference equation (7.77) with asymptotic estimate in (7.78) is the same as the solution to the following summation equation with the same asymptotic estimate:

$$\delta_n^p(x) = \sum_{j=n+1}^\infty \frac{[s_n^p(x)r_j^p(x) - r_n^p(x)s_j^p(x)]S_j^p(x)}{[s_{n+1}^p(x)r_n^p(x) - s_n^p(x)r_{n+1}^p(x)](j+\tau_0)^{p+2}}$$
$$+ \sum_{j=n+1}^\infty \frac{[s_n^p(x)R_j^p(x) - r_n^p(x)S_j^p(x)]\delta_j^p(x)}{[s_{n+1}^p(x)r_n^p(x) - s_n^p(x)r_{n+1}^p(x)](j+\tau_0)^{p+2}}. \tag{7.79}$$

We will construct such a solution by successive approximation. Starting with $\delta^p_{n,0} = 0$, we define $\delta^p_{n,k}(x)$ recursively by

$$\delta^p_{n,k}(x) = \sum_{j=n+1}^{\infty} \frac{[s^p_n(x)r^p_j(x) - r^p_n(x)s^p_j(x)]S^p_j(x)}{[s^p_{n+1}(x)r^p_n(x) - s^p_n(x)r^p_{n+1}(x)](j+\tau_0)^{p+2}} + \sum_{j=n+1}^{\infty} \frac{[s^p_n(x)R^p_j(x) - r^p_n(x)S^p_j(x)]\delta^p_{j,k-1}(x)}{[s^p_{n+1}(x)r^p_n(x) - s^p_n(x)r^p_{n+1}(x)](j+\tau_0)^{p+2}}. \tag{7.80}$$

Since $\tilde{A}_s(\eta)$ and $\tilde{B}_s(\eta)$ are bounded, it follows from (7.61) and (7.62) that

$$|r^p_n(x)| \le C_p\alpha_n(x), \qquad |s^p_n(x)| \le C_p\beta_n(x), \tag{7.81}$$

for some constant C_p. Moreover, in view of $|e^{iN\eta}| \cdot |e^{-iN\eta}| = 1$, we have from (7.72), (7.75) and (7.81) that

$$\begin{aligned}
&|[s^p_n(x)r^p_j(x) - r^p_n(x)s^p_j(x)]S^p_j(x)| \\
\le&(C_p)^2C'_p(1+|t|)[\beta_n(x)\alpha_j(x) + \alpha_n(x)\beta_j(x)]\beta_j(x) \\
\le&2(C_p)^2C'_p(1+|t|)\beta_n(x)\alpha_j(x)\beta_j(x) \le M'_p\beta_n(x),
\end{aligned} \tag{7.82}$$

for some constant M'_p. Note from (7.61) and (7.62) that (exercise)

$$\lim_{n\to\infty}[r^p_{n+1}(x)s^p_n(x) - s^p_{n+1}r^p_n(x)] = \lim_{n\to\infty}\frac{2i\sin(u^+)}{H_0} = 2i. \tag{7.83}$$

Especially,

$$|r^p_{n+1}(x)s^p_n(x) - r^p_n(x)s^p_{n+1}(x)| > 1, \tag{7.84}$$

for all sufficiently large n. A combination of (7.80)-(7.82) and (7.84) yields

$$|\delta^p_{n,1}(x)| \le \sum_{j=n+1}^{\infty}\frac{M'_p\beta_n(x)}{(j+\tau_0)^{p+2}} \le \frac{M''_p}{N^{p+1}}\beta_n(x), \tag{7.85}$$

for some positive constant M''_p. Similarly, we can prove by induction that

$$|\delta^p_{n,k}(x) - \delta^p_{n,k-1}(x)| \le \left(\frac{M''_p}{N^{p+1}}\right)^k \beta_n(x), \tag{7.86}$$

from which it follows that, for all sufficiently large n,

$$\delta^p_{n,k}(x) = \sum_{j=1}^{k}[\delta^p_{n,j}(x) - \delta^p_{n,j-1}(x)] \tag{7.87}$$

converges as $k \to \infty$. Clearly, the limit function $\delta^p_n(x)$ satisfies (7.79) as well as (7.77). Moreover, we obtain from (7.85)-(7.87)

$$|\delta^p_n(x)| \le \frac{M_p}{N^{p+1}}\beta_n(x), \tag{7.88}$$

for some positive constant M_p. In a similar manner, we can show that

$$|\varepsilon^p_n(x)| \le \frac{M_p}{N^{p+1}}\alpha_n(x), \tag{7.89}$$

thus completing the proof. □

For each $p \geq 0$, we define $P_n^p(x) := r_n^p(x) + \varepsilon_n^p(x)$ and $Q_n^p(x) := s_n^p(x) + \delta_n^p(x)$. It is obvious that $P_n^p(x)$ and $Q_n^p(x)$ are two solutions to the difference equation (7.1). Making use of the fact that $\varepsilon_n^p(x) \to 0$ and $\delta_n^p(x) \to 0$ as $n \to \infty$, one can prove that (exercise)

$$P_{n+1}^p(x)Q_n^p(x) - P_n^p(x)Q_{n+1}^p(x) = 2i. \tag{7.90}$$

Especially, $P_n^p(x)$ and $Q_n^p(x)$ are two linearly independent solutions to (7.1). For any $p \neq q$, it can be shown that (exercise) $P_n^p(x) = P_n^q(x)$ and $Q_n^p(x) = Q_n^q(x)$. Hence, $P_n(x) = P_n^p(x)$ and $Q_n(x) = Q_n^p(x)$ are independent of p. Now, we obtain the following theorem.

Theorem 7.1. *Assume that the coefficients $C_1(n)$ and $C_0(n)$ in the linear difference equation (7.1) have the asymptotic expansions of the form given in (7.2) with $a_0 < 0$, $b_0 = 2$ and $b_1 = 0$. Set $N = n + \tau_0$ with $\tau_0 = -a_1/(2a_0)$. Let $\eta(t)$ be given as in (7.40) and $\Lambda(t) = (H_0)^{1/2} = [1-(a_0t/2+1)^2)]^{1/4}$. Denote $\tilde{A}_s(\eta) = \Lambda(t)A_s(\eta)$ and $\tilde{B}_s(\eta) = \Lambda(t)B_s(\eta)$, where $A_s(\eta)$ and $B_s(\eta)$ are successively defined by (7.42)-(7.43) and (7.48)-(7.49) with $\tilde{A}_0(\eta) = \Lambda(t)A_0(\eta) = 1$ and $\tilde{B}_0(\eta) = \Lambda(t)B_0(\eta) = 0$. Then equation (7.1) has, for each integer $p \geq 0$, a pair of linearly independent solutions $P_n(x)$ and $Q_n(x)$ such that*

$$P_n(N^2t) = \frac{e^{iN\eta}}{[-a_0t(1+\frac{a_0t}{4})]^{1/4}}\left[\sum_{s=0}^{p}\frac{\tilde{A}_s(\eta) + i\tilde{B}_s(\eta)}{N^s} + \varepsilon_p(t,N)\right], \tag{7.91}$$

$$Q_n(N^2t) = \frac{e^{-iN\eta}}{[-a_0t(1+\frac{a_0t}{4})]^{1/4}}\left[\sum_{s=0}^{p}\frac{\tilde{A}_s(\eta) - i\tilde{B}_s(\eta)}{N^s} + \delta_p(t,N)\right], \tag{7.92}$$

where $x = N^2t$, $t < t_* - \sigma$, $t_* = -2/a_0 > 0$, and $\sigma > 0$ is an arbitrarily small positive constant. Moreover,

$$|\varepsilon_p(t,N)| \leq \frac{M_p}{N^{p+1}}, \qquad |\delta_p(t,N)| \leq \frac{M_p}{N^{p+1}}, \tag{7.93}$$

for some positive constant M_p.

Proof. Note that $\varepsilon_n^p(x) = (H_0)^{-1/2}e^{iN\eta}\varepsilon_p(t,N)$ and $\delta_n^p(x) = (H_0)^{-1/2}e^{-iN\eta}\delta_p(t,N)$ with $(H_0)^{-1/2} = [1-(a_0t/2+1)^2)]^{-1/4}$. The result is a direct consequence of Lemma 7.4. □

7.6 Exercises

1. Prove Proposition 7.1.
2. Prove (7.25)-(7.27).

3. Let $P_n(x)$ in (7.28) be a solution to (7.1) with series expansions in (7.29). Assume $A_s = B_s = 0$ for all $s < p$ and either A_p or B_p is nonzero. Show that (7.37)-(7.38) become $(\psi_0 - G_0)A_p = (\psi_0 - G_0)B_p = 0$, from which we have $\cos u_0 = \psi_0$.
4. Verify that $\eta(t)$ defined in (7.40) is a solution to the differential equation (7.39). Show that any general solution of (7.39) can be expressed as $\eta(t) + c_0\sqrt{t}$ for some constant c_0.
5. Verify (7.44) and (7.45).
6. Let u_j, $u_{m,l}$ and Φ_l be given as in Proposition 7.1. Show that $u_j = u_j(t)$ are analytic functions for $t \in \mathbf{C}\backslash(-\infty, 0]\cup[t_*, \infty)$. Prove that $\Phi_l(u_0)u_{m,l}$ and $\Phi_{l+1}(u_0)u_{m,l}$ can be analytically continued to $(-\infty, 0]$.
7. Verify (7.53) and (7.54).
8. Prove (7.71).
9. Verify (7.74)
10. Verify (7.83).
11. Let $r_n^p(x)$ and $s_n^p(x)$ with $p \geq 0$ be defined as in (7.61) and (7.62). If $P_n(x)$ and $Q_n(x)$ are two solutions to (7.1) such that $P_n(x) - r_n^0(x) \to 0$ and $Q_n(x) - s_n^0(x) \to 0$ as $n \to \infty$, prove that $P_{n+1}(x)Q_n(x) - P_n(x)Q_{n+1}(x) = 2i$. Moreover, show that such pair of solutions are unique; namely, if $P_n^*(x)$ and $Q_n^*(x)$ are two solutions to (7.1) such that $P_n^*(x) - r_n^0(x) \to 0$ and $Q_n^*(x) - s_n^0(x) \to 0$ as $n \to \infty$, then $P_n^*(x) = P_n(x)$ and $Q_n^*(x) = Q_n(x)$.

Part 2

Riemann-Hilbert Problems

Chapter 8

Riemann-Hilbert Problem

8.1 Introduction

Riemann-Hilbert problem is an inverse problem of finding an analytic function with given jump conditions across contours in the complex plane. It has been widely applied in the asymptotic theory of differential equations [Deift and Zhou (1993)] and orthogonal polynomials [Deift *et al.* (1999)]. One significant application of Riemann-Hilbert problem is the asymptotic analysis of orthogonal polynomials with respect to the Freud weight. Unlike the classical orthogonal polynomials, there is no integral representation or differential equation for these orthogonal polynomials, and thus, the classical steepest descent method (for integrals) or WKB method (for differential equations) is not applicable. In this chapter, we sketch the standard procedure of deriving asymptotic formulas of orthogonal polynomials via Riemann-Hilbert problem.

Let Σ be an oriented contour in the complex plane and $f(z)$ a function that is analytic near Σ with one-sided limits on Σ. For each $z \in \Sigma$, we denote by $f_+(z)$ (resp. $f_-(z)$) the limit of $f(\zeta)$ as ζ approaches z from the left (resp. right) of Σ. Especially, if Σ is the real line $\mathbf{R}$ oriented to the right, the notations become

$$f_\pm(x) = \lim_{\varepsilon\to 0^+} f(x \pm i\varepsilon), \quad x \in \mathbf{R}. \tag{8.1}$$

As an illustration, we consider the following scalar Riemann-Hilbert problem [Wang and Wong (2011), (3.20)-(3.21)] on $\Sigma = (-i\infty, i\infty)$ with upwards orientation.

(D1) $D(z)$ is analytic in $\mathbf{C} \setminus (-i\infty, i\infty)$.
(D2) $D(z)$ satisfies the jump condition

$$D_+(z) = D_-(z)J_D(z), \quad z \in (-i\infty, i\infty), \tag{8.2}$$

where $J_D(z) = 1 - e^{-2\pi|z|}$ for $z \in (-i\infty, i\infty)$.

(D3) For $z \in \mathbf{C} \setminus (-i\infty, i\infty)$, $D(z) = 1 + O(|z|^{-1})$ as $z \to \infty$.

By Plemelj formula, we can write $D(z)$ in terms of Cauchy integral:

$$\begin{aligned} D(z) &= \exp\left\{\frac{1}{2\pi i}\int_{-i\infty}^{i\infty} \ln J_D(\zeta)\frac{d\zeta}{\zeta - z}\right\} \\ &= \exp\left\{\frac{1}{2\pi i}\int_0^{\infty}\left[\frac{\ln(1-e^{-2\pi s})}{s+iz} - \frac{\ln(1-e^{-2\pi s})}{s-iz}\right]ds\right\}. \end{aligned} \tag{8.3}$$

A simple calculation (exercise) yields

$$D(z) = \begin{cases} \dfrac{\Gamma(z+1)}{\sqrt{2\pi z}(z/e)^z}, & \operatorname{Re} z > 0, \\ \\ \dfrac{\sqrt{-2\pi z}(-z/e)^{-z}}{\Gamma(-z+1)}, & \operatorname{Re} z < 0. \end{cases} \tag{8.4}$$

Throughout this chapter, we take the principal component of the logarithmic function with a branch cut on the negative real line. The power function z^α is defined as $e^{\alpha \ln z}$, which is analytic in $\mathbf{C} \setminus (-\infty, 0]$ such that

$$\lim_{\varepsilon \to 0^+} (x \pm i\varepsilon)^\alpha = |x|^\alpha e^{\pm \alpha\pi i}, \quad x < 0.$$

For convenience, given any two real numbers $a < b$, we use $\sqrt{(z-a)(z-b)}$ to indicate the product of power functions $(z-a)^{1/2}(z-b)^{1/2}$, which is analytic in $\mathbf{C} \setminus (-\infty, b]$ and can be analytically continued to $(-\infty, a)$. Especially,

$$\lim_{z \to x} \sqrt{(z-a)(z-b)} = \lim_{z \to x} (z-a)^{1/2}(z-b)^{1/2} = -(a-x)^{1/2}(b-x)^{1/2} < 0$$

for $x < a$. To avoid possible confusion, we never use the notation $\sqrt{(a-z)(b-z)}$. Instead, we use $\sqrt{a-z} \cdot \sqrt{b-z}$ to indicate the product of power functions $(a-z)^{1/2}(b-z)^{1/2}$, which is analytic in $\mathbf{C} \setminus [a, \infty)$ and can be analytically continued to (b, ∞). Especially,

$$\lim_{z \to x} \sqrt{a-z} \cdot \sqrt{b-z} = \lim_{z \to x} (a-z)^{1/2}(b-z)^{1/2} = -(x-a)^{1/2}(x-b)^{1/2} < 0$$

for $x > b$. It is easily seen that $\sqrt{(z-a)(z-b)} = e^{\pm\pi i}\sqrt{a-z} \cdot \sqrt{b-z}$ for $z \in \mathbf{C}_\pm$. For the special case $a = -b$ with $b > 0$, we use $\sqrt{z^2 - b^2}$ to denote the analytic function in $\mathbf{C} \setminus [-b, b]$ that behaves like z as $z \to \infty$; we use $\sqrt{-b-z} \cdot \sqrt{b-z}$ to denote the analytic function in $\mathbf{C} \setminus [-b, b]$ that behaves like $-z$ as $z \to \infty$.

The Pauli matrices

$$\sigma_1 = \begin{pmatrix} 0 & 1 \\ 1 & 0 \end{pmatrix}, \quad \sigma_2 = \begin{pmatrix} 0 & -i \\ i & 0 \end{pmatrix}, \quad \sigma_3 = \begin{pmatrix} 1 & 0 \\ 0 & -1 \end{pmatrix} \tag{8.5}$$

will be used frequently. Given any function $f(z)$ and real number $\alpha \in \mathbf{R}$, we denote for the sake of convenience

$$f(z)^{\alpha\sigma_3} := \begin{pmatrix} f(z)^{\alpha} & 0 \\ 0 & f(z)^{-\alpha} \end{pmatrix},$$

provided both $f(z)^{\alpha}$ and $f(z)^{-\alpha}$ are well-defined.

8.2 Some standard parametrices

Given $a < b$, we consider the following matrix-valued Riemann-Hilbert problem on $[a, b]$ with right orientation.

(M1) $M(z)$ is analytic in $\mathbf{C} \setminus [a, b]$.
(M2) $M(z)$ satisfies the jump condition

$$M_+(x) = M_-(x) \begin{pmatrix} 0 & 1 \\ -1 & 0 \end{pmatrix}, \tag{8.6}$$

for $x \in (a, b)$.
(M3) $M(z) = I + O(|z|^{-1})$, as $z \to \infty$.

In view of the factorization

$$\begin{pmatrix} 0 & 1 \\ -1 & 0 \end{pmatrix} = \frac{1}{2} \begin{pmatrix} 1 & -1 \\ -i & -i \end{pmatrix} \begin{pmatrix} -i & 0 \\ 0 & i \end{pmatrix} \begin{pmatrix} 1 & i \\ -1 & i \end{pmatrix},$$

one can easily verify (exercise) that

$$\begin{aligned} M(z) &= \begin{pmatrix} \dfrac{\sqrt{z-a} + \sqrt{z-b}}{2(z-a)^{1/4}(z-b)^{1/4}} & \dfrac{i(\sqrt{z-a} - \sqrt{z-b})}{2(z-a)^{1/4}(z-b)^{1/4}} \\ \\ \dfrac{-i(\sqrt{z-a} - \sqrt{z-b})}{2(z-a)^{1/4}(z-b)^{1/4}} & \dfrac{\sqrt{z-a} + \sqrt{z-b}}{2(z-a)^{1/4}(z-b)^{1/4}} \end{pmatrix} \\ &= \begin{pmatrix} \dfrac{\sqrt{a-z} + \sqrt{b-z}}{2(a-z)^{1/4}(b-z)^{1/4}} & \dfrac{i(\sqrt{a-z} - \sqrt{b-z})}{2(a-z)^{1/4}(b-z)^{1/4}} \\ \\ \dfrac{-i(\sqrt{a-z} - \sqrt{b-z})}{2(a-z)^{1/4}(b-z)^{1/4}} & \dfrac{\sqrt{a-z} + \sqrt{b-z}}{2(a-z)^{1/4}(b-z)^{1/4}} \end{pmatrix} \end{aligned} \tag{8.7}$$

satisfies the Riemann-Hilbert problem (M1)-(M3). The special case with $a = -1$ and $b = 1$ coincides with the Riemann-Hilbert problem considered in Deift *et al.* (1999, (6.13)-(6.15)).

Next, we introduce the *Airy parametrix* which was defined in Deift *et al.* (1999) and later modified in Wong and Zhang (2007); Wen *et al.* (2018):

$$A(z) := \begin{cases} \begin{pmatrix} \mathrm{Ai}(z) & -\omega^2 \mathrm{Ai}(\omega^2 z) \\ \mathrm{Ai}'(z) & -\omega \mathrm{Ai}'(\omega^2 z) \end{pmatrix}, & z \in \mathbf{C}_+, \\ \\ \begin{pmatrix} \mathrm{Ai}(z) & \omega \mathrm{Ai}(\omega z) \\ \mathrm{Ai}'(z) & \omega^2 \mathrm{Ai}'(\omega z) \end{pmatrix}, & z \in \mathbf{C}_-, \end{cases} \tag{8.8}$$

where $\omega = e^{2\pi i/3}$ and $\mathbf{C}_\pm$ denotes the upper/lower half complex plane. In view of the connection formula for the Airy function [Olver *et al.* (2010), (9.2.12)]

$$\mathrm{Ai}(z) + \omega \mathrm{Ai}(\omega z) + \omega^2 \mathrm{Ai}(\omega^2 z) = 0, \tag{8.9}$$

we obtain the jump condition for $A(z)$ on the real line with right orientation:

$$A_+(x) = A_-(x) \begin{pmatrix} 1 & 1 \\ 0 & 1 \end{pmatrix}, \quad x \in \mathbf{R}. \tag{8.10}$$

From the asymptotic behavior of the Airy function [Olver (1974a), p. 392] (or [Olver *et al.* (2010), (9.7.5)-(9.7.6)]), we have as $z \to \infty$,

$$\mathrm{Ai}(z) \sim \frac{z^{-1/4}}{2\sqrt{\pi}} e^{-\frac{2}{3} z^{3/2}}, \quad \mathrm{Ai}'(z) \sim \frac{-z^{1/4}}{2\sqrt{\pi}} e^{-\frac{2}{3} z^{3/2}} \tag{8.11}$$

for $|\arg z| < \pi - \delta$ with any $\delta > 0$, and

$$-\omega^2 \mathrm{Ai}(\omega^2 z) \sim \frac{iz^{-1/4}}{2\sqrt{\pi}} e^{\frac{2}{3} z^{3/2}}, \quad -\omega \mathrm{Ai}'(\omega^2 z) \sim \frac{iz^{1/4}}{2\sqrt{\pi}} e^{\frac{2}{3} z^{3/2}} \tag{8.12}$$

for $z \in \mathbf{C}_+$, and

$$\omega \mathrm{Ai}(\omega z) \sim \frac{iz^{-1/4}}{2\sqrt{\pi}} e^{\frac{2}{3} z^{3/2}}, \quad \omega^2 \mathrm{Ai}'(\omega z) \sim \frac{iz^{1/4}}{2\sqrt{\pi}} e^{\frac{2}{3} z^{3/2}} \tag{8.13}$$

for $z \in \mathbf{C}_-$. Substituting the above asymptotic formulas into (8.8) yields

$$2\sqrt{\pi} \begin{pmatrix} z^{1/4} & 0 \\ 0 & z^{-1/4} \end{pmatrix} A(z) \sim \begin{pmatrix} e^{-\frac{2}{3} z^{3/2}} & ie^{\frac{2}{3} z^{3/2}} \\ -e^{-\frac{2}{3} z^{3/2}} & ie^{\frac{2}{3} z^{3/2}} \end{pmatrix}, \tag{8.14}$$

as $z \to \infty$. Moreover, the determinant of $A(z)$ is continuous (since its jump is 1) on the real line and it approaches $i/(2\pi)$ as $z \to \infty$. Hence, it is a constant:

$$\det A(z) = \frac{i}{2\pi}. \tag{8.15}$$

8.3 Orthogonal polynomials with continuous weights

Let $\pi_n(x)$ be the monic orthogonal polynomials with respect to the weight function $w(x)$ on $\mathbf{R}$; namely,

$$\int_{\mathbf{R}} \pi_m(x)\pi_n(x)w(x)dx = \frac{\delta_{mn}}{\gamma_n^2}, \tag{8.16}$$

where γ_n is the leading coefficient of orthonormal polynomial $p_n(x) = \gamma_n\pi_n(x)$ and δ_{mn} is the Kronecker delta symbol such that $\delta_{nm} = 1$ if $n = m$ and $\delta_{mn} = 0$ if $n \neq m$. Following Deift (1999), we assume that, for any $k \geq 0$, both $x^k w(x)$ and $x^k w'(x)$ are in $L^2(\mathbf{R})$, and consider the following Riemann-Hilbert problem on $\mathbf{R}$ with right orientation.

(Y1) $Y(z)$ is analytic in $\mathbf{C} \setminus \mathbf{R}$.
(Y2) $Y(z)$ satisfies the jump condition

$$Y_+(x) = Y_-(x)\begin{pmatrix} 1 & w(x) \\ 0 & 1 \end{pmatrix}, \tag{8.17}$$

for $x \in \mathbf{R}$.
(Y3) As $z \to \infty$, $Y(z)z^{-n\sigma_3} = I + O(|z|^{-1})$.

The asymptotic condition (Y3) implies that the matrix-valued function $Y(z)$ also depends on n. For convenience, we drop the dependence of $Y(z)$ on n.

Proposition 8.1. *The Riemann-Hilbert problem (Y1)-(Y3) has a unique solution given by*

$$Y(z) = \begin{pmatrix} \pi_n(z) & \dfrac{1}{2\pi i}\displaystyle\int_{\mathbf{R}} \frac{\pi_n(x)w(x)}{x-z}dx \\ -2\pi i\gamma_{n-1}^2\pi_{n-1}(z) & -\gamma_{n-1}^2\displaystyle\int_{\mathbf{R}} \frac{\pi_{n-1}(x)w(x)}{x-z}dx \end{pmatrix} \tag{8.18}$$

for $z \in \mathbf{C} \setminus \mathbf{R}$.

Proof. We first show that the solution $Y(z)$ (if exists) is invertible with $\det Y(z) = 1$. Note from (Y1) that $\det Y(z)$ is analytic in $\mathbf{C} \setminus \mathbf{R}$. The jump condition (Y2) implies that $\det Y_+(x) = \det Y_-(x)$ for $x \in \mathbf{R}$, and hence $\det Y(z)$ is an entire function. By (Y3), we have $\det Y(z) \to 1$ as $z \to \infty$. Consequently, $\det Y(z) = 1$ for all $z \in \mathbf{C}$. Especially, $Y(z)$ is invertible.

Next, we prove that the solution of the Riemann-Hilbert problem (Y1)-(Y3) is unique. Assume the contrary, and let $Y(z)$ and $\tilde{Y}(z)$ be two distinct solutions. Denote $K(z) = \tilde{Y}(z)Y(z)^{-1}$, which is also analytic in $\mathbf{C} \setminus \mathbf{R}$ by (Y1). In view of (Y2), we obtain $K_+(x) = K_-(x)$ for $x \in \mathbf{R}$. Consequently, $K(z)$ is an entire function. Moreover, (Y3) implies that $K(z) \to I$ as $z \to \infty$, and thus, $K(z) = I$. This proves the uniqueness of the solution.

Finally, we verify that $Y(z)$ constructed in (8.18) is indeed a solution. It is obvious that $Y(z)$ is analytic in $\mathbf{C} \setminus \mathbf{R}$, which gives (Y1). For any $x \in \mathbf{R}$, we note from Plemelj formula that

$$Y_{12,+}(x) - Y_{12,-}(x) = \pi_n(x)w(x) = Y_{11}(x)w(x),$$

and

$$Y_{22,+}(x) - Y_{22,-}(x) = -2\pi i\gamma_{n-1}^2\pi_{n-1}(x)w(x) = Y_{21}(x)w(x).$$

Hence, (Y2) is proved. As $z \to \infty$, we have

$$z^{-n}Y_{11}(z) = z^{-n}\pi_n(z) = 1 + O(|z|^{-1}),$$

and

$$z^{-n}Y_{21}(z) = -2\pi i\gamma_{n-1}^2 z^{-n}\pi_{n-1}(z) = O(|z|^{-1}).$$

Moreover, since

$$\frac{1}{z-x} = \frac{1}{z} + \frac{x}{z(z-x)} = \sum_{k=0}^{n-2} \frac{x^k}{z^{k+1}} + \frac{x^{n-1}}{z^{n-1}(z-x)} = \sum_{k=0}^{n-1} \frac{x^k}{z^{k+1}} + \frac{x^n}{z^n(z-x)},$$

we obtain from the orthogonality relation (8.16) that

$$\begin{aligned} z^n Y_{12}(z) =& \frac{1}{2\pi i}\int_{\mathbf{R}} \frac{x^n\pi_n(x)w(x)}{x-z}dx \sim \frac{-1}{2\pi i\gamma_n^2 z} = O(|z|^{-1}), \\ z^n Y_{22}(z) =& \gamma_{n-1}^2 \int_{\mathbf{R}} \frac{x^{n-1}\pi_{n-1}(x)w(x)z}{z-x}dx \to 1, \end{aligned}$$

as $z \to \infty$. Thus, (Y3) is verified. □

8.4 Deift-Zhou nonlinear steepest descent method

In this section, we sketch an outline of nonlinear steepest descent method invented by Deift and Zhou (1993) and later developed for orthogonal polynomials [Deift *et al.* (1999)]. First, we shall make a sequence of transformations: scaling ($Y \to U$), normalization ($U \to T$), and factorization ($T \to V$). Next, we use suitable parametrices to construct a matrix-valued function U^* which is expected to be an approximation of the scaled matrix-valued function U. Finally, we prove that U is indeed asymptotically approximated by U^* as the polynomial degree tends to infinity.

As in the study of difference equation, we usually scale the variable as $z \to (\tau_1 N)^\theta z$, where $N = n + \tau_0$. The values of constants θ, τ_1 and τ_0 vary for different orthogonal polynomials. The scaling transformation is given by

$$U(z) := (\tau_1 N)^{-n\theta\sigma_3} Y((\tau_1 N)^\theta z), \tag{8.19}$$

where σ_3 is the Pauli matrix defined in (8.5). It is readily seen from (Y1)-(Y3) that $U(z)$ satisfies the following Riemann-Hilbert problem on $\mathbf{R}$ with right orientation.

(U1) $U(z)$ is analytic in $\mathbf{C} \setminus \mathbf{R}$.

(U2) $U(z)$ satisfies the jump condition

$$U_+(x) = U_-(x) \begin{pmatrix} 1 & w((\tau_1 N)^\theta x) \\ 0 & 1 \end{pmatrix}, \tag{8.20}$$

for $x \in \mathbf{R}$.

(U3) As $z \to \infty$, $U(z) z^{-n\sigma_3} = I + O(|z|^{-1})$.

Note from (8.18) and (8.19) that the $(1,1)$ entry of matrix U is

$$U_{11}(z) = (\tau_1 N)^{-n\theta} Y_{11}((\tau_1 N)^\theta z) = (\tau_1 N)^{-n\theta} \pi_n((\tau_1 N)^\theta z). \tag{8.21}$$

The normalization transformation involves the equilibrium measure of a minimal energy problem related to the potential function

$$v(z) = -\frac{1}{N} \ln w((\tau_1 N)^\theta z). \tag{8.22}$$

In general, $v(z)$ may depend on n. As we did for $Y(z)$ and $U(z)$, we drop the dependence on n for the sake of convenience. The equilibrium measure is the unique minimizer of the energy functional

$$\mathbb{E}(d\mu) = -\int_{\mathbf{R}^2} \ln|x-y| d\mu(x) d\mu(y) + \int_{\mathbf{R}} v(x) d\mu(x). \tag{8.23}$$

Assume that the equilibrium measure has the form $\rho(x)dx$ and it is supported on an interval $[a,b]$, where a and b are called the Mhaskar-Rakhmanov-Saff (MRS) numbers. In the asymptotic theory of orthogonal polynomials, we also call a and b the turning points (or transition points). Usually, the weight function (taking Freud weight as an example) is real analytic with possible integrable singularities. Hence, it is natural to assume that the potential function $v(z)$ is analytic for $z \in \mathbf{C} \setminus (-\infty, b]$. Now, we define the g-function and ϕ-function as

$$g(z) = \int_a^b \ln(z-x) \rho(x) dx, \tag{8.24}$$

$$\phi(z) = \frac{v(z) + l}{2} - g(z), \tag{8.25}$$

for $z \in \mathbf{C} \setminus (-\infty, b]$, where

$$l = 2g(b) - v(b) \tag{8.26}$$

is the Lagrange multiplier. An important property of the g-function is that it satisfies the following jump conditions on the real line.

$$g_+(x) + g_-(x) - v(x) - l = \begin{cases} 0, & x \in (a,b), \\ -2\phi(x), & x > b, \\ -2\phi_+(x) - 2\pi i, & x < a, \end{cases} \tag{8.27}$$

$$g_+(x) - g_-(x) = \begin{cases} -2\phi_+(x) = 2\phi_-(x), & x \in (a,b), \\ 0, & x > b, \\ 2\pi i, & x < a. \end{cases} \tag{8.28}$$

On account of (8.25), the two equations (8.27) and (8.28) are equivalent (exercise). It is obvious from (8.24) that $g_+(x) - g_-(x) = 0$ for $x > b$, and

$$g_+(x) - g_-(x) = \int_a^b 2\pi i \rho(y) dy = 2\pi i$$

for $x < a$. The remaining equation $g_+(x) + g_-(x) - v(x) - l = 0$ for $x \in (a,b)$ is called the Euler-Lagrange equation, the proof of which can be found in Saff and Totik (1997, Chapter I). From the definition of the Lagrange multiplier in (8.26), the Euler-Lagrange equation is equivalent to

$$g'_+(x) + g'_-(x) = v'(x), \quad x \in (a,b). \tag{8.29}$$

By Plemelj formula, we have the following integral representation (exercise)

$$g'(z) = \frac{(z-a)^{1/2}(z-b)^{1/2}}{2\pi} \int_a^b \frac{v'(x)}{\sqrt{(x-a)(b-x)}} \frac{dx}{z-x}, \tag{8.30}$$

for $z \in \mathbf{C} \setminus [a,b]$. Note from (8.24) that

$$g'(z) = \int_a^b \frac{\rho(x)}{z-x} dx \sim \frac{1}{z}, \tag{8.31}$$

as $z \to \infty$. Let $z \to \infty$ in (8.30). We obtain

$$\int_a^b \frac{v'(x)}{\sqrt{(x-a)(b-x)}} dx = 0. \tag{8.32}$$

Since $z/(z-x) = 1 + x/(z-x)$, we further obtain from (8.30), (8.31) and (8.32) that

$$\int_a^b \frac{xv'(x)}{\sqrt{(x-a)(b-x)}} dx = 2\pi. \tag{8.33}$$

Actually, one could solve the MRS numbers a and b from the two equations (8.32) and (8.33). In view of (8.29) and (8.31), we obtain from Cauchy integral formula (or Plemelj formula) that

$$\rho(x) = \frac{i}{2\pi}[g'_+(x) - g'_-(x)] = \mathrm{Re}\,\frac{ig'_+(x)}{\pi}, \quad x \in (a,b). \tag{8.34}$$

Hence, to solve the minimization problem for the energy functional in (8.23), one could use the above formula to calculate the equilibrium measure $d\mu(x) = \rho(x)dx$, where $g'(z)$ has the integral representation in (8.30), and the MRS numbers a and b can be solved from the system (8.32)-(8.33).

Now, we are ready to make the normalization transformation:

$$T(z) = e^{-\frac{1}{2}l_n\sigma_3}U(z)e^{[-Ng(z)+\tau_0 \ln \frac{\varphi(z)}{2} + \frac{1}{2}l_n]\sigma_3}, \tag{8.35}$$

where

$$\varphi(z) = \frac{(\sqrt{z-a}+\sqrt{z-b})^2}{2} = z - \frac{a+b}{2} + \sqrt{(z-a)(z-b)}, \tag{8.36}$$

for $z \in \mathbf{C} \setminus (-\infty, b]$, and

$$l_n = Nl - 2\tau_0 \ln \frac{b-a}{4} \tag{8.37}$$

is the modified Lagrange multiplier. Recall that $N = n + \tau_0$. It follows from (U3) and (8.24) that $U(z)e^{-Ng(z)\sigma_3} \sim z^{-\tau_0\sigma_3}$ as $z \to \infty$. The purpose of introducing $\varphi(z)$ is to cancel the term $z^{-\tau_0\sigma_3}$ and ensure $T(z) = I + O(|z|^{-1})$ as $z \to \infty$. For convenience, we also define

$$\psi(z) := \ln \frac{b-a}{4} - \ln \frac{\varphi(z)}{2} = -2\ln \frac{\sqrt{z-a}+\sqrt{z-b}}{\sqrt{b-a}}, \tag{8.38}$$

for $z \in \mathbf{C} \setminus (-\infty, b]$. One can easily verify that (exercise)

$$\ln \frac{\varphi_+(x)}{2} + \ln \frac{\varphi_-(x)}{2} - 2\ln \frac{b-a}{4} = \begin{cases} 0, & a < x < b, \\ -2\psi(x), & x > b, \\ -2\psi_+(x) - 2\pi i, & x < a, \end{cases} \tag{8.39}$$

and

$$\ln \frac{\varphi_+(x)}{2} - \ln \frac{\varphi_-(x)}{2} = \begin{cases} -2\psi_+ = 2\psi_-, & a < x < b, \\ 0, & x > b, \\ 2\pi i, & x < a. \end{cases} \tag{8.40}$$

It is readily seen from (U1)-(U3) that $T(z)$ satisfies the following Riemann-Hilbert problem on $\mathbf{R}$ with right orientation:

(T1) $T(z)$ is analytic in $\mathbf{C} \setminus \mathbf{R}$.
(T2) $T(z)$ satisfies the jump condition

$$T_+(x) = T_-(x) J_T(x) = T_-(x) \begin{pmatrix} J_{T,11}(x) & J_{T,12}(x) \\ J_{T,21}(x) & J_{T,22}(x) \end{pmatrix}, \tag{8.41}$$

for $x \in \mathbf{R}$, where

$$J_{T,11}(x) = e^{N[g_-(x)-g_+(x)]+\tau_0[\ln \varphi_+(x) - \ln \varphi_-(x)]}, \tag{8.42}$$

$$J_{T,12}(x) = e^{N[g_+(x)+g_-(x)-v(x)-l]-\tau_0[\ln \frac{\varphi_+(x)\varphi_-(x)}{4} - 2\ln \frac{b-a}{4}]}, \tag{8.43}$$

$$J_{T,22}(x) = e^{N[g_+(x)-g_-(x)]-\tau_0[\ln \varphi_+(x) - \ln \varphi_-(x)]}, \tag{8.44}$$

and $J_{T,21}(x) = 0$.
(T3) As $z \to \infty$, $T(z) = I + O(|z|^{-1})$.

In view of (8.22), (8.27), (8.28), (8.39) and (8.40), we express the jump matrix as

$$J_T(x) = \begin{pmatrix} e^{2N\phi_+(x) - 2\tau_0\psi_+(x)} & 1 \\ 0 & e^{2N\phi_-(x) - 2\tau_0\psi_-(x)} \end{pmatrix} \tag{8.45a}$$

for $x \in (a, b)$, and

$$J_T(x) = \begin{pmatrix} 1 & e^{-2N\phi(x)+2\tau_0\psi(x)} \\ 0 & 1 \end{pmatrix} \tag{8.45b}$$

for $x \in (b, \infty)$, and

$$J_T(x) = \begin{pmatrix} 1 & e^{-2N\phi_+(x)+2\tau_0\psi_+(x)} \\ 0 & 1 \end{pmatrix} \tag{8.45c}$$

for $x \in (-\infty, a)$. According to the characterization of the equilibrium measure [Saff and Totik (1997), Remark I.1.5], we have $\phi(x) > 0$ for $x > b$ and $\operatorname{Re} \phi_+(x) > 0$ for $x < a$. Hence, as $N \to \infty$, the jump matrix $J_T(x)$ converges exponentially to the identity matrix for $x \in (-\infty, a) \cup (b, \infty)$. For $x \in (a, b)$, we factorize the jump matrix as

$$J_T(x) = \begin{pmatrix} 1 & 0 \\ e^{2N\phi_-(x) - 2\tau_0\psi_-(x)} & 1 \end{pmatrix} \begin{pmatrix} 0 & 1 \\ -1 & 0 \end{pmatrix} \begin{pmatrix} 1 & 0 \\ e^{2N\phi_+(x) - 2\tau_0\psi_+(x)} & 1 \end{pmatrix}. \tag{8.46}$$

This suggests us to add two contours $\Sigma = \Sigma_+ \cup \Sigma_-$ from a to b in the upper and lower complex planes, respectively, and then perform the following factorization transformation.

$$V(z) = \begin{cases} T(z), & z \in \mathbf{C} \setminus (\bar{\Omega} \cup \mathbf{R}), \\ T(z) \begin{pmatrix} 1 & 0 \\ -e^{2N\phi(z) - 2\tau_0 \psi(z)} & 1 \end{pmatrix}, & z \in \Omega_+ \\ T(z) \begin{pmatrix} 1 & 0 \\ e^{2N\phi(z) - 2\tau_0 \psi(z)} & 1 \end{pmatrix}, & z \in \Omega_-; \end{cases} \tag{8.47}$$

see Figure 8.1 for an illustration of the contours $\Sigma = \Sigma_+ \cup \Sigma_-$ and the upper and lower lens regions $\Omega = \Omega_+ \cup \Omega_-$.

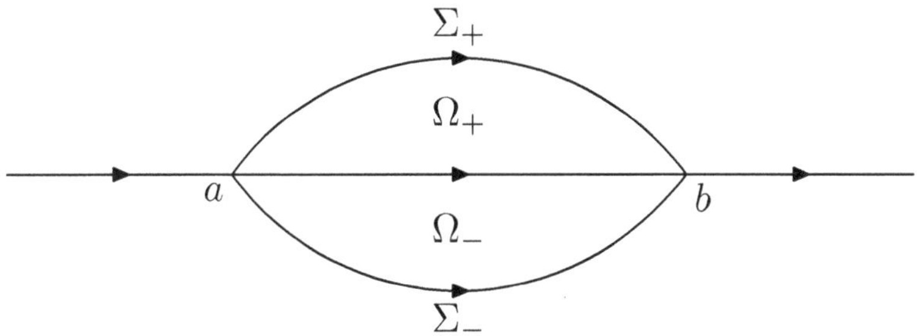

Fig. 8.1: The lens-shaped regions $\Omega = \Omega_+ \cup \Omega_-$ and the contours $\Sigma = \Sigma_+ \cup \Sigma_-$.

We then obtain from (T1)-(T3) with the jump conditions (8.45) and the factorization (8.46) that $V(z)$ satisfies the following Riemann-Hilbert problem on $\Sigma \cup \mathbf{R}$ with right orientation:

(V1) $V(z)$ is analytic in $\mathbf{C} \setminus (\Sigma \cup \mathbf{R})$.
(V2) $V(z)$ satisfies the jump conditions

$$V_+(z) = V_-(z) \begin{pmatrix} 1 & 0 \\ e^{2N\phi(z) - 2\tau_0 \psi(z)} & 1 \end{pmatrix} \tag{8.48}$$

for $z \in \Sigma$, and

$$V_+(x) = V_-(x) \begin{pmatrix} 0 & 1 \\ -1 & 0 \end{pmatrix} \tag{8.49}$$

for $x \in (a, b)$, and

$$V_+(x) = V_-(x) \begin{pmatrix} 1 & e^{-2N\phi(x) + 2\tau_0 \psi(x)} \\ 0 & 1 \end{pmatrix} \tag{8.50}$$

for $x \in (b, \infty)$, and

$$V_+(x) = V_-(x) \begin{pmatrix} 1 & e^{-2N\phi_+(x)+2\tau_0\psi_+(x)} \\ 0 & 1 \end{pmatrix} \tag{8.51}$$

for $x \in (-\infty, a)$.

(V3) As $z \to \infty$, $V(z) = I + O(|z|^{-1})$.

When the variable z is bounded away from the MRS numbers a and b, one can show that all jump matrices other than the one on (a, b) tend to the identity matrix as $N \to \infty$. In view of the Riemann-Hilbert problem (M1)-(M3) for the parametrix $M(z)$ explicitly given in (8.7), we can formally approximate $V(z)$, and thus $T(z)$, by $M(z)$; namely, $T(z) \sim V(z) \sim M(z)$. Tracing back to $U(z)$ by (8.35) and making use of (8.22), (8.25), (8.37) and (8.38), we have the following formal approximation

$$\begin{aligned} U(z) \sim & e^{(l_n/2)\sigma_3} M(z) e^{[Ng(z)-l_n/2-\tau_0 \ln(\varphi(z)/2)]\sigma_3} \\ = & e^{(l_n/2)\sigma_3} M(z) e^{[-N\phi(z)+\tau_0\psi(z)]\sigma_3} w((\tau_1 N)^\theta z)^{-\sigma_3/2} \\ = & \frac{1}{2} e^{(l_n/2)\sigma_3} M(z) e^{\tau_0\psi(z)\sigma_3} \begin{pmatrix} 1 & -1 \\ -i & -i \end{pmatrix} \\ & \times \begin{pmatrix} e^{-N\phi(z)} & ie^{N\phi(z)} \\ -e^{-N\phi(z)} & ie^{N\phi(z)} \end{pmatrix} w((\tau_1 N)^\theta z)^{-\sigma_3/2} \\ \sim & \sqrt{\pi} e^{(l_n/2)\sigma_3} M(z) e^{\tau_0\psi(z)\sigma_3} \begin{pmatrix} 1 & -1 \\ -i & -i \end{pmatrix} \\ & \times f(z)^{\sigma_3/4} A(f(z)) w((\tau_1 N)^\theta z)^{-\sigma_3/2}, \end{aligned} \tag{8.52}$$

where, in the last step, we have defined

$$f(z) := \left[\frac{3}{2} N\phi(z)\right]^{2/3}, \quad z \in \mathbf{C} \setminus (-\infty, b], \tag{8.53}$$

and made use of the Airy parametrix (8.8) and its asymptotic formula (8.14). Assume that $f(z)$ can be analytically continued to the interval $(a, b]$. The right-hand side of (8.52) suggests a parametrix in a large complex region including b. To find a parametrix in a neighborhood of a, we shall introduce the $\tilde{\phi}$-function

$$\tilde{\phi}(z) := \begin{cases} \phi(z) + \pi i, & z \in \mathbf{C}_+, \\ \phi(z) - \pi i, & z \in \mathbf{C}_-, \end{cases} \tag{8.54}$$

and define

$$\tilde{f}(z) := \left[\frac{3}{2} N\tilde{\phi}(z)\right]^{2/3}, \quad z \in \mathbf{C} \setminus [a, \infty). \tag{8.55}$$

Here, we have made use of (8.25) and (8.28) to obtain that $\phi_+(x)-\phi_-(x) = -2\pi i$ for $x < a$, and hence, $\tilde{\phi}(z)$ is analytic for $z \in \mathbf{C} \setminus [a, \infty)$. We further assume that $\tilde{f}(z)$ can be analytically continued to the interval $[a, b)$. Set

$$\tilde{\psi}(z) := -2 \ln \frac{\sqrt{a-z}+\sqrt{b-z}}{\sqrt{b-a}} = \begin{cases} \psi(z) + \pi i, & z \in \mathbf{C}_+, \\ \psi(z) - \pi i, & z \in \mathbf{C}_-. \end{cases} \tag{8.56}$$

It then follows from (8.54), (8.56), and $N = n + \tau_0$ that $e^{-N\phi(z)+\tau_0\psi(z)} = (-1)^n e^{-N\tilde{\phi}(z)+\tau_0\tilde{\psi}(z)}$ for $z \in \mathbf{C}_\pm$. Thus, (8.52) can be rewritten as

$$\begin{aligned} U(z) \sim & (-1)^n e^{(l_n/2)\sigma_3} M(z) e^{[-N\tilde{\phi}(z)+\tau_0\tilde{\psi}(z)]\sigma_3} w((\tau_1 N)^\theta z)^{-\sigma_3/2} \\ = & \frac{(-1)^n}{2} e^{(l_n/2)\sigma_3} M(z) e^{\tau_0\tilde{\psi}(z)\sigma_3} \sigma_3 \begin{pmatrix} 1 & -1 \\ -i & -i \end{pmatrix} \\ & \times \begin{pmatrix} e^{-N\tilde{\phi}(z)} & ie^{N\tilde{\phi}(z)} \\ -e^{-N\tilde{\phi}(z)} & ie^{N\tilde{\phi}(z)} \end{pmatrix} \sigma_3 w((\tau_1 N)^\theta z)^{-\sigma_3/2} \\ \sim & (-1)^n \sqrt{\pi} e^{(l_n/2)\sigma_3} M(z) e^{\tau_0\tilde{\psi}(z)\sigma_3} \sigma_3 \begin{pmatrix} 1 & -1 \\ -i & -i \end{pmatrix} \\ & \times \tilde{f}(z)^{\sigma_3/4} A(\tilde{f}(z)) \sigma_3 w((\tau_1 N)^\theta z)^{-\sigma_3/2}, \end{aligned} \tag{8.57}$$

where, we have again made use of (8.8) and (8.14). Comparing with (8.52), we have introduced two additional σ_3 in second step of (8.57) and applied the identity $e^{-N\tilde{\phi}(z)\sigma_3} = \sigma_3 e^{-N\tilde{\phi}(z)\sigma_3}\sigma_3$. As we shall see later, this is because the jump matrix for $A(f(z))$ on the interval $x > a$ is $\begin{pmatrix} 1 & 1 \\ 0 & 1 \end{pmatrix}$, while the jump matrix for $A(\tilde{f}(z))$ on the interval $x < b$ is $\begin{pmatrix} 1 & -1 \\ 0 & 1 \end{pmatrix}$. After a right multiplication by σ_3, we notice that the jump matrix for $A(\tilde{f}(z))\sigma_3$ coincides with the jump matrix for $A(f(z))$ on the interval (a, b). The two formal asymptotic formulas (8.52) and (8.57) suggest us to define the following parametrix

$$U^*(z) := \begin{cases} E(z) A(f(z)) w((\tau_1 N)^\theta z)^{-\sigma_3/2}, & \operatorname{Re} z > c, \\ (-1)^n \sigma_3 \tilde{E}(z) A(\tilde{f}(z)) w((\tau_1 N)^\theta z)^{-\sigma_3/2} \sigma_3, & \operatorname{Re} z < c, \end{cases} \tag{8.58}$$

where $c \in (a, b)$ is any fixed constant, and

$$E(z) := \sqrt{\pi} e^{(l_n/2)\sigma_3} M(z) e^{\tau_0\psi(z)\sigma_3} \begin{pmatrix} 1 & -1 \\ -i & -i \end{pmatrix} f(z)^{\sigma_3/4}, \tag{8.59}$$

$$\tilde{E}(z) := \sqrt{\pi} e^{(l_n/2)\sigma_3} \sigma_3 M(z) \sigma_3 e^{\tau_0\tilde{\psi}(z)\sigma_3} \begin{pmatrix} 1 & -1 \\ -i & -i \end{pmatrix} \tilde{f}(z)^{\sigma_3/4}. \tag{8.60}$$

We assume that $f'(x) > 0$ for $x > a$ and $\tilde{f}'(x) < 0$ for $x < b$. Hence, when z approaches $x \in (a, \infty)$ from the upper (lower) half-plane, $f(z)$ approaches $f(x)$ from the upper (lower) half-plane; when z approaches $x \in (-\infty, b)$ from the upper (lower) half-plane, $\tilde{f}(z)$ approaches $\tilde{f}(x)$ from the lower (upper) half-plane. It is readily seen from the jump condition (8.10) that

$$U_+^*(x) = U_-^*(x) \begin{pmatrix} 1 & w((\tau_1 N)^\theta x) \\ 0 & 1 \end{pmatrix}, \quad x \in (-\infty, c) \cup (c, \infty). \tag{8.61}$$

As $z \to \infty$, we obtain (exercise) from the above formulas and (8.14) that

$$U^*(z) \sim e^{\frac{1}{2}l_n \sigma_3} M(z) e^{[Ng(z) - \tau_0 \ln \frac{\varphi(z)}{2} - \frac{1}{2}l_n]\sigma_3} = [I + O(|z|^{-1})] z^{n\sigma_3}, \tag{8.62}$$

where in the last step, we have made use of $M(z) = I + O(|z|^{-1})$ in (M3), $g(z) \sim \ln z$ in (8.24) and $\varphi(z) \sim 2z$.

Finally, we define

$$S(z) := e^{-(l_n/2)\sigma_3} U(z) [U^*(z)]^{-1} e^{(l_n/2)\sigma_3}, \tag{8.63}$$

and note that $S(z)$ satisfies the following Riemann-Hilbert problem on the vertical line $\operatorname{Re} z = c$ with upward orientation (exercise).

(S1) $S(z)$ is analytic for $\operatorname{Re} z \neq c$.

(S2) $S(z)$ satisfies the jump condition

$$S_+(z) = S_-(z) J_S(z) \tag{8.64}$$

for $\operatorname{Re} z = c$, where

$$J_S(z) = (-1)^n e^{-\frac{1}{2}l_n \sigma_3} E(z) A(f(z)) \sigma_3 [A(\tilde{f}(z))]^{-1} \tilde{E}(z)^{-1} \sigma_3 e^{\frac{1}{2}l_n \sigma_3}. \tag{8.65}$$

(S3) As $z \to \infty$, $S(z) = I + O(|z|^{-1})$.

Making use of asymptotic expansions for Airy function and its derivative, one can prove that

$$J_S(z) \sim I + \sum_{k=1}^{\infty} \frac{J_{S,k}(z)}{N^k}, \tag{8.66}$$

as $N \to \infty$. Consequently, the solution of the Riemann-Hilbert problem (S1)-(S3) has the asymptotic expansion

$$S(z) \sim I + \sum_{k=1}^{\infty} \frac{S_k(z)}{N^k}, \tag{8.67}$$

as $N \to \infty$, where $S_k(z)$ can be determined recursively by

$$S_k(z) = \frac{1}{2\pi i} \int_{\operatorname{Re} \zeta = c} \sum_{j=1}^{k} (S_{k-j})_-(\zeta) J_{S,j}(\zeta) \frac{d\zeta}{\zeta - z}, \quad k = 1, 2, \cdots, \tag{8.68}$$

for $\operatorname{Re} z \neq c$.

8.5 An example: Hermite polynomials

In this section, we use the Hermite polynomials as a simple example to illustrate the application of Riemann-Hilbert problem and the standard sequence of transformations.

The material in this section is taken from Wong and Zhang (2007). The Hermite polynomials have the following hypergeometric representation [Koekoek *et al.* (2010), (9.15.1)]

$$H_n(x) = (2x)^n {}_2F_0\left(\begin{matrix} -n/2 \quad (1-n)/2 \\ - \end{matrix} \middle| -\frac{1}{x^2}\right), \tag{8.69}$$

and they satisfy the following orthogonality relation [Koekoek *et al.* (2010), (9.15.2)]

$$\int_{\mathbf{R}} H_m(x) H_n(x) e^{-x^2} dx = 2^n n! \sqrt{\pi} \delta_{mn}. \tag{8.70}$$

It is obvious that the leading coefficient of $H_n(x)$ is 2^n and the monic Hermite polynomials $\pi_n(x) = 2^{-n} H_n(x)$ satisfy the orthogonality relation (8.16) with weight function

$$w(x) = e^{-x^2}, \quad x \in \mathbf{R}, \tag{8.71}$$

and

$$\gamma_n^2 = 2^n / (n! \sqrt{\pi}). \tag{8.72}$$

Moreover, the matrix-valued function $Y(z)$ defined in (8.18) satisfies the Riemann-Hilbert problem (Y1)-(Y3). Now, we set

$$\theta = 1/2, \quad \tau_1 = 2, \quad \tau_0 = 1/2, \quad N = n + \tau_0 = n + 1/2. \tag{8.73}$$

The scaling transformation (8.19) gives a scaled matrix-valued function $U(z)$ which satisfies the Riemann-Hilbert problem (U1)-(U3). The potential function for the Hermite polynomials is

$$v(z) = -\frac{1}{N} \ln w((\tau_1 N)^\theta z) = -\frac{1}{N} \ln e^{-2Nz^2} = 2z^2. \tag{8.74}$$

The system (8.32)-(8.33) reads

$$\int_a^b \frac{4x}{\sqrt{(x-a)(b-x)}} dx = 0, \tag{8.75}$$

$$\int_a^b \frac{4x^2}{\sqrt{(x-a)(b-x)}} dx = 2\pi, \tag{8.76}$$

from which one could solve the MRS numbers as (exercise)

$$a = -1, \quad b = 1. \tag{8.77}$$

The integral representation (8.30) can be evaluated as (exercise)

$$g'(z) = \frac{\sqrt{z^2-1}}{2\pi} \int_{-1}^{1} \frac{4x}{\sqrt{1-x^2}} \frac{dx}{z-x} = 2z - 2\sqrt{z^2-1}. \tag{8.78}$$

Since $g'_+(x) = 2x - 2i\sqrt{1-x^2}$ for $x \in (-1,1)$, we obtain from (8.34)

$$\rho(x) = \frac{2\sqrt{1-x^2}}{\pi}, \quad x \in (-1,1). \tag{8.79}$$

The g-function in (8.24) is now calculated as (exercise)

$$g(z) = z(z - \sqrt{z^2-1}) - \frac{1}{2} + \ln \frac{z+\sqrt{z^2-1}}{2}. \tag{8.80}$$

Here, instead of a direct and tedious evaluation of the integral in (8.24), one could actually find the antiderivative of $g'(z) = 2z - 2\sqrt{z^2-1}$ and then make use of the asymptotic behavior $g(z) \sim \ln z$ as $z \to \infty$. Consequently, the Lagrange multipliers in (8.26) and (8.37) are

$$l = 2g(1) - v(1) = -1 - 2\ln 2, \tag{8.81}$$

$$l_n = Nl - \ln\frac{1}{2} = -N - (2N-1)\ln 2. \tag{8.82}$$

The ϕ-function in (8.25) and $\tilde{\phi}$-function are explicitly given as

$$\phi(z) = z\sqrt{z^2-1} - \ln(z + \sqrt{z^2-1}) \tag{8.83}$$

for $z \in \mathbf{C} \setminus (-\infty, 1]$, and

$$\tilde{\phi}(z) = -z\sqrt{-1-z}\cdot\sqrt{1-z} - \ln(-z + \sqrt{-1-z}\cdot\sqrt{1-z}), \tag{8.84}$$

for $z \in \mathbf{C} \setminus [-1, \infty)$. Since $\phi'(z) = 2\sqrt{z^2-1}$, we have $\phi(z) \sim (4\sqrt{2}/3)(z-1)^{3/2}$ as $z \to 1$. Based on the conformal mapping analysis (see Figure 8.2), we note that $\arg\phi_\pm(x) = \pm 3\pi/2$ for $x \in (-1,1)$ and $f(z)$ defined in (8.53) can be analytically continued to the interval $(-1,1]$. Similarly, one could show (exercise) that $\tilde{\phi}(z) \sim (4\sqrt{2}/3)(-1-z)^{3/2}$ as $z \to -1$, $\arg\tilde{\phi}_\pm(x) = \mp 3\pi/2$ for $x \in (-1,1)$, and $\tilde{f}(z)$ defined in (8.53) can be analytically continued to the interval $(-1,1]$.

After the normalization transform $(U \to T)$ in (8.35) and the factorization transform $(T \to V)$ in (8.47), we construct the parametrix $U^*(z)$ as in (8.58) with $c = 0$. One can verify (exercise) that $f'(x) > 0$ for $x > -1$ and $\tilde{f}'(x) < 0$ for $x < 1$. On account of (8.7), (8.38) and (8.56), the

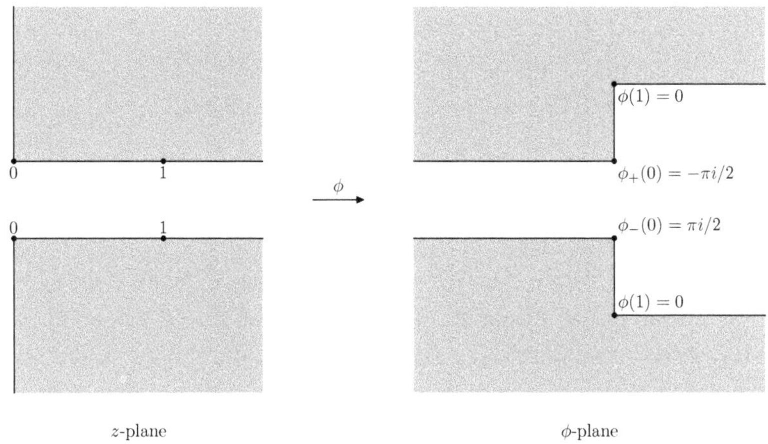

Fig. 8.2: The conformal mapping under $\phi(z)$ from right-half z-plane to ϕ-plane.

matrix-valued functions $E(z)$ and $\tilde{E}(z)$ in (8.59) and (8.60) can be explicitly calculated as (exercise)

$$E(z) = \sqrt{2\pi} e^{\frac{1}{2} l_n \sigma_3} \begin{pmatrix} 1 & 0 \\ -zi & -i \end{pmatrix} \left(\frac{f(z)^{1/4}}{(z-1)^{1/4}(z+1)^{1/4}} \right)^{\sigma_3}, \tag{8.85}$$

$$\tilde{E}(z) = \sqrt{2\pi} e^{\frac{1}{2} l_n \sigma_3} \begin{pmatrix} 1 & 0 \\ zi & -i \end{pmatrix} \left(\frac{\tilde{f}(z)^{1/4}}{(-1-z)^{1/4}(1-z)^{1/4}} \right)^{\sigma_3}. \tag{8.86}$$

It is interesting to notice the symmetry relations (exercise)

$$\tilde{\phi}(z) = \phi(-z), \quad \tilde{f}(z) = f(-z), \quad \tilde{E}(z) = E(-z), \quad \operatorname{Re} z \leq 0. \tag{8.87}$$

Let $S(z)$ be the matrix-valued function defined in (8.63). We calculate the jump matrix $J_S(z)$ in (8.65) as (exercise)

$$\begin{aligned} J_S(z) = &(-1)^{n+1} 2\pi i \begin{pmatrix} 1 & 0 \\ -zi & -i \end{pmatrix} (z-1)^{-\sigma_3/4} (z+1)^{-\sigma_3/4} \begin{pmatrix} m_{11} & m_{12} \\ m_{21} & m_{22} \end{pmatrix} \\ &\times (-1-z)^{\sigma_3/4} (1-z)^{\sigma_3/4} \begin{pmatrix} 1 & 0 \\ z & i \end{pmatrix} \sigma_3, \quad \operatorname{Re} z = 0, \end{aligned} \tag{8.88}$$

where

$$m_{11} = (f/\tilde{f})^{1/4} \omega^2 [\operatorname{Ai}(f) \operatorname{Ai}'(\omega \tilde{f}) - \operatorname{Ai}(\omega^2 f) \operatorname{Ai}'(\tilde{f})], \tag{8.89}$$

$$m_{12} = (f\tilde{f})^{1/4} \omega [\omega \operatorname{Ai}(\omega^2 f) \operatorname{Ai}(\tilde{f}) - \operatorname{Ai}(f) \operatorname{Ai}(\omega \tilde{f})], \tag{8.90}$$

$$m_{21} = (f\tilde{f})^{-1/4} \omega [\omega \operatorname{Ai}'(f) \operatorname{Ai}'(\omega \tilde{f}) - \operatorname{Ai}'(\omega^2 f) \operatorname{Ai}'(\tilde{f})], \tag{8.91}$$

$$m_{22} = (\tilde{f}/f)^{1/4} \omega [\operatorname{Ai}'(\omega^2 f) \operatorname{Ai}(\tilde{f}) - \operatorname{Ai}'(f) \operatorname{Ai}(\omega \tilde{f})], \tag{8.92}$$

for $\operatorname{Im} z > 0$, and

$$m_{11} = (f/\tilde{f})^{1/4}\omega[\operatorname{Ai}(\omega f)\operatorname{Ai}'(\tilde{f}) - \operatorname{Ai}(f)\operatorname{Ai}'(\omega^2\tilde{f})], \tag{8.93}$$

$$m_{12} = (f\tilde{f})^{1/4}\omega[\omega\operatorname{Ai}(f)\operatorname{Ai}(\omega^2\tilde{f}) - \operatorname{Ai}(\omega f)\operatorname{Ai}(\tilde{f})], \tag{8.94}$$

$$m_{21} = (f\tilde{f})^{-1/4}\omega[\omega\operatorname{Ai}'(\omega f)\operatorname{Ai}'(\tilde{f}) - \operatorname{Ai}'(f)\operatorname{Ai}'(\omega^2\tilde{f})], \tag{8.95}$$

$$m_{22} = (\tilde{f}/f)^{1/4}\omega^2[\operatorname{Ai}'(f)\operatorname{Ai}(\omega^2\tilde{f}) - \operatorname{Ai}'(\omega f)\operatorname{Ai}(\tilde{f})], \tag{8.96}$$

for $\operatorname{Im} z < 0$. In view of the symmetry relations (8.87), we have (exercise) $J_S(-z) = \sigma_3 J_S(z)^{-1}\sigma_3$. Hence, it suffices to prove (8.66) for $\operatorname{Im} z > 0$. From the asymptotic expansion of the Airy function [Olver (1974a), p. 392] (or [Olver *et al.* (2010), (9.7.5)-(9.7.6)]), we have as $N \to \infty$,

$$f^{1/4}\operatorname{Ai}(f) \sim \frac{e^{-N\phi}}{2\sqrt{\pi}}\sum_{k=0}^{\infty} c_k(-N\phi)^{-k}, \tag{8.97}$$

$$f^{-1/4}\operatorname{Ai}'(f) \sim -\frac{e^{-N\phi}}{2\sqrt{\pi}}\sum_{k=0}^{\infty} d_k(-N\phi)^{-k}, \tag{8.98}$$

and

$$(\omega^{-1}f)^{1/4}\operatorname{Ai}(\omega^2 f) \sim \frac{e^{N\phi}}{2\sqrt{\pi}}\sum_{k=0}^{\infty} c_k(N\phi)^{-k}, \tag{8.99}$$

$$(\omega^{-1}f)^{-1/4}\operatorname{Ai}'(\omega^2 f) \sim -\frac{e^{N\phi}}{2\sqrt{\pi}}\sum_{k=0}^{\infty} d_k(N\phi)^{-k}, \tag{8.100}$$

and

$$\tilde{f}^{1/4}\operatorname{Ai}(\tilde{f}) \sim \frac{e^{-N\tilde{\phi}}}{2\sqrt{\pi}}\sum_{k=0}^{\infty} c_k(-N\tilde{\phi})^{-k}, \tag{8.101}$$

$$\tilde{f}^{-1/4}\operatorname{Ai}'(\tilde{f}) \sim -\frac{e^{-N\tilde{\phi}}}{2\sqrt{\pi}}\sum_{k=0}^{\infty} d_k(-N\tilde{\phi})^{-k}, \tag{8.102}$$

and

$$(\omega\tilde{f})^{1/4}\operatorname{Ai}(\omega\tilde{f}) \sim \frac{e^{N\tilde{\phi}}}{2\sqrt{\pi}}\sum_{k=0}^{\infty} c_k(N\tilde{\phi})^{-k}, \tag{8.103}$$

$$(\omega\tilde{f})^{-1/4}\operatorname{Ai}'(\omega\tilde{f}) \sim -\frac{e^{N\tilde{\phi}}}{2\sqrt{\pi}}\sum_{k=0}^{\infty} d_k(N\tilde{\phi})^{-k}, \tag{8.104}$$

where $c_0 = d_0 = 1$ and

$$c_k = \frac{\Gamma(3k+1)}{54^k k!\Gamma(k+1/2)}, \quad d_k = -\frac{6k+1}{6k-1}c_k, \quad k = 1, 2, \cdots. \tag{8.105}$$

A simple though tedious verification gives the asymptotic expansion (8.66) for $J_S(z)$. From the asymptotic expansion (8.67) for $S(z)$, we have the following main theorem for the Hermite polynomials.

Theorem 8.1. *Denote* $N = n + 1/2$, $l_n = -N - (2N-1)\ln 2$ *and*

$$f(z) = \left[\frac{3}{2}N\phi(z)\right]^{2/3}, \qquad \tilde{f}(z) = \left[\frac{3}{2}N\tilde{\phi}(z)\right]^{2/3}, \tag{8.106}$$

where $\phi(z)$ *and* $\tilde{\phi}(z)$ *are given in* (8.83) *and* (8.84), *respectively. Let* $\pi_n(z) = 2^{-n}H_n(z)$ *be the monic Hermite polynomial of degree* n. *We have as* $n \to \infty$,

$$\begin{aligned}\pi_n(\sqrt{2n+1}z) &\sim \sqrt{2\pi}(2n+1)^{n/2}e^{l_n/2+(n+1/2)z^2}\\ &\times\left\{\frac{f(z)^{1/4}}{(z-1)^{1/4}(z+1)^{1/4}}\operatorname{Ai}(f(z))\sum_{k=0}^{\infty}\frac{A_k(z)}{(n+1/2)^k}\right.\\ &\left.+\frac{(z-1)^{1/4}(z+1)^{1/4}}{f(z)^{1/4}}\operatorname{Ai}'(f(z))\sum_{k=1}^{\infty}\frac{B_k(z)}{(n+1/2)^k}\right\},\end{aligned} \tag{8.107}$$

for $\operatorname{Re} z > 0$, *and*

$$\begin{aligned}\pi_n(\sqrt{2n+1}z) &\sim (-1)^n\sqrt{2\pi}(2n+1)^{n/2}e^{l_n/2+(n+1/2)z^2}\\ &\times\left\{\frac{\tilde{f}(z)^{1/4}}{(-1-z)^{1/4}(1-z)^{1/4}}\operatorname{Ai}(\tilde{f}(z))\sum_{k=0}^{\infty}\frac{\tilde{A}_k(z)}{(n+1/2)^k}\right.\\ &\left.+\frac{(-1-z)^{1/4}(1-z)^{1/4}}{\tilde{f}(z)^{1/4}}\operatorname{Ai}'(\tilde{f}(z))\sum_{k=1}^{\infty}\frac{\tilde{B}_k(z)}{(n+1/2)^k}\right\},\end{aligned} \tag{8.108}$$

for $\operatorname{Re} z < 0$, *where* $A_0(z) = \tilde{A}_0(z) = 1$. *The asymptotic formula for* $\operatorname{Re} z = 0$ *can be obtained by taking limit in either* (8.107) *or* (8.108). *Actually, the above two formulas are equivalent due to the symmetry relation* $\pi_n(-\sqrt{2n+1}z) = (-1)^n\pi_n(\sqrt{2n+1}z)$ *for all* $z \in \mathbf{C}$.

Proof. Note from (8.63) that $U(z) = e^{l_n\sigma_3/2}S(z)e^{-l_n\sigma_3/2}U^*(z)$. Hence, we obtain

$$U_{11}(z) = S_{11}(z)U_{11}^*(z) + S_{12}(z)U_{21}^*(z)e^{l_n}.$$

When $\operatorname{Re} z > 0$, we have from (8.8), (8.58) and (8.85)

$$\begin{aligned}U_{11}^*(z) =&\sqrt{2\pi}e^{l_n/2}\frac{f(z)^{1/4}}{(z-1)^{1/4}(z+1)^{1/4}}\operatorname{Ai}(f(z))e^{(n+1/2)z^2},\\ U_{21}^*(z) =&-i\sqrt{2\pi}e^{-l_n/2}\left[\frac{zf(z)^{1/4}}{(z-1)^{1/4}(z+1)^{1/4}}\operatorname{Ai}(f(z))\right.\\ &\left.+\frac{(z-1)^{1/4}(z+1)^{1/4}}{f(z)^{1/4}}\operatorname{Ai}'(f(z))\right]e^{(n+1/2)z^2}.\end{aligned}$$

Recall from (8.21) that $\pi_n(\sqrt{2n+1}z) = (2n+1)^{n/2}U_{11}(z)$. The formula (8.107) follows from the asymptotic expansion for $S(z)$ in (8.67) and the above equations. If $\operatorname{Re} z < 0$, it follows from (8.8), (8.58) and (8.86) that

$$\begin{aligned}
U_{11}^*(z) =&(-1)^n\sqrt{2\pi}e^{l_n/2}\frac{\tilde{f}(z)^{1/4}}{(-1-z)^{1/4}(1-z)^{1/4}}\operatorname{Ai}(\tilde{f}(z))e^{(n+1/2)z^2},\\
U_{21}^*(z) =&(-1)^n i\sqrt{2\pi}e^{-l_n/2}\left[\frac{-z\tilde{f}(z)^{1/4}}{(-1-z)^{1/4}(1-z)^{1/4}}\operatorname{Ai}(\tilde{f}(z))\right.\\
&\left.+\frac{(-1-z)^{1/4}(1-z)^{1/4}}{\tilde{f}(z)^{1/4}}\operatorname{Ai}'(\tilde{f}(z))\right]e^{(n+1/2)z^2}.
\end{aligned}$$

In view of (8.67), we obtain (8.108). □

8.6 Exercises

1. Verify (8.4).
2. Verify (8.7).
3. Prove that (8.27) and (8.28) are equivalent.
4. Derive (8.30).
5. Verify (8.39) and (8.40).
6. Prove (8.62).
7. Verify that $S(z)$ defined in (8.63) satisfies the Riemann-Hilbert problem (S1)-(S3).
8. Prove (8.77).
9. Verify (8.78).
10. Prove (8.80).
11. Let $\tilde{\phi}(z)$ be given in (8.84). Prove that $\tilde{\phi}(z) \sim (4\sqrt{2}/3)(-1-z)^{3/2}$ as $z \to -1$, $\arg\tilde{\phi}_\pm(x) = \mp 3\pi/2$ for $x \in (-1,1)$, and $\tilde{f}(z)$ defined in (8.55) can be analytically continued to the interval $(-1,1]$.
12. Let $f(z)$ and $\tilde{f}(z)$ be defined in (8.106). Prove that $f'(x) > 0$ for $x > -1$ and $\tilde{f}'(x) < 0$ for $x < 1$.
13. Verify (8.85) and (8.86).
14. Prove the symmetry relation (8.87).
15. Prove (8.88).
16. Let $J_S(z)$ be given in (8.88). Show that $J_S(-z) = \sigma_3 J_S(z)^{-1}\sigma_3$.

Chapter 9

Global Asymptotics of Generalized Freud Orthogonal Polynomials

9.1 Introduction

The material in this chapter is taken from Wen *et al.* (2018). Given $\alpha \in \mathbf{R}$, $t \in \mathbf{R}$ and $m > 0$, we consider the following Riemann-Hilbert problem on $\mathbf{R}$ with right orientation.

(Y1) $Y(z)$ is analytic in $\mathbf{C} \setminus \mathbf{R}$.
(Y2) $Y(z)$ satisfies the jump condition

$$Y_+(x) = Y_-(x) \begin{pmatrix} 1 & |x|^{2\alpha} e^{-m(x^4+tx^2)} \\ 0 & 1 \end{pmatrix}, \tag{9.1}$$

for $x \in \mathbf{R}$.
(Y3) As $z \to \infty$, $Y(z) z^{-n\sigma_3} = I + O(|z|^{-1})$.
(Y4) As $z \to 0$, $Y(z) = \begin{cases} \begin{pmatrix} O(1) & O(|z|^{2\alpha}) \\ O(1) & O(|z|^{2\alpha}) \end{pmatrix}, & \alpha \le 0, \\ \\ O(1), & \alpha \ge 0. \end{cases}$

For the special case with $m = 1$ and $\alpha > -1/2$, it was shown in Clarkson *et al.* (2016) that the recurrence coefficient for the monic orthogonal polynomials $Y_{11}(z)$ satisfies the first discrete Painlevé equation, and is related to solutions of the fourth Painlevé equation. Asymptotic properties of the orthogonal polynomials and the recurrence coefficient as n or t tends to infinity were investigated in Clarkson and Jordaan (2018). Another special case with $\alpha = 0$ and $t < 0$ was investigated in Bleher and Its (1999), where asymptotic formulas of $Y_{11}(z)$ as both n and m tend to infinity with $\varepsilon < n/m < 1/\varepsilon$ were obtained.

Denote $N = n + \alpha$ and $c = (m/N)^2 t$. We introduce the scaling transformation

$$U(z) := (N/m)^{-(n+\alpha)\sigma_3/4} Y((N/m)^{1/4} z)(N/m)^{\alpha\sigma_3/4}. \tag{9.2}$$

It is readily seen that $U(z)$ satisfies the following Riemann-Hilbert problem on $\mathbf{R}$ with right orientation.

(U1) $U(z)$ is analytic in $\mathbf{C} \setminus \mathbf{R}$.
(U2) $U(z)$ satisfies the jump condition

$$U_+(x) = U_-(x) \begin{pmatrix} 1 & |x|^{2\alpha} e^{-N(x^4 + cx^2)} \\ 0 & 1 \end{pmatrix}, \tag{9.3}$$

for $x \in \mathbf{R}$.
(U3) As $z \to \infty$, $U(z) z^{-n\sigma_3} = I + O(|z|^{-1})$.
(U4) As $z \to 0$, $U(z) = \begin{cases} \begin{pmatrix} O(1) & O(|z|^{2\alpha}) \\ O(1) & O(|z|^{2\alpha}) \end{pmatrix}, & \alpha \le 0, \\ O(1), & \alpha \ge 0. \end{cases}$

9.2 Equilibrium measure

The potential function corresponding to the jump matrix in (9.3) is a quartic polynomial $v(z) = z^4 + cz^2$. We first consider the case $c \ge -2$. By symmetry, (8.32) is automatically satisfied when $a = -b$. The equilibrium measure is supported on $(-b, b)$, where $b > 0$ is determined by (8.32) with $a = -b$; namely,

$$2\pi = \int_{-b}^{b} \frac{xv'(x)}{\sqrt{b^2 - x^2}} dx = \int_{-b}^{b} \frac{4x^4 + 2cx^2}{\sqrt{b^2 - x^2}} dx.$$

The integral on the right-hand side of the above equation is (exercise)

$$\int_{-b}^{b} \frac{4x^4 + 2cx^2}{\sqrt{b^2 - x^2}} dx = \frac{3\pi b^4}{2} + c\pi b^2. \tag{9.4}$$

Hence, we obtain

$$b = \sqrt{\frac{\sqrt{c^2 + 12} - c}{3}}. \tag{9.5}$$

It then follows from (8.30) with $a = -b$ that

$$g'(z) = \frac{\sqrt{z^2 - b^2}}{2\pi} \int_{-b}^{b} \frac{4x^3 + 2cx}{\sqrt{b^2 - x^2}} \frac{dx}{z - x}.$$

A simple calculation (exercise) gives

$$g'(z) = 2z^3 + cz - (2z^2 + c + b^2)\sqrt{z^2 - b^2}. \tag{9.6}$$

From (8.34), we obtain the equilibrium measure

$$\rho(x) = \operatorname{Re} \frac{ig'_+(x)}{\pi} = \frac{1}{\pi}(2x^2 + c + b^2)\sqrt{b^2 - x^2}, \quad x \in (-b, b). \tag{9.7}$$

Recall from (8.24) and (8.25) that

$$g(z) = \int_{-b}^{b} \ln(z - x)\rho(x)dx, \tag{9.8}$$

$$\phi(z) = \frac{v(z) + l}{2} - g(z), \tag{9.9}$$

for $z \in \mathbf{C} \setminus (-\infty, b]$, where

$$l = 2g(b) - v(b) \tag{9.10}$$

is the Lagrange multiplier. It is readily seen that $\phi(b) = 0$ and

$$\phi'(z) = v'(z)/2 - g'(z) = (2z^2 + c + b^2)\sqrt{z^2 - b^2}. \tag{9.11}$$

Thus, we have (exercise)

$$\phi(z) = \frac{(2z^2 + 2c + b^2)z\sqrt{z^2 - b^2}}{4} - \frac{3b^4 + 2cb^2}{4} \ln \frac{z + \sqrt{z^2 - b^2}}{b}, \tag{9.12}$$

for $z \in \mathbf{C} \setminus (-\infty, b]$. One can further calculate from (9.7) and (9.8) that (exercise)

$$\begin{aligned} g(b) &= \frac{1}{\pi} \int_{-b}^{b} \ln(b - x)(2x^2 + c + b^2)\sqrt{b^2 - x^2}dx \\ &= \frac{3b^4 + 2cb^2}{4} \ln \frac{b}{2} + \frac{5b^4 + 4cb^2}{16}. \end{aligned} \tag{9.13}$$

Recall from (9.5) that $3b^4 + 2cb^2 = 4$. It then follows from (9.10) that

$$l = 2 \ln \frac{b}{2} - \frac{3b^4 + 4cb^2}{8} = \ln \frac{b^2}{4} - 1 + \frac{3b^4}{8}. \tag{9.14}$$

Finally, we have an explicit expression for the g-function.

$$\begin{aligned} g(z) =& \frac{z^4 + cz^2}{2} + \ln \frac{b}{2} - \frac{1}{2} + \frac{3b^4}{16} - \frac{(2z^2 + 2c + b^2)z\sqrt{z^2 - b^2}}{4} \\ &+ \frac{3b^4 + 2cb^2}{4} \ln \frac{z + \sqrt{z^2 - b^2}}{b}, \end{aligned} \tag{9.15}$$

for $z \in \mathbf{C} \setminus (-\infty, b]$. We recall the property of the g-function from (8.27)-(8.28).

$$g_+(x) + g_-(x) - v(x) - l = \begin{cases} 0, & x \in (-b, b), \\ -2\phi(x), & x > b, \\ -2\phi_+(x) - 2\pi i, & x < -b, \end{cases} \tag{9.16}$$

$$g_+(x) - g_-(x) = \begin{cases} -2\phi_+(x) = 2\phi_-(x), & x \in (-b, b), \\ 0, & x > b, \\ 2\pi i, & x < -b. \end{cases} \tag{9.17}$$

Next, we consider the case when $c \le -2$. It was demonstrated in Bleher and Its (1999); Claeys *et al.* (2008) that the equilibrium measure $d\mu(x) = \rho(x)dx$ is supported on two intervals which are symmetric about the origin: $E := [-b, -a] \cup [a, b]$. As we shall see later, these two intervals are disconnected (i.e., $a > 0$) if $c < -2$ and merged into one interval (i.e., $a = 0$) if $c = -2$. We define the g-function and ϕ-function as

$$g(z) = \int_E \ln(z - x)\rho(x)dx, \tag{9.18}$$

$$\phi(z) = \frac{v(z) + l}{2} - g(z), \tag{9.19}$$

for $z \in \mathbf{C} \setminus (-\infty, b]$, where

$$l = 2g(b) - v(b) \tag{9.20}$$

is the Lagrange multiplier. We have the following jump conditions on the real line.

$$g_+(x) + g_-(x) - v(x) - l = \begin{cases} 0, & x \in (-b, -a) \cup (a, b), \\ -2\phi_+(x) - \pi i, & x \in (-a, a), \\ -2\phi(x), & x > b, \\ -2\phi_+(x) - 2\pi i, & x < -b, \end{cases} \tag{9.21}$$

$$g_+(x) - g_-(x) = \begin{cases} -2\phi_+(x) = 2\phi_-(x), & x \in (-b, -a) \cup (a, b), \\ \pi i, & x \in (-a, a), \\ 0, & x > b, \\ 2\pi i, & x < -b. \end{cases} \tag{9.22}$$

On account of (9.19), the two equations (9.21) and (9.22) are equivalent (exercise). It is obvious from (9.18) that $g_+(x) - g_-(x) = 0$ for $x > b$, and

$$g_+(x) - g_-(x) = \int_E 2\pi i\rho(y)dy = 2\pi i$$

for $x < -b$. By symmetry, we have $\rho(x) = \rho(-x)$ for $x \in (-b,-a) \cup (a,b)$, and hence

$$g_+(x) - g_-(x) = \int_a^b 2\pi i\rho(y)dy = \pi i$$

for $x \in (-a,a)$. The remaining equation $g_+(x)+g_-(x)-v(x)-l = 0$ for $x \in (-b,-a) \cup (a,b)$ is called the *Euler-Lagrange equation*, the proof of which can be found in Saff and Totik (1997, Chapter I). From the definition of the Lagrange multiplier in (9.20), the Euler-Lagrange equation is equivalent to

$$g'_+(x) + g'_-(x) = v'(x), \quad x \in (-b,-a) \cup (a,b). \tag{9.23}$$

By Plemelj formula, we have the following integral representation (exercise)

$$g'(z) = \frac{z(z^2-a^2)^{1/2}(z^2-b^2)^{1/2}}{\pi} \int_a^b \frac{4x^3+2cx}{\sqrt{(x^2-a^2)(b^2-x^2)}} \frac{dx}{z^2-x^2}, \tag{9.24}$$

for $z \in \mathbf{C} \setminus [-b,-a] \cup [a,b]$, where we have made use of the fact that $v'(x) = 4x^3 + 2cx$ is an odd function. A further calculation yields (exercise)

$$g'(z) = z[c + 2z^2 - 2(z^2-a^2)^{1/2}(z^2-b^2)^{1/2}]. \tag{9.25}$$

Note from (9.18) that

$$g'(z) = \int_E \frac{\rho(x)}{z-x} dx \sim \frac{1}{z}, \tag{9.26}$$

as $z \to \infty$. Since $z^2/(z^2-x^2) = 1 + x^2/(z^2-x^2)$, we obtain from (9.24) and (9.26) that (exercise)

$$0 = \int_a^b \frac{4x^3+2cx}{\sqrt{(x^2-a^2)(b^2-x^2)}} dx = \pi(b^2+a^2+c), \tag{9.27}$$

$$\pi = \int_a^b \frac{4x^5+2cx^3}{\sqrt{(x^2-a^2)(b^2-x^2)}} dx = \frac{\pi[(b^2-a^2)^2 + 2(b^2+a^2)(b^2+a^2+c)]}{4}. \tag{9.28}$$

By solving the above two equations we find explicit expressions for the MRS numbers

$$a = \sqrt{-1-c/2}, \quad b = \sqrt{1-c/2}. \tag{9.29}$$

In view of (9.23), (9.25) and (9.26), we obtain from Cauchy integral formula (or Plemelj formula) that

$$\rho(x) = \frac{i}{2\pi}[g'_+(x) - g'_-(x)] = \operatorname{Re}\frac{ig'_+(x)}{\pi} = \frac{2|x|\sqrt{(x^2-a^2)(b^2-x^2)}}{\pi}, \tag{9.30}$$

for $x \in (-b,-a) \cup (a,b)$. Substituting this into (9.18) yields (exercise)

$$\begin{aligned} g(b) =& \int_a^b \ln(b^2-x^2)\frac{2x\sqrt{(x^2-a^2)(b^2-x^2)}}{\pi}dx \\ =& \frac{(b^2-a^2)^2}{8}(\ln\frac{b^2-a^2}{4}+\frac{1}{2}). \end{aligned} \tag{9.31}$$

Note from (9.29) that $b^2 - a^2 = 2$. We then have $g(b) = (1-\ln 4)/4$. This together with (9.20) and (9.29) gives

$$l = -\ln 2 - \frac{1}{2} + \frac{c^2}{4}. \tag{9.32}$$

To calculate $\phi(z)$, we observe from (9.19), (9.20), $v'(z) = 4z^3 + 2cz$, and (9.25) that $\phi(b) = 0$ and

$$\phi'(z) = 2z(z^2-a^2)^{1/2}(z^2-b^2)^{1/2}. \tag{9.33}$$

A simple integration yields (exercise)

$$\begin{aligned} \phi(z) =& \frac{2z^2-a^2-b^2}{4}\sqrt{(z^2-a^2)(z^2-b^2)} \\ &- \frac{(b^2-a^2)^2}{8}\ln\frac{2z^2-a^2-b^2+2\sqrt{(z^2-a^2)(z^2-b^2)}}{b^2-a^2}, \end{aligned} \tag{9.34}$$

for $z \in \mathbf{C}\setminus(-\infty, b]$. Note from (9.29) that $b^2 - a^2 = 2$ and $b^2 + a^2 = -c$. We finally obtain from (9.19) and (9.32) that

$$\begin{aligned} g(z) =& \frac{z^4+cz^2-\ln 2-1/2+c^2/4}{2} - \frac{2z^2+c}{4}\sqrt{(z^2-a^2)(z^2-b^2)} \\ &+ \frac{1}{2}\ln[z^2+c/2+\sqrt{(z^2-a^2)(z^2-b^2)}], \end{aligned} \tag{9.35}$$

for $z \in \mathbf{C}\setminus(-\infty, b]$.

9.3 Contour deformation

We make the normalization transformation:

$$T(z) = e^{-Nl\sigma_3/2}U(z)e^{[-Ng(z)+Nl/2+\alpha\ln z]\sigma_3}. \tag{9.36}$$

It is readily seen from (U1)-(U4) that $T(z)$ satisfies the following Riemann-Hilbert problem on $\mathbf{R}$ with right orientation.

(T1) $T(z)$ is analytic in $\mathbf{C} \setminus \mathbf{R}$.
(T2) $T(z)$ satisfies the jump condition

$$T_+(x) = T_-(x) J_T(x) = T_-(x) \begin{pmatrix} J_{T,11}(x) & J_{T,12}(x) \\ J_{T,21}(x) & J_{T,22}(x) \end{pmatrix}, \tag{9.37}$$

for $x \in \mathbf{R} \setminus \{0\}$, where

$$J_{T,11}(x) = e^{N[g_-(x) - g_+(x)] + \alpha[(\ln x)_+ - (\ln x)_-]}, \tag{9.38}$$

$$J_{T,12}(x) = e^{N[g_+(x) + g_-(x) - v(x) - l]}, \tag{9.39}$$

$$J_{T,22}(x) = e^{N[g_+(x) - g_-(x)] - \alpha[(\ln x)_+ - (\ln x)_-]}, \tag{9.40}$$

and $J_{T,21}(x) = 0$.
(T3) As $z \to \infty$, $T(z) = I + O(|z|^{-1})$.
(T4) As $z \to 0$, $T(z) = \begin{cases} O(|z|^\alpha), & \alpha \le 0, \\ \begin{pmatrix} O(|z|^\alpha) & O(|z|^{-\alpha}) \\ O(|z|^\alpha) & O(|z|^{-\alpha}) \end{pmatrix}, & \alpha \ge 0. \end{cases}$

First, we consider the case $c \ge -2$. For convenience, we define

$$\tilde{\phi}(z) = \phi(z) \pm \pi i, \quad z \in \mathbf{C}_\pm; \tag{9.41}$$

see (8.54). It is readily seen from (9.9) and (9.17) that

$$\tilde{\phi}_+(x) - \tilde{\phi}_-(x) = \phi_+(x) - \phi_-(x) + 2\pi i = g_-(x) - g_+(x) + 2\pi i = 0$$

for $x \in (-\infty, -b)$. Hence, $\tilde{\phi}(z)$ is analytic in $\mathbf{C} \setminus [-b, \infty)$. In view of (9.16) and (9.17), we express the jump matrix as

$$J_T(x) = \begin{cases} \begin{pmatrix} e^{2N\phi_+(x)} & 1 \\ 0 & e^{2N\phi_-(x)} \end{pmatrix}, & x \in (0, b), \\ \begin{pmatrix} e^{2N\tilde{\phi}_+(x)} & 1 \\ 0 & e^{2N\tilde{\phi}_-(x)} \end{pmatrix}, & x \in (-b, 0), \end{cases} \tag{9.42a}$$

and

$$J_T(x) = \begin{cases} \begin{pmatrix} 1 & e^{-2N\phi(x)} \\ 0 & 1 \end{pmatrix}, & x \in (b, \infty), \\ \begin{pmatrix} 1 & e^{-2N\tilde{\phi}(x)} \\ 0 & 1 \end{pmatrix}, & x \in (-\infty, -b). \end{cases} \tag{9.42b}$$

Note from (9.12) and (9.41) that $\phi(x) > 0$ for $x > b$ and $\mathrm{Re}\,\tilde{\phi}(x) = \mathrm{Re}\,\phi_+(x) > 0$ for $x < -b$. Hence, as $N \to \infty$, the jump matrix $J_T(x)$ converges exponentially to the identity matrix for $x \in (-\infty, -b) \cup (b, \infty)$. For $x \in (-b, b)$, we factorize the jump matrix as

$$J_T(x) = \begin{cases} \begin{pmatrix} 1 & 0 \\ e^{2N\phi_-(x)} & 1 \end{pmatrix} \begin{pmatrix} 0 & 1 \\ -1 & 0 \end{pmatrix} \begin{pmatrix} 1 & 0 \\ e^{2N\phi_+(x)} & 1 \end{pmatrix}, & x \in (0, b), \\ \begin{pmatrix} 1 & 0 \\ e^{2N\tilde{\phi}_-(x)} & 1 \end{pmatrix} \begin{pmatrix} 0 & 1 \\ -1 & 0 \end{pmatrix} \begin{pmatrix} 1 & 0 \\ e^{2N\tilde{\phi}_+(x)} & 1 \end{pmatrix}, & x \in (-b, 0). \end{cases}$$

This suggests us to add four contours $\Sigma = \Sigma_1 \cup \Sigma_2 \cup \Sigma_3 \cup \Sigma_4$ encircling $(-b, 0) \cup (0, b)$ in the complex plane, and then perform the following factorization transformation.

$$V(z) = \begin{cases} T(z), & z \in \mathbf{C} \setminus (\bar{\Omega} \cup \mathbf{R}), \\ T(z) \begin{pmatrix} 1 & 0 \\ -e^{2N\phi(z)} & 1 \end{pmatrix}, & z \in \Omega_1, \\ T(z) \begin{pmatrix} 1 & 0 \\ e^{2N\phi(z)} & 1 \end{pmatrix}, & z \in \Omega_2, \\ T(z) \begin{pmatrix} 1 & 0 \\ -e^{2N\tilde{\phi}(z)} & 1 \end{pmatrix}, & z \in \Omega_3, \\ T(z) \begin{pmatrix} 1 & 0 \\ e^{2N\tilde{\phi}(z)} & 1 \end{pmatrix}, & z \in \Omega_4; \end{cases} \tag{9.43}$$

see Figure 9.1 for an illustration of the contours $\Sigma = \Sigma_1 \cup \Sigma_2 \cup \Sigma_3 \cup \Sigma_4$ and the upper and lower lens regions $\Omega = \Omega_1 \cup \Omega_2 \cup \Omega_3 \cup \Omega_4$.

We then obtain from (T1)-(T4) with the jump conditions (9.42) that $V(z)$ satisfies the following Riemann-Hilbert problem on $\Sigma \cup \mathbf{R}$ with right orientation.

(V1) $V(z)$ is analytic in $\mathbf{C} \setminus (\Sigma \cup \mathbf{R})$.
(V2) $V(z)$ satisfies the jump conditions $V_+(z) = V_-(z) J_V(z)$ on $\Sigma \cup \mathbf{R}$,

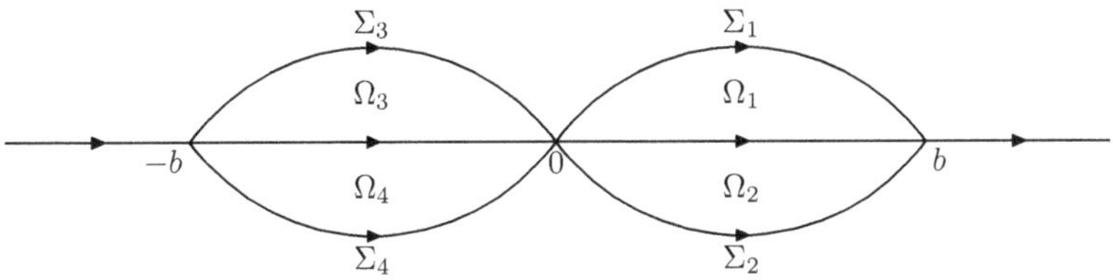

Fig. 9.1: The lens-shaped regions $\Omega = \Omega_1 \cup \Omega_2 \cup \Omega_3 \cup \Omega_4$ and the contours $\Sigma = \Sigma_1 \cup \Sigma_2 \cup \Sigma_3 \cup \Sigma_4$.

where

$$J_V(z) = \begin{cases} \begin{pmatrix} 1 & 0 \\ e^{2N\phi(z)} & 1 \end{pmatrix}, & z \in \Sigma_1 \cup \Sigma_2, \\ \\ \begin{pmatrix} 1 & 0 \\ e^{2N\tilde{\phi}(z)} & 1 \end{pmatrix}, & z \in \Sigma_3 \cup \Sigma_4, \end{cases} \tag{9.44}$$

and

$$J_V(x) = \begin{cases} \begin{pmatrix} 1 & e^{-2N\phi(x)} \\ 0 & 1 \end{pmatrix}, & x \in (b, \infty), \\ \\ \begin{pmatrix} 1 & e^{-2N\tilde{\phi}(x)} \\ 0 & 1 \end{pmatrix}, & x \in (-\infty, -b), \end{cases} \tag{9.45}$$

and

$$J_V(x) = \begin{pmatrix} 0 & 1 \\ -1 & 0 \end{pmatrix} \tag{9.46}$$

for $x \in (-b, 0) \cup (0, b)$.

(V3) As $z \to \infty$, $V(z) = I + O(|z|^{-1})$.

(V4) If $\alpha \leq 0$, $T(z) = O(|z|^{\alpha})$ as $z \to 0$. If $\alpha \geq 0$, $V(z) = O(|z|^{-\alpha})$ as z approaches 0 from inside of the lens regions; and

$$V(z) = \begin{pmatrix} O(|z|^{\alpha}) & O(|z|^{-\alpha}) \\ O(|z|^{\alpha}) & O(|z|^{-\alpha}) \end{pmatrix}$$

as z approaches 0 from outside of the lens regions.

Next, we consider the case $c \leq -2$. For convenience, we define

$$\tilde{\phi}(z) = \phi(z) \pm \pi i, \quad \bar{\phi}(z) = \phi(z) \pm \pi i/2, \quad z \in \mathbf{C}_{\pm}. \tag{9.47}$$

It is readily seen from (9.19) and (9.22) that

$$\tilde{\phi}_+(x) - \tilde{\phi}_-(x) = \phi_+(x) - \phi_-(x) + 2\pi i = g_-(x) - g_+(x) + 2\pi i = 0$$

for $x \in (-\infty, -b)$, and

$$\bar{\phi}_+(x) - \bar{\phi}_-(x) = \phi_+(x) - \phi_-(x) + \pi i = g_-(x) - g_+(x) + \pi i = 0$$

for $x \in (-a, a)$. Hence, $\tilde{\phi}(z)$ is analytic in $\mathbf{C} \setminus [-b, \infty)$ and $\bar{\phi}(z)$ is analytic in $\mathbf{C} \setminus (-\infty, -a] \cup [a, \infty)$. In view of (9.21) and (9.22), we express the jump matrix as

$$J_T(x) = \begin{cases} \begin{pmatrix} e^{2N\phi_+(x)} & 1 \\ 0 & e^{2N\phi_-(x)} \end{pmatrix}, & x \in (a, b), \\ \\ \begin{pmatrix} e^{2N\tilde{\phi}_+(x)} & 1 \\ 0 & e^{2N\tilde{\phi}_-(x)} \end{pmatrix}, & x \in (-b, -a), \end{cases} \tag{9.48a}$$

and

$$J_T(x) = \begin{cases} \begin{pmatrix} 1 & e^{-2N\phi(x)} \\ 0 & 1 \end{pmatrix}, & x \in (b, \infty), \\ \\ \begin{pmatrix} 1 & e^{-2N\tilde{\phi}(x)} \\ 0 & 1 \end{pmatrix}, & x \in (-\infty, -b), \end{cases} \tag{9.48b}$$

and

$$J_T(x) = \begin{cases} \begin{pmatrix} e^{-(n+\alpha)\pi i} & e^{-2N\bar{\phi}(x)} \\ 0 & e^{(n+\alpha)\pi i} \end{pmatrix}, & x \in (0, a), \\ \\ \begin{pmatrix} e^{(n+\alpha)\pi i} & e^{-2N\bar{\phi}(x)} \\ 0 & e^{-(n+\alpha)\pi i} \end{pmatrix}, & x \in (-a, 0). \end{cases} \tag{9.48c}$$

Similar to the previous case, the following factorization of the jump matrix

$$J_T(x) = \begin{cases} \begin{pmatrix} 1 & 0 \\ e^{2N\phi_-(x)} & 1 \end{pmatrix} \begin{pmatrix} 0 & 1 \\ -1 & 0 \end{pmatrix} \begin{pmatrix} 1 & 0 \\ e^{2N\phi_+(x)} & 1 \end{pmatrix}, & x \in (a, b), \\ \\ \begin{pmatrix} 1 & 0 \\ e^{2N\tilde{\phi}_-(x)} & 1 \end{pmatrix} \begin{pmatrix} 0 & 1 \\ -1 & 0 \end{pmatrix} \begin{pmatrix} 1 & 0 \\ e^{2N\tilde{\phi}_+(x)} & 1 \end{pmatrix}, & x \in (-b, -a) \end{cases}$$

suggests us to add four contours $\Sigma = \Sigma_1 \cup \Sigma_2 \cup \Sigma_3 \cup \Sigma_4$ encircling $(-b, -a) \cup (a, b)$ in the complex plane, and then perform the following factorization transformation.

$$V(z) = \begin{cases} T(z), & z \in \mathbf{C} \setminus (\bar{\Omega} \cup \mathbf{R}), \\ T(z) \begin{pmatrix} 1 & 0 \\ -e^{2N\phi(z)} & 1 \end{pmatrix}, & z \in \Omega_1, \\ T(z) \begin{pmatrix} 1 & 0 \\ e^{2N\phi(z)} & 1 \end{pmatrix}, & z \in \Omega_2, \\ T(z) \begin{pmatrix} 1 & 0 \\ -e^{2N\tilde{\phi}(z)} & 1 \end{pmatrix}, & z \in \Omega_3, \\ T(z) \begin{pmatrix} 1 & 0 \\ e^{2N\tilde{\phi}(z)} & 1 \end{pmatrix}, & z \in \Omega_4; \end{cases} \tag{9.49}$$

see Figure 9.2 for an illustration of the contours $\Sigma = \Sigma_1 \cup \Sigma_2 \cup \Sigma_3 \cup \Sigma_4$ and the upper and lower lens regions $\Omega = \Omega_1 \cup \Omega_2 \cup \Omega_3 \cup \Omega_4$.

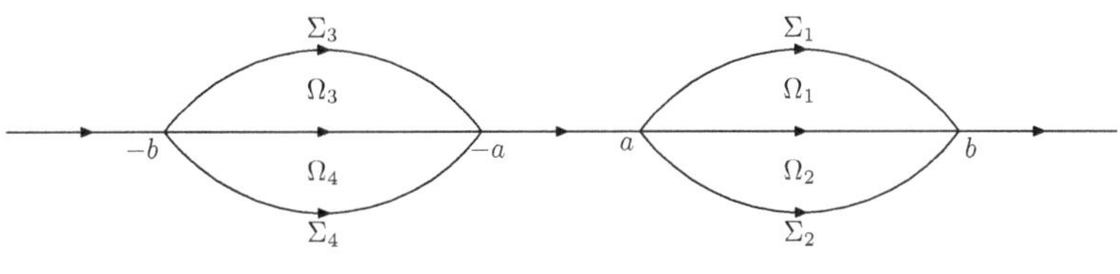

Fig. 9.2: The lens-shaped regions $\Omega = \Omega_1 \cup \Omega_2 \cup \Omega_3 \cup \Omega_4$ and the contours $\Sigma = \Sigma_1 \cup \Sigma_2 \cup \Sigma_3 \cup \Sigma_4$.

We then obtain from (T1)-(T4) with the jump conditions (9.42) that $V(z)$ satisfies the following Riemann-Hilbert problem on $\Sigma \cup \mathbf{R}$ with right orientation.

(V1) $V(z)$ is analytic in $\mathbf{C} \setminus (\Sigma \cup \mathbf{R})$.
(V2) $V(z)$ satisfies the jump conditions $V_+(z) = V_-(z) J_V(z)$ on $\Sigma \cup \mathbf{R}$,

where

$$J_V(z) = \begin{cases} \begin{pmatrix} 1 & 0 \\ e^{2N\phi(z)} & 1 \end{pmatrix}, & z \in \Sigma_1 \cup \Sigma_2, \\ \\ \begin{pmatrix} 1 & 0 \\ e^{2N\tilde{\phi}(z)} & 1 \end{pmatrix}, & z \in \Sigma_3 \cup \Sigma_4, \end{cases} \tag{9.50a}$$

and

$$J_V(x) = \begin{cases} \begin{pmatrix} 1 & e^{-2N\phi(x)} \\ 0 & 1 \end{pmatrix}, & x \in (b, \infty), \\ \\ \begin{pmatrix} 1 & e^{-2N\tilde{\phi}(x)} \\ 0 & 1 \end{pmatrix}, & x \in (-\infty, -b), \end{cases} \tag{9.50b}$$

and

$$J_V(x) = \begin{cases} \begin{pmatrix} e^{-(n+\alpha)\pi i} & e^{-2N\bar{\phi}(x)} \\ 0 & e^{(n+\alpha)\pi i} \end{pmatrix}, & x \in (0, a), \\ \\ \begin{pmatrix} e^{(n+\alpha)\pi i} & e^{-2N\bar{\phi}(x)} \\ 0 & e^{-(n+\alpha)\pi i} \end{pmatrix}, & x \in (-a, 0), \end{cases} \tag{9.50c}$$

and

$$J_V(x) = \begin{pmatrix} 0 & 1 \\ -1 & 0 \end{pmatrix} \tag{9.50d}$$

for $x \in (-b, -a) \cup (a, b)$.

(V3) As $z \to \infty$, $V(z) = I + O(|z|^{-1})$.

(V4) If $\alpha \leq 0$, then $V(z) = O(|z|^{\alpha})$ as $z \to 0$. If $\alpha \geq 0$ and $c < -2$, then

$$V(z) = \begin{pmatrix} O(|z|^{\alpha}) & O(|z|^{-\alpha}) \\ O(|z|^{\alpha}) & O(|z|^{-\alpha}) \end{pmatrix}$$

as $z \to 0$. If $\alpha \geq 0$ and $c = -2$, then $V(z) = O(|z|^{-\alpha})$ as z approaches 0 from inside of the lens regions; and

$$V(z) = \begin{pmatrix} O(|z|^{\alpha}) & O(|z|^{-\alpha}) \\ O(|z|^{\alpha}) & O(|z|^{-\alpha}) \end{pmatrix}$$

as z approaches 0 from outside of the lens regions.

9.4 Construction of parametrices

First, we consider the case $c > -2$. When the variable z is bounded away from the MRS numbers $(-b, b)$, one can show that the jump matrix $J_V(z)$ tends to the identity matrix as $N \to \infty$ (exercise). In view of Riemann-Hilbert problem (M1)-(M3) for the parametrix $M(z)$ explicitly given in (8.7) (with $a = -b$), we can formally approximate $V(z)$, and thus $T(z)$, by $M(z)$; namely,

$$T(z) \sim V(z) \sim M(z) = \frac{1}{2} \begin{pmatrix} \beta(z) + \beta(z)^{-1} & -i[\beta(z) - \beta(z)^{-1}] \\ i[\beta(z) - \beta(z)^{-1}] & \beta(z) + \beta(z)^{-1} \end{pmatrix},$$

where

$$\beta(z) := \frac{(z-b)^{1/4}}{(z+b)^{1/4}}. \tag{9.51}$$

Tracing back to $U(z)$ by (9.36) and making use of (9.9) and (9.10), we have the following formal approximation

$$U(z) \sim e^{Nl\sigma_3/2} M(z) e^{[Ng(z) - Nl/2 - \alpha \ln z]\sigma_3}. \tag{9.52}$$

To cancel the singularity of $|x|^{2\alpha}$ in the weight function, we shall make use of the *Bessel parametrix* defined as

$$B^{(\alpha)}(z) := z^{1/2} \begin{pmatrix} I_{\alpha+1/2}(ze^{-\pi i/2}) & -K_{\alpha+1/2}(ze^{-\pi i/2}) \\ -iI_{\alpha-1/2}(ze^{-\pi i/2}) & -iK_{\alpha-1/2}(ze^{-\pi i/2}) \end{pmatrix} (\sqrt{\pi} e^{\alpha\pi i/2})^{\sigma_3}, \tag{9.53a}$$

for $z \in \mathbf{C}_+$, and

$$B^{(\alpha)}(z) := z^{1/2} \begin{pmatrix} -iI_{\alpha+1/2}(ze^{\pi i/2}) & -iK_{\alpha+1/2}(ze^{\pi i/2}) \\ I_{\alpha-1/2}(ze^{\pi i/2}) & -K_{\alpha-1/2}(ze^{\pi i/2}) \end{pmatrix} (\sqrt{\pi} e^{-\alpha\pi i/2})^{\sigma_3}, \tag{9.53b}$$

for $z \in \mathbf{C}_-$, where $I_{\alpha\pm1/2}$ and $K_{\alpha\pm1/2}$ are modified Bessel functions defined in $\mathbf{C} \setminus (-\infty, 0]$. On account of the identities

$$I_\nu(ze^{\pi i}) = e^{\nu\pi i} I_\nu(z), \quad K_\nu(ze^{\pi i}) = e^{-\nu\pi i} K_\nu(z) - \pi i I_\nu(z), \quad \arg z \in (-\pi, 0),$$

we obtain $B_+^{(\alpha)}(x) = B_-^{(\alpha)}(x) J_B^{(\alpha)}(x)$ for $x \in \mathbf{R}$, where

$$J_B^{(\alpha)}(x) = \begin{cases} \begin{pmatrix} 1 & 1 \\ 0 & 1 \end{pmatrix}, & x > 0, \\ \begin{pmatrix} e^{2\alpha\pi i} & 1 \\ 0 & e^{-2\alpha\pi i} \end{pmatrix}, & x < 0. \end{cases} \tag{9.54}$$

Furthermore, from the asymptotic formulas of modified Bessel functions

$$I_\nu(z) \sim \frac{e^z}{\sqrt{2\pi z}}, \quad K_\nu(z) \sim \sqrt{\frac{\pi}{2z}} e^{-z}, \quad z \to \infty, \ |\arg z| < \pi/2,$$

we have as $z \to \infty$,

$$B^{(\alpha)}(z) \sim \frac{1}{\sqrt{2}} \begin{pmatrix} 1 & -i \\ -i & 1 \end{pmatrix} e^{i(\pi/4 - z + \alpha\pi/2)\sigma_3} \tag{9.55}$$

for $z \in \mathbf{C}_+$, and

$$B^{(\alpha)}(z) \sim \frac{1}{\sqrt{2}} \begin{pmatrix} 1 & -i \\ -i & 1 \end{pmatrix} e^{i(\pi/4 - z + \alpha\pi/2)\sigma_3} \begin{pmatrix} 0 & -1 \\ 1 & 0 \end{pmatrix} \tag{9.56}$$

for $z \in \mathbf{C}_-$. Now, we introduce

$$\eta(z) := \begin{cases} -i\phi(z) + i\phi_+(0), & z \in \mathbf{C}_+, \\ i\phi(z) + i\phi_+(0), & z \in \mathbf{C}_-, \end{cases} \tag{9.57}$$

and

$$E_0(z) := \begin{cases} M(z) e^{(-\alpha\pi i/2 + N\phi_+(0) + \pi i/4)\sigma_3}, & z \in \mathbf{C}_+, \\ M(z) \begin{pmatrix} 0 & 1 \\ -1 & 0 \end{pmatrix} e^{(-\alpha\pi i/2 + N\phi_+(0) + \pi i/4)\sigma_3}, & z \in \mathbf{C}_-. \end{cases} \tag{9.58}$$

Since $\phi_+(x) = -\phi_-(x)$ for $x \in (-b, b)$ by (9.17), the function $\eta(z)$ is analytic for $z \in \mathbf{C} \setminus (-\infty, -b] \cup [b, \infty)$. It also follows from the jump condition of $M(z)$ on $(-b, b)$ that $E_0(z)$ is analytic for $z \in \mathbf{C} \setminus (-\infty, -b] \cup [b, \infty)$. The parametrix in a neighborhood of the origin is defined as

$$P(z) := \frac{1}{\sqrt{2}} e^{Nl\sigma_3/2} E_0(z) \begin{pmatrix} 1 & i \\ i & 1 \end{pmatrix} B^{(\alpha)}(N\eta(z)) e^{N(z^4 + cz^2)\sigma_3/2} z^{-\alpha\sigma_3}, \tag{9.59}$$

for $z \in B_\delta(0)$, where $B_\delta(0)$ denotes the disk centered at the origin with a fixed small radius $\delta > 0$. It is easy to verify that $P(z)$ satisfies the following conditions of a local Riemann-Hilbert problem (exercise).

(P1) $P(z)$ is analytic in $B_\delta(0) \setminus [-\delta, \delta]$.

(P2) For $x \in (-\delta, 0) \cup (0, \delta)$, we have

$$P_+(x) = P_-(x) \begin{pmatrix} 1 & |x|^{2\alpha} e^{-N(x^4 + cx^2)} \\ 0 & 1 \end{pmatrix}.$$

(P3) For $z \in \partial B_\delta(0) \setminus \{\pm\delta\}$, $P(z) \sim e^{Nl\sigma_3/2} M(z) e^{[Ng(z) - Nl/2 - \alpha \ln z]\sigma_3}$, as $n \to \infty$.

The parametrix away from the origin can be constructed via Airy parametrix defined in (8.8) and $M(z)$ defined in (8.7). Similar as in (8.53), (8.55) and (8.58), we define

$$f(z) := \left[\frac{3}{2}N\phi(z)\right]^{2/3}, \quad z \in \mathbf{C} \setminus (-\infty, -b], \tag{9.60}$$

$$\tilde{f}(z) := \left[\frac{3}{2}N\tilde{\phi}(z)\right]^{2/3}, \quad z \in \mathbf{C} \setminus [b, \infty), \tag{9.61}$$

and

$$P^*(z) := \begin{cases} E(z)Q(z)e^{[Nv(z)/2-\alpha \ln z]\sigma_3}, & |\arg z| \le \pi/4, \\ \tilde{E}(z)Q(z)e^{[Nv(z)/2-\alpha \ln(-z)]\sigma_3}, & |\arg z| \in [3\pi/4, \pi], \\ e^{Nl\sigma_3/2}M(z)e^{[Ng(z)-Nl/2-\alpha \ln z]\sigma_3}, & |\arg z| \in [\pi/4, 3\pi/4], \end{cases} \tag{9.62}$$

where

$$Q(z) = \begin{cases} A(f(z)), & \operatorname{Re} z > 0, \\ A(\tilde{f}(z))\sigma_3, & \operatorname{Re} z < 0, \end{cases} \tag{9.63}$$

and

$$E(z) := \sqrt{\pi}e^{Nl\sigma_3/2}M(z)\begin{pmatrix} 1 & -1 \\ -i & -i \end{pmatrix} f(z)^{\sigma_3/4}, \tag{9.64}$$

$$\tilde{E}(z) := (-1)^n\sqrt{\pi}e^{Nl\sigma_3/2}M(z)\sigma_3\begin{pmatrix} 1 & -1 \\ -i & -i \end{pmatrix} \tilde{f}(z)^{\sigma_3/4}. \tag{9.65}$$

Unlike the Hermite polynomials where $\phi(z) \sim z^2$ as $z \to \infty$ (see (8.83)), here, we have from (9.9) that $\phi(z) \sim z^4/2$ as $z \to \infty$. Hence, the Stokes line $\arg f(z) = \pi$ is asymptotic to the ray $\arg z = 3\pi/8$ as $z \to \infty$. We choose the regions of Airy parametrix as $|\arg z| \le \pi/4$ and $|\arg z| \in [3\pi/4, \pi]$. Outside of these regions; namely, when $|\arg z| \in [\pi/4, 3\pi/4]$, we make use of $M(z)$ to define the parametrix. It can be shown from (9.11) that $\operatorname{Re}\phi(z) < 0$ and $\arg f(z) < \pi$ for $\arg z = \pi/4$ (exercise). The global parametrix for the case $c > -2$ is defined as

$$U^*(z) = \begin{cases} P(z), & z \in B_\delta(0), \\ P^*(z), & z \in \mathbf{C} \setminus B_\delta(0). \end{cases} \tag{9.66}$$

Next, we consider the critical case $c = -2$. Note that the equilibrium measure $\rho(x)$, the ϕ-function and the g-function in (9.7), (9.12) and (9.15) coincide with those in (9.30), (9.34) and (9.35), respectively. The parametrix away from the origin is defined the same as in (9.62). However, the parametrix in the neighborhood of the origin requires a special treatment. The second Painlevé equation $u''(s) = su(s) + 2u(s)^3 - \alpha$ with $v(s) = u'(s)$ is equivalent to the compatibility condition of the Lax pair

$$\partial_\zeta \Psi(\zeta, s) = [-i(4\zeta^2 + s + 2u^2)\sigma_3 + (4\zeta u + \alpha/\zeta)\sigma_1 - 2v\sigma_2]\Psi, \tag{9.67}$$

$$\partial_s \Psi(\zeta, s) = (-i\zeta\sigma_3 + u\sigma_1)\Psi. \tag{9.68}$$

For each $k = 1, \cdots, 6$, denote by Ψ_k the canonical solution of (9.67) with asymptotic behavior

$$\Psi_k(\zeta, s) = (I + O(|\zeta|^{-1}))e^{-i(4\zeta^3/3+s\zeta)\sigma_3} \tag{9.69}$$

as $\zeta \to \infty$ with $(k-2)\pi/3 < \arg\zeta < k\pi/3$. The corresponding Stokes matrices S_k are defined as $\Psi_{k+1}(\zeta, s) = \Psi_k(\zeta, s)S_k$ for $k = 1, \cdots, 5$, and $\Psi_1(\zeta, s) = \Psi_6(\zeta, s)S_6$. It is shown in Fokas *et al.* (2006, p. 164) that S_{k+3} is the transpose of S_k for $k = 1, 2, 3$, and

$$S_1 = \begin{pmatrix} 1 & 0 \\ s_1 & 1 \end{pmatrix}, \quad S_2 = \begin{pmatrix} 1 & s_2 \\ 0 & 1 \end{pmatrix}, \quad S_3 = \begin{pmatrix} 1 & 0 \\ s_3 & 1 \end{pmatrix}, \tag{9.70}$$

with $s_1 + s_2 + s_3 + s_1 s_2 s_3 = -2i\sin(\alpha\pi)$. Moreover, it is proved in Claeys *et al.* (2008) that if

$$s_1 = e^{-\alpha\pi i}, \quad s_2 = 0, \quad s_3 = -e^{\alpha\pi i},$$

then $u(s)$ is the Painlevé II solution satisfying the asymptotic boundary conditions

$$u(s) \sim \begin{cases} \alpha/s, & s \to \infty, \\ \sqrt{-s/2}, & s \to -\infty. \end{cases} \tag{9.71}$$

In this case, $\Psi_2(\zeta, s) = \Psi_3(\zeta, s) \sim e^{-i(4\zeta^3/3+s\zeta)\sigma_3}$ and $\Psi_5(\zeta, s) = \Psi_6(\zeta, s) \sim e^{-i(4\zeta^3/3+s\zeta)\sigma_3}$ as $\zeta \in \infty$ in $\mathbf{C}_+$ and $\mathbf{C}_-$, respectively. Now, we construct the parametrix near the origin as

$$R(z) = M(z)e^{n\pi i\sigma_3/2}\Psi_3(\zeta, 0)e^{i(4\zeta^3/3-n\pi/2)\sigma_3} \tag{9.72a}$$

for $z \in \mathbf{C}_+$, and

$$R(z) = M(z)\begin{pmatrix} 0 & 1 \\ -1 & 0 \end{pmatrix} e^{n\pi i\sigma_3/2}\Psi_6(\zeta, 0)e^{i(4\zeta^3/3-n\pi/2)\sigma_3}\begin{pmatrix} 0 & -1 \\ 1 & 0 \end{pmatrix} \tag{9.72b}$$

for $z \in \mathbf{C}_-$, where

$$\zeta = \zeta(z) := \left\{\frac{3}{4}N[\pi/2 \mp i\phi(z)]\right\}^{1/3}, \quad z \in \mathbf{C}_\pm \tag{9.73}$$

is defined as an analytic function (exercise) in a neighborhood of the origin such that $\zeta(z) \sim (N/\sqrt{2})^{1/3}z$ as $z \to 0$. One can easily verify that $R(z)$ satisfies the following local Riemann-Hilbert problem (exercise).

(R1) $R(z)$ is analytic in $B_\delta(0) \setminus [-\delta, \delta]$.
(R2) For $x \in (-\delta, 0) \cup (0, \delta)$, we have $R_+(x) = R_-(x) J_R(x)$, where

$$J_R(x) = \begin{cases} \begin{pmatrix} e^{2N\phi_+} & 1 \\ 0 & e^{2N\phi_-} \end{pmatrix}, & x \in (0, \delta), \\ \\ \begin{pmatrix} e^{2N\tilde{\phi}_+} & 1 \\ 0 & e^{2N\tilde{\phi}_-} \end{pmatrix}, & x \in (-\delta, 0). \end{cases} \tag{9.74}$$

(R3) For $z \in \partial B_\delta(0) \setminus \{\pm\delta\}$, $R(z) \sim M(z)$, as $n \to \infty$.

For simplicity, we denote $\varphi_1(\zeta) = \Psi_{3,11}(\zeta, 0) = \Psi_{2,11}(\zeta, 0)$ and $\varphi_2(\zeta) := \Psi_{3,21}(\zeta, 0) = \Psi_{2,21}(\zeta, 0)$ for $\zeta \in \mathbf{C}_+$. It is readily seen that the pair $\varphi_1(\zeta)$ and $\varphi_2(\zeta)$ is the unique solution to the linear system

$$\begin{pmatrix} \varphi_1'(\zeta) \\ \varphi_2'(\zeta) \end{pmatrix} = [-i(4\zeta^2 + 2u_0^2)\sigma_3 + (4\zeta u_0 + \alpha/\zeta)\sigma_1 - 2v_0\sigma_2] \begin{pmatrix} \varphi_1(\zeta) \\ \varphi_2(\zeta) \end{pmatrix} \tag{9.75}$$

with the asymptotic boundary condition

$$e^{4\zeta^3/3}\varphi_1(\zeta) = 1 + O(|\zeta|^{-1}), \quad e^{4\zeta^3/3}\varphi_2(\zeta) = O(|\zeta|^{-1}), \quad \zeta \to \infty, \ \zeta \in \mathbf{C}_+, \tag{9.76}$$

where $u_0 = u(0)$ and $v_0 = u'(0)$. One can further prove (exercise) that $\varphi_1(\zeta) = \Psi_{6,22}(-\zeta, 0) = \Psi_{5,22}(-\zeta, 0)$ and $\varphi_2(\zeta) = \Psi_{6,12}(-\zeta, 0) = \Psi_{5,12}(-\zeta, 0)$ for $\zeta \in \mathbf{C}_+$. Consequently, we obtain (exercise) from (8.7), (9.72) and (9.73) that

$$R_{11}(z) e^{-N\phi(z) - \alpha \ln z} = (-1)^n R_{11}(-z) e^{-N\phi(-z) - \alpha \ln(-z)}, \quad z \in \mathbf{C}_-. \tag{9.77}$$

The global parametrix for the critical case $c = -2$ is defined as

$$U^*(z) = \begin{cases} e^{Nl\sigma_3/2} R(z) e^{[Ng(z) - Nl/2 - \alpha \ln z]\sigma_3}, & z \in B_\delta(0), \\ P^*(z), & z \in \mathbf{C} \setminus B_\delta(0), \end{cases} \tag{9.78}$$

where $P^*(z)$ and $R(z)$ are given in (9.62) and (9.72), respectively.

Finally, we consider the case $c < -2$. The jump conditions in (9.50) motivate us to consider the following limiting Riemann-Hilbert problem.

(K1) $K(z)$ is analytic in $\mathbf{C} \setminus [-b, b]$.

(K2) For $|x| \in (0,a) \cup (a,b)$, we have $K_+(x) = K_-(x)J_K(x)$, where

$$J_K(x) = \begin{cases} \begin{pmatrix} 0 & 1 \\ -1 & 0 \end{pmatrix}, & x \in (-b,-a) \cup (a,b), \\ \begin{pmatrix} e^{-(n+\alpha)\pi i} & 0 \\ 0 & e^{(n+\alpha)\pi i} \end{pmatrix}, & x \in (0,a), \\ \begin{pmatrix} e^{(n+\alpha)\pi i} & 0 \\ 0 & e^{-(n+\alpha)\pi i} \end{pmatrix}, & x \in (-a,0). \end{cases}$$

(K3) As $z \to \infty$, $K(z) = I + O(|z|^{-1})$.

To solve this Riemann-Hilbert problem, we shall introduce the Szegö function

$$D(z) = \exp\left[-\frac{\alpha\sqrt{(z^2-a^2)(z^2-b^2)}}{2}\int_0^{a^2} \frac{1}{\sqrt{(x^2-a^2)(b^2-x^2)}}\frac{dx}{x-z^2}\right] \tag{9.79}$$

which is analytic in $\mathbf{C} \setminus [-b,b]$ and satisfies the jump conditions (exercise)

$$\begin{aligned} D_+(x)D_-(x) &= 1, \quad x \in (-b,-a) \cup (a,b), \\ D_+(x) &= D_-(x)e^{\alpha\pi i}, \quad x \in [0,a], \\ D_+(x) &= D_-(x)e^{-\alpha\pi i}, \quad x \in [-a,0]. \end{aligned} \tag{9.80}$$

Moreover,

$$D(\infty) = [(b+a)/(b-a)]^{\alpha/2}. \tag{9.81}$$

It is readily seen that

$$\tilde{K}(z) := D(\infty)^{-\sigma_3}K(z)D(z)^{\sigma_3} \tag{9.82}$$

satisfies the following Riemann-Hilbert problem.

($\tilde{K}1$) $\tilde{K}(z)$ is analytic in $\mathbf{C} \setminus [-b,b]$.
($\tilde{K}2$) For $|x| \in (0,a) \cup (a,b)$, we have $\tilde{K}_+(x) = \tilde{K}_-(x)J_{\tilde{K}}(x)$, where

$$J_{\tilde{K}}(x) = \begin{cases} \begin{pmatrix} 0 & 1 \\ -1 & 0 \end{pmatrix}, & x \in (-b,-a) \cup (a,b), \\ \begin{pmatrix} e^{-n\pi i} & 0 \\ 0 & e^{n\pi i} \end{pmatrix}, & x \in (-a,a). \end{cases}$$

($\tilde{K}3$) As $z \to \infty$, $\tilde{K}(z) = I + O(|z|^{-1})$.

One can verify (exercise) that

$$\tilde{K}(z) = \frac{1}{2}\begin{pmatrix} \gamma(z) + \gamma(z)^{-1} & i[\gamma(z) - \gamma(z)^{-1}] \\ -i[\gamma(z) - \gamma(z)^{-1}] & \gamma(z) + \gamma(z)^{-1} \end{pmatrix} \tag{9.83}$$

solves ($\tilde{K}1$)-($\tilde{K}3$), where

$$\gamma(z) := \frac{(z+b)^{1/4}(z-a)^{(-1)^n/4}}{(z-b)^{1/4}(z+a)^{(-1)^n/4}}. \tag{9.84}$$

Now, we fix $c_0 \in (a,b)$, for example, $c_0 = \sqrt{-c/2}$. Similar as in the case $c \geq -2$, we define

$$f(z) := \left[\frac{3}{2}N\phi(z)\right]^{2/3}, \quad z \in \mathbf{C} \setminus (-\infty, a], \tag{9.85}$$

$$\tilde{f}(z) := \left[\frac{3}{2}N\tilde{\phi}(z)\right]^{2/3}, \quad z \in \mathbf{C} \setminus [-a, \infty). \tag{9.86}$$

By symmetry of the generalized Freud weight function, we have $\tilde{\phi}(z) = \phi(-z)$ and $\tilde{f}(z) = f(-z)$. Note from (9.33), (9.34) and (9.47) that (exercise)

$$\bar{\phi}(z) = \int_a^z 2w\sqrt{(w^2-a^2)(w^2-b^2)}dw = \int_{-a}^z 2w\sqrt{(w^2-a^2)(w^2-b^2)}dw, \tag{9.87}$$

which is analytic in $\mathbf{C} \setminus (-\infty, -a] \cup [a, \infty)$. We then have

$$\bar{\phi}(z) \sim \begin{cases} (4a/3)\sqrt{(2a)(b^2-a^2)}(a-z)^{3/2}, & z \to a, \\ e^{\pm 2\pi i}(4a/3)\sqrt{(2a)(b^2-a^2)}(z+a)^{3/2}, & z \to -a, \ z \in \mathbf{C}_\pm. \end{cases}$$

Define

$$\bar{f}(z) := \left[\frac{3}{2}N\bar{\phi}(z)\right]^{2/3} \sim 2a(b^2-a^2)^{1/3}(a-z), \quad z \to a, \tag{9.88}$$

and

$$\hat{f}(z) := \left[\frac{3}{2}N\bar{\phi}(z)e^{\mp 2\pi i}\right]^{2/3} \sim 2a(b^2-a^2)^{1/3}(z+a), \quad z \to -a, \ z \in \mathbf{C}_\pm. \tag{9.89}$$

By analytic continuation, $\bar{f}(z)$ is analytic in $\mathbf{C} \setminus (-\infty, -a] \cup [b, \infty)$ and $\hat{f}(z)$ is analytic in $\mathbf{C} \setminus (-\infty, -b] \cup [a, \infty)$. We also have $\hat{f}(z) = \bar{f}(-z)$ by symmetry of the generalized Freud weight function. Making use of the Airy parametrix in (8.8), we define

$$Q(z) := \begin{cases} A(f(z)), & |\arg z| \le \pi/4,\ |z| \ge c_0, \\ A(\tilde{f}(z))\sigma_3, & 3\pi/4 \le |\arg z| \le \pi,\ |z| \ge c_0, \\ A(\bar{f}(z))\sigma_3, & \operatorname{Re} z \ge 0,\ |z| \le c_0, \\ A(\hat{f}(z)), & \operatorname{Re} z \le 0,\ |z| \le c_0. \end{cases} \tag{9.90}$$

The global parametrix is constructed as follows.

$$U^*(z) := \begin{cases} E(z)Q(z)e^{[\frac{Nv(z)}{2} - \alpha \ln z]\sigma_3}, & |\arg z| \le \frac{\pi}{4},\ |z| \ge c_0, \\ \tilde{E}(z)Q(z)e^{[\frac{Nv(z)}{2} - \alpha \ln(-z)]\sigma_3}, & |\arg z| \in [\frac{3\pi}{4}, \pi],\ |z| \ge c_0, \\ \bar{E}(z)Q(z)e^{[\frac{Nv(z)}{2} - \alpha \ln z]\sigma_3}, & \operatorname{Re} z \ge 0,\ |z| \le c_0, \\ \hat{E}(z)Q(z)e^{[\frac{Nv(z)}{2} - \alpha \ln(-z)]\sigma_3}, & \operatorname{Re} z \le 0,\ |z| \le c_0, \\ e^{\frac{Nl}{2}\sigma_3} K(z) e^{[Ng(z) - \frac{Nl}{2} - \alpha \ln z]\sigma_3}, & |\arg z| \in [\frac{\pi}{4}, \frac{3\pi}{4}],\ |z| \ge c_0, \end{cases} \tag{9.91}$$

where

$$E(z) := \sqrt{\pi} e^{Nl\sigma_3/2} K(z) \begin{pmatrix} 1 & -1 \\ -i & -i \end{pmatrix} f(z)^{\sigma_3/4}, \tag{9.92}$$

$$\tilde{E}(z) := (-1)^n \sqrt{\pi} e^{Nl\sigma_3/2} K(z)\sigma_3 \begin{pmatrix} 1 & -1 \\ -i & -i \end{pmatrix} \tilde{f}(z)^{\sigma_3/4}, \tag{9.93}$$

$$\bar{E}(z) := \sqrt{\pi} e^{Nl\sigma_3/2} K(z)\sigma_3 e^{\pm N\pi i\sigma_3/2} \begin{pmatrix} 1 & -1 \\ -i & -i \end{pmatrix} \bar{f}(z)^{\sigma_3/4}, \tag{9.94}$$

$$\hat{E}(z) := (-1)^n \sqrt{\pi} e^{Nl\sigma_3/2} K(z) e^{\mp N\pi i\sigma_3/2} \begin{pmatrix} 1 & -1 \\ -i & -i \end{pmatrix} \hat{f}(z)^{\sigma_3/4}. \tag{9.95}$$

9.5 Uniform asymptotic expansions

It is readily seen from the construction of the global parametrix $U^*(z)$ in the previous section that (exercise)

$$U^*_+(x) = U^*_-(x) \begin{pmatrix} 1 & |x|^{2\alpha} e^{-Nv(x)} \\ 0 & 1 \end{pmatrix}, \tag{9.96}$$

for $x \in \mathbf{R} \setminus \{0\}$, and $U^*(z)$ has the same large z behavior as $U(z)$ in (U3). In addition to the real line, the global parametrix $U^*(z)$ has jumps on the contour

$$\Gamma := \begin{cases} \partial B_\delta(0) \cup \{||\arg z| - \pi/2| = \pi/4 :\ |z| \ge \delta\}, & c \ge -2, \\ \partial B_{c_0}(0) \cup \{||\arg z| - \pi/2| = \pi/4 :\ |z| \ge c_0\}, & c < -2. \end{cases} \tag{9.97}$$

Now, we define

$$S(z) := e^{-Nl\sigma_3/2}U(z)[U^*(z)]^{-1}e^{Nl\sigma_3/2}, \tag{9.98}$$

which satisfies the following Riemann-Hilbert problem.

(S1) $S(z)$ is analytic in $\mathbf{C}\setminus\Gamma$.
(S2) $S(z)$ satisfies the jump condition $S_+(z) = S_-(z)J_S(z)$ for $z\in\Gamma$, where

$$J_S(z) = e^{-Nl\sigma_3/2}U_-^*(z)[U_+^*(z)]^{-1}e^{Nl\sigma_3/2}. \tag{9.99}$$

(S3) As $z\to\infty$, $S(z) = I + O(|z|^{-1})$.

Similar as in the proof of (8.66), we obtain

$$J_S(z) \sim \begin{cases} I+\sum_{k=1}^{\infty}\frac{J_{S,k}(z)}{N^k}, & c\neq -2,\\ I+\sum_{k=1}^{\infty}\frac{J_{S,k}(z)}{N^{k/3}}, & c=-2, \end{cases} \tag{9.100}$$

as $N\to\infty$. Consequently, the solution of the Riemann-Hilbert problem (S1)-(S3) has the asymptotic expansion

$$S(z) \sim \begin{cases} I+\sum_{k=1}^{\infty}\frac{S_k(z)}{N^k}, & c\neq -2,\\ I+\sum_{k=1}^{\infty}\frac{S_k(z)}{N^{k/3}}, & c=-2, \end{cases} \tag{9.101}$$

as $N\to\infty$, where $S_k(z)$ can be determined recursively by

$$S_k(z) = \frac{1}{2\pi i}\int_\Gamma \sum_{j=1}^{k} S_{k-j}(\zeta)J_{S,j}(\zeta)\frac{d\zeta}{\zeta - z}, \quad k=1,2,\cdots, \tag{9.102}$$

for $z\in\mathbf{C}\setminus\Gamma$.

Similar as in the proof of Theorem 8.1, we obtain our main results as stated in the following theorem.

Theorem 9.1. *Let $Y(z)$ be the solution of the Riemann-Hilbert problem (Y1)-(Y4) and $Y_{11}(z)$ be the 11-entry of the matrix $Y(z)$. Denote $N=n+\alpha$ and $c=(m/N)^2t$. If $c>-2$, we have as $n\to\infty$,*

$$\begin{aligned} Y_{11}\left(\frac{N^{1/4}z}{m^{1/4}}\right) &\sim \sqrt{\pi}\left(\frac{N}{m}\right)^{n/4} z^{-\alpha}e^{\frac{N}{2}[v(z)+l]}\\ &\times\left\{\frac{f(z)^{1/4}}{\beta(z)}\,\mathrm{Ai}(f(z))\left[1+\sum_{k=1}^{\infty}\frac{A_{2,k}(z)}{N^k}\right]\right.\\ &\left.-\frac{\beta(z)}{f(z)^{1/4}}\,\mathrm{Ai}'(f(z))\left[1+\sum_{k=1}^{\infty}\frac{B_{2,k}(z)}{N^k}\right]\right\}, \end{aligned} \tag{9.103}$$

for $|\arg z| \le \pi/4$ *and* $|z| \ge \delta$, *and*

$$Y_{11}\left(\frac{N^{1/4}z}{m^{1/4}}\right) \sim \left(\frac{N}{m}\right)^{n/4} z^{-\alpha} e^{\frac{N}{2}[v(z)+l-2\phi(z)]} \times \frac{\beta(z)+\beta(z)^{-1}}{2}\left\{1+\sum_{k=1}^{\infty}\frac{C_{2,k}(z)}{N^k}\right\}, \tag{9.104}$$

for $|\arg z| \in [\pi/4, 3\pi/4]$ *and* $|z| \ge \delta$, *where* $v(z) = z^4 + cz^2$, $l = \ln(b^2/4) - 1 + 3b^4/8$ *with* $b = \sqrt{(\sqrt{c^2+12}-c)/3}$, $\beta(z) = (z-b)^{1/4}/(z+b)^{1/4}$, $\phi(z)$ *is given in* (9.9), $f(z) = [3N\phi(z)/2]^{2/3}$, *and* $\delta \in (0, b)$ *is any fixed small number. The asymptotic expansion for* $|\arg z| \in [3\pi/4, \pi]$ *and* $|z| \ge \delta$ *can be obtained by the symmetry relation* $Y_{11}(-z) = (-1)^n Y_{11}(z)$. *Let* $\eta(z)$ *and* $E_0(z)$ *be defined in* (9.57) *and* (9.58), *respectively. For* $c > -2$ *and* $|z| \le \delta$, *we have as* $n \to \infty$,

$$Y_{11}\left(\frac{N^{1/4}z}{m^{1/4}}\right) \sim \sqrt{\frac{\pi N\eta(z)}{2i}}\left(\frac{N}{m}\right)^{n/4} z^{-\alpha} e^{\frac{N}{2}[v(z)+l]} \times \left\{J_{\nu+1/2}(N\eta(z))\sum_{k=0}^{\infty}\frac{A_{0,k}(z)}{N^k} + J_{\nu-1/2}(N\eta(z))\sum_{k=0}^{\infty}\frac{B_{0,k}(z)}{N^k}\right\}, \tag{9.105}$$

where $A_{0,0}(z) = E_{0,11}(z) + iE_{0,12}(z)$ *and* $B_{0,0}(z) = iE_{0,11}(z) + E_{0,12}(z)$.

If $c = -2$, *the asymptotic expansions for* $|z| \ge \delta$ *are similar to those for the case* $c > -2$ *with higher-order terms* N^{-k} *replaced with* $N^{-k/3}$. *Let* $\varphi_1(\zeta)$ *and* $\varphi_2(\zeta)$ *be uniquely determined by the linear system* (9.75) *with the asymptotic boundary condition* (9.76). *For* $|z| \le \delta$ *with* $z \in \mathbf{C}_+$, *we have as* $n \to \infty$,

$$Y_{11}\left(\frac{N^{1/4}z}{m^{1/4}}\right) \sim \left(\frac{N}{m}\right)^{n/4} z^{-\alpha} e^{\frac{N}{2}[v(z)+l]+\frac{\alpha\pi i}{2}} \times \left\{\varphi_1(\zeta(z))\sum_{k=0}^{\infty}\frac{A_{1,k}(z)}{N^{k/3}} + \varphi_2(\zeta(z))\sum_{k=0}^{\infty}\frac{B_{1,k}(z)}{N^{k/3}}\right\}, \tag{9.106}$$

where

$$A_{1,0}(z) = \frac{e^{n\pi i/2}}{2}[\beta(z)+\beta(z)^{-1}], \quad B_{1,0}(z) = \frac{e^{-n\pi i/2}}{2i}[\beta(z)-\beta(z)^{-1}].$$

The asymptotic expansion for $|z| \le \delta$ *with* $z \in \mathbf{C}_-$ *can be obtained via the reflection formula* $Y_{11}(-z) = (-1)^n Y_{11}(z)$.

If $c < -2$, *the MRS numbers are given by* $a = \sqrt{-1-c/2}$ *and* $b = \sqrt{1-c/2}$. *Denote* $l = -\ln 2 - 1/2 - c^2/4$ *and* $c_0 = \sqrt{-c/2} \in (a, b)$. *Let the functions* $\phi(z)$, $f(z)$, $\tilde{f}(z)$, $\bar{f}(z)$, *and* $\hat{f}(z)$ *be defined as in* (9.34), (9.85),

(9.86), (9.88), *and* (9.89), *respectively. Recall from* (9.82) *that* $K(z) = D(\infty)^{\sigma_3}\tilde{K}(z)D(z)^{-\sigma_3}$, *where* $D(z)$, $D(\infty)$, *and* $\tilde{K}(z)$ *are given in* (9.79), (9.81), *and* (9.83), *respectively. As* $n \to \infty$, *we have*

$$Y_{11}\left(\frac{N^{1/4}z}{m^{1/4}}\right) \sim \sqrt{\pi}\left(\frac{N}{m}\right)^{n/4} z^{-\alpha} e^{\frac{N}{2}[v(z)+l]} \times \left\{ \frac{K_{11}(z) - iK_{12}(z)}{f(z)^{-1/4}} \operatorname{Ai}(f(z)) \left[1 + \sum_{k=1}^{\infty} \frac{A_k(z)}{N^k}\right] - \frac{K_{11}(z) + iK_{12}(z)}{f(z)^{1/4}} \operatorname{Ai}'(f(z)) \left[1 + \sum_{k=1}^{\infty} \frac{B_k(z)}{N^k}\right] \right\}, \tag{9.107}$$

for $|\arg z| \le \pi/4$ *and* $|z| \ge c_0$, *and*

$$Y_{11}\left(\frac{N^{1/4}z}{m^{1/4}}\right) \sim (-1)^n\sqrt{\pi}\left(\frac{N}{m}\right)^{n/4} (-z)^{-\alpha} e^{\frac{N}{2}[v(z)+l]} \times \left\{ \frac{K_{11}(z) + iK_{12}(z)}{\tilde{f}(z)^{-1/4}} \operatorname{Ai}(\tilde{f}(z)) \left[1 + \sum_{k=1}^{\infty} \frac{\tilde{A}_k(z)}{N^k}\right] - \frac{K_{11}(z) - iK_{12}(z)}{\tilde{f}(z)^{1/4}} \operatorname{Ai}'(\tilde{f}(z)) \left[1 + \sum_{k=1}^{\infty} \frac{\tilde{B}_k(z)}{N^k}\right] \right\}, \tag{9.108}$$

for $|\arg z| \in [3\pi/4, \pi]$ *and* $|z| \ge c_0$, *and*

$$Y_{11}\left(\frac{N^{1/4}z}{m^{1/4}}\right) \sim \sqrt{\pi}\left(\frac{N}{m}\right)^{n/4} z^{-\alpha} e^{\frac{N}{2}[v(z)+l]} \times \left\{ \frac{K_{11}(z)e^{N\pi i/2} + iK_{12}(z)e^{-N\pi i/2}}{\bar{f}(z)^{-1/4}} \operatorname{Ai}(\bar{f}(z)) \left[1 + \sum_{k=1}^{\infty} \frac{\bar{A}_k(z)}{N^k}\right] - \frac{K_{11}(z)e^{N\pi i/2} - iK_{12}(z)e^{-N\pi i/2}}{\bar{f}(z)^{1/4}} \operatorname{Ai}'(\bar{f}(z)) \left[1 + \sum_{k=1}^{\infty} \frac{\bar{B}_k(z)}{N^k}\right] \right\}, \tag{9.109}$$

for $\operatorname{Re} z \ge 0$ *and* $|z| \le c_0$, *and*

$$Y_{11}\left(\frac{N^{1/4}z}{m^{1/4}}\right) \sim (-1)^n\sqrt{\pi}\left(\frac{N}{m}\right)^{n/4} (-z)^{-\alpha} e^{\frac{N}{2}[v(z)+l]} \times \left\{ \frac{K_{11}(z)e^{-N\pi i/2} - iK_{12}(z)e^{N\pi i/2}}{\hat{f}(z)^{-1/4}} \operatorname{Ai}(\hat{f}(z)) \left[1 + \sum_{k=1}^{\infty} \frac{\hat{A}_k(z)}{N^k}\right] - \frac{K_{11}(z)e^{-N\pi i/2} + iK_{12}(z)e^{N\pi i/2}}{\hat{f}(z)^{1/4}} \operatorname{Ai}'(\hat{f}(z)) \left[1 + \sum_{k=1}^{\infty} \frac{\hat{B}_k(z)}{N^k}\right] \right\}, \tag{9.110}$$

for $\operatorname{Re} z \le 0$ *and* $|z| \le c_0$, *and*

$$Y_{11}\left(\frac{N^{1/4}z}{m^{1/4}}\right) \sim \left(\frac{N}{m}\right)^{n/4} z^{-\alpha} e^{\frac{N}{2}[v(z)+l-2\phi(z)]} K_{11}(z) \left\{1 + \sum_{k=1}^{\infty} \frac{C_k(z)}{N^k}\right\}, \tag{9.111}$$

for $|\arg z| \in [\pi/4, 3\pi/4]$ *and* $|z| \ge c_0$. *Actually, the asymptotic expansions for* $\operatorname{Re} z \ge 0$ *and* $\operatorname{Re} z \le 0$ *are equivalent because of the symmetry relation* $Y_{11}(-z) = (-1)^n Y_{11}(z)$.

Proof. We only give a proof of (9.103). The other asymptotic expansions can be verified in a similar manner (exercise). Note from (9.2) and (9.98) that

$$\begin{aligned} Y((N/m)^{1/4}z) =& (N/m)^{(n+\alpha)\sigma_3/4} U(z) (N/m)^{-\alpha\sigma_3/4} \\ =& (N/m)^{(n+\alpha)\sigma_3/4} e^{Nl\sigma_3/2} S(z) e^{-Nl\sigma_3/2} U^*(z) (N/m)^{-\alpha\sigma_3/4}. \end{aligned}$$

On account of (9.62)-(9.66), we obtain

$$\begin{aligned} Y((N/m)^{1/4}z) =& \sqrt{\pi}(N/m)^{(n+\alpha)\sigma_3/4} e^{Nl\sigma_3/2} S(z) M(z) \begin{pmatrix} 1 & -1 \\ -i & -i \end{pmatrix} f(z)^{\sigma_3/4} \\ & \times A(f(z)) e^{[Nv(z)/2 - \alpha \ln z]\sigma_3} (N/m)^{-\alpha\sigma_3/4}. \end{aligned}$$

Recall from (8.7) (with $a = -b$) that

$$\begin{aligned} M(z) =& \frac{1}{2}\begin{pmatrix} \beta(z) + \beta(z)^{-1} & -i[\beta(z) - \beta(z)^{-1}] \\ i[\beta(z) - \beta(z)^{-1}] & \beta(z) + \beta(z)^{-1} \end{pmatrix} \\ =& \frac{1}{2}\begin{pmatrix} 1 & -1 \\ -i & -i \end{pmatrix} \beta(z)^{-\sigma_3} \begin{pmatrix} 1 & i \\ -1 & i \end{pmatrix}, \end{aligned}$$

where

$$\beta(z) := \frac{(z-b)^{1/4}}{(z+b)^{1/4}}.$$

We then have

$$\begin{aligned} Y((N/m)^{1/4}z) =& \sqrt{\pi}(N/m)^{(n+\alpha)\sigma_3/4} e^{Nl\sigma_3/2} S(z) \begin{pmatrix} 1 & -1 \\ -i & -i \end{pmatrix} \left[\frac{f(z)^{1/4}}{\beta(z)}\right]^{\sigma_3} \\ & \times A(f(z)) e^{[Nv(z)/2 - \alpha \ln z]\sigma_3} (N/m)^{-\alpha\sigma_3/4}. \end{aligned}$$

Consequently,

$$\begin{aligned} Y_{11}\left(\frac{N^{1/4}z}{m^{1/4}}\right) =& \sqrt{\pi}\left(\frac{N}{m}\right)^{n/4} z^{-\alpha} e^{\frac{N}{2}[v(z)+l]} \\ & \times \left\{ S_{11}(z)\left[A_{11}(f(z))\frac{f(z)^{1/4}}{\beta(z)} - A_{21}(f(z))\frac{\beta(z)}{f(z)^{1/4}}\right] \right. \\ & \left. - iS_{12}(z)\left[A_{12}(f(z))\frac{f(z)^{1/4}}{\beta(z)} + A_{22}(f(z))\frac{\beta(z)}{f(z)^{1/4}}\right]\right\}. \end{aligned}$$

Now, (9.103) follows from the asymptotic expansion of $S(z)$ in (9.101) and the definition of Airy parametrix in (8.8). □

9.6 Exercises

1. Verify (9.4).
2. Verify (9.6).
3. Verify (9.12) and (9.13).
4. Show that the two equations (9.21) and (9.22) are equivalent.
5. Verify (9.24) and (9.25).
6. Verify (9.27) and (9.28).
7. Verify (9.31) and (9.34).
8. Let $c > -2$. If z is bounded away from $[-b, b]$, show that the jump matrix $J_V(z)$ tends to the identity matrix as $N \to \infty$.
9. Show that $P(z)$ defined in (9.59) satisfies the Riemann-Hilbert problem (P1)-(P3).
10. Let $\phi(z)$ and $f(z)$ be defined as in (9.9) and (9.60), respectively. Prove that $\operatorname{Re}\phi(z) < 0$ and $\arg f(z) < \pi$ for $\arg z = \pi/4$
11. Prove that the function $\zeta(z)$ defined in (9.73) is an analytic function in a neighborhood of the origin with $\zeta(z) \sim (N/\sqrt{2})^{1/3} z$ as $z \to 0$.
12. Show that $R(z)$ defined in (9.72) satisfies the Riemann-Hilbert problem (R1)-(R3).
13. Let $\varphi_1(\zeta)$ and $\varphi_2(\zeta)$ be uniquely determined by the linear system (9.75) with the asymptotic boundary condition (9.76). Let $\Psi_k(\zeta, s)$ $(k = 1, \cdots, 6)$ be uniquely determined by the linear system (9.67) with the asymptotic boundary condition (9.69). Prove that $\varphi_1(\zeta) = \Psi_{3,11}(\zeta, 0) = \Psi_{2,11}(\zeta, 0) = \Psi_{6,22}(-\zeta, 0) = \Psi_{5,22}(-\zeta, 0)$ and $\varphi_2(\zeta) = \Psi_{3,21}(\zeta, 0) = \Psi_{2,21}(\zeta, 0) = \Psi_{6,12}(-\zeta, 0) = \Psi_{5,12}(-\zeta, 0)$ for $\zeta \in \mathbf{C}_+$.
14. Verify (9.77).
15. Show that the function $D(z)$ defined in (9.79) is analytic in $\mathbf{C} \setminus [-b, b]$ and satisfies the jump conditions (9.80).
16. Show that $\tilde{K}(z)$ defined in (9.83) solves Riemann-Hilbert problem ($\tilde{K}1$)-($\tilde{K}3$).
17. Verify (9.87).
18. Verify (9.96).
19. Verify (9.104), (9.105), (9.106), (9.107), (9.108), (9.109), (9.110), and (9.111).

Chapter 10

Freud Weight Revisited: New Special Functions

10.1 Introduction

Kriecherbauer and McLaughlin (1999) considered the Riemann-Hilbert problem of the orthogonal polynomials with respect to the Freud weight

$$w(x) = \exp\left\{-\frac{\Gamma(\beta/2)\Gamma(1/2)}{\Gamma(\beta/2+1/2)}|x|^{\beta}\right\}, \quad x \in \mathbf{R}, \tag{10.1}$$

where $\beta > 0$ is a given constant. If $\beta \in (0,1)$, the asymptotic formulas of the Freud orthogonal polynomials are given in terms of the unique solution of following Riemann-Hilbert problem.

($\tilde{L}1$) $\tilde{L}(z)$ is analytic in $\mathbf{C} \setminus \mathbf{R}$.
($\tilde{L}2$) $\tilde{L}(z)$ satisfies the jump condition

$$\tilde{L}_+(x) = \tilde{L}_-(x)\begin{pmatrix} 1 & -\eta(x) \\ \eta(-x) & 1 \end{pmatrix}, \tag{10.2}$$

for $x \in \mathbf{R}$.
($\tilde{L}3$) As $z \to \infty$, $\tilde{L}(z) \to I$.
($\tilde{L}4$) As $z \to 0$, $\tilde{L}(z) = O(\ln|z|)$.

Here, the function $\eta(x)$ in the jump matrix is defined as

$$\eta(x) = \begin{cases} 2i(-1)^{n+1}e^{-x^{\beta}\sin(\pi\beta/2)}\sin[x^{\beta}\cos(\pi\beta/2)], & x \geq 0, \\ 0 & , \ x \leq 0, \end{cases} \tag{10.3}$$

where n is the polynomial degree; see Kriecherbauer and McLaughlin (1999, (6.41)-(6.43)).

The material in this chapter is taken from Wong and Zhao (2016). In this chapter, we will introduce a pair of special functions defined as solutions of integral equations, and then solve the Riemann-Hilbert problems

($\tilde{L}1$)-($\tilde{L}4$) in terms of the new special functions. First, we will replace the condition ($\tilde{L}4$) with a more natural description of the asymptotic behavior at the origin, and consider the following Riemann-Hilbert problem.

(L1) $L(z)$ is analytic in $\mathbf{C} \setminus \mathbf{R}$.
(L2) $L(z)$ satisfies the jump condition

$$L_+(x) = L_-(x) \begin{pmatrix} 1 & -\eta(x) \\ \eta(-x) & 1 \end{pmatrix}, \tag{10.4}$$

for $x \in \mathbf{R}$.
(L3) As $z \to \infty$, $L(z) \to I$.
(L4) As $z \to 0$, $L(z) = O(\varepsilon_\beta(z))$, where

$$\varepsilon_\beta(z) := \begin{cases} 1 & 1/2 < \beta < 1, \\ \ln|z| & \beta = 1/2, \\ |z|^{\beta - 1/2}, & 0 < \beta < 1/2. \end{cases} \tag{10.5}$$

By Liouville's theorem, the solutions to the Riemann-Hilbert problems ($\tilde{L}1$)-($\tilde{L}4$) and (L1)-(L4) are unique (exercise).

Lemma 10.1. *The solutions to the Riemann-Hilbert problems ($\tilde{L}1$)-($\tilde{L}4$) and (L1)-(L4) are the same: $\tilde{L}(z) = L(z)$.*

Proof. The function $\tilde{L}(z)L(z)^{-1}$ has no jump on the real line, and has only an isolated singularity at the origin. Since the singularity at the origin is not a pole or an essential singularity, it must be removable. By ($\tilde{L}3$) and (L3), we have $\tilde{L}(z)L(z)^{-1} \to I$ as $z \to \infty$. It follows from Liouville's theorem that $\tilde{L}(z)L(z)^{-1} = I$ for all z. This completes the proof. □

10.2 Integral equations and special functions

Let $f(z) = L_{11}(z)$ and $g(z) = L_{12}(z)$ be the $(1,1)$-entry and $(1,2)$-entry of $L(z)$, respectively. It is readily seen from (L1)-(L4) that $f(z)$ is analytic for $z \in \mathbf{C} \setminus (-\infty, 0]$ and satisfies the jump condition (exercise)

$$f_+(x) - f_-(x) = \eta(-x) g(x), \quad x \in (-\infty, 0). \tag{10.6}$$

Moreover, $f(z) \to 1$ as $z \to \infty$ and $f(z) = O(\varepsilon_\beta(z))$ as $z \to 0$. In a similar manner, we obtain that $g(z)$ is analytic for $z \in \mathbf{C} \setminus [0, \infty)$ and satisfies the jump condition (exercise)

$$g_+(x) - g_-(x) = -\eta(x) f(x), \quad x \in (0, \infty). \tag{10.7}$$

Moreover, $g(z) \to 0$ as $z \to \infty$ and $g(z) = O(\varepsilon_\beta(z))$ as $z \to 0$. Solving the Riemann-Hilbert problems (10.6) and (10.7) respectively, we have (exercise)

$$f(z) = 1 - \frac{1}{2\pi i}\int_0^\infty \frac{\eta(x)g(-x)}{x+z}dx, \; z \in \mathbf{C} \setminus (-\infty, 0], \tag{10.8}$$

$$g(z) = -\frac{1}{2\pi i}\int_0^\infty \frac{\eta(x)f(x)}{x-z}dx, \; z \in \mathbf{C} \setminus [0, \infty). \tag{10.9}$$

Define two special functions

$$u(x) := f(x) + (-1)^n g(-x), \quad v(x) = f(x) - (-1)^n g(-x), \tag{10.10}$$

for $x \in (0, \infty)$. It follows from (10.8) and (10.9) that (exercise)

$$u(x) = 1 + \int_0^\infty \frac{K(t)u(t)}{t+x}dt, \quad x \in (0, \infty), \tag{10.11}$$

$$v(x) = 1 - \int_0^\infty \frac{K(t)v(t)}{t+x}dt, \quad x \in (0, \infty), \tag{10.12}$$

where the kernel function $K(t)$ is given by

$$K(t) = \frac{(-1)^{n+1}\eta(t)}{2\pi i} = \frac{1}{\pi}e^{-t^\beta \sin(\pi\beta/2)} \sin[t^\beta \cos(\pi\beta/2)], \tag{10.13}$$

for $t \in (0, \infty)$.

10.3 Unique solvability of the integral equations

In this section, we will prove that the integral equations (10.11) and (10.12) are uniquely solvable. First, we denote by $\mathbf{R}_+ = (0, \infty)$ the positive real line, and $C_\pm = \{z \in \mathbf{C} : \pm \operatorname{Im} z > 0\}$ the upper and lower half complex plane, respectively. For any square integrable function $\phi \in L^2(\mathbf{R}_+)$, we define $T\phi \in L^2(\mathbf{R}_+)$ as

$$(T\phi)(x) = \int_0^\infty \frac{K(t)\phi(t)}{t+x}dt, \quad x \in \mathbf{R}_+, \tag{10.14}$$

where $K(t)$ is the kernel function defined in (10.13). Since $K(t) = O(e^{-t^\beta \sin(\pi\beta/2)})$ as $t \to \infty$ and $K(t) = O(t^\beta)$ as $t \to 0^+$, the function $K(t)/(t+x)$ is square integrable in $\mathbf{R}_+ \times \mathbf{R}_+$ (exercise):

$$\int_0^\infty \int_0^\infty \frac{[K(t)]^2}{(t+x)^2}dxdt < \infty. \tag{10.15}$$

It then follows from Yosida (1980, p. 277) that the linear operator is compact on $L^2(\mathbf{R}_+)$. By using a similar argument as in Kriecherbauer and McLaughlin (1999); Wong and Zhao (2016); Zhou (1989), we will prove that the null space of $I - T$ is trivial.

Lemma 10.2. *Let T be the linear compact operator defined in* (10.14). *If $T\phi = \phi$ for some $\phi \in L^2(\mathbf{R}_+)$, then $\phi(x) \equiv 0$ for $x \in \mathbf{R}_+$.*

Proof. By analytic continuation, the function

$$\phi(z) = \int_0^\infty \frac{K(t)\phi(t)}{t+z}dt \tag{10.16}$$

is analytic for $z \in \mathbf{C} \setminus (-\infty, 0]$ (exercise). It also follows from (10.13) that $K(t)$ can be analytically continued as an analytic function in $\mathbf{C} \setminus (-\infty, 0]$. By Plemelj formula, we have

$$\phi_+(-x) - \phi_-(-x) = -2i\pi K(x)\phi(x), \quad x > 0. \tag{10.17}$$

It is obvious from (10.16) that $\overline{\phi(z)} = \phi(\bar{z})$ for $z \in \mathbf{C} \setminus (-\infty, 0]$. Hence, $\phi_-(-x) = \overline{\phi_+(-x)}$ for $x > 0$. This together with (10.17) implies that

$$\phi_\pm(-x) = \psi(x) \mp i\pi K(x)\phi(x), \quad x > 0, \tag{10.18}$$

where $\psi(x) = \operatorname{Re}\phi_\pm(-x)$ is a real-valued function. Now, we define

$$\Phi(z) := \phi(z)^2 + \phi(-z)^2, \tag{10.19}$$

which is analytic for $z \in \mathbf{C}_\pm$. It is readily seen that

$$\Phi_\pm(x) = \begin{cases} \phi(x)^2 + \phi_\mp(-x)^2, & x > 0, \\ \phi_\pm(x)^2 + \phi(-x)^2, & x < 0. \end{cases} \tag{10.20}$$

We also obtain from (10.16) and (10.19) that $\Phi(z) = O(1/|z|^2)$ as $z \to \infty$. By Cauchy integral formula, we have

$$\begin{aligned} 0 &= \int_{-\infty}^{\infty} \Phi_+(x)dx \\ &= \int_{-\infty}^{0} [\phi_+(x)^2 + \phi(-x)^2]dx + \int_0^\infty [\phi(x)^2 + \phi_-(-x)^2]dx \\ &= \int_0^\infty [2\phi(x)^2 + \phi_-(-x)^2 + \phi_+(-x)^2]dx. \end{aligned}$$

However, we note from (10.13) and (10.18) that the integrand is always nonnegative:

$$\begin{aligned} 2\phi(x)^2 + \phi_-(-x)^2 + \phi_+(-x)^2 &= 2\phi(x)^2 + 2\psi(x)^2 - 2[\pi K(x)\phi(x)]^2 \\ &= 2[1 - |\pi K(x)|^2]\phi(x)^2 + 2\psi(x)^2 \geq 0. \end{aligned}$$

Therefore, $\phi(x) = 0$ almost everywhere for $x > 0$. By the integral representation in (10.16), we have $\phi(z) \equiv 0$. This completes the proof. □

By Riesz-Schauder theory of linear compact operators [Yosida (1980), p. 283], the operator $I - T$ is invertible in $L^2(\mathbf{R}_+)$. Define

$$h(x) := \int_0^\infty \frac{K(t)}{t+x} dt, \quad x \in \mathbf{R}_+. \tag{10.21}$$

It is readily seen from (10.13) that $h \in L^2(\mathbf{R}_+)$. Hence, the functions $u = 1 + (I - T)^{-1}h$ and $v = 1 - (I + T)^{-1}h$ are the unique solutions of the integral equations (10.11) and (10.12), respectively. By analytic continuation, we can extend u and v as analytic functions in $\mathbf{C} \setminus (-\infty, 0]$. Finally, we have the following result.

Theorem 10.1. *There exists a unique pair of special functions $u(z)$ and $v(z)$, which are analytic for $z \in \mathbf{C} \setminus (-\infty, 0]$ and satisfy the integral equations (10.11) and (10.12). Define*

$$f(z) = [u(z) + v(z)]/2 \tag{10.22}$$

for $z \in \mathbf{C} \setminus (-\infty, 0]$, and

$$g(z) = (-1)^n [u(-z) - v(-z)]/2 \tag{10.23}$$

for $z \in \mathbf{C} \setminus [0, \infty)$. The solution of Riemann-Hilbert problem (L1)-(L4) is given by

$$L(z) = \begin{pmatrix} f(z) & g(-z) \\ g(z) & f(-z) \end{pmatrix}, \quad z \in \mathbf{C}_\pm. \tag{10.24}$$

10.4 Boundedness of the special functions

In the previous section, we have shown that the special functions u and v are square integrable on $\mathbf{R}_+$ and analytic in $\mathbf{C} \setminus (-\infty, 0]$. In this section, we will further prove that u and v are bounded on $\mathbf{R}_+$.

Theorem 10.2. *Let $u(z)$ and $v(z)$ be the unique pair of special functions, which are analytic for $z \in \mathbf{C} \setminus (-\infty, 0]$ and satisfy the integral equations (10.11) and (10.12). There exists a constant $C > 0$ such that $|u(x)| + |v(x)| \leq C$ for all $x \in \mathbf{R}_+$.*

Proof. We only prove that $u \in L^\infty(\mathbf{R}_+)$. The proof of boundedness of v can be given in a similar manner. Recall from the previous section that $u - 1 = (I - T)^{-1}h \in L^2(\mathbf{R}_+)$, we have $u(x) \to 1$ as $x \to \infty$. It suffices to show that $u(x) = O(1)$ as $x \to 0^+$. Since $u - 1 \in L^2(\mathbf{R}_+)$, there exists

$C_1 > 0$ such that $|u(x)| \le C_1 x^{-1/2}$ for all $x \in (0,1)$. On account of (10.13), there exists $C_2 > 0$ such that $|K(t)| \le C_2 t^{\beta}$ for all $t \in (0,1)$. Hence,

$$\begin{aligned}|u(x)| &\le 1 + \int_0^1 \frac{|K(t)u(t)|}{t+x}dt + \int_1^\infty \frac{|K(t)u(t)|}{t+x}dt \\ &\le C_1C_2 \int_0^1 \frac{t^{\beta-1/2}}{t+x}dt + C_3 + 1,\end{aligned}$$

where

$$C_3 := \sup_{t\in[1,\infty)} |u(t)| \cdot \int_1^\infty |K(t)|dt < \infty.$$

If $\beta > 1/2$, then

$$\int_0^1 \frac{t^{\beta-1/2}}{t+x}dt \le \int_0^1 t^{\beta-3/2}dt = \frac{1}{\beta - 1/2}, \quad x \in (0,1).$$

If $\beta = 1/2$, then (exercise)

$$\int_0^1 \frac{t^{\beta-1/2}}{t+x}dt = -\ln x \le \frac{\beta}{2e}x^{-\beta/2}, \quad x \in (0,1). \tag{10.25}$$

If $\beta < 1/2$, then

$$\int_0^1 \frac{t^{\beta-1/2}}{t+x}dt \le \int_0^x \frac{t^{\beta-1/2}}{x}dt + \int_x^1 \frac{t^{\beta-1/2}}{t}dt = \frac{x^{\beta-1/2}}{\beta+1/2} + \frac{x^{\beta-1/2}-1}{1/2-\beta},$$

for $x \in (0,1)$. Combining the above three cases, we obtain that as $x \to 0^+$,

$$u(x) = \begin{cases} O(1), & \beta > 1/2, \\ O(x^{-\beta/2}), & \beta = 1/2, \\ O(x^{\beta-1/2}), & \beta < 1/2. \end{cases}$$

In general, if $u(x) = O(x^{-\alpha})$ as $x \to 0^+$, for some $\alpha \in [0,1/2]$, then (exercise)

$$u(x) = \begin{cases} O(1), & \beta > \alpha, \\ O(x^{-\beta/2}), & \beta = \alpha, \\ O(x^{\beta-\alpha}), & \beta < \alpha, \end{cases} \tag{10.26}$$

as $x \to 0^+$. If $\beta > 1/2$, then we apply the above argument once to obtain $u(x) = O(1)$ as $x \to 0^+$. If $2\beta > 1/2 \ge \beta$, then we apply the above argument once to obtain either $u(x) = O(x^{\beta-1/2})$ (when $\beta < 1/2$) or $u(x) = O(x^{-1/4})$ (when $\beta = 1/2$) as $x \to 0^+$. Applying the above argument one more time gives $u(x) = O(1)$ as $x \to 0^+$. In general, by repeating the above argument k times, where $k\beta > 1/2$, we obtain $u(x) = O(1)$ as $x \to 0^+$. This completes the proof. □

10.5 Some open problems on the special functions

The graphs of $u(x)$ and $v(x)$ with different values of β are illustrated in Figures 10.1 and 10.2, respectively. It is conjectured that $u(x)$ is decreasing and $v(x)$ is increasing for $x \in \mathbf{R}_+$. We also observe that the initial values $u(0)$ and $v(0)$ are decreasing and increasing functions of $\beta \in (0,1)$. The proof of the monotonicity properties of the special functions is proposed as open problems.

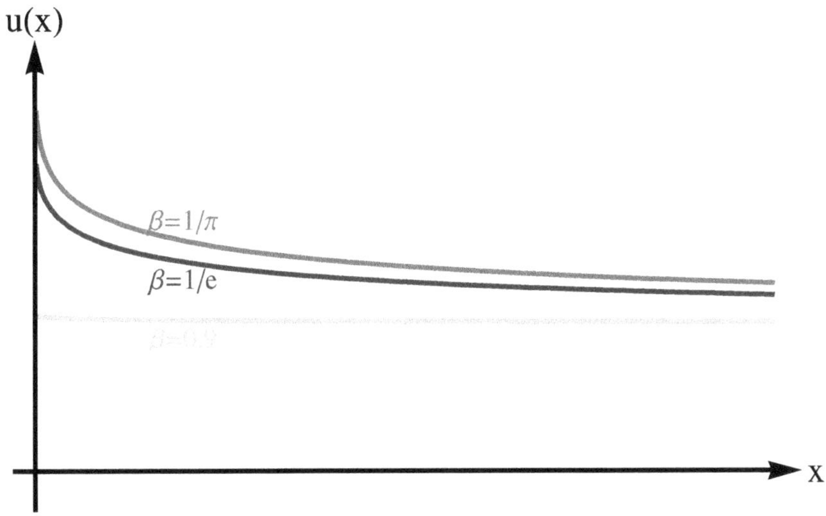

Fig. 10.1: The graph of $u(x)$ with different values of β.

It is easily seen from the integral equations (10.11) and (10.12) that (exercise)

$$u(z) \sim 1 + \sum_{k=1}^{\infty} c_k z^{-k}, \tag{10.27}$$

$$v(z) \sim 1 + \sum_{k=1}^{\infty} d_k z^{-k}, \tag{10.28}$$

as $z \to \infty$, where the coefficients are given as the moments

$$c_k = (-1)^{k-1} \int_0^{\infty} t^{k-1} K(t) u(t) dt, \tag{10.29}$$

$$d_k = (-1)^{k-1} \int_0^{\infty} t^{k-1} K(t) v(t) dt. \tag{10.30}$$

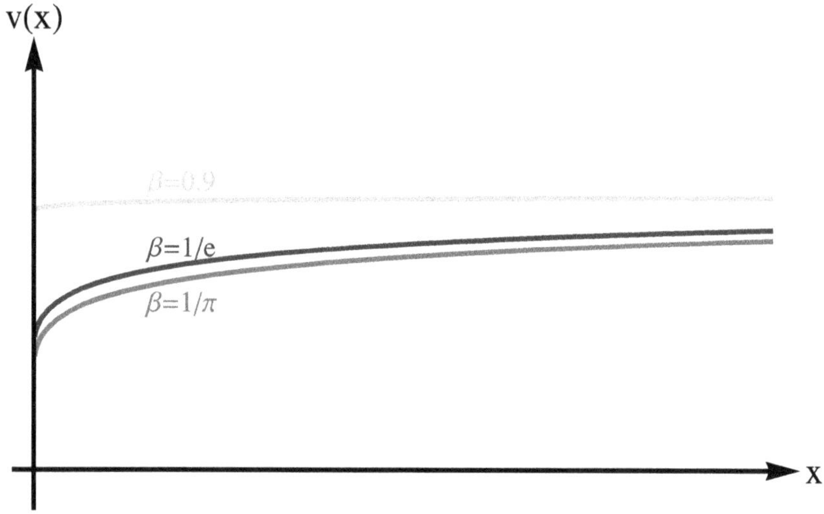

Fig. 10.2: The graph of $v(x)$ with different values of β.

It is left as an open problem to derive explicit expressions of the coefficients c_k and d_k in the asymptotic expansions (10.27) and (10.28).

The kernel function $K(t)$ in (10.13) suggests us to consider the following Stieltjes integral

$$G_\beta(z) := \int_0^\infty \frac{e^{-t^\beta}}{t+z}dt, \tag{10.31}$$

which is analytic for $z \in \mathbf{C} \setminus (-\infty, 0]$ and has the asymptotic expansion

$$G_\beta(z) \sim \sum_{k=1}^\infty \frac{(-1)^{k-1}\Gamma(k/\beta)}{\beta} z^{-k}, \quad z \to \infty. \tag{10.32}$$

Next, we investigate the asymptotic behavior of $G_\beta(z)$ as $z \to 0$. We shall consider three different cases depending on whether β is an irrational number, an integer, or a rational number.

If $\beta > 0$ is irrational, then we have

$$G_\beta(z) \sim -\ln z + \sum_{k=1}^\infty \frac{(-1)^{k+1}\pi}{k!\sin(\beta k\pi)} z^{\beta k} + \sum_{k=1}^\infty c_{\beta,k}(-z)^k, \quad z \to 0, \tag{10.33}$$

where

$$c_{\beta,0} = \sum_{j=1}^\infty \frac{(-1)^j}{j!\beta j} + \frac{\Gamma(0,1)}{\beta} = \frac{1}{\beta}\left[\int_0^1 \frac{e^{-t}-1}{t}dt + \int_1^\infty \frac{e^{-t}}{t}dt\right] = -\frac{\gamma}{\beta}, \tag{10.34}$$

and

$$c_{\beta,k} = \sum_{j=0}^{\infty} \frac{(-1)^j}{j! \cdot (\beta j - k)} + \frac{\Gamma(-k/\beta, 1)}{\beta}, \quad k = 1, 2, \cdots. \tag{10.35}$$

Here,

$$\Gamma(a, z) = \int_z^{\infty} t^{a-1} e^{-t} dt$$

is the incomplete Gamma function, γ is the Euler constant, and the last equality in the expression of $c_{\beta,0}$ in (10.34) follows from Olver *et al.* (2010, (6.2.1),(6.2.3)-(6.2.4)).

If β is a positive integer, then $G_\beta(z) = G_{\beta,1}(z) + G_{\beta,2}(z)$, where

$$G_{\beta,1}(z) := \int_0^1 \frac{e^{-t^\beta}}{t+z} dt = \sum_{j=0}^{\infty} \frac{(-1)^j}{j!} \int_0^1 \frac{t^{\beta j}}{t+z} dt,$$

and

$$G_{\beta,2}(z) := \int_1^{\infty} \frac{e^{-t^\beta}}{t+z} dt = \sum_{k=0}^{\infty} (-z)^k \int_1^{\infty} \frac{e^{-t^\beta}}{t^{k+1}} dt.$$

Since

$$\begin{aligned} \int_0^1 \frac{t^{\beta j}}{t+z} dt &= \sum_{k=0}^{\beta j - 1} (-z)^k \int_0^1 t^{\beta j - k - 1} dt + (-z)^{\beta j} \int_0^1 \frac{1}{t+z} dt \\ &= \sum_{k=0}^{\beta j - 1} \frac{(-z)^k}{\beta j - k} + (-z)^{\beta j} \ln \frac{1+z}{z}, \end{aligned}$$

and

$$\int_1^{\infty} \frac{e^{-t^\beta}}{t^{k+1}} dt = \int_1^{\infty} \frac{e^{-s}}{\beta s^{k/\beta+1}} ds = \frac{\Gamma(-k/\beta, 1)}{\beta},$$

we have

$$\begin{aligned} G_\beta(z) \sim & \sum_{j=0}^{\infty} \frac{(-1)^j}{j!} \sum_{k=0}^{\beta j - 1} \frac{(-z)^k}{\beta j - k} + \sum_{j=0}^{\infty} \frac{(-1)^j}{j!} (-z)^{\beta j} \ln \frac{1+z}{z} \\ & + \sum_{k=0}^{\infty} (-z)^k \frac{\Gamma(-k/\beta, 1)}{\beta}, \quad z \to 0. \end{aligned}$$

Making use of the Taylor expansion

$$\ln(1+z) = -\sum_{l=1}^{\infty} \frac{(-z)^l}{l},$$

we obtain by rearranging the like terms that

$$G_\beta(z) \sim \sum_{j=0}^{\infty} \frac{(-1)^{j+1}}{j!}(-z)^{\beta j} \ln z + \sum_{k=0}^{\infty} d_{\beta,k}(-z)^k, \quad z \to 0, \tag{10.36}$$

where

$$d_{\beta,k} = \sum_{j=0,\beta j \neq k}^{\infty} \frac{(-1)^j}{j!(\beta j - k)} + \frac{\Gamma(-k/\beta, 1)}{\beta}. \tag{10.37}$$

If $\beta = p/q$ is rational, where p and q are relatively prime positive integers. We make a change of variable $t^{1/q} = s$ and rewrite (10.31) as

$$G_\beta(z) = \int_0^\infty \frac{qs^{q-1}e^{-s^p}}{s^q + \zeta^q} ds, \quad \zeta^q = z. \tag{10.38}$$

Let $\omega_l = e^{(2l+1)\pi i/q}$ for $l = 0, \cdots, q-1$ be the roots of the equation $\omega^q + 1 = 0$. We then have the factorization

$$s^q + \zeta^q = \prod_{l=0}^{q-1}(s - \omega_l \zeta), \tag{10.39}$$

and the partial fractions

$$\frac{qs^{q-1}}{s^q + \zeta^q} = \sum_{l=0}^{q-1} \frac{1}{s - \omega_l \zeta}. \tag{10.40}$$

Substituting this into (10.38) gives

$$G_\beta(z) = \sum_{l=0}^{q-1} G_p(e^{(2l+1-q)\pi i/q} z^{1/q}). \tag{10.41}$$

Since p is a positive integer, for each $l = 0, \cdots, q-1$, the asymptotic expansion of $G_p(e^{(2l+1-q)\pi i/q} z^{1/q})$ as $z \to 0$ can be derived from (10.36). Substituting these expansions into the above equation gives the asymptotic expansion of $G_\beta(z)$ with $\beta = p/q$ being a rational number. Relating the special functions $u(z)$ and $v(z)$ to the function $G_\beta(z)$ is left as an open problem. In particular, we expect that $u(z)$ and $v(z)$ have asymptotic expansions in terms of z^k and $z^{\beta k}$ (with nonnegative integers k) as $z \to 0$. To determine the coefficients of z^k and $z^{\beta k}$ in the asymptotic expansions is an interesting open problem.

10.6 Exercises

1. Prove that the solution of the Riemann-Hilbert problems ($\tilde{L}1$)-($\tilde{L}4$) is unique.
2. Prove that the solution of the Riemann-Hilbert problems (L1)-(L4) is unique.
3. Verify (10.6) and (10.7).
4. Show that $f(z)$ given in (10.8) satisfies the jump condition (10.6) and the asymptotic behavior $f(z) \to 1$ as $z \to \infty$.
5. Show that $g(z)$ given in (10.9) satisfies the jump condition (10.7) and the asymptotic behavior $g(z) \to 0$ as $z \to \infty$.
6. Verify (10.11) and (10.12).
7. Verify (10.15).
8. Show that the function $\phi(z)$ in (10.16) is analytic for $z \in \mathbf{C} \setminus (-\infty, 0]$.
9. Show that $L(z)$ defined in (10.24) satisfies the Riemann-Hilbert problem (L1)-(L4).
10. Verify (10.25).
11. Prove (10.26).
12. Verify (10.27) and (10.28).
13. Prove that the Stieltjes integral $G_\beta(z)$ defined in (10.31) is analytic for $z \in \mathbf{C}\setminus(-\infty, 0]$ and has the asymptotic expansion in (10.32) as $z \to \infty$.
14. Verify (10.40).
15. Verify (10.36).

Chapter 11

Global Asymptotics of the Meixner Polynomials

11.1 Introduction

In this chapter, we investigate the asymptotic behavior of the Meixner polynomials. These polynomials have many applications in statistical physics. For instance, they are used in the study of the shape fluctuations in a certain two-dimensional random growth model; see Johansson (2000) and the references therein.

The material in this chapter is taken from Wang and Wong (2011). For $\beta > 0$ and $0 < c < 1$, the Meixner polynomials are given by

$$m_n(z; \beta, c) = (\beta)_n \cdot {}_2F_1(-n, -z; \beta; 1 - c^{-1}). \tag{11.1}$$

They satisfy the discrete orthogonality relation

$$\sum_{k=0}^{\infty} m_n(k; \beta, c) m_p(k; \beta, c) \frac{c^k (\beta)_k}{k!} = (1 - c)^{-\beta} c^{-n} n! (\beta)_n \delta_{np}. \tag{11.2}$$

This notation is adopted in Erdélyi *et al.* (1953, Section 10.24) and also in Jin and Wong (1998).

Using probabilistic arguments, Maejima and Van Assche (1985) have given an asymptotic formula for $m_n(n\alpha; \beta, c)$ when $\alpha < 0$ and β is a positive integer. Their result is in terms of elementary functions. By using the steepest descent method for integrals, Jin and Wong (1998) have derived two infinite asymptotic expansions for $m_n(n\alpha; \beta, c)$; one holds uniformly for $0 < \varepsilon \le \alpha \le 1 + \varepsilon$, and the other holds uniformly for $1 - \varepsilon \le \alpha \le M < \infty$. Both expansions involve the parabolic cylinder function and its derivative.

In view of Gauss's contiguous relations for hypergeometric functions [Abramowitz and Stegun (1970), Section 15.2] and the connection formula [Jin and Wong (1998)]

$$m_n(-x-\beta;\beta,c^{-1}) = c^n m_n(x;\beta,c),$$

we may restrict our study to the case $1 \le \beta < 2$ and $0 < c < 1$. Fixing $0 < c < 1$ and $1 \le \beta < 2$, we intend to investigate the large-n behavior of $m_n(nz - \beta/2; \beta, c)$ for z in the whole complex plane, including a neighborhood of the origin and regions extending to infinity. Our approach is based on the nonlinear steepest descent method for oscillatory Riemann-Hilbert problems, first introduced by Deift and Zhou (1993) for nonlinear partial differential equations, later developed in Deift *et al.* (1999) for orthogonal polynomials with respect to exponential weights, and further extended in Baik *et al.* (2003, 2007) to a general class of discrete orthogonal polynomials.

A direct application of the method in Baik *et al.* (2003, 2007) would, however, only give local asymptotics. For instance, in the case of Meixner polynomials, one would have to divide the complex plane into at least six regions (one near the origin, two near the two turning points and three in between, including an unbounded one), and give correspondingly six different asymptotic formulas. To reduce the number of these regions, we shall make some modifications to the method in Baik *et al.* (2003, 2007). Our approach is motivated by the previous work in Dai and Wong (2007); Ou and Wong (2010), and the main idea is to extend, as large as possible, the two regions of validity of the two asymptotic formulas near the two turning points. There have already been several examples in which we only need two regions with appropriate asymptotic formulas to cover the entire plane; the Hermite polynomial [Wong and Zhang (2007)] is one of such examples. However, for discrete orthogonal polynomials, there might be cuts starting from the finite endpoints of the intervals of orthogonality. For instance, in the case of Krawtchouk polynomials considered in Dai and Wong (2007), there are two cuts $(-\infty, 0]$ and $[1, \infty)$ where no asymptotic formulas are given. In the present chapter, we shall give two asymptotic formulas for the Meixner polynomial $m_n(nz - \beta/2; \beta, c)$, one valid inside a rectangle with two vertical boundary lines passing through $z = 0$ and $z = 1$, and the other valid outside the rectangle. Both formulas can be extended slightly beyond the boundary of the rectangle, and they are asymptotically equal to each other in the overlap region.

11.2 Standard formulation of Riemann-Hilbert problem

From (11.1), we note that the leading coefficient of $m_n(z;\beta,c)$ is $(1-c^{-1})^n$. Thus, the monic Meixner polynomials are given by

$$\pi_n(z) := (1-\frac{1}{c})^{-n} m_n(z;\beta,c). \tag{11.3}$$

For convenience, in (11.3) we have suppressed the dependence of $\pi_n(z)$ on c and β. Furthermore, throughout the chapter we shall fix the parameters $c \in (0,1)$ and $\beta \in [1,2)$. The orthogonality property of $\pi_n(z)$ can be easily derived from (11.2), and we have

$$\sum_{k=0}^{\infty} \pi_n(k)\pi_p(k)w(k) = \delta_{np}/\gamma_n^2, \tag{11.4}$$

where

$$\gamma_n^2 = \frac{(1-c)^{2n+\beta}c^{-n}}{\Gamma(n+\beta)\Gamma(n+1)} \tag{11.5}$$

and

$$w(z) := \frac{\Gamma(z+\beta)}{\Gamma(z+1)}c^z. \tag{11.6}$$

Let $P(z)$ be the 2×2 matrix defined by

$$P(z) := \begin{pmatrix} \pi_n(z) & \sum\limits_{k=0}^{\infty} \dfrac{\pi_n(k)w(k)}{z-k} \\ \gamma_{n-1}^2\pi_{n-1}(z) & \sum\limits_{k=0}^{\infty} \dfrac{\gamma_{n-1}^2\pi_{n-1}(k)w(k)}{z-k} \end{pmatrix}. \tag{11.7}$$

A proof of the following result can be found in Baik *et al.* (2007, Section 1.5.1). The only difference is that their N should be replaced by ∞.

Proposition 11.1. *The matrix-valued function $P(z)$ defined in (11.7) is the unique solution of the following interpolation problem:*

(P1) $P(z)$ is analytic in $\mathbf{C}\setminus\mathbf{N}$;

(P2) at each $z=k\in\mathbf{N}$, the first column of $P(z)$ is analytic and the second column of $P(z)$ has a simple pole with residue

$$\operatorname*{Res}_{z=k} P(z) = \lim_{z\to k} P(z)\begin{pmatrix} 0 & w(z) \\ 0 & 0 \end{pmatrix} = \begin{pmatrix} 0 & w(k)P_{11}(k) \\ 0 & w(k)P_{21}(k) \end{pmatrix}; \tag{11.8}$$

(P3) for z bounded away from $\mathbf{N}$, $P(z)\begin{pmatrix} z^{-n} & 0 \\ 0 & z^n \end{pmatrix} = I+O(|z|^{-1})$ *as* $z \to \infty$.

Let $\mathbb{X}$ denote the set defined by

$$\mathbb{X} := \{X_k\}_{k=0}^{\infty}, \qquad \text{where} \qquad X_k := \frac{k+\beta/2}{n}; \tag{11.9}$$

cf. Baik *et al.* (2003, 2007); Ou and Wong (2010). The X_k's are called *nodes*. For the sake of simplicity, we put

$$B(z) := \prod_{j=0}^{n-1}(z - X_j). \tag{11.10}$$

Our first transformation is given by

$$\begin{aligned} Q(z) &:= n^{-n\sigma_3}P(nz-\beta/2)B(z)^{-\sigma_3} \\ &= \begin{pmatrix} n^{-n} & 0 \\ 0 & n^n \end{pmatrix} P(nz-\beta/2) \begin{pmatrix} B(z)^{-1} & 0 \\ 0 & B(z) \end{pmatrix}, \end{aligned} \tag{11.11}$$

where $\sigma_3 := \begin{pmatrix} 1 & 0 \\ 0 & -1 \end{pmatrix}$ is a Pauli matrix. In this chapter, we shall also make use of another Pauli matrix, namely $\sigma_1 := \begin{pmatrix} 0 & 1 \\ 1 & 0 \end{pmatrix}$.

Proposition 11.2. *The matrix-valued function* $Q(z)$ *defined in (11.11) is the unique solution of the following interpolation problem:*

(Q1) $Q(z)$ *is analytic in* $\mathbb{C} \setminus \mathbb{X}$;

(Q2) at each node X_k *with* $k \in \mathbb{N}$ *and* $k \geq n$, *the first column of* $Q(z)$ *is analytic and the second column of* $Q(z)$ *has a simple pole with residue*

$$\operatorname*{Res}_{z=X_k} Q(z) = \lim_{z\to X_k} Q(z)\begin{pmatrix} 0 & w(nz-\beta/2)B(z)^2 \\ 0 & 0 \end{pmatrix}; \tag{11.12}$$

at each node X_k *with* $k \in \mathbb{N}$ *and* $k < n$, *the second column of* $Q(z)$ *is analytic and the first column of* $Q(z)$ *has a simple pole with residue*

$$\operatorname*{Res}_{z=X_k} Q(z) = \lim_{z\to X_k} Q(z)\begin{pmatrix} 0 & 0 \\ \dfrac{(z-X_k)^2}{w(nz-\beta/2)B(z)^2} & 0 \end{pmatrix}; \tag{11.13}$$

(Q3) for z bounded away from $\mathbb{X}$, $Q(z) = I + O(|z|^{-1})$ *as* $z \to \infty$.

Proof. This is obvious from Proposition 11.1 and the definition of $Q(z)$ in (11.11). □

The purpose of our next transformation is to remove the poles in the interpolation problem for $Q(z)$ (cf. Baik *et al.* (2007, Section 4.2)). Let $\delta > 0$ be a sufficiently small number. We define (see Figure 11.1 below)

$$R(z) := Q(z)\begin{pmatrix} 1 & 0 \\ -\Delta_\pm(z) & 1 \end{pmatrix} \tag{11.14a}$$

for $\operatorname{Re} z \in (0,1)$ and $\pm \operatorname{Im} z \in (0,\delta)$, and

$$R(z) := Q(z)\begin{pmatrix} 1 & -\nabla_\pm(z) \\ 0 & 1 \end{pmatrix} \tag{11.14b}$$

for $\operatorname{Re} z \in (1,\infty)$ and $\pm \operatorname{Im} z \in (0,\delta)$, and

$$R(z) := Q(z) \tag{11.14c}$$

for $\operatorname{Re} z \notin [0,\infty)$ or $\operatorname{Im} z \notin [-\delta,\delta]$, where

$$\nabla_\pm(z) := \frac{n\pi w(nz-\beta/2)B(z)^2}{e^{\mp i\pi(nz-\beta/2)}\sin(n\pi z-\beta\pi/2)}, \tag{11.15}$$

$$\Delta_\pm(z) := \frac{e^{\pm i\pi(nz-\beta/2)}\sin(n\pi z-\beta\pi/2)}{n\pi w(nz-\beta/2)B(z)^2}. \tag{11.16}$$

Lemma 11.1. *For each $k \in \mathbf{N}$, the singularity of $R(z)$ at the node $X_k = \frac{k+\beta/2}{n}$ is removable, that is,* $\operatorname*{Res}_{z=X_k} R(z) = 0$.

Proof. For any $k \in \mathbf{N}$ with $k \geq n$, we have $X_k = \frac{k+\beta/2}{n} > 1$ since $1 \leq \beta < 2$. From (11.15), it is evident that the residue of $\nabla_\pm(z)$ at $z = X_k$ is

$$\operatorname*{Res}_{z=X_k} \nabla_\pm(z) = w(nX_k-\beta/2)B(X_k)^2.$$

From (11.12), we also note that the residue of $Q_{12}(z)$ at $z = X_k$ is $Q_{11}(X_k)$ multiplied by $w(nX_k-\beta/2)B(X_k)^2$. Thus, it follows from (11.14b) that the residue of $R_{12}(z) = Q_{12}(z) - \nabla_\pm(z)Q_{11}(z)$ at $z = X_k$ is zero. Similarly, one can show that $\operatorname*{Res}_{z=X_k} R_{22}(z) = 0$. Since $R_{11}(z) = Q_{11}(z)$ and $R_{21}(z) = Q_{21}(z)$, and since $Q_{11}(z)$ and $Q_{21}(z)$ are analytic by Proposition 11.2, the residues of $R_{11}(z)$ and $R_{21}(z)$ at X_k are zero. For any $k \in \mathbf{N}$ with $k < n$, we have $X_k = \frac{k+\beta/2}{n} < 1$ since $1 \leq \beta < 2$. From (11.10), (11.13) and (11.16), we observe that

$$\operatorname*{Res}_{z=X_k} \Delta_\pm(z) = \frac{1}{w(nX_k-\beta/2)}\prod_{\substack{j=0\\ j\neq k}}^{n-1}(X_k-X_j)^{-2},$$

$$\operatorname*{Res}_{z=X_k} Q_{11}(z) = \frac{Q_{12}(X_k)}{w(nX_k-\beta/2)}\prod_{\substack{j=0\\ j\neq k}}^{n-1}(X_k-X_j)^{-2}.$$

Thus, the residue of $R_{11}(z) = Q_{11}(z) - \Delta_\pm(z)Q_{12}(z)$ at $z = X_k$ is zero. Similarly, one can prove that the residue of $R_{21}(z)$ at $z = X_k$ is also zero. Since $R_{12}(z) = Q_{12}(z)$ and $R_{22}(z)=Q_{22}(z)$, and since $Q_{12}(z)$ and $Q_{22}(z)$ are analytic by Proposition 11.2, the residues of $R_{12}(z)$ and $R_{22}(z)$ at X_k are zero. This completes the proof. □

From the definition in (11.14) and Lemma 11.1, the jump conditions of $R(z)$ given in the following proposition are easily verified (exercise).

Proposition 11.3. *Let Σ_R be the oriented contour shown in Figure 11.1. Denote by $R_+(z)$ and $R_-(z)$, respectively, the limiting values of $R(z)$ on Σ_R taken from the left and from the right of the contour. The jump matrix $J_R(z) := R_-(z)^{-1}R_+(z)$ has the following explicit expressions. For* $\operatorname{Re} z = 1$ *and* $\operatorname{Im} z \in (-\delta, \delta)$, *we have*

$$J_R(z) = \begin{pmatrix} 1 - \Delta_\pm(z)\nabla_\pm(z) & \nabla_\pm(z) \\ -\Delta_\pm(z) & 1 \end{pmatrix}. \tag{11.17}$$

On the positive real line, we have

$$J_R(x) = \begin{pmatrix} 1 & 0 \\ \Delta_-(x) - \Delta_+(x) & 1 \end{pmatrix} \tag{11.18a}$$

for $x \in (0, 1)$, and

$$J_R(x) = \begin{pmatrix} 1 & \nabla_-(x) - \nabla_+(x) \\ 0 & 1 \end{pmatrix} \tag{11.18b}$$

for $x \in (1, \infty)$. Furthermore, we have

$$J_R(z) = \begin{pmatrix} 1 & 0 \\ \Delta_\pm(z) & 1 \end{pmatrix} \tag{11.19a}$$

for $z = i$ Im z with Im $z \in (-\delta, \delta)$ *and* $z =$ Re $z \pm i\delta$ *with* Re $z \in (0, 1)$, *and*

$$J_R(z) = \begin{pmatrix} 1 & \nabla_\pm(z) \\ 0 & 1 \end{pmatrix} \tag{11.19b}$$

for $z = \operatorname{Re} z \pm i\delta$ *with* $\operatorname{Re} z \in (1, \infty)$.

For simplicity, we define

$$\theta(z) := n\pi z - \beta\pi/2, \tag{11.20}$$

$$v(z) := -z \log c, \tag{11.21}$$

$$C := 2i\pi c^{-\beta/2} n^\beta, \tag{11.22}$$

$$W(z) := \frac{(nz)^{1-\beta}\Gamma(nz + \beta/2)}{\Gamma(nz + 1 - \beta/2)}. \tag{11.23}$$

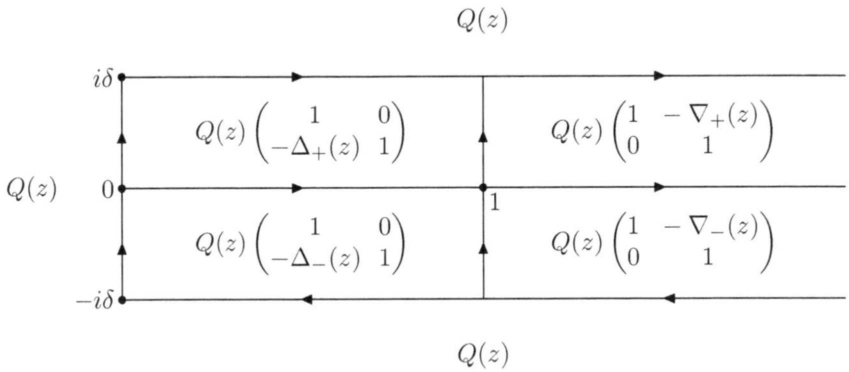

Fig. 11.1: The transformation $Q \to R$ and the oriented contour Σ_R.

In view of (11.6) and the above notations, the functions defined in (11.15) and (11.16) become (exercise)

$$\nabla_\pm = \frac{Cz^{\beta-1}WB^2}{2i\sin\theta e^{\mp i\theta+nv}} \qquad \text{and} \qquad \Delta_\pm = \frac{2i\sin\theta e^{\pm i\theta+nv}}{Cz^{\beta-1}WB^2}. \tag{11.24}$$

It is easy to see that

$$\Delta_\pm\nabla_\pm = e^{\pm 2i\theta}$$

for $z \in \mathbf{C}_\pm$. Also,

$$\nabla_- - \nabla_+ = -Cx^{\beta-1}e^{-nv}WB^2$$

for $z = x \in (1, \infty)$, and

$$\Delta_- - \Delta_+ = \frac{4\sin^2\theta}{Cx^{\beta-1}e^{-nv}WB^2}$$

for $z = x \in (0, 1)$.

The following proposition follows readily (exercise) from Proposition 11.2 and the definition of $R(z)$ in (11.14).

Proposition 11.4. *The matrix-valued function $R(z)$ defined in (11.14) is the unique solution of the following Riemann-Hilbert problem:*

(R1) $R(z)$ is analytic in $\mathbb{C} \setminus \Sigma_R$;
(R2) for $z \in \Sigma_R$, $R_+(z) = R_-(z)J_R(z)$, where the jump matrix $J_R(z)$ is given in Proposition 11.3;
(R3) for $z \in \mathbb{C} \setminus \Sigma_R$, $R(z) = I + O(|z|^{-1})$ as $z \to \infty$.

11.3 Some auxiliary functions

To construct our parametrix, we should introduce some auxiliary functions. First, define the two constants

$$a := \frac{1-\sqrt{c}}{1+\sqrt{c}} \qquad \text{and} \qquad b := \frac{1+\sqrt{c}}{1-\sqrt{c}}. \tag{11.25}$$

These constants are the two turning points for the Meixner polynomials; see Jin and Wong (1998, (2.6)). Let

$$\phi(z) := z \log \frac{\sqrt{bz-1}+\sqrt{az-1}}{\sqrt{bz-1}-\sqrt{az-1}} - \log \frac{\sqrt{z-a}+\sqrt{z-b}}{\sqrt{z-a}-\sqrt{z-b}} \tag{11.26a}$$

for $z \in \mathbf{C} \setminus (-\infty, b]$ and

$$\widetilde{\phi}(z) := z \log \frac{\sqrt{1-az}+\sqrt{1-bz}}{\sqrt{1-az}-\sqrt{1-bz}} - \log \frac{\sqrt{b-z}+\sqrt{a-z}}{\sqrt{b-z}-\sqrt{a-z}} \tag{11.26b}$$

for $z \in \mathbf{C} \setminus (-\infty, 0] \cup [a, \infty)$. These two functions are analogues of the ϕ-function and $\widetilde{\phi}$-function in Chapter 8. It is clear from the definitions that (exercise)

$$\widetilde{\phi}(z) = \phi(z) \pm i\pi(1-z) \tag{11.27}$$

for $z \in \mathbf{C}_\pm$. As $z \to \infty$, we have

$$\phi(z) = z \log \frac{\sqrt{b}+\sqrt{a}}{\sqrt{b}-\sqrt{a}} - \log z + \log \frac{b-a}{4} - 1 + O(\frac{1}{z}).$$

Here we have used the fact that $ab = 1$. Put

$$l := 2 \log \frac{b-a}{4} - 2, \tag{11.28}$$

and recall the definition of $v(z)$ in (11.21). Since $(\sqrt{b}+\sqrt{a})/(\sqrt{b}-\sqrt{a}) = 1/\sqrt{c}$ by (11.25), it follows from the above two equations that (exercise)

$$-\phi(z) + v(z)/2 + l/2 = \log z + O(\frac{1}{z}) \tag{11.29}$$

as $z \to \infty$. For convenience, we define

$$F(z) := \left[\frac{3}{2} n\phi(z)\right]^{2/3} \qquad \text{and} \qquad \widetilde{F}(z) := \left[-\frac{3}{2} n\widetilde{\phi}(z)\right]^{2/3}. \tag{11.30}$$

Note by (11.26) that $\phi(b) = 0$ and (exercise)

$$\phi'(z) = \log \frac{\sqrt{bz-1}+\sqrt{az-1}}{\sqrt{bz-1}-\sqrt{az-1}} = \log \frac{\sqrt{1-bz}+\sqrt{1-az}}{\sqrt{1-bz}-\sqrt{1-az}}. \tag{11.31}$$

Using (11.27), it is readily seen that $\tilde{\phi}(a) = 0$ and $\tilde{\phi}(0) = \frac{1}{2}\log c$. The mapping properties of the functions $\phi(z)$ and $\widetilde{\phi}(z)$ are illustrated in Figure 11.2. From this figure and the definitions (11.26) and (11.30), we have the following proposition.

Proposition 11.5. *For $z \in \mathbf{C} \setminus [a,b]$, we have*

$$\pm \arg F(z) \in (-\pi, \pi) \qquad \text{and} \qquad \arg \widetilde{F}(z) \in (-\pi, \pi). \tag{11.32}$$

For $\operatorname{Re} z \in (a,b)$ and $\pm \operatorname{Im} z \in [0,\delta]$, we have

$$\pm \arg F(z) \in (\pi/3, \pi] \qquad \text{and} \qquad \mp \arg \widetilde{F}(z) \in (\pi/3, \pi]. \tag{11.33}$$

For $x \geq 0$ and δ sufficient small, we have

$$\operatorname{Re} \phi(x \pm i\delta) \sim \begin{cases} \phi(x), & x \geq b; \\ -2\delta \arctan \sqrt{\dfrac{1-ax}{bx-1}}, & a < x < b; \\ \widetilde{\phi}(x) - \pi\delta, & 0 \leq x \leq a. \end{cases} \tag{11.34}$$

Proof. It is easy to prove (11.32) and (11.33) by using (11.26), (11.30) and Figure 11.2. For small $\delta > 0$, we have from a two-term Taylor expansion

$$\operatorname{Re} \phi(x \pm i\delta) \sim \operatorname{Re} \phi_\pm(x) \mp \delta \operatorname{Im} \phi'_\pm(x).$$

Note by (11.31) that

$$\operatorname{Im} \phi'_\pm(x) = \begin{cases} 0, & x \geq b; \\ \pm 2 \arctan \sqrt{\dfrac{1-ax}{bx-1}}, & a < x < b; \\ \pm\pi, & 0 \leq x \leq a. \end{cases}$$

Moreover, (11.26) and (11.27) imply

$$\operatorname{Re} \phi_\pm(x) = \begin{cases} \phi(x), & x \geq b; \\ 0, & a < x < b; \\ \widetilde{\phi}(x), & 0 \leq x \leq a. \end{cases}$$

Thus, (11.34) follows from the above three equations. □

Define

$$N(z) := \begin{pmatrix} \dfrac{(z-1)^{\frac{1-\beta}{2}} (\frac{\sqrt{z-a}+\sqrt{z-b}}{2})^\beta}{(z-a)^{1/4}(z-b)^{1/4}} & \dfrac{-i(z-1)^{\frac{\beta-1}{2}} (\frac{\sqrt{z-a}-\sqrt{z-b}}{2})^\beta}{(z-a)^{1/4}(z-b)^{1/4}} \\ \\ \dfrac{i(z-1)^{\frac{1-\beta}{2}} (\frac{\sqrt{z-a}-\sqrt{z-b}}{2})^{2-\beta}}{(z-a)^{1/4}(z-b)^{1/4}} & \dfrac{(z-1)^{\frac{\beta-1}{2}} (\frac{\sqrt{z-a}+\sqrt{z-b}}{2})^{2-\beta}}{(z-a)^{1/4}(z-b)^{1/4}} \end{pmatrix}. \tag{11.35}$$

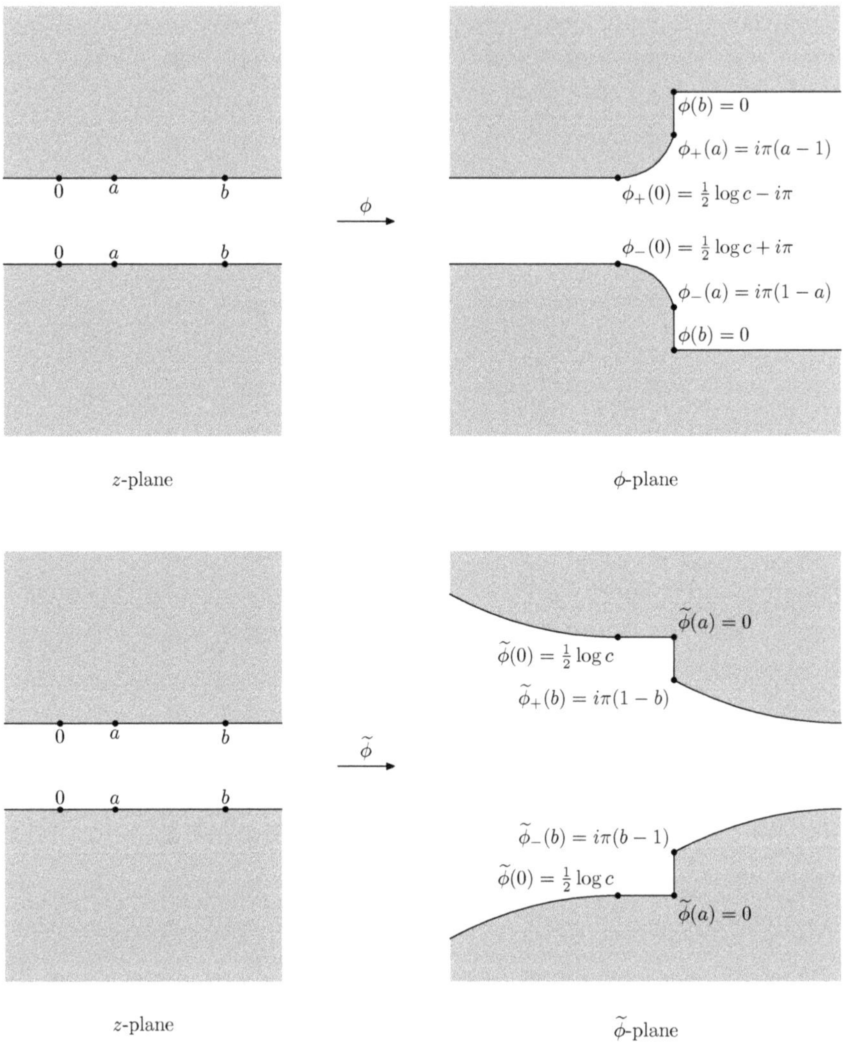

Fig. 11.2: The z-plane under the mappings $\phi(z)$ and $\widetilde{\phi}(z)$.

It is easy to verify that $N(z)$ is analytic in $\mathbf{C} \setminus [a, b]$ and (exercise)

$$N_+(x) = N_-(x) \begin{pmatrix} 0 & -|x-1|^{\beta-1} \\ |x-1|^{1-\beta} & 0 \end{pmatrix}, \quad x \in (a, b). \quad (11.36)$$

The matrix $N(z)$ is analogous to the matrix $N(z)$ in Deift *et al.* (1999). Now we introduce the Airy parametrix which is also similar to the one in

Deift *et al.* (1999). For $z \in \mathbf{C}_\pm$, define

$$\mathbb{A}(z) := \begin{pmatrix} \operatorname{Ai}(z) & -i\operatorname{Bi}(z) \\ i\operatorname{Ai}'(z) & \operatorname{Bi}'(z) \end{pmatrix} \begin{pmatrix} 1 & \mp 1/2 \\ 0 & 1/2 \end{pmatrix}. \tag{11.37}$$

It is clear that

$$\mathbb{A}_+(x) = \mathbb{A}_-(x) \begin{pmatrix} 1 & -1 \\ 0 & 1 \end{pmatrix} \tag{11.38}$$

on the real line. For convenience, set $\omega = e^{2\pi i/3}$. Note that (cf. Abramowitz and Stegun (1970, (10.4.9)))

$$2\omega \operatorname{Ai}(\omega z) = -\operatorname{Ai}(z) + i\operatorname{Bi}(z) \qquad \text{and} \qquad 2\omega^2 \operatorname{Ai}(\omega^2 z) = -\operatorname{Ai}(z) - i\operatorname{Bi}(z).$$

From (11.37), we obtain (exercise)

$$\mathbb{A}(z) = \begin{cases} \begin{pmatrix} \operatorname{Ai}(z) & \omega^2 \operatorname{Ai}(\omega^2 z) \\ i\operatorname{Ai}'(z) & i\omega \operatorname{Ai}'(\omega^2 z) \end{pmatrix} & z \in \mathbf{C}_+; \\[2ex] \begin{pmatrix} \operatorname{Ai}(z) & -\omega \operatorname{Ai}(\omega z) \\ i\operatorname{Ai}'(z) & -i\omega^2 \operatorname{Ai}'(\omega z) \end{pmatrix} & z \in \mathbf{C}_-. \end{cases} \tag{11.39}$$

Furthermore, in view of (cf. Abramowitz and Stegun (1970, (10.4.7)))

$$\operatorname{Ai}(z) + \omega \operatorname{Ai}(\omega z) + \omega^2 \operatorname{Ai}(\omega^2 z) = 0,$$

we have (exercise)

$$\mathbb{A}(z) \begin{pmatrix} 1 & 0 \\ \pm 1 & 1 \end{pmatrix} = \begin{cases} \begin{pmatrix} -\omega \operatorname{Ai}(\omega z) & \omega^2 \operatorname{Ai}(\omega^2 z) \\ -i\omega^2 \operatorname{Ai}'(\omega z) & i\omega \operatorname{Ai}'(\omega^2 z) \end{pmatrix} & z \in \mathbf{C}_+; \\[2ex] \begin{pmatrix} -\omega^2 \operatorname{Ai}(\omega^2 z) & -\omega \operatorname{Ai}(\omega z) \\ -i\omega \operatorname{Ai}'(\omega^2 z) & -i\omega^2 \operatorname{Ai}'(\omega z) \end{pmatrix} & z \in \mathbf{C}_-. \end{cases} \tag{11.40}$$

Recall the asymptotic expansions of the Airy function and its derivative (cf. Olver (1974a, p. 392) or Wong (1989, p. 47))

$$\begin{aligned} \operatorname{Ai}(z) &\sim \frac{z^{-1/4}}{2\sqrt{\pi}} e^{-\frac{2}{3}z^{3/2}} \sum_{s=0}^{\infty} \frac{(-1)^s u_s}{(\frac{2}{3}z^{3/2})^s}, \\ \operatorname{Ai}'(z) &\sim -\frac{z^{1/4}}{2\sqrt{\pi}} e^{-\frac{2}{3}z^{3/2}} \sum_{s=0}^{\infty} \frac{(-1)^s v_s}{(\frac{2}{3}z^{3/2})^s}, \end{aligned} \tag{11.41}$$

as $z \to \infty$ with $|\arg z| < \pi$, where u_s, v_s are constants with $u_0 = v_0 = 1$. For $\arg z \in (-\pi, 0]$, we have $\arg(\omega z) \in (-\pi/3, 2\pi/3]$. Applying (11.41) to (11.39) gives (exercise)

$$\mathbb{A}(z) = \frac{z^{-\sigma_3/4}}{2\sqrt{\pi}} \begin{pmatrix} 1 & -i \\ -i & 1 \end{pmatrix} (I + O(|z|^{-3/2}))e^{-\frac{2}{3}z^{3/2}\sigma_3} \tag{11.42}$$

as $z \to \infty$ with $\arg z \in (-\pi, \pi)$. Applying (11.41) to (11.40) gives (exercise)

$$\mathbb{A}(z) \begin{pmatrix} 1 & 0 \\ \pm 1 & 1 \end{pmatrix} = \frac{z^{-\sigma_3/4}}{2\sqrt{\pi}} \begin{pmatrix} 1 & -i \\ -i & 1 \end{pmatrix} (I + O(|z|^{-3/2}))e^{-\frac{2}{3}z^{3/2}\sigma_3} \tag{11.43}$$

as $z \to \infty$ with $|\arg z| \in (\pi/3, \pi]$. Here the sign $\pm$ means plus sign when z is in the upper half-plane, and minus sign when z is in the lower half-plane. Finally, we introduce a crucial function which enables us to obtain global asymptotic formulas without any cut in the complex plane. For z not on the imaginary line, we define

$$D(z) := \begin{cases} \dfrac{e^{nz}\Gamma(nz - \beta/2 + 1)}{\sqrt{2\pi}(nz)^{nz+(1-\beta)/2}} & \operatorname{Re} z > 0; \\[2ex] \dfrac{\sqrt{2\pi}(-nz)^{-nz+(\beta-1)/2}}{e^{-nz}\Gamma(-nz + \beta/2)} & \operatorname{Re} z < 0. \end{cases} \tag{11.44}$$

The jump of $D(z)$ on the imaginary line is given by (exercise)

$$J_D(z) := D_-(z)^{-1}D_+(z) = 1 - e^{\pm 2i\pi(nz-\beta/2)}, \tag{11.45}$$

where $D_+(z)$ ($D_-(z)$) is the limiting value of $D(z)$ taken from the left (right) of the imaginary axis. It will be seen in the proof of Lemma 11.2 that the usage of $D(z)$ is to cancel the jump $1 - e^{\pm 2i\pi(nz-\beta/2)}$ across the imaginary axis. The explicit formula of $D(z)$ is obtained by solving a one-dimensional Riemann-Hilbert problem and calculating a Cauchy integral. As $n \to \infty$, applying Stirling's formula (cf. Abramowitz and Stegun (1970, (6.1.40))) to (11.44), we have (exercise)

$$D(z) = 1 + O(1/n) \tag{11.46}$$

uniformly for z bounded away from the origin.

11.4 Cyclic conditions

In this section, we introduce the multiplicative cyclic condition in Qiu and Wong (2008) for the jump matrix $J_K(z)$ of a general Riemann-Hilbert problem, and state a theorem in Qiu and Wong (2008) about the asymptotic expansion of the solution to the Riemann-Hilbert problem.

Let $\Gamma = \Gamma_1 \cup \cdots \cup \Gamma_m$ be the union of simple and smooth curves with a certain orientation. For each pair of indices j and k, the intersection $\Gamma_j \cap \Gamma_k$ is either empty or a set of finitely many points. Without loss of generality, we assume the intersection points are the endpoints of Γ_j and Γ_k; otherwise, we can break Γ_j and Γ_k into several curves such that all of the intersection points are endpoints. Let Γ_j^0 be the interior of Γ_j (namely, Γ_j with the endpoints removed) and denote $\Gamma^0 = \Gamma_1^0 \cup \cdots \cup \Gamma_m^0$. The finite set $\Gamma \setminus \Gamma^0$ consists of two kinds of points: (1) the *point of self-intersection* of Γ which belongs to $\Gamma_j \cap \Gamma_k$ for some j and k; and (2) the *endpoint* of Γ which lies in $\Gamma \setminus \Gamma^0$ but does not belong to any $\Gamma_j \cap \Gamma_k$. We also call each curve Γ_j a *branch* of Γ. The orientation of each branch Γ_j induces left $(+)$ side and right $(-)$ side on Γ_j. Let $J_K(z)$ be a matrix-valued function piecewisely defined on Γ^0 and denote by $J_K^j(z)$ the restriction of J_K on Γ_j^0. We make the following assumptions.

(A1) There exists a small $\delta > 0$ such that for each $j = 1, \cdots, M$, the restriction $J_K^j(z)$ is invertible and has an analytic continuation to the δ-neighborhood of Γ^j defined as $\Gamma_j(\delta) := \{z \in \mathbf{C} : d(z, \Gamma_j) < \delta\}$. Here, $d(z, \Gamma_j) = \inf_{w \in \Gamma_j} |z - w|$ is the distance from z to Γ_j.

(A2) There exists an $\alpha > 0$ such that if Γ_j is unbounded for some j, then

$$J_K^j(z) = I + O(|z|^{-\alpha}) \tag{11.47}$$

as $z \to \infty$ for $z \in \Gamma_j(\delta)$.

(A3) Let ζ be any point of self-intersection of Γ. Collect all the branches connected by ζ and order them counterclockwise as $\Gamma_{j_1}, \cdots, \Gamma_{j_q}$. We assume the following multiplicative *cyclic condition*

$$(J_K^{j_q})^{\varepsilon_q}(z) \cdot (J_K^{j_{q-1}})^{\varepsilon_{q-1}}(z) \cdots (J_K^{j_1})^{\varepsilon_1}(z) = I, \tag{11.48}$$

for all $z \in \mathbf{C}$ such that $|z - \zeta| < \delta$, where $\varepsilon_k = 1$ (resp. $\varepsilon_k = -1$) if Γ_{j_k} is directed towards (resp. away from) ζ. When $\varepsilon_k = -1$, the matrix $(J_K^{j_k})^{-1}(z)$ means the inverse of $(J_K^{j_k})(z)$.

Now, we consider the following matrix Riemann-Hilbert problem for $K : \mathbf{C} \setminus \Gamma \to \mathbf{C}^2 \times \mathbf{C}^2$ with a parameter N.

(B1) $K(z) = K(z; N)$ is analytic for $z \in \mathbf{C} \setminus \Gamma$;

(B2) $K_+(z) = K_-(z) J_K(z)$ for $z \in \Gamma^0$;

(B3) fixing any N, we have $K(z) = I + o(1)$ as $z \to \infty$ with $z \in \mathbf{C} \setminus \Gamma$;

(B4) fixing any N, we have $\lim_{z \to \zeta} (z - \zeta) K(z) = 0$ at every endpoint or point of self-intersection $\zeta \in \Gamma \setminus \Gamma^0$.

The following theorem comes from Qiu and Wong (2008, Theorem 3.8).

Theorem 11.1. *Assume that* Γ *has no endpoint and the jump matrix* $J_K(z) = J_K(z; N)$ *satisfies the assumptions (A1)-(A3), where* (11.47) *is uniform in* N. *Assume further that*

$$J_K^j(z) \sim I + \sum_{k=1}^{\infty} \frac{J_{K,k}^j(z)}{N^k}, \tag{11.49}$$

as $N \to \infty$, *uniformly for* $z \in \Gamma_j(\delta)$, *where for each* $k \geq 1$, *the function* $J_{K,k}^j(z)$ *is analytic in* $\Gamma_j(\delta)$. *If* Γ_j *is unbounded, then the above expansion is uniformly for all large* z *in the sense that*

$$\|J_K^j(z) - I - \sum_{k=1}^{p} \frac{J_{K,k}^j(z)}{N^k}\| \leq \frac{C_p}{N^{p+1}|z|^\alpha} \tag{11.50}$$

for all large $z \in \Gamma_j(\delta)$, *where* $C_p > 0$ *is a constant independent of* N *and* z, *and* $\alpha > 0$ *is given in (A2). If* $K(z) = K(z; N)$ *is the solution of the Riemann-Hilbert problem (B1)-(B4), then we have the following asymptotic expansion*

$$K(z) \sim I + \sum_{k=1}^{\infty} \frac{K_k(z)}{N^k}, \tag{11.51}$$

as $N \to \infty$, *uniformly for* $z \in \mathbf{C} \setminus \Gamma$, *where the coefficients* $K_k(z)$ *are analytic in* $\mathbf{C} \setminus \Gamma$. *The above asymptotic expansion is uniform in the neighborhood of infinity in the sense that for each nonnegative integer* p,

$$\|K(z) - I - \sum_{k=1}^{p} \frac{K_k(z)}{N^k}\| \leq \frac{c_p r_\alpha(z)}{N^{p+1}}, \tag{11.52}$$

where $c_p > 0$ *is a constant independent of* N *and* z, *and*

$$r_\alpha(z) := \begin{cases} (\log|z|)/|z|^\alpha, & 0 < \alpha \leq 1, \\ 1/|z|, & \alpha > 1. \end{cases} \tag{11.53}$$

We refer to Qiu and Wong (2008, Theorem 3.8) for the proof of the above theorem.

11.5 Construction of parametrices

For $\operatorname{Re} z \notin [0,1]$ or $\operatorname{Im} z \notin [-\delta, \delta]$, we define

$$\widetilde{R}(z) := \sqrt{\pi}[Ce^{nl}]^{\sigma_3/2} N(z)(z-1)^{\frac{\beta-1}{2}\sigma_3} \begin{pmatrix} 1 & i \\ i & 1 \end{pmatrix} F(z)^{\sigma_3/4} \mathbb{A}(F) \times \left[\frac{D(z)^2 e^{nv(z)}}{Cz^{\beta-1}B(z)^2}\right]^{\sigma_3/2}. \tag{11.54a}$$

For $\operatorname{Re} z \in (0,1)$ and $\operatorname{Im} z \in (-\delta, \delta)$, we define

$$\widetilde{R}(z) := (-1)^{n+1}\sqrt{\pi}[Ce^{nl}]^{\sigma_3/2}N(z)(1-z)^{\frac{\beta-1}{2}\sigma_3}\begin{pmatrix}1 & i\\ i & 1\end{pmatrix}\widetilde{F}(z)^{-\sigma_3/4}$$
$$\times\sigma_1\mathbb{A}(\widetilde{F})\sigma_1\left[\frac{4\sin^2\theta(z)D(z)^2e^{nv(z)}}{Cz^{\beta-1}B(z)^2}\right]^{\sigma_3/2}. \tag{11.54b}$$

Note that $\tilde{R}(z)$ has jumps across the negative real axis and the imaginary axis; they are caused by the functions $z^{\beta-1}$ and $D(z)$ respectively. Our parametrix $\tilde{R}(z)$ is analogous to that in Dai and Wong (2007, (4.54)). The main difference is that the parametrix here has a factor involving the auxiliary function $D(z)$ defined in (11.44). This factor will make our asymptotic formulas valid in much bigger regions, one of which includes the cut $(-\infty, 0]$. As $z \to \infty$, a combination of (11.29), (11.30) and (11.42) gives

$$\sqrt{\pi}\begin{pmatrix}1 & i\\ i & 1\end{pmatrix}F(z)^{\sigma_3/4}\mathbb{A}(F) = e^{-n\phi\sigma_3}[I+O(\frac{1}{z})] = e^{(-\frac{nv}{2}-\frac{nl}{2})\sigma_3}z^{n\sigma_3}[I+O(\frac{1}{z})].$$

Furthermore, it is easily seen from the definitions (11.10), (11.35) and (11.44) that we have respectively $B(z) \sim z^n$, $N(z) \sim I$ and $D(z) \sim 1$ as $z \to \infty$. Thus we obtain from (11.54a) and the above formula that (exercise)

$$\widetilde{R}(z) = I + O(\frac{1}{z}) \tag{11.55}$$

as $z \to \infty$. Define

$$K(z) := (Ce^{nl})^{-\sigma_3/2}R(z)\widetilde{R}(z)^{-1}(Ce^{nl})^{\sigma_3/2}. \tag{11.56}$$

It is clear from (11.55) and Proposition 11.4 that

$$K(z) = I + O(\frac{1}{z}) \tag{11.57}$$

as $z \to \infty$. Let Σ_K denote the oriented contour consisting of Σ_R in Figure 11.1, the negative real axis, and the two infinite lines from $z = \pm i\delta$ to $z = \pm i\infty$ on the imaginary axis. The jump matrix of $K(z)$ is given by

$$J_K(z) = (Ce^{nl})^{-\sigma_3/2}\widetilde{R}_-(z)J_R(z)\widetilde{R}_+(z)^{-1}(Ce^{nl})^{\sigma_3/2}. \tag{11.58}$$

Lemma 11.2. *$J_K(z) = I + O(1/n)$ and $K(z) = I + O(1/n)$ as $n \to \infty$.*

Proof. In view of the structure of the contour Σ_K, we divide our discussion into eight cases and consider each case separately.

Case I. For $\operatorname{Re} z = 1$ and $\pm \operatorname{Im} z \in [0, \delta]$, we have from (11.17) and (11.24) that

$$J_R(z) = \begin{pmatrix} 1 - e^{\pm 2i\theta} & \dfrac{Cz^{\beta-1}B^2W}{2i\sin\theta e^{\mp i\theta + nv}} \\ \\ \dfrac{-2i\sin\theta e^{\pm i\theta+nv}}{Cz^{\beta-1}B^2W} & 1 \end{pmatrix}.$$

This together with (11.54) and (11.58) gives (exercise)

$$\begin{aligned} J_K(z) = N(z)(z-1)^{\frac{\beta-1}{2}\sigma_3} & \left[\begin{pmatrix} 1 & i \\ i & 1 \end{pmatrix} F(z)^{\sigma_3/4}\mathbb{A}(F)\right] \\ & \times \begin{pmatrix} \dfrac{1-e^{\pm 2i\theta}}{2\sin\theta} & -iD^2We^{\pm i\theta} \\ \\ \dfrac{-ie^{\pm i\theta}}{D^2W} & 2\sin\theta \end{pmatrix} \left[\begin{pmatrix} 1 & i \\ i & 1 \end{pmatrix} \widetilde{F}(z)^{-\sigma_3/4}\sigma_1\mathbb{A}(\widetilde{F})\sigma_1\right]^{-1} \\ & \times (1-z)^{\frac{1-\beta}{2}\sigma_3} N(z)^{-1}(-1)^{n+1}. \end{aligned} \tag{11.59}$$

On account of (11.30), we obtain from (11.43) that as $n \to \infty$,

$$\begin{pmatrix} 1 & i \\ i & 1 \end{pmatrix} F(z)^{\sigma_3/4}\mathbb{A}(F) = [I + O(\frac{1}{n})]\frac{e^{-n\phi\sigma_3}}{\sqrt{\pi}} \begin{pmatrix} 1 & 0 \\ \mp 1 & 1 \end{pmatrix}$$

and

$$\left[\begin{pmatrix} 1 & i \\ i & 1 \end{pmatrix} \widetilde{F}(z)^{-\sigma_3/4}\sigma_1\mathbb{A}(\widetilde{F})\sigma_1\right]^{-1} = \sqrt{\pi}\begin{pmatrix} 1 & \mp 1 \\ 0 & 1 \end{pmatrix} e^{n\widetilde{\phi}\sigma_3}[I + O(\frac{1}{n})].$$

From (11.20) and (11.27) we have

$$e^{n\widetilde{\phi}} = (-1)^n e^{n\phi \mp i\theta \mp i\pi\beta/2}.$$

As $n \to \infty$, applying Stirling's formula (cf. Abramowitz and Stegun (1970, (6.1.40))) to (11.23) yields

$$W(z) = 1 + O(1/n) \tag{11.60}$$

uniformly for z bounded away from the negative real axis. Applying the last four equations and (11.46) to (11.59) gives

$$\begin{aligned} J_K(z) &= N(z)(z-1)^{\frac{\beta-1}{2}\sigma_3} \\ & \quad \times \begin{pmatrix} 1 & \pm(1 - D^2W)e^{-2n\widetilde{\phi} \mp i\pi\beta} \\ \\ \mp(1 - D^{-2}W^{-1})e^{2n\phi} & 1 - e^{\pm 2i\theta}(2 - D^2W - D^{-2}W^{-1}) \end{pmatrix} \\ & \quad \times (z-1)^{\frac{1-\beta}{2}\sigma_3} N(z)^{-1}[I + O(\frac{1}{n})] \\ &= I + O(\frac{1}{n}). \end{aligned}$$

Here we have used the fact that in the present case, $\operatorname{Re}\phi(z) \leq 0$ and $\operatorname{Re}\widetilde{\phi}(z) \geq 0$; see Figure 11.2.

Case II. For $z = x \in [1, \infty)$, we have from (11.18) and (11.24)

$$J_R(x) = \begin{pmatrix} 1 & -Cx^{\beta-1}B^2We^{-nv} \\ 0 & 1 \end{pmatrix}.$$

This together with (11.38), (11.54) and (11.58) gives (exercise)

$$J_K(x) = \left[N(x)(x-1)^{\frac{\beta-1}{2}\sigma_3}\begin{pmatrix} 1 & i \\ i & 1 \end{pmatrix} F(x)^{\sigma_3/4}\mathbb{A}_-(F)\right]\begin{pmatrix} 1 & 1-D^2W \\ 0 & 1 \end{pmatrix}$$
$$\times\left[N(x)(x-1)^{\frac{\beta-1}{2}\sigma_3}\begin{pmatrix} 1 & i \\ i & 1 \end{pmatrix} F(x)^{\sigma_3/4}\mathbb{A}_-(F)\right]^{-1}. \qquad (11.61)$$

Note that the matrices $N(z)$ and $F(z)^{\sigma_3/4}$ are both discontinuous across the interval $[1, b)$. But a combination of them makes the jumps vanish. Observe from Figure 11.2 and (11.30) that $\arg\phi_\pm(x) = \pm 3\pi/2$ and $\arg F_\pm(x) = \pm\pi$ for $x \in [1, b)$. Thus, we have $F_+(x)^{\sigma_3/4} = F_-(x)^{\sigma_3/4}e^{i\pi\sigma_3/2}$. It then follows from (11.36) that the matrix

$$N(z)(z-1)^{\frac{\beta-1}{2}\sigma_3}\begin{pmatrix} 1 & i \\ i & 1 \end{pmatrix} F(z)^{\sigma_3/4}$$

has no jump on the interval $[1, b)$. Applying (11.46) and (11.60) to (11.61) gives $J_K(x) = I + O(1/n)$.

Case III. For $z = x \in [0, 1]$, we can proceed in a similar manner as in Case II and obtain $J_K(x) = I + O(1/n)$.

Case IV. For $z = \operatorname{Re} z \pm i\delta$ with $\operatorname{Re} z \in (1, \infty)$, we have from (11.19b) and (11.24)

$$J_R(z) = \begin{pmatrix} 1 & \dfrac{Cz^{\beta-1}B^2W}{2i\sin\theta e^{\mp i\theta+nv}} \\ 0 & 1 \end{pmatrix}.$$

This together with (11.54) and (11.58) gives (exercise)

$$J_K(z) = N(z)(z-1)^{\frac{\beta-1}{2}\sigma_3}\left[\begin{pmatrix} 1 & i \\ i & 1 \end{pmatrix} F(z)^{\sigma_3/4}\mathbb{A}(F)e^{n\phi\sigma_3}\right]$$
$$\times\begin{pmatrix} 1 & \dfrac{D^2We^{-2n\phi\pm 2i\theta}}{2i\sin\theta e^{\pm i\theta}} \\ 0 & 1 \end{pmatrix}\left[\begin{pmatrix} 1 & i \\ i & 1 \end{pmatrix} F(z)^{\sigma_3/4}\mathbb{A}(F)e^{n\phi\sigma_3}\right]^{-1}$$
$$\times(z-1)^{\frac{1-\beta}{2}\sigma_3}N(z)^{-1}. \qquad (11.62)$$

Note from (11.32) that in this case, we have $\arg F(z) \in (-\pi, \pi)$. Thus, coupling (11.30) and (11.42), we obtain

$$\begin{pmatrix} 1 & i \\ i & 1 \end{pmatrix} F(z)^{\sigma_3/4}\mathbb{A}(F)e^{n\phi\sigma_3} = \frac{1}{\sqrt{\pi}}[I + O(\frac{1}{n})].$$

Applying this to (11.62) yields

$$J_K(z) = I + O(\frac{1}{n}).$$

Here we have used the facts that $\mp 2i\sin\theta e^{\pm i\theta} = 1 - e^{\pm 2i\theta} \sim 1$ as $n \to \infty$ and $\mathrm{Re}(-n\phi \pm i\theta) = -n(\mathrm{Re}\,\phi + \pi\delta) < 0$; see (11.20) and (11.34).

Case V. For $z = x \pm i\delta$ with $x \in (0,1)$, we have from (11.19a) and (11.24)

$$J_R(z) = \begin{pmatrix} 1 & 0 \\ \dfrac{2i\sin\theta e^{\pm i\theta + nv}}{Cz^{\beta-1}B^2W} & 1 \end{pmatrix}.$$

This together with (11.54) and (11.58) gives (exercise)

$$\begin{aligned} J_K(z) = (-1)^{n+1}N(z)(1-z)^{\frac{\beta-1}{2}\sigma_3}\left[\begin{pmatrix} 1 & i \\ i & 1 \end{pmatrix} \widetilde{F}(z)^{-\sigma_3/4}\sigma_1\mathbb{A}(\widetilde{F})\sigma_1\right] \\ \times \begin{pmatrix} 2\sin\theta & 0 \\ iD^{-2}W^{-1}e^{\pm i\theta} & (2\sin\theta)^{-1} \end{pmatrix} \\ \times \left[\begin{pmatrix} 1 & i \\ i & 1 \end{pmatrix} F(z)^{\sigma_3/4}\mathbb{A}(F)\right]^{-1}(z-1)^{\frac{1-\beta}{2}\sigma_3}N(z)^{-1}. \end{aligned} \tag{11.63}$$

Note from (11.32) that in this case, $\arg F(z) \in (-\pi, \pi)$ and $\arg \widetilde{F}(z) \in (-\pi, \pi)$. Thus, we obtain from (11.30) and (11.42)

$$\begin{pmatrix} 1 & i \\ i & 1 \end{pmatrix} \widetilde{F}(z)^{-\sigma_3/4}\sigma_1\mathbb{A}(\widetilde{F})\sigma_1 = [I + O(\frac{1}{n})]\frac{e^{-n\widetilde{\phi}\sigma_3}}{\sqrt{\pi}}$$

and

$$\left[\begin{pmatrix} 1 & i \\ i & 1 \end{pmatrix} F(z)^{\sigma_3/4}\mathbb{A}(F)\right]^{-1} = \sqrt{\pi}e^{n\phi\sigma_3}[I + O(\frac{1}{n})]$$

Applying the last two equations to (11.63) yields

$$\begin{aligned} J_K(z) &= N(z)(z-1)^{\frac{\beta-1}{2}\sigma_3} \\ &\times \begin{pmatrix} 2(-1)^{n+1}\sin\theta e^{n\phi - n\widetilde{\phi} \pm i\pi(1-\beta)/2} & 0 \\ i(-1)^{n+1}D^{-2}W^{-1}e^{n\phi + n\widetilde{\phi} \pm i\theta \mp i\pi(1-\beta)/2} & \dfrac{e^{n\widetilde{\phi} - n\phi \mp i\pi(1-\beta)/2}}{2(-1)^{n+1}\sin\theta} \end{pmatrix} \\ &\times (z-1)^{\frac{1-\beta}{2}\sigma_3}N(z)^{-1}[I + O(\frac{1}{n})]. \end{aligned}$$

Using (11.20), (11.26), (11.27) and (11.34), one can show that

$$\mathrm{Re}\{n\phi + n\widetilde{\phi} \pm i\theta\} = \mathrm{Re}\{2n\phi\} < 0$$

and

$$2(-1)^{n+1}\sin\theta e^{n\phi - n\widetilde{\phi} \pm i\pi(1-\beta)/2} = \mp 2i\sin\theta e^{\pm i\theta} = 1 - e^{\pm 2i\theta} \sim 1.$$

Thus, we again have

$$J_K(z) = I + O(\frac{1}{n}), \qquad \text{as} \quad n \to \infty.$$

Case VI. For $z = \pm iy$ with $y \in (0, \delta)$, we have from (11.19) and (11.24)

$$J_R(z) = \begin{pmatrix} 1 & 0 \\ \dfrac{2i\sin\theta e^{\pm i\theta + nv}}{Cz^{\beta-1}B^2W} & 1 \end{pmatrix}.$$

This together with (11.54) and (11.58) gives (exercise)

$$\begin{aligned} J_K(z) = (-1)^{n+1}N(z)(1-z)^{\frac{\beta-1}{2}\sigma_3}\left[\begin{pmatrix} 1 & i \\ i & 1 \end{pmatrix}\widetilde{F}(z)^{-\sigma_3/4}\sigma_1\mathbb{A}(\widetilde{F})\sigma_1\right] \\ \times \begin{pmatrix} \dfrac{2D_-\sin\theta}{D_+} & 0 \\ \dfrac{ie^{\pm i\theta}}{D_+D_-W} & \dfrac{D_+}{2D_-\sin\theta} \end{pmatrix} \\ \times \left[\begin{pmatrix} 1 & i \\ i & 1 \end{pmatrix} F(z)^{\sigma_3/4}\mathbb{A}(F)\right]^{-1}(z-1)^{\frac{1-\beta}{2}\sigma_3}N(z)^{-1}. \end{aligned} \tag{11.64}$$

Note from (11.32) that in this case, $\arg F(z) \in (-\pi, \pi)$ and $\arg \widetilde{F}(z) \in (-\pi, \pi)$. Thus, as in Case IV we have from (11.30) and (11.42)

$$\begin{pmatrix} 1 & i \\ i & 1 \end{pmatrix}\widetilde{F}(z)^{-\sigma_3/4}\sigma_1\mathbb{A}(\widetilde{F})\sigma_1 = [I + O(\frac{1}{n})]\frac{e^{-n\widetilde{\phi}\sigma_3}}{\sqrt{\pi}}$$

and

$$\left[\begin{pmatrix} 1 & i \\ i & 1 \end{pmatrix} F(z)^{\sigma_3/4}\mathbb{A}(F)\right]^{-1} = \sqrt{\pi}e^{n\phi\sigma_3}[I + O(\frac{1}{n})].$$

Applying the above two equations to (11.64) yields

$$\begin{aligned} J_K(z) = N(z)(z-1)^{\frac{\beta-1}{2}\sigma_3}\begin{pmatrix} \dfrac{2(-1)^{n+1}\sin\theta}{J_D e^{n\widetilde{\phi} - n\phi \mp i\pi(1-\beta)/2}} & 0 \\ O(e^{2n\,\mathrm{Re}\,\phi}) & \dfrac{J_D e^{n\widetilde{\phi} - n\phi \mp i\pi(1-\beta)/2}}{2(-1)^{n+1}\sin\theta} \end{pmatrix} \\ \times (z-1)^{\frac{1-\beta}{2}\sigma_3}N(z)^{-1}[I + O(\frac{1}{n})]. \end{aligned}$$

Here we have used (11.23), (11.44) and the asymptotic formula for $\Gamma(x \pm iy)$ as $y \to +\infty$. Since

$$2(-1)^{n+1} \sin\theta e^{n\phi - n\widetilde{\phi} \pm i\pi(1-\beta)/2} = 1 - e^{\pm 2i\theta} = J_D$$

and $\operatorname{Re}\phi(z) < 0$, as before we again have $J_K(z) = I + O(1/n)$ as $n \to \infty$. As mentioned in a statement following (11.45), the usage of $D(z)$ defined in (11.44) is to cancel the jump $1 - e^{\pm 2i\theta} = 1 - e^{\pm 2i\pi(nz - \beta/2)}$. Without this function, the jump matrix $J_K(z)$ is not asymptotically equal to the identity matrix in this case.

Case VII. For $\operatorname{Re} z = 0$ and $|\operatorname{Im} z| \geq \delta$, we have $J_R(z) = I$; see Figure 11.1. Thus, (11.54) and (11.58) imply (exercise)

$$\begin{aligned} J_K(z) = &\left[N(z)(z-1)^{\frac{\beta-1}{2}\sigma_3} \begin{pmatrix} 1 & i \\ i & 1 \end{pmatrix} F(z)^{\sigma_3/4} \mathbb{A}(F) \right] J_D^{-\sigma_3} \\ &\times \left[N(z)(z-1)^{\frac{\beta-1}{2}\sigma_3} \begin{pmatrix} 1 & i \\ i & 1 \end{pmatrix} F(z)^{\sigma_3/4} \mathbb{A}(F) \right]^{-1}. \end{aligned} \tag{11.65}$$

Note that by (11.45), $J_D = 1 - e^{\pm 2i\pi(nz - \beta/2)}$ is exponentially small for $|\operatorname{Im} z| \geq \delta$. From (11.65), it again follows that $J_K = I + O(1/n)$ in this case.

Case VIII. For $z = x \in (-\infty, 0)$, we have $J_R(x) = I$. Thus, (11.54) and (11.58) imply (exercise)

$$\begin{aligned} J_K(z) = N(x)(1-x)^{\frac{\beta-1}{2}\sigma_3} e^{\frac{i\pi(1-\beta)}{2}\sigma_3} &\left[\begin{pmatrix} 1 & i \\ i & 1 \end{pmatrix} F_-(x)^{\sigma_3/4} \mathbb{A}(F_-) e^{n\phi_- \sigma_3} \right] \\ \times e^{i\pi(\beta-1)\sigma_3} &\left[\begin{pmatrix} 1 & i \\ i & 1 \end{pmatrix} F_+(x)^{\sigma_3/4} \mathbb{A}(F_+) e^{n\phi_+ \sigma_3} \right]^{-1} \\ \times (1-x)^{\frac{1-\beta}{2}\sigma_3} e^{\frac{i\pi(1-\beta)}{2}\sigma_3} N(x)^{-1}. \end{aligned} \tag{11.66}$$

Here we have used the fact that $e^{n(\phi_+ - \phi_-)} = e^{2ni\pi} = 1$ for $x < 0$; see (11.26). Note from (11.32) that $\arg F_\pm \in (-\pi, \pi)$ in this case. Hence, we obtain from (11.30) and (11.42)

$$\begin{pmatrix} 1 & i \\ i & 1 \end{pmatrix} F_\pm(x)^{\sigma_3/4} \mathbb{A}(F_\pm) e^{n\phi_\pm \sigma_3} = \frac{1}{\sqrt{\pi}}[I + O(\frac{1}{n})].$$

Applying the last equation to (11.66) yields

$$\begin{aligned} J_K(z) &= N(x)(1-x)^{\frac{\beta-1}{2}\sigma_3} e^{\frac{i\pi(1-\beta)}{2}\sigma_3} e^{i\pi(\beta-1)\sigma_3} e^{\frac{i\pi(1-\beta)}{2}\sigma_3} \\ &\quad \times (1-x)^{\frac{1-\beta}{2}\sigma_3} N(x)^{-1}[I + O(\frac{1}{n})] \\ &= I + O(\frac{1}{n}). \end{aligned}$$

In conclusion, we have shown that $J_K(z) = I + O(1/n)$ on the contour of $K(z)$. It is not difficult to verify that the multiplicative cyclic condition (11.48) holds for the jump matrix $J_K(z)$. An application of Theorem 11.1 then gives $K(z) = I + O(1/n)$ as $n \to \infty$. $\square$

Combining Lemma 11.2 with (11.7), (11.10), (11.11), (11.14) and (11.56), we obtain (exercise)

$$\pi_n(nz - \beta/2) = n^n B(z)\widetilde{R}_{11}(z)[I + O(\frac{1}{n})] \tag{11.67a}$$

for $\operatorname{Re} z \notin [0,1]$ or $\operatorname{Im} z \notin [-\delta, \delta]$, and

$$\pi_n(nz - \beta/2) = n^n B(z)[\widetilde{R}_{11}(z) + \Delta_\pm(z)\widetilde{R}_{12}(z)][I + O(\frac{1}{n})] \tag{11.67b}$$

for $\operatorname{Re} z \in (0,1)$ and $\operatorname{Im} z \in (0, \pm\delta)$.

11.6 Main results

Theorem 11.2. *As $n \to \infty$, we have*

$$\begin{aligned}\pi_n(nz - \beta/2) &= n^n \sqrt{\pi} D(z) e^{nv(z)/2 + nl/2} \\ &\times \left\{ \frac{(\frac{\sqrt{z-a}+\sqrt{z-b}}{2})^\beta + (\frac{\sqrt{z-a}-\sqrt{z-b}}{2})^\beta}{z^{(\beta-1)/2}(z-a)^{1/4}(z-b)^{1/4}F(z)^{-1/4}} \operatorname{Ai}(F) \right. \\ &\left. - \frac{(\frac{\sqrt{z-a}+\sqrt{z-b}}{2})^\beta - (\frac{\sqrt{z-a}-\sqrt{z-b}}{2})^\beta}{z^{(\beta-1)/2}(z-a)^{1/4}(z-b)^{1/4}F(z)^{1/4}} \operatorname{Ai}'(F) \right\} \\ &\times [1 + O(\frac{1}{n})]\end{aligned} \tag{11.68}$$

for $\operatorname{Re} z \notin [0,1]$ *or* $\operatorname{Im} z \notin [-\delta, \delta]$, *and*

$$\begin{aligned}\pi_n(nz - \beta/2) &= (-n)^n \sqrt{\pi} D(z) e^{nv(z)/2 + nl/2} \\ &\times \left\{ \frac{(\frac{\sqrt{b-z}+\sqrt{a-z}}{2})^\beta + (\frac{\sqrt{b-z}-\sqrt{a-z}}{2})^\beta}{z^{(\beta-1)/2}(b-z)^{1/4}(a-z)^{1/4}\widetilde{F}(z)^{-1/4}} \right. \\ &\times [\cos\theta \operatorname{Ai}(\widetilde{F}) - \sin\theta \operatorname{Bi}(\widetilde{F})] \\ &+ \frac{(\frac{\sqrt{b-z}+\sqrt{a-z}}{2})^\beta - (\frac{\sqrt{b-z}-\sqrt{a-z}}{2})^\beta}{z^{(\beta-1)/2}(b-z)^{1/4}(a-z)^{1/4}\widetilde{F}(z)^{1/4}} \\ &\left. \times [\cos\theta \operatorname{Ai}'(\widetilde{F}) - \sin\theta \operatorname{Bi}'(\widetilde{F})] \right\} [I + O(\frac{1}{n})]\end{aligned} \tag{11.69}$$

for $\operatorname{Re} z \in (0,1)$ *and* $\operatorname{Im} z \in (-\delta, \delta)$, *where the constant l is given in (11.28) and the functions $v(z), F(z)$ and $D(z)$ are respectively given in (11.21), (11.30) and (11.44). The asymptotic formula on the boundary of the two regions can be obtained by taking limits from either side.*

Proof. From (11.35), (11.37), (11.54) and (11.67a), it is easy to obtain (11.68). We now prove (11.69). Define

$$\widetilde{Q}(z) := \widetilde{R}(z)\begin{pmatrix} 1 & 0 \\ \Delta_{\pm}(z) & 1 \end{pmatrix}. \tag{11.70}$$

From (11.67b), we have (exercise)

$$\pi_n(nz - \beta/2) = n^n B(z)\widetilde{Q}_{11}(z)[I + O(\frac{1}{n})]. \tag{11.71}$$

Thus, we only need to calculate $\widetilde{Q}_{11}(z)$. First, we observe from (11.24), (11.54) and (11.70) that

$$\widetilde{Q}(z) = (-1)^{n+1}\sqrt{\pi}[Ce^{nl}]^{\sigma_3/2}N(z)(1-z)^{\frac{\beta-1}{2}\sigma_3}\begin{pmatrix} 1 & i \\ i & 1 \end{pmatrix}\widetilde{F}(z)^{-\sigma_3/4}$$
$$\times\sigma_1\mathbb{A}(\widetilde{F})\sigma_1\begin{pmatrix} 1 & 0 \\ \dfrac{ie^{\pm i\theta(z)}}{2D^2(z)W(z)\sin\theta(z)} & 1 \end{pmatrix}$$
$$\times\left[\frac{4\sin^2\theta(z)D(z)^2e^{nv(z)}}{Cz^{\beta-1}B(z)^2}\right]^{\sigma_3/2}.$$

Second, (11.35) gives

$$N(z)(1-z)^{\frac{\beta-1}{2}\sigma_3}\begin{pmatrix} 1 & i \\ i & 1 \end{pmatrix}$$
$$= \begin{pmatrix} \dfrac{\varphi_+(z)^\beta - \varphi_-(z)^\beta}{(b-z)^{1/4}(a-z)^{1/4}} & \dfrac{\varphi_+(z)^\beta + \varphi_-(z)^\beta}{-i(b-z)^{1/4}(a-z)^{1/4}} \\ \dfrac{\varphi_+(z)^{2-\beta} - \varphi_-(z)^{2-\beta}}{-i(b-z)^{1/4}(a-z)^{1/4}} & \dfrac{\varphi_+(z)^{2-\beta} + \varphi_-(z)^{2-\beta}}{(b-z)^{1/4}(a-z)^{1/4}} \end{pmatrix},$$

where we have denoted $\varphi_{\pm}(z) := (\sqrt{b-z} \pm \sqrt{a-z})/2$ for sake of simplicity.

Finally, (11.37) implies

$$\sigma_1 \mathbb{A}(\widetilde{F})\sigma_1 \begin{pmatrix} 1 & 0 \\ \dfrac{ie^{\pm i\theta}}{2D^2W\sin\theta} & 1 \end{pmatrix} (2\sin\theta)^{\sigma_3}$$

$$= \begin{pmatrix} \mathrm{Bi}'(\widetilde{F}) & i\,\mathrm{Ai}'(\widetilde{F}) \\ -i\,\mathrm{Bi}(\widetilde{F}) & \mathrm{Ai}(\widetilde{F}) \end{pmatrix} \begin{pmatrix} 1/2 & 0 \\ \pm 1/2 & 1 \end{pmatrix} \begin{pmatrix} 2\sin\theta & 0 \\ iD^{-2}W^{-1}e^{\pm i\theta} & (2\sin\theta)^{-1} \end{pmatrix}$$

$$= \begin{pmatrix} \sin\theta\,\mathrm{Bi}'(\widetilde{F}) - [\cos\theta + (D^{-2}W^{-1}-1)e^{\pm i\theta}]\,\mathrm{Ai}'(\widetilde{F}) & \dfrac{i\,\mathrm{Ai}'(\widetilde{F})}{2\sin\theta} \\ -i\{\sin\theta\,\mathrm{Bi}(\widetilde{F}) - [\cos\theta + (D^{-2}W^{-1}-1)e^{\pm i\theta}]\,\mathrm{Ai}(\widetilde{F})\} & \dfrac{\mathrm{Ai}(\widetilde{F})}{2\sin\theta} \end{pmatrix}.$$

Applying the last three equations to (11.71) gives

$$\begin{aligned} &\pi_n(nz-\beta/2) \\ &= (-n)^n\sqrt{\pi}D(z)e^{nv(z)/2+nl/2}\Bigg\{\frac{\varphi_+(z)^\beta + \varphi_-(z)^\beta}{z^{(\beta-1)/2}(b-z)^{1/4}(a-z)^{1/4}\widetilde{F}(z)^{-1/4}} \\ &\times[\cos\theta\,\mathrm{Ai}(\widetilde{F}) - \sin\theta\,\mathrm{Bi}(\widetilde{F}) + e^{\pm i\theta}(D^{-2}W^{-1}-1)\,\mathrm{Ai}(\widetilde{F})] \\ &+\frac{\varphi_+(z)^\beta - \varphi_-(z)^\beta}{z^{(\beta-1)/2}(b-z)^{1/4}(a-z)^{1/4}\widetilde{F}(z)^{1/4}} \\ &\times[\cos\theta\,\mathrm{Ai}'(\widetilde{F}) - \sin\theta\,\mathrm{Bi}'(\widetilde{F}) + e^{\pm i\theta}(D^{-2}W^{-1}-1)\,\mathrm{Ai}'(\widetilde{F})]\Bigg\}[I+O(\frac{1}{n})]. \end{aligned}$$

On account of (11.26), (11.30), and (11.41), $\mathrm{Ai}(\widetilde{F})$ and $\mathrm{Ai}'(\widetilde{F})$ are exponentially small when z approaches the origin; by (11.46) and (11.60), we also have $D^{-2}W^{-1}-1 = O(1/n)$ for $z \neq 0$. Since we can always neglect the terms $(D^{-2}W^{-1}-1)\,\mathrm{Ai}(\widetilde{F})$ and $(D^{-2}W^{-1}-1)\,\mathrm{Ai}'(\widetilde{F})$, formula (11.69) is proved.

To justify that the asymptotic formula on the curve separating the two regions can be obtain by taking limits from either side, we just note that the regions of validity of both formulas (11.68) and (11.69) can be slightly extended beyond their boundaries, and that in the overlap region these two formulas are asymptotically equal. □

Remark 11.1. The formulas (6.9) in Jin and Wong (1998) and (2.35) in Jin and Wong (1999) are asymptotically equal to (11.68) in the present chapter, while the formulas (6.27) in Jin and Wong (1998) and (4.19) in Jin and Wong (1999) are asymptotically equal to (11.69).

11.7 Exercises

1. Prove Proposition 11.3.
2. Verify (11.24).
3. Prove Proposition 11.4.
4. Verify (11.27) and (11.29).
5. Verify (11.31).
6. Verify (11.36).
7. Verify (11.39) and (11.40).
8. Prove (11.42) and (11.43).
9. Verify (11.45) and (11.46).
10. Verify (11.55).
11. Prove (11.59), (11.61), (11.62), (11.63), (11.64), (11.65), and (11.66).
12. Verify (11.67a) and (11.67b).
13. Verify (11.71).

Chapter 12

Global Asymptotics of the Stieltjes-Wigert Polynomials

12.1 Introduction

In this chapter, we study the Stieltjes-Wigert polynomials, a typical example in the class of q-orthogonal polynomials. We first introduce the notation for the q-factorial:

$$(a;q)_n := \prod_{j=0}^{n-1}(1-aq^j),$$

and q-binomial:

$$\begin{bmatrix} n \\ j \end{bmatrix}_q := \frac{(q;q)_n}{(q;q)_j(q;q)_{n-j}},$$

where by default an empty product equals 1. The Stieltjes-Wigert polynomials are defined by

$$p_n(x;q) = \frac{(-1)^n q^{n/2+1/4}}{\sqrt{(q;q)_n}} \sum_{j=0}^{n} \begin{bmatrix} n \\ j \end{bmatrix}_q q^{j^2}(-q^{1/2}x)^j; \tag{12.1}$$

see Szegö (1975, (2.7.4)). Let

$$q = \exp\{-(2k^2)^{-1}\}. \tag{12.2}$$

The Stieltjes-Wigert polynomials are orthogonal with respect to the weight function

$$w(x) = k\pi^{-1/2}\exp\{-k^2(\ln x)^2\}, \quad x > 0; \tag{12.3}$$

see Szegö (1975, (2.7.1)). For fixed k, an asymptotic formula of $p_n(x;q)$ as $n \to \infty$ was derived in Wigert (1923), and the formula can be expressed in terms of the q-Airy function; see Ismail (2005). The cases when $k = n/2$ and $k = (n-1)/2$ were considered in Baik and Suidan (2007) to investigate the asymptotics of a non-intersecting Brownian bridge process. The monic Stieltjes-Wigert polynomials are denoted by

$$\pi_n(x) := p_n(x;q)/\gamma_n, \tag{12.4}$$

where

$$\gamma_n := \frac{q^{n^2+n+1/4}}{\sqrt{(q;q)_n}} \tag{12.5}$$

is the leading coefficient of $p_n(x;q)$.

The material in this chapter is taken from Wang and Wong (2006). In this chapter, we will consider the following Riemann-Hilbert problem (RHP) on $[0,\infty)$ with right orientation.

(Y1) $Y(z)$ is analytic in $\mathbf{C}\setminus[0,\infty)$.
(Y2) $Y(z)$ satisfies the jump condition

$$Y_+(x) = Y_-(x)\begin{pmatrix}1 & w(x)\\ 0 & 1\end{pmatrix}, \tag{12.6}$$

for $x\in(0,\infty)$, where $w(x)$ is the weight function given in (12.3).
(Y3) As $z\to\infty$, $Y(z)z^{-n\sigma_3} = I + O(|z|^{-1})$.
(Y4) As $z\to 0$, $Y(z) = O(1)$.

According to a well-known theorem of Fokas *et al.* (1991), the unique solution to the above RHP is given by

$$Y(z) = \begin{pmatrix}\pi_n(z) & C[\pi_n w](z)\\ -2\pi i\gamma_{n-1}^2\pi_{n-1}(z) & -2\pi i\gamma_{n-1}^2 C[\pi_{n-1}w](z)\end{pmatrix}, \quad z\in\mathbf{C}\setminus[0,\infty),$$

where $C[f](z)$ is the Cauchy transform

$$C[f](z) := \frac{1}{2\pi i}\int_0^\infty \frac{f(x)}{x-z}dx, \quad z\in\mathbf{C}\setminus[0,\infty).$$

In Baik and Suidan (2007), by setting $k=(n-1)/2$ and applying the equilibrium measure calculated in Wang and Wong (2006), the authors derived an asymptotic formula for $Y(z)$. Throughout this chapter, we assume

$$\frac{k}{n} = O(1), \quad \frac{n}{k} = O(1), \tag{12.7}$$

as $n\to\infty$. We will follow the procedure in Wang and Wong (2006) to calculate the equilibrium measure and derive global asymptotics of the monic Stieltjes-Wigert polynomials $\pi_n(z)$ as $n\to\infty$.

12.2 Equilibrium measure

We choose $\tau_1 = 1$, $\theta = 0$, and $N = n + 1/2$ in (8.22). The potential function for the weight function in (12.3) becomes

$$v(z) = -\frac{1}{N}\ln w(z) = \frac{k^2}{N}(\ln x)^2 - \frac{\ln(k/\sqrt{\pi})}{N}. \tag{12.8}$$

For simplicity, we denote

$$\kappa := 2k^2/N, \quad \eta = 1/\kappa = N/(2k^2). \tag{12.9}$$

It is easy to calculate $v'(x) = \kappa(\ln x)/x$. Hence, the equations for the MRS numbers (8.32)-(8.33) can be written as

$$\int_a^b \frac{\ln x}{x\sqrt{(x-a)(b-x)}}dx = 0, \tag{12.10}$$

$$\int_a^b \frac{\ln x}{\sqrt{(x-a)(b-x)}}dx = 2\pi\eta. \tag{12.11}$$

One can prove that (exercise) for any $0 < a < b$,

$$\int_a^b \frac{\ln x}{x\sqrt{(x-a)(b-x)}}dx = -\frac{2\pi}{\sqrt{ab}}\ln\frac{\sqrt{a}+\sqrt{b}}{2\sqrt{ab}}, \tag{12.12}$$

$$\int_a^b \frac{\ln x}{\sqrt{(x-a)(b-x)}}dx = 2\pi\ln\frac{\sqrt{a}+\sqrt{b}}{2}. \tag{12.13}$$

Hence, we obtain from (12.10)-(12.11) that

$$2\sqrt{ab} = \sqrt{a}+\sqrt{b} = 2e^{\eta}.$$

Solving these equations gives (exercise)

$$a = (e^{\eta} - \sqrt{e^{2\eta}-e^{\eta}})^2 = 2e^{2\eta} - e^{\eta} - 2e^{\eta}\sqrt{e^{2\eta}-e^{\eta}}, \tag{12.14}$$

$$b = (e^{\eta} + \sqrt{e^{2\eta}-e^{\eta}})^2 = 2e^{2\eta} - e^{\eta} + 2e^{\eta}\sqrt{e^{2\eta}-e^{\eta}}. \tag{12.15}$$

Next, we set $v'(x) = \kappa(\ln x)/x$ in (8.30) to find

$$g'(z) = \frac{(z-a)^{1/2}(z-b)^{1/2}}{2\pi}\int_a^b \frac{\kappa\ln x}{x\sqrt{(x-a)(b-x)}}\frac{dx}{z-x}, \tag{12.16}$$

for $z \in \mathbf{C}\setminus[a,b]$. Since

$$\frac{z}{x(z-x)} = \frac{1}{x} + \frac{1}{z-x},$$

we obtain

$$g'(z) = \frac{\kappa(z-a)^{1/2}(z-b)^{1/2}}{2\pi z}[I(z) - I(0)], \tag{12.17}$$

where

$$I(z) := \int_a^b \frac{\ln x}{\sqrt{(x-a)(b-x)}} \frac{dx}{z-x}. \tag{12.18}$$

First, we introduce two complex contours surrounding the negative real line $(-\infty, 0]$ and the interval $[a, b]$, respectively:

$$\begin{aligned}\Gamma_{0,\varepsilon} :=& \{u + i\varepsilon : \ u \le 0\} \cup \{\varepsilon e^{i\theta} : \ \theta \in [-\pi/2, \pi/2]\} \cup \{u - i\varepsilon : \ u \le 0\}, \\ \Gamma_{1,\varepsilon} :=& \{u + i\varepsilon : \ u \in [a, b]\} \cup \{b + \varepsilon e^{i\theta} : \ \theta \in [-\pi/2, \pi/2]\} \\ & \cup \{u - i\varepsilon : \ u \in [a, b]\} \cup \{a + \varepsilon e^{i\theta} : \ \theta \in [\pi/2, 3\pi/2]\}.\end{aligned}$$

If we define the distance from a point $z \in \mathbf{C}$ to a given closed subset $C \in \mathbf{C}$ as

$$\mathrm{dist}(z, C) := \inf_{w \in C} |z - w|,$$

then the two contours can be rewritten as

$$\begin{aligned}\Gamma_{0,\varepsilon} :=& \{z \in \mathbf{C} : \ \mathrm{dist}(z, (-\infty, 0]) = \varepsilon\}, \\ \Gamma_{1,\varepsilon} :=& \{z \in \mathbf{C} : \ \mathrm{dist}(z, [a, b]) = \varepsilon\}.\end{aligned}$$

The contours $\Gamma_{0,\varepsilon}$ and $\Gamma_{1,\varepsilon}$ are oriented such that, if z moves along each contour, then $\mathrm{Re}\, z$ is increasing (resp. decreasing) when $\mathrm{Im}\, z$ is positive (resp. negative); see Figure 12.1.

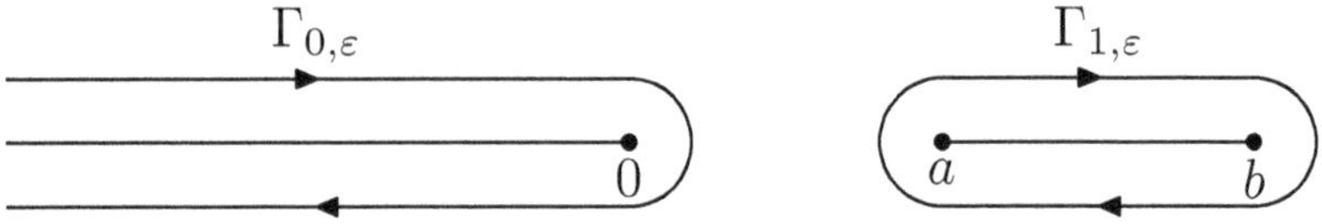

Fig. 12.1: The contours $\Gamma_{0,\varepsilon}$ and $\Gamma_{1,\varepsilon}$.

Next, we denote

$$\Gamma_0 := \lim_{\varepsilon \to 0^+} \Gamma_{0,\varepsilon}, \quad \Gamma_1 := \lim_{\varepsilon \to 0^+} \Gamma_{1,\varepsilon},$$

and define

$$u(\zeta) := \frac{\ln \zeta}{(\zeta - a)^{1/2} (\zeta - b)^{1/2} (\zeta - z)}.$$

Since $u(\zeta) = O(|\ln \zeta| / |\zeta|^2)$ as $\zeta \to \infty$, we obtain from Cauchy's theorem that

$$\int_{\Gamma_0 \cup \Gamma_1} u(\zeta) d\zeta = \int_{\Gamma_0 \cup \Gamma_1} \frac{\ln \zeta}{(\zeta - a)^{1/2} (\zeta - b)^{1/2}} \frac{d\zeta}{(\zeta - z)} = \frac{2\pi i \ln z}{(z - a)^{1/2} (z - b)^{1/2}}.$$

The integral on the left-hand side can be expressed as the sum of

$$\int_{\Gamma_1} u(\zeta)d\zeta = \int_a^b \frac{2\ln x}{i\sqrt{(x-a)(b-x)}}\frac{dx}{x-z} = 2iI(z),$$

and

$$\int_{\Gamma_0} u(\zeta)d\zeta = \int_0^\infty \frac{2\pi i}{\sqrt{(x+a)(x+b)}}\frac{dx}{x+z}. \tag{12.19}$$

A further computation (exercise) gives

$$\int_0^\infty \frac{(z-a)^{1/2}(z-b)^{1/2}}{\sqrt{(x+a)(x+b)}}\frac{dx}{x+z} = \ln\frac{[z+\sqrt{ab}+(z-a)^{1/2}(z-b)^{1/2}]^2}{(\sqrt{a}+\sqrt{b})^2 z}. \tag{12.20}$$

A combination of the above formulas gives

$$I(z) = \frac{-2\pi}{(z-a)^{1/2}(z-b)^{1/2}}\ln\frac{z+\sqrt{ab}+(z-a)^{1/2}(z-b)^{1/2}}{(\sqrt{a}+\sqrt{b})z}.$$

In the calculation, we assume $z \notin (-\infty, 0] \cup [a, b]$. By analytic continuation, the above formula is valid for $z \in \mathbf{C} \setminus [a, b]$. In particular,

$$\begin{aligned} I(0) &= \lim_{\varepsilon\to 0^+} I(-\varepsilon) \\ &= \lim_{\varepsilon\to 0^+} \frac{2\pi}{\sqrt{(a+\varepsilon)(b+\varepsilon)}}\ln\frac{\varepsilon-\sqrt{ab}+\sqrt{(a+\varepsilon)(b+\varepsilon)}}{(\sqrt{a}+\sqrt{b})\varepsilon} \\ &= \frac{2\pi}{\sqrt{ab}}\ln\frac{\sqrt{b}+\sqrt{a}}{2\sqrt{ab}} \\ &= 0, \end{aligned} \tag{12.21}$$

where we have made use of (12.14)-(12.15) in the last equality. Substituting the above two formulas into (12.17) yields

$$g'(z) = -\frac{\kappa}{z}\ln\frac{z+\sqrt{ab}+(z-a)^{1/2}(z-b)^{1/2}}{(\sqrt{a}+\sqrt{b})z}. \tag{12.22}$$

This together with (8.34) gives the equilibrium measure

$$\rho(x) = \operatorname{Re}\frac{ig'_+(x)}{\pi} = \frac{\kappa}{\pi x}\arctan\frac{\sqrt{(x-a)(b-x)}}{x+\sqrt{ab}}, \quad x \in (a, b). \tag{12.23}$$

The g-function in (8.24) and the ϕ-function in (8.25) are given by

$$g(z) = \int_a^b \ln(z-x)\rho(x)dx, \tag{12.24}$$

$$\phi(z) = \frac{v(z)+l}{2} - g(z), \tag{12.25}$$

for $z \in \mathbf{C} \setminus (-\infty, b]$, where

$$l = 2g(b) - v(b) \tag{12.26}$$

is the Lagrange multiplier. Similar as in (8.27)-(8.28), one can prove that (exercise)

$$g_+(x) + g_-(x) - v(x) - l = \begin{cases} 0, & x \in (a,b), \\ -2\phi(x), & x > b, \\ -2\phi_+(x) - 2\pi i, & 0 < x < a, \end{cases} \tag{12.27}$$

$$g_+(x) - g_-(x) = \begin{cases} -2\phi_+(x) = 2\phi_-(x), & x \in (a,b), \\ 0, & x > b, \\ 2\pi i, & x < a. \end{cases} \tag{12.28}$$

Note that $\tau_0 = N - n = 1/2$. It follows from (12.8) that the modified Lagrange multiplier in (8.37) can be expressed as (exercise)

$$l_n = 2Ng(b) + \ln w(b) - \ln \frac{b-a}{4}. \tag{12.29}$$

In view of (12.22), (12.25) and (12.26), we also obtain (exercise)

$$\phi(z) = \int_b^z \frac{\kappa}{\zeta} \ln \frac{\zeta + \sqrt{ab} + (\zeta - a)^{1/2}(\zeta - b)^{1/2}}{(\sqrt{a} + \sqrt{b})\zeta^{1/2}} d\zeta, \tag{12.30}$$

for $z \in \mathbf{C} \setminus (-\infty, b]$. If $x \in (a,b)$, then we have (exercise)

$$\phi_\pm(x) = \pm\pi i \int_b^x \rho(t)dt, \tag{12.31}$$

where ρ is the density function of the equilibrium measure explicitly given in (12.23).

12.3 Contour deformation

Since we do not scale the variable; namely, we set $\theta = 0$ in (8.19) and obtain $U(z) = Y(z)$, the normalization transformation in (8.35) becomes

$$T(z) = e^{-\frac{1}{2}l_n\sigma_3} Y(z) e^{[-Ng(z) + \frac{1}{2}\ln\frac{\varphi(z)}{2} + \frac{1}{2}l_n]\sigma_3}, \tag{12.32}$$

where

$$\varphi(z) = \frac{(\sqrt{z-a} + \sqrt{z-b})^2}{2} = z - \frac{a+b}{2} + \sqrt{(z-a)(z-b)}. \tag{12.33}$$

It is readily seen from (Y1)-(Y4) that $T(z)$ satisfies the following Riemann-Hilbert problem on $[0, \infty)$ with right orientation.

(T1) $T(z)$ is analytic in $\mathbf{C} \setminus [0, \infty)$.
(T2) $T_+(x) = T_-(x) J_T(x)$ for $x \in (0, \infty)$,
(T3) As $z \to \infty$, $T(z) = I + O(|z|^{-1})$.
(T4) As $z \to 0$, $T(z) = O(1)$.

Recall from (8.38) that

$$\psi(z) := \ln \frac{b-a}{4} - \ln \frac{\varphi(z)}{2} = -2 \ln \frac{\sqrt{z-a} + \sqrt{z-b}}{\sqrt{b-a}}, \tag{12.34}$$

for $z \in \mathbf{C} \setminus (-\infty, b]$. Similar as in (8.45), the jump matrix of $T(z)$ can be expressed as

$$J_T(x) = \begin{pmatrix} e^{2N\phi_+(x) - \psi_+(x)} & 1 \\ 0 & e^{2N\phi_-(x) - \psi_-(x)} \end{pmatrix} \tag{12.35a}$$

for $x \in (a, b)$, and

$$J_T(x) = \begin{pmatrix} 1 & e^{-2N\phi(x) + \psi(x)} \\ 0 & 1 \end{pmatrix} \tag{12.35b}$$

for $x \in (b, \infty)$, and

$$J_T(x) = \begin{pmatrix} 1 & e^{-2N\phi_+(x) + \psi_+(x)} \\ 0 & 1 \end{pmatrix} \tag{12.35c}$$

for $x \in (0, a)$. Next, we add two contours $\Sigma = \Sigma_+ \cup \Sigma_-$ from a to b in the upper and lower complex planes, respectively, and then make a factorization transformation.

$$V(z) = \begin{cases} T(z), & z \in \mathbf{C} \setminus (\bar{\Omega} \cup [0, \infty)), \\ T(z) \begin{pmatrix} 1 & 0 \\ -e^{2N\phi(z) - \psi(z)} & 1 \end{pmatrix}, & z \in \Omega_+ \\ T(z) \begin{pmatrix} 1 & 0 \\ e^{2N\phi(z) - \psi(z)} & 1 \end{pmatrix}, & z \in \Omega_-; \end{cases} \tag{12.36}$$

see Figure 12.2 for an illustration of the contours $\Sigma = \Sigma_+ \cup \Sigma_-$ and the upper and lower lens regions $\Omega = \Omega_+ \cup \Omega_-$. We can easily prove that $V(z)$ satisfies the following Riemann-Hilbert problem on $\Sigma \cup [0, \infty)$ with right orientation.

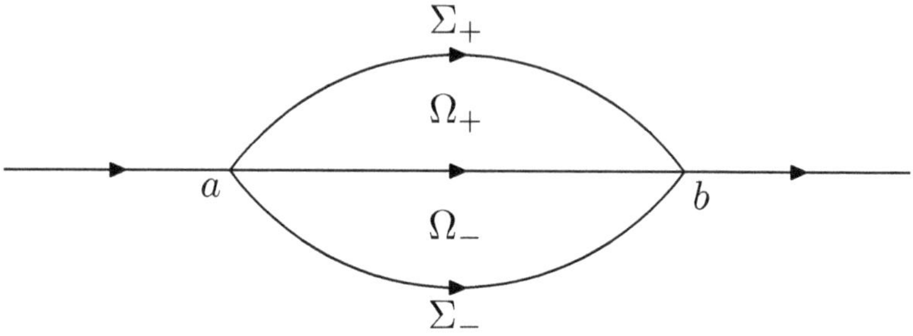

Fig. 12.2: The lens-shaped regions $\Omega = \Omega_+ \cup \Omega_-$ and the contours $\Sigma = \Sigma_+ \cup \Sigma_-$.

(V1) $V(z)$ is analytic in $\mathbf{C} \setminus (\Sigma \cup [0,\infty))$.
(V2) $V(z)$ satisfies the jump conditions

$$V_+(z) = V_-(z) \begin{pmatrix} 1 & 0 \\ e^{2N\phi(z)-\psi(z)} & 1 \end{pmatrix} \tag{12.37}$$

for $z \in \Sigma$, and

$$V_+(x) = V_-(x) \begin{pmatrix} 0 & 1 \\ -1 & 0 \end{pmatrix} \tag{12.38}$$

for $x \in (a,b)$, and

$$V_+(x) = V_-(x) \begin{pmatrix} 1 & e^{-2N\phi(x)+\psi(x)} \\ 0 & 1 \end{pmatrix} \tag{12.39}$$

for $x \in (b,\infty)$, and

$$V_+(x) = V_-(x) \begin{pmatrix} 1 & e^{-2N\phi_+(x)+\psi_+(x)} \\ 0 & 1 \end{pmatrix} \tag{12.40}$$

for $x \in (0,a)$.
(V3) As $z \to \infty$, $V(z) = I + O(|z|^{-1})$.
(V4) As $z \to 0$, $V(z) = O(1)$.

12.4 Construction of parametrices

In view of Riemann-Hilbert problem (M1)-(M3) for the parametrix $M(z)$ explicitly given in (8.7), we can formally approximate $T(z) \sim V(z) \sim M(z)$. Recall that $\tau_0 = 1/2$. The formal approximation in (8.52) is rewritten as

$$\begin{aligned} Y(z) \sim & \sqrt{\pi} e^{(l_n/2)\sigma_3} M(z) e^{\psi(z)\sigma_3/2} \begin{pmatrix} 1 & -1 \\ -i & -i \end{pmatrix} \\ & \times f(z)^{\sigma_3/4} A(f(z)) w(z)^{-\sigma_3/2}, \end{aligned} \tag{12.41}$$

where

$$f(z) := \left[\frac{3}{2}N\phi(z)\right]^{2/3}, \quad z \in \mathbf{C} \setminus (-\infty, b], \tag{12.42}$$

and A is the Airy parametrix defined in (8.8). One can prove that $f(z)$ can be analytically continued to the interval $(a, b]$ (exercise). The right-hand side of (12.41) suggests a parametrix in a large complex region including b. To find a parametrix in a neighborhood of a, we shall introduce the $\tilde{\phi}$-function

$$\tilde{\phi}(z) := \begin{cases} \phi(z) + \pi i, & z \in \mathbf{C}_+, \\ \phi(z) - \pi i, & z \in \mathbf{C}_-, \end{cases} \tag{12.43}$$

and define

$$\tilde{f}(z) := \left[\frac{3}{2}N\tilde{\phi}(z)\right]^{2/3}, \quad z \in \mathbf{C} \setminus (-\infty, 0] \cup [a, \infty). \tag{12.44}$$

Here, we have made use of (12.25) and (12.28) to show that $\phi_+(x) - \phi_-(x) = -2\pi i$ for $0 < x < a$, and hence, $\tilde{\phi}(z)$ is analytic for $z \in \mathbf{C} \setminus (-\infty, 0] \cup [a, \infty)$. Similar as in (12.30), we obtain the following integral representation (exercise)

$$\tilde{\phi}(z) = \int_a^z \frac{\kappa}{\zeta} \ln \frac{\zeta + \sqrt{ab} + (\zeta - a)^{1/2}(\zeta - b)^{1/2}}{(\sqrt{a} + \sqrt{b})\zeta^{1/2}} d\zeta, \tag{12.45}$$

for $z \in \mathbf{C} \setminus (-\infty, 0] \cup [a, \infty)$. If $x \in (a, b)$, then we have (exercise)

$$\tilde{\phi}_\pm(x) = \pm\pi i \int_a^x \rho(t)dt, \tag{12.46}$$

where ρ is the density function of the equilibrium measure explicitly given in (12.23). Therefore, $\tilde{f}(z)$ can be analytically continued to the interval $[a, b)$ (exercise). Set

$$\tilde{\psi}(z) := -2 \ln \frac{\sqrt{a - z} + \sqrt{b - z}}{\sqrt{b - a}} = \begin{cases} \psi(z) + \pi i, & z \in \mathbf{C}_+, \\ \psi(z) - \pi i, & z \in \mathbf{C}_-. \end{cases} \tag{12.47}$$

The formal asymptotic formula (8.57) can be rewritten as

$$\begin{aligned} Y(z) \sim &(-1)^n \sqrt{\pi} e^{(l_n/2)\sigma_3} M(z) e^{\tilde{\psi}(z)\sigma_3/2} \sigma_3 \begin{pmatrix} 1 & -1 \\ -i & -i \end{pmatrix} \\ &\times \tilde{f}(z)^{\sigma_3/4} A(\tilde{f}(z)) \sigma_3 w(z)^{-\sigma_3/2}. \end{aligned} \tag{12.48}$$

As in (8.58), the two formal asymptotic formulas (12.41) and (12.48) suggest us to define the following parametrix

$$Y^*(z) := \begin{cases} E(z)A(f(z))w(z)^{-\sigma_3/2}, & \operatorname{Re} z > c, \\ (-1)^n\sigma_3\tilde{E}(z)A(\tilde{f}(z))w(z)^{-\sigma_3/2}\sigma_3, & \operatorname{Re} z < c, \end{cases} \tag{12.49}$$

where $c = \sqrt{ab} \in (a, b)$, and

$$E(z) :=\sqrt{\pi}e^{(l_n/2)\sigma_3}M(z)e^{\psi(z)\sigma_3/2}\begin{pmatrix}1 & -1 \\ -i & -i\end{pmatrix}f(z)^{\sigma_3/4}, \tag{12.50}$$

$$\tilde{E}(z) :=\sqrt{\pi}e^{(l_n/2)\sigma_3}\sigma_3M(z)\sigma_3e^{\tilde{\psi}(z)\sigma_3/2}\begin{pmatrix}1 & -1 \\ -i & -i\end{pmatrix}\tilde{f}(z)^{\sigma_3/4}. \tag{12.51}$$

One can prove that $f'(x) > 0$ for $x > a$ and $\tilde{f}'(x) < 0$ for $0 < x < b$ (exercise). Hence, when z approaches $x \in (a, \infty)$ from the upper (lower) half-plane, $f(z)$ approaches $f(x)$ from the upper (lower) half-plane; when z approaches $x \in (0, b)$ from the upper (lower) half-plane, $\tilde{f}(z)$ approaches $\tilde{f}(x)$ from the lower (upper) half-plane. It is readily seen from the jump condition (8.10) that

$$Y^*_+(x) = Y^*_-(x)\begin{pmatrix}1 & w(x) \\ 0 & 1\end{pmatrix}, \quad x \in (0, c) \cup (c, \infty). \tag{12.52}$$

Finally, we define

$$S(z) := e^{-(l_n/2)\sigma_3}Y(z)[Y^*(z)]^{-1}e^{(l_n/2)\sigma_3}, \tag{12.53}$$

and note that $S(z)$ satisfies the following Riemann-Hilbert problem on the vertical line $\operatorname{Re} z = c$ with upward orientation and on the negative real line with right orientation (exercise).

(S1) $S(z)$ is analytic in $\operatorname{Re} z \neq c$ and $z \notin (-\infty, 0]$.

(S2) $S(z)$ satisfies the jump condition

$$S_+(z) = S_-(z)J_S(z) \tag{12.54}$$

for $\operatorname{Re} z = c$ or $z \in (-\infty, 0)$, where

$$J_S(z) = (-1)^n e^{-\frac{1}{2}l_n\sigma_3}E(z)A(f(z))\sigma_3[A(\tilde{f}(z))]^{-1}\tilde{E}(z)^{-1}\sigma_3e^{\frac{1}{2}l_n\sigma_3} \tag{12.55}$$

for $\operatorname{Re} z = c$, and

$$J_S(x) =M(x)\sigma_3e^{\tilde{\psi}(x)\sigma_3/2}\begin{pmatrix}1 & -1 \\ -i & -i\end{pmatrix}K(x) \times \begin{pmatrix}1 & -1 \\ -i & -i\end{pmatrix}^{-1}e^{-\tilde{\psi}(x)\sigma_3/2}\sigma_3[M(x)]^{-1} \tag{12.56}$$

with

$$K(x) := \tilde{f}_+(x)^{\sigma_3/4} A(\tilde{f}_+(x)) \left[\frac{w_-(x)}{w_+(x)}\right]^{\sigma_3/2} [A(\tilde{f}_-(x))]^{-1} \tilde{f}_-(x)^{-\sigma_3/4} \tag{12.57}$$

for $x \in (-\infty, 0)$.

(S3) As $z \to \infty$, $S(z) = I + O(|z|^{-1})$.

(S4) As $z \to 0$, $S(z) = O(1)$.

The jump matrix $J_S(z)$ is asymptotically equal to the identity matrix as the polynomial degree n tends to infinity.

Lemma 12.1. *For* $\operatorname{Re} z = c$ *or* $z \in (-\infty, 0)$, *we have* $\lim_{n\to\infty} J_S(z) = I$.

Proof. Assume $\operatorname{Re} z = c$ and $n \to \infty$. We obtain from (8.14) and (12.42) that

$$2\sqrt{\pi} f(z)^{\sigma_3/4} A(f(z)) \sim \begin{pmatrix} 1 & i \\ -1 & i \end{pmatrix} e^{-N\phi(z)\sigma_3}.$$

This together with (12.50) yields

$$e^{-\frac{1}{2}l_n\sigma_3} E(z) A(f(z)) \sim M(z) e^{[\psi(z)/2 - N\phi(z)]\sigma_3},$$

where we have made use of the identity

$$\begin{pmatrix} 1 & -1 \\ -i & -i \end{pmatrix} \begin{pmatrix} 1 & i \\ -1 & i \end{pmatrix} = \begin{pmatrix} 2 & 0 \\ 0 & 2 \end{pmatrix}.$$

In a similar manner, we obtain from (8.14), (12.44) and (12.51) that

$$e^{-\frac{1}{2}l_n\sigma_3} \tilde{E}(z) A(\tilde{f}(z)) \sim \sigma_3 M(z) \sigma_3 e^{[\tilde{\psi}(z)/2 - N\tilde{\phi}(z)]\sigma_3}.$$

Substituting the above asymptotic formulas into (12.55) gives

$$J_S(z) \sim (-1)^n M(z) e^{[\psi(z)/2 - N\phi(z)]\sigma_3} e^{-[\tilde{\psi}(z)/2 - N\tilde{\phi}(z)]\sigma_3} [M(z)]^{-1}.$$

It then follows from $N = n + 1/2$, (12.43) and (12.47) that

$$[\frac{\psi(z)}{2} - N\phi(z)] - [\frac{\tilde{\psi}(z)}{2} - N\tilde{\phi}(z)] = N[\tilde{\phi}(z) - \phi(z)] - \frac{1}{2}[\tilde{\psi}(z) - \psi(z)] = \pm n\pi i,$$

and

$$J_S(z) \sim (-1)^n M(z) e^{\pm n\pi i\sigma_3} [M(z)]^{-1} = I.$$

Now, we assume $z = x \in (-\infty, 0)$ and $n \to \infty$. It again follows from (8.14) and (12.44) that

$$2\sqrt{\pi} \tilde{f}_\pm(x)^{\sigma_3/4} A(\tilde{f}_\pm(x)) \sim \begin{pmatrix} 1 & i \\ -1 & i \end{pmatrix} e^{-N\tilde{\phi}_\pm(x)\sigma_3}.$$

Note from (12.25) and (12.43) that

$$\tilde{\phi}_\pm(x) = \phi_\pm(x) \pm \pi i = \frac{v_\pm(x) + l}{2} - g_\pm(x) \pm \pi i.$$

This together with (12.8) implies

$$e^{-N\tilde{\phi}_\pm(x)}[w_\pm(x)]^{-1/2} = e^{-Nl/2 + Ng_\pm(x) \mp N\pi i}.$$

Recall from (12.28) that $g_+(x) - g_-(x) = 2\pi i$. We further obtain

$$e^{-N\tilde{\phi}_+(x)} \left[\frac{w_-(x)}{w_+(x)}\right]^{1/2} e^{N\tilde{\phi}_+(x)} = e^{N[g_+(x) - g_-(x)] - 2N\pi i} = 1.$$

Substituting the above formulas into (12.57) gives

$$K(x) \sim \begin{pmatrix} 1 & i \\ -1 & i \end{pmatrix} e^{-N\tilde{\phi}_+(x)\sigma_3} \left[\frac{w_-(x)}{w_+(x)}\right]^{\sigma_3/2} e^{N\tilde{\phi}_+(x)\sigma_3} \begin{pmatrix} 1 & i \\ -1 & i \end{pmatrix}^{-1} = I.$$

We then obtain from (12.56) that $J_S(x) \to I$ as $n \to \infty$. This completes the proof. □

An extension of the above lemma gives the following asymptotic expansion

$$J_S(z) \sim I + \sum_{k=1}^{\infty} \frac{J_{S,k}(z)}{N^k}, \tag{12.58}$$

as $n \to \infty$. Consequently, the solution of the Riemann-Hilbert problem (S1)-(S4) has the asymptotic expansion

$$S(z) \sim I + \sum_{k=1}^{\infty} \frac{S_k(z)}{N^k}, \tag{12.59}$$

as $n \to \infty$, where $S_k(z)$ can be determined recursively by

$$S_k(z) = \frac{1}{2\pi i} \int_{\{\operatorname{Re}\zeta = c\} \cup \mathbf{R}_-} \sum_{j=1}^{k} (S_{k-j})_-(\zeta) J_{S,j}(\zeta) \frac{d\zeta}{\zeta - z}, \quad k = 1, 2, \cdots, \tag{12.60}$$

for $\operatorname{Re} z \neq c$ and $z \notin \mathbf{R}_-$.

12.5 Uniform asymptotic expansions

In view of (8.7), (12.34) and (12.47), we may rewrite (12.50) and (12.51) as

$$E(z) = \sqrt{\pi} e^{(l_n/2)\sigma_3} \begin{pmatrix} 1 & 0 \\ \frac{-i(2z-a-b)}{b-a} & -2i \end{pmatrix} \left[\frac{(b-a)^2 f(z)}{(z-a)(z-b)}\right]^{\sigma_3/4}, \tag{12.61}$$

$$\tilde{E}(z) = \sqrt{\pi} e^{(l_n/2)\sigma_3} \begin{pmatrix} 1 & 0 \\ \frac{-i(a+b-2z)}{b-a} & -2i \end{pmatrix} \left[\frac{(b-a)^2 \tilde{f}(z)}{(a-z)(b-z)}\right]^{\sigma_3/4}. \tag{12.62}$$

We state our main theorem below.

Theorem 12.1. *Let $S(z)$ be the solution to the Riemann-Hilbert problem (S1)-(S4) with the asymptotic expansion in* (12.59). *Let $\pi_n(z) = Y_{11}(z)$ be the monic Stieltjes-Wigert polynomial. If* $\operatorname{Re} z > c$, *we have*

$$\pi_n(z) = \sqrt{\frac{\pi e^{l_n}}{w(z)}}\bigg\{ \operatorname{Ai}(f) \left[\frac{(b-a)^2 f(z)}{(z-a)(z-b)}\right]^{1/4} \left[S_{11}(z) - \frac{i(2z-a-b)}{b-a} S_{12}(z)\right] \\ - 2i \operatorname{Ai}'(f) \left[\frac{(z-a)(z-b)}{(b-a)^2 f(z)}\right]^{1/4} S_{12}(z)\bigg\}. \tag{12.63}$$

If $\operatorname{Re} z < c$ *and* $z \notin (-\infty, 0]$, *we have*

$$\pi_n(z) = (-1)^n \sqrt{\frac{\pi e^{l_n}}{w(z)}}\bigg\{ \operatorname{Ai}(\tilde{f}) \left[\frac{(b-a)^2 \tilde{f}(z)}{(a-z)(b-z)}\right]^{1/4} \\ \times \left[S_{11}(z) + \frac{i(a+b-2z)}{b-a} S_{12}(z)\right] \\ + 2i \operatorname{Ai}'(\tilde{f}) \left[\frac{(a-z)(b-z)}{(b-a)^2 \tilde{f}(z)}\right]^{1/4} S_{12}(z)\bigg\}. \tag{12.64}$$

Here, l_n is the modified Lagrange multiplier defined in (12.29), *and f and $\tilde{f}$ are defined in* (12.42) *and* (12.44), *respectively.*

Proof. We only prove the case $\operatorname{Re} z > c$. The other case is left as an exercise. Note from (12.49) and (12.53) that

$$Y(z) = e^{(l_n/2)\sigma_3} S(z) e^{-(l_n/2)\sigma_3} E(z) A(f(z)) w(z)^{-\sigma_3/2}.$$

Substituting (12.61) into the above formula gives

$$Y(z) = \sqrt{\pi} e^{(l_n/2)\sigma_3} S(z) \begin{pmatrix} 1 & 0 \\ \frac{-i(2z-a-b)}{b-a} & -2i \end{pmatrix} \left[\frac{(b-a)^2 f(z)}{(z-a)(z-b)}\right]^{\sigma_3/4} \\ \times A(f(z)) w(z)^{-\sigma_3/2}.$$

Denote by

$$B := S(z) \begin{pmatrix} 1 & 0 \\ \frac{-i(2z-a-b)}{b-a} & -2i \end{pmatrix} \left[\frac{(b-a)^2 f(z)}{(z-a)(z-b)}\right]^{\sigma_3/4}.$$

We then have

$$\pi_n(z) = Y_{11}(z) = \sqrt{\pi} e^{l_n/2} [B_{11} A_{11}(f) + B_{12} A_{21}(f)] w(z)^{-1/2}.$$

Recall from (8.8) that $A_{11}(f) = \mathrm{Ai}(f)$ and $A_{21}(f) = \mathrm{Ai}'(f)$. On the other hand, it is easy to calculate

$$B_{11} = \left[\frac{(b-a)^2 f(z)}{(z-a)(z-b)}\right]^{1/4} \left[S_{11}(z) - \frac{i(2z-a-b)}{b-a} S_{12}(z)\right],$$

and

$$B_{12} = -2i \left[\frac{(z-a)(z-b)}{(b-a)^2 f(z)}\right]^{1/4} S_{12}(z).$$

A combination of the above formulas gives (12.63). □

We remark that the above result is valid when k is unbounded as $n \to \infty$; see (12.7). The case when k is fixed and $n \to \infty$ has been studied in the literature; see [Wigert (1923); Ismail (2005); Wang and Wong (2007); Li and Wong (2013)], where q-Airy function was used in the asymptotic formula. Furthermore, Li and Wong (2013) derived a limiting relation between Airy function and q-Airy function as $q \to 1$.

12.6 Exercises

1. Prove (12.12) and (12.13).
2. Verify (12.14) and (12.15).
3. Prove (12.19) and (12.20).
4. Verify (12.21).
5. Verify (12.23).
6. Prove (12.27) and (12.28).
7. Prove (12.29).
8. Verify (12.30).
9. Verify (12.31).
10. Prove that $f(z)$ defined in (12.42) can be analytically continued to the interval $(a, b]$.
11. Verify (12.45) and (12.46).
12. Prove that $\tilde{f}(z)$ defined in (12.44) can be analytically continued to the interval $[a, b)$.
13. Let f and $\tilde{f}$ be defined in (12.42) and (12.44), respectively. Prove that $f'(x) > 0$ for $x > a$ and $\tilde{f}'(x) < 0$ for $0 < x < b$.
14. Verify that $S(z)$ defined in (12.53) is the solution to the Riemann-Hilbert problem (S1)-(S4).
15. Verify (12.61) and (12.62).
16. Prove (12.64).

Part 3
Differential Equations

Chapter 13

Asymptotics of Differential Equations with Irregular Singularities

13.1 Classification of singularities

The material in this chapter is taken from Olver (1974a, Chapters 5 and 7). We consider the linear differential equation

$$\frac{d^2w}{dz^2} + f(z)\frac{dw}{dz} + g(z)w = 0, \tag{13.1}$$

where $f(z)$ and $g(z)$ are analytic functions in a neighborhood of infinity; namely, the following series expansions

$$f(z) = \sum_{s=0}^{\infty} f_s z^{-s}, \quad g(z) = \sum_{s=0}^{\infty} g_s z^{-s} \tag{13.2}$$

converge for $|z| > R$, where $R \geq 0$ is a fixed number. A simple change of variable $z = 1/t$ transforms the equation (13.1) into (exercise)

$$\frac{d^2w}{dt^2} + p(t)\frac{dw}{dt} + q(t)w = 0, \tag{13.3}$$

where

$$p(t) = 2z - z^2 f(z), \quad q(t) = z^4 g(z). \tag{13.4}$$

The singularity of (13.1) at infinity (resp. (13.3) at $t = 0$) can be classified as follows.

(1) If both $p(t)$ and $q(t)$ are analytic at $t = 0$, then $t = 0$ (resp. $z = \infty$) is an *ordinary point* for (13.3) (resp. (13.1)). Otherwise, $t = 0$ (resp. $z = \infty$) is a *singular point* for (13.3) (resp. (13.1)).

(2) Assume $t = 0$ is a singular point for (13.3) (i.e., either $p(t)$ or $q(t)$ is not analytic at $t = 0$). If both $tp(t)$ and $t^2q(t)$ are analytic at $t = 0$ (resp. both $zf(z)$ and $z^2g(z)$ are analytic at $z = \infty$), then the singularity of (13.3) at $t = 0$ (resp. (13.1) at $z = \infty$) is *regular*.

(3) If either $tp(t)$ or $t^2q(t)$ is not analytic at $t=0$ (resp. either $zf(z)$ or $z^2g(z)$ is not analytic at $z=\infty$), then we say (13.3) (resp. (13.1)) has an *irregular singularity* at $t=0$ (resp. $z=\infty$).

The first two cases (no singularity or regular singularity) are easier to investigate because one can express the solutions in terms of convergent series expansions. The main purpose of this chapter is to consider the third case and derive two linearly independent asymptotic solutions of (13.1) as $z\to\infty$. However, in the rest of this section, we shall divert ourselves to the first two cases and briefly explain how to obtain the convergent series expansions of the solutions. For the first case when both $p(t)$ and $q(t)$ are analytic at $t=0$. We let

$$p(t)=\sum_{s=0}^{\infty}p_st^s, \qquad q(t)=\sum_{s=0}^{\infty}q_st^s, \tag{13.5}$$

and find the solution

$$w(t)=\sum_{s=0}^{\infty}w_st^s. \tag{13.6}$$

Substituting (13.5) and (13.6) into (13.3) yields the recurrence relation (exercise)

$$(s+2)(s+1)w_{s+2}+\sum_{k=0}^{s}(k+1)w_{k+1}p_{s-k}+\sum_{k=0}^{s}w_kq_{s-k}=0, \tag{13.7}$$

for $s=0,1,\cdots$, where w_0 and w_1 are arbitrary constants. The following lemma will be used to prove the convergence of the series in (13.6).

Lemma 13.1. *Let w_s with $s=0,1,2,\cdots$ be an infinite sequence satisfying the recurrence inequalities*

$$Q(s)|w_{s+1}|\le\sum_{k=0}^{s}L(k)|w_k|$$

for $s\ge n$ and some given integer $n\ge 0$, where $Q(s)>0$ for all $s\ge n$, and $L(k)\ge 0$ for all $k\ge 0$. If $Q(s)/Q(s-1)\to\rho>0$ and $L(s)/Q(s)\to 0$ as $s\to\infty$, then the series expansion

$$\sum_{s=0}^{\infty}w_st^s$$

converges for all $|t|<\rho$.

Proof. Let W_s be defined as $W_0 = |w_0|, \cdots, W_n = |w_n|$, and

$$Q(s)W_{s+1} = \sum_{k=0}^{s} L(k)W_k, \quad s \geq n.$$

By induction, we obtain $|w_s| \leq W_s$ for all $s \geq 0$. On the other hand, it follows from the above recurrence relation that (exercise)

$$Q(s)W_{s+1} - Q(s-1)W_s = L(s)W_s, \quad s \geq n+1. \tag{13.8}$$

In particular,

$$\frac{W_{s+1}}{W_s} = \frac{Q(s-1) + L(s)}{Q(s)} \to \rho^{-1}$$

as $s \to \infty$. This together with $|w_s| \leq W_s$ completes the proof. □

Proposition 13.1. *Assume $p(t)$ and $q(t)$ are analytic for $|t| < r$ for some $r > 0$. For any given w_0 and w_1, the series expansion of $w(t)$ in* (13.6) *with coefficients defined recursively by* (13.7) *is also analytic for $|t| < r$.*

Proof. Given any fixed $\rho \in (0, r)$, there exists $M > 0$ such that $|p(t)| \leq M$ and $|tq(t)| \leq M$ for all $|t| \leq \rho$. By Cauchy integral formula, we obtain (exercise)

$$|p_s| \leq M\rho^{-s}, \quad |q_s| \leq M\rho^{-s-1}, \quad s = 0, 1, \cdots. \tag{13.9}$$

It then follows from (13.7) that (exercise)

$$(s+2)(s+1)|w_{s+2}| \leq M \sum_{k=0}^{s+1} (k+1)|w_k|\rho^{-s+k-1}, \quad s = 0, 1, \cdots. \tag{13.10}$$

By choosing $n = 1$, $Q(s) = s(s+1)\rho^s$ and $L(k) = M(k+1)\rho^k$ in Lemma 13.1, one can show that the series expansion of $w(t)$ in (13.6) converges for all $|t| < \rho$. Since $\rho \in (0, r)$ is arbitrarily chosen, $w(t)$ is analytic for $|t| < r$. □

Next, we consider the second case when (13.3) has a regular singularity at $t = 0$; namely,

$$p(t) = \sum_{s=0}^{\infty} p_s t^{s-1}, \quad q(t) = \sum_{s=0}^{\infty} q_s t^{s-2}, \tag{13.11}$$

where at least one of p_0, q_0, and q_1 is nonzero. Ignoring the higher-order terms in the above expansions, we arrive at the Cauchy-Euler equidimensional equation:

$$\frac{d^2w}{dt^2} + \frac{p_0}{t}\frac{dw}{dt} + \frac{q_0}{t^2}w = 0. \tag{13.12}$$

Substituting $w = t^\alpha$ into this equation gives the following *indicial equation*:

$$\alpha(\alpha - 1) + p_0\alpha + q_0 = 0. \tag{13.13}$$

The solutions of the indicial equation α and $\tilde{\alpha}$ are called the *indices (exponents)* of the singularity. Without loss of generality, we assume $\mathrm{Re}(\alpha - \tilde{\alpha}) \geq 0$.

Proposition 13.2. *Let $tp(t)$ and $t^2q(t)$ be analytic functions in $|t| < r$ with some $r > 0$; namely $p(t)$ and $q(t)$ have the series expansions in* (13.11). *Assume that at least one of p_0, q_0, and q_1 is nonzero. Denote by α and $\tilde{\alpha}$ the roots of the indicial equation* (13.13) *with* $\mathrm{Re}(\alpha - \tilde{\alpha}) \geq 0$. *The differential equation* (13.3) *has a solution with the following series expansion*

$$w(t) = \sum_{s=0}^{\infty} w_s t^{\alpha+s}, \tag{13.14}$$

where $w_0 = 1$ and $w_1, w_2, \cdots$ are recursively determined by

$$(s+1)(2\alpha + s + p_0)w_{s+1} + \sum_{k=0}^{s} [(\alpha + k)p_{s+1-k} + q_{s+1-k}]w_k = 0, \tag{13.15}$$

for $s = 0, 1, \cdots$. Moreover, $t^{-\alpha}w(t)$ is analytic in $|t| < r$.

Proof. Substituting (13.14) into (13.3) gives (13.15). It suffices to show that for any given $\rho \in (0, r)$ the function $t^{-\alpha}w(t)$ is analytic in $|t| < \rho$. Similar to the proof of Proposition 13.1, we can find a large constant $M > 0$ such that $|p_k| \leq M\rho^{-k+1}$ and $|q_k| \leq M\rho^{-k+1}$ for any $k = 0, 1, \cdots$. For any $s > 2|\alpha| + |p_0|$, it follows from (13.15) that

$$(s+1)(s - 2|\alpha| - |p_0|)|w_{s+1}| \leq M \sum_{k=0}^{s} (|\alpha| + k + 1)|w_k|\rho^{-s+k}.$$

By choosing n sufficiently large, $Q(s) = (s+1)(s - 2|\alpha| - |p_0|)\rho^s$ and $L(k) = M(|\alpha| + k + 1)\rho^k$, we obtain from Lemma 13.1 that $t^{-\alpha}w(t)$ is analytic for $|t| < \rho$. The proof is complete, since $\rho \in (0, r)$ is arbitrary. □

To find another linearly independent solution of (13.3), we need to consider two subcases: (i) $\alpha - \tilde{\alpha} \notin \mathbf{Z}$, and (ii) $\alpha - \tilde{\alpha} \in \mathbf{Z}$. For the first subcase, one can show that $2\tilde{\alpha} + s + p_0 \neq 0$ for all nonnegative integers s and then use the same argument as in the proof of Proposition 13.2 to find another solution

$$\tilde{w}(t) = \sum_{s=0}^{\infty} \tilde{w}_s t^{\tilde{\alpha}+s},$$

where $\tilde{w}_0 = 1$ and $\tilde{w}_1, \tilde{w}_2, \cdots$ are recursively determined by

$$(s+1)(2\tilde{\alpha}+s+p_0)\tilde{w}_{s+1} + \sum_{k=0}^{s}[(\tilde{\alpha}+k)p_{s+1-k} + q_{s+1-k}]\tilde{w}_k = 0,$$

for $s = 0, 1, \cdots$. Moreover, $t^{-\tilde{\alpha}}\tilde{w}(t)$ is analytic for $|t| < r$.

Finally, we consider the second subcase when $\alpha - \tilde{\alpha} \in \mathbf{Z}$. Recall that we have assumed $\operatorname{Re}(\alpha - \tilde{\alpha}) \geq 0$. Hence, $n := \alpha - \tilde{\alpha}$ is a nonnegative integer. Let $w(t)$ be the solution given in Proposition 13.2. We shall find another linearly independent solution as follows:

$$\tilde{w}(t) = \sum_{s=0}^{\infty} \tilde{w}_s t^{\tilde{\alpha}+s} + cw(t)\ln t. \tag{13.16}$$

Substituting this into (13.3) and making use of the fact that $w(t)$ is a solution of (13.3), we obtain after a simple calculation (exercise)

$$\begin{aligned} &\sum_{s=0}^{\infty}\left\{s(s-n)\tilde{w}_s + \sum_{k=0}^{s-1}[(\tilde{\alpha}+k)p_{s-k} + q_{s-k}]\tilde{w}_k\right\}t^{\tilde{\alpha}+s-2} \\ &\qquad + c\sum_{s=0}^{\infty}\left\{(2s+n)w_s + \sum_{k=0}^{s-1}p_{s-k}w_k\right\}t^{\alpha+s-2} = 0. \end{aligned} \tag{13.17}$$

Comparing the coefficients of $t^{\tilde{\alpha}+s-2}$ gives the recurrence relations

$$s(s-n)\tilde{w}_s + \sum_{k=0}^{s-1}[(\tilde{\alpha}+k)p_{s-k} + q_{s-k}]\tilde{w}_k = 0, \tag{13.18}$$

for $s = 0, 1, \cdots, n-1$, and

$$\begin{aligned} &s(s-n)\tilde{w}_s + \sum_{k=0}^{s-1}[(\tilde{\alpha}+k)p_{s-k} + q_{s-k}]\tilde{w}_k \\ &\qquad + c\left\{(2s-n)w_{s-n} + \sum_{k=0}^{s-n-1}p_{s-n-k}w_k\right\} = 0, \end{aligned} \tag{13.19}$$

for $s = n, n+1, \cdots$. If $n = 0$, then $\tilde{\alpha} = \alpha = (1-p_0)/2$. We choose $\tilde{w}_0 = 0$ and $c = 1$. The coefficients $\tilde{w}_1, \tilde{w}_2, \cdots$ can be determined recursively from (13.19) with $s = 1, 2, \cdots$, respectively. If $n > 0$, then we choose $\tilde{w}_0 = 1$ and $\tilde{w}_n = 0$. The coefficients $\tilde{w}_1, \cdots, \tilde{w}_{n-1}$ can be determined recursively from (13.18) with $s = 1, \cdots, n-1$, respectively. Next, we solve c from the equation (13.19) with $s = n$; namely, we have (exercise)

$$c = -\frac{1}{n}\sum_{k=0}^{n-1}[(\tilde{\alpha}+k)p_{n-k} + q_{n-k}]\tilde{w}_k. \tag{13.20}$$

Then, the coefficients $\tilde{w}_{n+1}, \cdots$ can be determined recursively from (13.19) with $s = n+1, \cdots$, respectively.

Proposition 13.3. *In addition to the conditions in Proposition 13.2, we assume that $n := \alpha - \tilde{\alpha}$ is a nonnegative integer. The differential equation* (13.3) *possesses two linearly independent solutions $w(t)$ and $\tilde{w}(t)$, where $w(t)$ is given as in Proposition 13.2, and $\tilde{w}(t)$ is defined in* (13.16) *with coefficients determined by the recurrence relations* (13.18)-(13.19) *and the following initial choices: $\tilde{w}_0 = 0$ and $c = 1$ if $n = 0$, and $\tilde{w}_0 = 1$ and $\tilde{w}_n = 0$ if $n > 0$. Moreover, the series expansion*

$$t^{-\tilde{\alpha}}[\tilde{w}(t) - cw(t)\ln t] = \sum_{s=0}^{\infty} \tilde{w}_s t^s$$

is analytic for $|t| < r$.

Proof. We fix any $\rho \in (0, r)$ and use Cauchy integral formula to find a large constant M such that the coefficients p_k, q_k, and w_k are bounded in absolute value by $M\rho^{-k}$ for all $k = 0, 1, \cdots$. Now, if $s > n$, we obtain from (13.19) that (exercise)

$$s(s-n)|\tilde{w}_s| \leq \sum_{k=0}^{s-1} M(|\tilde{\alpha}|+k+1)|\tilde{w}_k|\rho^{-s+k} + cM\rho^{-s+n}[(2s-n)+(s-n)M]. \tag{13.21}$$

Let $\tilde{W}_s = |\tilde{w}_s|\rho^s$ for $s = 0, \cdots, n$, and

$$s(s-n)\tilde{W}_s = \sum_{k=0}^{s-1} M(|\tilde{\alpha}| + k + 1)\tilde{W}_k + cM(M+2)\rho^n s$$

for $s = n+1, \cdots$. By induction, we obtain $|\tilde{w}_s|\rho^s \leq \tilde{W}_s$ for all $s \geq 0$. On the other hand, the above recurrence relation implies

$$s(s-n)\tilde{W}_s - (s-1)(s-n-1)\tilde{W}_{s-1} = M(|\tilde{\alpha}| + s)\tilde{W}_{s-1} + cM(M+2)\rho^n.$$

Denote $V_s := s(s-n)\tilde{W}_s$. For any $\varepsilon > 0$, there exists $N > 0$ such that

$$V_s - V_{s-1} = \frac{M(|\tilde{\alpha}| + s)V_{s-1} + cM(M+2)\rho^n}{(s-1)(s-n-1)} \leq \varepsilon V_{s-1} + \varepsilon^2,$$

for all $s \geq N$. Hence, we obtain

$$V_s + \varepsilon \leq (1+\varepsilon)(V_{s-1} + \varepsilon), \quad s \geq N.$$

Since $\varepsilon > 0$ is arbitrary, we then have

$$\limsup_{s\to\infty} \frac{\tilde{W}_s}{\tilde{W}_{s-1}} = \limsup_{s\to\infty} \frac{V_s}{V_{s-1}} \leq 1,$$

which implies that the series expansion $W_0 + W_1 t + W_2 t^2 + \cdots$ converges for all $|t| < 1$. This together with $|\tilde{w}_s|\rho^s \leq W_s$ for all $s \geq 0$ yields the desired result. $\square$

13.2 Formal asymptotic expansions

Throughout this section, we assume that (13.1) has an irregular singularity at $z = \infty$; namely, either $zf(z)$ or $z^2g(z)$ is not analytic at $z = \infty$. Recall the series expansions of $f(z)$ and $g(z)$ in (13.2). The coefficients f_0, g_0, and g_1 cannot all be zero. We look for solutions with the following asymptotic expansions

$$w(z) \sim e^{\lambda z} \sum_{s=0}^{\infty} w_s z^{\mu - s}, \quad w_0 = 1. \tag{13.22}$$

Substituting this and (13.2) into (13.1), we obtain the following recurrence relations by equating the coefficients of $e^{\lambda z} z^{\mu - s}$ (exercise):

$$\begin{aligned} \lambda^2 w_s &+ 2\lambda(\mu - s + 1)w_{s-1} + (\mu - s + 2)(\mu - s + 1)w_{s-2} \\ &+ \sum_{k=0}^{s} (\lambda f_{s-k} + g_{s-k})w_k + \sum_{k=0}^{s} (\mu - k + 1) f_{s-k} w_{k-1} = 0, \end{aligned} \tag{13.23}$$

for $s = 0, 1, \cdots$, where we have set $w_{-1} = w_{-2} = 0$ for convenience. When $s = 0$, the above equation gives the *characteristic equation*:

$$\lambda^2 + f_0\lambda + g_0 = 0. \tag{13.24}$$

Temporarily, we assume that this equation has two distinct *characteristic roots*: λ and $\tilde{\lambda}$. Setting $s = 1$ in (13.23) gives

$$(2\lambda + f_0)\mu = -(\lambda f_1 + g_1), \tag{13.25}$$

which determines μ because the coefficient $2\lambda + f_0 \neq 0$. The coefficients $w_1, w_2, \cdots$ can be determined recursively from (13.23) with $s = 2, 3, \cdots$. In a similar manner, we find another linearly independent solution with the asymptotic expansion

$$\tilde{w}(z) \sim e^{\tilde{\lambda} z} \sum_{s=0}^{\infty} \tilde{w}_s z^{\tilde{\mu} - s}, \quad \tilde{w}_0 = 1, \tag{13.26}$$

where $\tilde{\mu} = -(\tilde{\lambda} f_1 + g_1)/(2\tilde{\lambda} + f_0)$ and

$$\begin{aligned} (2\tilde{\lambda} + f_0)(s-1)\tilde{w}_{s-1} =& (\tilde{\mu} - s + 2)(\tilde{\mu} - s + 1)\tilde{w}_{s-2} + \sum_{k=0}^{s-2} (\tilde{\lambda} f_{s-k} + g_{s-k})\tilde{w}_k \\ &+ \sum_{k=0}^{s-1} (\tilde{\mu} - k + 1) f_{s-k} \tilde{w}_{k-1}, \end{aligned} \tag{13.27}$$

for $s = 2, 3, \cdots$.

Now, we consider the critical case when the two characteristic roots of (13.24) are the same: $\lambda = \tilde{\lambda} = -f_0/2$. In this case, we have $f_0^2 = 4g_0$. We shall make use of Fabry's transformation:

$$Z = z^{1/2}, \quad W(Z) = e^{f_0 z/2} w(z). \tag{13.28}$$

Equation (13.1) becomes (exercise)

$$\frac{d^2W}{dZ^2} + F(Z)\frac{dW}{dZ} + G(Z)W = 0, \tag{13.29}$$

where

$$F(Z) = (2f_1 - 1)Z^{-1} + \sum_{s=2}^{\infty} 2f_s Z^{1-2s}, \tag{13.30}$$

$$G(Z) = 4g_1 - 2f_0 f_1 + \sum_{s=2}^{\infty} (4g_s - 2f_0 f_s) Z^{2-2s}. \tag{13.31}$$

If $4g_1 - 2f_0 f_1 = 0$, then (13.29) has at most a regular singularity at $Z = \infty$ and we can find two linearly independent solutions as in the previous section. If $4g_1 - 2f_0 f_1 \neq 0$, then the characteristic equation for (13.29) has two distinct roots. Hence, the critical case does not require a special theory after Fabry's transformation. From now on, we assume without loss of generality that the characteristic equation (13.24) has two distinct characteristic roots: λ and $\tilde{\lambda}$.

13.3 Error bounds

We assume $f_0^2 \neq 4g_0$ and denote by λ and $\tilde{\lambda}$ the two distinct characteristic roots of the characteristic equation (13.24). The main objective of this section is to provide a rigorous error bound for the formal asymptotic solution $w(z)$ in (13.22). For each large integer n, we write $w(z) = L_n(z) + \varepsilon_n(z)$, where

$$L_n(z) = e^{\lambda z} \sum_{s=0}^{n-1} w_s z^{\mu - s}. \tag{13.32}$$

Recall that $w_0 = 1$ and $w_1, w_2, \cdots$ are determined recursively from (13.23). It is easily verified that (exercise)

$$\begin{aligned} R_n(z) :=& L_n''(z) + f(z)L_n'(z) + g(z)L_n(z) \\ =& e^{\lambda z}(\mu - n + 1)(\mu - n) w_{n-1} z^{\mu-n-1} \\ &+ e^{\lambda z} \sum_{s=n+1}^{\infty} \sum_{k=0}^{n-1} [\lambda f_{s-k} + (\mu - k) f_{s-k-1} + g_{s-k}] w_k z^{\mu - s}. \end{aligned} \tag{13.33}$$

In particular,

$$\lim_{z\to\infty}\frac{R_n(z)}{e^{\lambda z}z^{\mu-n-1}} =(\mu-n+1)(\mu-n)w_{n-1} + \sum_{k=0}^{n-1}[\lambda f_{n+1-k}+(\mu-k)f_{n-k}+g_{n+1-k}]w_k.$$

Therefore, $w(z) = L_n(z) + \varepsilon_n(z)$ is a solution to (13.1) if and only if

$$\begin{aligned}\varepsilon_n''(z)+f_0\varepsilon_n'(z)+g_0\varepsilon_n(z) =&R_n(z)+[f_0-f(z)]\varepsilon_n'(z)+[g_0-g(z)]\varepsilon_n(z)\\ =:&h_n(z,\varepsilon_n(z),\varepsilon_n'(z)).\end{aligned} \tag{13.34}$$

We intend to show that the above equation possesses a solution such that $\varepsilon_n(z)/(e^{\lambda z}z^{\mu-n}) = O(1)$ as $z \to \infty$. To this end, we shall consider the integral equation

$$\varepsilon_n(z) = \mathcal{K}_n(\varepsilon_n)(z) := \int_{\gamma_z} K(z,t)h_n(t,\varepsilon_n(t),\varepsilon_n'(t))dt, \tag{13.35}$$

where

$$K(z,t) = \frac{e^{\lambda(z-t)}-e^{\tilde{\lambda}(z-t)}}{\lambda-\tilde{\lambda}}, \tag{13.36}$$

and the path of integration γ_z connects z to ∞ in the sense that $\mathrm{Re}[(\tilde{\lambda}-\lambda)(t-z)] = 0$ for all $t \in \gamma_z$. To describe this integration path, we first introduce for each $c \geq 0$ and $\alpha \in \mathbf{R}$ the open complex domain

$$\Omega_c^\alpha := \{z = re^{i\theta}: \quad r > c, \quad -\pi < \theta + \alpha < \pi\}. \tag{13.37}$$

Denote $\phi := \arg(\tilde{\lambda} - \lambda) \in [-\pi, \pi]$. Recall that $f(z)$ and $g(z)$ are analytic for $|z| > R$. For any $z = re^{i\theta} \in \Omega_R^\phi$, we have $r > R$ and $\theta + \phi \in (-\pi, \pi)$. Now, we choose the integration path γ_z as

$$\gamma_z = \begin{cases}\{z+ie^{-i\phi}y: \quad y\geq 0\}, & \theta+\phi\in[0,\pi)\\ \{z-ie^{-i\phi}y: \quad y\geq 0\}, & \theta+\phi\in(-\pi,0).\end{cases} \tag{13.38}$$

It is readily seen (exercise) that $t \in \gamma_z$ implies $t \in \Omega_R^\phi$ and $\mathrm{Re}[e^{i\phi}(t-z)] = 0$. Furthermore, for any $a \geq 0$, we have (exercise)

$$\int_{\gamma_z} |t|^{-a-1}|dt| \leq \int_0^\infty (|z|^2+y^2)^{-(a+1)/2}dy = C_a|z|^{-a}, \tag{13.39}$$

where

$$C_a := \frac{\Gamma(a/2)\Gamma(1/2)}{2\Gamma(a/2+1/2)} = O(a^{-1/2}), \tag{13.40}$$

as $a \to \infty$.

Now, we set $\varepsilon_{n,0}(z) = 0$ and define $\varepsilon_{n,k}(z)$ recursively as

$$\varepsilon_{n,k}(z) = \mathcal{K}_n(\varepsilon_{n,k-1})(z), \quad k = 1, 2, \cdots.$$

We shall prove that

$$\varepsilon_n(z) := \sum_{k=1}^{\infty} [\varepsilon_{n,k}(z) - \varepsilon_{n,k-1}(z)] \tag{13.41}$$

converges in Ω_R^ϕ. Fix any $\rho > R$. It follows from (13.2) and (13.33) that $|z[f_0 - f(z)]| + |z[g_0 - g(z)]| \le A$ and $|R_n(z)/(e^{\lambda z} z^{\mu-n-1})| \le B_n$ for $z \in \Omega_\rho^\phi$. Moreover, we have from (13.36) that

$$|e^{\lambda(t-z)} K(z,t)| \le \frac{2}{|\lambda - \tilde{\lambda}|}, \quad |e^{\lambda(t-z)} \partial_z K(z,t)| \le \frac{|\lambda| + |\tilde{\lambda}|}{|\lambda - \tilde{\lambda}|}, \tag{13.42}$$

for $z \in \Omega_R^\phi$ and $t \in \gamma_z$. Hence, we can find a large constant $M > 0$ independent of n and z such that (exercise)

$$\begin{aligned} |\varepsilon_{n,1}(z)| + |\varepsilon_{n,1}'(z)| &\le B_n M |e^{\lambda z}| \int_{\gamma_z} |t|^{\operatorname{Re}\mu - n - 1} |dt| \\ &\le B_n M C_{n-\operatorname{Re}\mu} |z|^{\operatorname{Re}\mu - n} |e^{\lambda z}|, \end{aligned} \tag{13.43}$$

where we have made use of (13.39). By induction, we further obtain (exercise)

$$\begin{aligned} |\varepsilon_{n,k}(z) - \varepsilon_{n,k-1}(z)| &+ |\varepsilon_{n,k}'(z) - \varepsilon_{n,k-1}'(z)| \\ &\le B_n (M C_{n-\operatorname{Re}\mu})^k |z|^{\operatorname{Re}\mu - n} |e^{\lambda z}|, \end{aligned} \tag{13.44}$$

for all $k \ge 1$. Recall from (13.40) that $C_{n-\operatorname{Re}\mu} = O(n^{1/2})$ as $n \to \infty$. Hence, for sufficiently large n such that $M C_{n-\operatorname{Re}\mu} < 1/2$, the series expansion of $\varepsilon_n(z)$ in (13.41) converges and

$$|\varepsilon_n(z)| + |\varepsilon_n'(z)| \le 2B_n |z|^{\operatorname{Re}\mu - n} |e^{\lambda z}|.$$

Moreover, $\varepsilon_n(z)$ is a fixed point of the integral operator $\mathcal{K}_n$ and satisfies the differential equation (13.34). Similarly, we can obtain error bounds for the truncated asymptotic expansions of $\tilde{w}(z)$ in (13.26) for $z \in \Omega_\rho^{\tilde{\phi}}$, where $\tilde{\phi} := \arg(\lambda - \tilde{\lambda}) \in [-\pi, \pi]$. These two asymptotic solutions $w(z)$ and $\tilde{w}(z)$ exist in the joint region $\Omega_\rho^\phi \cap \Omega_\rho^{\tilde{\phi}}$. For any complex number z, the arguments of z and $-z$ differ by π (modulo 2π). Hence, we have $|\phi - \tilde{\phi}| = \pi$. One can show that (exercise)

$$\Omega_\rho^\phi \cap \Omega_\rho^{\tilde{\phi}} = \{z = re^{i\theta} : \ r > \rho, \ -\max\{\phi, \tilde{\phi}\} < \theta < -\min\{\phi, \tilde{\phi}\}\}. \tag{13.45}$$

In general, $w(z)$ and $\tilde{w}(z)$ constructed above may depend on n. However, since $\tilde{w}(z)/w(z) \sim e^{(\tilde{\lambda} - \lambda) z} z^{\tilde{\mu} - \mu}$ is exponentially large when $\arg z$ is close to

$-\phi$ and exponentially small when $\arg z$ is close to $-\tilde{\phi}$, both solutions are independent of n. We summarize our results in the following theorem.

Theorem 13.1. *Assume f and g are analytic for $|z| > R$ with $f_0^2 \neq 4g_0$. Let λ and $\tilde{\lambda}$ be two distinct characteristic roots satisfying the characteristic equation* (13.24)*. Denote $\phi := \arg(\tilde{\lambda} - \lambda) \in [-\pi, \pi]$ and $\tilde{\phi} := \arg(\lambda - \tilde{\lambda}) \in [-\pi, \pi]$. For any $\rho > R$, the differential equation* (13.1) *possesses two linearly independent solutions $w(z)$ and $\tilde{w}(z)$ with asymptotic expansions in* (13.22) *and* (13.26)*, respectively, as $z \to \infty$ and z lies in the region $\Omega_\rho^\phi \cap \Omega_\rho^{\tilde{\phi}}$ given in* (13.45).

13.4 Exercises

1. Show that the change of variable $z = 1/t$ converts (13.1) into (13.3) with (13.4).
2. Verify (13.7).
3. Verify (13.8).
4. Verify (13.9).
5. Verify (13.10).
6. Let α be the root of the indicial equation (13.13) with the largest real part. Verify that $w(t)$ in (13.14) with coefficients determined recursively by (13.15) is a solution to (13.3). In particular, prove that $2\alpha+s+p_0 \neq 0$ for all $s = 0, 1, \cdots$.
7. Prove (13.17).
8. Verify (13.20).
9. Verify (13.21).
10. Verify (13.23) and (13.27).
11. Prove that Fabry's transformation (13.28) transforms the equation (13.1) into (13.29).
12. Verify (13.33).
13. Let γ_z be defined as in (13.38), where $z = re^{i\theta} \in \Omega_R^\phi$. Prove that $t \in \gamma_z$ implies $t \in \Omega_R^\phi$ and $\mathrm{Re}[e^{i\phi}(t - z)] = 0$. Verify (13.39).
14. Verify (13.43) and (13.44).
15. Prove (13.45).

Chapter 14

Asymptotics of Differential Equations with a Turning Point

14.1 Liouville transformation

We consider the linear differential equation

$$\frac{d^2w}{dx^2} = [u^2 f(x) + g(x)]w, \quad x \in (a,b), \tag{14.1}$$

where $u > 0$ is a large parameter, (a,b) is a finite or infinite real interval, $g(x)$ is a continuous real or complex function in (a,b), and $f(x)$ has a simple zero $x_0 \in (a,b)$ such that $f(x)/(x-x_0)$ is positive and twice continuously differentiable in (a,b). Here, x_0 is called the turning point.

The material in this chapter is taken from Olver (1974a, Chapter 11). For the special case when $f(x) = x - x_0$ and $g(x) = 0$, the equation (14.1) reduces to the Airy equation. In this chapter, we will show that as $u \to \infty$, the solution of (14.1) is close to the solution of the Airy equation. To this end, we first introduce the Liouville transformation:

$$\frac{d\zeta}{dx} = \sqrt{\frac{f(x)}{\zeta}} > 0, \quad W = \sqrt{\frac{d\zeta}{dx}}w = (f/\zeta)^{1/4}w. \tag{14.2}$$

Note that $f(x)/(x-x_0) > 0$ for $x \in (a,b)$. The positivity of $f(x)/\zeta$ implies that $\zeta(x_0) = 0$ and $\zeta/(x-x_0) > 0$ for $x \in (a,b)$. A simple integration of the first equation in (14.2) gives (exercise)

$$\begin{cases} \frac{2}{3}[\zeta(x)]^{3/2} = \int_{x_0}^{x} [f(t)]^{1/2}dt, & \text{if } x \geq x_0, \\ \frac{2}{3}[-\zeta(x)]^{3/2} = \int_{x}^{x_0} [-f(t)]^{1/2}dt, & \text{if } x \leq x_0. \end{cases} \tag{14.3}$$

We also obtain from the chain rule that (exercise)

$$\frac{dw}{dx} = \frac{1}{4}f^{1/4}\zeta^{-5/4}W - \frac{1}{4}f^{-5/4}f'\zeta^{1/4}W + f^{1/4}\zeta^{-1/4}\frac{dW}{d\zeta}, \tag{14.4}$$

$$\frac{d^2w}{dx^2} = \left[-\frac{5f^{3/4}}{16\zeta^{11/4}} + \frac{5\zeta^{1/4}(f')^2}{f^{9/4}} - \frac{\zeta^{1/4}f''}{f^{5/4}}\right]W + (f/\zeta)^{3/4}\frac{d^2W}{d\zeta^2}. \tag{14.5}$$

Substituting the above two formulas into (14.1) yields the following model problem (exercise)

$$\frac{d^2W}{d\zeta^2} = [u^2\zeta + \psi(\zeta)]W, \quad \zeta \in (\alpha, \beta), \tag{14.6}$$

where $\alpha = \zeta(a)$, $\beta = \zeta(b)$, and

$$\psi(\zeta) := \frac{5}{16\zeta^2} + \frac{f''\zeta}{4f^2} - \frac{5(f')^2\zeta}{16f^3} + \frac{\zeta g}{f}. \tag{14.7}$$

Since $u > 0$ is large, we may formally approximate $\psi(\zeta)$ by 0 and hence find two approximate solutions to the equation (14.6) in terms of the Airy functions: $\mathrm{Ai}(u^{2/3}\zeta)$ and $\mathrm{Bi}(u^{2/3}\zeta)$. In the following three sections, we will state some basic properties of the Airy functions, derive formal asymptotic expansions of two linearly independent solutions, and provide error estimates for the truncated asymptotic expansions.

We conclude this section with the following lemma.

Lemma 14.1. *Assume $p(x) := [f(x)/(x - x_0)]^{1/2}$ is positive and twice continuously differentiable for $x \in (a, b)$. Let ζ and $\psi(\zeta)$ be defined as in* (14.3) *and* (14.7), *respectively. The function $\zeta(x)/(x - x_0)$ is positive and twice continuously differentiable for $x \in (a, b)$. The function $\psi(\zeta)$ is continuous for $\zeta \in (\zeta(a), \zeta(b))$.*

Proof. Since $p(x) := [f(x)/(x - x_0)]^{1/2}$ is positive and twice continuously differentiable for $x \in (a, b)$, it is readily seen from (14.3) that $\zeta(x)/(x - x_0)$ is positive and twice continuously differentiable for $x \in (a, x_0) \cup (x_0, b)$. Now, we define

$$q(x) = \begin{cases} [\zeta(x)/(x - x_0)]^{3/2}, & x > x_0, \\ [-\zeta(x)/(x_0 - x)]^{3/2}, & x < x_0. \end{cases}$$

If $x > x_0$, we have from (14.3) and integration by parts that

$$\begin{aligned} q(x) &= \frac{3}{2(x - x_0)^{3/2}} \int_{x_0}^{x} (t - x_0)^{1/2} p(t) dt \\ &= p(x) - \frac{1}{(x - x_0)^{3/2}} \int_{x_0}^{x} (t - x_0)^{3/2} p'(t) dt \\ &= p(x) - \frac{2}{5}(x - x_0)p'(x) + \frac{2}{5(x - x_0)^{3/2}} \int_{x_0}^{x} (t - x_0)^{5/2} p''(t) dt. \end{aligned}$$

Similarly, we obtain

$$\begin{aligned} q(x) &= \frac{3}{2(x_0 - x)^{3/2}} \int_x^{x_0} (x_0 - t)^{1/2} p(t) dt \\ &= p(x) + \frac{1}{(x_0 - x)^{3/2}} \int_x^{x_0} (x_0 - t)^{3/2} p'(t) dt \\ &= p(x) + \frac{2}{5}(x_0 - x) p'(x) + \frac{2}{5(x_0 - x)^{3/2}} \int_x^{x_0} (x_0 - t)^{5/2} p''(t) dt. \end{aligned}$$

for $x < x_0$. Coupling the above two formulas implies that $q(x)$ is twice continuously differentiable at $x = x_0$. In particular, $q(x_0) = p(x_0) > 0$. Hence, $\zeta(x)/(x - x_0) = q(x)^{2/3}$ is positive and twice continuously differentiable for $x \in (a, b)$.

Finally, we note that $\eta(x) := [\zeta(x)/f(x)]^{1/4}$ is positive and twice continuously differentiable for $x \in (a, b)$. It follows from (14.2) and (14.7) that (exercise)

$$\psi(\zeta) = g(x)[\eta(x)]^4 - [\eta(x)]^3 \eta''(x) \tag{14.8}$$

is continuous for $x \in (a, b)$, and hence continuous for $\zeta \in (\zeta(a), \zeta(b))$. □

14.2 Airy functions and auxiliary functions

When $f(x) = x$, and $g(x) = 0$, the equation (14.1) becomes $w''(x) = u^2 x w(x)$. Upon a scaling, we may assume further $u = 1$. Now, we look for a solution with series representation $w(x) = \sum_{n=0}^{\infty} w_n x^n$. Since $w''(x) = \sum_{n=0}^{\infty} w_{n+2}(n+2)(n+1)x^n$ and $xw(x) = \sum_{n=1}^{\infty} w_{n-1} x^n$, we obtain from $w''(x) = xw(x)$ that $w_2 = 0$ and $w_{n+2}(n+2)(n+1) = w_{n-1}$ for $n = 1, 2, \cdots$. It is readily seen that (exercise)

$$w_{3k+2} = 0, \quad w_{3k+1} = \frac{w_1 \Gamma(4/3)}{3^{2k} k! \Gamma(k + 4/3)}, \quad w_{3k} = \frac{w_0 \Gamma(2/3)}{3^{2k} k! \Gamma(k + 2/3)}, \quad k \geq 0. \tag{14.9}$$

We will study two special solutions with specific values of $w_0 = w(0)$ and $w_1 = w'(0)$. First, we introduce the Airy function defined by the integral

$$\mathrm{Ai}(x) := \frac{1}{\pi} \int_0^{\infty} \cos(xt + t^3/3) dt = \frac{1}{2\pi i} \int_{-i\infty}^{i\infty} e^{xs - s^3/3} ds. \tag{14.10}$$

It is obvious that $\mathrm{Ai}''(x) = x\,\mathrm{Ai}(x)$. By Cauchy's theorem, we can replace the integration path of the imaginary axis with $\gamma_2 \cup \gamma_1$, where $\gamma_2 := \{re^{4i\pi/3} : r > 0\}$ and $\gamma_1 := \{re^{2i\pi/3} : r > 0\}$ oriented along the increasing direction of imaginary parts. For convenience, we denote $\omega := e^{2i\pi/3}$. A simple calculation gives

$$\begin{aligned}\mathrm{Ai}(x) =&\frac{1}{2\pi i}[\int_{\infty}^{0} e^{\omega^2 xr - r^3/3}\omega^2 dr + \int_{0}^{\infty} e^{\omega xr - r^3/3}\omega dr]\\ =&\sum_{n=0}^{\infty} \frac{x^n(\omega^{n+1} - \omega^{2n+2})}{2\pi i n!} \int_0^{\infty} r^n e^{-r^3/3} dr\\ =&\sum_{n=0}^{\infty} \frac{x^n \sin[2(n+1)\pi/3]}{\pi n!} 3^{n/3-2/3}\Gamma(n/3+1/3).\end{aligned}$$

In particular, $\mathrm{Ai}(0) = 3^{-1/6}\Gamma(1/3)/(2\pi) = 3^{-2/3}/\Gamma(2/3)$ and $\mathrm{Ai}'(0) = -3^{1/6}\Gamma(2/3)/(2\pi) = -3^{-1/3}/\Gamma(1/3)$. By analytic continuation, we can regard $\mathrm{Ai}(x)$ as an entire function on the complex plane. Furthermore, we have the following connection formula (exercise)

$$\mathrm{Ai}(x) + \omega\,\mathrm{Ai}(\omega x) + \omega^2\,\mathrm{Ai}(\omega^2 x) = 0. \tag{14.11}$$

Next, we define

$$\mathrm{Bi}(x) = i\omega^2\,\mathrm{Ai}(\omega^2 x) - i\omega\,\mathrm{Ai}(\omega x) \tag{14.12}$$

It is easy to verify that $\mathrm{Bi}''(x) = x\,\mathrm{Bi}(x)$ with $\mathrm{Bi}(0) = 3^{-1/6}/\Gamma(2/3)$ and $\mathrm{Bi}'(0) = 3^{1/6}/\Gamma(1/3)$. Hence, we have $\mathrm{Bi}(x) \in \mathbf{R}$ when $x \in \mathbf{R}$. The Wronskian of $\mathrm{Ai}(x)$ and $\mathrm{Bi}(x)$ is calculated as

$$\mathrm{Ai}(x)\,\mathrm{Bi}'(x) - \mathrm{Ai}'(x)\,\mathrm{Bi}(x) = \frac{1}{\pi}. \tag{14.13}$$

One can prove that (exercise) $\mathrm{Ai}(x)$, $\mathrm{Bi}(x)$, and $\mathrm{Bi}(x) - \mathrm{Ai}(x)$ are positive when $x > 0$. For convenience, we let $a_1 < 0$, $b_1 < 0$, and $c_1 < 0$ be the largest zeros of $\mathrm{Ai}(x)$, $\mathrm{Bi}(x)$, and $\mathrm{Bi}(x) - \mathrm{Ai}(x)$, respectively. It is easily seen that (exercise)

$$a_1 < b_1 < c_1 < 0. \tag{14.14}$$

The graphs of $\mathrm{Ai}(x)$, $\mathrm{Bi}(x)$, and $\mathrm{Bi}(x) - \mathrm{Ai}(x)$ are plotted in Figure 14.1 for illustration.

Now, we define three auxiliary functions

$$E(x) := \begin{cases} 1, & x \le c_1, \\ \sqrt{\mathrm{Bi}(x)/\mathrm{Ai}(x)}, & x \ge c_1, \end{cases} \tag{14.15}$$

$$M(x) := \begin{cases} \sqrt{[\mathrm{Ai}(x)]^2 + [\mathrm{Bi}(x)]^2}, & x \le c_1, \\ \sqrt{2\,\mathrm{Ai}(x)\,\mathrm{Bi}(x)}, & x \ge c_1, \end{cases} \tag{14.16}$$

$$\theta(x) := \begin{cases} \arctan[\mathrm{Ai}(x)/\mathrm{Bi}(x)], & x \le c_1, \\ \pi/4, & x \ge c_1. \end{cases} \tag{14.17}$$

It is readily seen that $E(x)\,\mathrm{Ai}(x) = M(x)\sin\theta(x)$ and $\mathrm{Bi}(x)/E(x) = M(x)\cos\theta(x)$.

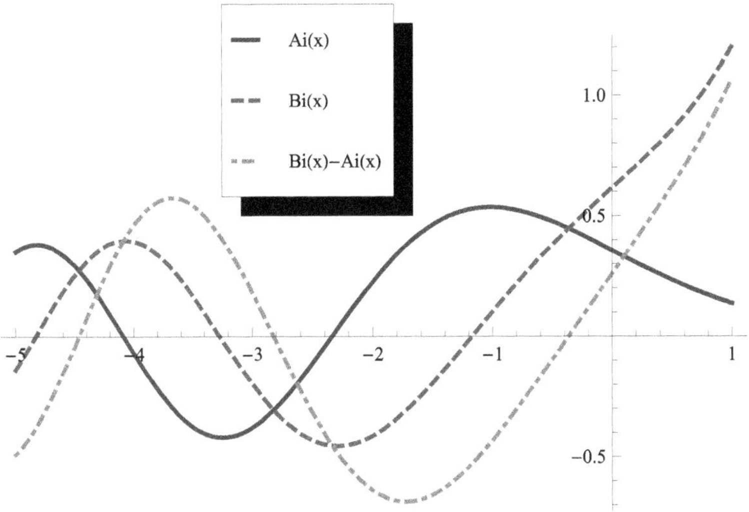

Fig. 14.1: The graphs of $\mathrm{Ai}(x)$, $\mathrm{Bi}(x)$, and $\mathrm{Bi}(x) - \mathrm{Ai}(x)$.

14.3 Formal asymptotic expansions

Note that the parameter u is large. If we ignore $\psi(\zeta)$ in (14.6), the equation becomes $d^2W/d\zeta^2 = u^2\zeta W$, which has two linearly independent solutions $\mathrm{Ai}(u^{2/3}\zeta)$ and $\mathrm{Bi}(u^{2/3}\zeta)$. This suggests us to try for a formal solution with asymptotic expansion

$$W(\zeta) \sim \mathrm{Ai}(u^{2/3}\zeta)\sum_{s=0}^{\infty}\frac{A_s(\zeta)}{u^{2s}} + \mathrm{Ai}'(u^{2/3}\zeta)\sum_{s=0}^{\infty}\frac{B_s(\zeta)}{u^{2s+4/3}}, \quad u \to \infty. \tag{14.18}$$

Taking formal derivatives and making use of the identity $\mathrm{Ai}''(u^{2/3}\zeta) = u^{2/3}\zeta\,\mathrm{Ai}(u^{2/3}\zeta)$, we obtain (exercise)

$$\frac{dW}{d\zeta} \sim \mathrm{Ai}(u^{2/3}\zeta)\sum_{s=0}^{\infty}\frac{A_s'(\zeta)+\zeta B_s(\zeta)}{u^{2s}} + \mathrm{Ai}'(u^{2/3}\zeta)\sum_{s=0}^{\infty}\frac{A_s(\zeta)+B_{s-1}'(\zeta)}{u^{2s-2/3}}, \tag{14.19}$$

$$\begin{aligned}\frac{d^2W}{d\zeta^2} \sim{}& \mathrm{Ai}(u^{2/3}\zeta)\sum_{s=0}^{\infty}\frac{A_{s-1}''(\zeta)+B_{s-1}(\zeta)+2\zeta B_{s-1}'(\zeta)+\zeta A_s(\zeta)}{u^{2s-2}}\\ &+ \mathrm{Ai}'(u^{2/3}\zeta)\sum_{s=0}^{\infty}\frac{2A_s'(\zeta)+\zeta B_s(\zeta)+B_{s-1}''(\zeta)}{u^{2s-2/3}},\end{aligned} \tag{14.20}$$

where we set $A_{-1}(\zeta) = B_{-1}(\zeta) = 0$ for convenience. Substituting the above expansions into (14.6) and comparing like powers of u, we obtain (exercise)

$$A_s''(\zeta) + B_s(\zeta) + 2\zeta B_s'(\zeta) = \psi(\zeta)A_s(\zeta), \quad 2A_s'(\zeta) + B_{s-1}''(\zeta) = \psi(\zeta)B_{s-1}(\zeta), \tag{14.21}$$

for $s \geq 0$. In particular, $A_0'(\zeta) = 0$. Without loss of generality, we may choose $A_0(\zeta) = 1$. The above equations can be solved recursively (exercise) as

$$B_s(\zeta) = \begin{cases} \frac{1}{\zeta^{1/2}} \int_0^\zeta \frac{\psi(z)A_s(z) - A_s''(z)}{2z^{1/2}} dz, & \zeta > 0, \\ \psi(\zeta)A_s(\zeta) - A_s''(\zeta), & \zeta = 0, \\ \frac{1}{(-\zeta)^{1/2}} \int_\zeta^0 \frac{\psi(z)A_s(z) - A_s''(z)}{2(-z)^{1/2}} dz, & \zeta < 0, \end{cases} \tag{14.22}$$

$$A_s(\zeta) = A_s(0) + \frac{1}{2}\int_0^\zeta [\psi(z)B_{s-1}(z) - B_{s-1}''(z)]dz. \tag{14.23}$$

In a similar manner, we find another linearly independent solution with formal asymptotic expansion as in (14.18), where Ai is replaced with Bi.

14.4 Error bounds

In this section, we provide rigorous error bounds to justify the asymptotic nature of (14.18). Our main theorem is stated below.

Theorem 14.1. *Assume $\psi \in C^\infty(I)$ with $I = (\alpha, \beta)$ being an open real interval containing 0, where α and β may be infinite. For each $u > 0$ and $n = 0, 1, 2, \cdots$, the equation (14.6) possesses a pair of solutions in $C^\infty(I)$, given by*

$$W_1 = \mathrm{Bi}(u^{2/3}\zeta)\sum_{s=0}^{n} \frac{A_s(\zeta)}{u^{2s}} + \mathrm{Bi}'(u^{2/3}\zeta)\sum_{s=0}^{n-1} \frac{B_s(\zeta)}{u^{2s+4/3}} + \varepsilon_1(\zeta), \tag{14.24}$$

$$W_2 = \mathrm{Ai}(u^{2/3}\zeta)\sum_{s=0}^{n} \frac{A_s(\zeta)}{u^{2s}} + \mathrm{Ai}'(u^{2/3}\zeta)\sum_{s=0}^{n-1} \frac{B_s(\zeta)}{u^{2s+4/3}} + \varepsilon_2(\zeta), \tag{14.25}$$

where $A_s(\zeta)$ and $B_s(\zeta)$ are determined recursively from (14.22) and (14.23). The error terms $\varepsilon_1(\zeta)$ and $\varepsilon_2(\zeta)$ are bounded by

$$|\varepsilon_1(\zeta)| \leq \frac{\lambda P(\zeta)\Phi_n(\zeta)e^{\lambda\Phi_0(\zeta)/u}}{u^{2n+1}}, \quad |\varepsilon_2(\zeta)| \leq \frac{\lambda Q(\zeta)\Phi_n(\zeta)e^{\lambda\Phi_0(\zeta)/u}}{u^{2n+1}}, \tag{14.26}$$

where $P(\zeta) := E(u^{2/3}\zeta)M(u^{2/3}\zeta)$, $Q(\zeta) := M(u^{2/3}\zeta)/E(u^{2/3}\zeta)$,

$$\begin{aligned} \lambda &= \sup_{x\in\mathbf{R}}[\pi u^{1/3}|x|^{1/2}P(x)Q(x)] = \sup_{x\in\mathbf{R}}[\pi u^{1/3}|x|^{1/2}(M(u^{2/3}x))^2] \\ &= \sup_{x\in\mathbf{R}}[\pi|x|^{1/2}(M(x))^2] \end{aligned} \tag{14.27}$$

is finite (exercise), and

$$\begin{aligned}\Phi_k(x) :=& 2\int_\alpha^x |\frac{d}{dy}[|y|^{1/2}B_k(y)]|dy = \int_\alpha^x |y^{-1/2}[2yB_k'(y) + B_k(y)]|dy \\ =& \int_\alpha^x |y^{-1/2}[\psi(y)A_k(y) - A_k''(y)]|dy \end{aligned} \tag{14.28}$$

for $k = 0, 1, \cdots$.

Proof. Let W_1 be given as in (14.24). We obtain (exercise)

$$\begin{aligned}\frac{d^2W_1}{d\zeta^2} - [u^2\zeta + \psi(\zeta)]W_1 =& \varepsilon_1''(\zeta) - [u^2\zeta + \psi(\zeta)]\varepsilon_1(\zeta) \\ & - \mathrm{Bi}(u^{2/3}\zeta)\frac{2\zeta B_n'(\zeta) + B_n(\zeta)}{u^{2n}}. \end{aligned} \tag{14.29}$$

Hence, W_1 is a solution to (14.6) if and only if

$$\varepsilon_1''(\zeta) - u^2\zeta\varepsilon_1(\zeta) = \psi(\zeta)\varepsilon_1(\zeta) + \mathrm{Bi}(u^{2/3}\zeta)\frac{2\zeta B_n'(\zeta) + B_n(\zeta)}{u^{2n}}.$$

Now, we define a kernel function

$$K(x, y) := \mathrm{Bi}(u^{2/3}x)\,\mathrm{Ai}(u^{2/3}y) - \mathrm{Ai}(u^{2/3}x)\,\mathrm{Bi}(u^{2/3}y), \tag{14.30}$$

for $\alpha < y \le x < \beta$, and an integral map

$$\mathcal{K}(h)(x) := \frac{\pi}{u^{2/3}}\int_\alpha^x K(x, y)[\psi(y)h(y) + \mathrm{Bi}(u^{2/3}y)\frac{2yB_n'(y) + B_n(y)}{u^{2n}}]dy, \tag{14.31}$$

for $x \in (\alpha, \beta)$ and $h \in C^\infty(I)$. It is easy to verify that (exercise) if $\varepsilon_1 \in C^\infty(I)$ is a fixed point of $\mathcal{K}$, then W_1 is a solution to (14.6). To find a fixed point of $\mathcal{K}$, we recursively define $\delta_{k+1} = \mathcal{K}(\delta_k)$ for $k \ge 0$ with $\delta_0(x) = 0$. For any $\alpha < y \le x < \beta$, we have (exercise)

$$|K(x, y)| \le \frac{E(u^{2/3}x)}{E(u^{2/3}y)}M(u^{2/3}x)M(u^{2/3}y) = P(x)Q(y). \tag{14.32}$$

It follows from (14.31) and (14.32) that

$$\begin{aligned}|\delta_1(x)| \le& \frac{\pi}{u^{2/3}}P(x)\int_\alpha^x Q(y)P(y)\frac{|2yB_n'(y) + B_n(y)|}{u^{2n}}dy, \\ |\delta_{k+1}(x) - \delta_k(x)| \le& \frac{\pi}{u^{2/3}}P(x)\int_\alpha^x Q(y)|\psi(y)[\delta_k(y) - \delta_{k-1}(y)]|dy.\end{aligned}$$

By induction, we obtain

$$|\delta_{k+1}(x) - \delta_k(x)| \le \frac{\lambda P(x)\Phi_n(x)}{u^{2n+1}} \cdot \frac{[\lambda\Phi_0(x)/u]^k}{k!}.$$

Therefore,

$$\varepsilon_1(x) := \lim_{k\to\infty} \delta_k(x) = \sum_{k=0}^{\infty}[\delta_{k+1}(x) - \delta_k(x)]$$

is well-defined and has an upper bound as given in (14.26). Moreover, it is a fixed point of the integral map $\mathcal{K}$, which implies that W_1 given in (14.24) is a solution to (14.6). The bound for ε_2 can be proved in a similar manner. □

14.5 Exercises

1. Verify (14.3).
2. Verify (14.4) and (14.5).
3. Verify (14.6).
4. Verify (14.8).
5. Verify (14.9).
6. Prove the identity (14.11).
7. Verify (14.13).
8. For each $x > 0$, prove that $\mathrm{Ai}(x) > 0$, $\mathrm{Bi}(x) > 0$, $\mathrm{Ai}'(x) < 0$, $\mathrm{Bi}'(x) > 0$, $\mathrm{Ai}''(x) > 0$, $\mathrm{Bi}''(x) > 0$, and $\mathrm{Ai}(x) - \mathrm{Bi}(x) < 0$.
9. Prove (14.14).
10. Verify (14.19) and (14.20).
11. Verify (14.21).
12. Show that the coefficients $A_s(\zeta)$ and $B_s(\zeta)$ defined recursively as in (14.22)-(14.23) satisfy the equations in (14.21). Moreover, if $\psi(\zeta)$ is infinitely differentiable, prove that both $A_s(\zeta)$ and $B_s(\zeta)$ are infinitely differentiable.
13. Prove that the constant λ defined in (14.27) is finite.
14. Verify (14.29).
15. Prove that if $\varepsilon_1 \in C^\infty(I)$ is a fixed point of the integral map $\mathcal{K}$ defined in (14.31), then W_1 given in (14.24) is a solution to (14.6).
16. Verify (14.32).

Chapter 15

Asymptotics of Differential Equations with a Simple Pole

15.1 Introduction

The material in this chapter is taken from Olver (1974a, Chapter 12). We consider the linear differential equation (14.1) with a large parameter $u > 0$. Again, we let (a, b) be a finite or infinite real interval. In this chapter, we consider the case when $f(x)$ has a simple pole $x_0 \in (a, b)$ such that $f(x)(x - x_0) > 0$ is positive. The function $g(x)$ may also have a simple or double pole at x_0. But we assume $(x - x_0)^2 g(x)$ is analytic. As in Chapter 14, we again first introduce the Liouville transformation:

$$\frac{d\zeta}{dx} = \sqrt{4\zeta f(x)} > 0, \quad W = \sqrt{\frac{d\zeta}{dx}}w = (4\zeta f)^{1/4}w. \tag{15.1}$$

A simple integration of the first equation in (15.1) gives (exercise)

$$\begin{cases} [\zeta(x)]^{1/2} = \int_{x_0}^{x} [f(t)]^{1/2} dt, & \text{if } x \geq x_0, \\ [-\zeta(x)]^{1/2} = \int_{x}^{x_0} [-f(t)]^{1/2} dt, & \text{if } x \leq x_0. \end{cases} \tag{15.2}$$

We also obtain from the chain rule that (exercise)

$$\frac{dw}{dx} = -4^{-3/4}\zeta^{-3/4}f^{1/4}W - 4^{-5/4}\zeta^{-1/4}f^{-5/4}f'W + 4^{1/4}\zeta^{1/4}f^{1/4}\frac{dW}{d\zeta}, \tag{15.3}$$

and

$$(4\zeta f)^{1/4}\frac{d^2w}{dx^2} = [\frac{3f}{4\zeta} - \frac{5(f')^2}{16f^2} - \frac{f''}{4f}]W + 4\zeta f\frac{d^2W}{d\zeta^2}. \tag{15.4}$$

Substituting the above two formulas into (14.1) yields the following model problem (exercise)

$$\frac{d^2W}{d\zeta^2} = [\frac{u^2}{4\zeta} + \varphi(\zeta)]W, \quad \zeta \in (\alpha, \beta), \tag{15.5}$$

where $\alpha = \zeta(a)$, $\beta = \zeta(b)$, and

$$\varphi(\zeta) := -\frac{3}{16\zeta^2} + \frac{f''}{16\zeta f^2} - \frac{5(f')^2}{64\zeta f^3} + \frac{g}{4f}. \tag{15.6}$$

Since $(x - x_0)^2 g(x)$ is analytic at $x = x_0$, the function $\zeta^2\varphi(\zeta)$ is analytic at $\zeta = 0$. We denote

$$\frac{\nu^2 - 1}{4} = \lim_{\zeta \to 0}[\zeta^2 \varphi(\zeta)] = \lim_{x \to x_0}[(x - x_0)^2 g(x)]. \tag{15.7}$$

The equation (15.5) is rewritten as

$$\frac{d^2W}{d\zeta^2} = [\frac{u^2}{4\zeta} + \frac{\nu^2 - 1}{4\zeta^2} + \frac{\psi(\zeta)}{\zeta}]W, \quad \zeta \in (\alpha, \beta), \tag{15.8}$$

where

$$\psi(\zeta) := \frac{1 - 4\nu^2}{16\zeta} + \frac{f''}{16f^2} - \frac{5(f')^2}{64f^3} + \frac{g}{4f}. \tag{15.9}$$

Since $u > 0$ is large, we may formally approximate $\psi(\zeta)$ by 0 and hence find two approximate solutions to the equation (15.8) in terms of the Bessel functions. In the next three sections, we will state some basic properties of Bessel functions and modified Bessel functions, derive formal asymptotic expansions of two linearly independent solutions, and provide error estimates for the truncated asymptotic expansions.

15.2 Bessel functions and modified Bessel functions

The Bessel function of the first kind is defined as

$$J_\nu(x) := \left(\frac{x}{2}\right)^\nu \sum_{k=0}^{\infty} \frac{(-x^2/4)^k}{k!\Gamma(\nu + k + 1)}. \tag{15.10}$$

The Bessel function of the second kind is defined as

$$Y_\nu(x) := \lim_{\mu \to \nu} \frac{J_\mu(x)\cos(\mu\pi) - J_{-\mu}(x)}{\sin(\mu\pi)}. \tag{15.11}$$

For convenience, we let $\mathcal{C}_\nu(x) = aJ_\nu(x) + bY_\nu(x)$ with $|a| + |b| \neq 0$ be any nontrivial linear combination of $J_\nu(x)$ and $Y_\nu(x)$. It can be verified that $\mathcal{C}_\nu(x)$ satisfies Bessel's equation (exercise)

$$\mathcal{C}_\nu''(x) + \mathcal{C}_\nu'(x)/x + [1 - (\nu/x)^2]\mathcal{C}_\nu(x) = 0. \tag{15.12}$$

The Wronskian of $J_\nu(x)$ and $Y_\nu(x)$ is

$$J_{\nu+1}(x)Y_\nu(x) - J_\nu(x)Y_{\nu+1}(x) = \frac{2}{\pi x}. \tag{15.13}$$

We further have

$$\mathcal{C}_{\nu-1}(x) + \mathcal{C}_{\nu+1}(x) = \frac{2\nu}{x}\mathcal{C}_\nu(x), \quad \mathcal{C}_{\nu-1}(x) - \mathcal{C}_{\nu+1}(x) = 2\mathcal{C}'_\nu(x), \tag{15.14}$$

and

$$\mathcal{C}'_\nu(x) = \mathcal{C}_{\nu-1}(x) - \frac{\nu}{x}\mathcal{C}_\nu(x) = -\mathcal{C}_{\nu+1}(x) + \frac{\nu}{x}\mathcal{C}_\nu(x). \tag{15.15}$$

The following two identities will be used later (exercise).

$$\frac{d}{dx}[2x^{1/2}\mathcal{C}_\nu(ux^{1/2})] = (1+\nu)x^{-1/2}\mathcal{C}_\nu(ux^{1/2}) - u\mathcal{C}_{\nu+1}(ux^{1/2}), \tag{15.16}$$

$$\frac{d}{dx}[2x\mathcal{C}_{\nu+1}(ux^{1/2})] = ux^{1/2}\mathcal{C}_\nu(ux^{1/2}) + (1-\nu)\mathcal{C}_{\nu+1}(ux^{1/2}). \tag{15.17}$$

Let $j_{\nu,s}$ and $y_{\nu,s}$ be the s-th positive zeros of $J_\nu(x)$ and $Y_\nu(x)$, one can prove that these zeros are interlacing (exercise):

$$y_{\nu,1} < j_{\nu,1} < y_{\nu,2} < j_{\nu,2} < \cdots \tag{15.18}$$

Let X_ν be the smallest positive zero of $J_\nu(x)+Y_\nu(x)$. It can be shown that (exercise)

$$0 < X_\nu < y_{\nu,1}. \tag{15.19}$$

The graphs of $J_1(x)$, $Y_1(x)$, and $J_1(x)+Y_1(x)$ are plotted in Figure 15.1 for illustration.

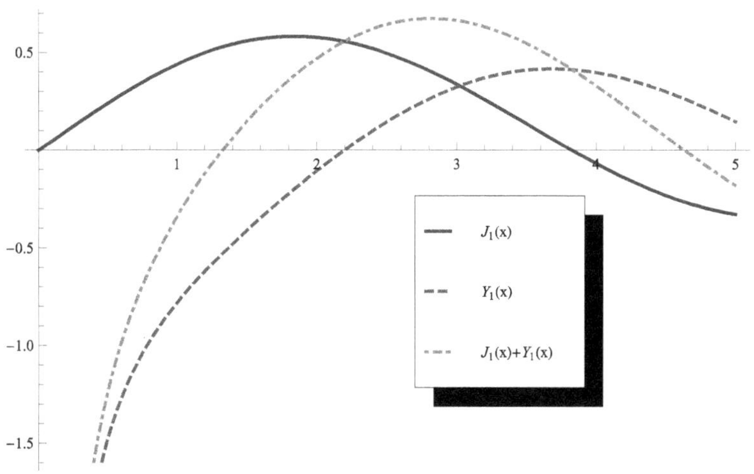

Fig. 15.1: The graphs of $J_1(x)$, $Y_1(x)$, and $J_1(x)+Y_1(x)$.

We introduce the following auxiliary functions

$$E_\nu(x) := \begin{cases} 1, & x \geq X_\nu, \\ \sqrt{-Y_\nu(x)/J_\nu(x)}, & 0 < x \leq X_\nu, \end{cases} \tag{15.20}$$

$$M_\nu(x) := \begin{cases} \sqrt{J_\nu^2(x) + Y_\nu^2(x)}, & x \geq X_\nu, \\ \sqrt{-2Y_\nu(x)J_\nu(x)}, & 0 < x \leq X_\nu, \end{cases} \tag{15.21}$$

$$\theta_\nu(x) := \begin{cases} \arctan[Y_\nu(x)/J_\nu(x)] + s\pi, & j_{\nu,s} \leq x < j_{\nu,s+1}, \\ (s - 1/2)\pi, & x = j_{\nu,s}, \\ \arctan[Y_\nu(x)/J_\nu(x)], & X_\nu \leq x < j_{\nu,1}, \\ -\pi/4, & 0 < x \leq X_\nu. \end{cases} \tag{15.22}$$

It is readily seen that $J_\nu(x)E_\nu(x) = M_\nu(x)\cos\theta_\nu(x)$ and $Y_\nu(x)/E_\nu(x) = M_\nu(x)\sin\theta_\nu(x)$. For any $x \geq y > 0$, we have (exercise)

$$E_\nu(x) \leq E_\nu(y), \quad |J_\nu(x)Y_\nu(y) - J_\nu(y)Y_\nu(x)| \leq M_\nu(x)M_\nu(y)\frac{E_\nu(y)}{E_\nu(x)}. \tag{15.23}$$

The modified Bessel function of the first kind is defined as

$$I_\nu(x) := e^{-\nu\pi i/2}J_\nu(ix) = \left(\frac{x}{2}\right)^\nu \sum_{k=0}^{\infty} \frac{(x^2/4)^k}{k!\Gamma(\nu+k+1)}. \tag{15.24}$$

The modified Bessel function of the second kind is defined as

$$K_\nu(x) := \lim_{\mu\to\nu} \frac{\pi[I_{-\mu}(x) - I_\mu(x)]}{2\sin(\mu\pi)}. \tag{15.25}$$

For convenience, we let $\mathcal{Z}_\nu(x) = aI_\nu(x) + be^{\nu\pi i}K_\nu(x)$ with $|a| + |b| \neq 0$ be any nontrivial linear combination of $I_\nu(x)$ and $e^{\nu\pi i}K_\nu(x)$. It can be verified that $\mathcal{Z}_\nu(x)$ satisfies modified Bessel's equation (exercise)

$$\mathcal{Z}_\nu''(x) + \mathcal{Z}_\nu'(x)/x - [1 + (\nu/x)^2]\mathcal{Z}_\nu(x) = 0. \tag{15.26}$$

The Wronskian of $I_\nu(x)$ and $K_\nu(x)$ is

$$I_{\nu+1}(x)K_\nu(x) + I_\nu(x)K_{\nu+1}(x) = \frac{1}{x}. \tag{15.27}$$

We further have

$$\mathcal{Z}_{\nu-1}(x) - \mathcal{Z}_{\nu+1}(x) = \frac{2\nu}{x}\mathcal{Z}_\nu(x), \quad \mathcal{Z}_{\nu-1}(x) + \mathcal{Z}_{\nu+1}(x) = 2\mathcal{Z}_\nu'(x), \tag{15.28}$$

and

$$\mathcal{Z}_\nu'(x) = \mathcal{Z}_{\nu-1}(x) - \frac{\nu}{x}\mathcal{Z}_\nu(x) = \mathcal{Z}_{\nu+1}(x) + \frac{\nu}{x}\mathcal{Z}_\nu(x). \tag{15.29}$$

The following two identities will be used later (exercise).

$$\frac{d}{d\zeta}[2\zeta^{1/2}\mathcal{Z}_\nu(u\zeta^{1/2})] = (1+\nu)\zeta^{-1/2}\mathcal{Z}_\nu(u\zeta^{1/2}) + u\mathcal{Z}_{\nu+1}(u\zeta^{1/2}), \tag{15.30}$$

$$\frac{d}{d\zeta}[2\zeta\mathcal{Z}_{\nu+1}(u\zeta^{1/2})] = u\zeta^{1/2}\mathcal{Z}_\nu(u\zeta^{1/2}) + (1-\nu)\mathcal{Z}_{\nu+1}(u\zeta^{1/2}). \tag{15.31}$$

15.3 Formal asymptotic expansions

Note that the parameter u is large. If we ignore $\psi(\zeta)/\zeta$ in (15.8), the equation becomes $d^2W/d\zeta^2 = [u^2/(4\zeta) + (\nu^2-1)/(4\zeta^2)]W$, which has a general solution $\zeta^{1/2}\mathcal{Z}_\nu(u\zeta^{1/2})$ when $\zeta > 0$ and $|\zeta|^{1/2}\mathcal{C}_\nu(u|\zeta|^{1/2})$ when $\zeta < 0$. When $\zeta > 0$, we consider a formal solution with asymptotic expansion

$$W(\zeta) \sim \zeta^{1/2}\mathcal{Z}_\nu(u\zeta^{1/2})\sum_{s=0}^{\infty}\frac{A_s(\zeta)}{u^{2s}} + \zeta\mathcal{Z}_{\nu+1}(u\zeta^{1/2})\sum_{s=0}^{\infty}\frac{B_s(\zeta)}{u^{2s+1}}, \tag{15.32}$$

as $u \to \infty$. Taking formal differentiations and making use of (15.30)-(15.31), we derive (exercise)

$$\begin{aligned}\frac{dW}{d\zeta} \sim&\zeta^{1/2}\mathcal{Z}_\nu(u\zeta^{1/2})\sum_{s=0}^{\infty}\frac{A_s'(\zeta)+(1+\nu)A_s(\zeta)/(2\zeta)+B_s(\zeta)/2}{u^{2s}}\\ &+\zeta\mathcal{Z}_{\nu+1}(u\zeta^{1/2})\sum_{s=0}^{\infty}\frac{B_{s-1}'(\zeta)+(1-\nu)B_{s-1}(\zeta)/(2\zeta)+A_s(\zeta)/(2\zeta)}{u^{2s-1}},\end{aligned} \tag{15.33}$$

and

$$\begin{aligned}\frac{d^2W}{d\zeta^2} \sim&\zeta^{1/2}\mathcal{Z}_\nu(u\zeta^{1/2})\sum_{s=0}^{\infty}\frac{A_s''(\zeta)+(1+\nu)A_s'(\zeta)/\zeta+B_s'(\zeta)+B_s(\zeta)/(2\zeta)}{u^{2s}}\\ &+\zeta^{1/2}\mathcal{Z}_\nu(u\zeta^{1/2})\sum_{s=-1}^{\infty}\frac{(\nu^2-1)A_s(\zeta)/(4\zeta^2)+A_{s+1}(\zeta)/(4\zeta)}{u^{2s}}\\ &+\zeta\mathcal{Z}_{\nu+1}(u\zeta^{1/2})\sum_{s=0}^{\infty}\frac{B_s''(\zeta)+(1-\nu)B_s'(\zeta)/\zeta+A_{s+1}'(\zeta)/\zeta}{u^{2s+1}}\\ &+\zeta\mathcal{Z}_{\nu+1}(u\zeta^{1/2})\sum_{s=-1}^{\infty}\frac{(\nu^2-1)B_s(\zeta)/(4\zeta^2)+B_{s+1}(\zeta)/(4\zeta)}{u^{2s+1}},\end{aligned} \tag{15.34}$$

where we set $A_{-1}(\zeta) = B_{-1}(\zeta) = 0$ for convenience. Substituting the above expansions into (15.8) and comparing the like powers of u, we obtain (exercise)

$$\zeta A_s''(\zeta) + (1+\nu)A_s'(\zeta) + \zeta B_s'(\zeta) + \frac{B_s(\zeta)}{2} =\psi(\zeta)A_s(\zeta), \tag{15.35}$$

$$A_s'(\zeta) + \zeta B_{s-1}''(\zeta) + (1-\nu)B_{s-1}'(\zeta) =\psi(\zeta)B_{s-1}(\zeta), \tag{15.36}$$

for $s \geq 0$. In particular, $A_0'(\zeta) = 0$. Without loss of generality, we may choose $A_0(\zeta) = 1$. The above equations can be solved recursively as

$$B_s(\zeta) = \frac{1}{\zeta^{1/2}} \int_0^\zeta \frac{\psi(z)A_s(z) - zA_s''(z) - (1+\nu)A_s'(z)}{z^{1/2}} dz, \tag{15.37}$$

$$A_s(\zeta) = A_s(0) + \int_0^\zeta [\psi(z)B_{s-1}(z) - zB_{s-1}''(z) - (1-\nu)B_{s-1}'(z)]dz. \tag{15.38}$$

When $\zeta < 0$, we consider a formal solution with asymptotic expansion

$$W(\zeta) \sim |\zeta|^{1/2}\mathcal{C}_\nu(u|\zeta|^{1/2}) \sum_{s=0}^\infty \frac{A_s(\zeta)}{u^{2s}} - |\zeta|\mathcal{C}_{\nu+1}(u|\zeta|^{1/2}) \sum_{s=0}^\infty \frac{B_s(\zeta)}{u^{2s+1}}, \tag{15.39}$$

as $u \to \infty$. Taking formal differentiations and making use of (15.16)-(15.17), we derive (exercise)

$$\begin{aligned}\frac{dW}{d\zeta} \sim & |\zeta|^{1/2}\mathcal{C}_\nu(u|\zeta|^{1/2}) \sum_{s=0}^\infty \frac{A_s'(\zeta) + (1+\nu)A_s(\zeta)/(2\zeta) + B_s(\zeta)/2}{u^{2s}} \\ & - |\zeta|\mathcal{C}_{\nu+1}(u|\zeta|^{1/2}) \sum_{s=0}^\infty \frac{B_{s-1}'(\zeta) + (1-\nu)B_{s-1}(\zeta)/(2\zeta) + A_s(\zeta)/(2\zeta)}{u^{2s-1}},\end{aligned} \tag{15.40}$$

and

$$\begin{aligned}\frac{d^2W}{d\zeta^2} \sim & |\zeta|^{1/2}\mathcal{C}_\nu(u|\zeta|^{1/2}) \sum_{s=0}^\infty \frac{A_s''(\zeta) + (1+\nu)A_s'(\zeta)/\zeta + B_s'(\zeta) + B_s(\zeta)/(2\zeta)}{u^{2s}} \\ & + |\zeta|^{1/2}\mathcal{C}_\nu(u|\zeta|^{1/2}) \sum_{s=-1}^\infty \frac{(\nu^2-1)A_s(\zeta)/(4\zeta^2) + A_{s+1}(\zeta)/(4\zeta)}{u^{2s}} \\ & - |\zeta|\mathcal{C}_{\nu+1}(u|\zeta|^{1/2}) \sum_{s=0}^\infty \frac{B_s''(\zeta) + (1-\nu)B_s'(\zeta)/\zeta + A_{s+1}'(\zeta)/\zeta}{u^{2s+1}} \\ & - |\zeta|\mathcal{C}_{\nu+1}(u|\zeta|^{1/2}) \sum_{s=-1}^\infty \frac{(\nu^2-1)B_s(\zeta)/(4\zeta^2) + B_{s+1}(\zeta)/(4\zeta)}{u^{2s+1}},\end{aligned} \tag{15.41}$$

where we set $A_{-1}(\zeta) = B_{-1}(\zeta) = 0$ for convenience. Substituting the above expansions into (15.8) and comparing the like powers of u, we again obtain (15.35) and (15.36) for $s \geq 0$. Without loss of generality, we may still choose $A_0(\zeta) = 1$. Note that $\zeta < 0$. The equations (15.35) and (15.36) can be solved recursively as

$$B_s(\zeta) = \frac{1}{(-\zeta)^{1/2}} \int_\zeta^0 \frac{\psi(z)A_s(z) - zA_s''(z) - (1+\nu)A_s'(z)}{(-z)^{1/2}} dz, \tag{15.42}$$

$$A_s(\zeta) = A_s(0) + \int_0^\zeta [\psi(z)B_{s-1}(z) - zB_{s-1}''(z) - (1-\nu)B_{s-1}'(z)]dz, \tag{15.43}$$

for $\zeta < 0$. It is easily seen that if $\psi \in C^\infty(I)$, where $I = (\alpha, \beta)$ is an open real interval containing 0, then both A_s and B_s are in $C^\infty(I)$.

15.4 Error bounds

In this section, we provide rigorous error bounds to justify the asymptotic nature of (15.32) and (15.39).

Theorem 15.1. *Assume $\psi \in C^\infty(0, \beta)$, where $\beta > 0$ may be infinite. For each $u > 0$ and $n = 0, 1, 2, \cdots$, the equation* (15.8) *possesses a pair of solutions in $C^\infty(0, \beta)$, given by*

$$W_1^+ = \zeta^{1/2} I_\nu(u\zeta^{1/2}) \sum_{s=0}^{n} \frac{A_s(\zeta)}{u^{2s}} + \zeta I_{\nu+1}(u\zeta^{1/2}) \sum_{s=0}^{n-1} \frac{B_s(\zeta)}{u^{2s+1}} + \varepsilon_1^+(\zeta), \tag{15.44}$$

$$W_2^+ = \zeta^{1/2} K_\nu(u\zeta^{1/2}) \sum_{s=0}^{n} \frac{A_s(\zeta)}{u^{2s}} - \zeta K_{\nu+1}(u\zeta^{1/2}) \sum_{s=0}^{n-1} \frac{B_s(\zeta)}{u^{2s+1}} + \varepsilon_2^+(\zeta), \tag{15.45}$$

where $A_s(\zeta)$ and $B_s(\zeta)$ are determined recursively from (15.37) *and* (15.38)*. The error terms $\varepsilon_1^+(\zeta)$ and $\varepsilon_2^+(\zeta)$ are bounded by*

$$|\varepsilon_1^+(\zeta)| \le \frac{\lambda P^+(\zeta)\Phi_n(\zeta)e^{\lambda\Phi_0(\zeta)/u}}{u^{2n+1}}, \quad |\varepsilon_2^+(\zeta)| \le \frac{\lambda Q^+(\zeta)\Phi_n(\zeta)e^{\lambda\Phi_0(\zeta)/u}}{u^{2n+1}}, \tag{15.46}$$

where $P^+(\zeta) := \zeta^{1/2} I_\nu(u\zeta^{1/2})$, $Q^+(\zeta) := \zeta^{1/2} K_\nu(u\zeta^{1/2})$,

$$\lambda = \sup_{x>0} [ux^{-1/2} P^+(x) Q^+(x)] = \sup_{x>0} [x I_\nu(x) K_\nu(x)] < \infty, \tag{15.47}$$

and

$$\Phi_k(x) := 2\int_0^x |\frac{d}{dy}[y^{1/2} B_k(y)]| dy \tag{15.48}$$

for $k = 0, 1, \cdots$.

Proof. Let W_1^+ be given as in (15.44). We obtain (exercise)

$$\frac{d^2W_1^+}{d\zeta^2} - [\frac{u^2}{4\zeta} + \frac{\nu^2-1}{4\zeta^2} + \frac{\psi(\zeta)}{\zeta}]W_1^+ = \frac{d^2\varepsilon_1^+}{d\zeta^2} - [\frac{u^2}{4\zeta} + \frac{\nu^2-1}{4\zeta^2} + \frac{\psi(\zeta)}{\zeta}]\varepsilon_1^+(\zeta)$$
$$- \zeta^{1/2}I_\nu(u\zeta^{1/2})\frac{B_n'(\zeta)+B_n(\zeta)/(2\zeta)}{u^{2n}}. \quad (15.49)$$

Hence, W_1^+ is a solution to (15.8) if and only if

$$\frac{d^2\varepsilon_1^+}{d\zeta^2} - [\frac{u^2}{4\zeta} + \frac{\nu^2-1}{4\zeta^2}]\varepsilon_1^+(\zeta) = \frac{\psi(\zeta)}{\zeta}\varepsilon_1^+(\zeta) + I_\nu(u\zeta^{1/2})\frac{[\zeta^{1/2}B_n(\zeta)]'}{u^{2n}}.$$

Now, we define a kernel function

$$K^+(x,y) := (xy)^{1/2}[I_\nu(ux^{1/2})K_\nu(uy^{1/2}) - K_\nu(ux^{1/2})I_\nu(uy^{1/2})], \quad (15.50)$$

for $0 < y \le x < \beta$, and an integral map

$$\mathcal{K}^+(h)(x) := 2\int_0^x K^+(x,y)[\frac{\psi(y)}{y}h(y) + I_\nu(uy^{1/2})\frac{[y^{1/2}B_n(y)]'}{u^{2n}}]dy, \quad (15.51)$$

for $x \in (0,\beta)$ and $h \in C^\infty(0,\beta)$. It is easy to verify that (exercise) if $\varepsilon_1^+ \in C^\infty(0,\beta)$ is a fixed point of $\mathcal{K}^+$, then W_1^+ is a solution to (15.8). To find a fixed point of $\mathcal{K}^+$, we recursively define $\delta_{k+1}^+ = \mathcal{K}^+(\delta_k^+)$ for $k \ge 0$ with $\delta_0^+(x) = 0$. For any $0 < y \le x < \beta$, we have

$$|K^+(x,y)| \le (xy)^{1/2}I_\nu(ux^{1/2})K_\nu(uy^{1/2}) = P^+(x)Q^+(y). \quad (15.52)$$

It follows from (15.51) and (15.52) that

$$|\delta_1^+(x)| \le 2P^+(x)\int_0^x y^{-1/2}Q^+(y)P^+(y)\frac{|[y^{1/2}B_n(y)]'|}{u^{2n}}dy,$$
$$|\delta_{k+1}^+(x) - \delta_k^+(x)| \le 2P^+(x)\int_0^x Q^+(y)|\frac{\psi(y)}{y}[\delta_k^+(y) - \delta_{k-1}^+(y)]|dy.$$

Since $\Phi_0(0) = 0$ and $\Phi_0'(x) = 2|x^{-1/2}\psi(x)|$ (exercise), we obtain by induction

$$|\delta_{k+1}^+(x) - \delta_k^+(x)| \le \frac{\lambda P^+(x)\Phi_n(x)}{u^{2n+1}} \cdot \frac{[\lambda\Phi_0(x)/u]^k}{k!}.$$

Therefore,

$$\varepsilon_1^+(x) := \lim_{k\to\infty}\delta_k^+(x) = \sum_{k=0}^\infty[\delta_{k+1}^+(x) - \delta_k^+(x)]$$

is well-defined and has an upper bound as given in (15.46). Moreover, it is a fixed point of the integral map $\mathcal{K}^+$, which implies that W_1^+ given in (15.44) is a solution to (15.8). The bound for ε_2^+ can be proved in a similar manner. □

Theorem 15.2. *Assume $\psi \in C^{\infty}(\alpha, 0)$, where $\alpha < 0$ may be negative infinity. For each $u > 0$ and $n = 0, 1, 2, \cdots$, the equation (15.8) possesses a pair of solutions in $C^{\infty}(\alpha, 0)$, given by*

$$W_1^- = |\zeta|^{1/2} J_\nu(u|\zeta|^{1/2}) \sum_{s=0}^{n} \frac{A_s(\zeta)}{u^{2s}} - |\zeta| J_{\nu+1}(u|\zeta|^{1/2}) \sum_{s=0}^{n-1} \frac{B_s(\zeta)}{u^{2s+1}} + \varepsilon_1^-(\zeta), \tag{15.53}$$

$$W_2^- = |\zeta|^{1/2} Y_\nu(u|\zeta|^{1/2}) \sum_{s=0}^{n} \frac{A_s(\zeta)}{u^{2s}} - |\zeta| Y_{\nu+1}(u|\zeta|^{1/2}) \sum_{s=0}^{n-1} \frac{B_s(\zeta)}{u^{2s+1}} + \varepsilon_2^-(\zeta), \tag{15.54}$$

where $A_s(\zeta)$ and $B_s(\zeta)$ are determined recursively from (15.42) and (15.43). The error terms $\varepsilon_1^-(\zeta)$ and $\varepsilon_2^-(\zeta)$ are bounded by

$$|\varepsilon_1^-(\zeta)| \le \frac{\lambda P^-(\zeta)\Psi_n(\zeta)e^{\lambda\Psi_0(\zeta)/u}}{u^{2n+1}}, \quad |\varepsilon_2^-(\zeta)| \le \frac{\lambda Q^-(\zeta)\Psi_n(\zeta)e^{\lambda\Psi_0(\zeta)/u}}{u^{2n+1}}, \tag{15.55}$$

where λ is defined in (15.47), $P^-(\zeta) := |\zeta|^{1/2} M_\nu(u|\zeta|^{1/2})/E_\nu(u|\zeta|^{1/2})$, $Q^-(\zeta) := |\zeta|^{1/2} M_\nu(u|\zeta|^{1/2}) E_\nu(u|\zeta|^{1/2})$, and

$$\Psi_k(x) := \pi \int_x^0 |\frac{d}{dy}[|y|^{1/2} B_k(y)]| dy \tag{15.56}$$

for $k = 0, 1, \cdots$.

Proof. Let W_1^- be given as in (15.53). We obtain (exercise)

$$\frac{d^2 W_1^-}{d\zeta^2} - [\frac{u^2}{4\zeta} + \frac{\nu^2 - 1}{4\zeta^2} + \frac{\psi(\zeta)}{\zeta}] W_1^- = \frac{d^2 \varepsilon_1^-}{d\zeta^2} - [\frac{u^2}{4\zeta} + \frac{\nu^2 - 1}{4\zeta^2} + \frac{\psi(\zeta)}{\zeta}] \varepsilon_1^-(\zeta)$$
$$- |\zeta|^{1/2} J_\nu(u|\zeta|^{1/2}) \frac{B_n'(\zeta) + B_n(\zeta)/(2\zeta)}{u^{2n}}. \tag{15.57}$$

Hence, W_1^- is a solution to (15.8) if and only if

$$\frac{d^2 \varepsilon_1^-}{d\zeta^2} - [\frac{u^2}{4\zeta} + \frac{\nu^2 - 1}{4\zeta^2}] \varepsilon_1^-(\zeta) = \frac{\psi(\zeta)}{\zeta} \varepsilon_1^-(\zeta) + J_\nu(u|\zeta|^{1/2}) \frac{[|\zeta|^{1/2} B_n(\zeta)]'}{u^{2n}}.$$

Now, we define a kernel function

$$K^-(x, y) := |xy|^{1/2} [J_\nu(u|x|^{1/2}) Y_\nu(u|y|^{1/2}) - Y_\nu(u|x|^{1/2}) J_\nu(u|y|^{1/2})], \tag{15.58}$$

for $\alpha < x \le y < 0$, and an integral map

$$\mathcal{K}^-(h)(x) := -\pi \int_x^0 K^-(x, y) [\frac{\psi(y)}{y} h(y) + J_\nu(u|y|^{1/2}) \frac{[|y|^{1/2} B_n(y)]'}{u^{2n}}] dy, \tag{15.59}$$

for $x \in (\alpha, 0)$ and $h \in C^\infty(\alpha, 0)$. It is easy to verify that (exercise) if $\varepsilon_1^- \in C^\infty(\alpha, 0)$ is a fixed point of $\mathcal{K}^-$, then W_1^- is a solution to (15.8). To find a fixed point of $\mathcal{K}^-$, we recursively define $\delta_{k+1}^- = \mathcal{K}^-(\delta_k^-)$ for $k \geq 0$ with $\delta_0^-(x) = 0$. For any $\alpha < x \leq y < 0$, it follows from (15.23) that

$$|K^-(x,y)| \leq P^-(x)Q^-(y). \tag{15.60}$$

Thus, by (15.59), we have

$$|\delta_1^-(x)| \leq \pi P^-(x) \int_x^0 y^{-1/2} Q^-(y) P^-(y) \frac{|[y^{1/2} B_n(y)]'|}{u^{2n}} dy,$$

$$|\delta_{k+1}^-(x) - \delta_k^-(x)| \leq \pi P^-(x) \int_x^0 Q^-(y) |\frac{\psi(y)}{y} [\delta_k^-(y) - \delta_{k-1}^-(y)]| dy.$$

Since $\Psi_0(0) = 0$ and $\Psi_0'(x) = -\pi |x|^{-1/2} |\psi(x)|$ (exercise), we obtain by induction

$$|\delta_{k+1}^-(x) - \delta_k^-(x)| \leq \frac{\lambda P^-(x) \Psi_n(x)}{u^{2n+1}} \cdot \frac{[\lambda \Psi_0(x)/u]^k}{k!}.$$

Therefore,

$$\varepsilon_1^-(x) := \lim_{k \to \infty} \delta_k^-(x) = \sum_{k=0}^{\infty} [\delta_{k+1}^-(x) - \delta_k^-(x)]$$

is well-defined and has an upper bound as given in (15.55). Moreover, it is a fixed point of the integral map $\mathcal{K}^-$, which implies that W_1^- given in (15.53) is a solution to (15.8). The bound for ε_2^- can be proved in a similar manner. □

15.5 Exercises

1. Verify (15.2).
2. Verify (15.3) and (15.4).
3. Verify (15.5).
4. Verify (15.12).
5. Verify (15.16) and (15.17).
6. Prove (15.18) and (15.19).
7. Verify (15.23).
8. Show that $\theta_\nu(x)$ defined in (15.22) is a continuous and nondecreasing function for $x > 0$.
9. Verify (15.26).
10. Verify (15.30) and (15.31).
11. Verify (15.33) and (15.34).

12. Verify (15.35) and (15.36).
13. Verify (15.40) and (15.41).
14. Prove that the constant λ defined in (15.47) is finite.
15. Verify (15.49).
16. Prove that if $\varepsilon_1^+ \in C^\infty(I)$ is a fixed point of the integral map $\mathcal{K}^+$ defined in (15.51), then W_1^+ given in (15.44) is a solution to (15.8).
17. Verify (15.52).
18. Prove that $\Phi_0(0) = 0$ and $\Phi_0'(x) = 2|x^{-1/2}\psi(x)|$; namely, $\Phi_0(x) = 2\int_0^x |y^{-1/2}\psi(y)|dy$.
19. Verify (15.57).
20. Prove that if $\varepsilon_1^- \in C^\infty(I)$ is a fixed point of the integral map $\mathcal{K}^-$ defined in (15.59), then W_1^- given in (15.53) is a solution to (15.8).
21. Verify (15.60).
22. Prove that $\Psi_0(0) = 0$ and $\Psi_0'(x) = -\pi|x|^{-1/2}|\psi(x)|$; namely, $\Psi_0(x) = \pi\int_x^0 |y|^{-1/2}|\psi(y)|dy$.

Chapter 16

Asymptotics of Differential Equations with Two Coalescing Turning Points

16.1 Introduction

The material in this chapter is primarily adapted from Olver (1975), while the final section is based on Dunster (2025). We are interested in the linear differential equation with a large parameter $u > 0$ and another bounded parameter $\kappa \in \mathbf{R}$:

$$\frac{d^2 w}{dx^2} = [u^2 f(u, \kappa, x) + g(u, \kappa, x)]w. \tag{16.1}$$

Assume there exists a certain critical value κ_0 such that $f(u, \kappa, x) = 0$ has two simple roots $z_1 < z_2$ for each $\kappa \neq \kappa_0$. Here, $z_1 = z_1(\kappa)$ and $z_2 = z_2(\kappa)$ are called turning points and depend on the parameter κ. We assume that as $\kappa \to \kappa_0$ these two turning points coalesce: $z_1(\kappa_0) = z_2(\kappa_0)$. Moreover, we assume that $f(u, \kappa, x)$ is negative for $x \in (z_1, z_2)$ and $\kappa \neq \kappa_0$. A typical example for this case is $f(u, \kappa, x) = x^2 - \kappa^2$. Actually, we can use the Liouville transformation

$$w = (\dot{x})^{1/2} W, \quad \dot{x} = dx/d\zeta \tag{16.2}$$

to convert (16.1) into (exercise)

$$\frac{d^2 W}{d\zeta^2} = [u^2 \dot{x}^2 f(u, \kappa, x) + \dot{x}^2 g(u, \kappa, x) + \dot{x}^{1/2} \frac{d^2}{d\zeta^2}(\dot{x}^{-1/2})]W. \tag{16.3}$$

We prescribe

$$\dot{x}^2 f(u, \kappa, x) = \zeta^2 - \alpha^2, \quad \zeta(z_1) = -\alpha, \quad \zeta(z_2) = \alpha. \tag{16.4}$$

A simple integration gives (exercise)

$$\int_{-\alpha}^{\zeta}(\alpha^2-\tau^2)^{1/2}d\tau=\int_{z_1}^{x}[-f(u,\kappa,t)]^{1/2}dt,\quad z_1\le x\le z_2, \tag{16.5}$$

$$\int_{\zeta}^{-\alpha}(\tau^2-\alpha^2)^{1/2}d\tau=\int_{x}^{z_1}[f(u,\kappa,t)]^{1/2}dt,\quad x\le z_1, \tag{16.6}$$

$$\int_{\alpha}^{\zeta}(\tau^2-\alpha^2)^{1/2}d\tau=\int_{z_2}^{x}[f(u,\kappa,t)]^{1/2}dt,\quad x\ge z_2. \tag{16.7}$$

In particular,

$$\frac{\pi}{2}\alpha^2=\int_{-\alpha}^{\alpha}(\alpha^2-\tau^2)^{1/2}d\tau=\int_{z_1}^{z_2}[-f(u,\kappa,t)]^{1/2}dt. \tag{16.8}$$

Now, we rewrite (16.3) as

$$\frac{d^2W}{d\zeta^2}=[u^2(\zeta^2-\alpha^2)+\psi(u,\alpha,\zeta)]W, \tag{16.9}$$

where

$$\psi(u,\alpha,\zeta):=\dot{x}^2g(u,\kappa,x)+\dot{x}^{1/2}\frac{d^2}{d\zeta^2}(\dot{x}^{-1/2}). \tag{16.10}$$

If $u>0$ is large, we may formally approximate $\psi(u,\alpha,\zeta)$ by 0 and hence find two approximate solutions to equation (16.9) in terms of the parabolic cylinder functions.

16.2 Parabolic cylinder functions

Consider the equation

$$y''(x)=(x^2/4+a)y(x). \tag{16.11}$$

A transformation $y(x)=e^{-x^2/4}Y(x)$ gives

$$Y''(x)-xY'(x)-(a+1/2)Y(x)=0.$$

The above equation possesses two linearly independent power series solutions (exercise):

$$Y_1(x)=\sum_{k=0}^{\infty}\frac{(a/2+1/4)_k}{2^k(1/2)_kk!}x^{2k},\quad Y_2(x)=\sum_{k=0}^{\infty}\frac{(a/2+3/4)_k}{2^k(3/2)_kk!}x^{2k+1}. \tag{16.12}$$

The Wronskian of $y_1(x):=e^{-x^2/4}Y_1(x)$ and $y_2(x):=e^{-x^2/4}Y_2(x)$ is (exercise)

$$y_1(x)y_2'(x)-y_1'(x)y_2(x)=1. \tag{16.13}$$

Now, we define

$$U(a,\pm x) := \frac{\sqrt{\pi}2^{-a/2-1/4}}{\Gamma(a/2+3/4)} y_1(x) \mp \frac{\sqrt{\pi}2^{-a/2+1/4}}{\Gamma(a/2+1/4)} y_2(x), \tag{16.14}$$

$$\begin{aligned}\bar{U}(a,\pm x) :=& \frac{\Gamma(1/4-a/2)\sin(3\pi/4-a\pi/2)}{\sqrt{\pi}2^{a/2+1/4}} y_1(x) \\ &\mp \frac{\Gamma(3/4-a/2)\sin(5\pi/4-a\pi/2)}{\sqrt{\pi}2^{a/2-1/4}} y_2(x).\end{aligned} \tag{16.15}$$

For any $a > -1/2$, it is easy to verify that (exercise)

$$U(a,x) = \frac{e^{-x^2/4}}{\Gamma(a+1/2)} \int_0^\infty t^{a-1/2} e^{-t^2/2-tx}\,dt, \tag{16.16}$$

from which we conclude $U(a,x) > 0$ for all $x \in \mathbf{R}$.

For each $a \le 0$, since the set of nonnegative roots of the equation $U(a,x) = \bar{U}(a,x)$ is finite and nonempty (exercise), we may denote by $X_a \ge 0$ the largest root. The graphs of $U(-3,x)$, $\bar{U}(-3,x)$, and $\bar{U}(-3,x) - U(-3,x)$ are plotted in Figure 16.1 for illustration.

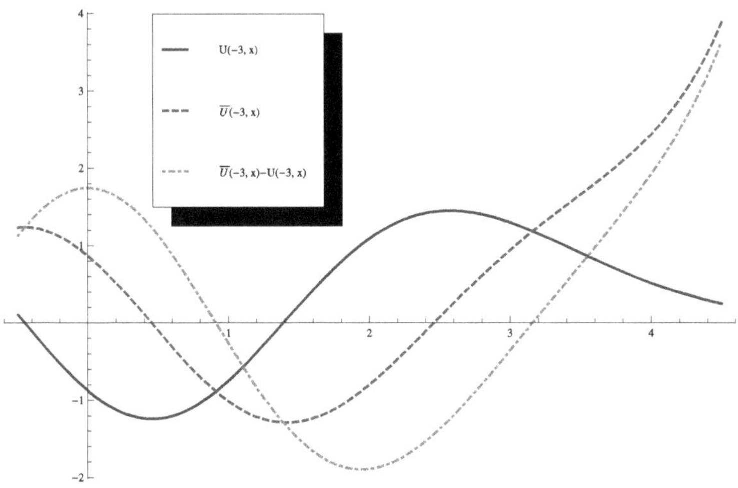

Fig. 16.1: The graphs of $U(-3,x)$, $\bar{U}(-3,x)$, and $\bar{U}(-3,x) - U(-3,x)$.

For all $a \le 0$, we introduce the following auxiliary functions

$$E_a(x) := \begin{cases} 1, & 0 \le x \le X_a, \\ \sqrt{\bar{U}(a,x)/U(a,x)}, & x \ge X_a, \end{cases} \tag{16.17}$$

$$M_a(x) := \begin{cases} \sqrt{U^2(a,x) + \bar{U}^2(a,x)}, & 0 \le x \le X_a, \\ \sqrt{2U(a,x)\bar{U}(a,x)}, & x \ge X_a, \end{cases} \tag{16.18}$$

$$\theta_a(x) := \begin{cases} \arctan[U(a,x)/\bar{U}(a,x)], & 0 \le x \le X_a, \\ \pi/4, & x \ge X_a. \end{cases} \tag{16.19}$$

It is readily seen that $U(a,x)E_a(x) = M_a(x)\sin\theta_a(x)$ and $\bar{U}(a,x)/E_a(x) = M_a(x)\cos\theta_a(x)$. For any $0 \le x \le y$, we have (exercise)

$$E_a(x) \le E_a(y), \quad |U(a,x)\bar{U}(a,y) - U(a,y)\bar{U}(a,x)| \le M_a(x)M_a(y)\frac{E_a(y)}{E_a(x)}. \tag{16.20}$$

16.3 Asymptotic solutions and error bounds

In this section, we shall find two linearly independent asymptotic solutions to (16.9) in terms of the parabolic cylinder functions. The error bounds of the asymptotic solutions will be given by the error-control function

$$\Phi(\zeta) = \Phi_{u,\alpha}(\zeta) := \int_0^{\zeta} \frac{|\psi(u,\alpha,t)|}{\Omega(t\sqrt{2u})}dt, \tag{16.21}$$

where $\Omega(x) \in C(\mathbf{R}, \mathbf{R}_+)$ is a conveniently chosen nonnegative function such that $\Omega(x) > 0$ for $x \ne 0$ and $\Omega(x) = O(|x|)$ as $|x| \to \infty$. Given two positive constants $A > 0$ and $\zeta_2 > 0$, we define a two-parameter region

$$B := [0, A] \times [0, \zeta_2) = \{(\alpha, \zeta) \in \mathbf{R}_+^2, \ \ 0 \le \alpha \le A, \ 0 \le \zeta < \zeta_2\}, \tag{16.22}$$

equipped with the Euclidean topology in $\mathbf{R}^2$. We say a two-variable function $W \in C^{0,2}(B, \mathbf{R})$ if W is continuous in B and has a continuous second partial derivative in ζ.

Theorem 16.1. *Assume that for each large $u > 0$, the function $\psi(u, \alpha, \zeta)$ is continuous for $(\alpha, \zeta) \in B := [0, A] \times [0, \zeta_2)$, and the integral definition of $\Phi(\zeta_2)$ in (16.21) converges uniformly with respect to α. Then for each large $u > 0$, the equation (16.9) has two solutions $W_1, W_2 \in C^{0,2}(B, \mathbf{R})$ given by*

$$W_1 = U(a,x) + \varepsilon_1, \quad W_2 = \bar{U}(a,x) + \varepsilon_2, \tag{16.23}$$

where $a = -u\alpha^2/2$, $x = \zeta\sqrt{2u}$, *and*

$$|\varepsilon_1| \le [M_a(x)/E_a(x)] \left\{ e^{\lambda[\Phi(\zeta_2)-\Phi(\zeta)]/\sqrt{u}} - 1 \right\}, \tag{16.24}$$

$$|\varepsilon_2| \le M_a(x)E_a(x) \left\{ e^{\lambda\Phi(\zeta)/\sqrt{u}} - 1 \right\}, \tag{16.25}$$

with

$$\lambda := \sup_{t\in[0,\zeta_2)} \frac{\sqrt{\pi}\Omega(t\sqrt{2u})M_a^2(t\sqrt{2u})}{2\Gamma(1/2-a)}. \tag{16.26}$$

Proof. First, we observe that $U(a,x)$ with $a = -u\alpha^2/2$ and $x = \zeta\sqrt{2u}$ satisfies the differential equation

$$\frac{d^2U}{d\zeta^2} = u^2(\zeta^2 - \alpha^2)U.$$

Hence, W_1 given in (16.23) is a solution to (16.9) if and only if

$$\frac{d^2\varepsilon_1}{d\zeta^2} - u^2(\zeta^2 - \alpha^2)\varepsilon_1 = \psi(u,\alpha,\zeta)[\varepsilon_1 + U(a,x)].$$

Now, we define a kernel function

$$K(x,y) := U(a,x)\bar{U}(a,y) - \bar{U}(a,x)U(a,y), \tag{16.27}$$

for $x = \zeta\sqrt{2u}$, $y = t\sqrt{2u}$, and $0 \le \zeta \le t < \zeta_2$, and an integral map

$$\mathcal{K}(h)(\alpha,\zeta) := \frac{\sqrt{\pi/u}}{2\Gamma(1/2-a)} \int_\zeta^{\zeta_2} K(x,y)\psi(u,\alpha,t)[h(\alpha,t) + U(a,y)]dt, \tag{16.28}$$

for $\alpha \in [0,A]$, $\zeta \in [0,\zeta_2)$ and $h \in C^{0,2}(B,\mathbf{R})$. Here again, we denote $x = \zeta\sqrt{2u}$ and $y = t\sqrt{2u}$ for simplicity. It is easy to verify that (exercise) if ε_1 is a fixed point of $\mathcal{K}$, then W_1 is a solution to (16.9). To find a fixed point of $\mathcal{K}$, we recursively define $\delta_{k+1} = \mathcal{K}(\delta_k)$ for $k \ge 0$ with $\delta_0 = 0$. For any $x = \zeta\sqrt{2u}$ and $y = t\sqrt{2u}$ with $0 \le \zeta \le t$, it follows from (16.20) that $|K(x,y)| \le P(x)Q(y)$, where

$$P(x) := M_a(x)/E_a(x), \ Q(y) := M_a(y)E_a(y). \tag{16.29}$$

Hence, from (16.28) we obtain $|\delta_1(\alpha,\zeta)| \le [\lambda P(x)/\sqrt{u}] \cdot [\Phi(\zeta_2) - \Phi(\zeta)]$ and

$$\begin{aligned}&|\delta_{k+1}(\alpha,\zeta) - \delta_k(\alpha,\zeta)| \\ \le& \frac{\sqrt{\pi/u}P(x)}{2\Gamma(1/2-a)} \int_\zeta^{\zeta_2} Q(y)|\psi(u,\alpha,t)[\delta_k(\alpha,t) - \delta_{k-1}(\alpha,t)]|dt.\end{aligned}$$

By induction, we have

$$|\delta_{k+1}(\alpha,\zeta) - \delta_k(\alpha,\zeta)| \le P(x) \cdot \frac{(\lambda/\sqrt{u})^{k+1}[\Phi(\zeta_2) - \Phi(\zeta)]^{k+1}}{(k+1)!},$$

Therefore,

$$\varepsilon_1 := \lim_{k\to\infty} \delta_k = \sum_{k=0}^{\infty}[\delta_{k+1} - \delta_k]$$

is well-defined and has an upper bound as given in (16.24). Moreover, it is a fixed point of the integral map $\mathcal{K}$, which implies that W_1 given in (16.23) is a solution to (16.9). The bound for ε_2 can be proved in a similar manner. □

16.4 More cases

In the previous sections, we assumed that for each $\kappa \neq \kappa_0$, the function $f(u,\kappa,x)$ has two simple real roots $z_1(\kappa) < z_2(\kappa)$ which coalesce as $\kappa \to \kappa_0$. We also assumed that $f(u,\kappa,x) < 0$ for $z_1(\kappa) < x < z_2(\kappa)$. Now, we consider a second case when $f(u,\kappa,x)$ is an even function that has two simple purely imaginary roots $iz_1(\kappa)$ and $iz_2(\kappa)$ with $z_1(\kappa) < z_2(\kappa)$ which coalesce as $\kappa \to \kappa_0$, and moreover, $f(u,\kappa,ix)$ is real for $x \in \mathbf{R}$ and positive for $z_1(\kappa) < x < z_2(\kappa)$. A typical example for the second case is $f(u,\kappa,x) = u^2(x^2+\kappa^2)$. Actually, by setting

$$\dot{x}^2 f(u,\kappa,x) = \zeta^2 + \alpha^2, \quad \zeta(iz_1) = -i\alpha, \quad \zeta(iz_2) = i\alpha, \tag{16.30}$$

we convert (16.3) into (exercise)

$$\frac{d^2W}{d\zeta^2} = [u^2(\zeta^2+\alpha^2) + \psi(u,\alpha,\zeta)]W, \tag{16.31}$$

where $\psi(u,\alpha,\zeta)$ is given by (16.10).

Theorem 16.2. *Fix $A > 0$ and $\zeta_1 < 0 < \zeta_2$. Assume that for each large $u > 0$, the function $\psi(u,\alpha,\zeta)$ is continuous for $(\alpha,\zeta) \in B_2 := [0,A] \times (\zeta_1,\zeta_2)$, and the integral definitions of $\Phi(\zeta_1)$ and $\Phi(\zeta_2)$ in (16.21) converge uniformly with respect to α. Then for each large $u > 0$, the equation (16.31) has two solutions $W_+, W_- \in C^{0,2}(B_2,\mathbf{R})$ given by*

$$W_+ = U(b,x) + \varepsilon_+, \quad W_- = U(b,-x) + \varepsilon_-, \tag{16.32}$$

where $b = u\alpha^2/2$, $x = \zeta\sqrt{2u}$, and

$$|\varepsilon_+| \le U(b,x)\left\{e^{\mu[\Phi(\zeta_2)-\Phi(\zeta)]/\sqrt{u}} - 1\right\}, \tag{16.33}$$

$$|\varepsilon_-| \le U(b,-x)\left\{e^{\mu[\Phi(\zeta)-\Phi(\zeta_1)]/\sqrt{u}} - 1\right\}, \tag{16.34}$$

with

$$\mu := \sup_{t\in(\zeta_1,\zeta_2)} \frac{\Gamma(1/2+b)\Omega(t\sqrt{2u})U(b,t\sqrt{2u})U(b,-t\sqrt{2u})}{2\sqrt{\pi}}. \tag{16.35}$$

Proof. First, we observe that $U(b,x)$ with $b = u\alpha^2/2$ and $x = \zeta\sqrt{2u}$ satisfies the differential equation

$$\frac{d^2U}{d\zeta^2} = u^2(\zeta^2 + \alpha^2)U.$$

Hence, W_+ given in (16.32) is a solution to (16.31) if and only if

$$\frac{d^2\varepsilon_+}{d\zeta^2} - u^2(\zeta^2 + \alpha^2)\varepsilon_+ = \psi(u,\alpha,\zeta)[\varepsilon_+ + U(b,x)].$$

Now, we define a kernel function

$$K_2(x,y) := U(b,x)U(b,-y) - U(b,-x)U(b,y), \tag{16.36}$$

for $x = \zeta\sqrt{2u}$, $y = t\sqrt{2u}$, and $\zeta_1 < \zeta \leq t < \zeta_2$. We also introduce an integral map

$$\mathcal{K}_2(h)(\alpha,\zeta) := \frac{\Gamma(1/2+b)}{2\sqrt{\pi u}} \int_\zeta^{\zeta_2} K_2(x,y)\psi(u,\alpha,t)[h(\alpha,t) + U(b,y)]dt, \tag{16.37}$$

for $\alpha \in [0,A]$, $\zeta \in (\zeta_1,\zeta_2)$ and $h \in C^{0,2}(B_2,\mathbf{R})$. Here again, we denote $x = \zeta\sqrt{2u}$ and $y = t\sqrt{2u}$ for simplicity. It is easy to verify that (exercise) if ε_+ is a fixed point of $\mathcal{K}_2$, then W_+ is a solution to (16.31). To find a fixed point of $\mathcal{K}_2$, we recursively define $\delta_{k+1} = \mathcal{K}_2(\delta_k)$ for $k \geq 0$ with $\delta_0 = 0$. For any $x = \zeta\sqrt{2u}$ and $y = t\sqrt{2u}$ with $\zeta_1 < \zeta \leq t < \zeta_2$, it follows from the monotonicity of $U(b,x)$ that $|K_2(x,y)| \leq U(b,x)U(b,-y)$. Hence, from (16.37) we obtain $|\delta_1(\alpha,\zeta)| \leq [\mu U(b,x)/\sqrt{u}] \cdot [\Phi(\zeta_2) - \Phi(\zeta)]$ and

$$\begin{aligned}&|\delta_{k+1}(\alpha,\zeta) - \delta_k(\alpha,\zeta)|\\ \leq&\frac{\Gamma(1/2+b)U(b,x)}{2\sqrt{\pi u}} \int_\zeta^{\zeta_2} U(b,-y)|\psi(u,\alpha,t)[\delta_k(\alpha,t) - \delta_{k-1}(\alpha,t)]|dt.\end{aligned}$$

We obtain by induction

$$|\delta_{k+1}(\alpha,\zeta) - \delta_k(\alpha,\zeta)| \leq U(b,x) \cdot \frac{(\mu/\sqrt{u})^{k+1}[\Phi(\zeta_2) - \Phi(\zeta)]^{k+1}}{(k+1)!}.$$

Therefore,

$$\varepsilon_+ := \lim_{k\to\infty} \delta_k = \sum_{k=0}^{\infty}[\delta_{k+1} - \delta_k]$$

is well-defined and has an upper bound as given in (16.33). Moreover, it is a fixed point of the integral map $\mathcal{K}_2$, which implies that W_+ given in (16.32) is a solution to (16.31). The bound for ε_- can be proved by symmetry. □

There are two more cases, Cases III and IV, which are similar to Cases I and II, respectively, except f is replaced with $-f$. The differential equations (16.9) and (16.31) are replaced by

$$\frac{d^2W}{d\zeta^2} = [u^2(\alpha^2 - \zeta^2) + \psi(u, \alpha, \zeta)]W, \tag{16.38}$$

and

$$\frac{d^2W}{d\zeta^2} = [-u^2(\zeta^2 + \alpha^2) + \psi(u, \alpha, \zeta)]W, \tag{16.39}$$

respectively. The asymptotic solutions shall be expressed in terms of the modified parabolic cylinder functions that satisfy the differential equation $y''(x) = (a - x^2/4)y(x)$. We refer the readers to Olver (1975) for more details concerning Cases III and IV.

16.5 Extension to complex and unbounded domains

Recently, Dunster (2025) extended Olver's analysis of two coalescing turning points to obtain asymptotic expansions with explicitly computable coefficients. These expansions remain valid when the argument is complex and unbounded, and even at certain singularities of the differential equation. The main result can be stated as follows. Consider

$$\frac{d^2w}{dz^2} = [u^2 f(\kappa, z) + g(z)]w, \tag{16.40}$$

where $f(\kappa, z)$ and $g(z)$ may depend on u and are analytic in an open domain Ω. For each $\kappa \in (\kappa_0, \kappa_1]$, the function $f(\kappa, z)$ has two simple real zeros $z_1(\kappa)$ and $z_2(\kappa)$. The closed interval $[z_1(\kappa), z_2(\kappa)]$ lies within Ω and shrinks to a point as $\kappa \to \kappa_0^+$. For any real $x \in \Omega$, $f(\kappa, x)$ is real, and moreover $f(\kappa, x) < 0$ for $x \in (z_1(\kappa), z_2(\kappa))$.

Same as in (16.8), we denote

$$\alpha = \sqrt{\frac{2}{\pi} \int_{z_1}^{z_2} [-f(\kappa, t)]^{1/2} dt}. \tag{16.41}$$

We further define two functions ξ and ζ by

$$\xi = \int_{z_2}^{z} [f(\kappa, t)]^{1/2} dt = \int_{\alpha}^{\zeta} (\tau^2 - \alpha^2)^{1/2} d\tau. \tag{16.42}$$

Let $z = x_1, x_2, x_3, x_4$ correspond to $\zeta = i\infty, \infty, -i\infty, -\infty$, respectively. Set

$$F_1 = \frac{1}{8f^2}\frac{d^2 f}{dz^2} - \frac{5}{32f^3}\left(\frac{df}{dz}\right)^2 + \frac{g}{2f}, \tag{16.43}$$

and

$$F_{k+1} = -\frac{1}{2f^{1/2}}\frac{dF_k}{dz} - \sum_{j=1}^{k-1}\frac{F_jF_{k-j}}{2}, \tag{16.44}$$

$$E_k = \int F_k f^{1/2} dz, \tag{16.45}$$

for $k \geq 1$, where the integration constants for E_k are arbitrarily chosen.

By using Liouville-Green approximation and asymptotic matching, Dunster (2025) showed that the differential equation (16.40) admits two linearly independent solutions

$$\begin{aligned} w_1 = e^{-(u\hat{\alpha}^2-1)\pi i/4}&[U(u\hat{\alpha}^2/2, -i\sqrt{2u}\zeta)A(u,\kappa,z) \\ &+ \partial_\zeta U(u\hat{\alpha}^2/2, -i\sqrt{2u}\zeta)B(u,\kappa,z)], \end{aligned} \tag{16.46}$$

$$\begin{aligned} w_3 = e^{(u\hat{\alpha}^2-1)\pi i/4}&[U(u\hat{\alpha}^2/2, i\sqrt{2u}\zeta)A(u,\kappa,z) \\ &+ \partial_\zeta U(u\hat{\alpha}^2/2, i\sqrt{2u}\zeta)B(u,\kappa,z)], \end{aligned} \tag{16.47}$$

and another pair of solutions

$$\begin{aligned} w_2 = \frac{\sqrt{2\pi}}{\Gamma(u\hat{\alpha}^2/2+1/2)}&[U(-u\hat{\alpha}^2/2, i\sqrt{2u}\zeta)A(u,\kappa,z) \\ &+ \partial_\zeta U(-u\hat{\alpha}^2/2, i\sqrt{2u}\zeta)B(u,\kappa,z)], \end{aligned} \tag{16.48}$$

$$\begin{aligned} w_4 = \frac{\sqrt{2\pi}}{\Gamma(u\hat{\alpha}^2/2+1/2)}&[U(-u\hat{\alpha}^2/2, -i\sqrt{2u}\zeta)A(u,\kappa,z) \\ &+ \partial_\zeta U(-u\hat{\alpha}^2/2, -i\sqrt{2u}\zeta)B(u,\kappa,z)], \end{aligned} \tag{16.49}$$

satisfying

$$w_2 = w_1 + w_3, \tag{16.50}$$

$$w_4 = e^{\lambda\pi i}w_1 + e^{-\lambda\pi i}w_3, \quad \lambda = (u\hat{\alpha}^2 - 1)/2. \tag{16.51}$$

The above relation defines the parameters λ and $\hat{\alpha}$. In fact, $\hat{\alpha}$ is a perturbation of α, with asymptotic expansion

$$\hat{\alpha}^2 \sim \alpha^2 + \frac{1}{\pi i}\sum_{k=0}^{\infty}\frac{E_{2k+1}^{-}(x_4) - E_{2k+1}^{+}(x_4)}{u^{2k+2}}, \quad u \to \infty, \tag{16.52}$$

where

$$E_{2k+1}^{\pm}(x_4) = \lim_{\varepsilon\to 0^+} E_{2k+1}(x_4 \pm i\varepsilon). \tag{16.53}$$

The coefficients $A(u,\kappa,z)$ and $B(u,\kappa,z)$ in (16.46)-(16.49) are slowly varying and have computable asymptotic expansions as $u \to \infty$. For further details, see Dunster (2025).

16.6 Exercises

1. Verify (16.3).
2. Verify (16.5)-(16.8).
3. Verify (16.12) and (16.13).
4. Verify (16.16).
5. Prove that for each $a \leq 0$, the set of nonnegative roots of the equation $U(a,x) = \bar{U}(a,x)$ is finite and nonempty.
6. Verify (16.20).
7. Prove that if ε_1 is a fixed point of the integral map $\mathcal{K}$ defined in (16.28), then W_1 given in (16.23) is a solution to (16.9).
8. Show that the transformation (16.30) converts (16.3) into (16.31) with $\psi(u, \alpha, \zeta)$ given by (16.10).
9. Prove that if ε_+ is a fixed point of the integral map $\mathcal{K}_2$ defined in (16.37), then W_+ given in (16.32) is a solution to (16.31).

Part 4

Integrals

Chapter 17

Classical Methods

17.1 Integration by parts

The material in this chapter is taken from Wong (1989). A simple yet powerful technique for deriving asymptotic expansions of definite integrals is the method of integration by parts. Each integration produces a term in the expansion, and the error term is given explicitly as an integral. This idea can be made clear by considering the following examples.

Example 1. The Exponential Integral.

This integral is defined by

$$\mathrm{Ei}(z) = \int_{-\infty}^{z} \frac{e^t}{t} dt, \qquad |\arg(-z)| < \pi, \tag{17.1}$$

where the integration can be taken along any path C in the complex t-plane with a cut along the positive real axis. A possible choice of C is the infinite line segment

$$-\infty < \mathrm{Re}\, t < \mathrm{Re}\, z, \quad \mathrm{Im}\, t = \mathrm{Im}\, z, \tag{17.2}$$

joining $t = -\infty$ to $t = z$ and parallel to the real axis. Repeated integration by parts gives

$$\mathrm{Ei}(z) = \frac{e^z}{z} + \int_{-\infty}^{z} \frac{e^t}{t^2} dt = \frac{e^z}{z} + \frac{e^z}{z^2} + 2\int_{-\infty}^{z} \frac{e^t}{t^3} dt = \frac{e^z}{z}\left[\sum_{k=0}^{n} \frac{k!}{z^k} + \varepsilon_n(z)\right], \tag{17.3}$$

where

$$\varepsilon_n(z) = (n+1)!ze^{-z} \int_{-\infty}^{z} \frac{e^t}{t^{n+2}} dt, \quad |\arg(-z)| < \pi. \tag{17.4}$$

To estimate the remainder $\varepsilon_n(z)$, we take C to be the line segment given in (17.2). Suppose $|\arg(-z)| \leq \pi - \delta$, where δ is any positive number in

$(0, \pi)$, and write $z = x + iy$. Along the segment $t = \tau + iy$, $-\infty < \tau \leq x$, we have $|e^t| = e^\tau$, $|t| \geq |z| \sin\delta$. Hence

$$|\varepsilon_n(z)| \leq \frac{(n+1)!}{|z|^{n+1}(\sin\delta)^{n+2}}. \tag{17.5}$$

Accordingly, as $z \to \infty$ in $|\arg(-z)| \leq \pi - \delta$,

$$\mathrm{Ei}(z) \sim \frac{e^z}{z}\sum_{k=0}^{\infty}\frac{k!}{z^k}. \tag{17.6}$$

If $\mathrm{Re}\, z \leq 0$, i.e., $|\arg(-z)| \leq \pi/2$, then we may take $\delta = \pi/2$ and (17.5) becomes

$$|\varepsilon_n(z)| \leq \frac{(n+1)!}{|z|^{n+1}}. \tag{17.7}$$

In this case, the error committed does not exceed the first neglected term in absolute value.

In this example, we are fortunate to have a sharp numerical upper bound for the absolute value of the remainder term. In general, it is often very difficult to find such an upper bound, and, even in cases when an upper bound can be obtained, it is quite useless for numerical calculation. This does not mean that the asymptotic expansion obtained is of no value, since for many purposes, particularly in applied mathematics, the asymptotic expansions obtained give much more information than we need to know. In fact, the knowledge of the first one or two terms of these expansions is usually sufficient for our purpose. However, to obtain results that are absolutely reliable for numerical computations, it is necessary to construct a sharp upper bound for the remainder term. Thus, if it is possible to obtain such an upper bound, then we must not neglect to do so.

Example 2. Fourier Integrals.

A second type of integrals to which the method of integration by parts can be applied is the Fourier integral

$$F(x) = \int_a^b f(t)e^{ixt}dt, \tag{17.8}$$

where (a, b) is a real finite interval and $f(t)$ is an N-times continuously differentiable function in $[a, b]$. By successive integration by parts, we have

$$F(x) = \sum_{n=0}^{N-1}\left(\frac{i}{x}\right)^{n+1}\left[e^{iax}f^{(n)}(a) - e^{ibx}f^{(n)}(b)\right] + \varepsilon_N(x), \tag{17.9}$$

where

$$\varepsilon_N(x) = \left(\frac{i}{x}\right)^N \int_a^b f^{(N)}(t)e^{ixt}dt. \tag{17.10}$$

The Riemann-Lebesgue lemma then gives $\varepsilon_N(x) = o(x^{-N})$ as $x \to +\infty$, thus showing that equation (17.9) is the asymptotic expansion of $F(x)$ as far as terms of order x^{-N}. If N is infinite, then (17.9) can be written as

$$F(x) \sim \sum_{n=0}^{\infty} \left(\frac{i}{x}\right)^{n+1} \left[e^{iax}f^{(n)}(a) - e^{ibx}f^{(n)}(b)\right], \tag{17.11}$$

as $x \to +\infty$.

A simple upper bound for the error term $\varepsilon_N(x)$ is given by

$$|\varepsilon_N(x)| \leq x^{-N} \int_a^b |f^{(N)}(t)|dt; \tag{17.12}$$

but this estimate falls short of the actual result $\varepsilon_N(x) = O(x^{-N-1})$. To improve this estimate, we integrate (17.10) by parts one more time and arrive at

$$\varepsilon_N(x) = \left(\frac{i}{x}\right)^{N+1} \left[e^{iax}f^{(N)}(a) - e^{ibx}f^{(N)}(b)\right] + \left(\frac{i}{x}\right)^{N+1} \int_a^b f^{(N+1)}(t)e^{ixt}dt, \tag{17.13}$$

which in turn gives

$$|\varepsilon_N(x)| \leq x^{-N-1} \left[|f^{(N)}(a)| + |f^{(N)}(b)| + \int_a^b |f^{(N+1)}(t)|dt\right]. \tag{17.14}$$

The foregoing results can easily be extended to semi-infinite intervals. For instance, if $f(t)$ is infinitely differentiable on $[a, \infty)$ and

$$f^{(s)}(t) = O(t^{-1-\varepsilon}), \tag{17.15}$$

as $t \to \infty$, for some $\varepsilon > 0$ and for every $s \geq 0$, then by letting $b \to \infty$ in (17.11) we establish

$$\int_a^{\infty} f(t)e^{ixt}dt \sim e^{iax} \sum_{n=0}^{\infty} f^{(n)}(a) \left(\frac{i}{x}\right)^{n+1}, \tag{17.16}$$

as $x \to +\infty$.

The above technique of integration by parts needs some modification when there are algebraic singularities at the end points. As a simple illustration, let us consider the infinite Fourier integral

$$F(x) = \int_0^{\infty} t^{\alpha-1}f(t)e^{ixt}dt, \tag{17.17}$$

where $0 < \alpha < 1$. Put

$$g_0(t) = t^{\alpha-1} e^{ixt} \tag{17.18}$$

and define

$$g_1(t) = -\int_t^{t+i\infty} \tau^{\alpha-1} e^{ix\tau} d\tau, \tag{17.19}$$

where the path of integration is the vertical line $\tau = t + iy$, $y \geq 0$. It is easily shown that $g_1'(t) = g_0(t)$ and

$$g_1(0) = -\int_0^{i\infty} \tau^{\alpha-1} e^{ix\tau} d\tau = -e^{i\pi\alpha/2}\Gamma(\alpha)x^{-\alpha}. \tag{17.20}$$

We assume that condition (17.15) again holds (with $\varepsilon = 0$). Upon integration by parts, we have from (17.17)

$$F(x) = f(0)e^{i\pi\alpha/2}\Gamma(\alpha)x^{-\alpha} - \int_0^{\infty} f'(t)g_1(t)dt. \tag{17.21}$$

To continue this process, we define

$$g_{n+1}(t) = -\int_t^{t+i\infty} g_n(\tau)d\tau = \frac{(-1)^{n+1}}{n!}\int_t^{t+i\infty} (z-t)^n z^{\alpha-1} e^{ixz} dz, \tag{17.22}$$

and observe that

$$g_{n+1}'(t) = g_n(t) \tag{17.23}$$

and

$$g_{n+1}(0) = \frac{(-1)^{n+1}\Gamma(n+\alpha)e^{i\pi(n+\alpha)/2}}{n!x^{n+\alpha}}. \tag{17.24}$$

Repeated integration by parts gives

$$F(x) = \sum_{n=0}^{N-1} \frac{f^{(n)}(0)\Gamma(n+\alpha)e^{i\pi(n+\alpha)/2}}{n!x^{n+\alpha}} + R_N, \tag{17.25}$$

where

$$R_N = (-1)^N \int_0^{\infty} f^{(N)}(t)g_N(t)dt. \tag{17.26}$$

To find a bound for R_N, we must first estimate $g_N(t)$. Note that along the path of integration in (17.22), we have

$$|g_{n+1}(t)| \leq \frac{t^{\alpha-1}}{n!}\int_0^{\infty} y^n e^{-xy} dy = t^{\alpha-1}x^{-n-1} \tag{17.27}$$

for $t > 0$, $x > 0$, $n = 0, 1, 2, \cdots$. From (17.26), it now follows that

$$|R_N| \leq x^{-N} \int_0^\infty t^{\alpha-1} f^{(N)}(t)| dt. \tag{17.28}$$

The clever technique given in this example is due to Erdélyi (1955). Although clever, the area of its application is rather limited. For a more powerful method in dealing with oscillatory integrals, we refer to Wong (1989, Chapter 4).

Example 3. Cases of Failure.

The above example clearly illustrates the basic idea behind the method of integration by parts. However, in applying this method, one often has the tendency to think that there is no need for the estimation of the remainder, since each integration yields a term in the asymptotic expansion and hence repeating this procedure infinitely many times should produce the entire expansion. This thought is incorrect, however, as we shall see in the present example. Let

$$S(x) = \int_0^\infty \frac{1}{\sqrt[3]{1+t}(x+t)} dt. \tag{17.29}$$

Repeated integration by parts gives

$$\begin{aligned} S(x) &= -\frac{3}{2}\frac{1}{x} + \frac{3 \cdot 1}{2} \int_0^\infty \frac{(1+t)^{2/3}}{(x+t)^2} dt \\ &= -\frac{3}{2}\frac{1}{x} - \frac{3^2 \cdot 1}{2 \cdot 5} + \frac{3^2 \cdot 2!}{2 \cdot 5} \int_0^\infty \frac{(1+t)^{5/3}}{(x+t)^3} dt \\ &= -\sum_{n=1}^{N-1} \frac{3^n (n-1)!}{2 \cdot 5 \cdots (3n-1)} \frac{1}{x^n} + \delta_N(x), \end{aligned} \tag{17.30}$$

where

$$\delta_N(x) = \frac{3^{N-1}(N-1)!}{2 \cdot 5 \cdots (3N-4)} \int_0^\infty \frac{(1+t)^{N-4/3}}{(x+t)^N} dt. \tag{17.31}$$

Thus, one might naturally conclude that

$$S(x) \sim -\sum_{n=1}^\infty \frac{3^n (n-1)!}{2 \cdot 5 \cdots (3n-1)} x^{-n}, \tag{17.32}$$

as $x \to +\infty$, which is obviously false, since the integral $S(x)$ is positive, whereas the terms in the series are all negative. Also, since $t \leq 1+t \leq x+t$ for $x \geq 1$, we have

$$\int_0^\infty \frac{t^{N-4/3}}{(x+t)^N} dt \leq \int_0^\infty \frac{(1+t)^{N-4/3}}{(x+t)^N} dt \leq \frac{3}{x^{1/3}}. \tag{17.33}$$

In view of the well-known identity (see Beals and Wong (2016, p. 18))

$$\int_0^\infty \frac{t^{\alpha-1}}{(1+t)^{\alpha+\beta}}dt = \frac{\Gamma(\alpha)\Gamma(\beta)}{\Gamma(\alpha+\beta)},$$

the integral on the extreme left of (17.33) can be explicitly evaluated to be $\Gamma(N-\frac{1}{3})\Gamma(\frac{1}{3})/[\Gamma(N)x^{1/3}]$. Therefore the remainder $\delta_N(x)$ is of higher order of magnitude than any of the terms in the series (17.30), violating the definition of an asymptotic expansion. The correct asymptotic expansion of $S(x)$ is given in Chapter 18.

17.2 Watson's Lemma

Another simple and useful technique for deriving asymptotic expansions of integrals is that of termwise integration of a series expansion of the integrand. This method is particularly easy to apply, when the integrand is exponentially decaying. To illustrate, we consider the integral

$$L(x) = \int_0^\infty \frac{t^{\lambda-1}}{1+t}e^{-xt}dt, \tag{17.34}$$

where $\lambda > 0$. Since

$$\frac{1}{1+t} = \sum_{n=0}^{N-1}(-1)^n t^n + (-1)^N \frac{t^N}{1+t}, \tag{17.35}$$

termwise integration gives

$$L(x) = \sum_{n=0}^{N-1}(-1)^n \frac{\Gamma(n+\lambda)}{x^{n+\lambda}} + R_N(x), \tag{17.36}$$

where

$$R_N(x) = (-1)^N \int_0^\infty \frac{t^{N+\lambda-1}}{1+t}e^{-xt}dt. \tag{17.37}$$

Thus

$$|R_N(x)| \le \frac{\Gamma(N+\lambda)}{x^{N+\lambda}}, \tag{17.38}$$

and

$$L(x) \sim \sum_{n=0}^{\infty}(-1)^n \frac{\Gamma(n+\lambda)}{x^{n+\lambda}}, \quad \text{as } x \to +\infty. \tag{17.39}$$

This is a special case of a much more general result due to Watson (1918, p. 133), now known as Watson's lemma. In view of its importance, Watson's statement and proof of the result are reproduced below.

Lemma 17.1 (Watson's Lemma). *If (i) $f(t)$ is analytic when $|t| \leq a + \delta$, where $a > 0, \delta > 0$, except at a branch point at the origin, and*

$$f(t) = \sum_{m=1}^{\infty} a_m t^{m/r-1} \tag{17.40}$$

when $|t| \leq a$, r being positive; (ii) $|f(t)| < Ke^{bt}$, where K and b are independent of t, when t is positive and $t \geq a$; (iii) $|\arg z| \leq \pi/2 - \Delta$, where $\Delta > 0$; and (iv) $|z|$ is sufficiently large; then there exists a complete asymptotic expansion given by the formula

$$F(z) = \int_0^{\infty} f(t)e^{-zt}dt \sim \sum_{m=1}^{\infty} a_m \Gamma(\frac{m}{r}) z^{-m/r}. \tag{17.41}$$

Proof. Let M be a fixed integer. Then we have

$$|f(t) - \sum_{m=1}^{M-1} a_m t^{m/r-1}| \leq K_1 t^{M/r-1} e^{bt} \tag{17.42}$$

for all $t > 0$, where K_1 is some number independent of t. Hence,

$$F(z) = \sum_{m=1}^{M-1} a_m \int_0^{\infty} t^{m/r-1} e^{-zt} dt + R_M(z),$$

where

$$\begin{aligned} |R_M(z)| &\leq K_1 \int_0^{\infty} t^{M/r-1} e^{bt} |e^{-zt}| dt \\ &< K_1 \Gamma(\frac{M}{r})(\operatorname{Re} z - b)^{-M/r}, \end{aligned} \tag{17.43}$$

provided that $\operatorname{Re} z > b$, which is the case when $|z|$ is sufficiently large. Since $(\operatorname{Re} z - b)^{-1} = O(1/z)$ for the range of values of z under consideration, we have

$$F(z) = \sum_{m=1}^{M-1} a_m \Gamma(\frac{m}{r}) z^{-m/r} + O(z^{-M/r}), \tag{17.44}$$

and so the integral possesses the complete asymptotic expansion. □

Among the known procedures, Watson's lemma is certainly one of the most frequently used methods for finding asymptotic expansions. However, its conditions and paths of integration are known to be more restrictive than necessary. In order to generalize, we consider the integral

$$F(z) = \int_0^{\infty e^{i\gamma}} f(t)e^{-zt}dt, \tag{17.45}$$

where γ is a fixed real number, and the path of integration is the straight line joining $t = 0$ to $t = \infty e^{i\gamma}$. The following result is given in Wyman (1964, p. 249).

Lemma 17.2 (Generalized Watson's Lemma). *Suppose that the integral in* (17.45) *exists for some fixed* $z = z_0$ *and that as* $t \to 0$ *along* $\arg t = \gamma$,

$$f(t) \sim \sum_{n=0}^{\infty} a_n t^{\lambda_n - 1}, \tag{17.46}$$

where $\operatorname{Re}\lambda_0 > 0$ *and* $\operatorname{Re}\lambda_{n+1} > \operatorname{Re}\lambda_n$. *Then*

$$F(z) \sim \sum_{n=0}^{\infty} a_n \Gamma(\lambda_n) z^{-\lambda_n} \tag{17.47}$$

as $z \to \infty$ *in* $|\arg(ze^{i\gamma})| \le \pi/2 - \Delta$, *for any real number* Δ *in the interval* $0 < \Delta \le \pi/2$.

Proof. The general properties of Laplace integrals (see Widder (1941, p. 37)) ensure that $F(z)$ exists as long as $\operatorname{Re}(ze^{i\gamma}) > \operatorname{Re}(z_0 e^{i\gamma})$. Put

$$g(t) = \int_c^t f(\tau) e^{-z_0 \tau} d\tau. \tag{17.48}$$

Then, by hypothesis, $|g(t)| \le M < \infty$ for all t on the line segment joining $t = c$ to $t = \infty e^{i\gamma}$. Thus, for any arbitrary fixed choice of $t = c = |c|e^{i\gamma}, 0 < |c| < \infty$, it is true that

$$\int_c^{\infty e^{i\gamma}} f(t) e^{-zt} dt = (z - z_0) \int_c^{\infty e^{i\gamma}} g(t) e^{-(z - z_0)t} dt. \tag{17.49}$$

Furthermore, the right-hand side is dominated by

$$\frac{M|z - z_0|}{\operatorname{Re}[(z - z_0)e^{i\gamma}]} \exp\{-\operatorname{Re}[(z - z_0)e^{i\gamma}]|c|\}. \tag{17.50}$$

Since

$$\begin{aligned} \operatorname{Re}[(z - z_0)e^{i\gamma}] &= \operatorname{Re}(ze^{i\gamma}) - \operatorname{Re}(z_0 e^{i\gamma}) \\ &\ge |z| \cos(\arg(ze^{i\gamma})) - |z_0| \ge |z| \sin\Delta - |z_0|, \end{aligned} \tag{17.51}$$

we have

$$\int_c^{\infty e^{i\gamma}} f(t) e^{-zt} dt = O(e^{-\delta |z||c|}), \tag{17.52}$$

for $|z| \ge 2|z_0| \csc\Delta$, where $\delta = \frac{1}{2}\sin\Delta$.

For every integer $N \ge 1$, there exist positive numbers K_N and r_N such that

$$f(t) = \sum_{n=0}^{N-1} a_n t^{\lambda_n - 1} + R_N \quad \text{for} \quad |t| \le r_N \text{ and } \arg t = \gamma, \tag{17.53}$$

where

$$|R_N| \le K_N |t^{\lambda_N - 1}|. \tag{17.54}$$

Hence,

$$\int_0^c f(t)e^{-zt}dt = \sum_{n=0}^{N-1} a_n \int_0^c t^{\lambda_n - 1}e^{-zt}dt + S_N \tag{17.55}$$

with $|c| \le r_N$ and

$$S_N = \int_0^c R_N e^{-zt}dt. \tag{17.56}$$

From (17.52), it follows that

$$\begin{aligned}\int_0^c t^{\lambda_n-1}e^{-zt}dt &= \int_0^{\infty e^{i\gamma}} t^{\lambda_n-1}e^{-zt}dt + O(e^{-\delta|c||z|}) \\ &= \Gamma(\lambda_n)z^{-\lambda_n} + O(e^{-\delta|c||z|}).\end{aligned} \tag{17.57}$$

Furthermore, (17.54) and (17.56) give

$$\begin{aligned}|S_N| &\le K_N \int_0^c |t^{\lambda_N-1}e^{-zt}dt| \le K_N \int_0^{\infty e^{i\gamma}} |t^{\lambda_N-1}e^{-zt}dt| \\ &= O(z^{-\lambda_N}), \text{ as } z \to \infty \text{ in } |\arg(ze^{i\gamma})| \le \pi/2 - \Delta.\end{aligned} \tag{17.58}$$

These results coupled together give

$$F(z) \sim \sum_{n=0}^{\infty} a_n \Gamma(\lambda_n) z^{-\lambda_n}, \tag{17.59}$$

uniformly in $\arg z$, as $z \to \infty$ in $|\arg(ze^{i\gamma})| \le \pi/2 - \Delta$. □

Example 4. The Bessel Function $J_\nu(x)$.

When $\operatorname{Re}\nu > -\frac{1}{2}$, the Bessel function of the first kind has the integral representation

$$\begin{aligned}J_\nu(x) &= \frac{(x/2)^\nu}{\Gamma(\frac{1}{2})\Gamma(\nu+\frac{1}{2})}\int_{-1}^{1}(1-t^2)^{\nu-1/2}\cos xt dt \\ &= \frac{(x/2)^\nu}{\Gamma(\frac{1}{2})\Gamma(\nu+\frac{1}{2})}\int_{-1}^{1}(1-t^2)^{\nu-1/2}e^{ixt}dt;\end{aligned} \tag{17.60}$$

see Olver *et al.* (2010, p. 224). By Cauchy's theorem, the last integral can be written as

$$\int_{-1}^{-1+i\infty}(1-t^2)^{\nu-1/2}e^{ixt}dt - \int_{1}^{1+i\infty}(1-t^2)^{\nu-1/2}e^{ixt}dt,$$

where the paths of integration are vertical lines through $t = -1$ and $t = +1$. In the first integral, we put $t = -1 + i2\tau$, and in the second integral, we put $t = 1 + i2\tau$. The resulting expression is

$$\int_{-1}^{1} (1-t^2)^{\nu-1/2} e^{ixt} dt = 2^{2\nu} \Big\{ e^{-i(x-\pi\nu/2-\pi/4)} I_-(x) + e^{i(x-\pi\nu/2-\pi/4)} I_+(x) \Big\}, \tag{17.61}$$

where

$$I_\pm(x) = \int_0^\infty \tau^{\nu-1/2}(1 \pm i\tau)^{\nu-1/2} e^{-2x\tau} d\tau. \tag{17.62}$$

Although this result is established when x is real and positive, by analytic continuation it also holds for complex x as long as $\mathrm{Re}\, x > 0$. Application of the generalized Watson's lemma to (17.62) with $\lambda_n = n + \nu + \frac{1}{2}$ and z replaced by $2x$ yields (exercise)

$$I_\pm(x) \sim \sum_{n=0}^{\infty} (-1)^n \frac{(\frac{1}{2}-\nu)_n}{n!} (\pm i)^n \Gamma(n+\nu+\frac{1}{2})(2x)^{-n-\nu-\frac{1}{2}}, \tag{17.63}$$

as $x \to \infty$ in $|\arg x| \le \pi/2 - \Delta$, where

$$(\lambda)_0 = 1, (\lambda)_k = \frac{\Gamma(\lambda+k)}{\Gamma(\lambda)} = \lambda(\lambda+1)\cdots(\lambda+k-1). \tag{17.64}$$

Substituting (17.63) in (17.61) and replacing x by z in the resulting expansion, we obtain

$$J_\nu(z) \sim \left(\frac{2}{\pi z}\right)^{1/2} \Bigg[\cos(z - \pi\nu/2 - \pi/4) \sum_{k=0}^{\infty} (-1)^k \frac{(\nu, 2k)}{(2z)^{2k}} - \sin(z - \pi\nu/2 - \pi/4) \sum_{k=0}^{\infty} (-1)^k \frac{(\nu, 2k+1)}{(2z)^{2k+1}} \Bigg] \tag{17.65}$$

as $z \to \infty$ in $|\arg z| \le \pi/2 - \Delta$. Here we have introduced the notation: $(\nu, 0) = 1$ and

$$(\nu, k) = \frac{(-1)^k}{k!}(\frac{1}{2}-\nu)_k(\frac{1}{2}+\nu)_k = \frac{(4\nu^2-1)(4\nu^2-3)\cdots(4\nu^2-(2k-1)^2)}{2^{2k}\cdot k!}. \tag{17.66}$$

To bound the error terms associated with the expansion (17.63), we assume that ν is real. From Taylor's theorem, we have

$$(1 \pm i\tau)^{\nu-\frac{1}{2}} = \sum_{k=0}^{n-1} (-1)^k \frac{(\frac{1}{2}-\nu)_k}{k!} (\pm i\tau)^k + r_n(\tau), \tag{17.67}$$

where

$$r_n(\tau) = (-1)^n \frac{(\frac{1}{2}-\nu)_n}{(n-1)!}(\pm i\tau)^n \int_0^1 (1-t)^{n-1}(1 \pm it)^{\nu-\frac{1}{2}-n}dt. \qquad (17.68)$$

Choose $n > \nu - \frac{1}{2}$ so that $|1 \pm it|^{\nu-n-1/2} \leq 1$ and

$$|r_n(\tau)| \leq \frac{|(\frac{1}{2}-\nu)_n|}{n!}\tau^n. \qquad (17.69)$$

From this it follows that the n-th error term of (17.63) does not exceed

$$\frac{(\frac{1}{2}-\nu)_n}{n!}\frac{\Gamma(\nu+n+\frac{1}{2})}{[2|x|\cos(\arg x)]^{\nu+n+\frac{1}{2}}}.$$

If x is positive, then this implies that the n-th error term of (17.63) is bounded in absolute value by the first neglected term of expansion. A combination of this result with (17.65) gives

$$J_\nu(x) = \sqrt{\frac{2}{\pi x}}\sum_{k=0}^{n-1}(-1)^k \frac{(\nu,k)}{2^k}\frac{\cos[x-(\nu+\frac{1}{2}+k)\pi/2]}{x^k} + \varepsilon_n(x), \qquad (17.70)$$

where

$$|\varepsilon_n(x)| \leq \sqrt{\frac{2}{\pi|x|}}\frac{|(\frac{1}{2}-\nu)_n|}{n!}\frac{\Gamma(\nu+n+\frac{1}{2})}{|\Gamma(\nu+\frac{1}{2})|}\frac{[\sec(\arg x)]^{\nu+n+\frac{1}{2}}}{(2|x|)^n}. \qquad (17.71)$$

If x is positive, then (17.71) simplifies to

$$|\varepsilon_n(x)| \leq \sqrt{\frac{2}{\pi x}}\frac{|(\nu,n)|}{2^n}x^{-n}. \qquad (17.72)$$

The sector of validity of (17.65) can be extended to $|\arg z| \leq \pi - \Delta$ by rotating the path of integration in (17.62) through an angle γ, where $-\pi/2 < \gamma < \pi/2$. To do this, we note that for $x > 0$, we have by Cauchy's theorem

$$I_\pm(x) = \int_0^{\infty e^{i\gamma}} \tau^{\nu-1/2}(1 \pm i\tau)^{\nu-1/2}e^{-2x\tau}d\tau. \qquad (17.73)$$

The integral on the right-hand side actually defines an analytic function in $-\pi/2-\gamma < \arg x < \pi/2-\gamma$. Thus, by varying γ in the interval $(-\pi/2, \pi/2)$, (17.73) provides an analytic continuation for $I_\pm(x)$ in the larger sector $|\arg x| \leq \pi - \Delta$. We shall denote the analytic continuation again by $I_\pm(x)$, with this understanding, the identity by (17.61) now holds for $|\arg x| \leq \pi - \Delta$. Applying the generalized Watson's lemma to (17.73) then shows that (17.63) is also valid for $-\pi/2 - \gamma + \Delta \leq \arg x \leq \pi/2 - \gamma - \Delta$. This, upon varying γ from $-\pi/2$ to $\pi/2$, would imply that (17.65) holds in the sector $|\arg z| \leq \pi - \Delta$.

17.3 Laplace's method

In probability theory, one frequently encounters integrals of the form

$$I_n = \int_a^b \varphi(x)[f(x)]^n dx, \tag{17.74}$$

where $\varphi(x)$ and $f(x)$ are continuous functions on the finite or infinite interval $[a, b]$, and $f(x)$ is positive there. Long ago, Laplace (1820) already made the observation that the major contribution to the integral I_n should come from the neighborhoods of the points where $f(x)$ attains its greatest value. For instance, if $f(x)$ has its maximum value only at the point ξ in (a, b) where $f'(\xi) = 0$ and $f''(\xi) < 0$, then Laplace's result is

$$I_n \sim \varphi(\xi)[f(\xi)]^{n+1/2} \left[\frac{-2\pi}{nf''(\xi)}\right]^{1/2} \tag{17.75}$$

as $n \to \infty$. An alternative form of this approximation is obtained by writing $f(x) = e^{h(x)}$; that is, if $h(x)$ attains its maximum only at ξ, where $h'(\xi) = 0$ and $h''(\xi) < 0$, then

$$I_n \sim \varphi(\xi)e^{nh(\xi)} \left[\frac{-2\pi}{nh''(\xi)}\right]^{1/2} \tag{17.76}$$

as $n \to \infty$. Formulas (17.75) and (17.76) are now known as Laplace's formula.

A heuristic argument for these formulas may proceed as follows. First, replace $\varphi(x)$ and $h(x)$ by the leading terms in their Taylor expansions at $x = \xi$, and then extend the integration limits to $-\infty$ and $+\infty$. Thus,

$$\begin{aligned}
\int_a^b \varphi(x)e^{nh(x)}dx &\approx \int_a^b \varphi(\xi)e^{n[h(\xi)+h''(\xi)(x-\xi)^2/2]}dx \\
&\approx \varphi(\xi)e^{nh(\xi)} \int_{-\infty}^{\infty} e^{n(x-\xi)^2h''(\xi)/2}dx \\
&= \varphi(\xi)e^{nh(\xi)} \left[\frac{-2\pi}{nh''(\xi)}\right]^{-1/2}.
\end{aligned} \tag{17.77}$$

To give a rigorous proof of the Laplace formula, we consider the equivalent integral

$$I(\lambda) = \int_a^b \varphi(x)e^{-\lambda h(x)}dx, \tag{17.78}$$

where λ is a large positive parameter. Our method is simply to reduce this integral to a Laplace transform and then apply Watson's lemma. First, we observe that by subdividing the range of integration at the minima and maxima of $h(x)$, and by reversing the sign of x whenever necessary, we may assume, without loss of generality, that $h(x)$ has only one minimum in $[a, b]$ which occurs at $x = a$. Next, we assume that

$$h(x) \sim h(a) + \sum_{s=0}^{\infty} a_s (x-a)^{s+\mu}, \tag{17.79}$$

and

$$\varphi(x) \sim \sum_{s=0}^{\infty} b_s (x-a)^{s+\alpha-1}, \tag{17.80}$$

as $x \to a^+$, and that the expansion (17.79) can be termwise differentiated, that is

$$h'(x) \sim \sum_{s=0}^{\infty} a_s (s+\mu)(x-a)^{s+\mu-1}, \tag{17.81}$$

as $x \to a^+$. Here μ is a positive constant, and α can be real or complex as long as $\operatorname{Re}\alpha > 0$. Also, we suppose, without loss of generality, that $a_0 \neq 0$ and $b_0 \neq 0$. The following result is an extension of Laplace's formula.

Theorem 17.1. *For the integral*

$$I(\lambda) = \int_a^b \varphi(x) e^{-\lambda h(x)} dx, \tag{17.82}$$

we assume that (i) $h(x) > h(a)$ for all $x \in (a, b)$, and for every $\delta > 0$ the infimum of $h(x) - h(a)$ in $[a+\delta, b)$ is positive; (ii)$h'(x)$ and $\varphi(x)$ are continuous in a neighborhood of $x = a$, except possibly at a; (iii) the expansions (17.79), (17.80) and (17.81) hold; and (iv) the integral $I(\lambda)$ converges absolutely for all sufficiently large λ. Then

$$I(\lambda) \sim e^{-\lambda h(a)} \sum_{s=0}^{\infty} \Gamma\left(\frac{s+\alpha}{\mu}\right) \frac{c_s}{\lambda^{(s+a)/\mu}}, \tag{17.83}$$

as $\lambda \to +\infty$, where the coefficients c_s are expressible in terms of a_s and b_s.

The first three coefficients are explicitly given by

$$c_0 = \frac{b_0}{\mu a_0^{\alpha/\mu}}, \quad c_1 = \left\{\frac{b_1}{\mu} - \frac{(\alpha+1)a_1 b_0}{\mu^2 a_0}\right\} \frac{1}{a_0^{(\alpha+1)/\mu}}, \tag{17.84}$$

and

$$c_2 = \left[\frac{b_2}{\mu} - \frac{(\alpha+2)a_1 b_1}{\mu^2 a_0} + \{(\alpha+\mu+2)a_1^2 - 2\mu a_0 a_2\} \frac{(\alpha+2)b_0}{2\mu^3 a_0^2}\right] \frac{1}{a_0^{(\alpha+2)/\mu}}. \tag{17.85}$$

For an explicit formula of the general coefficient c_s, see Nemes (2013).

Proof. Conditions (*ii*) and (*iii*) imply the existence of a number c in (a, b) such that $h'(x)$ and $\varphi(x)$ are continuous in $(a, c]$, and $h'(x)$ is also positive there. Put $T = h(c) - h(a)$, and introduce the new variable

$$t = h(x) - h(a). \tag{17.86}$$

Since $h(x)$ is increasing in (a, c), we may write

$$e^{\lambda h(a)} \int_a^c \varphi(x) e^{-\lambda h(x)} dx = \int_0^T f(t) e^{-\lambda t} dt \tag{17.87}$$

with $f(t)$ being the continuous function on $(0, T]$ given by

$$f(t) = \varphi(x) \frac{dx}{dt} = \frac{\varphi(x)}{h'(x)}. \tag{17.88}$$

Now substitute (17.79) into equation (17.86). Reversion of the resulting expression yields an asymptotic expansion of the form (exercise)

$$x - a \sim \sum_{s=1}^{\infty} \alpha_s t^{s/\mu}, \quad \text{as} \quad t \to 0^+; \tag{17.89}$$

The first three coefficients are given by

$$\alpha_1 = \frac{1}{a_0^{1/\mu}}, \; \alpha_2 = -\frac{a_1}{\mu a_0^{1+2/\mu}}, \tag{17.90}$$

and

$$\alpha_3 = \frac{(\mu + 3)a_1^2 - 2\mu a_0 a_2}{2\mu^2 a_0^{2+3/\mu}}. \tag{17.91}$$

Substitution of (17.89) in (17.88) then gives

$$f(t) \sim \sum_{s=0}^{\infty} c_s t^{(s+\alpha-\mu)/\mu}, \quad \text{as} \quad t \to 0^+. \tag{17.92}$$

To the integral on the right-hand side of (17.87) we can apply Watson's lemma. Thus

$$\int_a^c \varphi(x) e^{-\lambda h(x)} dx \sim e^{-\lambda h(a)} \sum_{s=0}^{\infty} \Gamma\left(\frac{s+\alpha}{\mu}\right) \frac{c_s}{\lambda^{(s+\alpha)/\mu}}, \tag{17.93}$$

as $\lambda \to +\infty$.

To complete the proof, we must also estimate the integral on the remaining range (c, b). Let λ_0 be a value of λ for which $I(\lambda)$ is absolutely convergent, and set

$$\varepsilon = \inf_{c \le x < b} \{h(x) - h(a)\}.$$

By condition (i), ε is positive. Now we compute, for $\lambda \geq \lambda_0$,

$$\begin{aligned}\lambda[h(x) - h(a)] =& (\lambda - \lambda_0)[h(x) - h(a)] + \lambda_0[h(x) - h(a)] \\ &\geq (\lambda - \lambda_0)\varepsilon + \lambda_0[h(x) - h(a)]\end{aligned}$$

and

$$|e^{\lambda h(a)} \int_c^b \varphi(x) e^{-\lambda h(x)} dx| \leq M e^{-\varepsilon\lambda}, \tag{17.94}$$

where M is a constant, namely,

$$M = e^{\lambda_0[\varepsilon + h(a)]} \int_c^b |\varphi(x)| e^{-\lambda_0 h(x)} dx.$$

The desired result (17.83) now follows from (17.93) and (17.94). □

Example 5. The Gamma function has the well-known integral representation

$$\Gamma(\lambda + 1) = \int_0^\infty u^\lambda e^{-u} du, \quad \lambda > 0. \tag{17.95}$$

The integrand has a maximum at $u = \lambda$ and its maximum value is $e^{-\lambda}\lambda^\lambda$. Since the location of the maximum is not fixed, we make the substitution $u = \lambda(1 + x)$. Then (17.95) takes the form

$$\begin{aligned}\Gamma(\lambda + 1) &= e^{-\lambda}\lambda^{\lambda+1} \int_{-1}^\infty [(1 + x)e^{-x}]^\lambda dx \\ &= e^{-\lambda}\lambda^{\lambda+1} \int_{-1}^\infty e^{-\lambda h(x)} dx,\end{aligned} \tag{17.96}$$

where

$$h(x) = x - \log(1 + x). \tag{17.97}$$

Splitting the interval of integration at $x = 0$, we have

$$e^\lambda \lambda^{\lambda+1} \Gamma(\lambda + 1) = \int_0^\infty e^{-\lambda h(x)} dx + \int_0^1 e^{-\lambda h(-x)} dx. \tag{17.98}$$

To each of the integrals on the right-hand side, we can now apply Theorem 17.1. Since

$$h(x) = \frac{1}{2}x^2 - \frac{1}{3}x^3 + \frac{1}{4}x^4 - \cdots, \quad x \in [0, 1),$$

we have, in the notations of (17.79) and (17.80), $\mu = 2, \alpha = 1, b_0 = 1, b_1 = b_2 = \cdots = 0$ and $a_s = (-1)^s/(s + 2)$. Thus, by (17.84) and (17.85), $c_0 = 1/\sqrt{2}, c_1 = 2/3$ and $c_2 = \sqrt{2}/12$. Expansion (17.83) then gives

$$\int_0^\infty e^{-\lambda h(x)} dx \sim \sqrt{\frac{\pi}{2}} \frac{1}{\lambda^{1/2}} + \frac{2}{3}\frac{1}{\lambda} + \frac{1}{12}\sqrt{\frac{\pi}{2}} \frac{1}{\lambda^{3/2}} + \cdots, \tag{17.99}$$

as $\lambda \to +\infty$. Similarly,

$$\int_0^1 e^{-\lambda h(x)} dx \sim \sqrt{\frac{\pi}{2}} \frac{1}{\lambda^{1/2}} - \frac{2}{3}\frac{1}{\lambda} + \frac{1}{12}\sqrt{\frac{\pi}{2}}\frac{1}{\lambda^{3/2}} - \cdots, \tag{17.100}$$

as $\lambda \to +\infty$. A combination of the last two expansions, together with (17.98) and the identity $\Gamma(\lambda+1) = \lambda\Gamma(\lambda)$, gives

$$\Gamma(\lambda) \sim e^{-\lambda}\lambda^{\lambda}\sqrt{\frac{2\pi}{\lambda}}\left[1 + \frac{1}{12\lambda} + \cdots\right], \quad \lambda \to +\infty. \tag{17.101}$$

To obtain higher order approximations, we repeat some of the arguments used in the proof of Theorem 17.1. Equation (17.86) suggests that in (17.99) we set

$$t = h(x) = \frac{1}{2}x^2 - \frac{1}{3}x^3 + \frac{1}{4}x^4 - \cdots. \tag{17.102}$$

Reversion of this expansion gives

$$x = \sum_{s=1}^{\infty} \alpha_s t^{s/2}. \tag{17.103}$$

An explicit expression for the coefficients α_s can be obtained from Lagrange's inversion formula (see Copson (1935, p. 125)), and a recurrence relation will be given later. From (17.88) and (17.92), it follows that

$$f(t) = \frac{dx}{dt} = \sum_{s=0}^{\infty} c_s t^{(s-1)/2}, \tag{17.104}$$

where

$$c_s = \frac{s+1}{2}\alpha_{s+1}. \tag{17.105}$$

By Watson's lemma, we have

$$\int_0^{\infty} e^{-\lambda h(x)} dx = \int_0^{\infty} e^{-\lambda x}\frac{dx}{dt} dt \sim \sum_{s=0}^{\infty} \Gamma(\frac{s+1}{2})\frac{c_s}{\lambda^{(s+1)/2}}, \tag{17.106}$$

In exactly the same manner, we set

$$t = h(-x) = \frac{1}{2}(-x)^2 - \frac{1}{3}(-x)^3 + \frac{1}{4}(-x)^4 - \cdots$$

in (17.100), and obtain, by reversion,

$$-x = \sum_{s=1}^{\infty} (-1)^s \alpha_s t^{s/2}.$$

Thus

$$\frac{dx}{dt} = \sum_{s=0}^{\infty} (-1)^s c_s t^{(s-1)/2},$$

and

$$\int_0^1 e^{-\lambda h(-x)} dx = \int_0^{\infty} e^{-\lambda t} \frac{dx}{dt} dt \sim \sum_{s=0}^{\infty} (-1)^s \Gamma(\frac{s+1}{2}) \frac{c_s}{\lambda^{(s+1)/2}}. \quad (17.107)$$

A combination of (17.98), (17.106) and (17.107) gives

$$\Gamma(\lambda) \sim e^{-\lambda} \lambda^{\lambda} \sum_{s=0}^{\infty} \Gamma(s + \frac{1}{2}) \frac{2c_{2s}}{\lambda^{s+1/2}}. \quad (17.108)$$

To determine the coefficients, we return to (17.102), i.e., $t = x - \log(1 + x)$. By differentiation, we have

$$x \frac{dx}{dt} = (1 + x). \quad (17.109)$$

Substitution of (17.103) in (17.109) yields $\alpha_1 = \sqrt{2}$ and the recurrence relation

$$\frac{s+2}{\sqrt{2}} \alpha_{s+1} = \alpha_s - \sum_{k=0}^{s-2} \frac{k+2}{2} \alpha_{k+2} \alpha_{s-k}, \quad (17.110)$$

$s = 1, 2, \cdots$. As usual, empty sums are understood to be zero. From (17.110), it follows readily that

$$\alpha_2 = \frac{2}{3}, \alpha_3 = \frac{\sqrt{2}}{18}, \alpha_4 = -\frac{2}{135}, \alpha_5 = \frac{\sqrt{2}}{1080},$$

and

$$c_2 = \frac{\sqrt{2}}{12}, \quad c_4 = \frac{\sqrt{2}}{432}.$$

Thus, we have finally

$$\Gamma(\lambda) \sim e^{-\lambda} \lambda^{\lambda} \left(\frac{2\pi}{\lambda} \right)^{1/2} \left[1 + \frac{1}{12\lambda} + \frac{1}{288\lambda^2} + \cdots \right]. \quad (17.111)$$

Taking just the first term gives what is known as Stirling's formula. An explicit formula for the general coefficient in the expansion (17.111) has been given by Nemes (2010b,a, 2013).

17.4 Principle of stationary phase

According to Watson (1944, pp. 229-230), the principle of stationary phase was first introduced by Thomson (1887), later known as Lord Kelvin, although the essence of this principle can be found in the works of Cauchy, Stokes, and Riemann. Kelvin's original problem was to find the asymptotic behavior of the integral

$$u(x) = \frac{1}{2\pi}\int_0^\infty \cos\{m[x - tf(m)]\}dm, \quad \text{as } x \to \infty,$$

where all variables are real. However, it is now more common to state the principle in terms of the integral

$$I(x) = \int_a^b g(t)e^{ixf(t)}dt, \tag{17.112}$$

where a, b and the function $f(t)$ are real, and x is a large real parameter.

The underlying principle of stationary phase is the assertion that the major asymptotic contribution to the integral (17.112) comes from points where the *phase function* $f(t)$ is *stationary*, i.e., where $f'(t)$ vanish. Suppose that $f(t)$ has only one such point, say c, in (a, b), and that $f''(c)$ is positive there, i.e., $f(c)$ is a simple minimum. Then, heuristically, we have

$$\begin{aligned} I(x) &\approx \int_{c-\varepsilon}^{c+\varepsilon} g(t)e^{ixf(t)}dt \quad (\varepsilon > 0) \\ &\approx g(c)\int_{c-\varepsilon}^{c+\varepsilon} e^{ix[f(c)+f''(c)(t-c)^2/2]}dt \\ &\approx g(c)e^{ixf(c)}\int_{-\infty}^{\infty} e^{ixf''(c)(t-c)^2/2}dt \\ &\approx g(c)\sqrt{\frac{2\pi}{xf''(c)}}e^{i[xf(c)+\pi/4]}. \end{aligned} \tag{17.113}$$

Similarly, if $f''(c) < 0$, i.e., $f(c)$ is a maximum, then

$$I(x) \sim g(c)\sqrt{\frac{2\pi}{-xf''(c)}}e^{i[xf(c)-\pi/4]}. \tag{17.114}$$

A rigorous proof of (17.113) has been given by Watson, but his discussion is rather restricted in scope. In particular, his argument does not seem capable of producing the higher-order terms in the asymptotic expansion. Considerably more satisfactory results have subsequently been obtained by van der Corput (1935, 1936); Erdélyi (1955); Olver (1974b). The argument which we shall present is due to Erdelyi.

First we assume that $f(t)$ has only a finite number of stationary points in the interval under consideration, i.e., $f'(t)$ has only a finite number of zeros there. Next, by subdividing the interval, we may assume that in each subinterval there is at most one stationary point, and that it is located at one of the endpoints. Without loss of generality, we may suppose that this point is the left endpoint. By changing the variable of integration from t to $-t$, or by changing $f(t)$ to $-f(t)$ and x to $-x$, we may also suppose that $f(t)$ is strictly increasing. Thus we shall consider the integral (17.112), in which $f(t)$ and $g(t)$ satisfy the following conditions:

(a) $f'(t) > 0$ in (a, b), and $f(t)$ has the form

$$f(t) = f(a) + (t-a)^{\rho} f_1(t), \quad f_1(a) \neq 0,$$

where $\rho \geq 1$ and $f_1(t)$ is infinitely differentiable in $[a, b]$.

(b) $g(t)$ is of the form

$$g(t) = (t-a)^{\lambda-1} g_1(t),$$

where $0 < \lambda \leq 1$ and $g_1(t)$ is also infinitely differentiable in $[a, b]$.

Our strategy is simply to reduce the integral $I(\lambda)$ in (17.112) to a Fourier-type integral, and then apply the integration-by-parts technique discussed in Sec.1. To this end, we introduce the variable u defined by

$$u^{\rho} = f(t) - f(a). \tag{17.115}$$

Since $f(t)$ is strictly increasing in $[a, b]$, there is a one-to-one relationship between t and u. Also, in consequence of condition (a), t is an infinitely differentiable function of u in $0 \leq u \leq B$, where $B^{\rho} = f(b) - f(a)$. In terms of the variable u, $I(x)$ becomes

$$I(x) = e^{ixf(a)} \int_0^B u^{\lambda-1} h(u) e^{ixu^{\rho}} du, \tag{17.116}$$

where

$$h(u) = u^{1-\lambda} g(t) \frac{dt}{du} = \left(\frac{t-a}{u} \right)^{\lambda-1} g_1(t) \frac{dt}{du}. \tag{17.117}$$

The smoothness conditions in (a) and (b) imply that $h(u)$ is a C^{∞}-function on $[0, B]$.

To proceed further, we extend $h(u)$ to a C^{∞}-function on $[0, \infty)$ with

$$h(u) \equiv 0 \text{ in a neighborhood of infinity,} \tag{17.118}$$

and write

$$
\begin{aligned}
I(x) &= e^{ixf(a)}\left(\int_0^\infty - \int_B^\infty\right) u^{\lambda-1}h(u)e^{ixu^\rho}\,du \\
&\equiv e^{ixf(a)}\left[I_1(x) - I_2(x)\right]. \qquad (17.119)
\end{aligned}
$$

The explicit extension of $h(u)$ on $[B,\infty)$ is not required; a construction of this extension is indicated in Rudin (1974, p. 418, Ex. 11 and Ex. 12). To apply successive integration by parts, we set

$$
k_0(u) = u^{\lambda-1}e^{ixu^\rho} \qquad (17.120)
$$

and define inductively

$$
k_{n+1}(u) = -\int_u^{u+\infty e^{i\pi/2\rho}} k_n(z)dz, \quad n = 0, 1, 2, \cdots, \qquad (17.121)
$$

where the path of integration is the ray $\arg(z-u) = \pi/2\rho$; cf. Sec. 1, equations (17.21)-(17.23). By interchanging the order of integration, it can be verified that

$$
k_{n+1}(u) = \frac{(-1)^{n+1}}{n!}\int_u^{u+\infty e^{i\pi/2\rho}} (z-u)^n z^{\lambda-1}e^{ixz^\rho}\,dz, \qquad (17.122)
$$

from which it follows that

$$
\begin{aligned}
k_{n+1}(0) &= \frac{(-1)^{n+1}}{n!}\int_0^{\infty e^{i\pi/2\rho}} z^{n+\lambda-1}e^{ixz^\rho}\,dz \\
&= \frac{(-1)^{n+1}}{n!\rho}\Gamma\left(\frac{n+\lambda}{\rho}\right)e^{i\pi(n+\lambda)/2\rho}x^{-(n+\lambda)/\rho}. \qquad (17.123)
\end{aligned}
$$

Since $dk_{n+1}/du = k_n(u)$, partial integration gives

$$
\begin{aligned}
I_1(x) &= \sum_{n=0}^{N-1}(-1)^{n+1}h^{(n)}(0)k_{n+1}(0) + R_N^{(1)}(x) \\
&= \sum_{n=0}^{N-1}\frac{1}{n!\rho}\Gamma\left(\frac{n+\lambda}{\rho}\right)h^{(n)}(0)e^{i\pi(n+\lambda)/2\rho}x^{-(n+\lambda)/\rho} + R_N^{(1)}(x)
\end{aligned}
$$

(17.124)

for any $N \geq 1$, the contributions from the upper limit of integration all vanishing in view of (17.118). The remainder $R_N^{(1)}(x)$ in (17.124) is given by

$$
R_N^{(1)}(x) = (-1)^N\int_0^\infty h^{(N)}(u)k_N(u)du. \qquad (17.125)
$$

The coefficients $h^{(n)}(0)$ in (17.124) can be expressed in terms of the original functions $f(t)$ and $g(t)$ via (17.115) and (17.117). In particular, we have

$$h(0) = g_1(a)f_1(a)^{-\lambda/\rho}. \tag{17.126}$$

To estimate $R_N^{(1)}(x)$, we first note that a further integration by parts gives

$$R_N^{(1)}(x) = (-1)^{N+1}\left[h^{(N)}(0)k_{N+1}(0) + \int_0^\infty h^{(N+1)}(u)k_{N+1}(u)du\right].$$

Hence, by (17.123),

$$\begin{aligned}|R_N^{(1)}(x)| \leq & \frac{1}{N!\rho}\Gamma\left(\frac{N+\lambda}{\rho}\right)|h^{(N)}(0)|x^{-(N+\lambda)/\rho} \\ & + \int_0^\infty |h^{(N+1)}(u)||k_{N+1}(u)|du. \end{aligned} \tag{17.127}$$

We next observe that along the path of integration in (17.122), we have $z = u + \zeta e^{i\pi/2\rho}$, where $u \geq 0$ and $\zeta \geq 0$. Clearly, $|z|^{\lambda-1} \leq u^{\lambda-1}$ for $\lambda \leq 1$, and

$$\rho\int_0^u (\eta + \zeta e^{i\pi/2\rho})^{\rho-1}d\eta = z^\rho - i\zeta^\rho.$$

Since the imaginary part of the integrand is positive, it follows that $\mathrm{Re}(ixz^\rho) + x\zeta^\rho \leq 0$, which in turn gives $|\exp(ixz^\rho)| \leq \exp(-x\zeta^\rho)$. Consequently,

$$\begin{aligned}|k_{N+1}(u)| &\leq \frac{u^{\lambda-1}}{N!}\int_0^\infty \zeta^N e^{-x\zeta^\rho}d\zeta \\ &= \frac{1}{N!\rho}\Gamma\left(\frac{N+1}{\rho}\right)u^{\lambda-1}x^{-(N+1)/\rho}. \end{aligned} \tag{17.128}$$

Inserting (17.128) in (17.127), we obtain

$$\begin{aligned}|R_N^{(1)}(x)| \leq & \frac{1}{N!\rho}\left[\Gamma\left(\frac{N+\lambda}{\rho}\right)|h^{(N)}(0)|x^{-(N+\lambda)/\rho}\right. \\ & \left. + \Gamma\left(\frac{N+1}{\rho}\right)\left\{\int_0^\infty u^{\lambda-1}|h^{(N+1)}(u)|du\right\}x^{-(N+1)/\rho}\right]. \end{aligned}$$

Since $0 < \lambda \leq 1$, this demonstrates that

$$R_N^{(1)}(x) = O(x^{-(N+\lambda)/\rho}) \quad \text{for} \quad x > 0. \tag{17.129}$$

We now turn to the second integral

$$I_2(x) = \int_B^\infty u^{\lambda-1}h(u)e^{ixu^\rho}du$$

in (17.119). The analysis for this integral is considerably simpler, since here the singular point $u = 0$ is outside the range of integration. The substitution $u^\rho = v$ gives

$$I_2(x) = \frac{1}{\rho}\int_{B^\rho}^{\infty} v^{\lambda/\rho-1}h(v^{1/\rho})e^{ixv}dv.$$

Since the integrand is infinitely differentiable, by repeated integration by parts we get

$$I_2(x) = \frac{1}{\rho}e^{ixB^\rho}\sum_{n=0}^{M-1} h_1^{(n)}(B^\rho)\left(\frac{i}{x}\right)^{n+1} + R_M^{(2)}(x) \tag{17.130}$$

for any $M \geq 1$, where

$$h_1(v) = v^{\lambda/\rho-1}h(v^{1/\rho}) \tag{17.131}$$

and

$$R_M^{(2)}(x) = \frac{1}{\rho}\left(\frac{i}{x}\right)^M \int_{B^\rho}^{\infty} h_1^{(M)}(v)e^{ixv}dv, \tag{17.132}$$

the contribution from the upper limit of integration again vanishing in view of (17.118); cf. Sec.1, equation (17.16). By the Riemann-Lebesgue lemma, $R_M^{(2)}(x) = o(x^{-M})$ as $x \to \infty$. Hence, a combination of (17.119), (17.124) and (17.130) gives

$$\begin{aligned} I(x) \sim & e^{ixf(a)}\sum_{n=0}^{\infty}\frac{1}{n!\rho}\Gamma\left(\frac{n+\lambda}{\rho}\right)h^{(n)}(0)e^{i\pi(n+\lambda)/2\rho}x^{-(n+\lambda)/\rho} \\ & - e^{ixf(b)}\sum_{n=0}^{\infty}\frac{1}{\rho}h_1^{(n)}(B^\rho)\left(\frac{i}{x}\right)^{n+1}, \end{aligned} \tag{17.133}$$

as $x \to +\infty$. The coefficients $(1/\rho)h_1^{(n)}(B^\rho)$ can be expressed in terms of the original functions $f(t)$ and $g(t)$. Indeed from (17.115), (17.117) and (17.131), we have

$$\frac{1}{\rho}h_1^{(n)}(B^\rho) = \left\{\frac{1}{f'(t)}\frac{d}{dt}\right\}^{(n)}\frac{g(t)}{f'(t)}\bigg|_{t=b};$$

cf. Olver (1974b, Eq. (2.3)).

If x is a negative parameter, or if $f'(t)$ is negative, then the asymptotic expansion of $I(x)$ can be obtained by taking complex conjugates or, equivalently, by changing the sign of i throughout.

To illustrate how the stationary phase approximation (17.113) follows from (17.133), we first split the integral (17.112) at $t = c$. Thus

$$I(x) = \int_a^c g(t)e^{ixf(t)}dt + \int_c^b g(t)e^{ixf(t)}dt.$$

We then replace t by $-t$ in the first integral on the right-hand side. By (17.133), we find that each integral on the right is asymptotically equal to half of the quantity

$$g(c)\sqrt{\frac{2\pi}{xf''(c)}}e^{i[xf(c)+\pi/4]}.$$

This of course establishes (17.113).

Example 6. The study of linear dispersive wave theory begins with the simple harmonic waves given by

$$u(x,t) = ae^{i(\kappa x-\omega t)}, \tag{17.134}$$

where t is time, a is the amplitude, κ is the wave number, and ω is the frequency. The *phase velocity* is the velocity dx/dt for which the phase function $\kappa x - \omega t$ is constant, i.e., $dx/dt = \omega/\kappa$. The term *dispersive* is referred to situation in which ω is a function of κ.

A continuous superposition of (17.134) yields

$$u(x,t) = \int_{-\infty}^{\infty} a(\kappa)e^{i\{\kappa x-\omega(\kappa)t\}}d\kappa. \tag{17.135}$$

To investigate the behavior of $u(x,t)$ for large positive t and fixed $\xi = x/t$, we apply the principle of stationary phase. The stationary points are determined from $\omega'(\kappa) = x/t$. The quantity $\omega'(\kappa)$ is referred to as the *group velocity*. If there is only one such point, then by (17.113) and (17.114)

$$u(x,t) \sim a(\kappa)\sqrt{\frac{2\pi}{t|\omega''(\kappa)|}}e^{i\kappa x-i\omega(\kappa)t-i(\pi/4)\mathrm{sgn}\omega''(\kappa)}, \tag{17.136}$$

as $t \to +\infty$. If there are more than one stationary point, then the asymptotic formula of $u(x,t)$ is obtained by adding up terms of the form (17.136).

To provide a concrete example of (17.135), we consider the Cauchy problem for the Klein-Gordon equation

$$u_{tt} - \gamma^2 u_{xx} + c^2 u = 0, \quad t > 0, \ -\infty < x < \infty, \tag{17.137}$$

with specified initial data $u(x,0)$ and $u_t(x,0)$. Let $U(\kappa,t)$ denote the Fourier transform of $u(x,t)$, i.e.,

$$U(\kappa,t) = \frac{1}{\sqrt{2\pi}}\int_{-\infty}^{\infty} e^{i\kappa x}u(x,t)dx.$$

The transformed equation is given by

$$\frac{\partial^2 U}{\partial t^2} + (\gamma^2\kappa^2 + c^2)U(\kappa,t) = 0,$$

and its solution is

$$U(\kappa, t) = a_+(\kappa)\exp(i\sqrt{\gamma^2\kappa^2 + c^2}t) + a_-(\kappa)\exp(-i\sqrt{\gamma^2\kappa^2 + c^2}t)$$

where $a_\pm(\kappa)$ are determined by the initial data. Taking the inverse Fourier transform, we obtain

$$\begin{aligned} u(x,t) =& \frac{1}{\sqrt{2\pi}} \int_{-\infty}^{\infty} a_+(\kappa)\exp\{i[\omega_+(\kappa)t - x\kappa]\}d\kappa \\ &+ \frac{1}{\sqrt{2\pi}} \int_{-\infty}^{\infty} a_-(\kappa)\exp\{i[\omega_-(\kappa)t - x\kappa]\}d\kappa, \end{aligned} \tag{17.138}$$

where $\omega_\pm(\kappa) = \pm\sqrt{\gamma^2\kappa^2 + c^2}$. Both integrals in (17.138) are of the form (17.135), and hence to each of them we can apply the asymptotic formula (17.136).

17.5 Method of steepest descents

This is a method for deriving asymptotic expansions of contour integrals of the form

$$I(\lambda) = \int_C g(z)e^{\lambda f(z)}dz, \tag{17.139}$$

where $f(z)$ and $g(z)$ are analytic functions, λ is again a large positive parameter and C is a contour in the z-plane. It was introduced by Debye (1909) in a paper concerning Bessel functions of large order. There are now so many references to this method in the literature that, by name, it is probably the best known procedure for finding asymptotic behavior of integrals. However, most descriptions of this procedure are so sketchy that it is probably also the most difficult to understand among all the classical methods. Debye's basic idea is to deform the contour C into a new path of integration C' so that the following conditions hold:

(i) C' passes through one or more zeros of $f'(z)$;
(ii) the imaginary part of $f(z)$ is constant on C'.

To obtain a geometrical interpretation of the new path of integration C', we write

$$f(z) = u(x, y) + iv(x, y), \tag{17.140}$$

where $z = x + iy$ and u and v are real. If u is treated as a third axis orthogonal to both x and y, then the equation $u = u(x, y)$ defines a surface S in the (x, y, u) space. Suppose that $z_0 = x_0 + iy_0$ is a zero of $f'(z)$. Then, by the Cauchy-Riemann equations, $f'(z) = u_x - iu_y$. Thus $f'(z_0) = 0$ implies

$$u_x(x_0, y_0) = u_y(x_0, y_0) = 0, \tag{17.141}$$

i.e., (x_0, y_0) is a critical point of $u(x, y)$. Since u is a harmonic function, $u(x, y)$ cannot have a maximum or a minimum at an interior point. Therefore, (x_0, y_0) must be a saddle point of $u(x, y)$; see Figure 17.1. For this reason, we call z_0 a *saddle point* of $f(z)$.

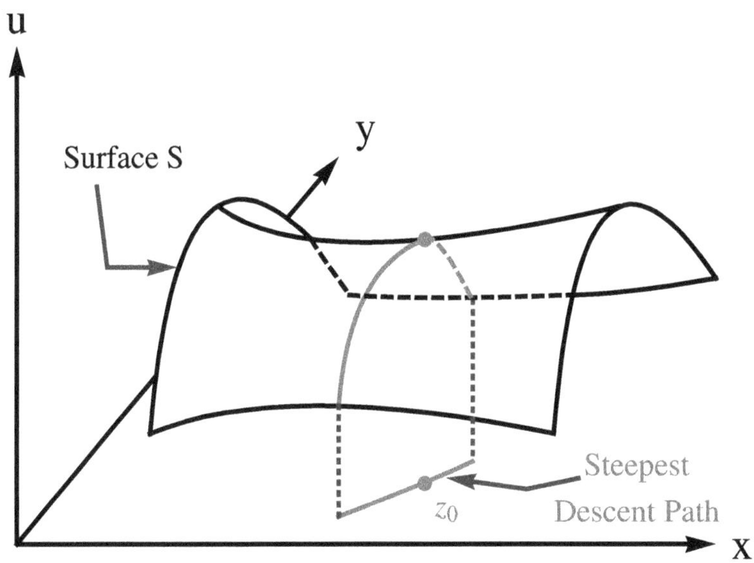

Fig. 17.1: Steepest descent path.

The shape of the surface S on the (x, y) plane can be represented by drawing the level curves on which u is constant. The curves v =constant are the orthogonal trajectories of the level curves, but so are the projections of the paths of steepest ascent or descent on the surface. The term *Steepest descent* stems from condition (ii) above.

We suppose that it is possible to deform the original path of integration C into a steepest path $v(x, y) = \text{constant} = \text{Im}f(z_0)$. On this path, we have

$$f(z) = f(z_0) - \tau, \tag{17.142}$$

where τ is real and is either monotonically increasing or monotonically decreasing. The integrand in (17.139) becomes $g(z)e^{\lambda f(z_0) - \lambda\tau}$. On a path where $\tau \to -\infty$, the integral may be divergent. For this reason we choose paths on which τ is positive and increasing. These are the paths of steepest descent from saddle points. We suppose that the contour C in (17.139) can be deformed into an equivalent contour consisting of paths of steepest descent through some of the saddle points. The saddle points involved in this deformation are called the *relevant saddle points*. Furthermore, if it is possible to express the original integral $I(\lambda)$ in (17.139) as a sum of integrals along paths of steepest descent from saddle points, then our problem is reduced to finding the asymptotic behavior of integrals of the form

$$e^{\lambda f(z_0)} \int_0^T g(\tau) \frac{dz}{d\tau} e^{-\lambda\tau} d\tau, \tag{17.143}$$

where T is a positive real number. In most cases, T is at $+\infty$, unless, for example, the integration path strikes another saddle point. The full contribution to the asymptotic expansion of the original integral $I(\lambda)$ is obtained by adding the contributions from all relevant saddle points. To the integral in (17.143), we can now apply Watson's lemma.

Example 7. The Airy integral defined by

$$\mathrm{Ai}(z) = \frac{1}{2\pi i} \int_L \exp\left(\frac{1}{3}t^3 - zt\right) dt, \tag{17.144}$$

where L is any contour which begins at infinity in the sector $-\pi/2 < \arg t < -\pi/6$ and ends at infinity in the sector $\pi/6 < \arg t < \pi/2$, plays an important role in "uniform asymptotic expansions" given later. For this reason, we shall give a complete discussion of the asymptotic behavior of this function and a summary of some of the analytical properties that it possesses.

From (17.144), it can be verified that $\mathrm{Ai}(z)$ has the Maclaurin expansion

$$\mathrm{Ai}(z) = 3^{-\frac{2}{3}} \left[\sum_{n=0}^{\infty} \frac{z^{3n}}{3^{2n} n! \Gamma(n + \frac{2}{3})} - 3^{-\frac{2}{3}} \sum_{n=0}^{\infty} \frac{z^{3n+1}}{3^{2n} n! \Gamma(n + \frac{4}{3})} \right]. \tag{17.145}$$

Thus, $\mathrm{Ai}(z)$ is an entire function. Furthermore, by integration by parts it can be verified that $\mathrm{Ai}(z)$ is a solution of

$$\frac{d^2 w}{dz^2} - zw = 0. \tag{17.146}$$

Clearly $\mathrm{Ai}(\omega z)$ and $\mathrm{Ai}(\omega^2 z)$ are all solutions of this differential equation, where $\omega = \exp(2\pi i/3)$. With the aid of Cauchy's theorem, we can show from (17.144) that these solutions are connected by the relation

$$\mathrm{Ai}(z) + \omega\, \mathrm{Ai}(\omega z) + \omega^2\, \mathrm{Ai}(\omega^2 z) = 0. \tag{17.147}$$

Instead of $\mathrm{Ai}(\omega z)$ and $\mathrm{Ai}(\omega^2 z)$, the function

$$\mathrm{Bi}(z) = i\omega^2\,\mathrm{Ai}(\omega^2 z) - i\omega\,\mathrm{Ai}(\omega z) \tag{17.148}$$

is often used as the second solution. It has the advantage of being real when z is real.

To put the integral (17.144) in the form of (17.139), we suppose for a moment that z is real and positive, and make the substitution $t = z^{1/2}u$. This yields

$$\mathrm{Ai}(z) = \frac{z^{1/2}}{2\pi i}\int_L e^{\lambda f(u)}\,du, \tag{17.149}$$

where $\lambda = z^{3/2}$ and $f(u) = \frac{1}{3}u^3 - u$. By analytic continuation, it follows that (17.149) is in fact valid for $-\pi/3 < \arg z < \pi/3$. The saddle points of the integrand in (17.149) are at $u = \pm 1$. Putting $u = s + it$, we have

$$f(u) = \frac{1}{3}s^3 - st^2 - s + i(s^2 t - t - \frac{1}{3}t^3). \tag{17.150}$$

Since $\mathrm{Im}\, f(\pm 1) = 0$, if $f(u) = U + iV$ then $V = 0$ implies either $t = 0$ or $t^2 - 3s^2 + 3 = 0$. We must now find the directions in which U decreases on the curve. On $t = 0$, we have $U = \frac{1}{3}s^3 - s$, which has a local minimum at $s = 1$ and a local maximum at $s = -1$. Thus, near $s = 1, t = 0$ is a steepest ascent curve. The other equation $t^2 - 3s^2 + 3 = 0$ represents two branches of a hyperbola with the asymptotes $t = \pm\sqrt{3}s$. On this hyperbola, $U = -\frac{8}{3}s^3 + 2s$, which decreases for $s > 1$. From Figure 17.2, it is clear that the branch of the hyperbola on the right half-plane is our desired path of steepest descent through the saddle point $u = 1$. (In Figure 17.2, arrows indicate the direction in which U decreases.)

We may now deform the original contour L into this path of steepest descent, and write

$$\mathrm{Ai}(z) = \frac{z^{1/2}}{2\pi i}\left(\int_1^{\infty e^{\pi i/3}} - \int_1^{\infty e^{-\pi i/3}}\right) e^{\lambda f(u)}\,du. \tag{17.151}$$

In both integrals above, $f(u) - f(1)$ is real and has a maximum at $u = 1$. Also, $f(u) - f(1)$ is decreasing as u moves away from $u = 1$. Put

$$\tau = f(1) - f(u) = -\frac{1}{3}u^3 + u - \frac{2}{3} = -(u-1)^2 - \frac{1}{3}(u-1)^3. \tag{17.152}$$

On our steepest descent path, τ is real; cf. Eq. (17.142). From (17.152), we have

$$\pm\, i\tau^{1/2} = (u-1)[1 + \frac{1}{3}(u-1)]^{1/2}, \tag{17.153}$$

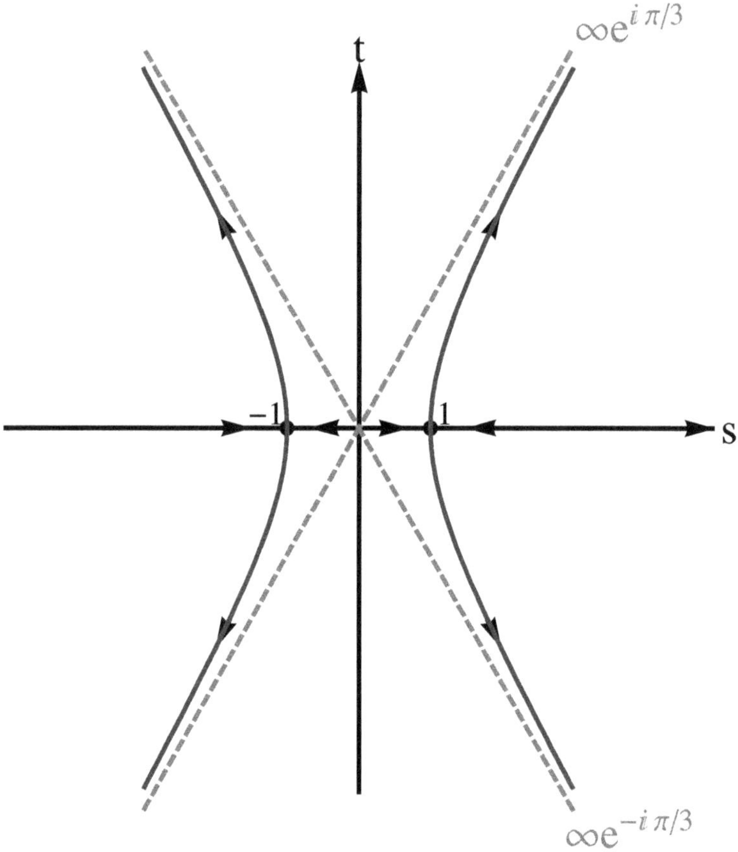

Fig. 17.2: Steepest paths for $\mathrm{Ai}(z)$.

where $[\cdots]^{1/2}$ is that branch which reduces to 1 at $u = 1$. By Lagrange's formula for the reversion of series (see Copson (1935, Section 6.23)),

$$u^{\pm} = 1 + \sum_{n=1}^{\infty} \alpha_n (\pm i\tau^{1/2})^n, \tag{17.154}$$

where

$$\alpha_n = \frac{1}{n!}\frac{d^{n-1}}{du^{n-1}}\left[1 + \frac{1}{3}(u-1)\right]^{-n/2}\Bigg|_{u=1}. \tag{17.155}$$

Since u^+ enters the first quadrant for increasing τ, we must take u^+ for the first integral in (17.151). Similarly, we take u^- for the second integral there, Equation (17.151) then becomes

$$\mathrm{Ai}(z) = \frac{z^{1/2}}{2\pi i} e^{-2\lambda/3} \int_0^\infty \left(\frac{du^+}{d\tau} - \frac{du^-}{d\tau} \right) e^{-\lambda\tau} d\tau. \tag{17.156}$$

Since $n\alpha_n$ is the coefficient of $(u-1)^{n-1}$ in the Taylor expansion of $[1 + \frac{1}{3}(u-1)]^{-n/2}$ at $u = 1$, it is easily shown that (exercise)

$$\alpha_n = \frac{(-1)^{n-1}\Gamma(\frac{3}{2}n - 1)}{n!\Gamma(\frac{1}{2}n)3^{n-1}}. \tag{17.157}$$

From this, it follows that

$$\frac{\partial u^+}{\partial \tau} - \frac{\partial u^-}{\partial \tau} = 2i \sum_{m=0}^\infty \frac{(-1)^m \Gamma(3m + \frac{1}{2})(m + \frac{1}{2})}{(2m+1)!\Gamma(m + \frac{1}{2})9^m} \tau^{m-1/2}. \tag{17.158}$$

By Watson's lemma,

$$\mathrm{Ai}(z) \sim \frac{1}{2\pi z^{1/4}} \exp(-\frac{2}{3} z^{3/2}) \sum_{m=0}^\infty \frac{(-1)^m \Gamma(3m + \frac{1}{2})}{(2m)!9^m} z^{-3m/2} \tag{17.159}$$

as $z \to \infty$, uniformly in $\arg z$ for $|\arg z| \le \pi/3 - \Delta, \Delta > 0$.

This example is purely illustrative, since a much simpler method is given in Exercise 1, by which the expansion (17.159) can be derived for the wider range of validity $|\arg z| \le \pi - \Delta, \Delta > 0$. To obtain an expansion in a sector that includes $\arg z = \pi$, one can use (17.147). Indeed, if $\pi/3 < \arg z < 5\pi/3$, then $\omega = e^{2\pi i/3}$ implies $-\pi/3 < \arg(\omega^{-1}z) < \pi$ and $-\pi < \arg(\omega^{-2}z) < \pi/3$. From (17.147), we have $\mathrm{Ai}(\omega^{-2}z) + \omega\,\mathrm{Ai}(\omega^{-1}z) + \omega^2\,\mathrm{Ai}(z) = 0$, which implies $\mathrm{Ai}(z) = -\omega^{-2}\,\mathrm{Ai}(\omega^{-2}z) - \omega^{-1}\,\mathrm{Ai}(\omega^{-1}z)$. Hence

$$\begin{aligned}\mathrm{Ai}(z) \sim{}& \frac{1}{2\pi z^{1/4}} \exp(-\frac{2}{3} z^{3/2}) \sum_{m=0}^\infty \frac{(-1)^m \Gamma(3m + \frac{1}{2})}{(2m)!9^m} z^{-3m/2} \\ &+ \frac{i}{2\pi z^{1/4}} \exp(\frac{2}{3} z^{3/2}) \sum_{m=0}^\infty \frac{\Gamma(3m + \frac{1}{2})}{(2m)!9^m} z^{-3m/2},\end{aligned} \tag{17.160}$$

an expansion that is uniformly valid in $\pi/3 + \Delta \le \arg z \le 5\pi/3 - \Delta$.

Observe that both expansions in (17.159) and (17.160) are valid in $\pi/3 + \Delta \le \arg z \le \pi - \Delta$, where Δ is positive but can be arbitrarily small. However, there is no inconsistency, since the first term on the right-hand side of (17.160) is *dominant* (exponentially increasing) and the second term is recessive (exponentially decreasing) in the sector $\pi/3 < \arg z < \pi$. The roles of these two terms are interchanged in the sector $-\pi/3 < \arg z < \pi/3$. The ray $\arg z = \pi/3$ is called a *Stokes line*. As $z \to \infty$ along this ray, the exponential factor in both terms are bounded, and bounded away from zero. Note also that there is a discontinuous change in the coefficients of the asymptotic expansion of $\mathrm{Ai}(z)$ when $\arg z$ changes continuously across $\arg z = \pi/3$; see equation (17.160). This circumstance was first discovered by Stokes, and is now known as *Stokes phenomenon*.

17.6 Exercises

1. Verify (17.63).
2. (a) Show that by restricting x to be real and positive, the Airy integral in (17.144) can be expressed as

$$\mathrm{Ai}(x) = \frac{1}{2\pi i}\int_{-i\infty}^{i\infty} \exp\left(\frac{1}{3}\tau^3 - x\tau\right) d\tau.$$

(b) By deforming the imaginary axis into the parallel vertical line $\mathrm{Re}\,\tau = \sqrt{x}$, show that we have

$$\mathrm{Ai}(x) = \frac{1}{\pi}e^{-\frac{2}{3}x^{3/2}}\int_0^\infty e^{-\sqrt{x}\rho^2}\cos(\frac{1}{3}\rho^3)d\rho.$$

(c) By replacing x by z and using analytic continuation, show that the above integral can be written as

$$\mathrm{Ai}(z) = \frac{1}{2\pi}e^{-\frac{2}{3}z^{3/2}}\int_0^\infty e^{-\sqrt{z}u}\cos(\frac{1}{3}u^{3/2})\frac{du}{\sqrt{u}}$$

for $|\arg z| < \pi$.

(d) Derive the asymptotic expansion

$$\mathrm{Ai}(z) \sim \frac{1}{2\pi z^{1/4}}e^{-\frac{2}{3}z^{3/2}}\sum_{n=0}^{\infty}\frac{\Gamma(3n+1/2)}{3^{2n}(2n)!}\frac{(-1)^n}{z^{3n/2}}$$

as $z \to \infty$ in $|\arg z| < \pi$; see Beals and Wong (2016, pp. 247-248).
3. Suppose that

$$h(x) \sim h(a) + \sum_{s=0}^{\infty} a_s(x-a)^{s+\mu}, \quad \text{as } x \to a^+,$$

and that $t = h(x) - h(a)$. Show that there exists an asymptotic representation of the form

$$x - a \sim \sum_{s=1}^{\infty} \alpha_s t^{s/\mu}, \quad \text{as } t \to 0^+,$$

and calculate the coefficients α_1, α_2 and α_3; cf. (17.90) and (17.91).
4. Verify (17.157).
5. Let

$$I(x) = \int_0^\infty K(t)K_0(xt)dt,$$

where $K_0(t)$ denotes the modified Bessel function of order 0. Assume that

$$f(t) = \sum_{s=0}^{\infty} a_s t^{s+\alpha-1} + f_n(t),$$

where $\alpha > 0$ and $|f_n(t)| \le M_n t^{n+\alpha-1} e^{\sigma t}$, for $t > 0$. Use the results

$$\int_0^\infty t^{\mu-1} K_0(t) dt = 2^{\mu-2} \Gamma^2(\mu/2), \quad \mu > 0,$$

and

$$0 \le K_0(t) \le \sqrt{\frac{\pi}{2t}} e^{-t}, \quad t > 0,$$

to derive the asymptotic expansion

$$I(x) = \frac{1}{4} \sum_{s=0}^{n-1} \Gamma^2\left(\frac{s+\alpha}{2}\right) \frac{a_s}{(x/2)^{s+\alpha}} + \delta_n(x).$$

Show that the remainder term satisfies

$$|\delta_n(t)| \le \sqrt{\frac{\pi}{2x}} M_n \Gamma(n+\alpha-1/2)(x-\sigma)^{-n-\alpha-1/2}.$$

6. Obtain a result, similar to the previous exercise, for the integral

$$I(x) = \int_0^\infty f(t) \operatorname{Ai}(xt) dt,$$

where $\operatorname{Ai}(x)$ is the Airy function which has the following properties:

$$\int_0^\infty t^{\mu-1} \operatorname{Ai}(t) dt = \frac{3^{2\mu/3-7/6}}{2\pi} \Gamma\left(\frac{\mu}{3}\right) \Gamma\left(\frac{\mu+1}{3}\right),$$

and

$$0 \le \operatorname{Ai}(t) \le \frac{1}{2\sqrt{\pi}} t^{-1/4} e^{-2t^{3/2}/3}, \quad t > 0.$$

7. Show that the integral

$$I(z) = \int_0^\infty e^{-z^2 t} \ln(1+t^{1/2}) dt$$

and its analytic continuation share the asymptotic expansion

$$I(z) \sim \sum_{s=1}^{\infty} \frac{(-1)^{s-1}\Gamma(s/2)}{2z^{s+2}}$$

as $z \to \infty$ in the sector $|\arg z| \le 5\pi/4 - \delta$ with any fixed small $\delta > 0$.

8. If L denotes the straight line joining $t = 0$ to $t = \pi(1+i)$, $(1+t)^{ix}$ has its principal value, and x is large and positive, show that

$$\int_L (1+t)^{ix} e^{ixe^t} dt = e^{ix} \left[\frac{i}{2x} + \frac{3i}{16x^3} + \frac{5}{32x^4} + O(x^{-5}) \right].$$

9. (a) Let $f(t)$ be m times continuously differentiable in $0 < t \leq b$, where m is a nonnegative integer and b is finite, and let

$$f(t) \sim \sum_{s=0}^{\infty} a_s t^{\lambda_s - 1},$$

as $t \to 0^+$, where $\lambda_0 > 0$ and $\lambda_{s+1} > \lambda_s$ for $s = 0, 1, 2, \cdots$. Suppose that this expansion can be differentiated term by term m times, and that n is a nonnegative integer such that $m < \lambda_n$. Show that

$$\int_0^b f(t)e^{ixt}dt = \sum_{s=0}^{n-1} a_s \Gamma(\lambda_s) e^{i\lambda_s \pi/2} x^{-\lambda_s} - e^{ibx} \sum_{s=0}^{m-1} f^{(s)}(b)(i/x)^{s+1} + \delta_{m,n}(x) - \varepsilon_{m,n}(x).$$

The error terms are given by

$$\delta_{m,n}(x) = \left(\frac{i}{x}\right)^m \int_0^b f_n^{(m)}(t)e^{ixt}dt,$$

and

$$\varepsilon_{m,n}(x) = \sum_{s=0}^{n-1} a_s e^{i\lambda_s \pi/2} \frac{\Gamma(\lambda_s)}{\Gamma(\lambda_s - m)} \Gamma(\lambda_s - m, -ixb) x^{-\lambda_s},$$

where the function $f_n(t)$ is defined by

$$f(t) = \sum_{s=0}^{n-1} a_s t^{\lambda_s - 1} + f_n(t)$$

and $\Gamma(a, z)$ is the complementary incomplete gamma function

$$\Gamma(a, z) = \int_z^{\infty} t^{\alpha-1} e^{-t} dt.$$

(b) Show that for any nonnegative integer m,

$$\Gamma(\alpha, z) = e^{-z} z^{\alpha-1} \sum_{j=0}^{m-1} \frac{\Gamma(\alpha)}{\Gamma(\alpha - j) z^j} + \frac{\Gamma(\alpha)\Gamma(\alpha - m, z)}{\Gamma(\alpha - m)},$$

and that by deforming the path of integration and then integrating by parts

$$|\Gamma(\alpha, \pm iy)| \leq 2y^{\alpha-1}, \quad \alpha < 1, \ y > 0.$$

(c) From (a) and (b) construct the error bounds

$$|\delta_{m,n}(x)| \leq \frac{1}{x^m} \int_0^b |f_n^{(m)}(t)| dt,$$

$$|\varepsilon_{m,n}(x)| \leq \frac{2}{x^{m+1}} \sum_{s=0}^{n-1} \frac{\Gamma(\lambda_s)}{\Gamma(\lambda_s - m)} |a_s| b^{\lambda_s - m - 1}.$$

If $\lambda_n \geq m+1$, show that $\delta_{m,n}(x)$ also satisfies

$$|\delta_{m,n}(x)| \leq \frac{1}{x^{m+1}}\left[|f_n^{(m)}(0)| + |f_n^{(m)}(b)| + \int_0^b |f_n^{(m+1)}(t)|dt\right].$$

10. Let $f(t)$ be an m times continuously differentiable function in $(0, \infty)$, m being a positive integer, and let

$$F(x) = \int_0^\infty f(t)e^{ixt}dt.$$

Suppose that

$$f(t) = \sum_{s=0}^{n-1} a_s t^{\lambda_s - 1}(\ln t) + f_n(t),$$

where n is the smallest nonnegative integer such that $\operatorname{Re}\lambda_n > m$, and $f_n^{(j)}(t) = O(t^{\lambda_n - j - 1}\ln t)$ as $t \to 0$, for $j = 0, 1, \cdots, m$. Also suppose that each of the integrals $\int_1^\infty f^{(j)}(t)e^{ixt}dt$, $j = 0, 1, \cdots, m$, converges uniformly for all sufficiently large x. Show that

$$F(x) = \sum_{s=0}^{n-1} a_s \Gamma(\lambda_s)e^{i\lambda_s\pi/2}[\psi(\lambda_s) + \frac{i\pi}{2} - \ln x]x^{-\lambda_s} + F_n(x),$$

where

$$F_n(x) = \left(\frac{i}{x}\right)^n \int_0^\infty f_n^{(m)}(t)e^{ixt}dt = o(x^{-n}),$$

as $x \to \infty$.

11. If $x > 0$ and n is any nonnegative integer, prove that

$$\left|\int_0^1 e^{-ixt}\ln(1+t)dt - \frac{ie^{-ix}\ln 2}{x} - \sum_{s=0}^{n-1} s!\left(1 - \frac{e^{-ix}}{2^{s+1}}\right)\left(\frac{i}{x}\right)^{s+2}\right| \leq \frac{2(n!)}{x^{n+2}}.$$

12. Let m be a positive integer. By expressing the integral

$$l(m) = \int_0^\pi \frac{\sin(mt)}{\sinh t}dt$$

in the form

$$\mathrm{Si}(m\pi) + \int_0^\pi \left(\frac{1}{\sinh t} - \frac{1}{t}\right)\sin(mt)dt,$$

where

$$\mathrm{Si}(z) := \int_0^z \frac{\sin t}{t}dt.$$

is known as a *sine integral*, show that

$$l(m) \sim \frac{\pi}{2} + \frac{(-1)^{m-1}}{m}\sum_{s=0}^\infty (-1)^s \frac{h_{2s}}{m^{2s}},$$

as $m \to \infty$, h_{2s} being the value of the $2s$-th derivative of $\operatorname{csch} t$ at $t = \pi$.

13. Let α and β be constants such that $0 < \alpha < 1$ and $\beta > 0$. Show that for large positive values of x,

$$\int_0^\infty e^{-t-xt^\alpha} t^{\beta-1} dt \sim \frac{\Gamma(\beta/\alpha)}{\alpha x^{\beta/\alpha}},$$

and

$$\int_0^\infty e^{-t+xt^\alpha} t^{\beta-1} dt \sim \sqrt{\frac{2\pi}{1-\alpha}} (\alpha x)^{(2\beta-1)/(2-2\alpha)} e^{(1-\alpha)(\alpha^\alpha x)^{1/(1-\alpha)}}.$$

14. Show that the modified Bessel function of order zero, i.e.,

$$I_0(x) = \frac{1}{\pi} \int_0^\pi e^{x\cos t} dt,$$

has the asymptotic expansion

$$I_0(x) \sim \frac{e^x}{\sqrt{2\pi x}} \sum_{s=0}^\infty \frac{1}{s!(8x)^s} \prod_{k=1}^s (2k-1)^2,$$

as $x \to \infty$.

15. Show that

$$\int_0^1 e^x \left(\frac{x}{1+x^2} \right)^n dx \sim \sqrt{\frac{\pi}{2n}} e 2^{-n},$$

as $n \to \infty$.

16. Prove the asymptotic equivalence

$$\int_0^\pi x^n \sin x dx \sim \pi^{n+1/2} n^{-2},$$

as $n \to \infty$.

17. Find the leading term in the asymptotic expansion of each of the following functions as $x \to \infty$,

$$(a) \quad \int_0^\pi \cos[x(\theta - \sin\theta)] d\theta,$$

$$(b) \quad \int_a^\infty \frac{1}{t} \exp[i(t + x/t)] dt, \ a > 0,$$

$$(c) \quad \int_a^\infty \frac{1}{t} \exp[i(t - x/t)] dt, \ a > 0.$$

18. Let N be a positive integer. Let $g(t)$ be N times, and $f(t)$ be $N+1$ times, continuously differentiable with $f'(t) \neq 0$ on $[a, b]$. Show that

$$F(x) = \int_a^b g(t) e^{ixf(t)} dt$$

has the expansion

$$F(x) = \sum_{n=0}^{N-1} \frac{(-1)^{n+1}}{(ix)^{n+1}} h^{(n)}(A) e^{if(a)x} + \sum_{n=0}^{N-1} \frac{(-1)^n}{(ix)^{n+1}} h^{(n)}(B) e^{if(b)x} + R_N,$$

where $A = f(a)$, $B = f(b)$, $h(u) = g(t)/f'(t)$, $u = f(t)$, and

$$R_N = \frac{(-1)^N}{(ix)^N} \int_a^b h^{(N)}(u) e^{ixu} du = o(x^{-N}),$$

as $x \to \infty$.

19. Show that for large positive x,

$$\int_0^\infty t e^{it^2(\ln t - \alpha)} dt \sim \sqrt{\frac{\pi}{e}} \exp[x - \frac{i}{2} e^{2x-1} + \frac{\pi i}{4}].$$

20. Show that

$$\int_0^{\pi/2} t \sin(x \cos t) dt \sim \frac{1}{x}(\frac{\pi}{2} - \cos x) + o(x^{-1}),$$

as $x \to \infty$.

21. Use the method of steepest descent to derive the asymptotic expansion of the Bessel function

$$J_\alpha(x) = \frac{1}{2\pi i} \int_{-\infty}^{(0^+)} t^{-\alpha-1} \exp[\frac{x}{2}(t - t^{-1})] dt,$$

as $x \to \infty$, where α is any complex number, and the loop contour starts from $-\infty$ on the real axis, passes around the origin in a counterclockwise direction and returns $-\infty$.

22. In the theory of wave pattern due to concentrated pressure moving over the free surface of water, there arose the integral

$$I(N, \theta) = \int_{-\infty - i\pi/4}^{\infty - i\pi/4} (\cosh u)^3 e^{Nf(u,\theta)} du,$$

where

$$f(u, \theta) = i[\cos\theta \cosh u - \frac{1}{2} \sin\theta \sinh(2u)].$$

Physically, N is the large non-dimensional radius measured from the disturbance, and $\theta > 0$ is the polar angle which appears as a parameter. (a) Show that the above integral is absolutely convergent for all $N > 0$. (b) Show that there exists a critical value θ_c such that (i) for $0 < \theta < \theta_c$, there are two real saddle points, (ii) for $\theta = \theta_c$, there is only one real saddle point which is of order 2, and (iii) for $\theta_c < \theta < \pi/2$, there are two conjugate complex saddle points in $|\operatorname{Im} u| < \pi/2$.

(c) Draw the steepest descent curve of $f(u,\theta)$ for each of the three cases mentioned in (b).

(d) Show that in case (i), we have

$$I(N,\theta) \sim \sqrt{\frac{2\pi}{N\sin\theta\sqrt{\cot^2\theta-8}}}\left\{\exp\left[\frac{iN\sin\theta\cosh^3\xi_-(\theta)}{\sinh\xi_-(\theta)}+\frac{i\pi}{4}\right]\right.$$
$$\left.+\exp\left[\frac{iN\sin\theta\cosh^3\xi_+(\theta)}{\sinh\xi_+(\theta)}-\frac{i\pi}{4}\right]\right\},$$

as $N\to\infty$, where $\xi_+(\theta)$ and $\xi_-(\theta)$ are the two real saddle points of $f(u,\theta)$ and $\xi_+(\theta)>\xi_-(\theta)$. Prove that the error in this approximation is $O((N\sin\theta)^{-3/2}(\cot^2\theta-8)^{-7/4})$.

(e) Show also that in case (ii), we have

$$I(n,\theta_c) = \frac{e^{i\sqrt{3}N/2}}{N^{1/3}}\left[\frac{3^{4/3}\Gamma(1/3)}{2^{3/2}} - \frac{5i}{N^{1/3}}\frac{3^{7/6}\Gamma(2/3)}{2^{5/2}} + O(1/N)\right],$$

as $N\to\infty$.

23. Consider the function

$$F(z) = \int_0^\infty e^{iz(t^3/3+t)}dt,$$

where $0<\arg z<\pi$. Show that, by rotating the path of integration through an angle ψ, $0<\psi<\pi/3$, $F(z)$ can be analytically continued to the region $-\pi/3<\arg z<\pi$. Furthermore, derive the asymptotic expansion

$$F(z) \sim i\sum_{m=0}^{\infty}\frac{(3m)!}{m!3^m}z^{-2m-1},$$

as $z\to\infty$, uniformly in the sector $-\pi/3+\delta\le\arg z\le\pi-\delta$, $\delta>0$. Note that this expansion holds, in particular, as $z\to\infty$ through the positive real values. Also, note that, since $F(z)=\overline{F(-z)}$, the domain of validity for the above expansion can be extended to $-\pi/3+\delta\le\arg z\le 4\pi/3-\delta$.

Chapter 18

Distribution Method

18.1 Introduction

The material in this chapter is taken from Wong (1989). In his Biographical Memoir of Arthur Erdélyi, Jones (1979) expressed that "by some odd mischance, he did not always discover the most important formulae." He also specifically mentioned two areas, where this had happened: (i) "he did not apply his theory for differential equations to find error estimates and so missed the theorems of most practical importance in asymptotic analysis"; (ii) "Another surprising omission from Erdélyi's investigation is the absence of an integral involving a generalized function, the asymptotic theory of which is both simpler and more general". From his own work Jones (1966, 1969, 1972, 1980, 1982a,b), it is evident that Jones himself is an expert in both asymptotic analysis and generalized functions. In this chapter, we will introduce an asymptotic method for integrals, which is based on the theory of generalized functions (also known as distribution theory).

We begin with an example. Consider the integral

$$F(x) = \int_0^\infty \frac{1}{\sqrt[3]{1+t}(x+t)} dt. \tag{18.1}$$

Using repeated integration by parts, we have

$$\begin{aligned} F(x) &= -\frac{3}{2x} + \frac{3 \cdot 1}{2} \int_0^\infty \frac{(1+t)^{2/3}}{(x+t)^2} dt \\ &= -\frac{3}{2x} - \frac{3^2}{2 \cdot 5x^2} + \frac{3^2 \cdot 2!}{2 \cdot 5} \int_0^\infty \frac{(1+t)^{5/3}}{(x+t)^3} dt \\ &= -\sum_{n=1}^{N-1} \frac{3^n \cdot (n-1)!}{2 \cdot 5 \cdots (3n-1) x^n} + \delta_N(x), \end{aligned} \tag{18.2}$$

where

$$\delta_N(x) = \frac{3^{N-1}(N-1)!}{2\cdot 5\cdots(3N-4)}\int_0^\infty \frac{(1+t)^{N-4/3}}{(x+t)^N}dt. \tag{18.3}$$

If we neglect the remainder $\delta_N(x)$ in (18.2), we are naturally led to conclude

$$F(x) \sim -\sum_{n=1}^{\infty}\frac{3^n\cdot(n-1)!}{2\cdot 5\cdots(3n-1)x^n} \tag{18.4}$$

as $x\to\infty$, which is obviously false, since the integral in (18.1) is positive but all terms in the series (18.4) are negative.

There is another formal approach to derive an expansion for the integral in (18.1); that is, to use the expansion

$$(1+t)^{-1/3} = \sum_{s=0}^{\infty}\binom{-1/3}{s}t^{-s-1/3}, \quad t>1. \tag{18.5}$$

Substituting (18.5) in (18.1), and integrating term by term, we obtain divergent integrals of the form

$$\int_0^\infty \frac{t^{-s-1/3}}{x+t}dt, \quad s=0,1,2,\cdots. \tag{18.6}$$

Can we give a meaning to these integrals? Yes, they can be interpreted in the theory of distributions. They are the limits of the distribution $t_+^{-s-1/3}$ acting on the test function $\varphi_\eta(t) = e^{-\eta t}/(x+t)$ as $\eta\to 0^+$; i.e., we set

$$\int_0^\infty \frac{t^{-s-1/3}}{x+t}dt = \frac{2\pi(-1)^s}{\sqrt{3}x^{s+1/3}}, \quad s=0,1,2,\cdots; \tag{18.7}$$

see Section 18.2 below. With this interpretation, termwise integration gives

$$F(x) \sim \frac{2\pi}{\sqrt{3}}\sum_{s=0}^{\infty}(-1)^s\binom{-1/3}{s}x^{-s-1/3} \tag{18.8}$$

as $x\to\infty$, which is again incorrect, in view of the correct result

$$F(x) \sim \frac{2\pi}{\sqrt{3}}\sum_{s=0}^{\infty}(-1)^s\binom{-1/3}{s}x^{-s-1/3} - \sum_{s=1}^{\infty}\frac{3^s(s-1)!}{2\cdot 5\cdots(3s-1)}x^{-s} \tag{18.9}$$

to be given in our next section.

The fact that expansion (18.8) misses all the terms in the second series in (18.9) raises the question what went wrong in the derivation of (18.7), knowing that the result (18.9) is correct (see Section 18.2 below). Note that the second series in (18.9) is exactly the incorrect result given in (18.4). Thus, this example illustrates the importance of error estimation; formal derivations often lead to incorrect results.

In the following sections, we shall show how to use the results established in distribution theory to derive asymptotic expansions of various integrals. Section 18.2 is devoted to a study of Stieltjes transform near infinity, and Section 18.3 deals with Laplace and Fourier transforms near the origin. In Section 18.4 we discuss the behavior of fractional integrals, and in Section 18.5 we use the method of regularization to derive asymptotic expansion for generalized Mellin convolution. Section 18.6 contains a discussion of a convolution integral on half line, and in Section 18.7 we present an asymptotic expansion for a Fourier transform on $\mathbf{R}^n$ near infinity.

18.2 Stieltjes transform

The Stieltjes transform of a locally integrable function $f(t)$ on $[0, \infty)$ is defined by

$$S_f(z) = \int_0^\infty \frac{f(t)}{t+z} dt, \tag{18.10}$$

where z is a complex variable in the cut plane $|\arg z| < \pi$. If $f(t)$ has an exponential rate of decay at ∞, then all the moments of f,

$$\mu_s = \int_0^\infty t^s f(t) dt, \quad s = 0, 1, 2, \cdots, \tag{18.11}$$

exist. Furthermore, if $z = x$ is real and positive, then it is easily shown that

$$S_f(x) = \sum_{s=0}^{n-1} (-1)^s \mu_s x^{-s-1} + \varepsilon_n(x), \tag{18.12}$$

where (exercise)

$$|\varepsilon_n(x)| \le x^{-n-1} \sup_{(0,\infty)} \left| \int_0^t \tau^n f(\tau) d\tau \right|. \tag{18.13}$$

However, if $f(t)$ decays only algebraically, then the above result is no longer valid.

In what follows, we shall assume that $f(t)$ possesses an asymptotic expansion of the form

$$f(t) \sim \sum_{s=0}^{\infty} a_s t^{-s-\alpha}, \tag{18.14}$$

as $t \to \infty$, where $0 < \alpha \le 1$. Substituting (18.14) in (18.10), and integrating term by term, we obtain divergent integrals of the form

$$\int_0^\infty \frac{t^{-s-\alpha}}{z+t} dt, \quad s = 0, 1, 2, \cdots. \tag{18.15}$$

To give a meaning to these integrals, we now begin to use some ideas from distribution theory.

Let $\mathcal{S}$ denote the space of infinitely differentiable functions that, together with their derivatives, approach zero more rapidly than any powers of $1/|t|$ as $|t| \to \infty$. More precisely, a function $\varphi(t)$ belongs to $\mathcal{S}$ or $\mathcal{S}(\mathbf{R})$ if and only if, given any $p, q = 0, 1, 2, \cdots$, there is a constant $C_{p,q}$ (depending on p, q and φ) such that

$$|t^p \varphi^{(q)}(t)| \leq C_{p,q}, \quad -\infty < t < \infty. \tag{18.16}$$

For example, the function $\exp(-t^2)$ belongs to $\mathcal{S}$ (exercise). The members of $\mathcal{S}$ are called *rapidly decreasing functions.* A sequence $\{\varphi_n\}$ of functions in $\mathcal{S}$ is said to converge to a function $\varphi \in \mathcal{S}$ if the sequence $\{\varphi_n^{(q)}\}$ converges uniformly to $\varphi^{(q)}$ on every finite interval and if the constants in the inequality

$$|t^p \varphi_n^{(q)}(t)| \leq C_{p,q} \tag{18.17}$$

can be chosen independently of n. For instance, the sequence $\varphi_n(t) = \frac{1}{n}\exp(-t^2)$ converges to zero in $\mathcal{S}$ as $n \to \infty$, whereas the sequence $\varphi_n(t) = \frac{1}{n}\exp(-t^2/n)$ does not (exercise). This is because the latter sequence does not satisfy (18.17) with $C_{p,q}$ independent of n. $\mathcal{S}$ is known as the *Schwartz space* and $\varphi(t) \in \mathcal{S}$ is called a *test function.* A continuous linear functional Λ on $\mathcal{S}$ is called a *(tempered) distribution*; that is, $\lim_{n\to\infty} \Lambda(\varphi_n) = \Lambda(\lim_{n\to\infty} \varphi_n)$ for any sequence $\{\varphi_n\}$ that is convergent in $\mathcal{S}$. In the following, we shall write $\langle \Lambda, \varphi \rangle$ for the action of the functional Λ on a test function φ. The set of tempered distributions on $\mathcal{S}$ is denoted by $\mathcal{S}'$.

A *distribution* is usually referred to as a linear functional on the space $\mathcal{D}$ of C^∞ functions on $(-\infty, \infty)$ with compact support. Since $\mathcal{D} \subset \mathcal{S}$ and convergence in $\mathcal{D}$ implies convergence in $\mathcal{S}$, every linear functional on $\mathcal{S}$ is a linear functional on $\mathcal{D}$; i.e., linear functionals on $\mathcal{S}$ are indeed distributions.

Let $\Lambda \in \mathcal{S}'(\mathbf{R})$. The distributional derivative Λ' of Λ is the functional defined by the formula

$$\langle \Lambda', \varphi \rangle = -\langle \Lambda, \varphi \rangle, \quad \varphi \in \mathcal{S}(\mathbf{R}). \tag{18.18}$$

Clearly the functional (18.18) is linear and continuous, and hence is itself a distribution. Higher order derivatives are defined in the same manner. Thus,

$$\langle \Lambda^{(k)}, \varphi \rangle = (-1)^k \langle \Lambda, \varphi^{(k)} \rangle, \quad k = 1, 2, \cdots . \tag{18.19}$$

If f is locally integrable on $[0, \infty)$ and satisfies $f(t) = O(t^{-\alpha})$ as $t \to \infty$ for some $\alpha > 0$, then f defines a distribution on $\mathcal{S}$ by

$$\langle f, \varphi \rangle = \int_0^\infty f(t)\varphi(t)dt. \tag{18.20}$$

Thus, if $t > 0$ and $0 < \alpha < 1$, then the function $t^{-\alpha}$ defines a distribution, supported on $[0, \infty)$, given by

$$\langle t^{-\alpha}, \varphi \rangle = \int_0^\infty t^{-\alpha}\varphi(t)dt. \tag{18.21}$$

Throughout the remaining portion of this section, we shall only consider distributions supported in $[0, \infty)$. Note that the integral in (18.21) cannot be used to define distributions associated with the functions $t^{-s-\alpha}$ with $s = 1, 2, \ldots$, in the same way as it is used to define the distribution associated with $t^{-\alpha}$. However, we can assign to these functions the distribution given by

$$\langle t^{-s-\alpha}, \varphi \rangle = \frac{1}{(\alpha)_s}\int_0^\infty t^{-\alpha}\varphi^{(s)}(t)dt, \tag{18.22}$$

where $(\alpha)_s = (\alpha)(\alpha+1)\cdots(\alpha+s-1)$ and $s = 1, 2, \cdots$. These are, by definition, the distributional derivative of the distribution given by (18.21). The functional δ defined by

$$\langle \delta, \varphi \rangle = \varphi(0) \tag{18.23}$$

is clearly a distribution on $\mathcal{S}$. The derivatives $\delta^{(s)}$ of the distribution δ are, by definition,

$$\langle \delta^{(s)}, \varphi \rangle = (-1)^s \varphi^{(s)}(0), \quad s = 1, 2, \cdots. \tag{18.24}$$

We now return to the expansion (18.14) and put

$$f(t) = \sum_{s=0}^{n-1} a_s t^{-s-\alpha} + f_n(t), \tag{18.25}$$

for each $n \geq 1$. To assign a distribution to the function $f_n(t)$, we first define inductively $f_{n,0}(t) = f_n(t)$ and

$$f_{n,j+1}(t) = -\int_t^\infty f_{n,j}(\tau)d\tau = \frac{(-1)^{j+1}}{j!}\int_t^\infty (\tau - t)^j f_n(\tau)d\tau, \tag{18.26}$$

for $j = 0, 1, \cdots, n-1$. For $0 < \alpha < 1$, it is easily seen that $f_{n,n}(t)$ is bounded on $[0, R]$ for any positive R and $O(t^{-\alpha})$ as $t \to \infty$. Therefore, we may define the distributions associated with $f_n(t)$ by

$$\langle f_n, \varphi \rangle = \langle f_{n,n}^{(n)}, \varphi \rangle = (-1)^n \langle f_{n,n}, \varphi^{(n)} \rangle = (-1)^n \int_0^\infty f_{n,n}(t)\varphi^{(n)}(t)dt. \tag{18.27}$$

We have now assigned a distribution to each function in (18.25). A natural question that arises is: "How are these distributions related?" The answer to this question is provided by the following lemma.

Lemma 18.1. *For $0 < \alpha < 1$ and $n \geq 1$, the identity*

$$\langle f, \varphi \rangle = \sum_{s=0}^{n-1} a_s \langle t^{-s-\alpha}, \varphi \rangle - \sum_{s=1}^{n} c_s \langle \delta^{(s-1)}, \varphi \rangle + \langle f_n, \varphi \rangle \tag{18.28}$$

holds for any rapidly decreasing function $\varphi \in \mathcal{S}$, where

$$c_s = \frac{(-1)^s}{(s-1)!} M[f; s] = \frac{(-1)^s}{(s-1)!} \int_0^\infty t^{s-1} f(t) dt \tag{18.29}$$

is the Mellin transform of $f(t)$, or its analytic continuation.

Proof. Let $f_0(t) = f(t)$. Then

$$f_{n+1}(t) = f_n(t) - a_n t^{-n-\alpha}$$

and

$$f_{n+1,n}(t) = f_{n,n}(t) - \frac{(-1)^n a_n}{(\alpha)_n} t^{-\alpha}, \tag{18.30}$$

for $n = 0, 1, 2, \cdots$. From this it follows, by integration by parts, that

$$\langle f_n, \varphi \rangle = -f_{n+1,n+1}(0) \langle \delta^{(n)}, \varphi \rangle + \langle f_{n+1}, \varphi \rangle + a_n \langle t^{-n-\alpha}, \varphi \rangle. \tag{18.31}$$

Note that $f_{n+1,n+1}(0)$ exists, by the remark following (18.26). Therefore,

$$\begin{aligned} \langle f_{n+1}, \varphi \rangle + \sum_{s=0}^{n} a_s \langle t^{-s-\alpha}, \varphi \rangle =& \langle f_n, \varphi \rangle + \sum_{s=0}^{n-1} a_s \langle t^{-s-\alpha}, \varphi \rangle \\ &+ f_{n+1,n+1}(0) \langle \delta^{(n)}, \varphi \rangle. \end{aligned}$$

Repeated application of this identity gives

$$\langle f_{n+1}, \varphi \rangle + \sum_{s=0}^{n} a_s \langle t^{-s-\alpha}, \varphi \rangle = \langle f, \varphi \rangle + \sum_{s=0}^{n} f_{s+1,s+1}(0) \langle \delta^{(s)}, \varphi \rangle.$$

Upon rearranging the terms in the above equation, we obtain the desired result (18.28). The formula of c_s in (18.29) is left as an exercise. □

To apply the above result to the Stieltjes transform (18.10), we take a specific function $\varphi_\varepsilon \in \mathcal{S}$ given by

$$\varphi_\varepsilon(t) = \frac{e^{-\varepsilon t}}{t + z}, \tag{18.32}$$

for $t \in (0, \infty)$. From (18.22), we have

$$\lim_{\varepsilon \to 0^+} \langle t^{-s-\alpha}, \varphi_\varepsilon \rangle = \frac{(-1)^s \pi}{\sin(\pi\alpha) z^{s+\alpha}}, \tag{18.33}$$

for $0 < \alpha < 1$ and $s = 0, 1, 2, \cdots$. Here we have made use of the fact that

$$\int_0^\infty \frac{t^{\lambda-1}}{(1+t)^{\lambda+\mu}} dt = \frac{\Gamma(\lambda)\Gamma(\mu)}{\Gamma(\lambda+\mu)}, \tag{18.34}$$

for $\lambda, \mu > 0$. From (18.20) and (18.27), we also have

$$\lim_{\varepsilon \to 0^+} \langle f, \varphi_\varepsilon \rangle = \lim_{\varepsilon \to 0^+} \int_0^\infty \frac{f(t)}{t+z} e^{-\varepsilon t} dt = S_f(z), \tag{18.35}$$

and

$$\lim_{\varepsilon \to 0^+} \langle f_n, \varphi_\varepsilon \rangle = \lim_{\varepsilon \to 0^+} (-1)^n \int_0^\infty f_{n,n}(t) \varphi_\varepsilon^{(n)}(t) dt = (-1)^n \int_0^\infty \frac{f_{n,n}(t)}{(t+z)^{n+1}} dt. \tag{18.36}$$

We are now ready to state and prove our first result for the Stieltjes transform.

Theorem 18.1. *Let f be a locally integrable function in $[0, \infty)$ which satisfies (18.14) with $0 < \alpha < 1$. Then for any $n \geq 1$,*

$$S_f(z) = \frac{\pi}{\sin(\pi\alpha)} \sum_{s=0}^{n-1} \frac{(-1)^s a_s}{z^{s+\alpha}} - \sum_{s=1}^{n} \frac{(s-1)! c_s}{z^s} + R_n(z), \tag{18.37}$$

where the coefficients are given in (18.29). The error term satisfies

$$R_n(z) = \frac{(-1)^n}{z^n} \int_0^\infty \frac{t^n f_n(t)}{t+z} dt. \tag{18.38}$$

Proof. By Lemma 18.1, we have

$$\langle f, \varphi_\varepsilon \rangle = \sum_{s=0}^{n-1} a_s \langle t^{-s-\alpha}, \varphi_\varepsilon \rangle - \sum_{s=1}^{n} c_s \langle \delta^{(s-1)}, \varphi_\varepsilon \rangle + \langle f_n, \varphi_\varepsilon \rangle. \tag{18.39}$$

Note that by (18.24),

$$\lim_{\varepsilon \to 0^+} \langle \delta^{(s-1)}, \varphi_\varepsilon \rangle = \frac{(s-1)!}{z^s}, \quad s = 1, 2, \cdots, \tag{18.40}$$

and that by partial integration,

$$n! \int_0^\infty \frac{f_{n,n}(t)}{(t+z)^{n+1}} dt = \frac{(-1)^n}{z^n} \int_0^\infty \frac{t^n f_n(t)}{t+z} dt. \tag{18.41}$$

Equation (18.37) now follows from (18.33), (18.35) and (18.36) by letting $\varepsilon \to 0^+$ in (18.39). □

Note that the integral in (18.38) is just the Stieltjes transform of $t^n f_n(t)$. Therefore, it is no more complicated than the original integral in (18.10), except that it is now multiplied by a decreasing factor $1/z^n$.

In the above discussion, we have restricted ourselves to the case $0 < \alpha < 1$. We next consider the case $\alpha = 1$. The distribution associated with t^{-s-1}, $s = 0, 1, 2, \cdots$, is defined by

$$\langle t^{-s-1}, \varphi \rangle = \frac{1}{s!} \int_0^\infty (\ln t) \varphi^{(s+1)}(t) dt. \tag{18.42}$$

Equation (18.25) now becomes

$$f(t) = \sum_{s=0}^{n-1} a_s t^{-s-1} + f_n(t). \tag{18.43}$$

We define $f_{n,j+1}$ with $j = 0, 1, \cdots, n-1$, as in (18.26). But, in this case, we have $f_{n,n}(t) = O(t^{-1})$ as $t \to \infty$ and $f_{n,n}(t) = O(\ln t)$ as $t \to 0^+$. Nevertheless, the distribution associated with $f_n(t)$ in (18.43) can still be defined by (18.27). The following is the analogue of Lemma 18.1.

Lemma 18.2. *If* $\alpha = 1$ *in* (18.25) *or equivalently* f *is given by* (18.43), *then for each* $n \geq 1$ *and for any function* $\varphi \in \mathcal{S}$ *we have*

$$\langle f, \varphi \rangle = \sum_{s=0}^{n-1} a_s \langle t^{-s-1}, \varphi \rangle - \sum_{s=1}^{n} d_s \langle \delta^{(s-1)}, \varphi \rangle + \langle f_n, \varphi \rangle, \tag{18.44}$$

where

$$d_s = \lim_{t \to 0^+} \left[f_{s,s}(t) + \frac{(-1)^{s-1}}{(s-1)!} a_{s-1} \ln t \right]. \tag{18.45}$$

For a more tractable expression for d_s, *see equations* (18.54) *and* (18.58) *below.*

Proof. Let $f_{0,0}(t) = f_0(t) = f(t)$. Then for any $n \geq 0$,

$$f_{n+1}(t) + a_n t^{-n-1} = f_n(t), \tag{18.46}$$

and hence

$$f_{n+1,n}(t) + \frac{(-1)^n}{n!} a_n t^{-1} = f_{n,n}(t). \tag{18.47}$$

It follows from (18.47) that

$$f_{n+1,n+1}(t) + \frac{(-1)^n}{n!} a_n \ln t$$

is an antiderivative of $f_{n,n}(t)$ on $(0,\infty)$, but so is the integral $\int_0^t f_{n,n}(\tau)d\tau$. Thus we have

$$\int_0^t f_{n,n}(\tau)d\tau = f_{n+1,n+1}(t) + \frac{(-1)^n}{n!}a_n \ln t + \text{constant}, \tag{18.48}$$

and the limit

$$d_{n+1} = \lim_{t\to 0^+}\left[f_{n+1,n+1}(t) + \frac{(-1)^n}{n!}a_n \ln t\right]$$

exists for each $n = 0, 1, 2, \cdots$. From (18.47) we also have, by integration by parts,

$$\langle f_n, \varphi\rangle = -d_{n+1}\langle \delta^{(n)}, \varphi\rangle + \langle f_{n+1}, \varphi\rangle + a_n\langle t^{-n-1}, \varphi\rangle.$$

Thus,

$$\begin{aligned}\langle f_{n+1}, \varphi\rangle + \sum_{s=0}^{n} a_s\langle t^{-s-1}, \varphi\rangle =& \langle f_n, \varphi\rangle + \sum_{s=0}^{n-1} a_s\langle t^{-s-1}, \varphi\rangle \\ &+ d_{n+1}\langle \delta^{(n)}, \varphi\rangle.\end{aligned}$$

Repeated application of this identity gives

$$\langle f_{n+1}, \varphi\rangle + \sum_{s=0}^{n} a_s\langle t^{-s-1}, \varphi\rangle = \langle f, \varphi\rangle + \sum_{s=0}^{n+1} d_s\langle \delta^{(s-1)}, \varphi\rangle,$$

which is of course equivalent to the statement of the lemma. □

Theorem 18.2. *Let f be a locally integrable function on $[0,\infty)$ and let f satisfy* (18.14) *with $\alpha = 1$. Then for any $n \geq 1$,*

$$S_f(z) = \ln z\sum_{s=0}^{n-1}(-1)^s a_s z^{-s-1} + \sum_{s=0}^{n-1}(-1)^s a_s^* z^{-s-1} + R_n(z), \tag{18.49}$$

where

$$a_s^* = \lim_{z\to s+1}\left\{M[f;z] + \frac{a_s}{z-s-1}\right\}, \tag{18.50}$$

and $M[f;z]$ is (the analytic continuation of) the Mellin transform of $f(t)$; see (18.29). *The remainder term satisfies*

$$R_n(z) = \frac{(-1)^n}{z^n}\int_0^\infty \frac{t^n f_n(t)}{t+z}dt. \tag{18.51}$$

Proof. Note that (exercise)

$$\int_0^\infty \frac{\ln t}{t+z} e^{-\varepsilon t} dt = O(\ln^2 \varepsilon)$$

as $\varepsilon \to 0^+$, and

$$\int_0^\infty \frac{\ln t}{(t+1)^{s+2}} dt = -\frac{1}{s+1} \sum_{k=1}^{s} \frac{1}{k}$$

for $s = 0, 1, 2, \cdots$, where an empty sum is understood to be zero. From (18.42), we have

$$\lim_{\varepsilon \to 0^+} \langle t^{-s-1}, \varphi_\varepsilon \rangle = \frac{(-1)^{s+1}}{z^{s+1}} \sum_{k=1}^{s} \frac{1}{k} + \frac{(-1)^s}{z^{s+1}} \ln z,$$

for $s = 0, 1, 2, \cdots$. Also note that

$$\lim_{\varepsilon \to 0^+} \langle \delta^{(s-1)}, \varphi_\varepsilon \rangle = \frac{(s-1)!}{z^s},$$

for $s = 1, 2, \cdots$. Equation (18.49) now follows from Lemma 18.2 by replacing φ by φ_ε and letting $\varepsilon \to 0^+$.

The coefficient a_s^* in (18.49) is first expressed in the form

$$(-1)^s a_s^* = (-1)^{s+1} a_s \sum_{k=1}^{s} \frac{1}{k} - s! d_{s+1}, \tag{18.52}$$

where d_{s+1} is as given in (18.45). Thus it remains to prove only the validity of (18.50).

Observe that equation (18.48) is equivalent to

$$\int_0^t f_{n,n}(\tau) d\tau = -\int_t^\infty f_{n+1,n}(\tau) d\tau + \frac{(-1)^n}{n!} a_n \ln t - d_{n+1}. \tag{18.53}$$

Taking $t = 1$ gives

$$d_{s+1} = -\int_1^\infty f_{s+1,s}(\tau) d\tau - \int_0^1 f_{s,s}(\tau) d\tau. \tag{18.54}$$

By partial integration,

$$\begin{aligned} \int_0^1 t^s f_s(t) dt =& f_{s,1}(1) + \sum_{k=2}^{s} (-1)^{k-1} s(s-1) \cdots (s-k+2) f_{s,k}(1) \\ &+ (-1)^s s! \int_0^1 f_{s,s}(t) dt, \end{aligned} \tag{18.55}$$

and

$$\int_1^\infty t^s f_{s+1}(t)dt = -f_{s+1,1}(1) + \sum_{k=2}^{s}(-1)^k s(s-1)\cdots(s-k+2)f_{s+1,k}(1)$$
$$+ (-1)^s s! \int_1^\infty f_{s+1,s}(t)dt. \tag{18.56}$$

From (18.47), we also have $f_{s+1,1}(1) = f_{s,1}(1) + a_s/s$ and

$$f_{s+1,k}(1) = f_{s,k}(1) - \frac{(-1)^k a_s}{s(s-1)\cdots(s-k+1)},$$

for $k = 2, \cdots, s$. Thus, coupling (18.55) and (18.56), we obtain

$$\int_0^1 t^s f_s(t)dt + \int_1^\infty t^s f_{s+1}(t)dt = -a_s \sum_{k=1}^{s} \frac{1}{k}$$
$$+ (-1)^s s! \left[\int_0^1 f_{s,s}(t)dt + \int_1^\infty f_{s+1,s}(t)dt\right]. \tag{18.57}$$

Equations (18.54) and (18.57) together yield

$$s!d_{s+1} = (-1)^{s+1} a_s \sum_{k=1}^{s} \frac{1}{k} + (-1)^{s+1}\left[\int_0^1 t^s f_s(t)dt + \int_1^\infty t^s f_{s+1}(t)dt\right]. \tag{18.58}$$

The quantity inside the square brackets in (18.58) is equal to (exercise)

$$\lim_{z\to s+1}\left\{M[f;z] + \frac{a_s}{z-s-1}\right\}.$$

Therefore, (18.50) follows immediately from (18.52) and (18.58). □

To construct numerical bounds for the remainder term $R_n(z)$ in (18.38) and (18.51), we normally assume that

$$M_n = \sup_{t>0} |t^{n+\alpha} f_n(t)| < \infty. \tag{18.59}$$

This condition does not follow from (18.14) and the local integrability of f, but will hold in most applications. From (18.59), it follows immediately that the function $f_n(t)$ satisfies

$$|f_n(t)| \le M_n t^{-n-\alpha}$$

for $t > 0$ and $0 < \alpha < 1$. If z is real and positive, say $z = x > 0$, then the remainder (18.38) is bounded by

$$|R_n(x)| \le \frac{\pi M_n}{\sin(\pi\alpha)x^{n+\alpha}}, \quad 0 < \alpha < 1. \tag{18.60}$$

If z is complex, then it can be shown that

$$|R_n(z)| \le \frac{M_n C_\alpha(\theta)}{|z|^{n+\alpha}}, \quad 0 < \alpha < 1, \tag{18.61}$$

where the constant $C_\alpha(\theta)$ depends on $\theta = \arg z \in (-\pi, \pi)$, but is independent of $|z|$. Inequalities (18.60) and (18.61) provide strict error bounds for the asymptotic expansion (18.37). Thus,

$$S_f(z) \sim \frac{\pi}{\sin(\pi\alpha)} \sum_{s=0}^{\infty} \frac{(-1)^s a_s}{z^{s+\alpha}} - \sum_{s=1}^{\infty} \frac{(s-1)! c_s}{z^s}, \tag{18.62}$$

as $|z| \to \infty$ in $|\arg z| \le \pi - \delta < \pi$. Note that the integral in (18.10) for $S_f(z)$ does not exist as $\arg z \to \pm\pi$.

In the case when f satisfies (18.14) with $\alpha = 1$, a natural way of extending the error analysis (18.60) and (18.61) is to assume that for some $\sigma \in (0, 1)$,

$$M_n(\sigma) = \sup_{t>0} |t^{n+\sigma} f_n(t)| < \infty. \tag{18.63}$$

The remainder in (18.51) then satisfies

$$|R_n(x)| \le \frac{\pi M_n(\sigma)}{\sin(\pi\sigma) x^{n+\sigma}} \tag{18.64}$$

if x is real and positive, and

$$|R_n(z)| \le \frac{M_n(\sigma) C_\sigma(\theta)}{|z|^{n+\sigma}} \tag{18.65}$$

if $0 \le |\arg z| < \pi$. Relations (18.64) and (18.65) confirm the asymptotic nature of the expansion (18.49)

$$S_f(z) \sim \ln z \sum_{s=0}^{\infty} (-1)^s a_s z^{-s-1} + \sum_{s=0}^{\infty} (-1)^s a_s^* z^{-s-1}, \tag{18.66}$$

as $z \to \infty$ in $|\arg z| \le \pi - \delta < \pi$. Note that the actual error $R_n(z)$ in the approximation (18.49) is $O(\ln z / z^{n+1})$. Hence the estimates (18.64) and (18.65) fall short of the actual result.

18.3 Laplace and Fourier transforms

As we have seen in Chapter 17, Watson's lemma gives the asymptotic expansion of the Laplace transform

$$L_f(x) = \int_0^{\infty} f(t) e^{-zt} dt, \quad \operatorname{Re} z > 0, \tag{18.67}$$

as $z \to \infty$. Also as we have seen in Chapter 17, the method of integration by parts gives the large-x behavior of the Fourier transform

$$F_f(x) = \int_0^\infty f(t)e^{ixt}dt. \tag{18.68}$$

However, the problem of deriving the asymptotic expansions of these two integral transforms for small values of x and z is not as simple. Fortunately, we have the results in Lemmas 18.1 and 18.2, from which we can essentially write out the expansions immediately.

We begin by choosing a test function $\varphi \in \mathcal{S}$ such that

$$\varphi(t) = e^{-zt}, \tag{18.69}$$

for $t \geq 0$. From (18.22) and (18.41), we have for $t > 0$

$$\langle t^{-s-\alpha}, \varphi\rangle = \Gamma(1-s-\alpha)z^{s+\alpha-1}, \quad 0 < \alpha < 1, \tag{18.70}$$

and

$$\langle t^{-s-1}, \varphi\rangle = \frac{(-1)^{s+1}}{s!}(\gamma + \ln z)z^s, \tag{18.71}$$

for $s = 0, 1, 2, \cdots$, where γ is the Euler constant.

Theorem 18.3. *Let $f(t)$ be a locally integrable function on $[0,\infty)$ and satisfy (18.14). (i) If $0 < \alpha < 1$ in (18.14) then for each $n \geq 1$*

$$L_f(z) = \sum_{s=0}^{n-1} a_s\Gamma(1-s-\alpha)z^{s+\alpha-1} - \sum_{s=1}^{n} c_s z^{s-1} + R_n(z), \tag{18.72}$$

where the coefficients c_s are given by (18.29). (ii) If $\alpha = 1$ then for each $n \geq 1$

$$L_f(z) = \ln z \sum_{s=0}^{n-1} \frac{(-1)^{s+1}}{s!}a_s z^s + \sum_{s=0}^{n-1} \gamma_s z^s + R_n(z), \tag{18.73}$$

where

$$\gamma_s = \frac{(-1)^{s+1}}{s!}\gamma a_s - d_{s+1}, \quad s = 0, 1, 2, \cdots, \tag{18.74}$$

and d_{s+1} is given by (18.45) or (18.50) and (18.52). The remainder in both cases is given by

$$R_n(z) = z^n \int_0^\infty f_{n,n}(t)e^{-zt}dt \tag{18.75}$$

with $f_{n,n}(t)$ being defined as before.

Proof. By Lemma 18.1, we have

$$L_f(z) = \sum_{s=0}^{n-1} a_s \langle t^{-\alpha-s}, \varphi\rangle - \sum_{s=1}^{n} c_s \langle \delta^{(s-1)}, \varphi\rangle + \langle f_n, \varphi\rangle,$$

where φ is given by (18.69). Since $\langle \delta^{(s)}, \varphi\rangle = z^s$, it follows from (18.70)

$$L_f(z) = \sum_{s=0}^{n-1} a_s \Gamma(1-s-\alpha) z^{s+\alpha-1} - \sum_{s=1}^{n} c_s z^{s-1} + \langle f_n, \varphi\rangle. \tag{18.76}$$

The last term is, by definition, given by

$$\langle f_n, \varphi\rangle = z^n \int_0^\infty f_{n,n}(t) e^{-zt} dt,$$

thus proving (i). To prove (ii), we use Lemma 18.2 and (18.71) instead of Lemma 18.1 and (18.70). The rest of the argument is similar. □

Corresponding expansions can be obtained for the Fourier transform

$$F_f(x) = \int_0^\infty f(t) e^{ixt} dt. \tag{18.77}$$

To this end, we put $z = \varepsilon - ix$ in (18.72) and (18.73). Upon letting $\varepsilon \to 0^+$, the final results follow from the lemma (exercise) below.

Lemma 18.3. *If the integral $\int_0^\infty f(t)dt$ exists as an improper Riemann integral, then*

$$\lim_{\varepsilon\to 0^+} \int_0^\infty f(t) e^{-\varepsilon t} dt = \int_0^\infty f(t) dt.$$

Theorem 18.4. *Let f be a locally integrable function on $[0,\infty)$ and satisfy* (18.14).

(i) If $0 < \alpha < 1$ in (18.14) *then*

$$F_f(x) = e^{-\alpha\pi i/2} \sum_{s=0}^{n-1} (-i)^{s-1} \Gamma(1-s-\alpha) a_s x^{\alpha+s-1} - \sum_{s=1}^{n} c_s (-ix)^{s-1} + R_n(x), \tag{18.78}$$

where the coefficients c_s are given by (18.29).

(ii) If $\alpha = 1$ then

$$F_f(x) = -\ln x \sum_{s=0}^{n-1} \frac{a_s}{s!} (ix)^s + \sum_{s=0}^{n-1} \gamma_s^* (-ix)^s + R_n(x), \tag{18.79}$$

where

$$\gamma_s^* = \frac{(-1)^{s+1}}{s!} \left(\gamma - \frac{i\pi}{2}\right) a_s - d_{s+1} \tag{18.80}$$

and d_{s+1} has the same meaning as given in (18.45) *or* (18.54) *and* (18.58). *The remainders in both cases are given by*

$$R_n(x) = (ix)^n \int_0^\infty f_{n,n}(t)e^{ixt}dt \tag{18.81}$$

with $f_{n,n}(t)$ being defined as before.

As an application of the above result, we consider the Kontorovich-Lebedev transform

$$I(x) = \int_0^\infty K_{it}(x)f(t)dt, \quad x > 0, \tag{18.82}$$

where $K_{it}(x)$ is the modified Bessel function of the third kind of imaginary order. This transform plays an important role in a number of problems concerning diffraction of waves by wedges. Here we shall assume that $f(t)$ is locally absolutely integrable on $(0, \infty)$ and satisfies $f(t) = O(t^b)$ as $t \to 0^+$, for some $b > -1$. Furthermore, we assume that $f(t)$ has an asymptotic expansion of the form (18.14):

$$f(t) \sim \sum_{s=0}^{\infty} a_s t^{-s-\alpha} \tag{18.83}$$

as $t \to \infty$, where $0 < \alpha < 1$. To apply the result of Theorem 18.4, we first observe the kernel function $K_{it}(x)$ has the integral representation

$$K_{it}(x) = \int_0^\infty \cos(ty)e^{-x\cosh y}dy. \tag{18.84}$$

Substituting (18.84) in (18.82) and reversing the order of integration, we have

$$I(x) = \int_0^\infty F_c(y)e^{-x\cosh y}dy, \tag{18.85}$$

where $F_c(y)$ is the Fourier cosine transform of $f(t)$; i.e.,

$$F_c(y) = \int_0^\infty f(t)\cos(ty)dt. \tag{18.86}$$

The interchange of the order of integration is justified by uniform convergence (exercise).

Following Laplace's method in Chapter 17, we set, in (18.85),

$$u^2 = \cosh y - 1 \tag{18.87}$$

and write $I(x)$ as

$$I(x) = e^{-x} \int_0^\infty \varphi(u)e^{-xu^2}du, \tag{18.88}$$

where

$$\varphi(u) = F_c(y(u))\frac{dy}{du}. \tag{18.89}$$

Equating the real parts in (18.78), we obtain

$$\begin{aligned} F_c(y) \sim & \sum_{s=0}^{\infty} a_s \Gamma(1-s-\alpha)\cos[(1-s-\alpha)\pi/2] y^{s+\alpha-1} \\ & + \sum_{s=0}^{\infty} \frac{(-1)^s}{(2s)!} M[f; 2s+1] y^{2s}, \end{aligned} \tag{18.90}$$

and from (18.87) it follows that

$$\frac{dy}{du} = \sqrt{2}(1+u^2/2)^{-1/2}, \tag{18.91}$$

and hence

$$y \sim \sqrt{2}u - \frac{\sqrt{2}}{12}u^3 + \frac{3\sqrt{2}}{160}u^5 + \cdots \tag{18.92}$$

as $u \to 0^+$. A combination of the results in (18.89)-(18.92) gives

$$\varphi(u) \sim \sqrt{2}\left[\sum_{s=0}^{\infty} \gamma_s u^{s+\alpha-1} + \sum_{s=0}^{\infty} \delta_s u^{2s}\right] \tag{18.93}$$

as $u \to 0^+$. The coefficients γ_s and δ_s are expressible in terms of a_s and $M[f; 2s+1]$, respectively; for example,

$$\begin{aligned} \gamma_0 =& 2^{(\alpha-1)/2} a_0 \Gamma(1-\alpha)\sin(\alpha\pi/2), \\ \gamma_1 =& 2^{\alpha/2} a_1 \Gamma(-\alpha)\cos(\alpha\pi/2), \\ \gamma_2 =& -2^{(\alpha-1)/2}(\alpha/12+1/6) a_0 \Gamma(1-\alpha)\sin(\alpha\pi/2) \\ & - 2^{(\alpha+1)/2} a_2 \Gamma(-1-\alpha)\sin(\alpha\pi/2), \end{aligned} \tag{18.94}$$

and

$$\begin{aligned} \delta_0 =& M[f;1], \quad \delta_1 = -M[f;3] - \frac{1}{4}M[f;1], \\ \delta_2 =& \frac{1}{6}M[f;5] + \frac{5}{12}M[f;3] + \frac{3}{32}M[f;1]. \end{aligned} \tag{18.95}$$

Substituting (18.93) into (18.88) and integrating term by term, we obtain

$$I(x) \sim \frac{e^{-x}}{\sqrt{2}}\left[\sum_{s=0}^{\infty} \Gamma\left(\frac{s+1}{2}\right)\frac{\gamma_s}{x^{(s+1)/2}} + \sum_{s=0}^{\infty} \Gamma\left(\frac{s+1}{2}\right)\frac{\delta_s}{x^{s+1/2}}\right] \tag{18.96}$$

as $x \to +\infty$. The last step is justified by Watson's lemma; see Chapter 17. For the case when $\alpha = 1$ in (18.83), we refer to Wong (1981), where the small x behavior of $I(x)$ is also given.

18.4 Fractional integrals

The Riemann-Liouville fractional integral of order μ is defined by

$$I^{\mu} f(x) = \frac{1}{\Gamma(\mu)} \int_0^x (x-t)^{\mu-1} f(t)dt, \quad \mu > 0, \tag{18.97}$$

where we again assume that $f(t)$ is locally integrable on $[0, \infty)$ and satisfies (18.14).

In terms of the convolution product

$$(f * g)(x) = \int_0^x f(x-t)g(t)dt \tag{18.98}$$

of two locally integrable functions on $[0, \infty)$, (18.97) can be written as

$$I^{\mu} f(x) = \frac{1}{\Gamma(\mu)} (t^{\mu-1} * f)(x). \tag{18.99}$$

Replacing $f(t)$ by its asymptotic expansion (18.14) and integrating term by term give rise to convolution integrals of the form

$$(t^{\mu-1} * t^{-s-\alpha})(x) = \int_0^x (x-t)^{\mu-1} t^{-s-\alpha} dt, \quad s = 0, 1, 2, \cdots. \tag{18.100}$$

Except for $s = 0$ and $0 < \alpha < 1$, none of these integrals exists in the usual sense. In view of what we have learned in Section 18.2 about distributions, it is natural to consider the left-hand side of (18.100) as the "convolution" of the distributions $t^{\mu-1}$ and $t^{-s-\alpha}$ for $0 < t < \infty$ and use the results in Lemma 18.1 and Lemma 18.2. Before proceeding further, we must first consider convolutions of distributions. Throughout the remaining portion of this section, we shall restrict ourselves to distributions supported on $[0, \infty)$, and this fact will not be mentioned further.

Let $L^+_{loc}(\mathbf{R})$ denote the class of functions which are locally integrable on $\mathbf{R}$ and which vanish on $(-\infty, 0)$. The distributions associated with functions in $L^+_{loc}(\mathbf{R})$ all have supports contained in $[0, \infty)$ and we have the following result.

Lemma 18.4. *Suppose $f \in L^+_{loc}(\mathbf{R})$. Then $Df \in L^+_{loc}(\mathbf{R})$ if and only if f is equal a.e. to a locally absolutely continuous function h on $\mathbf{R}$. In that case, $Df = h'$ a.e. and $h(0) = 0$.*

Proof. Suppose $g := Df \in L^+_{loc}(\mathbf{R})$. Then we have

$$\langle Df, \varphi \rangle = \langle g, \varphi \rangle = \int_0^\infty g(x)\varphi(x)dx, \quad \varphi \in \mathcal{D}(\mathbf{R}). \tag{18.101}$$

By the definition of distributional derivative (18.18),

$$\langle Df, \varphi\rangle = -\langle f, \varphi'\rangle = -\int_0^\infty f(x)\varphi'(x)dx. \tag{18.102}$$

Therefore,

$$\int_0^\infty f(x)\varphi'(x)dx = -\int_0^\infty g(x)\varphi(x)dx.$$

The member on the right-hand side can be written as

$$\int_0^\infty g(x)\int_x^\infty \varphi'(t)dtdx,$$

which, by Fubini's theorem, is in turn equal to

$$\int_0^\infty \varphi'(t)\int_0^t g(x)dxdt.$$

It now follows that for almost all $x \in (0,\infty)$,

$$f(x) = \int_0^x g(t)dx. \tag{18.103}$$

The integral on the right-hand side is obviously an absolutely continuous function; we denote it by $h(x)$. Thus, $h(0) = 0$ and $h'(x) = g(x) = Df(x)$ almost everywhere in $(0,\infty)$.

To prove the converse, we insert (18.103) in (18.102) and apply Fubini's theorem. The result is $Df = g = h'$ a.e. on $(0,\infty)$. □

Remark 18.1. It should be mentioned that an everywhere differentiable function with a locally integrable derivative is automatically locally absolutely continuous; see Rudin (1974, Theorem 8.21).

There is one further fact concerning differentiation which we need in the sequel. With "absolutely continuous" replaced by "continuously differentiable", this result may be found in most books on Advanced Calculus.

Lemma 18.5. *Suppose that $f(t,x)$ is integrable in $[a,b]\times[c,d]$, and absolutely continuous as a function of x, for each $t \in [a,b]$. Suppose also that $\partial f(t,x)/\partial x$ is integrable in $[a,b]\times[c,d]$. Then the function $F(x)$ defined by $F(x) = \int_a^b f(t,x)dt$ is absolutely continuous in $[c,d]$, and for almost all $x \in [c,d]$*

$$F'(x) = \int_a^b \frac{\partial}{\partial x} f(t,x)dt.$$

Proof. We first recall that a function $h(x)$ is absolutely continuous on an interval $[c, d]$ if and only if h' is absolutely integrable and

$$h(x) = \int_c^x h'(u)du + h(c)$$

for all $x \in [c, d]$. Thus

$$F(x_i') - F(x_i) = \int_a^b \int_{x_i}^{x_i'} \frac{\partial}{\partial \tau} f(t, \tau) d\tau dt. \tag{18.104}$$

Since $\partial f(t, x)\partial x$ is integrable in $[a, b] \times [c, d]$, if $\sum_{i=1}^n |x_i' - x_i|$ is small then it follows from (18.104) that $\sum_{i=1}^n |F(x_i') - F(x_i)|$ is also small; see Rudin (1974, p. 33, Ex. 12). This proves that $F(x)$ is absolutely continuous. Now write

$$F(x) = \int_a^b \left[f(t, c) + \int_c^x \frac{\partial}{\partial \tau} f(t, \tau) d\tau \right] dt.$$

By Fubini's theorem,

$$F(x) = \int_a^b f(t, c)dt + \int_c^x \int_a^b \frac{\partial}{\partial \tau} f(t, \tau) dt d\tau,$$

and also the inner integral on the right-hand side is an integrable function of τ. Thus

$$F'(x) = \int_a^b \frac{\partial}{\partial x} f(t, x) dt$$

almost everywhere in $[c, d]$. This proves the lemma. □

If f and g belong to $L_{loc}^+(\mathbf{R})$, then

$$(f * g)(x) = \int_0^x f(x - t)g(t)dt \tag{18.105}$$

defines $f * g \in L_{loc}^+(\mathbf{R})$. This product, together with the usual addition and scalar multiplication, makes $L_{loc}^+(\mathbf{R})$ a commutative, associative linear algebra over the complex field. Note that $f * g$ is continuous at 0, and at the points of continuity of either f or g.

Equation (18.105) can be considered as the definition of the convolution of two distributions, each associated with a function in $L_{loc}^+(\mathbf{R})$. We now wish to extend this definition of a convolution product in a space of distributions.

Definition 18.1. Let K^+ denote the space of all distributions of the form $D^n f$, where n is a nonnegative integer and $f \in L_{loc}^+(\mathbf{R})$. For F and G in K^+, say $F = D^n f$ and $G = D^m g$, we define

$$F * G = D^{n+m}(f * g). \tag{18.106}$$

The consistency equation whether

$$D(f * g) = Df * g \tag{18.107}$$

is answered in the following lemma, which also serves as the motivation for Definition 18.1.

Lemma 18.6. *If f, g and Df all belong to $L^+_{loc}(\mathbf{R})$, then $f * g$ is locally absolutely continuous in $\mathbf{R}$, and* (18.107) *holds.*

Proof. By Lemma 18.4, there exists a locally absolutely continuous function $h(x)$ in $\mathbf{R}$ such that $f(x) = h(x)$ a.e. in $\mathbf{R}$, and $h(0) = 0$. Therefore,

$$f(x) = \int_0^x h'(t)dt \tag{18.108}$$

almost everywhere for $x \geq 0$, and $f'(x) = h'(x)$ a.e. in $\mathbf{R}$. Equation (18.108) can be written as $f = h' * H$, where H is the Heaviside function. Thus for $x \geq 0$, we have a.e. $f * g = (h' * H) * g = (h' * g) * H$ and $(f * g)(x) = \int_0^x (h' * g)(t)dt$. This implies that $f * g$ is equal a.e. to a locally absolutely continuous function, and $(f * g)' = h' * g$ a.e. in $[0, \infty)$. By Lemma 18.4, $D(f * g) = (f * g)'$, and $Df = f' = h'$. This proves the lemma. □

The following theorem follows immediately from the definition in (18.106).

Theorem 18.5. *K^+ is a commutative, associative linear algebra. Thus, for all F, G and H in K^+, we have*

$$F * G = G * F, \tag{18.109}$$

$$(F * G) * H = F * (G * H), \tag{18.110}$$

$$F * (G + H) = F * G + F * H, \tag{18.111}$$

$$(F + G) * H = F * H + G * H. \tag{18.112}$$

The algebra K^+ clearly contains the distributions H, δ and x^λ for any real (or complex) number λ and $x > 0$. The last mentioned distribution was discussed in (18.21), (18.22) and (18.42). Using the fact that $\delta = DH$, we can derive the following equation. For any nonnegative integer n and $F \in K^+$,

$$D^n\delta * F = D^n F. \tag{18.113}$$

In particular,

$$D^n\delta * x^\lambda = \lambda(\lambda - 1)\cdots(\lambda - n + 1)x^{\lambda - n} \tag{18.114}$$

for all n and all $x > 0$, if λ is not a positive integer or zero. If λ is a nonnegative integer, then (18.114) remains valid for $n \leq \lambda$ (understood $x_+^0 = H$), and is replaced by

$$D^n \delta * x_+^\lambda = \lambda! D^{n-\lambda} H = \lambda! D^{n-\lambda-1} \delta \tag{18.115}$$

for $n > \lambda$.

Using the fact that

$$\int_0^1 (1-x)^\mu x^\lambda dx = \frac{\Gamma(\mu+1)\Gamma(\lambda+1)}{\Gamma(\mu+\lambda+2)} \tag{18.116}$$

whenever $\mu > -1$ and $\lambda > -1$, one can also show that

$$x^\mu * x^\lambda = \frac{\Gamma(\mu+1)\Gamma(\lambda+1)}{\Gamma(\mu+\lambda+2)} x^{\mu+\lambda+1}, \quad x > 0, \tag{18.117}$$

provided that μ, λ and $\mu+\lambda$ are not negative integers. If μ and λ are not negative integers, but $\mu+\lambda$ is a negative integer (≤ -1), then (18.117) must be replaced by

$$x^\mu * x^\lambda = \Gamma(\mu+1)\Gamma(\lambda+1) D^{-\mu-\lambda-1} H, \quad x > 0. \tag{18.118}$$

Finally, if $0 < \sigma < 1$, then

$$x^{-\sigma} \ln x = \frac{x^{1-\sigma}}{1-\sigma}[\ln x - \gamma - \psi(2-\sigma)], \quad x > 0, \tag{18.119}$$

where $\gamma = 0.57721\cdots$ is the Euler constant and ψ is the logarithmic derivative of the Gamma function. This leads to the formula

$$\begin{aligned} x^{-n-\sigma} * x^{-m} &= \frac{(-1)^{m-1}\Gamma(1-\sigma-n)}{(m-1)!\Gamma(1-\sigma)} \\ &\quad \times D^{m+n} \left\{ \frac{x^{1-\sigma}}{1-\sigma} [\ln x - \gamma - \psi(2-\sigma)] \right\} \end{aligned} \tag{18.120}$$

for $x > 0$. We are now ready to state our first main result of the section.

Theorem 18.6. *Let f be a locally integrable function on $[0,\infty)$ and satisfy (18.14) with $0 < \alpha < 1$. Then for any $n \geq 1$*

$$I^\mu f(x) = \sum_{s=0}^{n-1} a_s \frac{\Gamma(1-s-\alpha)}{\Gamma(\mu+1-s-\alpha)} x^{\mu-s-\alpha} - \sum_{s=1}^{n} \frac{c_s}{\Gamma(\mu+1-s)} x^{\mu-s} + \frac{\delta_n(x)}{x^n}, \tag{18.121}$$

where $c_s = (-1)^s M[f;s]/(s-1)!$ and the error term is given by

$$\delta_n(s) = \sum_{j=0}^{m} \binom{n}{j} \frac{\Gamma(\mu+1)}{\Gamma(\mu+1-j)} I^\mu[t^{n-j} f_{n,j}](x), \tag{18.122}$$

$f_{n,j}(t)$ being defined by (18.26).

Proof. Since $\mathcal{D} \subset \mathcal{S}$, Lemma 18.1 holds with $\mathcal{S}$ replaced by $\mathcal{D}$. Thus we have, as distributions in $\mathcal{D}'$,

$$f = \sum_{s=0}^{n-1} a_s t^{-s-\alpha} - \sum_{s=1}^{n} c_s \delta^{(s-1)} + f_n,$$

which, by the associative law, gives

$$t^{\mu-1} * f = \sum_{s=0}^{n-1} a_s t^{\mu-1} * t^{-s-\alpha} - \sum_{s=1}^{n} c_s t^{\mu-1} * \delta^{(s-1)} + t^{\mu-1} * f_n. \quad (18.123)$$

Since $f_n = f_{n,n}^{(n)}$, by the definition of a convolution of two distributions

$$t^{\mu-1} * f_n = D^n(t^{\mu-1} * f_{n,n}), \quad (18.124)$$

where D denotes the distributional derivative; see (18.106). Now, $t^{\mu-1} * f_{n,n}$ is a regular distribution on $(0,\infty)$ and is given by the locally integrable function

$$(t^{\mu-1} * f_{n,n})(x) = \int_0^x (x-t)^{\mu-1} f_{n,n}(t)dt = x^\mu \int_0^1 (1-u)^{\mu-1} f_{n,n}(xu)du.$$

By Lemma 18.4, the last integral is a locally absolutely continuous function on $(0,\infty)$, i.e., absolutely continuous on each compact subinterval of $(0,\infty)$. Furthermore,

$$\left[x^\mu \int_0^1 (1-u)^{\mu-1} f_{n,n}(xu)du\right]' = \mu x^{\mu-1} \int_0^1 (1-u)^{\mu-1} f_{n,n}(xu)du + x^\mu \int_0^1 (1-u)^{\mu-1} u f_{n,n-1}(xu)du$$

almost everywhere in $(0,\infty)$. Since $f_{n,j}(t)$ is absolutely continuous in $(0,\infty)$, for $j = 1, \cdots, n-1$, this process can be repeated. The result is

$$(t^{\mu-1} * f_{n,n})^{(n)}(x) = \sum_{j=0}^{n} \binom{n}{j} \frac{\Gamma(\mu+1)}{\Gamma(\mu+1-j)} x^{\mu-j} \times \int_0^1 (1-u)^{\mu-1} u^{n-j} f_{n,j}(xu)du = \Gamma(\mu) x^{-n} \delta_n(x)$$

almost everywhere in $(0,\infty)$. Since each function that has been differentiated is locally absolutely integrable, the result in Lemma 18.4 implies

$$D^n(t^{\mu-1} * f_{n,n}) = (t^{\mu-1} * f_{n,n})^{(n)}(x) = \Gamma(\mu) x^{-n} \delta_n(x) \quad (18.125)$$

for $t > 0$. Coupling (18.124) and (18.125) gives

$$t^{\mu-1} * f_n = \Gamma(\mu) x^{-n} \delta_n(x). \quad (18.126)$$

We now return to (18.123) and suppose for a moment that $\mu \neq 1, 2, \cdots$ and $\mu - \alpha$ is not a nonnegative integer. From (18.114) and (18.117), we have

$$t^{\mu-1} * \delta^{(s-1)} = \frac{\Gamma(\mu)}{\Gamma(\mu+1-s)} t^{\mu-s}, \quad t > 0, \tag{18.127}$$

and

$$t^{\mu-1} * t^{-s-\alpha} = \frac{\Gamma(\mu)\Gamma(1-s-\alpha)}{\Gamma(\mu+1-s-\alpha)} t^{\mu-s-\alpha}, \quad t > 0. \tag{18.128}$$

Inserting (18.127) and (18.128) into (18.123) yields

$$\begin{aligned} t^{\mu-1} * f = &\sum_{s=0}^{n-1} a_s \frac{\Gamma(\mu)\Gamma(1-s-\alpha)}{\Gamma(\mu+1-s-\alpha)} t^{\mu-s-\alpha} - \sum_{s=1}^{n} \frac{c_s\Gamma(\mu)}{\Gamma(\mu+1-s)} t^{\mu-s} \\ &+ t^{\mu-1} * f_n \end{aligned} \tag{18.129}$$

for $t > 0$. Note that each distribution in (18.129) is determined by a locally integrable function in $(0, \infty)$. By replacing these distributions by their corresponding functions, we have from (18.126)

$$\begin{aligned} (t^{\mu-1} * f)(x) = &\sum_{s=0}^{n-1} a_s \frac{\Gamma(\mu)\Gamma(1-s-\alpha)}{\Gamma(\mu+1-s-\alpha)} x^{\mu-s-\alpha} - \sum_{s=1}^{n} \frac{c_s\Gamma(\mu)}{\Gamma(\mu+1-s)} x^{\mu-s} \\ &+ \Gamma(\alpha) x^{-n} \delta_n(x), \end{aligned} \tag{18.130}$$

holding in the sense of distributions in $(0, \infty)$. It is straightforward to show that equation (18.130) in fact holds pointwise almost everywhere in $(0, \infty)$. Since every convolution in (18.129) has at least one factor which is continuous in $(0, \infty)$, we further conclude that (18.130), or equivalently (18.121), holds for each x in $(0, \infty)$. The assumption that $\mu \neq 1, 2, \cdots$ and $\mu - \alpha \neq 0, 1, \cdots$ can be removed by analytic continuation. This completes the proof of the theorem. □

Theorem 18.7. *Let f be a locally integrable function on $[0, \infty)$ and let f satisfy* (18.14) *with $\alpha = 1$. Then for any $n \geq 1$*

$$\begin{aligned} I^\mu f(x) = &\sum_{s=0}^{n-1} \frac{(-1)^s a_s}{\Gamma(\mu+1)s!} \left\{ x^\mu [\ln x - \gamma - \psi(\mu+1)] \right\}^{(s+1)} \\ &- \sum_{s=1}^{n} \frac{d_s}{\Gamma(\mu-s+1)} x^{\mu-s} + \frac{\delta_n(x)}{x^n}, \end{aligned} \tag{18.131}$$

where γ is the Euler constant, ψ is the logarithmic derivative of the gamma function, and $\delta_n(x)$ satisfies (18.122). *The coefficient d_s has the same meaning as in* (18.45) *or equivalently in* (18.54) *and* (18.58).

Proof. We again assume that $\mu \neq 1, 2, \cdots$. From Lemma 18.2, we have

$$f = \sum_{s=0}^{n-1} a_s t^{-s-1} - \sum_{s=1}^{n} d_s \delta^{(s-1)} + f_n$$

in $\mathcal{D}'$, and hence,

$$t^{\mu-1} * f = \sum_{s=0}^{n-1} a_s t^{\mu-1} * t^{-s-1} - \sum_{s=1}^{n} d_s t^{\mu-1} * \delta^{(s-1)} + t^{\mu-1} * f_n. \quad (18.132)$$

The results in (18.114) and (18.120) give

$$t^{\mu-1} * f = \sum_{s=0}^{n-1} \frac{(-1)^s a_s}{\mu \cdot s!} D^{s+1} \{t^\mu [\ln t - \gamma - \psi(\mu+1)]\}$$
$$- \sum_{s=1}^{n} \frac{d_s \Gamma(\mu)}{\Gamma(\mu - s + 1)} t^{\mu-s} + D^n (t^{\mu-1} * f_{n,n}). \quad (18.133)$$

Note that the function $t^\mu[\ln t - \gamma - \psi(\mu+1)]$ and all its derivatives are locally absolutely continuous in $(0, \infty)$. Thus the distributional derivatives in the first sum can be replaced by the corresponding ordinary derivatives. A similar remark applies to the last term in (18.133); see (18.125). Replacing each distribution in (18.133) by its corresponding function gives

$$(t^{\mu-1} * f)(x) = \sum_{s=0}^{n-1} \frac{(-1)^s a_s}{\mu \cdot s!} \{x^\mu [\ln x - \gamma - \psi(\mu+1)]\}^{(s+1)}$$
$$- \sum_{s=1}^{n} \frac{d_s \Gamma(\mu)}{\Gamma(\mu - s + 1)} x^{\mu-s} + (t^{\mu-1} * f_{n,n})^{(n)}(x),$$

valid for x in $(0, \infty)$. From here on the argument parallels that in the proof of Theorem 18.6, and the theorem is proved. □

The error term $\delta_n(x)$ can be bounded as follows. Let f satisfy (18.14) with $0 < \alpha < 1$, and assume

$$M_n = \sup_{t>0} |t^{n+\alpha} f_n(t)| < \infty; \quad (18.134)$$

cf. (18.59). A typical term in the sum (18.122) is given by the integral

$$I^\mu[t^{n-j} f_{n,j}] = \frac{1}{\Gamma(\mu)} \int_0^x (x-t)^{\mu-1} t^{n-j} f_{n,j}(t) dt. \quad (18.135)$$

From (18.26) and (18.134), it is easy to see that the function $f_{n,j}(t)$ satisfies

$$|f_{n,j}(t)| \leq \frac{\Gamma(n+\alpha-j)}{\Gamma(n+\alpha)} M_n t^{j-n-\alpha},$$

and hence,

$$|I^{\mu}[t^{n-j}f_{n,j}](x)| \leq \frac{\Gamma(n+\alpha-j)\Gamma(1-\alpha)}{\Gamma(\mu+1-\alpha)\Gamma(n+\alpha)} M_n x^{\mu-\alpha}.$$

This together with (18.122) gives

$$|\delta_n(x)| \leq M_n \left[\frac{\Gamma(\mu+1)\Gamma(1-\alpha)}{\Gamma(\mu+1-\alpha)\Gamma(n+\alpha)} \sum_{j=0}^{n} \binom{n}{j} \frac{\Gamma(n+\alpha-j)}{|\Gamma(\mu+1-j)|} \right] x^{\mu-\alpha} \tag{18.136}$$

The estimate (18.136) establishes the asymptotic nature of the expansion (18.121):

$$I^{\mu}f(x) \sim \sum_{s=0}^{\infty} a_s \frac{\Gamma(1-s-\alpha)}{\Gamma(\mu+1-s-\alpha)} x^{\mu-s-\alpha} - \sum_{s=1}^{\infty} \frac{c_s}{\Gamma(\mu+1-s)} x^{\mu-s}, \tag{18.137}$$

as $x \to +\infty$.

In the case when f satisfies (18.14) with $\alpha = 1$, we assume as in (18.63) that for some $\rho \in (0,1)$

$$M_n(\rho) = \sup_{t>0} |t^{n+\rho} f_n(t)| < \infty. \tag{18.138}$$

The remainder in (18.131) then satisfies

$$|\delta_n(x)| \leq M_n(\rho) \left[\frac{\Gamma(\mu+1)\Gamma(1-\rho)}{\Gamma(\mu+1-\rho)\Gamma(n+\rho)} \sum_{j=0}^{n} \binom{n}{j} \frac{\Gamma(n+\rho-j)}{\Gamma(\mu+1-j)} \right] x^{\mu-\rho}. \tag{18.139}$$

Example. We illustrate the above results with the integral

$$I(x) = \int_0^1 \frac{\tau}{\sqrt{1-\tau^2}} e^{-2ix\tau^2} F(2\tau\sqrt{x/\pi}) d\tau, \tag{18.140}$$

where F is the Fresnel integral defined by

$$F(t) = \int_t^{\infty} e^{i\pi\xi^2/2} d\xi. \tag{18.141}$$

The change of variable $x\tau^2 = t$ puts $I(x)$ into the form of a fractional integral

$$I(x) = \frac{1}{2\sqrt{x}} \int_0^x (x-t)^{-1/2} f(t) dt, \tag{18.142}$$

where

$$f(t) = e^{-2it} F(2\sqrt{t/\pi}). \tag{18.143}$$

In terms of the complementary error function, $f(t)$ can be written as

$$f(t) = \frac{1}{\sqrt{2}} e^{-2it+i\pi/4} \operatorname{erfc}(e^{-i\pi/4}\sqrt{2t}). \tag{18.144}$$

Thus it follows from the asymptotic expansion of this function (exercise)

$$f(t) = \frac{it^{-1/2}}{2\sqrt{\pi}} \left[1 + \sum_{s=1}^{n-1} (-i)^s \frac{1 \cdot 3 \cdots (2s-1)}{2^{2s} t^s} + r_n(t) \right], \tag{18.145}$$

where

$$|r_n(t)| \le \sqrt{2} \frac{1 \cdot 3 \cdots (2n-1)}{2^{2n} t^n}. \tag{18.146}$$

Let $a_0 = i/(2\sqrt{\pi})$ and

$$a_s = (-1)^s i^{s+1} \frac{1 \cdot 3 \cdots (2s-1)}{2^{2s+1}\sqrt{\pi}}, \quad s = 1, 2, \cdots. \tag{18.147}$$

With $\mu = \alpha = 1/2$, Theorem 18.6 gives at once

$$I(x) = \frac{\sqrt{\pi}}{2\sqrt{x}} \left[\sum_{s=0}^{n-1} \frac{a_s \Gamma(1/2 - s)}{\Gamma(1-s)} x^{-s} - \sum_{s=1}^{n} \frac{c_s}{\Gamma(3/2 - s)} x^{1/2-s} + \frac{\delta_n(x)}{x^n} \right]. \tag{18.148}$$

The coefficients c_s can be calculated from the Mellin transform of f (see Erdélyi *et al.* (1954, p. 325)), and is given by

$$c_s = \frac{(i/2)^s e^{i\pi/4}}{\sqrt{2}}. \tag{18.149}$$

Since the terms in the first sum are all zero except for the first one, (18.148) reduces to

$$I(x) \sim \frac{i\sqrt{\pi}}{4} x^{-1/2} + \frac{e^{-i\pi/4}}{4\sqrt{2}} \sum_{s=0}^{\infty} \frac{(1/2)_s}{(2i)^s} x^{-s-1} \tag{18.150}$$

as $x \to \infty$, where $(1/2)_s = \Gamma(s+1/2)/\Gamma(1/2)$.

The similarity of the two series in (18.145) and (18.150) indicates that the integral $I(x)$ can be evaluated in closed form in terms of the Fresnel integral (exercise).

18.5 The method of regularization

The idea of using distributions can further be extended to derive asymptotic expansions for the convolution integral

$$I(x) = \int_0^\infty f(t) h(xt) dt. \tag{18.151}$$

It is easy to see that this integral is equivalent to the more symmetric convolution defined by

$$(f * g)(x) = \int_0^\infty f(t)g(x/t)\frac{dt}{t}. \tag{18.152}$$

Here we shall assume that both f and g are members of the following class of functions.

Let $\mathcal{F}$ be the family of locally integrable functions $f(t)$ on $(0, \infty)$, with asymptotic expansions of the forms

$$f(t) \sim \sum_{s=0}^{\infty} a_s t^{s+\alpha} \tag{18.153}$$

as $t \to 0^+$, and

$$f(t) \sim \sum_{s=0}^{\infty} b_s t^{-s-\beta} \tag{18.154}$$

as $t \to +\infty$, where α and β are complex numbers. It is easily verified that (i) if f and g belong to $\mathcal{F}$ then fg also belongs to $\mathcal{F}$, and (ii) if f belongs to $\mathcal{F}$ then so does $\tilde{f}$, where $\tilde{f}(t) = f(1/t)$.

Without restrictions on the exponents α and β in (18.153) and (18.154), the integral

$$\int_0^\infty f(t)dt \tag{18.155}$$

will normally diverge. We shall now introduce the concept of "regularization" to give a meaningful definition to the infinite integral in (18.155).

Consider a function $f(x)$ that has only one non-integrable singularity at x_0, for example, $f(x) = 1/(x - x_0)$ in $\mathbf{R}$. In general, the integral

$$\int_0^\infty f(x)\varphi(x)dx, \tag{18.156}$$

where $\varphi(x)$ belongs to $\mathcal{D}(\mathbf{R})$ diverges. However, it will converge if $\varphi(x)$ vanishes in a neighborhood of x_0. A distribution Λ which has the property that for all $\varphi \in \mathcal{D}(\mathbf{R})$ vanishing in a neighborhood of x_0, $\langle \Lambda, \varphi \rangle$ has the value given by (18.156) is called a *regularization of the function* $f(x)$, and, when φ does not vanish near x_0, $\langle \Lambda, \varphi \rangle$ is called the *regularization of the integral* (18.156). For example, $\Lambda = D \ln |x|$ is a regularization of $1/x$. So is $\Lambda + P(D)\delta$, where P is any polynomial.

Here we shall first construct regularizations of some particularly important functions. Let us begin with the function

$$x_+^\lambda = \begin{cases} 0, & x \leq 0, \\ x^\lambda, & x > 0, \end{cases}$$

where λ is a complex number. For $\operatorname{Re}\lambda > -1$, this function defines the regular distribution

$$\langle x_+^\lambda, \varphi \rangle = \int_0^\infty x^\lambda \varphi(x)dx, \quad \varphi \in \mathcal{D}(\mathbf{R}). \tag{18.157}$$

The right-hand side is obviously an analytic function of λ for $\operatorname{Re}\lambda > -1$. To define a regularization of x_+^λ for $\operatorname{Re}\lambda \leq -1$, we rewrite (18.157) in the form

$$\int_0^1 x^\lambda [\varphi(x) - \varphi(0)]dx + \int_1^\infty x^\lambda \varphi(x)dx + \frac{\varphi(0)}{\lambda+1}.$$

The first term here is defined for $\operatorname{Re}\lambda > -2$, the second for all λ, and the third for $\lambda \neq -1$. Thus, the function in (18.157) can be analytically continued to $\operatorname{Re}\lambda > -2$ with $\lambda \neq -1$. Proceeding similarly, we can continue x_+^λ into the region $\operatorname{Re}\lambda > -n-1$ with $\lambda \neq -1, -2, \cdots, -n$, by using

$$\begin{aligned}\int_0^\infty x^\lambda \varphi(x)dx &= \int_0^1 x^\lambda [\varphi(x) - \varphi(0) - x\varphi'(0) - \cdots - \frac{x^{n-1}}{(n-1)!}\varphi^{(n-1)}(0)]dx \\ &\quad + \int_1^\infty x^\lambda \varphi(x)dx + \sum_{k=1}^n \frac{\varphi^{(k-1)}(0)}{(k-1)!(\lambda+k)}.\end{aligned} \tag{18.158}$$

Note that for $\operatorname{Re}\lambda > -n-1$ with $\lambda \neq -1, -2, \cdots, -n$, the right-hand side exists and defines a regularization of the integral on the left. Also, if $-n-1 < \operatorname{Re}\lambda \leq -n$ with $\lambda \neq -n$, and if the test function vanishes in a neighborhood of the origin, what remains on the right is $\int_0^\infty x^\lambda \varphi(x)dx$. Furthermore, in any strip of the form $-n-1 < \operatorname{Re}\lambda < -n$, equation (18.158) can be written in the simple form

$$\langle x_+^\lambda, \varphi \rangle = \int_0^1 x^\lambda [\varphi(x) - \varphi(0) - x\varphi'(0) - \cdots - \frac{x^{n-1}}{(n-1)!}\varphi^{(n-1)}(0)]dx \tag{18.159}$$

since for $1 \leq k \leq n$

$$\int_1^\infty x^{\lambda+k-1}dx = -\frac{1}{\lambda+k}. \tag{18.160}$$

Equation (18.158) shows that when $\langle x_+^\lambda, \varphi \rangle$ is considered as a function of λ, it has simple poles at $\lambda = -1, -2, \cdots$, and that its residue at $\lambda = -k$ is $\varphi^{(k-1)}(0)/(k-1)!$. This may be expressed by saying that the distribution x_+^λ itself has a simple pole at $\lambda = -k$ with residue

$$\frac{(-1)^k}{(k-1)!}\delta^{(k-1)}(x), \quad k = 1, 2, \cdots.$$

For $\operatorname{Re}\lambda > 0$, we have by integration by parts

$$\langle x_+^\lambda, \varphi' \rangle = -\langle \lambda x_+^{\lambda-1}, \varphi \rangle, \quad \varphi \in \mathcal{D}(\mathbf{R}).$$

Since both sides of this equation can be analytically continued to the entire λ-plane except for the points $0, -1, -2, \cdots$, by uniqueness of analytic continuation, this equation holds in the whole λ-plane except for these points. Thus $Dx_+^\lambda = \lambda x_+^{\lambda-1}$, in agreement with (18.22) with $s = 1$ and $\alpha = -\lambda$.

Let us now consider the problem of regularizing the function

$$x_{0\le x\le b}^\lambda = \begin{cases} x^\lambda, & 0 \le x \le b, \\ 0, & \text{otherwise}, \end{cases}$$

or equivalently the finite integral

$$\int_0^b x^\lambda \varphi(x) dx, \quad 0 < b < \infty, \ \varphi \in \mathcal{D}(\mathbf{R}). \tag{18.161}$$

For $\operatorname{Re}\lambda > -1$, this integral converges and is an analytic function of λ. Furthermore, it can be continued analytically to the entire λ-plane via the identity

$$\int_0^b x^\lambda \varphi(x) dx = \int_0^b x^\lambda [\varphi(x) - \varphi(0) - \varphi'(0)x - \cdots - \varphi^{(n-1)}(0)\frac{x^{n-1}}{(n-1)!}] dx$$
$$+ \varphi(0)\frac{b^{\lambda+1}}{\lambda+1} + \cdots + \varphi^{(n-1)}(0)\frac{b^{\lambda+n}}{(n-1)!(\lambda+n)}, \tag{18.162}$$

which is valid for all λ with $\operatorname{Re}\lambda > -n-1$ except for $\lambda = -1, -2, \cdots, -n$. Note that n can be arbitrarily large. We shall use (18.162) to define the distribution which regularizes the integral (18.161) for $\lambda \ne -1, -2, \cdots$. That is, we define

$$\langle x_{0\le x\le b}^\lambda, \varphi \rangle = \int_0^b x^\lambda [\varphi(x) - \varphi(0) - \varphi'(0)x - \cdots - \varphi^{(n-1)}(0)\frac{x^{n-1}}{(n-1)!}] dx$$
$$+ \varphi(0)\frac{b^{\lambda+1}}{\lambda+1} + \cdots + \varphi^{(n-1)}(0)\frac{b^{\lambda+n}}{(n-1)!(\lambda+n)}, \tag{18.163}$$

for $\operatorname{Re}\lambda > -n-1$ and $\lambda \ne -1, -2, \cdots, -n$.

Note that if $\varphi(x)$ vanishes in a neighborhood of $x = 0$, then (18.163) shows that the value $\langle x_{0\le x\le b}^\lambda, \varphi \rangle$ is indeed the same as that given by the integral (18.161). If we take $\varphi^*(x)$ to be a test function in $\mathcal{D}(\mathbf{R})$ which equals "1' on the interval $[0, b]$, then we have

$$\langle x_{0\le x\le b}^\lambda, \varphi^* \rangle = \frac{b^{\lambda+1}}{\lambda+1}. \tag{18.164}$$

Interpreting the integral $\int_0^b t^\lambda dt$ as the result of applying the distribution $x^\lambda_{0\le x\le b}$ to the test function $\varphi^*(x)$ gives

$$\int_0^b x^\lambda dx = \frac{b^{\lambda+1}}{\lambda+1}, \tag{18.165}$$

valid for all $\lambda \neq -1, -2, \cdots$. Since the right-hand side has only one singularity at $\lambda = -1$, we use (18.165) to define this integral also for $\lambda = -2, -3, \cdots$.

Having given a meaning to the integral $\int_0^b x^\lambda dx$, one now naturally wishes to give a meaning to the integral

$$\int_b^\infty x^\lambda dx, \quad 0 < b < \infty, \tag{18.166}$$

in a similar manner. However, there is unfortunately no test function in $\mathcal{D}(\mathbf{R})$ which is equal to "1" on the infinite interval $[b,\infty)$. Therefore, we must proceed in a different manner. Let $\hat{\mathcal{D}}(b,\infty)$ denote the space of all C^∞ functions $\psi(x)$ on $[b,\infty)$ such that for $x \in (0, 1/b)$, we have $\psi(1/x) = \varphi(x)$ for some $\varphi \in \mathcal{D}(\mathbf{R})$. Define a functional $x^\lambda_{b\le x<\infty}$ on $\hat{\mathcal{D}}(b,\infty)$ by

$$\langle x^\lambda_{b\le x<\infty}, \psi\rangle = \int_0^{1/b} y^{-\lambda-2}\varphi(y)dy, \tag{18.167}$$

where $\psi \in \hat{\mathcal{D}}(b,\infty)$, $\varphi(y) = \psi(1/y)$ and the integral on right is of course interpreted in the sense of regularization. Equation (18.167) makes sense for all $\lambda \neq -1, 0, 1, \cdots$. We now interpret the integral $\int_b^\infty x^\lambda dx$ as the result of the action of $x^\lambda_{b\le x<\infty}$ on a test function $\psi^*(x)$ such that $\psi^*(1/x) = \varphi^*(x) = 1$ for $0 \le x \le 1/b$. Thus

$$\int_b^\infty x^\lambda dx = -\frac{b^{\lambda+1}}{\lambda+1}, \tag{18.168}$$

for $\lambda \neq -1, 0, 1, \cdots$. As before, we use (18.168) to define $\int_b^\infty x^\lambda dx$ for $\lambda = 0, 1, \cdots$. Therefore, the only singularity is again $\lambda = -1$.

In view of (18.165) and (18.168), we put

$$\int_0^\infty x^\lambda dx = \int_0^b x^\lambda dx + \int_b^\infty x^\lambda dx = 0. \tag{18.169}$$

The last equality is valid for all λ except $\lambda = -1$.

The case $\lambda = -1$ needs a separate treatment. Observe that for all test functions φ whose support does not contain 0, we have

$$\int_0^b t^{-1}\varphi(t)dt = \varphi(b)\ln b - \int_0^b \varphi'(t)\ln t dt. \tag{18.170}$$

This identity suggests that we may define the distribution $t^{-1}_{0<t\le b}$ by

$$\langle t^{-1}_{0<t\le b}, \varphi\rangle = \varphi(b)\ln b - \int_0^b \varphi'(t)\ln t dt, \tag{18.171}$$

that is,

$$t^{-1}_{0<t\le b} = (\ln b)\delta(t-b) + [(\ln t)\chi_{[0,b]}]', \tag{18.172}$$

where $\chi_{[0,b]}$ is the characteristic function which is equal to 1 on $[0,b]$, and where the derivative on the right-hand side is taken in the distributional sense. As in (18.164), we again let $\varphi^*(t)$ denote a test function in $\mathcal{D}(\mathbf{R})$ which equals "1" on $[0,b]$. From (18.164) it follows that

$$\langle t^{-1}_{0<t\le b}, \varphi^*\rangle = \ln b. \tag{18.173}$$

We shall regard the integral $\int_0^b t^{-1}dt$ as the action of the distribution $t^{-1}_{0<t\le b}$ on φ^*. Thus (18.173) gives

$$\int_0^b t^{-1}dt = \ln b. \tag{18.174}$$

By transforming the interval $[b,\infty)$ to $(0,1/b]$ via the substitution $\tau = 1/t$, we also have

$$\int_b^\infty t^{-1}dt = -\ln b; \tag{18.175}$$

cf. (18.167) and (18.168). Formulas (18.174) and (18.175) suggest that equation (18.169) should remain valid when $\lambda = -1$. That is, we shall assign the meaning

$$\int_0^\infty x^\lambda dx = \int_0^b x^\lambda dx + \int_b^\infty x^\lambda dx = 0 \tag{18.176}$$

for all complex λ.

The above result will play an important role in the derivation of the asymptotic expansions of the Mellin-type convolution given in (18.151) or its equivalent and more symmetric convolution defined by (18.152). Without restrictions on the exponents α and β in the expansions (18.153) and (18.154), the integral

$$\int_0^\infty f(t)dt \tag{18.177}$$

will not normally converge. We shall now use the concept of regularization to give a meaningful definition to the infinite integral in (18.177). Suggested by (18.176), we split the interval of integration in (18.177) at $t = b$ and consider first the integral $\int_0^b f(t)dt$. Choose n such that $n+\operatorname{Re}\alpha > -1$ and write (18.153) as

$$f(t) = \sum_{s=0}^{n-1} a_s t^{s+\alpha} + f_{0,n}(t) \tag{18.178}$$

with

$$f_{0,n}(t) = O(t^{n+\alpha}) \tag{18.179}$$

as $t \to 0^+$. We define

$$\int_0^b f(t)dt := \sum_{s=0}^{n-1} a_s \int_0^b t^{s+\alpha}dt + \int_0^b f_{0,n}(t)dt, \tag{18.180}$$

where the integrals under the summation sign are understood in the sense of either (18.165) or (18.174), depending upon whether $s+\alpha$ is not or is equal to -1, and where the last integral on the right-hand side exists and is hence taken in the ordinary sense.

Similarly, we choose m such that $m + \operatorname{Re}\beta > 1$ and write (18.154) in the form

$$f(t) = \sum_{s=0}^{m-1} b_s t^{-s-\beta} + f_{\infty,m}(t) \tag{18.181}$$

with

$$f_{\infty,m}(t) = O(t^{-m-\beta}) \tag{18.182}$$

as $t \to +\infty$. The integral $\int_b^\infty f(t)dt$ is then defined by

$$\int_b^\infty f(t)dt := \sum_{s=0}^{m-1} b_s \int_b^\infty t^{-s-\beta}dt + \int_b^\infty f_{\infty,m}(t)dt, \tag{18.183}$$

where the integrals in the sum are understood in the sense of either (18.168) or (18.175), depending upon whether $s+\beta$ is not or is equal to 1, and where the last integral on the right-hand side exists and is taken in the ordinary sense.

Definition 18.2. If f is in $\mathcal{F}$ then the integral $\int_0^\infty f(t)dt$ is given the meaning

$$\int_0^\infty f(t)dt = \int_0^b f(t)dt + \int_b^\infty f(t)dt, \tag{18.184}$$

where the two integrals on the right are defined by the formulas (18.180) and (18.183). The value of the right-hand side is called the *regularization of the (formal) integral* on the left.

We shall also call the right-hand sides of equations (18.165), (18.168), (18.174), (18.175) and (18.176), respectively, the regularization of the integrals on the corresponding left-hand sides of these equations.

A few comments are now in order. First, it is easily shown that the definitions of $\int_0^b f(t)dt$ and $\int_b^\infty f(t)dt$ given in (18.180)-(18.183) are independent of the choices of n and m as long as $n + \operatorname{Re}\alpha > -1$ and $m + \operatorname{Re}\beta > 1$. Second, it can also be shown that the definition of $\int_0^\infty f(t)dt$ given in (18.184) is independent of the choice of b. Third, the new definition of $\int_0^\infty f(t)dt$ agrees with the usual one when the integral converges in the ordinary sense. Finally, for the functions of $\mathcal{F}$, we have the linearity property

$$\int_0^\infty [c_1 f_1(t) + c_2 f_2(t)]dt = c_1 \int_0^\infty f_1(t)dt + c_2 \int_0^\infty f_2(t)dt, \qquad (18.185)$$

where all these integrals are understood in the sense of Definition 18.2.

Definition 18.3. If $f \in \mathcal{F}$, then we define the *(generalized) Mellin transform* of f by

$$M[f;z] = \int_0^\infty t^{z-1} f(t)dt, \qquad (18.186)$$

where the integral on the right is understood in the sense of Definition 18.2.

The above definition is easily verified to be equivalent to the definition of Mellin transform given in (18.29), when $z \neq -s-\alpha$ and $z \neq s+\beta$, where $s+\alpha$ and $s+\beta$ are the exponents in the asymptotic expansions of f. Thus, $M[f;z]$ determines a meromorphic function with poles at $-s-\alpha$ and $s+\beta$. Furthermore, the present definition allows one to assign value to $M[f;-s-\alpha]$ and $M[f;s+\beta]$. The following lemma shows that these values are closely related to the meromorphic functions.

Lemma 18.7. *Let f be in $\mathcal{F}$ with asymptotic expansions* (18.153) *and* (18.154), *and let $M[f;z]$ be the Mellin transform defined in Definition 18.3.*

(i) If $-i-\alpha \neq s+\beta$ for all s, then

$$M[f;-i-\alpha] = \lim_{z\to -i-\alpha} \left\{ M[f;z] - \frac{a_i}{z+i+\alpha} \right\}.$$

(ii) If $j+\beta \neq -s-\alpha$ for all s, then

$$M[f;j+\beta] = \lim_{z\to j+\beta} \left\{ M[f;z] + \frac{b_j}{z-j-\beta} \right\}.$$

(iii) If $-i-\alpha = j+\beta$ for some i and j, then

$$\begin{aligned} M[f;-i-\alpha] &= M[f;j+\beta] \\ &= \lim_{z\to -i-\alpha} \left\{ M[f;z] - \frac{a_i}{z+i+\alpha} + \frac{b_j}{z-j-\beta} \right\}. \end{aligned}$$

Proof. By Definitions 18.2 and 18.3,

$$M[f;z] = \int_0^1 t^{z-1} f(t)dt + \int_1^\infty t^{z-1} f(t)dt. \tag{18.187}$$

For convenience, we have taken $b = 1$ in (18.184). From (18.165), (18.168), (18.178) and (18.181), it follows that the first integral in (18.187) is given by

$$\sum_{s=0}^{n-1} \frac{a_s}{z+s+\alpha} + \int_0^1 t^{z-1} f_{0,n}(t)dt$$

and the second integral by

$$-\sum_{s=0}^{m-1} \frac{b_s}{z-s-\beta} + \int_1^\infty t^{z-1} f_{\infty,m}(t)dt.$$

The last two integrals converge, and each defines an analytic function of z in the strip $-\operatorname{Re}\alpha - n < \operatorname{Re} z < m + \operatorname{Re}\beta$. Choose $n > i$, and write

$$\begin{aligned} M[f;z] - \frac{a_i}{z+i+\alpha} = \sum_{s=0,s\neq i}^{n-1} \frac{a_s}{z+s+\alpha} + \int_0^1 t^{z-1} f_{0,n}(t)dt \\ - \sum_{s=0}^{m-1} \frac{b_s}{z-s-\beta} + \int_1^\infty t^{z-1} f_{\infty,m}(t)dt. \end{aligned} \tag{18.188}$$

From this, it follows that

$$\begin{aligned} \lim_{z\to -i-\alpha} \left\{ M[f;z] - \frac{a_i}{z+i+\alpha} \right\} = \sum_{s=0,s\neq i}^{n-1} \frac{a_s}{s-i} + \sum_{s=0}^{m-1} \frac{b_s}{i+\alpha+s+\beta} \\ + \int_1^\infty t^{-i-\alpha-1} f_{\infty,m}(t)dt. \end{aligned}$$

Direct calculation, using (18.174), shows that the right-hand side of the above equation is equal to $M[f; -i-\alpha]$, thus proving (i).

Statements (ii) and (iii) are proved in a similar manner, using (18.175) where necessary. □

As remarked earlier, divergent integrals must be handled with care since standard rules do not usually apply. The next result shows the change-of-variable formula is no longer legitimate.

Lemma 18.8. *Let f be in $\mathcal{F}$ with asymptotic expansions* (18.153) *and* (18.154), *and let $f_x(t) = f(x/t)$. Then*

(i) $M[f_x; z] = x^z M[f; -z]$ *if* $z \neq s + \alpha$ *and* $z \neq -s - \beta$ *for all* s;

(ii) $M[f_x; i + \alpha] = x^{i+\alpha}\{M[f; -i - \alpha] - a_i \ln x\}$ *if* $i + \alpha \neq -s - \beta$ *for all* s;

(iii) $M[f_x; -j - \beta] = x^{-j-\beta}\{M[f; j + \beta] + b_j \ln x\}$ *if* $j + \beta \neq -s - \alpha$ *for all* s;

(iv) $M[f_x; z] = x^z\{M[f; -z] - (a_i - b_j) \ln x\}$ *if* $z = i + \alpha = -j - \beta$ *for some* i *and* j.

Proof. The statement (i) can be proved as in the proof of Lemma 18.7, except that instead of (18.187) we begin with

$$M[f_x; z] = \int_0^x t^{z-1} f(x/t)dt + \int_x^\infty t^{z-1} f(x/t)dt. \tag{18.189}$$

The two integrals on the right are given by

$$x^z \sum_{s=0}^{m-1} \frac{b_s}{z + s + \beta} + \int_0^x t^{z-1} f_{\infty,m}(x/t)dt$$

and

$$-x^z \sum_{s=0}^{n-1} \frac{a_s}{z - s - \alpha} + \int_x^\infty t^{z-1} f_{0,n}(x/t)dt,$$

respectively. For $-m - \operatorname{Re}\beta < \operatorname{Re} z < n + \operatorname{Re}\alpha$, the two remainder integrals exist as ordinary integrals, and hence by a change of variable they are equal to

$$x^z \int_1^\infty u^{-z-1} f_{\infty,m}(u)du$$

and

$$x^z \int_0^1 u^{-z-1} f_{0,n}(u)du.$$

A combination of these results gives

$$\begin{aligned} M[f_x; z] =x^z \bigg[&\sum_{s=0}^{m-1} \frac{b_s}{z + s + \beta} + \int_1^\infty u^{-z-1} f_{\infty,m}(u)du \\ &- \sum_{s=0}^{n-1} \frac{a_s}{z - s - \alpha} + \int_0^1 u^{-z-1} f_{0,n}(u)du \bigg]. \end{aligned}$$

The quantity inside the square brackets is exactly $M[f; z]$; see (18.188). This completes the proof of statement (i).

Statements (ii), (iii) and (iv) can be proved in a similar manner by using (18.174) and (18.175) when it is appropriate. □

Returning to the convolution integral (18.152), we assume that $f(t)$ has the asymptotic expansion (18.153) and (18.154), and that $g(t)$ has the corresponding expansions

$$g(t) \sim \sum_{s=0}^{\infty} c_s t^{s+\mu}, \quad \text{as } t \to 0^+, \tag{18.190}$$

$$g(t) \sim \sum_{s=0}^{\infty} d_s t^{-s-\nu}, \quad \text{as } t \to \infty, \tag{18.191}$$

where μ and ν are complex numbers. For the convergence of the integral (18.152), we also impose the conditions that the exponents in these expansions satisfy

$$\mathrm{Re}(\alpha+\nu) > 0, \quad \mathrm{Re}(\beta+\mu) > 0. \tag{18.192}$$

From (18.176) and (18.178) it is easily seen that

$$M[f;z] = M[f_{0,n};z] \tag{18.193}$$

for all values of z. Similarly, if (18.190) is written as

$$g(t) = \sum_{s=0}^{m-1} c_s t^{s+\mu} + g_{0,m}(t) \tag{18.194}$$

with $g_{0,m}(t) = O(t^{m+\mu})$ as $t \to 0^+$, then we also have

$$M[g_{0,m}(x/t);z] = M[g_x;z], \tag{18.195}$$

where $g_x(t) = g(x/t)$. Since $\mathrm{Re}(\alpha+\nu) > 0$ by (18.192), Lemma 18.8 (i) gives

$$M[g_x; i+\alpha] = x^{i+\alpha} M[g; -i-\alpha] \tag{18.196}$$

as long as $i+\alpha \neq s+\mu$ for all s and Lemma 18.8 (ii) gives

$$M[g_x; i+\alpha] = x^{i+\alpha}\{M[g;-i-\alpha] - c_{i+\alpha-\mu} \ln x\} \tag{18.197}$$

as long as $i+\alpha = j+\mu$ for some j.

Multiplying (18.178) and (18.194) together gives

$$f(t)g(x/t) = \sum_{i=0}^{n-1}\sum_{j=0}^{m-1} a_i c_j x^{j+\mu} t^{i+\alpha-j-\mu} + \sum_{i=0}^{n-1} a_i t^{i+\alpha} g_{0,m}(x/t)$$
$$+ \sum_{j=0}^{m-1} c_j x^{j+\mu} t^{-j-\mu} f_{0,n}(t) + f_{0,n}(t) g_{0,m}(x/t). \tag{18.198}$$

Inserting this into (18.152) and applying the linearity property mentioned above, we obtain from (18.176), (18.193) and (18.195)

$$(f*g)(x) = \sum_{i=0}^{n-1} a_i M[g_x; i+\alpha] + \sum_{j=0}^{m-1} c_j M[f; -j-\mu]x^{j+\mu} + \delta_{m,n}(x), \tag{18.199}$$

where

$$\delta_{m,n}(x) = \int_0^\infty f_{0,n}(t) g_{0,m}(x/t) t^{-1} dt. \tag{18.200}$$

Choose m and n so that

$$m \geq 1 - \mathrm{Re}(\mu+\nu), \quad n \geq 1 - \mathrm{Re}(\alpha+\beta) \tag{18.201}$$

and

$$m + \mathrm{Re}(\mu-\alpha) - 1 < n < m + \mathrm{Re}(\mu-\alpha) + 1. \tag{18.202}$$

Under these conditions, the integral in (18.200) exists as an ordinary integral. If it happens that $0 < \alpha \leq 1$ and $0 < \mu \leq 1$, then (18.202) is satisfied with $n = m$.

Theorem 18.8. *Let f and g belong to $\mathcal{F}$, and let their asymptotic expansions be given by* (18.153), (18.154), (18.190) *and* (18.191). *Suppose that* (18.192) *holds and $s+\alpha \neq r+\mu$ for all r and s. Then*

$$(f*g)(x) = \sum_{i=0}^{n-1} a_i M[g; -i-\alpha]x^{i+\alpha} + \sum_{j=0}^{m-1} c_j M[f; -j-\mu]x^{j+\mu} + O(x^{\rho_{m,n}}) \tag{18.203}$$

as $x \to 0^+$, where $\rho_{m,n} = \min\{n + \mathrm{Re}\,\alpha, m + \mathrm{Re}\,\mu\}$.

Proof. Without loss of generality, we may assume that n and m are sufficiently large so that (18.201) and (18.202) hold. Then, from (18.196) and (18.199), it follows that

$$(f*g)(x) = \sum_{i=0}^{n-1} a_i M[g; -i-\alpha]x^{i+\alpha} + \sum_{j=0}^{m-1} c_j M[f; -j-\mu]x^{j+\mu} + \delta_{m,n}(x),$$

where $\delta_{m,n}(x)$ is given in (18.200). Hence it suffices to show

$$\delta_{m,n}(x) = O(x^{\rho_{m,n}}) \tag{18.204}$$

as $x \to 0^+$. For $0 < x < 1$, it is easy to see that

$$\int_0^x f_{0,n}(t)g_{0,m}(x/t)t^{-1}dt = O(x^{n+\alpha}),$$

$$\int_x^1 f_{0,n}(t)g_{0,m}(x/t)t^{-1}dt = O(x^{\rho_{m,n}}),$$

$$\int_1^\infty f_{0,n}(t)g_{0,m}(x/t)t^{-1}dt = O(x^{m+\mu});$$

thus proving (18.204) and the theorem. □

Theorem 18.9. *Let f and g be given as in Theorem 18.8, and let the conditions in* (18.192) *hold. If $s + \alpha = r + \mu$ for some s and r, then*

$$\begin{aligned}(f * g)(x) = &\sum_{i=0}^{\mu-\alpha-1} a_i M[g; -i-\alpha]x^{i+\alpha} + \sum_{i=\mu-\alpha}^{n-1} a_i[c_i^* - c_{i+\alpha-\mu}\ln x]x^{i+\alpha} \\ &+ \sum_{j=0}^{\alpha-\mu-1} c_j M[f; -j-\mu]x^{j+\mu} + \sum_{j=\alpha-\mu}^{m-1} a_j^* c_j x^{j+\mu} \\ &+ O(x^{\rho_{m,n}}\ln x)\end{aligned} \tag{18.205}$$

as $x \to 0^+$, where $\rho_{m,n} = \min\{n + \operatorname{Re}\alpha, m + \operatorname{Re}\mu\}$ and

$$c_i^* = \lim_{z \to -i-\alpha}\left\{M[g; z] - \frac{c_{i+\alpha-\mu}}{z+i+\alpha}\right\}, \tag{18.206}$$

$$a_j^* = \lim_{z \to -j-\mu}\left\{M[f; z] - \frac{a_{j+\mu-\alpha}}{z+j+\mu}\right\}. \tag{18.207}$$

The coefficients a_i and c_j with negative subscripts are understood to be zero.

Proof. We again assume without loss of generality that m and n are sufficiently large so that (18.201) and (18.202) hold. Observe that since $s + \alpha = r + \mu$, $\alpha - \mu$ is an integer. Hence, if $i \geq 0$ and $i \geq \mu - \alpha$ then $i = \mu - \alpha + j$ for some $j \geq 0$. Consequently, $i + \alpha = j + \mu$ for all $i \geq \mu - \alpha$ and $j \geq \alpha - \mu$. Therefore, from (18.196), (18.197) and (18.199) we obtain

$$\begin{aligned}(f * g)(x) = &\sum_{i=0}^{\mu-\alpha-1} a_i M[g; -i-\alpha]x^{i+\alpha} \\ &+ \sum_{i=\mu-\alpha}^{n-1} a_i\left\{M[g; -i-\alpha] - c_{i+\alpha-\mu}\ln x\right\}x^{i+\alpha} \\ &+ \sum_{j=0}^{m-1} c_j M[f; -j-\mu]x^{j+\mu} + \delta_{m,n}(x).\end{aligned} \tag{18.208}$$

In view of Lemma 18.7(i) and (18.192)

$$\begin{aligned} M[g;-i-\alpha] =& M[g;-j-\mu] = \lim_{z\to -j-\mu}\left\{M[g;z]-\frac{c_j}{z+j+\mu}\right\}\\ =& \lim_{z\to -i-\alpha}\left\{M[g;z]-\frac{c_{i+\alpha-\mu}}{z+i+\alpha}\right\}=c_i^*\end{aligned}$$

for $i \geq \mu-\alpha$, and similarly $M[f;-j-\mu]=a_j^*$ for $j\geq \alpha-\mu$. Thus (18.205) will follow immediately from (18.208), provided we can show

$$\delta_{m,n}(x) = O(x^{\rho_{m,n}}\ln x) \tag{18.209}$$

as $x\to 0^+$.

For $0<x<1$, it is easily shown that

$$\begin{aligned}\int_0^x f_{0,n}(t)g_{0,m}(x/t)t^{-1}dt =& O(x^{n+\alpha}),\\ \int_1^\infty f_{0,n}(t)g_{0,m}(x/t)t^{-1}dt =& O(x^{m+\mu}).\end{aligned}$$

Since $n+\alpha=m+\mu$, we also have

$$\int_x^1 f_{0,n}(t)g_{0,m}(x/t)t^{-1}dt = O(x^{m+\mu}\ln x).$$

Therefore (18.209) holds and the theorem is proved. □

Note that if $\alpha=\mu$ then (18.205) simplifies to

$$(f*g)(x)\sim\sum_{i=0}^\infty A_i x^{i+\alpha}-\sum_{i=0}^\infty a_i c_i x^{i+\alpha}\ln x \tag{18.210}$$

as $x\to 0^+$, where $A_i=a_ic_i^*+a_i^*c_i$. Analogous results for large x behavior can be found in Wong and McClure (1984).

We now return to the error term $\delta_{m,n}(x)$ in (18.200), and assume that m and n satisfy (18.201) and (18.202). For simplicity, we also assume that the exponents α and μ are real. To construct numerical bounds for $\delta_{m,n}(x)$, we set

$$M_n = \sup_{t>0}|t^{-n-\alpha}f_{0,n}(t)| \tag{18.211}$$

$$N_m = \sup_{t>0}|t^{-m-\mu}g_{0,m}(t)|. \tag{18.212}$$

If M_n is finite and $m+\mu-1<n+\alpha<m+\mu$ then from (18.200) it follows immediately that

$$|\delta_{m,n}(x)|\leq x^{n+\alpha}M_n\int_0^\infty \tau^{-n-\alpha-1}|g_{0,m}(\tau)|d\tau. \tag{18.213}$$

On the other hand, if N_m is finite and $m+\mu < n+\alpha < m+\mu+1$ then

$$|\delta_{m,n}(x)| \leq x^{m+\mu} N_m \int_0^\infty \tau^{-m-\mu-1} |f_{0,n}(\tau)| d\tau. \tag{18.214}$$

In the case when both results (18.213) and (18.214) do not hold, then we may proceed as follows. In view of the assumptions that $f_{0,n}(t) = O(t^{n+\alpha})$ and $g_{0,m}(t) = O(t^{m+\mu})$ as $t \to 0^+$, it is not unreasonable to suppose that there exist numbers M_n^* and N_m^* such that

$$|f_{0,n}(t)| \leq M_n^* t^{n+\alpha}, \tag{18.215}$$

$$|g_{0,m}(t)| \leq N_m^* t^{m+\mu}, \tag{18.216}$$

for $0 < t < 1$. Put

$$F_{m,n} = \int_1^\infty t^{-m-\mu-1} |f_{0,n}(t)| dt, \tag{18.217}$$

$$G_{m,n} = \int_1^\infty t^{-n-\alpha-1} |g_{0,m}(t)| dt. \tag{18.218}$$

Under the conditions in (18.201) and (18.202), both integrals are convergent. A simple estimation then yields

$$\begin{aligned} |\delta_{m,n}(x)| \leq & x^{n+\alpha} M_n^* \left(G_{m,n} + \frac{N_m^*}{m+\mu-n-\alpha} \right) \\ & + x^{m+\mu} N_m^* \left(F_{m,n} + \frac{M_n^*}{n+\alpha-m-\mu} \right) \end{aligned} \tag{18.219}$$

if $n+\alpha \neq m+\mu$, and

$$|\delta_{m,n}(x)| \leq x^{n+\alpha} (A_{m,n}^* - M_n^* N_m^* \ln x) \tag{18.220}$$

if $n+\alpha = m+\mu$, for $0 < x < 1$, where $A_{m,n}^* = M_n^* G_{m,n} + N_m^* F_{m,n}$.

Example. The above results can be applied to the function

$$R_F(x,y,z) = \frac{1}{2} \int_0^\infty \frac{d\tau}{\sqrt{(\tau+x)(\tau+y)(\tau+z)}}, \tag{18.221}$$

which is called by Carlson (1977, p. 264) the elliptic integral of the first kind. In our notations, this function can be written as

$$2\sqrt{z} R_F(x,y,z) = (f * g)(1/z) \tag{18.222}$$

with

$$f(t) = \frac{1}{\sqrt{(1+xt)(1+yt)}}, \quad g(t) = \frac{1}{\sqrt{1+t}}.$$

Clearly $f(t) \sim \sum_{s=0}^{\infty} a_s t^s$ as $t \to 0^+$, where a_s is a polynomial in x and y of degree s, and $g(t) \sim \sum_{s=0}^{\infty} c_s t^s$ as $t \to 0^+$, where $c_s = (-1)^s \Gamma(s + 1/2)/(s!\sqrt{\pi})$. Thus, with $\alpha = \mu = 0$, Theorem 18.9 gives

$$2\sqrt{z} R_F(x, y, z) \sim \sum_{i=0}^{\infty} a_i c_i z^{-i} \ln z + \sum_{i=0}^{\infty} A_i z^{-i}, \tag{18.223}$$

as $z \to \infty$. It can be shown (exercise) that

$$A_0 = 4 \ln 2 - \ln x - 2 \ln(1 + \sqrt{y/x}), \tag{18.224}$$

but higher coefficients are difficult to calculate. Since $a_0 = c_0 = 1$, coupling (18.223) and (18.224) gives

$$R_F(x, y, z) \sim \frac{1}{\sqrt{z}} \ln \frac{4\sqrt{z}}{\sqrt{x} + \sqrt{y}}, \tag{18.225}$$

as $z \to \infty$. For a more detailed discussion of this example, see Carlson and Gustafson (1985).

18.6 Laplace convolutions

The material in this section is taken from Li and Wong (1994). Let $f(t)$ and $g(t)$ be locally integrable functions on $[0, \infty)$. The convolution integral

$$(f * g)(x) = \int_0^x f(x - t) g(t) dt, \quad x > 0, \tag{18.226}$$

occurs frequently in Laplace transform theory [Doetsch (1974); Sneddon (1972)] and Volterra integral equations [Miller (1971)]. Asymptotic behavior of this integral, as $x \to \infty$, has been investigated by Riekstiņš (1970) and Handelsman and Lew (1974), under the condition that $f(t)$ and $g(t)$ have asymptotic representations in descending powers of t near $t = \infty$. For simplicity, we assume that

$$f(t) \sim \sum_{s=0}^{\infty} a_s t^{-s-\alpha}, \quad 0 < \alpha \leq 1, \tag{18.227}$$

$$g(t) \sim \sum_{s=0}^{\infty} b_s t^{-s-\beta}, \quad 0 < \beta \leq 1, \tag{18.228}$$

as $t \to \infty$. The results of Riekstiņš (1970) and Handelsman and Lew (1974) establish the existence of asymptotic expansions of $(f * g)(x)$ in certain forms, but these expansions involve exponents and coefficients that are not explicitly known. To be of any practical use, these quantities should be determined explicitly. The purpose of this section is to provide formulas for these quantities, and to construct computable bounds for the error terms associated with these expansions. For instance, in the case $0 < \alpha, \beta < 1$, it will be shown that

$$(f * g)(x) = \sum_{s=0}^{n-1} A_s x^{-s-\alpha} + \sum_{s=0}^{n-1} B_s x^{-s-\beta} + \sum_{s=0}^{n-1} C_s x^{1-s-\alpha-\beta} + R_n(x), \tag{18.229}$$

where the remainder satisfies

$$|R_n(x)| \leq Q_n x^{1-\alpha-\beta-n} \tag{18.230}$$

for $x > 1$. The coefficients A_s, B_s and C_s in (18.229) are given explicitly by (18.278)-(18.280) below, and the constant Q_n is also known; see the paragraph containing (18.287).

Our approach is based on the distribution method presented in the previous sections. Here we first introduce the noncommutative convolution product

$$(f \odot g)(x) := \int_0^{x/2} f(x-t)g(t)dt, \tag{18.231}$$

and explain why we need such a product. Then, we give a derivation of (18.229) and a proof of (18.230). Note that the validity of (18.229) and (18.230) is under the assumption that $0 < \alpha, \beta < 1$. The cases when one of the exponents in (18.227)-(18.228) is 1 and when both exponents are equal to 1 are discussed later in the section. At the end of the section, we show how the results obtained for the convolution product (18.226) can be used to derive the asymptotic expansions of the convolution integrals

$$G^+(x) = \int_0^\infty f(t)g(x+t)dt, \tag{18.232}$$

$$G^-(x) = \int_0^\infty f(t)g(x-t)dt, \tag{18.233}$$

for $x > 0$. In (18.233), it is assumed that $g(-t) = g(t)$ for every $t \neq 0$. Asymptotic behavior of these two integrals has been considered previously by Muki and Sternberg (1970) in a study of an integral equation. Our investigation was motivated by a study of the asymptotic behavior of solutions to the renewal equation [Wong and Wong (1976)]

$$u(t) = g(t) + \int_0^t u(t-\tau)f(\tau)d\tau, \tag{18.234}$$

where $f(t)$ is a probability density function on $[0, \infty)$, and the Wiener-Hopf equation [Pipkin (1991)]

$$u(t) = g(t) + \int_0^\infty f(t-\tau)u(\tau)d\tau, \tag{18.235}$$

where $f(-t) = f(t)$; see also Beals and Wong (2020). In both (18.234) and (18.235), f and g satisfy (18.227) and (18.228), respectively. For a more recent investigation of (18.235), we refer to Li and Wong (2021).

Returning to (18.227) and (18.228), we write

$$f(t) = \sum_{s=0}^{n-1} a_s t^{-s-\alpha} + f_n(t), \tag{18.236}$$

$$g(t) = \sum_{s=0}^{n-1} b_s t^{-s-\beta} + g_n(t), \tag{18.237}$$

for each $n \geq 1$. As in Section 18.2, we may consider each function in (18.236) as a distribution, and from Lemma 18.1 we have

$$f = \sum_{s=0}^{n-1} a_s t^{-s-\alpha} - \sum_{s=1}^{n} c_s D^{s-1}\delta + f_n, \tag{18.238}$$

where $0 < \alpha < 1$ and

$$c_s = \frac{(-1)^s}{(s-1)!} M[f; s]. \tag{18.239}$$

If $0 < \beta < 1$, then the corresponding result for (18.237) is

$$g = \sum_{s=0}^{n-1} b_s t^{-s-\beta} - \sum_{s=1}^{n} d_s D^{s-1}\delta + g_n, \tag{18.240}$$

where

$$d_s = \frac{(-1)^s}{(s-1)!} M[g; s]. \tag{18.241}$$

Since the convolution product $*$ in (18.226) (i.e., (18.105)) is distributive, taking the convolution of f and g given in (18.238) and (18.240) leads to the distribution $f_n * g_n$. From (18.27), we have $D^n(f_{n,n}) = f_n$ and $D^n(g_{n,n}) = g_n$, by Definition 18.1

$$f_n * g_n = D^n(f_{n,n}) * D^n(g_{n,n}) = D^{2n}(f_{n,n} * g_{n,n}). \tag{18.242}$$

Now $f_{n,n} * g_{n,n}$ is a distribution in $(0, \infty)$ given by the locally integrable function

$$(f_{n,n} * g_{n,n})(x) = \int_0^x f_{n,n}(x-t)g_{n,n}(t)dt = x\int_0^1 f_{n,n}(x-xu)g_{n,n}(xu)du. \tag{18.243}$$

To differentiate this function under the integral sign, we use the result stated in Lemma 18.5.

Since $f_{n,j}$ and $g_{n,j}$ are locally absolutely continuous in $(0,\infty)$ for $j = 1,\cdots,n$, the function $(f_{n,n}*g_{n,n})(x)$ in (18.243) can be differentiated n (and only n) times under integral signs. But equation (18.242) requires $2n$ times differentiation. Therefore, a straightforward extension of the method given in Section 18.4 does not work, and we shall make the following necessary modification.

The integral in (18.226) can be written as

$$(f*g)(x) := \int_0^{x/2} f(x-t)g(t)dt + \int_0^{x/2} g(x-t)f(t)dt. \tag{18.244}$$

This device was given by Riekstiņš (1970). Let $L^+_{loc}(\mathbf{R})$ denote the class of functions which are locally integrable on $\mathbf{R}$ and vanish on $(-\infty,0)$. The formula in (18.244) holds for any f and g in $L^+_{loc}(\mathbf{R})$. The advantage of the convolution $\odot$ over the convolution $*$ is that the function f in (18.231) needs to be locally integrable only on the open interval $(0,\infty)$, and not on the half-closed interval $[0,\infty)$ as required in (18.226). Thus, equation (18.231) may be used to also define the convolution $f\odot g$, where f is a distribution in $L_{loc}(0,\infty)$ and g is a distribution in $L^+_{loc}(\mathbf{R})$. We wish to extend this to a definition of convolutions of the form $f\odot D^m g$, where m is a nonnegative integer.

Let f and g be two $(m-1)$-times differentiable functions in $(0,\infty)$, and let $g\in L^+_{loc}(\mathbf{R})$. For $x>0$, we define

$$f\odot D^m g = D^m(f\odot g) + \frac{1}{2}\sum_{l=0}^{m-1}\left[f(x/2)g^{(l)}(x/2)\right]^{(m-1-l)}. \tag{18.245}$$

The motivation of this defintion, and the answer to the obvious consistency question, are provided by the following lemma.

Lemma 18.9. *If f, g and Dg all belong to $L^+_{loc}(\mathbf{R})$, then $f\odot g$ and $g\odot f$ are both locally absolutely continuous, and we have*

$$D(f\odot g) = f\odot Dg - \frac{1}{2}f(x/2)g(x/2), \tag{18.246}$$

$$D(g\odot f) = Dg\odot f + \frac{1}{2}f(x/2)g(x/2). \tag{18.247}$$

Proof. Since $Dg\in L^+_{loc}(\mathbf{R})$, by Lemma 18.4 there exists a locally absolutely continuous function $h(x)$ such that $g(x) = h(x)$ a.e. in $\mathbf{R}$ and

$h(0) = 0$. By Lemma 18.4, we also have $Dg = g' = h'$ and $h' \in L^+_{loc}(\mathbf{R})$. Thus, for $x \geq 0$ we have

$$\int_0^x (f \odot h')(t)dt = \int_0^{x/2} h'(\tau) \int_{2\tau}^x f(t-\tau)dtd\tau$$
$$= \int_0^{x/2} [f(x-\tau) + f(\tau)]g(\tau)d\tau.$$

The last equality is obtained by integration by parts. Equation (18.231) then gives

$$\int_0^x (f \odot h')(t)dt = (f \odot g)(x) + \int_0^{x/2} f(\tau)g(\tau)d\tau. \qquad (18.248)$$

Since h (and hence g) is locally absolutely continuous, g is locally bounded. Furthermore, since f is locally integrable, the integral on the right exists. Equation (18.248) implies that $f \odot g$ is locally absolutely continuous, and that

$$f \odot g' = f \odot h' = (f \odot g)' + \frac{1}{2} f(x/2)g(x/2).$$

By Lemma 18.4, $D(f \odot g) = (f \odot g)'$ and $Dg = g'$. This proves equation (18.246).

Using a similar argument, it can be shown that $g \odot f$ is locally absolutely continuous. Since $Dg \in L^+_{loc}(\mathbf{R})$, replacing f and g by Dg and f in (18.244), respectively, gives

$$Dg * f = Dg \odot f + f \odot Dg,$$

which, in view of the definition of $*$ in (18.106), is equivalent to

$$D(g * f) = Dg \odot f + f \odot Dg. \qquad (18.249)$$

Coupling (18.246) and (18.249), we obtain

$$\begin{aligned} Dg \odot f =& D(g * f) - f \odot Dg \\ =& D(g * f) - D(f \odot g) - \frac{1}{2} f(x/2)g(x/2) \\ =& D(g \odot f) - \frac{1}{2} f(x/2)g(x/2). \end{aligned}$$

The last equality follows from (18.231) and (18.244). This proves (18.247).
□

Using the fact that $\delta = DH$ with H being the Heaviside function, it can be shown directly from (18.245) that

$$f \odot D^m \delta = D^m f \tag{18.250}$$

for any $f(x)$ which vanishes on $(-\infty, 0)$ and has a locally absolutely continuous mth derivative on $(0, \infty)$. In particular, we have

$$t^{-s-\alpha} \odot D^j \delta = (-1)^j (\alpha + s)_j t^{-s-j-\alpha}, \tag{18.251}$$

where s and j are nonnegative integers and $0 < \alpha \leq 1$. In (18.251) we have used the Pochhammer notation $(\gamma)_0 = 1$ and $(\gamma)_n = \gamma(\gamma+1)\cdots(\gamma+n-1)$ for $n = 1, 2, \cdots$.

The incomplete Beta integral is defined by

$$B_x(a, b) = \int_0^x t^{a-1}(1-t)^{b-1} dt, \tag{18.252}$$

where $a > 0$, $b > 0$ if $x \geq 1$ and b can be negative or zero if $x < 1$. In terms of this integral, it is evident for $0 < \alpha \leq 1$, $0 < \beta < 1$ and s any nonnegative integer, we have

$$t^{-a-s} \odot t^{-\beta} = B_{1/2}(1-\beta, 1-\alpha-s) t^{1-\alpha-\beta-s}.$$

Using the fact that

$$t^{-\beta-j} = \frac{(-1)^j}{(\beta)_j} D^j t^{-\beta},$$

it can be shown from the definition (18.245) that

$$t^{-\alpha-s} \odot t^{-\beta-j} = e_{s,j}(\alpha, \beta) t^{1-\alpha-\beta-s-j} \tag{18.253}$$

in $(0, \infty)$, where

$$\begin{aligned} e_{s,j}(\alpha, \beta) =& \frac{(\alpha+\beta+s-1)_j}{(\beta)_j} B_{1/2}(1-\beta, 1-\alpha-s) \\ & - \sum_{l=0}^{j-1} 2^{\alpha+\beta+s+l-1} \frac{(\alpha+\beta+s+l)_{j-1-l}}{(\beta+l)_{j-l}}. \end{aligned} \tag{18.254}$$

If $\beta = 1$, then we use the facts that

$$t^{-1-j} = \frac{(-1)^j}{j!} D^{j+1} \ln t$$

and

$$t^{-\alpha-s} \odot \ln t = [\xi_s(\alpha) \ln t + \eta_s(\alpha)]^{1-\alpha-s},$$

where $\xi_s(1) = 2$,

$$\xi_s(\alpha) = \frac{1}{\alpha - 1 + s}(2^{\alpha-1+s} - 1) \tag{18.255}$$

for $0 < \alpha < 1$, $s = 0, 1, \cdots$, or $\alpha = 1$, $s = 1, 2, \cdots$, and

$$\eta_s(\alpha) = \int_0^{1/2} (1-u)^{-\alpha-s} \ln u du = \frac{d}{da} B_{1/2}(a, 1-\alpha-s)\big|_{a=1}, \tag{18.256}$$

for $0 < \alpha \leq 1$ and $s = 0, 1, \cdots$. The definition (18.245) then gives

$$t^{-\alpha-s} \odot t^{-1-j} = [h_{s,j}(\alpha) \ln t + k_{s,j}(\alpha)] t^{-\alpha-s-j} \tag{18.257}$$

in $(0, \infty)$, where

$$h_{s,j}(\alpha) = \frac{(\alpha+s)_j}{j!} 2^{\alpha-1-s} - \frac{(\alpha-1+s)_{j+1}}{j!} \xi_s(\alpha), \tag{18.258}$$

and

$$\begin{aligned} k_{s,j}(\alpha) =& (j+1)\frac{(\alpha-1+s)_j}{j!}\xi_s(\alpha) - \frac{(\alpha-1+s)_{j+1}}{j!}\eta_s(\alpha) \\ &- \frac{(\alpha+s)_j}{j!} 2^{\alpha-1+s} \ln 2 \\ &+ \sum_{l=0}^{j-1} \left[\binom{j+1}{l} \frac{(\alpha-1+s)_l}{(j+1-l)_l} \xi_s(\alpha) - \frac{(\alpha+1+s+l)_{j-l-1}}{(l+1)_{j-l}} 2^{\alpha+s+l} \right. \\ &\left. - \binom{j}{l} \frac{(\alpha+s)_l}{(j-l)_{l+1}} 2^{\alpha-1+s} \right], \quad 0 < \alpha \leq 1. \end{aligned} \tag{18.259}$$

To derive the expansion in (18.229), we need one further preliminary result.

Lemma 18.10. *Let g_1 and g_2 be in $L^+_{loc}(\mathbf{R})$, and suppose that g_1 is n-times differentiable and g_2 is m-times differentiable in $(0, \infty)$. Let $N = \max\{n, m\}$, and suppose that f is N-times differentiable in $(0, \infty)$. Then we have*

$$f \odot (D^n g_1 + D^m g_2) = f \odot D^n g_1 + f \odot D^m g_2, \tag{18.260}$$

in $(0, \infty)$, i.e., the convolution $\odot$ is left-distributive.

Proof. Without loss of generality, we assume that $n \geq m$, and write $n = m + k$, $k \geq 0$. Then

$$D^n g_1 + D^m g_2 = D^n (g_1 + g_2^{(-k)}),$$

where $g_2^{(-k)}$ is the kth iterated integral of $g_2(x)$; i.e.,

$$g_2^{(-k)} = \frac{1}{(k-1)!}\int_0^x (x-t)^{k-1} g_2(t)dt.$$

Let $h = g_1 + g_2^{(-k)}$. Then $f \odot (D^n g_1 + D^m g_2) = f \odot D^n h$, and $h \in L^+_{loc}(\mathbf{R})$. By definition (18.245)

$$f \odot (D^n g_1 + D^m g_2) = D^n(f \odot h) + \frac{1}{2}\sum_{l=0}^{n-1}[f(x/2)h^{(l)}(x/2)]^{(n-1-l)}.$$

In terms of g_1 and $g_2^{(-k)}$, this becomes

$$\begin{aligned} f \odot (D^n g_1 + D^m g_2) =& D^n(f \odot g_1) + D^n(f \odot g_2^{(-k)}) \\ &+ \frac{1}{2}\sum_{l=0}^{n-1}[f(x/2)g_1^{(l)}(x/2)]^{(n-1-l)} \\ &+ \frac{1}{2}\sum_{l=0}^{n-1}[f(x/2)g_2^{(l-k)}(x/2)]^{(n-1-l)}. \end{aligned} \tag{18.261}$$

By re-indexing, we have

$$\begin{aligned} \sum_{l=0}^{n-1}[f(x/2)g_2^{(l-k)}(x/2)]^{(n-1-l)} =& \sum_{j=0}^{m-1}[f(x/2)g_2^{(j)}(x/2)]^{(m-1-j)} \\ &+ \sum_{i=1}^{k}[f(x/2)g_2^{(-i)}(x/2)]^{(m-1+i)}. \end{aligned} \tag{18.262}$$

The Leibnitz rule gives

$$D^n(f \odot g_2^{(-k)}) = D^n(f \odot g_2) - \frac{1}{2}\sum_{i=1}^{k}[f(x/2)g_2^{(-i)}(x/2)]^{(m-1+i)}. \tag{18.263}$$

In (18.262) and (18.263), empty sums are understood to be zero. Inserting (18.262) and (18.263) in (18.261), the final result now follows from (18.245). □

We are now ready to derive the expansion in (18.229). From (18.236), we have

$$(f \odot g)(x) = \sum_{s=0}^{n-1} a_s(t^{-s-\alpha} \odot g)(x) + (f_n \odot g) \tag{18.264}$$

for $0 < x < \infty$. Each of the convolutions in (18.264) exists as an ordinary integral in the form (18.231), and defines a locally integrable function in $(0, \infty)$. By Lemma 18.10, the convolution product $\odot$ in (18.231) is left-distributive. Hence for each $s = 0, 1, \cdots, n-1$, we have from (18.240)

$$t^{-s-\alpha} \odot g = \sum_{j=0}^{n-1} b_j t^{-s-\alpha} \odot t^{-\beta-j} - \sum_{j=0}^{n-1} d_{j+1} t^{-s-\alpha} \odot D^j \delta + t^{-s-\alpha} \odot g_n, \tag{18.265}$$

which, combined with (18.251) and (18.253), gives

$$\begin{aligned} t^{-s-\alpha} \odot g = &\sum_{j=0}^{n-1} b_j e_{s,j}(\alpha, \beta) t^{1-\alpha-\beta-s-j} + \sum_{j=0}^{n-1} (-1)^{j+1} d_{j+1} (\alpha + s)_j t^{-s-j-\alpha} \\ &+ t^{-s-\alpha} \odot g_n. \end{aligned} \tag{18.266}$$

Since the distribution defined by g_n is $D^n g_{n,n}$, by (18.245)

$$t^{-s-\alpha} \odot g_n = D^n(t^{-s-\alpha} \odot g_{n,n}) + \frac{1}{2} \sum_{l=0}^{n-1} [(t/2)^{-s-\alpha} g_{n,n-l}(t/2)]^{(n-1-l)}. \tag{18.267}$$

Note that $g_{n,n}$ is locally integrable on $[0, \infty)$. Thus, from (18.231) we have

$$(t^{-s-\alpha} \odot g_{n,n})(x) = x^{1-\alpha-s} \int_0^{1/2} (1-u)^{-\alpha-s} g_{n,n}(xu) du. \tag{18.268}$$

Since $g_{n,j}$ is locally absolutely continuous in $(0, \infty)$ for $j = 1, 2, \cdots, n$, by Lemma 18.5 we can differentiate (18.268) under the integral sign n times to obtain

$$\begin{aligned} &(t^{-s-\alpha} \odot g_{n,n})^{(n)}(x) \\ =&\sum_{j=0}^{n} \binom{n}{j} (-1)^j (\alpha - 1 + s)_j x^{1-\alpha-s-j} \int_0^{1/2} (1-u)^{-\alpha-s} u^{n-j} g_{n,j}(xu) du \\ =:&x^{-n} \varepsilon_s^{(1)}(x), \end{aligned} \tag{18.269}$$

almost everywhere in $(0, \infty)$. We have written the last expression in the form $x^{-n}\varepsilon_s^{(1)}(x)$ in order to indicate that the rate of decay of $\varepsilon_s^{(1)}(x)$ is independent of n; see below. Furthermore, since each function that has been differentiated is locally absolutely continuous, by Lemma 18.4

$$D^n(t^{-s-\alpha} \odot g_{n,n}) = (t^{-s-\alpha} \odot g_{n,n})^{(n)}(x) = x^{-n} \varepsilon_s^{(1)}(x). \tag{18.270}$$

Now each distribution in (18.266) is determined by a locally integrable function in $(0, \infty)$. By replacing these distributions by their corresponding functions, we obtain from (18.267) and (18.270)

$$\begin{aligned}(t^{-s-\alpha} \odot g)(x) = &\sum_{j=0}^{n-1} b_j e_{s,j}(\alpha, \beta) x^{1-\alpha-\beta-s-j} \\ &+ \sum_{j=0}^{n-1} (-1)^{j+1} d_{j+1} (\alpha+s)_j x^{-s-j-\alpha} \\ &+ \frac{1}{2} \sum_{j=0}^{n-1} [(x/2)^{-s-\alpha} g_{n,n-j}(x/2)]^{(n-1-j)} + x^{-n} \varepsilon_s^{(1)}(x)\end{aligned} \tag{18.271}$$

holding in the sense of distributions in $(0, \infty)$. It is straightforward to show that this equation in fact holds pointwise almost everywhere in $(0, \infty)$. Since every convolution in (18.271) and (18.269) has at least one factor which is continuous in $(0, \infty)$, we conclude that (18.271) holds for each x in $(0, \infty)$. Substituting (18.271) in (18.264) gives

$$\begin{aligned}(f \odot g)(x) = &\sum_{s=0}^{n-1} \sum_{j=0}^{n-1} a_s b_j e_{s,j}(\alpha, \beta) x^{1-\alpha-\beta-s-j} \\ &+ \sum_{s=0}^{n-1} \sum_{j=0}^{n-1} (-1)^{j+1} a_s d_{j+1} (\alpha+s)_j x^{-s-j-\alpha} \\ &+ \frac{1}{2} \sum_{s=0}^{n-1} \sum_{j=0}^{n-1} a_s [(x/2)^{-s-\alpha} g_{n,n-j}(x/2)]^{(n-1-j)} \\ &+ \sum_{s=0}^{n-1} a_s x^{-n} \varepsilon_s^{(1)}(x) + (f_n \odot g)(x)\end{aligned} \tag{18.272}$$

for $0 < \alpha \leq 1$ and $0 < \beta < 1$. Reversing the roles of f and g, we obtain a corresponding result for $(g \odot f)(x)$. Adding the two results together yields

$$\begin{aligned}(f \odot g)(x) = &\sum_{s=0}^{n-1} \sum_{j=0}^{n-1} a_s b_j [e_{s,j}(\alpha, \beta) + e_{j,s}(\beta, \alpha)] x^{1-\alpha-\beta-s-j} \\ &+ \sum_{s=0}^{n-1} \sum_{j=0}^{n-1} (-1)^{j+1} a_s d_{j+1} (\alpha+s)_j x^{-s-j-\alpha}\end{aligned}$$

$$+\sum_{s=0}^{n-1}\sum_{j=0}^{n-1}(-1)^{j+1}b_s c_{j+1}(\beta+s)_j x^{-s-j-\beta}$$
$$+R_{n,4}(x)+R_{n,5}(x)+R_{n,6}(x), \qquad (18.273)$$

where

$$R_{n,4}(x)=\frac{1}{2}\sum_{s=0}^{n-1}\sum_{j=0}^{n-1}[a_s(x/2)^{-s-\alpha}g_{n,n-j}(x/2)$$
$$+b_s(x/2)^{-s-\beta}f_{n,n-j}(x/2)]^{(n-1-j)}, \qquad (18.274)$$

$$R_{n,5}(x)=\sum_{s=0}^{n-1}[a_s\varepsilon_s^{(1)}(x)+b_s\varepsilon_s^{(2)}(x)]x^{-n}, \qquad (18.275)$$

$$R_{n,6}(x)=(f_n\odot g)(x)+(g_n\odot f)(x), \qquad (18.276)$$

and $\varepsilon_s^{(2)}(x)$ is given by

$$-x^{-n}\varepsilon_s^{(2)}(x)=\sum_{j=0}^{n}\binom{n}{j}(-1)^j(\beta-1+s)_j x^{1-\beta-s-j}$$
$$\times\int_0^{1/2}(1-u)^{-\beta-s}u^{n-j}f_{n,j}(xu)du. \qquad (18.277)$$

Each of these double sums in (18.273) can be rearranged to give a simple sum, and truncating each of these single sums after n terms gives the expansion (18.229) with

$$A_n=\sum_{j=0}^{s}(-1)^{j+1}a_{s-j}d_{j+1}(\alpha+s-j)_j, \qquad (18.278)$$

$$B_n=\sum_{j=0}^{s}(-1)^{j+1}b_{s-j}c_{j+1}(\beta+s-j)_j, \qquad (18.279)$$

$$C_n=\sum_{j=0}^{s}a_{s-j}b_j[e_{s-j,j}(\alpha,\beta)+e_{j,s-j}(\beta,\alpha)]. \qquad (18.280)$$

The remainder $R_n(x)$ in (18.229) can be written as

$$R_n(x)=\sum_{i=1}^{6}R_{n,i}(x), \qquad (18.281)$$

where

$$R_{n,1}(x) = \sum_{s=n}^{2n-2} \sum_{j=s-n+1}^{n-1} a_{s-j} b_j [e_{s-j,j}(\alpha,\beta) + e_{j,s-j}(\beta,\alpha)] x^{1-\alpha-\beta-s}, \tag{18.282}$$

$$R_{n,2}(x) = \sum_{s=n}^{2n-2} \sum_{j=s-n+1}^{n-1} (-1)^{j+1} a_{s-j} d_{j+1} (\alpha+s-j)_j x^{-s-\alpha}, \tag{18.283}$$

$$R_{n,3}(x) = \sum_{s=n}^{2n-2} \sum_{j=s-n+1}^{n-1} (-1)^{j+1} b_{s-j} c_{j+1} (\beta+s-j)_j x^{-s-\beta}. \tag{18.284}$$

The last three terms on the right-hand side of (18.281) are given in (18.274)-(18.276).

Next, we establish the inequality in (18.230). To do that we first estimate the terms $R_{n,1}(x)$, $R_{n,2}(x)$ and $R_{n,3}(x)$. By re-indexing, we have from (18.282)

$$\begin{aligned} R_{n,1}(x) = &\sum_{s=0}^{n-2} \sum_{j=0}^{n-2-s} a_{n-j-1} b_{s+j+1} [e_{n-j-1,s+j+1}(\alpha,\beta) \\ &+ e_{s+j+1,n-j-1}(\beta,\alpha)] x^{1-\alpha-\beta-n-s}. \end{aligned}$$

Hence

$$|R_{n,1}(x)| \le \gamma_{n,1} x^{1-\alpha-\beta-n},$$

where

$$\gamma_{n,1} = \sum_{s=0}^{n-2} \sum_{j=0}^{n-2-s} |a_{n-j-1} b_{s+j+1} [e_{n-j-1,s+j+1}(\alpha,\beta) + e_{s+j+1,n-j-1}(\beta,\alpha)]|.$$

Similarly, we have

$$|R_{n,2}(x)| \le \bar{A}_n x^{-\alpha-n}, \quad |R_{n,3}(x)| \le \bar{B}_n x^{-\beta-n},$$

where

$$\bar{A}_n = \sum_{s=0}^{n-2} \sum_{j=0}^{n-2-s} |a_{n-j-1} d_{s+j+2} (\alpha+n-j-1)_{s+j+1}|,$$

$$\bar{B}_n = \sum_{s=0}^{n-2} \sum_{j=0}^{n-2-s} |b_{n-j-1} c_{s+j+2} (\beta+n-j-1)_{s+j+1}|.$$

Next we consider the terms $R_{n,4}(x)$, $R_{n,5}(x)$ and $R_{n,6}(x)$. As in (18.134), we assume

$$M_n(\alpha) = \sup_{t>0} |t^{n+\alpha} f_n(t)| < \infty,$$
$$N_n(\beta) = \sup_{t>0} |t^{n+\beta} g_n(t)| < \infty.$$

These conditions do not follow from (18.227), (18.228), and the local integrability of f and g, but will be true in most applications. From (18.26), it is easily seen that for $j = 0, \cdots, n$,

$$|f_{n,j}(t)| \leq \frac{M_n(\alpha)}{(\alpha+n-j)_j} t^{-\alpha-n+j}. \tag{18.285}$$

Similarly, for $j = 0, 1, \cdots, n$,

$$|g_{n,j}(t)| \leq \frac{N_n(\alpha)}{(\beta+n-j)_j} t^{-\beta-n+j}. \tag{18.286}$$

Hence, from (18.269) and (18.277), we have

$$|\varepsilon_s^{(1)}(x)| \leq N_n(\beta) B_{1/2}(1-\beta, 1-\alpha-s) \sum_{j=0}^{n} \binom{n}{j} \frac{|(\alpha+s-1)_j|}{|(\beta+n-j)_j|} x^{1-\alpha-\beta-s},$$

$$|\varepsilon_s^{(2)}(x)| \leq M_n(\alpha) B_{1/2}(1-\alpha, 1-\beta-s) \sum_{j=0}^{n} \binom{n}{j} \frac{|(\beta+s-1)_j|}{|(\alpha+n-j)_j|} x^{1-\alpha-\beta-s}.$$

Coupling the last two estimates gives

$$|R_{n,5}(x)| \leq \gamma_{n,5} x^{1-\alpha-\beta-n}$$

for $x > 1$, where

$$\gamma_{n,5} = \sum_{s=0}^{n-1} \sum_{j=0}^{n} \binom{n}{j} \left\{ N_n(\beta) |a_s| B_{1/2}(1-\beta, 1-\alpha-s) \frac{|(\alpha+s-1)_j|}{|(\beta+n-j)_j|} \right.$$
$$\left. + M_n(\alpha) |b_s| B_{1/2}(1-\alpha, 1-\beta-s) \frac{|(\beta+s-1)_j|}{|(\alpha+n-j)_j|} \right\}.$$

To estimate $R_{n,4}(x)$, we carry out the differentiation in (18.274). The result is

$$R_{n,4}(x) = \sum_{s=0}^{n-1} \left\{ \sum_{l=0}^{n-1} \sum_{j=0}^{n-1-l} (-1)^j 2^{-n+l} \binom{n-l-1}{j} \right.$$
$$\times \left[a_s (\alpha+s)_j (x/2)^{-\alpha-j} g_{n,j+1}(x/2) \right.$$
$$\left. \left. + b_s (\beta+s)_j (x/2)^{-\beta-j} f_{n,j+1}(x/2) \right] \right\} (x/2)^{-s}.$$

From (18.285) and (18.286), it follows that

$$|R_{n,4}(x)| \le \gamma_{n,4}x^{1-\alpha-\beta-n},$$

where

$$\gamma_{n,4} = \sum_{s=0}^{n-1}\left\{\sum_{l=0}^{n-1}\sum_{j=0}^{n-1-l} 2^{\alpha+\beta+l+s-1}\binom{n-l-1}{j}\right.$$
$$\left.\times\left[\frac{|a_s|(\alpha+s)_j}{(\beta+n-j-1)_{j+1}}N_n(\beta)+\frac{|b_s|(\beta+s)_j}{(\alpha+n-j-1)_{j+1}}M_n(\alpha)\right]\right\}.$$

Since the convolution $f_n \odot g_n$ and $g_n \odot f_n$ exist as ordinary integrals, their estimations are rater easy, and we have from (18.276)

$$|R_{n,6}(x)| \le \gamma_{n,6}x^{1-\alpha-\beta-n},$$

where

$$\begin{aligned}\gamma_{n,6} =& M_n(\alpha)N_0(\beta)B_{1/2}(1-\beta, 1-\alpha-n)\\ &+ M_0(\alpha)N_n(\beta)B_{1/2}(1-\alpha, 1-\beta-n).\end{aligned}$$

A combination of the estimates for $R_{n,i}(x)$ with $i = 1, \cdots, 6$ gives

$$|R_n(x)| \le \bar{A}_n x^{-\alpha-n} + \bar{B}_n x^{-\beta-n} + \bar{C}_n x^{1-\alpha-\beta-n}, \tag{18.287}$$

where

$$\bar{C}_n = \gamma_{n,1} + \gamma_{n,4} + \gamma_{n,5} + \gamma_{n,6}.$$

The result (18.230) now follows from (18.287) by letting $Q_n = \bar{A}_n + \bar{B}_n + \bar{C}_n$.

We are now ready to consider the case $0 < \alpha < 1$ and $\beta = 0$. If one of the exponents α and β in (18.227) and (18.228) is equal to 1, then we may assume, without loss of generality, that $0 < \alpha < 1$ and $\beta = 1$. In this case, the incomplete Beta integral in (18.252) does not exist, and hence the expansion in (18.229) no longer holds. To derive the asymptotic expansion of $f * g$ in this case, we shall make use of (18.257) instead of (18.253). Since the argument here is similar to that for the derivation of (18.229), we shall keep our discussion brief. The analogue of (18.240) is

$$g = \sum_{s=0}^{n-1} b_s t^{-s-1} - \sum_{s=1}^{n} d_s^* D^{s-1}\delta + g_n,$$

where

$$d_s^* = \lim_{t\to 0}\left[g_{s,s}(t) + \frac{(-1)^{s-1}}{(s-1)!}b_{s-1}\ln t\right]; \tag{18.288}$$

see Lemma 18.2. (For a more tractable expansion of d_s^*; see the proof of Theorem 18.2.) From (18.257) and (18.251), it follows that

$$t^{-s-\alpha} \odot g = \sum_{j=0}^{n-1} b_j [h_{s,j}(\alpha) \ln t + k_{s,j}(\alpha)] t^{-s-j-\alpha}$$
$$+ \sum_{j=0}^{n-1} (-1)^{j+1} (\alpha+s)_j d_{j+1}^* t^{-s-j-\alpha} + t^{-s-\alpha} \odot g_n.$$

Since the results (18.267) and (18.270) continue to hold when $\beta = 1$, by the same argument used for (18.272) we obtain

$$(f \odot g)(x) = \sum_{s=0}^{n-1} \sum_{j=0}^{n-1} a_s b_j [h_{s,j}(\alpha) \ln x + k_{s,j}(\alpha)] x^{-s-j-\alpha}$$
$$+ \sum_{s=0}^{n-1} \sum_{j=0}^{n-1} (-1)^{j+1} (\alpha+s)_j a_s d_{j+1}^* x^{-s-j-\alpha}$$
$$+ \frac{1}{2} \sum_{s=0}^{n-1} \sum_{j=0}^{n-1} a_s [(x/2)^{-\alpha-s} g_{n,n-j}(x/2)]^{(n-1-j)}$$
$$+ \sum_{s=0}^{n-1} a_s \varepsilon_s^{(1)}(x) x^{-n} + (f_n \odot g)(x) \tag{18.289}$$

for all x in $(0, \infty)$. Furthermore, since (18.272) holds even when $\alpha = 1$, reversing the roles of f and g (and hence α and β) in this equation gives an expansion for $(g \odot f)(x)$. We now add the expansions for $f \odot g$ and $g \odot f$ together, and rearrange the terms as in the derivation for (18.229). The final result is

$$(f * g)(x) = \sum_{s=0}^{n-1} (D_s \ln x + E_s) x^{-\alpha-s} + \sum_{s=0}^{n-1} F_s x^{-1-s} + R_n(x), \tag{18.290}$$

where

$$D_s = \sum_{j=0}^{s} a_{s-j} b_j h_{s-j,j}(\alpha), \tag{18.291}$$

$$E_s = \sum_{j=0}^{s} a_{s-j} \{ b_{s-j} [k_{s-j,j}(\alpha) + e_{j,s-j}(1,\alpha)] + (-1)^{j+1} (\alpha+s-j)_j d_{j+1}^* \}, \tag{18.292}$$

$$F_s = \sum_{j=0}^{s} (-1)^{j+1} (1+s-j)_j b_{s-j} c_{j+1}. \tag{18.293}$$

To estimate the remainder $R_n(x)$, we assume, again as in (18.138), that there exists $\beta' \in (0,1)$ such that

$$N_n(\beta') = \sup_{t>0} |t^{n+\beta'} g_n(t)| < \infty.$$

For $x > 1$, it can be shown as before that there are computable constants P_n and Q_n such that

$$|R_n(x)| \le (P_n \ln x + Q_n) x^{1-\beta'-\alpha-n}. \tag{18.294}$$

Note that the actual error $R_n(x)$ in the expansion (18.290) is $O(x^{-\alpha-n}\ln x)$. Hence the estimate (18.294) is slightly short of the actual result.

Finally we come to the case $\alpha = \beta = 1$. This case turns out to be simpler than one might have expected. This is mainly due to split of the convolution integral $(f * g)(x)$ in (18.226) into two convolution integrals $f(\odot g)(x)$ and $(g \odot f)(x)$ of the form (18.231). The expansion for $(f \odot g)(x)$ given in (18.289) allows the possibility of $\alpha = \beta = 1$; cf. (18.257)-(18.259). By reversing the roles of f and g, the corresponding expansion for $(g \odot f)(x)$ also allows this possibility. For convenience, we set $h_{s,j} = h_{s,j}(1)$ and $k_{s,j} = k_{s,j}(1)$.

Now let $\alpha = 1$ in (18.289), and write down the corresponding expansion for $(g \odot f)(x)$. Adding up the two expansions and rearranging the terms as before, we obtain

$$(f * g)(x) = \sum_{s=0}^{n-1} (G_s \ln x + H_s) x^{-s-1} + R_n(x), \tag{18.295}$$

where

$$G_s = \sum_{j=0}^{s} a_{s-j} b_j (h_{s-j,j} + h_{j,s-j}), \tag{18.296}$$

$$\begin{aligned} H_s = &\sum_{j=0}^{s} [a_{s-j} b_j (k_{s-j,j} + k_{j,s-j}) \\ &+ (-1)^{j+1} (1+s-j)_j (a_{s-j} d^*_{j+1} + b_{s-j} c^*_{j+1})], \end{aligned} \tag{18.297}$$

and c^*_{j+1} has the same meaning as d^*_{j+1} given in equation (18.288) except that $g_{s,s}(t)$ and b_{s-1} are now replaced by $f_{s,s}(t)$ and a_{s-1}.

To obtain an error bound, one may assume that there exists $\rho \in (0,1)$ such that

$$\sup_{t>0} |t^{n+\rho} f_n(t)| + \sup_{t>0} |t^{n+\rho} g_n(t)| < \infty.$$

Under this condition, it can again be shown that there are computable constants P_n and Q_n such that

$$|R_n(x)| \leq (P_n \ln x + Q_n)x^{1-2\rho-n}$$

for $x > 1$.

As the last topic in this section, we discuss the behavior of the two infinite convolution integrals $G^+(x)$ and $G^-(x)$ in (18.232) and (18.233). In the case of $G^-(x)$, it is also assumed that $g(-t) = g(t)$ for all $t \neq 0$. Muki and Sternberg (1970) constructed one-term asymptotic approximation for the integrals $G^+(x)$ and $G^-(x)$ under the conditions

$$g(t) = \frac{a}{t^m} + O(t^{-m-1}), \quad a \neq 0, \ 1 < m < \infty, \tag{18.298}$$

$$f(t) = \frac{b}{t^\mu} + O(t^{-\mu}), \quad b \neq 0, \ 1 \leq \mu < \infty. \tag{18.299}$$

If f and g have the asymptotic expansions (18.227) and (18.228), then it is natural to expect that infinite asymptotic expansions can be derived for these integrals. This is indeed the case, and we shall show how this can be done. The asymptotic expansions of $G^+(x)$ can be easily obtained from the corresponding results for the generalized Stieltjes transform

$$S_f(x) = \int_0^\infty \frac{f(t)}{(t+x)^\rho} dt. \tag{18.300}$$

The asymptotic expansions of this integral can be derived in the same manner as those for the integral in (18.10). For an alternative method via the Mellin convolution (18.151), see Wong (1989, pp. 185-186 and Ex. 3).

We now suppose that f and g satisfy (18.227) and (18.228), respectively. When $0 < \alpha < 1$, we can show that

$$G^+(x) = \sum_{s=0}^\infty A_s^* x^{1-s-\alpha-\beta} + \sum_{s=0}^\infty B_s^* x^{-s-\beta}, \tag{18.301}$$

as $x \to \infty$, where

$$A_s^* = \Gamma(s+\alpha+\beta-1) \sum_{i=0}^s a_i b_{s-i} \frac{\Gamma(1-i-\alpha)}{\Gamma(s-i+\beta)}, \tag{18.302}$$

$$B_s^* = \Gamma(s+\beta) \sum_{i=0}^s (-1)^i \frac{b_{s-i}}{\Gamma(s-i+\beta)} M[f; i+1], \tag{18.303}$$

and $M[f; z]$ is the Mellin transform defined in (18.29).

If $\alpha = 1$ then we have

$$G^+(x) = \sum_{s=0}^\infty (C_s^* \ln x + D_s^*) x^{-s-\beta}, \tag{18.304}$$

where

$$C_s^* = \Gamma(s+\beta)\sum_{i=0}^{s}\frac{(-1)^i}{i!\Gamma(s-i+\beta)}a_i b_{s-i}, \tag{18.305}$$

$$D_s^* = \sum_{i=0}^{s} c_i(s-i+\rho)b_{s-i}. \tag{18.306}$$

The constants $c_i(\rho)$ are given by

$$c_s(\rho) = \frac{(-1)^s\Gamma(\rho+s)}{s!\Gamma(\rho)}\{[\psi(s+1)-\psi(s+\rho)]a_s + a_s^*(\rho)\}, \tag{18.307}$$

where $\psi(z) = \Gamma'(z)/\Gamma(z)$, and

$$a_s^*(\rho) = \lim_{z\to s+\rho}\left\{M[f;z-\rho+1] + \frac{a_s}{z-s-\rho}\right\}. \tag{18.308}$$

Although no explicit bounds can be given for the error terms associated with the asymptotic expansion (18.301) and (18.304), it is evident from our analysis such bounds can indeed be constructed.

In view of the condition $g(t) = g(-t)$ for every $t \neq 0$, the integral $G^-(x)$ in (18.233) can be written as

$$G^-(x) = (f * g)(x) + \int_0^\infty g(t)f(x+t)dt. \tag{18.309}$$

The second term on the right-hand side is an integral of the form (18.232) which has just been considered. Hence, the asymptotic expansion of $G^-(x)$ can be obtained by adding up the corresponding results for the two integrals on the right of (18.309).

18.7 Fourier transform in $\mathbf{R}^n$

In this final section, we will use the distribution method to derive the asymptotic expansions of the multi-dimensional Fourier transform

$$\hat{f}(t) = \frac{1}{(2\pi)^{n/2}}\int_{\mathbf{R}^n} f(x)e^{it\cdot x}dx, \tag{18.310}$$

where $x = (x_1, \cdots, x_n)$, $t = (t_1, \cdots, t_n)$, $t\cdot x = t_1x_1 + \cdots + t_nx_n$ and f is a complex-valued locally integrable function on $\mathbf{R}^n$. In (18.310), it is only assumed that the Cauchy limit

$$\lim_{\rho\to\infty}\int_{|x|<\rho} f(x)e^{it\cdot x}dx$$

exists for every $t \in \mathbf{R}^n$, where $|x| = \sqrt{x_1^2 + \cdots + x_n^2}$. Thus, the function f need not belong to $L^1(\mathbf{R}^n)$.

To proceed further, we first recall some standard notations. If $p = (p_1, \cdots, p_n)$ is a multi-index of non-negative integers, then we put $|p| = p_1 + \cdots + p_n$, $x^p = x_1^{p_1} \cdots x_n^{p_n}$, $D^p = D_1^{p_1} \cdots D_n^{p_n}$, where $D_j = \partial/\partial x_j$ and

$$D_p = (i)^{-|p|} D^p = \left(\frac{\partial}{i\partial x_1}\right)^{p_1} \cdots \left(\frac{\partial}{i\partial x_n}\right)^{p_n}. \tag{18.311}$$

For any open set $\Omega \in \mathbf{R}^n$ and for each nonnegative integer m, the set $C^m(\Omega)$ consists of all complex functions f in Ω whose derivative $D^p f$ exists and is a continuous function in Ω for each multi-index p with $|p| \le m$. The Laplacian is denoted by

$$\Delta = \frac{\partial^2}{\partial x_1^2} + \cdots + \frac{\partial^2}{\partial x_n^2}.$$

The followings are all standard assumptions for our problem.

(A1) $f \in C^{2m}(\mathbf{R}^n \backslash \{0\})$, m being a nonnegative integer. $f(x)$ is expressible in the form

$$f(x) = \sum C_{p,q} x^p r^q + \varphi(x), \tag{18.312}$$

where the q's are real numbers, the p's are multi-indices, $r = \sqrt{x_1^2 + \cdots + x_n^2}$ and the sum is finite. For all p and q under the summation, we require $q + |p| > -n$, and if $Q = \max\{q + |p|\}$ then $2m - 1 \le Q + n < 2m + 1$.

(A2) As $r \to 0^+$, the remainder $\varphi(x)$ in (18.312) satisfies

$$(\Delta^j \varphi)(x) = O(r^{Q-2j+1}), \quad j = 0, 1, \cdots, m,$$

if $Q + n \neq 2m - 1$, and

$$(\Delta^j \varphi)(x) = O(r^{Q-2j+2}), \quad j = 0, 1, \cdots, m,$$

if $Q + n = 2m - 1$.

(A3) For some $\rho > 0$, each of the integrals

$$\int_{|x|>\rho} (\Delta^j f)(x) e^{it\cdot x} dx, \quad j = 0, 1, \cdots, m,$$

converges uniformly for all sufficiently large $|t|$.

Remark 18.2. Under the above assumptions, the Fourier transform $\hat{f}(t)$, as given in (18.310), and the Fourier transform

$$(\Delta^m \varphi)^n(t) = \frac{1}{(2\pi)^{n/2}} \int_{\mathbf{R}^n} (\Delta^m \varphi)(x) e^{it\cdot x} dx \tag{18.313}$$

both exist uniformly for all sufficiently large values of $|t|$, see Shivakumar and Wong (1979, Remark (iii)). From this it follows that $\hat{f}(t)$ and $(\Delta^m \varphi)^n(t)$ are continuous functions of $t \in \mathbf{R}^n$. Also, it is easy to see that for each $j = 0, 1, \cdots, m$, $(\Delta^j \varphi)(x)$ is locally integrable on $\mathbf{R}^n$ and of at most finite algebraic growth at infinity.

Let $\mathcal{S}$ denote the space of rapidly decreasing functions and let $\mathcal{S}'$ be the space of tempered distributions (see Section 18.2). If p is a multi-index and $u \in \mathcal{S}'$ then the formula

$$\langle D^p u, \varphi \rangle = (-1)^{|p|} \langle u, D^p \varphi \rangle \tag{18.314}$$

defines a distribution $D^p u \in \mathcal{S}'$; see Rudin (1973, Section 6.12). If f and $D^p f$ are both locally integrable in $\mathbf{R}^n$ with at most finite algebraic growth at infinity then, by integration by parts, one can show that

$$\int_{\mathbf{R}^n} (D^p f)(x)\eta(x)dx = (-1)^{|p|} \int_{\mathbf{R}^n} f(x)(D^p \eta)(x)dx \tag{18.315}$$

for every $\eta \in \mathcal{S}$, under the hypothesis that f has continuous partial derivatives of all orders up to $|p| - 1$, all these derivatives have at most finite algebraic growth at infinity, and all partial derivatives of order $|p|$ are locally absolutely continuous at each variable $x_1, \cdots, x_n$. Note that these conditions are weaker than those given in Rudin (1973, p. 144). A rigorous proof of (18.315) can be provided easily by using the argument given in Schwartz (1966a, p. 80) and a remark made in Gel'fand and Shilov (1964, p. 21, Section 2.2, paragraph 2). The validity of equation (18.315) shows that the distributional derivative and the ordinary derivative are equal (in the sense of distribution).

What we need is a variant of (18.315), namely,

$$\int_{\mathbf{R}^n} f(x)(\Delta\eta)(x)dx = \int_{\mathbf{R}^n} (\Delta f)(x)\eta(x)dx \tag{18.316}$$

for every $\eta \in \mathcal{S}$. This will imply that the distributional and the ordinary Laplacian are equal also. For equation (18.316) to hold we require that f has continuous second partial derivatives in $\mathbf{R}^n$ except possibly at the origin, f grows at most algebraically at infinity, and $f(x) = O(r^\alpha)$ as $r \to 0^+$, with $\alpha + n > 0$. The result (18.316) follows easily from an application of Green's theorem [Schwartz (1966a), p. 88]. To be more rigorous, Green's theorem should be applied to the shell domain $S = \{x \in \mathbf{R}^n;\ \varepsilon \le |x| \le \rho\}$. Relation (18.316) then results as a limit when $\varepsilon \to 0^+$ and $\rho \to +\infty$.

If $\eta \in \mathcal{S}$ then $\hat{\eta}$, as defined in (18.310), exists and belongs to $\mathcal{S}$; see Rudin (1973, p. 168). If $u \in \mathcal{S}'$ then the Fourier transform $\hat{u}$ is defined by the equation

$$\langle \hat{u}, \eta \rangle = \langle u, \hat{\eta} \rangle, \quad \eta \in \mathcal{S}. \tag{18.317}$$

It is well-known that $\hat{u} \in \mathcal{S}'$ and the following identities [Rudin (1973), p. 176] hold: if P is a polynomial in n variables, say $P(x) = \sum c_p x^p$, and if the differential operator $P(D)$ is defined by $P(D) = \sum c_p D_p$, then

$$\widehat{P(D)u} = P(-t)\hat{u}, \quad \widehat{Pu} = P(D)\hat{u}. \tag{18.318}$$

If f is locally integrable in $\mathbf{R}^n$ and increases no more rapidly than an algebraic function, and if the Cauchy limit in the integral of (18.310) exists uniformly in $t \in \mathbf{R}^n$, then the distributional Fourier transform of f given in (18.317) and the ordinary Fourier transform of f given in (18.310) are equal as functionals; see Gel'fand and Shilov (1964, p. 200). The following result is crucial to our main theorem given below.

Lemma 18.11. *Let $|p| + q + n > 0$. If $q + n \neq -2l$, $l = 0, 1, 2, \cdots$, then*

$$\widehat{x^p r^q} = L(q) D_p(|t|^{-q-n}) \quad in \quad \mathcal{S}', \tag{18.319}$$

where $L(q) = 2^{q+n/2}\Gamma((q+n)/2)/\Gamma(-q/2)$. If $q + n = -2l$, $l = 0, 1, 2, \cdots$, then

$$\widehat{x^p r^q} = L^*(q) D_p(|t|^{2l} \ln |t|) \quad in \quad \mathcal{S}', \tag{18.320}$$

where $L^*(q) = (-1)^{l+1}/2^{n/2+2l-1} l! \Gamma(n/2+1)$.

Note that the value of $L^*(q)$ given in Shivakumar and Wong (1979) is slightly in error.

Proof. It is known that

$$\widehat{r^q} = L(q)|t|^{-q-n} \quad \text{in} \quad \mathcal{S}' \tag{18.321}$$

if $q + n \neq -2l$, $l = 0, 1, 2, \cdots$, and

$$\widehat{r^q} = L^*(q)|t|^{2l} \ln |t| + K|t|^{2l} \quad \text{in} \quad \mathcal{S}' \tag{18.322}$$

if $q + n = -2l$, $l = 0, 1, 2, \cdots$, where K is some constant; see Schwartz (1966b, p. 257-258) and Gel'fand and Shilov (1964, p. 194-195). Thus (18.319) follows immediately from the second equation in (18.318). Since $D_p(|t|^{2l}) = 0$ for $|p| > 2l$, (18.320) also follows immediately from the second equation in (18.318), thus proving the lemma. □

We now return to equation (18.312). Note that each function in this equation generates a distribution in $\mathcal{S}'$ by means of the definition (18.20), and that these distributions are again related via this equation, i.e.,

$$f = \sum C_{p,q} x^p r^q + \varphi \ \text{ in } \ \mathcal{S}'. \tag{18.323}$$

Taking (distributional) Fourier transform on both sides, we have by our lemma

$$\hat{f} = \sum{}' C_{p,q} L(q) D_p(|t|^{-q-n}) + \sum{}'' C_{p,q} L^*(q) D_p(|t|^{2l} \ln |t|) + \hat{\varphi} \tag{18.324}$$

in $\mathcal{S}'$, where $\sum'$ excludes those q's for which $q+n$ is a negative even integer, and $\sum''$ includes only those q's for which $q+n=-2l$, a negative even integer. Applying the first equation in (18.318) m times gives

$$\widehat{\Delta^m \varphi} = (-1)^m |t|^{2m} \hat{\varphi} \ \text{ in } \ \mathcal{S}'. \tag{18.325}$$

Inserting (18.325) in (18.324), we have as functionals

$$\begin{aligned} |t|^{2m} \hat{f} =& \sum{}' C_{p,q} L(q) |t|^{2m} D_p(|t|^{-q-n}) \\ &+ \sum{}'' C_{p,q} L^*(q) |t|^{2m} D_p(|t|^{2l} \ln |t|) + (-1)^m \widehat{\Delta^m \varphi}. \end{aligned} \tag{18.326}$$

In view of Remark 18.2 and the remark following (18.318) and concerning the ordinary and distributional Fourier transforms, we see that each distribution in (18.326) is defined by a locally integrable function in $\mathbf{R}^n$, and hence that equation (18.326) holds not only as functionals but also pointwise almost everywhere in $\mathbf{R}^n$. Since all functions involved are in fact continuous (see again the remark following (18.318)), equation (18.326) is valid for every $t \in \mathbf{R}^n$.

Therefore, we have established the following theorem, which is the main result in Shivakumar and Wong (1979).

Theorem 18.10. *Under the conditions (A1), (A2) and (A3), we have*

$$\hat{f}(t) = \sum{}' C_{p,q} L(q) D_p(|t|^{-q-n}) + \sum{}'' C_{p,q} L^*(q) D_p(|t|^{2l} \ln |t|) + \delta_m(t), \tag{18.327}$$

where the sums $\sum'$ *and* $\sum''$ *are given in* (18.324) *and the remainder satisfies*

$$\delta_m(t) = (-1)^m |t|^{-2m} \widehat{\Delta^m \varphi}(t). \tag{18.328}$$

Remark 18.3. As already pointed out in Remark 18.2, the Fourier transform $\widehat{\Delta^m \varphi}(t)$ given in (18.313) exists uniformly for all sufficiently large value of $t \in \mathbf{R}^n$. Hence, by a generalization of the Riemann-Lebesgue lemma, we have

$$\delta_m(t) = o(|t|^{-2m}) \tag{18.329}$$

as $|t| \to \infty$. Thus, (18.327) is indeed an asymptotic expansion.

18.8 Exercises

1. Verify (18.13).
2. Prove that $\varphi(t) = \exp(-t^2)$ belongs to $\mathcal{S}$.
3. Show that the sequence $\varphi_n(t) = \frac{1}{n}\exp(-t^2)$ converges to zero in $\mathcal{S}$ as $n \to \infty$, whereas the sequence $\varphi_n(t) = \frac{1}{n}\exp(-t^2/n)$ does not.
4. Verify (18.29).
5. Prove that

$$\int_0^\infty \frac{\ln t}{t+z} e^{-\varepsilon t} dt = O(\ln^2 \varepsilon)$$

as $\varepsilon \to 0^+$, and

$$\int_0^\infty \frac{\ln t}{(t+1)^{s+2}} dt = -\frac{1}{s+1} \sum_{k=1}^{s} \frac{1}{k},$$

for $s = 0, 1, 2, \cdots$, where an empty sum is understood to be zero.

6. Let

$$\varphi_\varepsilon(t) = \frac{e^{-\varepsilon t}}{t+z}.$$

Show that

$$\lim_{\varepsilon \to 0^+} \langle \delta^{(s-1)}, \varphi_\varepsilon \rangle = \frac{(s-1)!}{z^s},$$

for $s = 1, 2, \cdots$.

7. For $n = 1$, derive (18.37) directly from (18.25) and (18.26). Use (18.30) to show that the remainder in (18.38) satisfies the recurrence relation

$$R_n(t) = (-1)^n \frac{\pi}{\sin(\alpha\pi)} \frac{a_n}{z^{n+\alpha}} - n! \frac{c_{n+1}}{z^{n+1}} + R_{n+1}(z).$$

Prove (18.37) by induction.

8. Show that the coefficient d_s in (18.45) has the alternative representation

$$d_{s+1} = -\int_0^\infty [f_{s,s}(t) - \frac{(-1)^s a_s}{s!(1+t)}] dt, \quad s = 0, 1, \cdots .$$

9. Verify that the quantity inside the square brackets in (18.58) is equal to

$$\lim_{z \to s+1} \left\{ M[f; z] + \frac{a_s}{z - s - 1} \right\}.$$

10. Prove Lemma 18.3 and Theorem 18.4.
11. Verify (18.85).

12. The complementary error function is defined by

$$\mathrm{erfc}(z) = \frac{2}{\sqrt{\pi}} \int_z^\infty e^{-t^2}\,dt.$$

Show that

$$\mathrm{erfc}(z) = \frac{e^{-z^2}}{z\sqrt{\pi}} \left[1 + \sum_{k=1}^{n} (-1)^k \frac{1\cdot 3\cdots(2k-1)}{(2z^2)^k} + r_n(z)\right],$$

where

$$r_n(z) = (-1)^{n+1} \frac{1\cdot 3\cdots(2n+1)}{2^n} z e^{z^2} \int_z^\infty \frac{e^{-t^2}}{t^{2n+2}}\,dt.$$

Also show that for $|\arg z| \le \pi/2 - \delta$,

$$|r_n(z)| \le \frac{1\cdot 3\cdots(2n+1)}{(2|z|^2)^{n+1}\sin\delta}.$$

13. Consider the integral $I(x)$ in (18.142) as a convolution in the Laplace transform theory, and let $f(t)$ denote the function defined in (18.143). Show by integration by parts that

$$L_f(p) = \frac{e^{i\pi/4}}{\sqrt{2}(p+2i)} - \frac{1}{p^{1/2}(p+2i)}.$$

By taking the inverse Laplace transform, prove that

$$I(x) = \frac{i\sqrt{\pi}}{4} x^{-1/2} - \frac{e^{i\pi/4-2ix}\sqrt{\pi}}{2\sqrt{2x}} F(2\sqrt{x/\pi}),$$

where F is the Fresnel integral defined in (18.141).

14. Prove the result in (18.145) and (18.146).
15. Derive an asymptotic expansion for the integral

$$\frac{1}{\Gamma(\alpha)} \int_0^x (x-t)^{\alpha-1} e^{-t} I_0(t)\,dt, \quad \alpha > 0,$$

as $x \to \infty$, where $I_0(t)$ is the modified Bessel function defined by

$$I_0(t) = \sum_{k=0}^{\infty} \frac{t^{2k}}{4^k (k!)^2}, \quad t > 0,$$

and has the asymptotic expansion

$$I_0(t) \sim \frac{e^t}{\sqrt{2\pi t}} \sum_{s=0}^{\infty} (-1)^s \frac{A_s}{t^s},$$

as $t \to \infty$, with

$$A_s = \frac{(-1)^s}{s!8^s} \prod_{j=1}^{s} (2j-1)^2, \quad s = 0, 1, \cdots. \tag{18.330}$$

16. Put

$$I(x) = \int_0^\infty \frac{t}{t^2 + x^2} e^t K_0(t) dt,$$

where $K_0(t)$ is the modified Bessel function defined by

$$K_0(t) = -[\ln(t/2) + \gamma] I_0(t) + \sum_{k=1}^{\infty} \sum_{j=1}^{k} \frac{(t^2/4)^k}{j(k!)^2},$$

and has the asymptotic expansion

$$K_0(t) \sim \sqrt{\frac{\pi}{2t}} e^{-t} \sum_{s=0}^{\infty} \frac{A_s}{t^s},$$

as $t \to \infty$, with A_s defined as in the previous exercise; see (18.330). Derive the asymptotic expansion

$$I(x) = \sum_{s=0}^{2m} a_s x^{-s-1/2} + \sum_{s=0}^{m-1} b_s x^{-2s-2} + \delta_{2m+1}(x),$$

as $x \to \infty$, where

$$a_s = \frac{(-1)^s (\pi/2)^{3/2}}{s! 8^s \sin[\pi(s+1)/2]} \prod_{k=1}^{s} (2k-1)^2$$

and

$$b_s = (-1)^s [(2s+1)!]^2 \prod_{k=1}^{2s+1} \frac{1}{2k+1}.$$

Show that

$$|\delta_{2m+1}(x)| \leq \frac{\pi^{3/2} x^{-2m-3/2}}{2^{6m+4}(2m+1)!} \prod_{k=0}^{4m} (2k+1)^2.$$

17. Construct an asymptotic expansion for the integral

$$\int_0^\infty t^n (\ln t)^m \frac{e^{-t}}{t+x} dt$$

as $x \to 0^+$, where n and m are nonnegative integers.

18. Use induction to show that in the sense of regularization

$$\int_0^\infty \frac{t^{-m}}{t+x} dt = (-1)^{m+1} \frac{\ln x}{x}, \quad m = 0, 1, 2, \cdots.$$

19. Let

$$f(t) = \frac{1}{\sqrt{(1+xt)(1+yt)}}.$$

Show that

$$M[f;z] = x^{-z}\Gamma(z)\Gamma(1-z)F(1/2, z; 1; 1-y/x),$$

where $F(a,b;c;z)$ is the hypergeometric function defined by the power series

$$F(a,b;c;z) = \sum_{s=0}^{\infty} \frac{(a)_s(b)_s}{(c)_s}\frac{z^s}{s!}$$

for $|z| < 1$. Prove that

$$a_0^* := \lim_{z\to 0}\{M[f;z] - \frac{1}{z}\} = \frac{\partial}{\partial b}F(1/2, b; 1; 1-y/x)\Big|_{b=0} - \ln x,$$

and

$$\frac{\partial}{\partial b}F(1/2, b; 1; z)\Big|_{b=0} = 2\ln 2 - 2\ln(1+\sqrt{1-z}).$$

Use these to verify the result stated in (18.224).

20. Consider the integral

$$I(x) = \int_0^1 \frac{\tau}{\sqrt{1-\tau^2}} e^{-2ix\tau^2} F(2\tau\sqrt{x/\pi})d\tau,$$

where F is the Fresnel integral defined by

$$F(t) = \int_t^{\infty} e^{i\pi\xi^2/2}d\xi.$$

Let

$$f(t) = e^{-2it}F(2\sqrt{t/\pi}).$$

Show that the change of variable $x\tau^2 = t$ puts $I(x)$ into the form of a fractional integral.

$$I(x) = \frac{1}{2\sqrt{x}}\int_0^x (x-t)^{-1/2}f(t)dt.$$

Show by integration by parts that

$$L_f(p) := \int_0^{\infty} f(t)e^{-pt}dt = \frac{e^{\pi i/4}}{\sqrt{2}(p+2i)} - \frac{1}{p^{1/2}(p+2i)}.$$

By taking the inverse Laplace transform, prove that

$$I(x) = \frac{i}{4}\sqrt{\frac{\pi}{x}} - \frac{e^{i\pi/4}}{2\sqrt{x}}e^{-2ix}\sqrt{\frac{\pi}{2}}F(2\sqrt{x/\pi}).$$

Chapter 19

Asymptotics of Integrals with a Coalescing Saddle Point

19.1 Coalescing saddle points

The material in this chapter is taken from Wong (1989). Consider contour integrals of the form

$$I(\lambda;\alpha) = \int_C g(z)e^{-\lambda f(z;\alpha)}dz, \tag{19.1}$$

where f and g are analytic functions of $z \in \mathbf{C}$, and f is also an analytic function of the parameter α. Suppose that there is a critical value of α, say $\alpha = \alpha_0$, such that for $\alpha \neq \alpha_0$, there are two distinct saddle points z_+ and z_- of multiplicity 1, but at $\alpha = \alpha_0$, these two points coincide and give a single saddle point z_0 of multiplicity 2. Thus,

$$f_z(z_+;\alpha) = f_z(z_-;\alpha) = 0, \quad f_{zz}(z_\pm;\alpha) \neq 0 \tag{19.2}$$

for $\alpha \neq \alpha_0$, and

$$f_z(z_0;\alpha_0) = f_{zz}(z_0;\alpha_0) = 0, \quad f_{zzz}(z_0;\alpha_0) \neq 0. \tag{19.3}$$

By the saddle point method (see Chapter 17), it can be verified that the asymptotic forms of $I(\lambda;\alpha)$ for large λ are

$$\begin{aligned} I(\lambda;\alpha) \sim & g(z_+)e^{-\lambda f(z_+;\alpha)}\left[\frac{2\pi}{\lambda f_{zz}(z_+;\alpha)}\right]^{1/2} \\ & + g(z_-)e^{-\lambda f(z_-;\alpha)}\left[\frac{2\pi}{\lambda f_{zz}(z_-;\alpha)}\right]^{1/2} \end{aligned} \tag{19.4}$$

when $\alpha \neq \alpha_0$, and

$$I(\lambda;\alpha) \sim \mu g(z_0)e^{-\lambda f(z_0;\alpha_0)}\Gamma(4/3)\left[\frac{3!}{\lambda f_{zzz}(z_0;\alpha_0)}\right]^{1/3} \tag{19.5}$$

when $\alpha = \alpha_0$, where μ is a constant whose value depends on the contour C. The behaviors of these approximations differ radically, since (19.4) becomes singular, as $\alpha \to \alpha_0$, and the order of λ changes (discontinuously) from 1/2 in (19.4) to 1/3 in (19.5).

Since the simplest phase function that exhibits the coalescing of two saddle points is in cubic polynomial, Chester *et al.* (1957) introduced, in what is now regarded as a classic paper, a change of variable of the form

$$f(z;\alpha) = \frac{1}{3}u^3 - \zeta u + \eta, \tag{19.6}$$

where the coefficients ζ and η are to be determined. In order to have an analytic function $z = z(u)$ from (19.6), neither $\partial z/\partial u$ nor $\partial u/\partial z$ can vanish in the relevant regions. Since

$$\frac{\partial z}{\partial u} = \frac{u^2 - \zeta}{f_z(z;\alpha)} \tag{19.7}$$

and $f_z(z;\alpha) = 0$ for $z = z_+$ and $z = z_-$, we must make z_- and z_+ correspond to $-\zeta^{1/2}$ and $\zeta^{1/2}$, respectively. This correspondence will make the derivative in (19.7) indeterminate, and from (19.6) we have

$$\zeta^{3/2} = \frac{3}{4}[f(z_-;\alpha) - f(z_+;\alpha)], \tag{19.8}$$

$$\eta = \frac{1}{2}[f(z_-;\alpha) + f(z_+;\alpha)]. \tag{19.9}$$

By rewriting (19.7) as

$$f_z(z;\alpha)\frac{dz}{du} = u^2 - \zeta, \tag{19.10}$$

and differentiating with respect to u, we obtain

$$\left(\left.\frac{dz}{du}\right|_{u=\pm\zeta^{1/2}}\right)^2 = \frac{\pm 2\zeta^{1/2}}{f_{zz}(z_\pm;\alpha)} \tag{19.11}$$

if $\alpha \neq \alpha_0$. If $\alpha = \alpha_0$, then a further differentiation gives

$$\left(\left.\frac{dz}{du}\right|_{u=0}\right)^3 = \frac{2}{f_{zzz}(z_0;\alpha_0)}. \tag{19.12}$$

The transformation (19.6) was shown by Chester *et al.* (1957) to be one-to-one and analytic for all α in a neighborhood of α_0. More precisely, we have the following theorem. For a proof of this result we refer to Chester *et al.* (1957) and Friedman (1959).

Theorem 19.1. *With ζ and η given by* (19.8) *and* (19.9)*, the transformation* (19.6) *has exactly one branch $u = u(z;\alpha)$ which can be expanded into a power series in z, with coefficients which are continuous in α for α near α_0. On this branch, the points $z = z_\pm$ correspond to $u = \pm\zeta^{1/2}$ respectively. Furthermore, for α near α_0, the correspondence $u \leftrightarrow z$ is one-to-one.*

It is obvious that the above theorem is only a local result. However, in practice, as we shall see later in examples, the transformation (19.6) often turns out to be one-to-one and analytic in a domain containing the path of integration. Assuming this to be case, we substitute (19.6) in (19.1) and obtain

$$I(\lambda;\alpha) = e^{-\lambda\eta}\int_{C'} e^{-\lambda(u^3/3-\zeta u)}\varphi_0(u)du, \tag{19.13}$$

where C' is the image of C and

$$\varphi_0(u) = g(z)\frac{dz}{du}. \tag{19.14}$$

Note that $\varphi_0(u)$ is an analytic function of u, and that it depends on ζ and hence on α. For simplicity of presentation, we shall not indicate the dependence on ζ explicitly. To derive an asymptotic expansion for the last integral, we follow Bleistein (1967) and write

$$\varphi_0(u) = a_0 + b_0 u + (u^2 - \zeta^2)\psi_0(u), \tag{19.15}$$

where the coefficients a_0 and b_0 are easily determined to be (exercise)

$$a_0 = \frac{1}{2}[\varphi_0(\zeta^{1/2}) + \varphi_0(-\zeta^{1/2})], \tag{19.16}$$

$$b_0 = \frac{1}{2\zeta^{1/2}}[\varphi_0(\zeta^{1/2}) - \varphi_0(-\zeta^{1/2})]. \tag{19.17}$$

Clearly,

$$\lim_{\sqrt{\zeta}\to 0} a_0 = \varphi_0(0), \qquad \lim_{\sqrt{\zeta}\to 0} b_0 = \varphi_0'(0). \tag{19.18}$$

Thus, b_0 is an analytic function of $\zeta^{1/2}$ with a removable singularity at the origin. With a_0 and b_0 so chosen, it is easily seen that the function

$$\psi_0(u) = \frac{\varphi_0(u) - a_0 - b_0 u}{u^2 - \zeta} \tag{19.19}$$

is analytic in u and has removable singularities at $u = \pm\zeta^{1/2}$. Indeed, we have

$$\lim_{u\to\pm\sqrt{\zeta}} \psi_0(u) = \frac{\pm 1}{2\zeta^{1/2}}[\varphi_0'(\pm\zeta^{1/2}) - b_0]. \tag{19.20}$$

Now we insert (19.15) in (19.13) and obtain

$$e^{\lambda\eta}I(\lambda;\alpha) = \frac{a_0}{\lambda^{1/3}}V(\lambda^{2/3}\zeta) + \frac{b_0}{\lambda^{2/3}}V'(\lambda^{2/3}\zeta) + I_1(\lambda;\alpha), \tag{19.21}$$

where

$$V(\lambda) = \int_{C'} e^{-v^3/3+\lambda v}dv, \tag{19.22}$$

and

$$I_1(\lambda;\alpha) = \int_{C'} (u^2 - \zeta)e^{-\lambda(u^3/3-\zeta u)}\psi_0(u)du. \tag{19.23}$$

To the integral $I_1(\lambda;\alpha)$ we apply an integration by parts, and the result is

$$I_1(\lambda;\alpha) = \frac{1}{\lambda}\int_{C'} \varphi_1(u)e^{-\lambda(u^3/3-\zeta u)}du, \tag{19.24}$$

where

$$\varphi_1(u) = \frac{d}{du}\psi_0(u). \tag{19.25}$$

Since $\varphi_1(u)$ is analytic in u, the last integral is exactly of the same form as the one on the right-hand side of (19.13). Hence the above procedure can be repeated. Defining inductively

$$\varphi_m(u) = a_m + b_m u + (u^2 - \zeta)\psi_m(u), \tag{19.26}$$

$$\varphi_{m+1}(u) = \psi'_m(u), \tag{19.27}$$

we have (exercise)

$$e^{\lambda\eta}I(\lambda;\alpha) = V(2^{2/3}\zeta)\sum_{m=0}^{n-1}\frac{a_m}{\lambda^{m+1/3}} + V'(2^{2/3}\zeta)\sum_{m=0}^{n-1}\frac{b_m}{\lambda^{m+2/3}} + I_n(\lambda;\alpha), \tag{19.28}$$

where

$$I_n(\lambda;\alpha) = \frac{1}{\lambda^n}\int_{C'} \varphi_n(u)e^{-\lambda(u^3/3-\zeta u)}du. \tag{19.29}$$

The integral $V(\lambda)$ in (19.22) can be expressed in terms of Airy functions, but which Airy function is appropriate depends on the path of integration C'. There are three choices for C': it can be any one of the contours γ_0, γ_1 and γ_2 shown in Figure 19.1.

If C' is the contour γ_j, $j = 0, 1, 2$, then it can be easily verified that

$$V(\lambda) = 2\pi i\omega^j \operatorname{Ai}(\lambda\omega^j), \tag{19.30}$$

where $\omega = e^{2\pi i/3}$ is the cube root of unity. For definiteness, we suppose that C' is the contour γ_0. Thus, (19.28) becomes

$$e^{\lambda\eta}I(\lambda;\alpha) = 2\pi i\left[\frac{\operatorname{Ai}(\lambda^{2/3}\zeta)}{\lambda^{1/3}}\sum_{m=0}^{n-1}\frac{a_m}{\lambda^m} + \frac{\operatorname{Ai}'(\lambda^{2/3}\zeta)}{\lambda^{2/3}}\sum_{m=0}^{n-1}\frac{b_m}{\lambda^m}\right] + I_n(\lambda;\alpha). \tag{19.31}$$

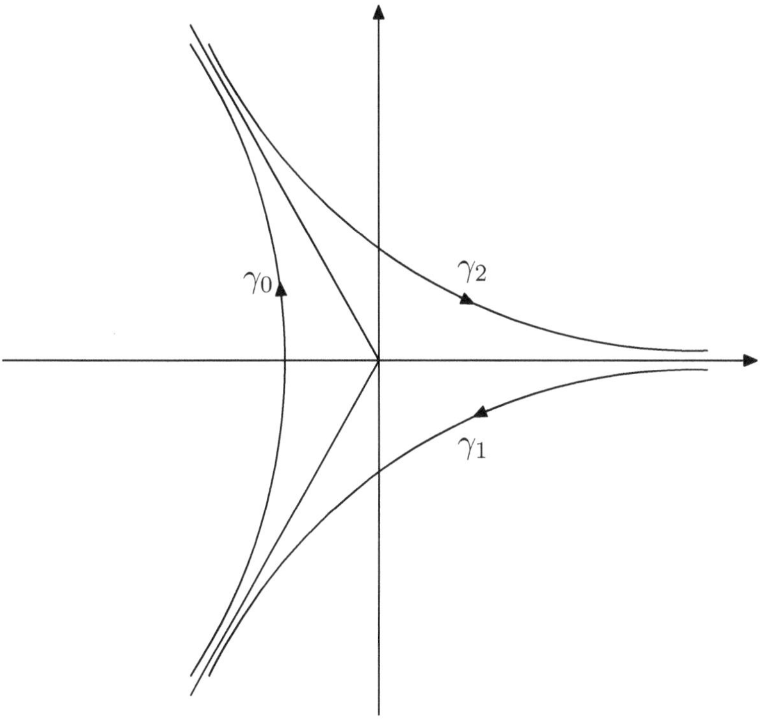

Fig. 19.1: Contour γ_j, $j = 0, 1, 2$.

For this to be a compound uniform asymptotic expansion, we must prove that there exist positive numbers C_n and D_n, independent of λ and α such that

$$|I_n(\lambda; \alpha)| \leq \frac{C_n}{\lambda^{n+1/3}} |\operatorname{Ai}(\lambda^{2/3}\zeta)| + \frac{D_n}{\lambda^{n+2/3}} |\operatorname{Ai}'(\lambda^{2/3}\zeta)|. \tag{19.32}$$

To achieve this, we first apply the recursive formula (19.26)-(19.27) once more. This gives

$$\begin{aligned} I_n(\lambda; \alpha) =& 2\pi i \left[\frac{a_n \operatorname{Ai}(\lambda^{2/3}\zeta)}{\lambda^{n+1/3}} + \frac{b_n \operatorname{Ai}'(\lambda^{2/3}\zeta)}{\lambda^{n+2/3}} \right] \\ &+ \frac{1}{\lambda^{n+1}} \int_{C'} \varphi_{n+1}(u) e^{-\lambda(u^3/3 - \zeta u)} du. \end{aligned}$$

Hence it suffices to show that the integral

$$I^* = \frac{1}{\lambda} \int_{C'} \varphi_{n+1}(u) e^{-\lambda(u^3/3 - \zeta u)} du \tag{19.33}$$

satisfies

$$|I^*| \le \frac{M_n}{\lambda^{1/3}} |\operatorname{Ai}(\lambda^{2/3}\zeta)| + \frac{N_n}{\lambda^{2/3}} |\operatorname{Ai}'(\lambda^{2/3}\zeta)| \tag{19.34}$$

for some constants M_n and N_n independent of λ and α. If $\lambda^{2/3}\zeta$ is bounded, say $|\lambda^{2/3}\zeta| \le \rho$, then we make the change of variable $u = \lambda^{-1/3}v$. This yields

$$I^* = \frac{1}{\lambda^{4/3}} \int_{C'} \varphi_{n+1}(\lambda^{-1/3}v) e^{-v^3/3+\lambda^{2/3}\zeta v} dv.$$

Since φ_n is an analytic function of both u and α, it is evident that there is a constant B_n, independent of λ and α, such that

$$|I^*| \le B_n \lambda^{-4/3}. \tag{19.35}$$

Furthermore, from differential equation theory, none of the zeros of $\operatorname{Ai}(z)$ and $\operatorname{Ai}'(z)$ coincide; hence (19.34) follows immediately from (19.35). If $|\lambda^{2/3}\zeta| \ge \rho$ then we suppose without loss of generality that λ is positive, and consider two separate cases: (i) $|\arg\zeta| < 2\pi/3$ and (ii) $|\arg(-\zeta)| < 2\pi/3$. In case (i), we make the change of variable $u = \zeta^{1/2}v$; and in case (ii), we put $u = (-\zeta)^{1/2}v$. The analysis for the two cases is similar, and we first consider case (i). The substitution $u = \zeta^{1/2}v$ gives

$$I^* = \frac{\zeta^{1/2}}{\lambda} \int_{\tilde{C}} \varphi_{n+1}(\zeta^{1/2}v) e^{-\lambda\zeta^{3/2}(v^3/3-v)} dv,$$

where $\tilde{C}$ is the image of C'. By temporarily assuming that ζ is positive, it is easily seen that $\tilde{C}$ retains the shape of C', i.e., that of γ_0 in Figure 19.1. The relevant saddle point for this integral is therefore at $v = -1$. The range of validity in ζ can now be extended by using analytic continuation. The ordinary saddle point method then gives

$$I^* \sim \frac{\sqrt{\pi}}{\lambda^{3/2}\zeta^{1/4}} \varphi_{n+1}(-\zeta^{1/2}) \exp(-\frac{2}{3}\lambda\zeta^{3/2}).$$

Since

$$\operatorname{Ai}(z) \sim \frac{1}{2\sqrt{\pi}} z^{-1/4} \exp(-\frac{2}{3} z^{3/2}),$$

$$\operatorname{Ai}'(z) \sim \frac{-1}{2\sqrt{\pi}} z^{1/4} \exp(-\frac{2}{3} z^{3/2}),$$

as $z \to \infty$ in $|\arg z| < 2\pi/3$ (see Olver *et al.* (2010, p. 198)), (19.34) holds for large λ, bounded ζ, and $|\lambda^{2/3}\zeta| \ge \rho > 0$. In case (ii), we have

$$I^* = \frac{\eta^{1/2}}{\lambda} \int_{\tilde{C}} \varphi_{n+1}(\eta^{1/2}v) e^{-\lambda\eta^{3/2}(v^3/3+v)} dv,$$

where $\eta = -\zeta$. We deform the contour $\tilde{C}$ to pass through the two saddle points $\pm i$. Using the saddle point method again and the fact that $\operatorname{Ai}(z)$ and $\operatorname{Ai}'(z)$ do not have the same zeros, it is easily seen that (19.34) also holds in this case, thus completing the proof of (19.32).

19.2 Bessel function

We investigate the integral

$$J_\nu(\nu a) = \frac{1}{2\pi i}\int_{\infty-\pi i}^{\infty+\pi i} e^{\nu(a\sinh z - z)}dz. \tag{19.36}$$

If we let C denote the above contour and

$$f(z;a) = z - a\sinh z, \tag{19.37}$$

then equation (19.36) can be written as

$$J_\nu(\nu a) = \frac{1}{2\pi i}\int_C e^{-\nu f(z;a)}dz. \tag{19.38}$$

The saddle points are given by the equation

$$\cosh z = \frac{1}{a}. \tag{19.39}$$

For $0 < a < 1$, it is convenient to write $a = \operatorname{sech}\alpha$ where $\alpha > 0$. Hence, the saddle points are $z = \pm\alpha + 2n\pi i$ where n is an integer. There are two saddle points in the strip $|\operatorname{Im} z| < \pi$ which concerns us, namely, $z = \pm\alpha$. For $a > 1$, we write $a = \sec\alpha$ where $0 < \alpha < \pi/2$. The relevant saddle points are now located on the imaginary axis; indeed, $z = \pm i\alpha$. When $a \to 1$, the corresponding parameter α approaches 0 and the two saddle points coalesce. For z near the origin, a Taylor expansion of the function $f(z;\alpha)$ in (19.37) gives

$$f(z;\alpha) = (a-1)z + \frac{1}{6}az^3 + \cdots, \tag{19.40}$$

which shows that as $a \to 1$ $(\alpha \to 0)$ the first two derivatives of f vanish at $z = 0$. This suggests the cubic transformation in (19.6):

$$f(z;\alpha) = \frac{1}{3}u^3 + \zeta u + \eta$$

used by Chester *et al.* (1957). The functions η and ζ in the cubic transformation can be found explicitly. Indeed, from (19.8)-(19.9) and (19.37), we have $\eta = 0$ and (exercise)

$$\begin{cases} \frac{2}{3}\zeta^{3/2} = \tanh\alpha - \alpha, & 0 < a < 1, \\ \frac{2}{3}(-\zeta)^{3/2} = \tan\alpha - \alpha, & a > 1. \end{cases} \tag{19.41}$$

With η and ζ given explicitly, replacing λ by ν in (19.31) gives

$$J_\nu(\nu a) \sim \frac{\operatorname{Ai}(\nu^{2/3}\zeta)}{\nu^{1/3}}\sum_{m=0}^{\infty}\frac{a_m}{\nu^m} + \frac{\operatorname{Ai}'(\nu^{2/3}\zeta)}{\nu^{2/3}}\sum_{m=0}^{\infty}\frac{b_m}{\nu^m}. \tag{19.42}$$

For the computation of the coefficients, we refer to Temme (2015, pp. 311-312) and Olver *et al.* (2010, pp. 232-233).

19.3 Hermite polynomials

From the generating function

$$\exp(2xh - h^2) = \sum_{n=0}^{\infty} H_n(x)\frac{h^n}{n!}, \tag{19.43}$$

we have by Cauchy integral formula

$$H_n(x) = \frac{n!}{2\pi i}\int_{C_0} \frac{e^{2xh-h^2}}{h^{n+1}}dh, \tag{19.44}$$

where C_0 is a circle oriented in the positive direction with radius $r > 0$ and center at the origin. By adding two infinite line segments along the negative real axis, one starts at $-\infty$ and ends at $-r$ in the lower half-plane, and the other starts at $-r$ and ends at $-\infty$ in the upper half-plane, the circle C_0 is then deformed into a Hankel loop Γ. Put $x = \sqrt{2n+1}t$; equation (19.44) becomes

$$H_n(\sqrt{2n+1}t) = \frac{n!}{2\pi i}\int_{\Gamma} h^{-n-1}e^{2\sqrt{2n+1}th-h^2}dh.$$

The change of variable $h = \sqrt{2n+1}z$ gives

$$H_n(\sqrt{2n+1}t) = \frac{n!}{2\pi i}(2n+1)^{-n/2}\int_{\Gamma} e^{(2n+1)(2tz-z^2-\frac{1}{2}\ln z)}\frac{dz}{z^{1/2}}. \tag{19.45}$$

Let $\mu = \sqrt{2n+1}$ and $x = \mu t$. The integral in (19.45) can be expressed as

$$I(\mu^2;t) = \int_{\Gamma} e^{\mu^2 f(z;t)}g(z)dz, \tag{19.46}$$

where

$$f(z;t) = 2tz - z^2 - \frac{1}{2}\ln z, \tag{19.47}$$

and $g(z) = z^{-1/2}$. From (19.43), it is readily seen that $H_n(-x) = (-1)^n H_n(x)$. Hence, without loss of generality, we may restrict ourselves to the case $t \in [0,\infty)$.

To find the saddle point of $f(z;t)$, we differentiate the function in (19.47) with respect to z and set it to zero; thus

$$\frac{\partial f}{\partial z} = 2t - 2z - \frac{1}{2z} = 0 \tag{19.48}$$

and the saddle points are at

$$z_{\pm} = \begin{cases} (t \pm \sqrt{t^2-1})/2, & t \geq 1, \\ (t \pm i\sqrt{1-t^2})/2, & 0 \leq t < 1. \end{cases} \tag{19.49}$$

Clearly the points z_+ and z_- coalesce at $z = 1$, when $t = 1$. We now make the cubic transformation (19.6):

$$f(z;t) = \frac{u^3}{3} - B^2(t)u + C(t). \tag{19.50}$$

A differentiation of equation (19.50) gives

$$\frac{\partial f}{\partial z}(z;t) = -\frac{2(z-z_+)(z-z_-)}{z} = (u^2 - B^2)\frac{\partial u}{\partial z}. \tag{19.51}$$

To ensure that the mapping $z \to u$ is conformal, $\partial u/\partial z$ must not vanish and hence $z_\pm$ must be mapped to the saddle points $u_\pm = \pm B(t)$ of the cubic function on the right-hand side of (19.50). The correspondences $z_\pm \leftrightarrow u_\pm$ imply that

$$2tz_\pm - z_\pm^2 - \frac{1}{2}\ln z_\pm = \mp\frac{2}{3}B^3(t) + C(t). \tag{19.52}$$

Adding these two equations yields

$$C(t) = \frac{t^2}{2} + \frac{1}{4} + \frac{\ln 2}{2}, \tag{19.53}$$

Moreover, we obtain

$$B(t) = -[\frac{3}{4}(t\sqrt{t^2-1} - \operatorname{arccosh} t)]^{1/3} \tag{19.54}$$

for $t > 1$, and

$$B(t) = -i[\frac{3}{4}(-t\sqrt{1-t^2} + \arccos t)]^{1/3} \tag{19.55}$$

for $0 < t < 1$.

Having found explicit formulas for the functions $B(t)$ and $C(t)$ in the cubic transformation (19.50), we are now ready to apply the result (19.31) for the integral $I(\lambda;\alpha)$ in (19.1) to the integral in (19.45). Before doing so, we first point out a few facts to avoid confusion which may arise in later discussion. The parameter α in (19.1) is now t in (19.45), the large variable λ in (19.1) is now $\mu^2 = 2n+1$ in (19.45), and the amplitude function in (19.1) is $z^{-1/2}$ in (19.45). Note the negative sign in front of the function $f(z;\alpha)$ in (19.1), which is absent in the phase function $f(z;t)$ in (19.45). This difference can be accommodated by changing u to $-u$ so that (19.50) becomes

$$\begin{aligned} f(z;t) &= -\,(z^2 - 2tz + \frac{1}{2}\ln z) = \frac{u^3}{3} - B^2(t)u + C(t) \\ &= -\,[(-u)^3/3 - B^2(t)(-u) - C(t)]. \end{aligned} \tag{19.56}$$

Furthermore, we note that in the present case the functions ζ and η in (19.6) correspond to $B^2(t)$ and $C(t)$ in (19.50).

From (19.45) and (19.46), we have

$$H_n(\mu t) = \frac{n!}{2\pi i \mu^n} \int_\Gamma e^{\mu^2 f(z;t)} g(z) dz, \tag{19.57}$$

where $g(z) = z^{-1/2}$. On account of (19.56) and (19.13), it follows

$$H_n(\mu t) = \frac{n! e^{\mu^2 C(t)}}{2\pi i \mu^n} \int_{\Gamma'} \varphi_0(u) e^{-\mu^2[(-u)^3/3 - B^2(t)(-u)]} du$$

where $\varphi_0(u) = z^{-1/2} dz/du$ and Γ' is the image of Γ under the mapping $z \to u$. Changing $-u$ to u gives

$$H_n(\mu t) = \frac{n! e^{\mu^2 C(t)}}{2\pi i \mu^n} \int_{\tilde{\Gamma}} \varphi_0(-u) e^{-\mu^2[u^3/3 - B^2(t)u]} du, \tag{19.58}$$

where $\tilde{\Gamma}$ is the image of Γ' under the change of variable from $-u$ to u and with the direction being reversed. The integral in (19.58) is now exactly in the form of (19.13). Hence, from (19.31) it follows that

$$H_n(\mu t) \sim \frac{n!}{\mu^n} \left[\mathrm{Ai}(\mu^{4/3} B^2) \sum_{m=0}^{\infty} \frac{a_m}{\mu^{2m+2/3}} + \mathrm{Ai}'(\mu^{4/3} B^2) \sum_{m=0}^{\infty} \frac{b_m}{\mu^{2m+4/3}} \right]. \tag{19.59}$$

The leading coefficients a_0 and b_0 are given by

$$a_0 = \frac{\varphi_0(-B) + \varphi_0(B)}{2}, \quad b_0 = \frac{\varphi_0(B) - \varphi_0(-B)}{2}; \tag{19.60}$$

noting that the function $\varphi_0(u)$ in (19.13) is now replaced by $\varphi_0(-u)$ in (19.59). To find exactly what are $\varphi_0(\pm B)$, we go back to (19.47) from which we have

$$\frac{\partial f}{\partial z} = \frac{-2(z - z_+)(z - z_-)}{z}; \tag{19.61}$$

see (19.48) and (19.49). By (19.50) and (19.61), we obtain

$$\frac{dz}{du} = \frac{\partial f/\partial u}{\partial f/\partial z} = -\frac{z(u^2 - B^2)}{2(z - z_+)(z - z_-)}. \tag{19.62}$$

Upon taking the limit as $u \to B$ and $u \to -B$ respectively, it follows

$$\left. \frac{dz}{du} \right|_{u = \pm B} = \sqrt{\frac{-B z_\pm}{z_+ - z_-}}.$$

Hence, we have from (19.60) and (19.54)

$$\left.\frac{dz}{du}\right|_{u=\pm B} = \begin{cases} [\frac{3(\sinh 2\beta - 2\beta)}{8}]^{1/6}\sqrt{\frac{e^{\pm\beta}}{2\sinh\beta}}, & t = \cosh\beta > 1, \\ [\frac{3(2\theta - \sin 2\theta)}{8}]^{1/6}\sqrt{\frac{e^{\pm i\theta}}{2\sin\theta}}, & t = \cos\theta \in (0,1), \\ 2^{-2/3}, & t = 1. \end{cases} \tag{19.63}$$

Consequently,

$$\varphi_0(\pm B(t)) = \begin{cases} [\frac{3(t\sqrt{t^2-1} - \operatorname{arccosh} t)}{4(t^2-1)^{3/2}}]^{1/6}, & t > 1, \\ [\frac{3(\arccos t - t\sqrt{1-t^2})}{4(1-t^2)^{3/2}}]^{1/6}, & 0 < t < 1, \\ 2^{-1/6}, & t = 1. \end{cases} \tag{19.64}$$

Finally, we calculate

$$a_0 = \begin{cases} [\frac{3(t\sqrt{t^2-1} - \operatorname{arccosh} t)}{4(t^2-1)^{3/2}}]^{1/6}, & t > 1, \\ [\frac{3(\arccos t - t\sqrt{1-t^2})}{4(1-t^2)^{3/2}}]^{1/6}, & 0 < t < 1, \\ 2^{-1/6}, & t = 1, \end{cases} \tag{19.65}$$

and $b_0 = 0$ for all $t \in [0, \infty)$.

To conclude this section, we now show how to find an explicit formula for the solution $u(x,t)$ of the cubic equation (19.50). Consider a cubic equation of the form

$$w^3 + \alpha w + \beta = 0, \tag{19.66}$$

where α and β are complex numbers. Suppose that $w = A\sin\theta$, where A and θ are complex numbers, is a solution of the cubic equation (19.66). Then

$$0 = A^3\sin^3\theta + \alpha A\sin\theta + \beta = \frac{\alpha A}{3}\sin 3\theta + \frac{A(4\alpha + 3A^2)}{3}\sin^3\theta + \beta.$$

If we choose $A = \sqrt{-4\alpha/3}$, then

$$\sin 3\theta = -\frac{3\beta}{\alpha A}. \tag{19.67}$$

Therefore, the three solutions of the cubic equation (19.66) are

$$w_1 = A\sin\frac{\varphi}{3}, \quad w_2 = A\sin\frac{\varphi + 2\pi}{3}, \quad w_3 = A\sin\frac{\varphi + 4\pi}{3}, \tag{19.68}$$

where

$$A = \sqrt{\frac{-4\alpha}{3}}, \tag{19.69}$$

and φ is a solution of

$$\sin\varphi = -\frac{3\beta}{\alpha A}. \tag{19.70}$$

Note that any possible values of A and φ satisfying (19.69) and (19.70) will work and give the same result. This is consistent with the fact that a cubic polynomial has exactly three complex roots counting multiplicities. In our case, the coefficients α and β in equation (19.66) are given by $\alpha = -3B^2(t)$ and $\beta = 3[C(t) - f(z;t)]$; see (19.56). Thus, it follows from (19.68) and (19.69) that $A = 2B(t)$ and

$$u_k(z;t) = 2B(t)\sin\frac{\varphi + 2(k-1)\pi}{3}, \quad k = 1, 2, 3, \tag{19.71}$$

where φ is a solution of

$$\sin\varphi = \frac{3[C(t) - f(z;t)]}{2B^3(t)};$$

see (19.67). Since $z_\pm \leftrightarrow u_\pm$, we have from (19.70)

$$\begin{aligned} &u_1(z_+;t) = B(t), \quad u_2(z_+;t) = B(t), \quad u_3(z_+;t) = -B(t), \\ &u_1(z_-;t) = -B(t), \quad u_2(z_-;t) = 2B(t), \quad u_3(z_-;t) = -B(t). \end{aligned}$$

Therefore, the desired solution is $u_1(z;t)$, and we simply call it $u(z;t)$.

Finally, we wish to point out that here we have established that the expansion (19.59) is uniformly valid only for t in any finite interval on the non-negative real line. The fact is that this expansion is actually uniformly valid for t in the infinite interval $[0,\infty)$. For a proof of the latter result, we refer to Shi (2008).

19.4 Exercises

1. Verify (19.16) and (19.17).
2. Prove (19.28).
3. Verify (19.41).
4. Verify (19.53), (19.54) and (19.55).
5. Prove (19.63), (19.64) and (19.65).
6. Let $f(t)$ be an analytic function in $|t| < |\alpha|$, and α be a pole of f of order p, which is not on the positive real t-axis. Suppose that

$$f(t) = \frac{b_{-p}}{(t-\alpha)^p} + \cdots + \frac{b_{-1}}{t-\alpha} + f^*(t),$$

where

$$f^*(t) = \sum_{n=0}^{\infty} a_n t^n,$$

for $|t| < \delta$ and $\delta > |\alpha|$. Let

$$I(\lambda) = \int_0^{\infty} t^{\mu} f(t) e^{-\lambda t} dt$$

be absolutely convergent for all large values of x. Show that

$$\begin{aligned} I(\lambda) = &\sum_{n=0}^{\infty} a_n \Gamma(\mu + n + 1) \lambda^{-\mu-n-1} \\ &+ \Gamma(\mu+1) \left(\frac{-\alpha}{\lambda}\right)^{\mu/2} e^{-\lambda\alpha/2} \\ &\times \sum_{k=1}^{p} b_{-k} \left(\frac{\lambda}{-\alpha}\right)^{k/2} W_{-k/2-\mu/2,\mu/2-k/2}(-\lambda x), \end{aligned}$$

as $\lambda \to \infty$, uniformly for α in a neighborhood of the origin, where $W_{k,m}(z)$ is the Whittaker function defined by

$$W_{k,m}(z) = \frac{e^{-z/2} z^{m+1/2}}{\Gamma(1/2 - k + m)} \int_0^{\infty} e^{-zt} t^{-1/2-k+m} (1+t)^{-1/2+k+m} dt$$

for $\mathrm{Re}(m-k) > -1/2$ and $\mathrm{Re}\, z > 0$.

7. Consider the integral

$$I(\lambda) = \int_0^{\infty} g(t) e^{i\lambda f(t)} dt,$$

where λ is a large positive parameter, $f(t)$ is a real-valued function, and $g(t)$ vanishes identically in a neighborhood of infinity. Assume that

$$f'(\alpha) = 0, \;\; f''(\alpha) = \pm 2a^2, \quad a > 0,$$

and $f'(t) \neq 0$ for $t \neq \alpha$, where α is some real number. Let $\alpha_- = \min\{\alpha, 0\}$, and suppose that both f and g are C^{∞} in $[\alpha_-, \infty)$. Note that as $\alpha \to 0$, the saddle point $t = \alpha$ coalesces with the endpoint $t = 0$. Put

$$u^2 = \pm[f(t) - f(\alpha)], \quad \mathrm{sgn}\, u = \mathrm{sgn}(t - \alpha).$$

Then

$$I(\lambda) = e^{i\lambda f(\alpha)} \int_{\eta}^{\infty} e^{\pm i\lambda u^2} h(u) du,$$

where

$$h(u) = \frac{2u}{\pm f'(t)} g(t)$$

and

$$\eta = \operatorname{sgn}(-\alpha)|f(0) - f(\alpha)|^{1/2}.$$

Define $h_0(u) = h(u)$ and

$$h_{s+1}(u) = \frac{d}{du}\left[\frac{h_s(u) - h_s(0)}{u}\right] = \int_0^1 v h_s''(uv)dv, \quad s \geq 0.$$

Derive the asymptotic expansion

$$\begin{aligned} I(\lambda) =& e^{i\lambda f(0)} \sum_{s=0}^{N} \left(\frac{\pm i}{2\lambda}\right)^s \left[\frac{h_s(0)}{\sqrt{\lambda}} F_{\pm}(\sqrt{\lambda}\eta) \pm \frac{i[h_s(\eta) - h_s(0)]}{2\lambda\eta}\right] \\ &+ \left(\frac{\pm i}{2\lambda}\right)^{N+1} \int_\eta^\infty e^{\pm i\lambda\mu^2} h_{N+1}(u)du, \end{aligned}$$

where $F_{\pm}(x)$ is the Fresnel integral defined by

$$F_{\pm}(x) = e^{\mp ix^2} \int_x^\infty e^{\pm iu^2} du.$$

8. In Chapter 17, Exercise 22, we have considered the integral

$$I(N, \theta) = \int_{-\infty - i\pi/4}^{\infty - i\pi/4} (\cosh u)^3 e^{Nf(u,\theta)} du,$$

where

$$f(u, \theta) = i[\cos\theta \cosh u - \sin\theta \sinh(2u)/2]$$

and θ is a real parameter. It was stated that there exists a critical value θ_c such that for $\theta \neq \theta_c$ there are two (relevant) saddle points of order 1, and for $\theta = \theta_c$ there is only one (relevant) saddle point (but) of order 2. Two asymptotic approximations were also given, one for $0 < \theta < \theta_c$ and one for $\theta = \theta_c$. It is easily seen that the amplitude of the approximation for $\theta < \theta_c$ approaches infinity when θ tends to θ_c from below. Thus two approximations do not join smoothly. A similar abrupt change occurs when θ tends θ_c from above. To derive an approximation which is uniform with respect to θ in a neighborhood of θ_c, we make the cubic change of variable

$$if(u, \theta) = \frac{v^3}{3} - \mu(\theta)v + \nu(\theta),$$

as suggested in Section 19.1.

(a) Determine the coefficients $\mu(\theta)$ and $\nu(\theta)$ so that the transformation $u \to v$ is analytic in a neighborhood of θ_c.

(b) Show that near $\theta = \theta_c$, $I(N,\theta)$ has an expansion of the form

$$e^{iN\nu(\theta)}I(N,\theta) \sim \frac{2\pi i \operatorname{Ai}(-N^{2/3}\mu(\theta))}{N^{1/3}} \sum_{m=0}^{\infty} \frac{A_m(\theta)}{N^m} + \frac{2\pi i \operatorname{Ai}'(-N^{2/3}\mu(\theta))}{N^{2/3}} \sum_{m=0}^{\infty} \frac{B_m(\theta)}{N^m}.$$

(c) Calculate the leading coefficients $A_0(\theta)$ and $B_0(\theta)$.

9. Consider the contour integral

$$I(\lambda) = \int_C g(t)e^{-\lambda f(t)}dt,$$

where λ is a large positive parameter and C is a continuous curve in the complex t-plane along which the functions $f(t)$ and $g(t)$ are defined. We assume that $f(t)$ has a single saddle point at $t = \alpha$ and $g(t)$ has a simple pole at $t = \beta$. We shall consider β near α, and in fact allow them to coalesce. In a neighborhood of α, we have

$$f(t) = f(\alpha) + \frac{f''(\alpha)}{2!}(t-\alpha)^2 + \frac{f'''(\alpha)}{3!}(t-\alpha)^3 + \cdots, \tag{19.72}$$

where $f''(\alpha) \neq 0$ for all α. If we set

$$u^2 = f(t) - f(\alpha), \tag{19.73}$$

then $I(\lambda)$ becomes

$$I(\lambda) = e^{-\lambda f(\alpha)} \int_{-\infty}^{\infty} q(u)e^{-\lambda u^2} du.$$

Here we have assumed that the image of C can be deformed into the real axis in the u-plane, and

$$q(u) = g(t)\frac{2u}{f'(t)}.$$

Note that $q(u)$ has a pole at $u = ib$, where $-b^2 = f(\beta) - f(\alpha)$. We can extract the pole by writing

$$q(u) = \frac{c_{-1}}{u - ib} + \psi(u). \tag{19.74}$$

Show that the constant c_{-1} is given by

$$c_{-1} = \lim_{u\to ib}[(u-ib)q(u)] = \lim_{t\to\beta}[(u-ib)g(t)\frac{2u}{f'(t)}] = -g_{-1}, \tag{19.75}$$

where g_{-1} is the residue of $g(t)$ at $t = \beta$. The function $\psi(u)$ has no pole at $u = 0$ or $u = -ib$. Show that from the last two equations, we have

$$\psi(0) = g(\alpha)\sqrt{\frac{2}{f''(\alpha)}} + \frac{g_{-1}}{\sqrt{f(\beta) - f(\alpha)}}$$
$$= g^*(\alpha)\sqrt{\frac{2}{f''(\alpha)}} - \frac{\sqrt{2}}{6} g_{-1} \frac{f'''(\alpha)}{[f''(\alpha)]^{1/2}} + O(\beta - \alpha), \qquad (19.76)$$

where $g^*(\alpha)$ is the "regular" part of $g(t)$, i.e.,

$$g(t) = \frac{g_{-1}}{t - \beta} + g^*(t).$$

To verify (19.76), first show that from (19.74) and (19.75), we have

$$\psi(0) = \lim_{u \to 0} [q(u) - \frac{c_{-1}}{u - ib}] = \lim_{t \to \alpha} [g(t) \frac{2u}{f'(t)} - \frac{g_{-1}}{u - ib}],$$

and from (19.72) and (19.73), we have

$$\lim_{t \to \alpha} \frac{2u}{f''(t)} = \sqrt{\frac{2}{f''(\alpha)}}.$$

Thus,

$$\psi(0) = g(\alpha)\sqrt{\frac{2}{f''(\alpha)}} + \frac{g_{-1}}{ib},$$

which is equivalent to the first equality in (19.76). The second equality in (19.76) follows from (19.72). The asymptotic expansion of the integral

$$\int_{-\infty}^{\infty} \psi(u) e^{-\lambda u^2} du$$

can be obtained by Watson's lemma (Lemma 17.1). We are thus left with the integral

$$K(\lambda) = \int_{-\infty}^{\infty} \frac{e^{-\lambda u^2}}{u - ib} du,$$

which can be expressed in terms of an incomplete gamma function or the complementary error function. To see this, show that $K(\lambda)$ satisfies the first-order linear differential equation

$$K'(\lambda) - b^2 K(\lambda) = -ib\sqrt{\frac{\pi}{\lambda}},$$

and show that the solution satisfying $K(\infty) = 0$ is given by

$$K(\lambda) = i\sqrt{\pi} e^{\lambda b^2} \Gamma(1/2, \lambda b^2) = i\pi e^{\lambda b^2} \operatorname{erfc}(\lambda^{1/2} b).$$

10. Use the Dirichlet-Mehler integral Szegö (1975, (4.8.6))

$$P_n(\cos\theta) = \frac{2}{\pi}\int_0^\theta \frac{\cos[(n+1/2)\varphi]}{(2\cos\varphi - 2\cos\theta)^{1/2}} d\varphi$$

to show that there exists analytic functions $a_\nu(\theta)$ in the interval $[0, \pi]$ such that

$$P_n(\cos\theta) = \sum_{\nu=0}^{m-1} a_\nu(\theta) \frac{J_\nu[(n+1/2)\theta]}{(n+1/2)^\nu} + O(n^{-m-1/2})$$

for every fixed $m \geq 1$, where the O-term is uniform with respect to $\theta \in [0, \pi - \varepsilon]$, $\varepsilon > 0$. Calculate the first two coefficients $a_0(\theta)$ and $a_1(\theta)$.

Chapter 20

Asymptotics of Jacobi Polynomials

20.1 History

The first asymptotic approximation of the Jacobi polynomials $P_n^{(\alpha,\beta)}(\cos\theta)$ as $n \to \infty$, which holds uniformly in the interval $0 \le \theta \le \pi - \varepsilon$ (with $\varepsilon > 0$), is probably the Hilb-type formula given by Szegö (1933a, p. 77): for $\alpha > -1$, $\beta \in \mathbf{R}$, and $N = n + (\alpha + \beta + 1)/2$, we have

$$\sin^\alpha \frac{\theta}{2} \cos^\beta \frac{\theta}{2} P_n^{(\alpha,\beta)}(\cos\theta) = \frac{N^{-\alpha}\Gamma(n+\alpha+1)}{n!}\sqrt{\frac{\theta}{\sin\theta}} J_\alpha(N\theta) + r_n(\theta), \tag{20.1}$$

where the remainder

$$r_n(\theta) = \begin{cases} \theta^{1/2} O(n^{-3/2}), & cn^{-1} \le \theta \le \pi - \varepsilon, \\ \theta^{\alpha+2} O(n^\alpha), & 0 \le \theta \le cn^{-1}, \end{cases}$$

with c and ε being fixed positive numbers; see also Szegö (1975, p. 197).

In fact, Szegö (1933a) gave an infinite asymptotic expansion for $P_n^{(\alpha,\beta)}(\cos\theta)$. The mth term in the expansion is of the form

$$\theta^m \sum_{\nu=0}^{m} f_{m,\nu}(\theta) J_{\alpha+\nu,m}(N\theta), \tag{20.2}$$

where

$$J_{\nu,m}(\xi) = \sum_{h=0}^{m} \binom{m}{h} J_{m-2h+\nu}(\xi). \tag{20.3}$$

The first few coefficient functions $f_{m,\nu}(\theta)$ can be calculated explicitly; see e.g., Frenzen and Wong (1985b). The remainder, associated with truncation after p terms, satisfies

$$R_p = \begin{cases} \theta^{p-\lambda_p-\alpha} O(n^{-\lambda_p}), & n\theta \ge c, \\ O(n^{\alpha-p}), & n\theta \le c, \end{cases} \tag{20.4}$$

where c is a positive constant, and

$$\lambda_p = \frac{1}{2} + [\frac{p+1}{2}]. \tag{20.5}$$

Szegö's paper is 79 pages long and was published in a journal that is not easily accessible. In view of the complicated nature of the general term given in (20.2)-(20.3), his result is probably not very useful.

Not knowing the above result of Szegö, Frenzen and Wong (1985a) gave a uniform asymptotic expansion for $P_n^{(\alpha,\beta)}(\cos\theta)$ involving $J_\alpha(N\theta)$, $J_{\alpha+1}(N\theta)$, $J_{\alpha+2}(N\theta), \cdots$. More precisely, they proved that for $\alpha > -1/2$, $\alpha - \beta > -2p$ and $\alpha + \beta \geq -1$, we have

$$\sin^\alpha \frac{\theta}{2} \cos^\beta \frac{\theta}{2} P_n^{(\alpha,\beta)}(\cos\theta) = \frac{\Gamma(n+\alpha+1)}{n!}\sqrt{\frac{\theta}{\sin\theta}} \times \left[\sum_{m=0}^{p-1} A_m(\theta)\frac{J_{\alpha+m}(N\theta)}{N^{\alpha+m}} + \sigma_p\right], \tag{20.6}$$

where $\sigma_p = \theta^\alpha O(N^{-p})$ and O-term is uniform with respect to $\theta \in [0, \pi - \varepsilon]$, $\varepsilon > 0$. Frenzen and Wong (1985b) also remarked that the condition $\alpha - \beta > -2p$ can be removed and the error estimate can be replaced by

$$\sigma_p = \theta^{\alpha+p} O(N^{-p}) \tag{20.7}$$

for $\theta \in [0, \pi - \varepsilon]$ with $\varepsilon > 0$. The first two coefficients in (20.6) are given by $A_0(\theta) = 1$ and

$$A_1(\theta) = \frac{(\alpha^2 - 1/4)(1 - \theta\cot\theta)}{2\theta} - \frac{\alpha^2 - \beta^2}{4}\tan\frac{\theta}{2}. \tag{20.8}$$

The method used by Frenzen and Wong begins with a Mehler-type integral for the Jacobi polynomial given by Gasper (1975). Note that the mth term in (20.6) clearly contains a decreasing factor $N^{-\alpha-m}$, whereas the corresponding term in (20.2) is not in strictly decreasing order. Furthermore, even if we multiply the series in (20.6) and the series

$$\frac{\Gamma(n+\alpha+1)}{n!} = n^\alpha \left[1 + \frac{\alpha(\alpha+1)}{2n} + \frac{\alpha(\alpha-1)(3\alpha^2+5\alpha+2)}{24n^2} + \cdots\right]$$

together, and regroup terms to form a new asymptotic expansion for $P_n^{(\alpha,\beta)}(\cos\theta)$, the resulting expansion still differs from that of Szegö.

A result, equivalent to (20.6), can be obtained by using Olver's differential equation theory in Olver (1974a, Chapter 12), and has been given explicitly by Baratella and Gatteschi (1988). They showed that for $\alpha > -1$ and β arbitrary, we have

$$\left(\sin\frac{\theta}{2}\right)^{\alpha+1/2}\left(\cos\frac{\theta}{2}\right)^{\beta+1/2} P_n^{(\alpha,\beta)}(\cos\theta) = 2^{-1/2}N^{-\alpha}\frac{\Gamma(n+\alpha+1)}{n!}$$
$$\times\left\{\theta^{1/2}J_\alpha(N\theta)\sum_{m=0}^{p}\frac{A_m(\theta)}{N^{2m}} + \theta^{3/2}J_{\alpha+1}(N\theta)\sum_{m=0}^{p-1}\frac{B_m(\theta)}{N^{2m+1}} + \varepsilon_p(N,\theta)\right\}, \tag{20.9}$$

where $N = n + (\alpha+\beta+1)/2$ and

$$\varepsilon_p(N,\theta) = \begin{cases}\theta O(N^{-2p-3/2}), & cN^{-1} \le \theta \le \pi - \varepsilon,\\ \theta^{\alpha+5/2}O(N^{-2p+\alpha}), & 0 \le \theta \le cN^{-1},\end{cases} \tag{20.10}$$

c and ε again being fixed positive constants. The first few coefficients are given by

$$A_0(\theta) = 1,\quad B_0(\theta) = \frac{g(\theta)}{4\theta},\quad A_1(\theta) = \frac{g'(\theta)}{8} - \frac{(1+2\alpha)g(\theta)}{8\theta} - \frac{g^2(\theta)}{32},$$

where

$$g(\theta) = \left(\frac{1}{4}-\alpha^2\right)\left(\cot\frac{\theta}{2} - \frac{2}{\theta}\right) - \left(\frac{1}{4}-\beta^2\right)\tan\frac{\theta}{2}.$$

The work of Frenzen and Wong (1985a) was motivated by a problem of L. Lorch concerning the Lebesgue constants for Jacobi series:

$$L_n(\alpha,\beta) = \frac{\Gamma(n+\alpha+\beta+1)}{\Gamma(\alpha+1)\Gamma(n+\beta+1)}$$
$$\times\int_0^\pi\left(\sin\frac{\theta}{2}\right)^{2\alpha+1}\left(\cos\frac{\theta}{2}\right)^{2\beta+1}|P_n^{(\alpha+1,\beta)}(\cos\theta)|d\theta. \tag{20.11}$$

Equation (20.11) gives $L_n(-1/2,-1/2) = \rho_n$, the Lebesgue constant commonly encountered in the study of Fourier series; see Zygmund (1935, p. 172). Many people have worked on the more general constants $L_n(\alpha,\beta)$; for example, Rau (1929); Szegö (1933a); Lorch (1959). It turns out that behind all this work, there was a conjecture of Szegö (1982) dated back to 1926; see Szegö (1933a, Footnote 6) and also the editor's comment in Askey (1982, p. 313). The conjecture states that the Lebesgue constants for the Legendre series at $x = 1$:

$$L_n := L_n(0,0) = \frac{n+1}{2}\int_{-1}^{1}|P_n^{(1,0)}(x)|dx$$
$$= (n+1)\int_0^\pi \sin\frac{\theta}{2}\cos\frac{\theta}{2}|P_n^{(1,0)}(\cos\theta)|d\theta, \tag{20.12}$$

is monotonically increasing in $n = 1, 2, \cdots$, i.e.,

$$L_{n+1} - L_n > 0 \tag{20.13}$$

for all $n \geq 1$.

The asymptotic result

$$L_n = \frac{2^{3/2}}{\sqrt{\pi}} n^{1/2} + o(n^{1/2}), \tag{20.14}$$

as $n \to \infty$, was first given by Gronwall (1913) and later by Szegö (1933b, 1934) with simpler proofs. In order to settle this conjecture for at least large values of n, Szegö posed to Lorch in the 1950s the problem of obtaining more refined results than that given in (20.14). Lorch (1959) showed that

$$L_n = \frac{2^{3/2}}{\sqrt{\pi}} n^{1/2} + B_0 + O(n^{-1/2}), \tag{20.15}$$

where B_0 is a computable constant. A simple calculation reveals that the result in (20.15) is insufficient to determine the asymptotic monotonicity of the sequence L_n. Lorch thus proposed to R. Wong in 1980 the problem of replacing the O-term in (20.15) by an explicitly determined expression plus terms of lower asymptotic order. The following result of Frenzen and Wong (1986) provides a solution to his problem:

$$L_n = \frac{2^{3/2}}{\sqrt{\pi}} n^{1/2} + B_0 + \sqrt{\frac{2}{\pi}} n^{-1/2} + D_0 + \varepsilon(n), \tag{20.16}$$

where D_0 is also a computable constant and the remainder $\varepsilon(n)$ satisfies

$$\varepsilon(n) = O(n^{-3/2}) \tag{20.17}$$

as $n \to \infty$. From (20.16), it can easily be verified that $\{L_n\}$ is an asymptotically increasing sequence. In the derivation of (20.16), Frenzen and Wong (1985a) made good use of their results for the uniform asymptotic expansion (20.6) of the Jacobi polynomials. To prove the conjecture of Szegö in (20.13), Qu and Wong (1988) showed that the remainder $\varepsilon(n)$ in (20.16) satisfies

$$|\varepsilon(n)| \leq \frac{15}{n^{3/2}} \tag{20.18}$$

for all $n \geq 49$. From (20.16) and (20.18), it can readily be shown that (20.13) is true for $n \geq 49$; see Qu and Wong (1988, p. 172). The first fifty L_n can be calculated numerically and an examination of the values shows that the sequence $\{L_n\}$ is indeed monotonically increasing. To achieve (20.18), Qu and Wong used the results of Baratella and Gatteschi mentioned earlier.

The objective of this chapter is to show that uniform problems can occur in other situations. For instance, consider the Maclaurin expansion

$$(e^{i\theta} - z)^{\alpha}(e^{-i\theta} - z)^{\beta} f(z) = \sum_{n=0}^{\infty} a_n(\theta) z^n, \tag{20.19}$$

where $f(z)$ is an analytic function in a domain containing the unit disk. If $\theta \neq 0$, then the function on the left of the equality has two singularities on the circle of convergence. If $\theta = 0$, then this function has only one singular point. In either case, the asymptotic expansion of $a_n(\theta)$, as $n \to \infty$, can be obtained by using the classical method of Darboux discussed in Wong (1989, pp. 116-117). The problem of allowing θ to vary in a closed interval $[0, \pi - \varepsilon]$ with $\varepsilon > 0$, is much more complicated. This problem was first addressed by Fields (1967) and dealt with more recently by Wong and Zhao (2005). The main objective of this chapter is to demonstrate how the technique used in Wong and Zhao (2005) can be applied directly to the special case of Jacobi polynomials. The material in this chapter is taken from the paper by Bai and Zhao (2007).

20.2 Statement of results

For $|w| < 1$, we let $R = \sqrt{1 - 2xw + w^2}$, which takes the value 1 for $w = 0$, while $(1 - w + R)^{-\alpha}$ and $(1 + w + R)^{-\beta}$ take the values $2^{-\alpha}$ and $2^{-\beta}$, respectively. The Jacobi polynomials have the generating function

$$2^{\alpha+\beta} R^{-1} (1 - w + R)^{-\alpha} (1 + w + R)^{-\beta} = \sum_{n=0}^{\infty} P_n^{(\alpha,\beta)}(x) w^n. \tag{20.20}$$

It is readily verified that R^{-1} has a pair of branch points $w = e^{\pm i\theta}$ in (20.20) for $\theta = 0, \pi$, where $\cos\theta = x$, while the factors $(1 - w + R)^{-\alpha}$ and $(1 + w + R)^{-\beta}$ possess no singularity other than the branch points $w = e^{\pm i\theta}$ (exercise). For $\theta = 0$, the pair $w = e^{\pm i\theta}$ coalesce into a single point $w = 1$, which is now a simple pole of R^{-1}, as well as a branch point of $(1 - w + R)^{-\alpha}$. A similar situation occurs when $\theta = \pi$. These bring up the uniform problem mentioned at the end of Section 20.1. In view of the reflection formula (see Szegö (1975, p. 59))

$$P_n^{(\alpha,\beta)}(-x) = (-1)^n P_n^{(\alpha,\beta)}(x), \tag{20.21}$$

we may, without loss of generality, restrict ourselves to the critical value $\theta = 0$.

By Cauchy integral formula, we have from (20.20)

$$P_n^{(\alpha,\beta)}(x) = \frac{1}{2\pi i}\int_{\Gamma_w} 2^{\alpha+\beta}R^{-1}(1-w+R)^{-\alpha}(1+w+R)^{-\beta}\frac{dw}{w^{n+1}}, \tag{20.22}$$

where Γ_w is an arbitrary simple closed curve encircling the origin but not the branch points $w = e^{\pm i\theta}$; here $\theta \in [0, \pi]$ is written as $x = \cos\theta$. Recall that when θ tends to its critical values 0 and π, the pair of branch points coalesce with each other.

For $n > -\alpha-\beta$, we can deform the integration path Γ_w so that it starts from $+\infty$ along the lower edge of the real line, encircling $e^{\pm i\theta}$ clockwise, and then goes to $+\infty$ along the upper edge of the real line. For the sake of convenience, we continue to denote the deformed path by Γ_w.

In (20.22), we now make the change of variable

$$w = e^{-\theta s}; \tag{20.23}$$

see Zygmund (1935, (2.2)). Note that this change of variable separates the coalescing branch points $w = e^{\pm i\theta}$, as θ tends to 0. The points $w = e^{\pm i\theta}$ correspond, respectively, to the points $s = \mp i$, and Γ_w is to a curve $-\Gamma$, where the minus sign indicates that the direction of the path has been reversed and Γ is a Hankel loop, which starts from $-\infty$, encircling both points $s = \pm i$ counterclockwise, and ends at $-\infty$.

Equation (20.23) defines a conformal mapping $w \leftrightarrow s$ from the complex w-plane to the strip $|\operatorname{Im} s| < \pi/\theta$ in the s-plane. Inserting (20.23) in (20.22) gives

$$P_n^{(\alpha,\beta)}(x) = \frac{2^{\alpha+\beta}\theta}{2\pi i}\int_{\Gamma}\frac{e^{N\theta s}ds}{R_1(e^{\theta s/2}-e^{-\theta s/2}+R_1)^{\alpha}(e^{\theta s/2}+e^{-\theta s/2}+R_1)^{\beta}}, \tag{20.24}$$

where $R_1 = (e^{\theta s}+e^{-\theta s}-2\cos\theta)^{1/2}$ and, instead of n, we have introduced a new large parameter $N = n + (\alpha+\beta+1)/2$. An advantage of the new variable is that it makes the integrand in (20.24) symmetrical.

We shall now carry out our formal derivation of the final asymptotic expansion on the basis of (20.24). First, we recall an integral representation for the Bessel function of the first kind (see Wong (1989, p. 112)).

$$J_\alpha(z) = \frac{1}{2\pi i}\int_{-\infty}^{(0^+)} e^{z(\tau-\tau^{-1})/2}\tau^{-\alpha-1}d\tau, \tag{20.25}$$

where $|\arg z| < \pi/2$. Put $s = (\tau-\tau^{-1})/2$, and we choose the branch $\tau = s+\sqrt{s^2+1}$ so that $\tau \sim 2s$ for large s. Then it follows that

$$J_\alpha(z) = \frac{1}{2\pi i}\int_{\Gamma} e^{zs}(s+\sqrt{s^2+1})^{-\alpha}\frac{ds}{\sqrt{s^2+1}}, \tag{20.26}$$

where Γ is the integration path in (20.24). Formula (20.26) is initially valid for $|\arg z| < \pi/2$; the sector of validity can be extended by analytic continuation.

Now we consider the small θ behavior of each factor in the integrand of (20.24) for fixed s. Straightforward calculation yields

$$R_1^{-1} \sim \theta^{-1}(s^2+1)^{-1/2},$$
$$(e^{\theta s/2} - e^{-\theta s/2} + R_1)^{-\alpha} \sim \theta^{-\alpha}(s+\sqrt{s^2+1})^{-\alpha},$$
$$(e^{\theta s/2} + e^{-\theta s/2} + R_1)^{-\beta} \sim 2^{-\beta}.$$

Substituting these into (20.24), and paying attention to (20.26), we have formally

$$P_n^{(\alpha,\beta)}(\cos\theta) \sim (\theta/2)^{-\alpha} J_\alpha(N\theta), \tag{20.27}$$

which reminds us of the Mehler-Heine type form (see Szegö (1975, p. 167))

$$\lim_{n\to\infty} n^{-\alpha} P_n^{(\alpha,\beta)}(\cos(z/n)) = (z/2)^{-\alpha} J_\alpha(z), \tag{20.28}$$

holding uniformly for $|z| \le \rho$ with ρ fixed. The approximation (20.27) also agrees with the Hilb-type formula (20.1) given by Szegö (1975, p. 197).

Inspired by the above formal derivation, we write

$$P_n^{(\alpha,\beta)}(\cos\theta) = \frac{(\theta/2)^{-\alpha}}{2\pi i}\int_\Gamma h_0(s,\theta) e^{N\theta s}(s+\sqrt{s^2+1})^{-\alpha}\frac{ds}{\sqrt{s^2+1}} \tag{20.29}$$

where

$$h_0(s,\theta) = \frac{2^\beta \theta^{1+\alpha}\sqrt{s^2+1}(s+\sqrt{s^2+1})^\alpha}{R_1(e^{\theta s/2} - e^{-\theta s/2} + R_1)^\alpha (e^{\theta s/2} + e^{-\theta s/2} + R_1)^\beta}. \tag{20.30}$$

For simplicity, we put

$$\Theta = \theta^2(s^2+1). \tag{20.31}$$

It will be shown in the next section that $h_0(s,\theta)$ has the following expression.

Lemma 20.1. *For $|\theta| < \pi$, we have*

$$h_0(s,\theta) = A_0 + \theta s B_0 + \theta\sqrt{s^2+1}C_0 + \theta^2 s\sqrt{s^2+1}D_0, \tag{20.32}$$

where

$$A_0 = \sum_{j=0}^\infty a_{0,j}(\theta^2)\Theta^j, \quad B_0 = \sum_{j=0}^\infty b_{0,j}(\theta^2)\Theta^j,$$
$$C_0 = \sum_{j=0}^\infty c_{0,j}(\theta^2)\Theta^j, \quad D_0 = \sum_{j=0}^\infty d_{0,j}(\theta^2)\Theta^j.$$

The coefficients $a_{0,j}(\theta^2)$, $b_{0,j}(\theta^2)$, $c_{0,j}(\theta^2)$ and $d_{0,j}(\theta^2)$, given in (20.48) below, are analytic functions in θ^2. All four expansions are absolutely convergent for $|s| < \pi/\theta$.

We leave the proof of this lemma to the next section. Here, we first introduce an iterative procedure: for $k = 0, 1, 2, \cdots$, we define

$$h_k(s,\theta) = a_{k,0}(\theta^2) + \theta s b_{k,0}(\theta^2) + \theta\sqrt{s^2+1}g_k(s,\theta) \tag{20.33}$$

and

$$h_{k+1}(s,\theta) = (-\sqrt{s^2+1}\frac{d}{ds} + \alpha)g_k(s,\theta). \tag{20.34}$$

The following result can readily be verified.

Lemma 20.2. *Each h_k has the expression*

$$h_k(s,\theta) = A_k + \theta s B_k + \theta\sqrt{s^2+1}C_k + \theta^2 s\sqrt{s^2+1}D_k, \tag{20.35}$$

where

$$A_k = \sum_{j=0}^{\infty} a_{k,j}(\theta^2)\Theta^j, \quad B_k = \sum_{j=0}^{\infty} b_{k,j}(\theta^2)\Theta^j,$$
$$C_k = \sum_{j=0}^{\infty} c_{k,j}(\theta^2)\Theta^j, \quad D_k = \sum_{j=0}^{\infty} d_{k,j}(\theta^2)\Theta^j,$$

with

$$\begin{aligned}
a_{k,j} =& \alpha c_{k-1,j} + (2j+1)\theta^2 b_{k-1,j+1} - 2j b_{k-1,j},\\
b_{k,j} =& \alpha d_{k-1,j} - (2j+1)a_{k-1,j+1},\\
c_{k,j} =& \alpha a_{k-1,j+1} + 2(j+1)\theta^2 d_{k-1,j+1} - (2j+1)d_{k-1,j},\\
d_{k,j} =& \alpha b_{k-1,j+1} - 2(j+1)c_{k-1,j+1},
\end{aligned}$$

for $k = 1, 2, \cdots$, all four series A_k, B_k, C_k and D_k being absolutely convergent in $\Theta = \theta^2(s^2+1)$ for $|s| < \pi/\theta$, with coefficients analytic in θ^2 for $|\theta| < \pi$.

From (20.26), it is readily seen that

$$\frac{1}{2\pi i}\int_\Gamma \frac{se^{zs}ds}{\sqrt{s^2+1}(s+\sqrt{s^2+1})^\alpha} = J'_\alpha(z) = -J_{\alpha+1}(z) + \frac{\alpha}{z}J_\alpha(z). \tag{20.36}$$

Appealing to (20.26) and (20.36), we can derive an expansion from the iteration procedure (20.32)-(20.34). We now state the main result of this chapter, the proof of which will again be given in next section.

Theorem 20.1. *For fixed $\alpha > -1$ and $\beta \in \mathbf{R}$, we have the asymptotic expansion*

$$(\theta/2)^\alpha P_n^{(\alpha,\beta)}(\cos\theta) = J_\alpha(N\theta)\sum_{k=0}^{m-1}\frac{a_{k,0}(\theta^2)}{N^k} + \theta J'_\alpha(N\theta)\sum_{k=0}^{m-1}\frac{b_{k,0}(\theta^2)}{N^k} + \varepsilon_m, \tag{20.37}$$

where

$$\varepsilon_m = \frac{N^{-m}}{2\pi i}\int_\Gamma h_m(s,\theta)e^{N\theta s}(s+\sqrt{s^2+1})^{-\alpha}\frac{ds}{\sqrt{s^2+1}}, \tag{20.38}$$

and there exists a constant M_m*, independent of* θ *and* n*, such that*

$$|\varepsilon_m| \leq M_m N^{-m}[|J_\alpha(N\theta)| + \theta|J'_\alpha(N\theta)|] \tag{20.39}$$

for $\theta \in [0, \pi - \varepsilon]$*, where* ε *is a small positive constant. The first few coefficients are given in* (20.47) *below.*

20.3 Technical details

We prove the identity in (20.32) by expanding each factor of $h_0(s,\theta)$ in (20.30) into an absolutely convergent series. To begin, we let

$$\eta(z) = \left(\frac{e^z - e^{-z}}{2z}\right)^{-1/2}. \tag{20.40}$$

From the Maclaurin expansion

$$\eta(z) = \sum_{j=0}^{\infty} \eta_j z^j$$

for $|z| < \pi$, it is readily seen that $\eta(z)$ is analytic in $|\operatorname{Im} z| < \pi$. Simple computation gives $\eta_0 = 1$, $\eta_1 = -1/12$ and $\eta_2 = 1/160$ (exercise). Since

$$\frac{\theta\sqrt{s^2+1}}{R_1} = \eta((s+i)\theta/2)\eta((s-i)\theta/2), \tag{20.41}$$

taking the obvious symmetry and the absolute convergence into account, by regrouping the series we have

$$\frac{\theta\sqrt{s^2+1}}{R_1} = \sum_{j=0}^{\infty} \gamma_j(\theta^2)\Theta^j. \tag{20.42}$$

The regrouped expansion is absolutely convergent so long as $|s| < 2\pi/\theta - 1$. Bring in the natural boundary $|\operatorname{Im} s| < \pi/\theta$ (cf. the paragraph between (20.23) and (20.24)), the coefficients $\gamma_j(\theta^2)$ are resulted from absolutely convergent series in θ^2, thus are analytic functions for $|\theta| < \pi$. The first few coefficients are

$$\begin{aligned}
\gamma_0(\theta^2) &= 1 + \frac{\theta^2}{12} + \frac{\theta^4}{160} + O(\theta^6),\\
\gamma_1(\theta^2) &= -\frac{1}{24} - \frac{\theta^2}{160} + O(\theta^4),\\
\gamma_2(\theta^2) &= \frac{7}{5760} + O(\theta^2).
\end{aligned}$$

Following similar steps, we also obtain

$$R_1 = \theta\sqrt{s^2+1}\sum_{j=0}^{\infty} r_{1,j}(\theta^2)\Theta^j, \tag{20.43}$$

with

$$\begin{aligned} r_{1,0}(\theta^2) =& 1 - \frac{\theta^2}{12} + \frac{\theta^4}{1440} + O(\theta^6), \\ r_{1,1}(\theta^2) =& \frac{1}{24} - \frac{\theta^2}{1440} + O(\theta^4), \\ r_{1,2}(\theta^2) =& \frac{1}{1920} + O(\theta^2). \end{aligned}$$

Hence $e^{\theta s/2} + e^{-\theta s/2} + R_1$ can be expanded in absolutely convergent series as

$$e^{\theta s/2} + e^{-\theta s/2} + R_1 = \sum_{j=0}^{\infty} r_{2,j}(\theta^2)\Theta^j + \sqrt{\Theta}\sum_{j=0}^{\infty} r_{1,j}(\theta^2)\Theta^j,$$

with $r_{2,0} = 2 - \theta^2/4 + O(\theta^4)$ and $r_{2,1} = 1/4 + O(\theta^2)$. The sum of the last two absolutely convergent series can of course be regarded as an analytic function in terms of $\sqrt{\Theta} = \theta\sqrt{s^2+1}$, convergent for $|s| < \pi/\theta$, $\theta \in (0, \pi)$. Note that $e^{\theta s/2} \pm e^{-\theta s/2} + R_1$ does not vanish for any s if $\theta \in (0, \pi)$.

Our argument works in this way: first we expand functions for small ranges of Θ, and then we find that the expansions should hold in larger domains. Thus we have

$$\left(\frac{e^{\theta s/2} + e^{-\theta s/2} + R_1}{2}\right)^{-\beta} = \sum_{j=0}^{\infty} \beta_{2,j}(\theta^2)\Theta^j + \sqrt{\Theta}\sum_{j=0}^{\infty} \beta_{1,j}(\theta^2)\Theta^j. \tag{20.44}$$

The first few coefficients are

$$\begin{aligned} \beta_{2,0} =& 1 + \frac{\beta\theta^2}{8} + O(\theta^4), \\ \beta_{2,1} =& \frac{\beta^2}{8} + O(\theta^2), \\ \beta_{1,0} =& -\frac{\beta}{2} - \frac{\beta(3\beta+1)}{48}\theta^2 + O(\theta^4). \end{aligned}$$

We now expand the remaining factors, and give the details. First, we write

$$e^{\theta s/2} - e^{-\theta s/2} = \theta s(1+G), \quad \sqrt{e^{\theta s} + e^{-\theta s} - 2\cos\theta} = \theta\sqrt{s^2+1}(1+H), \tag{20.45}$$

where

$$G = \sum_{l=1}^{\infty} \frac{(s\theta/2)^{2l}}{(2l+1)!} = \sum_{j=0}^{\infty} g_j(\theta^2)\Theta^j,$$

and the coefficients $g_j(\theta^2)$ are analytic functions in θ^2 for $|\theta| < \pi$, with

$$g_0 = -\frac{\theta^2}{24} + \frac{\theta^4}{1920} + O(\theta^6),\ g_1 = \frac{1}{24} - \frac{\theta^2}{960} + O(\theta^4),\ g_2 = \frac{1}{1920} + O(\theta^2).$$

Since the left-hand side of the second equation in (20.45) is R_1 (see the line following (20.24)), it readily follows from (20.43) that H is also of the power series form given for G. Indeed, we have

$$H = \sum_{j=0}^{\infty} h_j(\theta^2)\Theta^j,$$

where

$$h_0 = -\frac{\theta^2}{12} + \frac{\theta^4}{1440} + O(\theta^6),\ h_1 = \frac{1}{24} - \frac{\theta^2}{1440} + O(\theta^4),\ h_2 = \frac{1}{1920} + O(\theta^2).$$

From (20.45), we have (exercise)

$$2(1 - \cos\theta) = (1+H)^2\theta^2 + \theta^2 s^2 (H - G)(2 + H + G).$$

Hence,

$$s^2(H - G) = \frac{2(1-\cos\theta)/\theta^2 - (1+H)^2}{2 + H + G},$$

which is also of the form $\sum_{j=0}^{\infty} f_j(\theta^2)\Theta^j$. Expanding both sides of this equation in power series of Θ^j and identifying the coefficients of like powers of Θ gives (exercise)

$$h_j - g_j = \theta^2 (f_{j+1} + h_{j+1} - g_{j+1}), \tag{20.46}$$

for $j = 0, 1, 2, \cdots$. Thus, we can write

$$G - H = \theta^2 \sum_{j=0}^{\infty} \hat{g}_j(\theta^2)\Theta^j,$$

where the $\hat{g}_j$'s are analytic functions of θ^2 with

$$\hat{g}_0(\theta^2) = \frac{1}{24} - \frac{\theta^2}{5760} + O(\theta^4),\quad \hat{g}_1(\theta^2) = -\frac{1}{2880} + O(\theta^2).$$

Hence, it is readily verified that (exercise)

$$\begin{aligned}\frac{e^{\theta s/2}-e^{-\theta s/2}+R_1}{\theta(s+\sqrt{s^2+1})} &= 1+[H+s^2(H-G)+s\sqrt{s^2+1}(G-H)]\\ =1+\Bigg\{&\Bigg[\Big(-\frac{\theta^2}{24}+\frac{\theta^4}{1920}+\cdots\Big)+\Big(-\frac{\theta^2}{1152}+\cdots\Big)\Theta+\cdots\Bigg]\\ &+\theta^2 s\sqrt{s^2+1}\Bigg[\Big(\frac{1}{24}-\frac{\theta^2}{5760}+\cdots\Big)+\Big(-\frac{1}{2880}+\cdots\Big)\Theta+\cdots\Bigg]\Bigg\}.\end{aligned}$$

In turn, we obtain (exercise)

$$\left(\frac{e^{\theta s/2}-e^{-\theta s/2}+R_1}{\theta(s+\sqrt{s^2+1})}\right)^{-\alpha}=\sum_{j=0}^{\infty}\alpha_{2,j}(\theta^2)\Theta^j+\theta^2 s\sqrt{s^2+1}\sum_{j=0}^{\infty}\alpha_{1,j}(\theta^2)\Theta^j, \tag{20.47}$$

where

$$\alpha_{2,0}=1+\frac{\alpha\theta^2}{24}+O(\theta^4),\ \alpha_{2,1}=\frac{\alpha\theta^2}{1152}+O(\theta^4),\ \alpha_{1,0}=-\frac{\alpha}{24}+O(\theta^2).$$

Substituting (20.42), (20.44) and (20.47) into (20.30) gives (20.32). The coefficients of A_0, B_0, C_0 and D_0 can be expanded as

$$\begin{aligned} a_{0,j} &= \sum_{l_1+l_2+l_3=j}\gamma_{l_1}\beta_{2,l_2}\alpha_{2,l_3},\ j=0,1,2,\cdots,\\ b_{0,j} &= \sum_{l_1+l_2+l_3=j-1}\gamma_{l_1}\beta_{1,l_2}\alpha_{1,l_3},\ j=1,2,3,\cdots,\\ c_{0,j} &= \sum_{l_1+l_2+l_3=j}\gamma_{l_1}\beta_{1,l_2}\alpha_{2,l_3},\ j=0,1,2,\cdots,\\ d_{0,j} &= \sum_{l_1+l_2+l_3=j}\gamma_{l_1}\beta_{2,l_2}\alpha_{1,l_3},\ j=0,1,2,\cdots, \end{aligned} \tag{20.48}$$

with

$$\begin{aligned} a_{0,0} &=1+\frac{2+\alpha+3\beta}{24}\theta^2+O(\theta^4),\quad a_{0,1}=\frac{3\beta^2-1}{24}+O(\theta^2),\\ b_{0,0} &=0,\quad b_{0,1}=\frac{\alpha\beta}{48}+O(\theta^2),\quad c_{0,0}=-\frac{\beta}{2}+O(\theta^2),\quad d_{0,0}=-\frac{\alpha}{24}+O(\theta^2). \end{aligned} \tag{20.49}$$

This completes the proof of Lemma 20.1.

To prove Theorem 20.1, we only need to establish the error estimate (20.39). This is not a difficult task, since from the iteration procedure (20.32)-(20.34) we know that each of the functions A_k, B_k, C_k and D_k in (20.35) is defined by an absolutely convergent series for $|s|<\pi/\theta$, when

$\theta \in (0, \pi)$. Thus we have h_m bounded for $|s\theta| < \sigma_0$ for some positive constant σ_0. For s not in this region, we can, as described right after (20.23), let $s \in (-\infty, -\sigma_0/\theta]$. Denote by Γ_I the portion of Γ in $|s| < \sigma_0/\theta$, and by Γ_O the upper and lower edges of $(-\infty, -\sigma_0/\theta]$. We can now rewrite (20.38) as

$$N^m \varepsilon_m := r_m^I + r_m^O, \tag{20.50}$$

where r_m^I and r_m^O denote, respectively, the integral in (20.38) along Γ_I and Γ_O.

To prove (20.39), we first recall that

$$J_\alpha(z) = (z/2)^\alpha \sum_{j=0}^{\infty} \frac{(-z^2/4)^j}{j!\Gamma(\alpha + j + 1)}, \tag{20.51}$$

where the branch of $(z/2)^\alpha$ is chosen so that it is real and positive when z is positive; see Olver *et al.* (2010, p. 217). In particular, we have

$$J_\alpha(N\theta) \sim \frac{(N\theta/2)^\alpha}{\Gamma(\alpha+1)}, \quad J'_\alpha(N\theta) \sim \frac{(N\theta/2)^{\alpha-1}}{2\Gamma(\alpha)} \tag{20.52}$$

for $N\theta$ bounded. The second approximation in (20.52) is valid for $\alpha \neq 0, -1, -2, \cdots$, whereas the first holds for $\alpha \neq -1, -2, \cdots$. Since we consider only the case $\alpha > -1$, thus the first approximation always holds. For unbounded $N\theta$, we have

$$J_\alpha(z) = \sqrt{\frac{2}{\pi z}}[\cos\chi + O(|z|^{-1})], \quad J'_\alpha(z) = \sqrt{\frac{2}{\pi z}}[-\sin\chi + O(|z|^{-1})], \tag{20.53}$$

where $\chi = z - \alpha\pi/2 - \pi/4$ and $z = N\theta$; see Olver *et al.* (2010, pp. 228-229).

Next, we note that along Γ_O, $s \leq -\sigma_0/\theta$; hence the factor $e^{N\theta s}$ in r_m^O is exponentially small and satisfies $e^{N\theta s} \leq e^{-\sigma_0 N}$. Also, observe that the orders of magnitude of all other factors in the integrand of (20.38) are at most algebraic and all these factors are independent of N. When $N\theta$ is bounded, by introducing a change of variable $u = \theta s$ and in view of (20.52), we have from (20.38)

$$|r_m^O| \leq M e^{-\sigma_0 N/2} |J_\alpha(N\theta)|, \tag{20.54}$$

where M is a constant independent of both N and θ. When $N\theta \to \infty$, (20.53) implies that (exercise)

$$\begin{aligned}
|J_\alpha(N\theta)| + |\theta J'_\alpha(N\theta)| &\geq \frac{\theta}{\pi}(|J_\alpha(N\theta)| + |J'_\alpha(N\theta)|) \\
&\geq \frac{\theta}{\pi}\sqrt{\frac{2}{\pi N\theta}}[|\cos\chi| + |\sin\chi| - K/(N\theta)] \geq \frac{\theta}{\pi}\sqrt{\frac{2}{\pi N\theta}}[1 - K/(N\theta)] \\
&\geq \frac{\theta}{2\pi}\sqrt{\frac{2}{\pi N\theta}} \geq \frac{1}{N},
\end{aligned} \tag{20.55}$$

where $K > 0$ is a constant and $N\theta > \max\{2K, 2\pi^3\}$. Notice that when $N\theta$ is bounded from below, r_m^O is still of order $e^{-\sigma_0 N}\theta^\alpha$. Also note that $\theta^\alpha \le \pi^\alpha + N^{-\alpha}$, which is obvious if $\alpha > 0$; if $\alpha \in (-1, 0)$, then we assume $N\theta \ge 1$ and have $(N\theta)^\alpha \le 1 \le 1 + (N\pi)^\alpha$. As a consequence, we again have $\theta^\alpha \le \pi^\alpha + N^{-\alpha}$. Thus, in the case when $N\theta$ is not bounded, we have

$$|r_m^O| \le Me^{-\sigma_0 N/2}[|J_\alpha(N\theta)| + \theta|J'_\alpha(N\theta)|], \tag{20.56}$$

where again M is a constant not depending on N and θ.

To estimate r_m^I, we again distinguish two cases: namely, $N\theta$ bounded and $N\theta$ unbounded. In the first case, we introduce a rescaling $u = N\theta s$, and obtain

$$r_m^I = \frac{(N\theta)^\alpha}{2\pi i}\int_{\tilde{\Gamma}_I} e^u\left(u + \sqrt{u^2 + (N\theta)^2}\right)^{-\alpha}\frac{du}{\sqrt{u^2 + (N\theta)^2}},$$

where $\tilde{\Gamma}_I = (N\theta)\Gamma_I$ is still compact. Since $\pm N\theta i$ keep fixed distances from $\tilde{\Gamma}_I$, in view of (20.52), we have, for bounded $N\theta$,

$$|r_m^I| \le M|J_\alpha(N\theta)|. \tag{20.57}$$

When $N\theta$ is not bounded, we have $N\theta \to \infty$. Picking up the contribution to r_m^I from $s = \pm i$, which are the singularities of the integrand in (20.38), we obtain a linear combination of $(N\theta)^{-1/2}e^{iN\theta}$ and $(N\theta)^{-1/2}e^{-iN\theta}$. This can be proved by first deforming the path of integration into two segments parallel to the real axis and then using Watson's lemma. In view of (20.53), we have $|r_m^I| \le M[|J_\alpha(N\theta)| + |J'_\alpha(N\theta)|]$. Furthermore, note that in this case we have $\theta > 1/N$ and $N > 1$; hence

$$\frac{|r_{m+1}^I|}{N} \le M[|J_\alpha(N\theta)| + |J'_\alpha(N\theta)|/N] \le M[|J_\alpha(N\theta)| + \theta|J'_\alpha(N\theta)|].$$

Now, from (20.37),

$$\varepsilon_m = J_\alpha(N\theta)\frac{a_{m,0}(\theta^2)}{N^m} + \theta J'_\alpha(N\theta)\frac{b_{m,0}(\theta^2)}{N^m} + \varepsilon_{m+1}.$$

In view of (20.50), one may write

$$r_m^I = -r_m^O + a_{m,0}(\theta^2)J_\alpha(N\theta) + b_{m,0}(\theta^2)\theta J'_\alpha(N\theta) + \frac{r_{m+1}^O + r_{m+1}^I}{N}.$$

Each term on the right-hand side is bounded by $M(|J_\alpha| + \theta|J'_\alpha|)$, with M being a generic constant. Therefore, we have (exercise)

$$|r_m^I| \le M[|J_\alpha(N\theta)| + \theta|J'_\alpha(N\theta)|]. \tag{20.58}$$

The desired result (20.39) now follows from a combination of (20.54), (20.56)-(20.58) and (20.50), thus proving Theorem 20.1.

20.4 Exercises

1. For $x = \cos\theta \in (-1, 1)$, let $R = \sqrt{1 - 2xw + w^2}$, which takes the value 1 for $w = 0$, while $(1 - w + R)^{-\alpha}$ and $(1 + w + R)^{-\beta}$ take the values $2^{-\alpha}$ and $2^{-\beta}$, respectively. Verify that $(1 - w + R)^{-\alpha}$ and $(1 + w + R)^{-\beta}$ possess no singularity other than the branch points $w = e^{\pm i\theta}$.
2. Let
$$\eta(z) = \left(\frac{e^z - e^{-z}}{2z}\right)^{-1/2}.$$
From the Maclaurin expansion
$$\eta(z) = \sum_{j=0}^{\infty} \eta_j z^j$$
for $|z| < \pi$, show that $\eta(z)$ is analytic in $|\operatorname{Im} z| < \pi$, and that $\eta_0 = 1$, $\eta_1 = -1/12$ and $\eta_2 = 1/160$.
3. Use (20.45) to show that
$$2(1 - \cos\theta) = (1 + H)^2\theta^2 + \theta^2 s^2 (H - G)(2 + H + G).$$
4. Verify (20.46).
5. Verify the expansion in (20.47).
6. Establish the inequality in (20.55).
7. Prove the inequality in (20.58).
8. The Jacobi polynomial has the integral representation
$$(1 - x)^\alpha (1 + x)^\beta P_n^{(\alpha,\beta)}(x) = \frac{1}{2\pi i}\int_C \left[\frac{z^2 - 1}{2(z - x)}\right]^n \frac{(1 - z)^\alpha (1 + z)^\beta}{z - x} dz, \tag{20.59}$$
where C is a closed contour enclosing the point $z = x$, but not the points $z = \pm 1$; see Copson (1965, p. 67).
(a) Put $\tau = z - x$. Show that equation (20.59) becomes
$$(1 - x)^\alpha (1 + x)^\beta P_n^{(\alpha,\beta)}(x) = \frac{1}{2\pi i}\int_{C'} \left[\frac{(\tau + x + 1)(\tau + x - 1)}{2\tau}\right]^n \times \frac{(1 - \tau - x)^\alpha (1 + \tau + x)^\beta}{\tau} d\tau, \tag{20.60}$$
where C' is the image of C under the transformation. Furthermore, show that the last equation can be written in the form
$$2^n (1 - x)^\alpha (1 + x)^\beta P_n^{(\alpha,\beta)}(x) = \frac{1}{2\pi i}\int_{C'} g(\tau) e^{N f(\tau, x)} d\tau, \tag{20.61}$$

where

$$f(\tau, x) = \ln \frac{(\tau + x + 1)(\tau + x - 1)}{\tau} \tag{20.62}$$

and

$$g(\tau) = \frac{(1 - \tau - x)^{\alpha}(1 + \tau + x)^{(-\alpha+\beta-1)/2}}{(\tau + x - 1)^{(\alpha+\beta+1)/2}\tau^{(-\alpha-\beta+1)/2}} \tag{20.63}$$

(b) In view of the well-known identity [Copson (1965), p. 59]

$$P_n^{(\alpha,\beta)}(x) = (-1)^n P_n^{(\alpha,\beta)}(-x),$$

we may without loss of generality assume that x belongs to the interval $[0, 1]$. To make the function $f(\tau, x)$ single-valued and analytic, we cut the τ-plane form 0 to $1-x$ along the positive real axis, and from $-\infty$ to $-(1+x)$ along the negative real axis. The integration path C' in (20.60) is taken to be the contour indicated in Figure 20.1 below. Note that the cuts for f differ from those indicated in Figure 20.1 for contour C'.

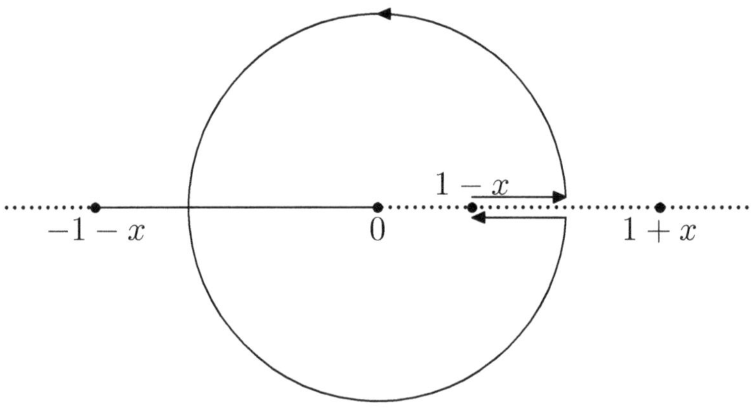

Fig. 20.1: Contour C'.

Show that the function $f(\tau, x)$ has two symmetrically located saddle points

$$\tau_{\pm} = \pm i\sqrt{1 - x^2}. \tag{20.64}$$

As $x \to \pm 1$, these two saddle points coalesce with each other and also with the logarithmic singularity at $\tau = 0$. Since x is restricted to $[0, 1]$, we are concerned with only the limit $x \to 1$. Show that in a neighborhood of $x = 1$, we have formally

$$\begin{aligned} f(\tau, x) &= \ln(\tau + x + 1) + \ln(\tau + x - 1) - \ln \tau \\ &= \ln(x+1) + \ln(1 + \frac{\tau}{x+1}) + \ln \tau + \ln(1 + \frac{x-1}{\tau}) - \ln \tau \\ &= \ln(x+1) + \frac{\tau}{1+x} - \frac{1-x}{\tau} + \cdots . \end{aligned}$$

This expansion and the behavior of the two saddle points suggest that we should compare $f(\tau, x)$ with the function

$$g(s, x) = s - \frac{A^2(x)}{s} + \gamma, \tag{20.65}$$

where γ and $A(x) \geq 0$ are to be determined. Thus we introduce the mapping $\tau = \tau(s)$ defined by

$$f(\tau, x) = g(s, x). \tag{20.66}$$

(c) The function $g(s, x)$ also has two saddle points

$$s_{\pm} = \pm i A(x). \tag{20.67}$$

In order for the transformation (20.66) to be analytic in the region of interest, we must have $d\tau/ds$ not equal to 0 or ∞. Now

$$f_\tau(\tau, x)\frac{d\tau}{ds} = g_s(s, x). \tag{20.68}$$

The left-hand side vanishes at $\tau = \tau_{\pm}$, while the right-hand side vanishes at $s = s_{\pm}$. Thus we must make $\tau = \tau_+$ correspond to $s = s_+$ and $\tau = \tau_-$ correspond to $s = s_-$. Show that this gives

$$\ln \frac{(\tau_+ + x + 1)(\tau_+ + x - 1)}{\tau_+} = iA(x) - \frac{A^2(x)}{iA(x)} + \gamma$$

and

$$\ln \frac{(\tau_- + x + 1)(\tau_- + x - 1)}{\tau_-} = -iA(x) + \frac{A^2(x)}{iA(x)} + \gamma,$$

and that the solution is

$$\gamma = \ln 2, \quad A(x) = \frac{1}{2} \arctan \frac{\sqrt{1 - x^2}}{x}. \tag{20.69}$$

Note that $A(x) \sim \sqrt{(1-x)/2}$ as $x \to 1$. Hence the saddle points of $g(s, x)$ also coalesce when x approaches 1. (It was shown by Wong and

Zhang (1996) that with this choice, the transformation (20.66) is one-to-one and analytic along the whole path of integration C'.) Show that by changing the integration variable in (20.61) to s, we obtain

$$(1-x)^\alpha(1+x)^\beta P_n^{(\alpha,\beta)}(x) = \frac{2^{(\alpha+\beta+1)/2}}{2\pi i}\int_\Gamma h(s)e^{N(s-A^2/s)}\frac{ds}{s^{\alpha+1}},$$

where

$$h(s) = \frac{(1-\tau-\alpha)^\alpha(\tau+x+1)^{(-\alpha+\beta-1)/2}}{(\tau+x-1)^{(\alpha+\beta+1)/2}\tau^{(-\alpha-\beta+1)/2}}s^{\alpha+1}\frac{d\tau}{ds},$$

and Γ is the image of C' under the transformation (20.66); see Figure 20.2. For a detailed study of the mapping $\tau \to s$, we refer to Wong and Zhang (1996). An improvement of the result in Wong and Zhang (1996) was given by Wong and Zhao (2003).

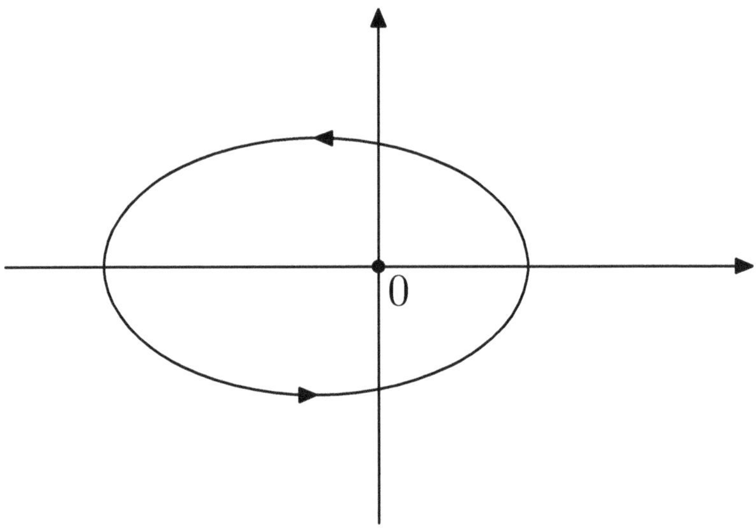

Fig. 20.2: Contour Γ.

9. The Laguerre polynomial has the integral representation

$$e^{-x/2}L_n^{(\alpha)}(x) = \frac{1}{2\pi i}\int_{-\infty}^{(0+)} \exp\left\{-\frac{x}{2}\frac{1+e^{-z}}{1-e^{-z}}\right\}(1-e^{-z})e^{nz}dz, \quad (20.70)$$

where the path of integration is the usual loop which begins and ends at $-\infty$ and encircles the origin in the positive direction; see [Szegö (1975),

p. 384]. Throughout this exercise we shall assume that $\alpha > -1$. Show that (20.70) can be written as

$$e^{-x/2}L_n^{(\alpha)}(x) = \frac{1}{2\pi i}\int_{-\infty}^{(0+)} \exp\left\{-\frac{x}{2}\coth\frac{z}{2} + Nz\right\} \times \left[\frac{\sinh(z/2)}{z/2}\right]^{-\alpha-1} z^{-\alpha-1}dz, \qquad (20.71)$$

with $N = n + (\alpha+1)/2$. If we replace z by $2z$ and let $\nu = 4N$ and $x = \nu t$, $-\infty < t < \infty$, show that (20.71) becomes

$$e^{-\nu t/2}L_n^{(\alpha)}(\nu t) = \frac{2^{-\alpha}}{2\pi i}\int_{-\infty}^{(0+)} \exp\left\{-\frac{\nu}{2}f(z,t)\right\}\left[\frac{\sinh z}{z}\right]^{-\alpha-1} z^{-\alpha-1}dz, \qquad (20.72)$$

where

$$f(z,t) = z - t\coth z. \qquad (20.73)$$

We shall represent $f(z,t)$ by the rational function

$$f(z,t) = u - \frac{A^2(t)}{u}. \qquad (20.74)$$

For this to be an analytic transformation in the region of interest, we must have dz/du not equal to 0 or ∞. Now

$$f_z(z,t)\frac{dz}{dt} = 1 + \frac{A^2(t)}{u^2} \qquad (20.75)$$

and $f_z(z,t)$ vanishes at $z = z_\pm$ in $|\operatorname{Im} z| < \pi/2$, where

$$z_\pm = \begin{cases} \pm i \arcsin\sqrt{t}, & 0 \le t < 1, \\ \mp \operatorname{arcsinh}\sqrt{-t}, & t < 0. \end{cases}$$

Since the right-hand side of (20.75) vanishes at $u = \pm iA(t)$, we must make $z = z_+$ correspond to $u = iA(t)$ and $z = z_-$ to $u = -iA(t)$. Show that this gives

$$A(t) = \begin{cases} [\arcsin\sqrt{t} + \sqrt{t(1-t)}]/2, & 0 \le t < 1, \\ i[\operatorname{arcsinh}\sqrt{-t} + \sqrt{t(t-1)}]/2, & t < 0. \end{cases} \qquad (20.76)$$

With this choice, the transformation (20.74) is one-to-one and analytic along the whole infinite loop in (20.72); for a proof, see Frenzen and Wong (1988). Show that changing the variable to u in (20.72) gives

$$e^{-\nu t/2}L_n^{(\alpha)}(\nu t) = \frac{2^{-\alpha}}{2\pi i}\int_{-\infty}^{(0+)} u^{-\alpha-1}h(u)\exp\left\{\frac{\nu}{2}[u - A^2(t)/u]\right\}du, \qquad (20.77)$$

where

$$h(u) = \left[\frac{\sinh z(u)}{z(u)}\right]^{-\alpha-1} \left[\frac{z(u)}{u}\right]^{-\alpha-1} \frac{dz}{du}. \tag{20.78}$$

From (20.73) and (20.74), we know that the mapping between z and u is determined by the equation

$$z - t\coth z = u - \frac{A^2(t)}{u}. \tag{20.79}$$

Hence, the function h in (20.78) depends also on the variable t; that is, $h = h(u,t)$. However, for simplicity, we shall not indicate this dependence explicitly.

To derive an asymptotic expansion from the integral in (20.77), we first deform the loop path of integration so that it consists of two straight lines along the negative real axis and a circle centered at the origin with radius ρ. The radius will be specified later, and depends on whether t is close to, or bounded away from, the origin. From (20.79), show that $z(u)$ is an odd function of u. From (20.78), show that $h(u)$ is an even function. Put $h_0(u) = h(u)$ and write

$$h_0(u) = \alpha_0 + \frac{\beta_0}{u} + \left[1 + \frac{A^2(t)}{u^2}\right] g_0(u), \tag{20.80}$$

where α_0, β_0 and $g_0(u)$ are to be determined. Since $h(u)$ is even, $h_0(iA) = h_0(-iA)$. Show that $\beta_0 = 0$, $\alpha_0 = h_0(iA)$ and

$$g_0(u) = u^2 \frac{h_0(u) - h_0(iA)}{u^2 - (iA)^2}. \tag{20.81}$$

Clearly, $g_0(u)$ is analytic for $u^2 \neq (iA)^2$ and has removable singularities at $u = \pm iA$. Near $u = \pm iA$,

$$g_0(u) = \frac{u^2}{u + (\pm iA)} \cdot \frac{h_0(u) - h_0(iA)}{u - (\pm iA)},$$

where either all "+" or all "−" signs are to be taken. From this, we conclude that $g_0(u)$ is analytic everywhere in the domain of $h_0(u)$. Now substitute (20.80) into (20.77), and show that

$$2^\alpha e^{-\nu t/2} L_n^{(\alpha)}(\nu t) = \frac{\alpha_0}{A^\alpha} J_\alpha(\nu A) - \frac{2}{\nu} \int_{-\infty}^{(0^+)} u^{-\alpha-1} h_1(u) e^{\nu[u - A^2(t)]/2} du, \tag{20.82}$$

where J_α is the Bessel function (see [Olver (1974a), p. 58]) and

$$h_1(u) = g_0'(u) - \frac{\alpha+1}{u} g_0(u).$$

We now digress to briefly discuss the asymptotic behavior of $z^{(n)}(u)$, for fixed t, as $u \to -\infty$, $n = 0, 1, 2, \cdots$. From (20.79), it is easily seen that $z(u) \sim u$ as $u \to -\infty$. Differentiating (20.79) with respect to u, we obtain

$$(1 + t\operatorname{csch}^2 z)z'(u) = 1 + \frac{A^2(t)}{u^2}, \tag{20.83}$$

which in turn yields $z'(u) = O(1)$ as $u \to -\infty$. Show that from (20.83), we also have, by Leibniz's rule,

$$\begin{aligned}(1 + t\operatorname{csch}^2 z)z^{(n+1)}(u) + \sum_{j=1}^{n} \binom{n}{j-1}(1 + t\operatorname{csch}^2 z)^{(n-j+1)} z^{(j)}(u) \\ =(-1)^n A^2(t)\frac{(n+1)!}{u^{n+1}}.\end{aligned} \tag{20.84}$$

Since every term in the above sum is exponentially decaying, show that (20.84) implies, by induction, that for fixed t,

$$\frac{d^{n+1}(z)}{du^{n+1}} = O(\frac{1}{u^{n+2}}), \quad n \geq 1, \tag{20.85}$$

as $u \to -\infty$.

To derive the form of an asymptotic expansion, we repeat the procedures indicated in (20.80) and (20.82). Thus, we define recursively

$$\begin{aligned} h_0(u) &= h(u), \\ h_n(u) &= \alpha_n + \frac{\beta_n}{u} + \left(1 + \frac{A^2}{u^2}\right) g_n(u), \\ h_{n+1}(u) &= g_n'(u) - \frac{\alpha+1}{u} g_n(u). \end{aligned}$$

Using induction, show that $h_n(u) = O(1)$, as $u \to -\infty$, for fixed t and for all $n \geq 0$. From this, show also that $g_n(u) = O(1)$, as $u \to -\infty$, for fixed t and for $n \geq$, $g_{2n}(u)$ and $h_{2n}(u)$ are even analytic functions, and that $g_{2n+1}(u)$ and $h_{2n+1}(u)$ are odd analytic functions. Consequently, we have

$$\alpha_{2n} = h_{2n}(iA), \quad \alpha_{2n+1} = 0$$

and

$$\beta_{2n} = 0, \quad \beta_{2n+1} = iAh_{2n+1}(iA).$$

Thus, for each n, both α_n and β_n are continuous in t for $-\infty < t < 1$. Show, by repeated application of integration by parts, that

$$\begin{aligned}2^\alpha e^{-\nu t/2} L_n^{(\alpha)}(\nu t) = &\frac{J_\alpha(\nu A)}{A^\alpha} \sum_{k=0}^{[(p-1)/2]} \alpha_{2k}(2/\nu)^{2k} \\ &- \frac{J_{\alpha+1}(\nu A)}{A^{\alpha+1}} \sum_{k=0}^{[p/2]-1} \beta_{2k+1}(2/\nu)^{2k+1} + \varepsilon_p. \quad (20.86)\end{aligned}$$

For an estimate of the error term ε_p, we refer to Frenzen and Wong (1988).

Chapter 21

Asymptotics of a Double Integral

21.1 Introduction

In this chapter, we consider double integrals of the form

$$I(\lambda;\alpha) = \iint_D g(x,y;\alpha)e^{i\lambda f(x,y;\alpha)}dxdy, \tag{21.1}$$

where λ is a large positive variable, α is an auxiliary parameter, D is a bounded domain in $\mathbf{R}^2$, f and g are C^∞ functions of (x,y) in D. These integrals arise frequently in diffraction theory; see, for instance, Wolf (1951); Bremmer (1955); Jones (1976). They also appear in asymptotic solutions of wave equations; see Poston and Stewart (1978, Chapter 12). When α is fixed, the behavior of $I(\lambda;\alpha)$, as $\lambda \to \infty$, has been studied extensively using a variety of methods (see, e.g., Jones and Kline (1958); Chako (1965); de Kok (1971)). In any case, it is now well-known that the main contributions to the asymptotic expansion of $I(\lambda;\alpha)$ come only from certain critical points. These include (i) stationary point of $f(x,y)$, i.e., points in D or on its boundary Γ which satisfy $f_x = f_y = 0$; (ii) points on the boundary at which a level curve of f is tangent to Γ; and (iii) points where Γ has a discontinuous turning tangent. Along these three kinds of critical points, the stationary points of $f(x,y)$ give the dominant behavior of $I(\lambda;\alpha)$. For a detailed discussion of this material, we refer to Wong (1989, Chapter VIII). If α is allowed to vary, then some of the stationary points may coalesce. As a consequence, the Hessian of $f(x,y)$ may vanish at these points, and the formula for the stationary phase approximation will fail to hold.

For a one-dimensional integral

$$I(\lambda;\alpha) = \int_a^b g(x;\alpha)e^{i\lambda f(x;\alpha)}dx, \tag{21.2}$$

asymptotic expansions that hold uniformly in α have been obtained by numerous investigators, including Chester *et al.* (1957); Bleistein (1966); Bleistein and Handelsman (1967); see also Wong (1989, Chapter VII). For double integrals of the form (21.1), there are only two papers dealing with this problem. The first one is by Fedoryuk (1964), who considered the case in which $f(x,y)$ has two stationary points which coalesce to an interior point of the domain D, as α approaches a critical value, say α_0. Fedoryuk, however, assumed that the Taylor expansion of $f(x,y;\alpha)$ is of the form

$$\begin{aligned} f(x,y;\alpha) =& f_0(\alpha) + \alpha(x-x_0)^2 + c_{02}(\alpha)(y-y_0)^2 \\ &+ \sum_{j+k\geq 3} c_{jk}(\alpha)(x-x_0)^j(y-y_0)^k \end{aligned} \tag{21.3}$$

or

$$\begin{aligned} f(x,y;\alpha) =& f_0(\alpha) + \alpha(x-x_0) + c_{02}(\alpha)(y-y_0)^2 \\ &+ \sum_{j+k\geq 3} c_{jk}(\alpha)(x-x_0)^j(y-y_0)^k, \end{aligned} \tag{21.4}$$

where α_0 was taken to be 0, $c_{02}(\alpha) \neq 0$ and the coefficient $c_{30}(\alpha)$ of $(x-x_0)^3$ also satisfied $c_{30}(0) \neq 0$. The other paper is by Bleistein and Handelsman (1969), who consider the case where the phase function $f(x,y;\alpha)$ has a stationary point in D which approaches the boundary Γ as $\alpha \to \alpha_0$.

In this chapter, we consider the following two cases: (1) $f(x,y;\alpha)$ has two interior stationary points $(x_\pm(\alpha), y_\pm(\alpha))$ which coalesce at an interior point $(x_0,y_0) \in D$ as $\alpha \to \alpha_0$; (2) $f(x,y;\alpha)$ has two interior stationary points $(x_\pm(\alpha), y_\pm(\alpha))$ which coalesce at a point $(x_0,y_0) \in \Gamma$ as $\alpha \to \alpha_0$, where Γ is the boundary of D. Note that we do not assume $f(x,y;\alpha)$ to be of the form (21.3) or (21.4). Our investigation was partly motivated by a paper of Chapman (1992) on propeller acoustics (see also Prentice (1994)).

For convenience, we shall assume that $f(x,y;\alpha)$ has only two stationary points $(x_\pm(\alpha), y_\pm(\alpha))$ in D, and that the boundary Γ of the domain is smooth. Let Δ_r denote the disk of radius r with its center at (x_0,y_0). Choose numbers δ and δ' such that $0 < \delta' < \delta$, and construct positive C^∞ functions $\mu(x,y)$ and $\nu(x,y)$ such that $\mu(x,y) = 1$ on $\Delta_{\delta'}$, $\mu(x,y) = 0$ outside Δ_δ, $\nu(x,y) = 0$ on $\Delta_{\delta'}$, $\nu(x,y) = 1$ outside Δ_δ, and

$$\mu(x,y) + \nu(x,y) = 1.$$

The set $\{\mu, \nu\}$ is called a partition of unity. For a construction of such a set, we refer to Wong (1989, p. 246). The integral in (21.1) can now be written as

$$\begin{aligned} I(\lambda;\alpha) =& \iint_{D\cap\Delta_\delta} g(x,y;\alpha)\mu(x,y)e^{i\lambda f(x,y;\alpha)}dxdy \\ &+ \iint_{D\setminus\Delta_{\delta'}} g(x,y;\alpha)\nu(x,y)e^{i\lambda f(x,y;\alpha)}dxdy \\ =& I_1(\lambda;\alpha) + I_2(\lambda;\alpha). \end{aligned} \tag{21.5}$$

If (x_0, y_0) is an interior point of D, then we shall take δ sufficiently small so that $\Delta_\delta \subset D$.

Without loss of generality, we may assume that α is so close to α_0 that both stationary points $(x_\pm(\alpha), y_\pm(\alpha))$ of $f(x,y;\alpha)$ lie inside $\Delta_{\delta'}$. Since $f(x,y;\alpha)$ has no stationary point in the interior of $D \subset \Delta_{\delta'}$, by using the divergence theorem, the second integral $I_2(\lambda;\alpha)$ can be reduced to a sum of one-dimensional integrals; see Wong (1989, p. 425, (2.9)). Since we know how to deal with one-dimensional integrals, we may consider only the first integral $I_1(\lambda;\alpha)$ in (21.5). For this reason, we shall simply write

$$I(\lambda;\alpha) = \iint_{D\cap\Delta_\delta} g(x,y;\alpha)e^{i\lambda f(x,y;\alpha)}dxdy. \tag{21.6}$$

where the amplitude function $g(x,y;\alpha)$ and all its derivatives vanish outside of Δ_δ.

The material presented in this chapter is taken from Qiu and Wong (2000). The material in the Exercise section is taken from Wong (1989, Chapter VIII).

21.2 Canonical form of the phase function

For simplicity, we shall write $f_\alpha(x,y) := f(x,y;\alpha)$. By assumption, $f_\alpha(x,y)$ has two simple stationary points $(x_\pm(\alpha), y_\pm(\alpha))$, which coalesce at (x_0, y_0) when α tends to α_0. Without loss of generality, we may take $\alpha_0 = 0$, $(x_0, y_0) = (0,0)$ and $f(0,0;\alpha) = 0$ for all small α. Under these assumptions, it is readily seen that $(0,0)$ is a degenerate stationary point of $f_0(x,y) = f(x,y;0)$; i.e., (21.8) holds. If we put

$$a_{ij} := \left.\frac{\partial^{i+j} f_0}{\partial x^i \partial y^j}\right|_{(0,0)}, \tag{21.7}$$

then the gradient of f_0 at $(0,0)$ is

$$\nabla f_0(0,0) = (a_{10}, a_{01}) = (0,0)$$

and the Hessian of f_0 at $(0,0)$ is

$$H_{f_0}(0,0) = a_{20}a_{02} - a_{11}^2 = 0. \tag{21.8}$$

The Taylor expansion of $f_0(x,y)$ at the origin is given by

$$f_0(x,y) = \frac{1}{2}(a_{20}x^2 + 2a_{11}xy + a_{02}y^2) + \frac{1}{6}(a_{30}x^3 + 3a_{21}x^2y + 3a_{12}xy^2 + a_{03}y^3) + \cdots. \tag{21.9}$$

In this chapter, a function (or mapping) is said to be *smooth* if it is a C^∞ function (or C^∞ mapping). All smooth functions (and mappings) here are defined in some neighborhood of the origin. We first recall the following result, known as the "splitting lemma" (see Poston and Stewart (1978, p. 95)).

Lemma 21.1. *Let $f(x,y;\alpha)$ be a smooth function of (x,y,α) defined in a neighborhood of the origin in $\mathbf{R}^2 \times \mathbf{R}$. Write $f_\alpha(x,y) = f(x,y;\alpha)$. Suppose that the Taylor expansion of $f_\alpha(x,y)$ at $(0,0)$ is of the form* (21.9)*, and that condition* (21.8) *holds with a_{20}, a_{11} and a_{02} not all zero. Then, for all small values of α, there exists a smooth and one-to-one mapping $(x,y) \to (s,t)$ given by $s = s_\alpha(x,y)$ and $t = t_\alpha(x,y)$ with $s_0(0,0) = t_0(0,0) = 0$, which depends also smoothly on α, such that*

$$f_\alpha(x,y) = \varepsilon s^2 + g_\alpha(t), \quad \varepsilon = \pm 1,$$

where $g_\alpha(t)$ is smooth in both t and α and $g_0(0) = 0$.

For our discussion of the case in which two stationary points of the phase function coalesce at a boundary point of the domain, we need a slightly different version of the splitting lemma.

Lemma 21.2. *Let $f_\alpha(x,y) = f(x,y;\alpha)$ be given as in Lemma 21.1. If the Taylor expansion of $f_0(x,y)$ at $(0,0)$ is of the form*

$$f_0(x,y) = \frac{1}{2}b_{20}x^2 + \frac{1}{6}(b_{30}x^3 + 3b_{21}x^2y + 3b_{12}xy^2 + b_{03}y^3) + \cdots, \tag{21.10}$$

where $b_{20} \neq 0$ and $b_{03} \neq 0$, then for all small α, there exists a smooth and one-to-one mapping $(x,y) \to (s,t)$ given by $s = s_\alpha(x,y)$ and $t = t_\alpha(x,y)$ with $s_0(0,0) = t_0(0,0) = 0$, which depends also smoothly on α, such that

$$f_\alpha(x,y) = \varepsilon s^2 + g_\alpha(t), \quad \varepsilon = \operatorname{sign} b_{20}, \tag{21.11}$$

where $g_\alpha(t)$ is smooth in both t and α and $g_0(t) = t^3/3$.

Note that in Lemma 21.2, the second variable t depends only on y, and not on x. Furthermore, we have also given the explicit form of the function $g_\alpha(t)$ when $\alpha = 0$, namely, $g_0(t) = t^3/3$. The proof of this result is just a modification of that for the splitting lemma given in Poston and Stewart (1978, p. 95). Hence, it will not be given here.

The next lemma is taken from Hörmander (1990, p. 204).

Lemma 21.3. *Let $g_\alpha(t) = g(t;\alpha)$ be a smooth function in a neighborhood of the origin in* $\mathbf{R}\times\mathbf{R}$ *which satisfies $g_0(t) = t^3/3$. Then there exists a smooth and one-to-one mapping $t \to v$ given by $v = v_\alpha(t)$ such that $v_0(t) = t$ and*

$$g_\alpha(t) = v^3/3 + \xi(\alpha)v + \eta(\alpha), \tag{21.12}$$

where $\xi(\alpha)$ and $\eta(\alpha)$ are smooth functions of α for all small values of α and $\xi(0) = \eta(0) = 0$.

From Lemma 21.2 and Lemma 21.3, we can see that if the phase function $f_\alpha(x,y)$, when $\alpha = 0$, has a Taylor expansion of the form (21.10) with $b_{20} \neq 0$ and $b_{03} \neq 0$, then $f_\alpha(x,y)$ can be mapped into the canonical form

$$f_\alpha(x,y) = \varepsilon u^2 + \frac{v^3}{3} + \xi(\alpha)v + \eta(\alpha), \tag{21.13}$$

where $\varepsilon = \operatorname{sign} b_{20}$ and $\xi(0) = \eta(0) = 0$. The mapping of the variables $(x,y) \to (u,v)$ is given by the equations of the form

$$u = u_\alpha(x,y), \quad v = v_\alpha(y). \tag{21.14}$$

We now turn to consider the phase function $f(x,y;\alpha)$, which has a Taylor expansion of the form (21.9) and satisfies (21.8) when $\alpha = 0$.

Theorem 21.1. *Let $f_\alpha(x,y) = f(x,y;\alpha)$ be a smooth function of (x,y,α) defined in a neighborhood of the origin of* $\mathbf{R}^2 \times \mathbf{R}$. *When $\alpha = 0$, $f_0(x,y)$ has a Taylor expansion of the form* (21.9) *and* (21.8) *holds. If $f_0(x,y)$ also satisfies the following two additional conditions:*

(A) the Hessian matrix of f_0 at $(0,0)$ has rank 1, i.e., not all a_{20}, a_{02} and a_{11} are zero;
(B) either

$$a_{03}a_{20}^2 - 3a_{12}a_{20}a_{11} + 3a_{21}a_{20}a_{02} - a_{30}a_{02}a_{11} \neq 0 \tag{21.15}$$

or

$$a_{30}a_{02}^2 - 3a_{21}a_{02}a_{11} + 3a_{12}a_{20}a_{02} - a_{03}a_{20}a_{11} \neq 0, \tag{21.16}$$

then for all small values of α, $f_\alpha(x,y)$ can be mapped into the canonical form

$$f_\alpha(x,y) = \varepsilon u^2 + \frac{v^3}{3} + \xi(\alpha)v + \eta(\alpha), \tag{21.17}$$

by a smooth and one-to-one mapping $(x, y) \to (u, v)$ *given by* $u = u_\alpha(x, y)$ *and* $v = v_\alpha(x, y)$*, where* $\varepsilon = \operatorname{sign} a_{20}$ *if* $a_{20} \neq 0$ *or* $\varepsilon = \operatorname{sign} a_{02}$ *if* $a_{02} \neq 0$*, and where* $\xi(\alpha)$ *and* $\eta(\alpha)$ *are smooth functions with* $\xi(0) = \eta(0) = 0$*. Moreover* $u_0(0, 0) = v_0(0, 0) = 0$*.*

Proof. As remarked earlier, if $f_0(x, y)$ has a Taylor expansion of the form (21.10) with $b_{20} \neq 0$ and $b_{03} \neq 0$, then by Lemmas 21.2 and 21.3, $f_\alpha(x, y)$ has the canonical form (21.17). So it is sufficient to prove that under the conditions of this theorem, $f_\alpha(x, y)$ can be mapped into a function F_α by a one-to-one and smooth transformation such that F_0 has a Taylor expansion of the form (21.10) with $b_{20} \neq 0$ and $b_{03} \neq 0$.

By (21.8) and condition (A), if $a_{11} = 0$ then one of a_{20} and a_{02} is not equal to 0; if $a_{11} \neq 0$ then both a_{20} and a_{02} are not equal to 0. By symmetry, we may assume without loss of generality that $a_{20} \neq 0$. Again by (21.8), the quadratic form of f_0 can be written as

$$Q_0(x, y) = \frac{a_{20}x^2 + 2a_{11}xy + a_{02}y^2}{2} = \frac{a_{20}}{2}\left(x + \frac{a_{11}}{a_{20}}y\right)^2.$$

Making the transformation $x_1 = x + (a_{11}/a_{20})y$ and $y_1 = y$, we obtain $Q_0(x, y) = a_{20}x_1^2/2$. After this transformation, let $F_\alpha(x_1, y_1) = f_\alpha(x, y)$. Then

$$F_0(x_1, y_1) = \frac{1}{2}a_{20}x_1^2 + \frac{1}{6}(b_{30}x_1^3 + 3b_{21}x_1^2y_1 + 3b_{12}x_1y_1^2 + b_{03}y_1^3) + \cdots,$$

where, by straightforward calculation, the coefficient of y_1^3 is given by

$$b_{03} = \frac{1}{a_{20}^2}\left(a_{03}a_{20}^2 - 3a_{12}a_{20}a_{11} + 3a_{21}a_{20}a_{02} - \frac{a_{30}a_{11}^2}{a_{20}}\right).$$

On account of (21.8), this can also be written as

$$b_{03} = \frac{1}{a_{20}^2}\left(a_{03}a_{20}^2 - 3a_{12}a_{20}a_{11} + 3a_{21}a_{20}a_{02} - a_{30}a_{02}a_{11}\right).$$

Hence, (21.15) is equivalent to $b_{03} \neq 0$. (If $a_{02} \neq 0$, then we can use (21.16) instead of (21.15); if both a_{20} and a_{02} are not zero, then we can use either (21.15) or (21.16).) This implies that $F_\alpha(x_1, y_1)$ satisfies the conditions of Lemma 21.2, and thus the proof is complete. □

Remark 21.1. Note that the additional conditions (A) and (B) in Theorem 21.1 are equivalent to $b_{20} \neq 0$ and $b_{03} \neq 0$ in Lemma 21.2. From the proofs of Lemmas 21.2 and 21.3, one can see that $b_{20} \neq 0$ and $b_{03} \neq 0$ are also the necessary conditions for f_α to have the canonical form (21.13). Therefore, our additional conditions in Theorem 21.1 are also necessary. Furthermore, we note that there are phase functions with two coalescing stationary points that cannot be mapped into the canonical form (21.17). For instance, $f_\alpha(x,y) = x^2 + y^5/5 - \alpha^2 y$ is such a function since $f_0(x,y) = x^2 + y^5/5$.

Now we turn to consider the case of two stationary points coalescing at a boundary point (x_0, y_0) of the domain D, and assume that the boundary point is located at the origin $(0,0)$. Let the boundary curve be given by $\Gamma : h(x,y) = 0$, where $h(x,y)$ is a smooth function defined in a neighborhood of the origin in $\mathbf{R}^2$. Then $h(0,0) = 0$ and $h(x,y)$ has the Taylor expansion

$$h(x,y) = h_{10}x + h_{01}y + \frac{h_{20}x^2 + 2h_{11}xy + h_{02}y^2}{2} + \cdots. \tag{21.18}$$

In view of the smoothness of Γ at $(0,0)$, we have $(h_{10}, h_{01}) \neq (0,0)$. Since a_{20}, a_{02} and a_{11} are not all zero and $a_{20}a_{02} = a_{11}^2 \geq 0$ by (21.8), we conclude that a_{20} and a_{02} must have the same sign if they are not zero. If $a_{11} \neq 0$, then both a_{20} and a_{02} are not zero. We set $\varepsilon = \operatorname{sign} a_{20} = \operatorname{sign} a_{02}$ and $\mu = \operatorname{sign}(\varepsilon a_{11})$; that is $\mu = 1$ if a_{11} has the same sign as a_{20}, and $\mu = -1$ if a_{11} has the opposite sign of a_{20}. If $a_{11} = 0$, then we define $\mu = 1$.

Lemma 21.4. *Let $f_\alpha(x,y)$ satisfy the conditions in Theorem 21.1. Let $h(x,y)$ be given as in* (21.18) *with $(h_{10}, h_{01}) \neq (0,0)$. Then the following conditions are equivalent:*

(1) With μ defined as above,

$$\mu|a_{02}|^{1/2}h_{10} - |a_{20}|^{1/2}h_{01} \neq 0. \tag{21.19}$$

(2) Let $(x(t), y(t))$ be a parametrized representation of Γ with $(x(0), y(0)) = (0,0)$ and $(x'(0), y'(0)) \neq (0,0)$; i.e., $h(x(t), y(t)) = 0$ for all small t. Then $t = 0$ is a simple stationary point of $f_0(x(t), y(t))$.

Moreover, the restriction of f_α to Γ, i.e., $f_\alpha(x(t), y(t))$, has a simple stationary point near $t = 0$ for sufficiently small α.

Proof. It is clear from (21.9) that $t = 0$ is always a stationary point of $f_0(x(t), y(t))$. Indeed, $f_0(x(0), y(0)) = f_0(0,0) = 0$ and

$$\left.\frac{d}{dt} f_0(x(t), y(t))\right|_{t=0} = \frac{\partial f_0}{\partial x}(0,0)x'(0) + \frac{\partial f_0}{\partial y}(0,0)y'(0) = 0.$$

So we need only to prove that (21.19) is equivalent to

$$\left.\frac{d^2}{dt^2} f_0(x(t), y(t))\right|_{t=0} \neq 0. \tag{21.20}$$

Direct calculation gives

$$\begin{aligned}\frac{d^2}{dt^2}f_0(x(t),y(t))\bigg|_{t=0} &= a_{20}x'(0)^2 + 2a_{11}x'(0)y'(0) + a_{02}y'(0)^2 \\ &= \varepsilon[|a_{20}|x'(0)^2 + 2\varepsilon a_{11}x'(0)y'(0) + |a_{02}|y'(0)^2].\end{aligned} \tag{21.21}$$

By virtue of (21.8), we can rewrite (21.21) as

$$\frac{d^2}{dt^2}f_0(x(t),y(t))\bigg|_{t=0} = \varepsilon[|a_{20}|^{1/2}x'(0) + \mu|a_{02}|^{1/2}y'(0)]^2.$$

Hence,

$$\frac{d^2}{dt^2}f_0(x(t),y(t))\bigg|_{t=0} = 0 \tag{21.22}$$

is equivalent to the vector $(x'(0),y'(0))$ and $(|a_{20}|^{1/2},\mu|a_{02}|^{1/2})$ being perpendicular, since both of these vectors are not equal to $(0,0)$.

On the other hand, $h(x(t),y(t)) = 0$ implies

$$h_{10}x'(0) + h_{01}y'(0) = 0.$$

That is, $(x'(0),y'(0))$ is perpendicular to (h_{10},h_{01}) since $(h_{10},h_{01}) \neq (0,0)$. From (21.22), it follows that (h_{10},h_{01}) is parallel to $(|a_{20}|^{1/2},\mu|a_{02}|^{1/2})$, or equivalently $|a_{20}|^{1/2}h_{01} = \mu|a_{02}|^{1/2}h_{10}$. This shows that (21.20) is equivalent to (21.19).

From (21.20) and the implicit function theorem, we see that the equation

$$\frac{d}{dt}f_\alpha(x(t),y(t)) = 0$$

has a unique solution $t(\alpha)$ for sufficiently small α and $t(0) = 0$. The solution $t(\alpha)$ is the simple stationary point of $f_\alpha(x(t),y(t))$. This completes the proof. □

From the canonical form (21.17), we can see that $f_0(x,y) = 0$ defines a curve with a cusp at $(0,0)$. Geometrically, condition (21.19) means that Γ is not tangent to this curve at $(0,0)$. In particular, if $f_0(x,y)$ is of the form (21.10), then (21.19) reduces to $h_{01} \neq 0$, i.e., Γ is not tangent to the y-axis at $(0,0)$.

Theorem 21.2. *Let $f_\alpha(x,y) = f(x,y;\alpha)$ satisfy the conditions of Theorem 21.1, and let Γ be a smooth curve defined by $h(x,y) = 0$, where $h(x,y)$ is a smooth function whose Taylor expansion at $(0,0)$ is of the form (21.18) with $(h_{10},h_{01}) \neq (0,0)$. If one of the conditions of Lemma 21.4 holds, then* *there exists a smooth and one-to-one mapping $u = u_\alpha(x, y)$, $v = v_\alpha(x, y)$ in a neighborhood of the origin in $\mathbf{R}^2$ such that $f_\alpha(x, y)$ is transformed into*

$$f_\alpha(x, y) = \varepsilon u^2 + \frac{v^3}{3} + \tilde{\xi}(\alpha)v + \tilde{\eta}(\alpha), \tag{21.23}$$

and Γ is mapped into a straight line defined by

$$v = \zeta(\alpha), \tag{21.24}$$

where ε is as in Theorem 21.1 and $\tilde{\xi}(\alpha)$, $\tilde{\eta}(\alpha)$ and $\zeta(\alpha)$ are all smooth functions with $\tilde{\xi}(0) = \tilde{\eta}(0) = \zeta(0) = 0$.

Proof. By Theorem 21.1, there exists a smooth and one-to-one change of variables $s = s_\alpha(x, y)$, $t = t_\alpha(x, y)$ with $s_0(0, 0) = t_0(0, 0) = 0$, such that $f_\alpha(x, y)$ is mapped into

$$f_\alpha(x, y) = \varepsilon s^2 + \frac{t^3}{3} + \xi(\alpha)t + \eta(\alpha) \tag{21.25}$$

for all small α. Here we have used variables (s, t), instead of (u, v), in Theorem 21.1. This change of variables also maps $h(x, y)$ into

$$h(x, y) = \tilde{h}_\alpha(x, y) = \tilde{h}_{00}(\alpha) + \tilde{h}_{10}(\alpha)s + \tilde{h}_{01}(\alpha)t + \cdots, \tag{21.26}$$

where $\tilde{h}_{ij}(\alpha)$ are all smooth functions of small α. Since $s_0(0, 0) = t_0(0, 0) = 0$, we have $\tilde{h}_{00}(0) = 0$. To prove that $\tilde{h}_{01}(\alpha) \neq 0$ for all small α, it suffices to show that $\tilde{h}_{01}(0) \neq 0$.

To this end, let

$$J = \begin{pmatrix} j_{11} & j_{12} \\ j_{21} & j_{22} \end{pmatrix}$$

be the Jacobian matrix of the mapping $s = s_0(x, y)$, $t = t_0(x, y)$ at $(0, 0)$. Then we have

$$\begin{aligned} s =& s_0(x, y) = j_{11}x + j_{12}y + \cdots, \\ t =& t_0(x, y) = j_{21}x + j_{22}y + \cdots. \end{aligned} \tag{21.27}$$

Substituting (21.27) into (21.25) with $\alpha = 0$, we obtain

$$f_0(x, y) = \varepsilon j_{11}^2 x^2 + 2\varepsilon j_{11} j_{12} xy + \varepsilon j_{12}^2 y^2 + \cdots.$$

Comparing it with (21.9) gives

$$\frac{a_{20}}{2} = \varepsilon j_{11}^2, \quad \frac{a_{02}}{2} = \varepsilon j_{12}^2, \quad a_{11} = 2\varepsilon j_{11} j_{12}. \tag{21.28}$$

The last equation in (21.28) shows that $\operatorname{sign}(j_{11} j_{12}) = \operatorname{sign}(\varepsilon a_{11}) = \mu$. So we have

$$j_{11} = \pm\sqrt{\frac{|a_{20}|}{2}}, \quad j_{12} = \pm\mu\sqrt{\frac{|a_{02}|}{2}}. \tag{21.29}$$

Substituting (21.27) into (21.26) with $\alpha = 0$, we also obtain

$$\begin{aligned}\tilde{h}_{10}(0)j_{11} + \tilde{h}_{01}(0)j_{21} &= h_{10},\\ \tilde{h}_{10}(0)j_{12} + \tilde{h}_{01}(0)j_{22} &= h_{01}.\end{aligned}$$

Solving these equations gives

$$\tilde{h}_{01}(0) = (j_{11}h_{01} - j_{12}h_{10})/|J|. \tag{21.30}$$

Inserting (21.29) into (21.30) yields

$$\tilde{h}_{01}(0) = \pm\frac{|a_{20}|^{1/2}h_{01} - \mu|a_{02}|^{1/2}h_{10}}{\sqrt{2}|J|}.$$

Thus, for all small α, $\tilde{h}_{01}(\alpha) \neq 0$ on account of (21.19).

Now (21.26) can be written as

$$h(x,y) = \tilde{h}_{00}(\alpha) + \tilde{h}_{01}(\alpha)\left[t + \frac{\tilde{h}_{10}(\alpha)}{\tilde{h}_{01}(\alpha)}s + \tilde{r}(s,t;\alpha)\right], \tag{21.31}$$

where $\tilde{r}(s,t;\alpha)$ is a smooth function of small (s,t,α) and satisfies

$$\tilde{r}(0,0;\alpha) = \frac{\partial}{\partial s}\tilde{r}(0,0;\alpha) = \frac{\partial}{\partial t}\tilde{r}(0,0;\alpha) = 0.$$

We make the change of variables

$$s_1 = s, \quad t_1 = t + \frac{\tilde{h}_{10}(\alpha)}{\tilde{h}_{01}(\alpha)}s + \tilde{r}(s,t;\alpha).$$

This is a one-to-one transformation, and its inverse mapping is given by

$$s = s_1, \quad t = t_1 - \frac{\tilde{h}_{10}(\alpha)}{\tilde{h}_{01}(\alpha)}s_1 + \tilde{r}_1(s_1,t_1;\alpha), \tag{21.32}$$

where $\tilde{r}_1(s_1,t_1;\alpha)$ also satisfies

$$\tilde{r}_1(0,0;\alpha) = \frac{\partial}{\partial s_1}\tilde{r}_1(0,0;\alpha) = \frac{\partial}{\partial t_1}\tilde{r}_1(0,0;\alpha) = 0. \tag{21.33}$$

Substituting (21.32) into (21.25) and (21.31) gives

$$h(x,y) = \tilde{h}_{00}(\alpha) + \tilde{h}_{01}(\alpha)t_1, \tag{21.34}$$

and

$$\begin{aligned}f_\alpha(x,y) =& \varepsilon s_1^2 + \frac{t_1^3}{3} - \frac{\tilde{h}_{10}(\alpha)}{\tilde{h}_{01}(\alpha)}s_1t_1^2 + \frac{\tilde{h}_{10}^2(\alpha)}{\tilde{h}_{01}^2(\alpha)}s_1^2t_1 - \frac{\tilde{h}_{10}^3(\alpha)s_1^3}{3\tilde{h}_{01}^3(\alpha)}\\ &+ \left(t_1 - \frac{\tilde{h}_{10}(\alpha)}{\tilde{h}_{01}(\alpha)}s_1\right)^2\tilde{r}_1(s_1,t_1,\alpha)\\ &+ \left(t_1 - \frac{\tilde{h}_{10}(\alpha)}{\tilde{h}_{01}(\alpha)}s_1\right)\tilde{r}_1(s_1,t_1,\alpha)^2 + \frac{\tilde{r}_1(s_1,t_1,\alpha)^3}{3}\\ &+ \xi(\alpha)\left(t_1 - \frac{\tilde{h}_{10}(\alpha)}{\tilde{h}_{01}(\alpha)}s_1 + \tilde{r}_1(s_1,t_1,\alpha)\right) + \eta(\alpha).\end{aligned} \tag{21.35}$$

Let us denote the function on the right-hand side of (21.35) by $\tilde{f}_\alpha(s_1,t_1)$. Then, since $\xi(0)=\eta(0)=0$, by (21.33) one can show that $\tilde{f}_\alpha(s_1,t_1)$ satisfies the conditions of Lemma 21.2, that is, the Taylor expansion of $\tilde{f}_0(s_1,t_1)$ at $(0,0)$ is of the form (21.10) with $b_{20}=2\varepsilon\neq 0$ and $b_{03}=2\neq 0$. By Lemmas 21.2 and 21.3, there exists a one-to-one change of variables $(s_1,t_1)\to(u,v)$ given by

$$u=u_\alpha(s_1,t_1),\quad v=v_\alpha(t_1) \tag{21.36}$$

with $u_0(0,0)=v_0(0)=0$ (see the statement following Lemma 21.3). This change of variables transforms $f_\alpha(x,y)=\tilde{f}_\alpha(s_1,t_1)$ into

$$f_\alpha(x,y)=\varepsilon u^2+\frac{v^3}{3}+\tilde{\xi}(\alpha)v+\tilde{\eta}(\alpha), \tag{21.37}$$

where $\tilde{\xi}(\alpha)$ and $\tilde{\eta}(\alpha)$ are smooth functions and $\tilde{\xi}(0)=\tilde{\eta}(0)=0$. Let us denote the inverse of the function v_α in the second equation in (21.36) by $t_1=\tilde{\tau}_\alpha(v)$. Then (21.34) becomes

$$h(x,y)=\tilde{h}_{00}(\alpha)+\tilde{h}_{01}(\alpha)\tilde{\tau}_\alpha(v),$$

and the equation $h(x,y)=0$ uniquely determines a function $v=\zeta(\alpha)$ by the implicit function theorem. This shows that the transformation (21.36) maps Γ into the line $v=\zeta(\alpha)$. Since $\tilde{h}_{00}(0)=0$ and $\tilde{\tau}_0(0)=0$, we obtain $\zeta(0)=0$. This completes the proof of the theorem. □

Note that the mapping $(x,y)\to(u,v)$ in (21.17) and (21.23) are one-to-one. Hence stationary points are mapped into stationary points. Using this fact, one can determine the coefficients ξ, η in (21.17) and $\tilde{\xi}$, $\tilde{\eta}$ in (21.23) by the values of $f_\alpha(x,y)$ at the stationary points (see (21.44) and (21.45) in Section 21.3). So, in fact, we have $\tilde{\xi}=\xi$ and $\tilde{\eta}=\eta$.

21.3 Derivation of asymptotic expansions: coalescence at an interior point

We now turn to the problem of deriving the asymptotic expansion of the integral (21.5). In this section, we consider only the case in which the phase function has two simple stationary points $(x_\pm(\alpha),y_\pm(\alpha))$ which coalesce at an interior point (x_0,y_0) of the domain D as α tends to α_0. In this case, we shall take δ so small that $\Delta_\delta\subset D$ and

$$I(\lambda;\alpha)=\iint_{\Delta_\delta} g(x,y;\alpha)e^{i\lambda f(x,y;\alpha)}dxdy. \tag{21.38}$$

Furthermore, by setting $g(x,y;\alpha) = 0$ outside Δ_δ, we can extend $g(x,y;\alpha)$ to a C^∞ function in the whole plane $\mathbf{R}^2$, which is supported on Δ_δ. Thus we have

$$I(\lambda;\alpha) = \iint_{\mathbf{R}^2} g(x,y;\alpha)e^{i\lambda f(x,y;\alpha)}dxdy. \tag{21.39}$$

We have already assumed that the phase function $f_\alpha(x,y) = f(x,y;\alpha)$ in (21.39) satisfies all conditions in Theorem 21.1. Hence, by Theorem 21.1, $f_\alpha(x,y)$ can be mapped into the canonical form

$$F_\alpha(u,v) = F(u,v;\alpha) = \varepsilon u^2 + \frac{v^3}{3} + \xi(\alpha)v + \eta(\alpha) \tag{21.40}$$

in Δ_δ if δ is taken sufficiently small. The integral (21.38) can therefore be written as

$$\begin{aligned} I(\lambda;\alpha) &= \iint_{\mathbf{R}^2} \varphi_0(u,v;\alpha)e^{i\lambda F(u,v;\alpha)}dudv \\ &= e^{i\lambda\eta(\alpha)} \iint_{\mathbf{R}^2} \varphi_0(u,v;\alpha)e^{i\lambda[\varepsilon u^2+v^3/3+\xi(\alpha)v]}dudv, \end{aligned} \tag{21.41}$$

where

$$\varphi_0(u,v;\alpha) = g(x,y;\alpha)\left|\frac{\partial(x,y)}{\partial(u,v)}\right| \tag{21.42}$$

and $|\partial(x,y)/\partial(u,v)|$ is the Jacobian of the mapping $(x,y) \to (u,v)$. Note that $\varphi_0(u,v;\alpha)$ is supported in a neighborhood $\tilde{\Delta}$ of the origin in $\mathbf{R}^2$, which is in fact the image of Δ_δ.

First, we shall determine the functions $\xi(\alpha)$ and $\eta(\alpha)$ in (21.41). By hypothesis, when α is sufficiently close to α_0, $f_\alpha(x,y)$ has two simple stationary points $(x_\pm(\alpha), y_\pm(\alpha))$ in Δ_δ. From the canonical form of $f_\alpha(x,y)$, we can see that one of the stationary points is a local extremum and the other is a saddle point. Note that the function $F(u,v;\alpha)$ also has two simple stationary points, namely, $(0, \pm\sqrt{-\xi(\alpha)})$, where $\xi(\alpha) < 0$. It is readily verified that for the mapping $(x,y) \to (u,v)$ to be one-to-one, stationary points must be mapped into stationary points. Furthermore, the correspondence between the stationary points must be carried out according to their classifications. For instance, if $(x_+(\alpha), y_+(\alpha))$ is a local extremum and $(x_-(\alpha), y_-(\alpha))$ is a saddle and also $\varepsilon = 1$, then $(x_+(\alpha), y_+(\alpha))$ must correspond to $(0, \sqrt{-\xi(\alpha)})$, and $(x_-(\alpha), y_-(\alpha))$ must correspond to $(0, -\sqrt{-\xi(\alpha)})$. As a result, from

$$f_\alpha(x,y) = F_\alpha(u,v), \tag{21.43}$$

we obtain (exercise)

$$\xi(\alpha) = -\left[\frac{3}{4}(f_\alpha(x_-(\alpha), y_-(\alpha)) - f_\alpha(x_+(\alpha), y_+(\alpha)))\right]^{2/3}, \tag{21.44}$$

$$\eta(\alpha) = \frac{1}{2}[f_\alpha(x_-(\alpha), y_-(\alpha)) + f_\alpha(x_+(\alpha), y_+(\alpha))]. \tag{21.45}$$

Next, we shall find the values of the Jacobian $|\partial(x, y)/\partial(u, v)|$ at the stationary points $(0, \pm\sqrt{-\xi(\alpha)})$. Differentiating both sides of (21.43) twice with respect to u and v, and evaluating the resulting expressions at $(u, v) = (0, \pm\sqrt{-\xi(\alpha)})$, we obtain

$$\begin{aligned}
\frac{\partial^2 f_\alpha}{\partial x^2}\left(\frac{\partial x}{\partial u}\right)^2 + 2\frac{\partial^2 f_\alpha}{\partial x \partial y}\frac{\partial x}{\partial u}\frac{\partial y}{\partial u} + \frac{\partial^2 f_\alpha}{\partial y^2}\left(\frac{\partial y}{\partial u}\right)^2 &= 2\varepsilon,\\
\frac{\partial^2 f_\alpha}{\partial x^2}\frac{\partial x}{\partial u}\frac{\partial x}{\partial v} + \frac{\partial^2 f_\alpha}{\partial x \partial y}\left(\frac{\partial x}{\partial u}\frac{\partial y}{\partial v} + \frac{\partial x}{\partial v}\frac{\partial y}{\partial u}\right) + \frac{\partial^2 f_\alpha}{\partial y^2}\frac{\partial y}{\partial u}\frac{\partial y}{\partial v} &= 0,\\
\frac{\partial^2 f_\alpha}{\partial x^2}\left(\frac{\partial x}{\partial v}\right)^2 + 2\frac{\partial^2 f_\alpha}{\partial x \partial y}\frac{\partial x}{\partial v}\frac{\partial y}{\partial v} + \frac{\partial^2 f_\alpha}{\partial y^2}\left(\frac{\partial y}{\partial v}\right)^2 &= \pm 2\sqrt{-\xi(\alpha)}.
\end{aligned}$$

Writing these equations in matrix form gives

$$\begin{pmatrix} \frac{\partial x}{\partial u} & \frac{\partial y}{\partial u} \\ \frac{\partial x}{\partial v} & \frac{\partial y}{\partial v} \end{pmatrix} \begin{pmatrix} \frac{\partial^2 f_\alpha}{\partial x^2} & \frac{\partial^2 f_\alpha}{\partial x \partial y} \\ \frac{\partial^2 f_\alpha}{\partial x \partial y} & \frac{\partial^2 f_\alpha}{\partial y^2} \end{pmatrix} \begin{pmatrix} \frac{\partial x}{\partial u} & \frac{\partial x}{\partial v} \\ \frac{\partial y}{\partial u} & \frac{\partial y}{\partial v} \end{pmatrix} = \begin{pmatrix} 2\varepsilon & 0 \\ 0 & \pm 2\sqrt{-\xi(\alpha)} \end{pmatrix}.$$

Hence we have (exercise)

$$\left|\frac{\partial(x, y)}{\partial(u, v)}\right|^2_{(0, \pm\sqrt{-\xi(\alpha)})} = \frac{H_{f_\alpha}(0, \pm\sqrt{-\xi(\alpha)})}{H_{f_\alpha}(x_\pm(\alpha), y_\pm(\alpha))} = \frac{\pm 4\varepsilon\sqrt{-\xi(\alpha)}}{H_{f_\alpha}(x_\pm(\alpha), y_\pm(\alpha))}, \tag{21.46}$$

where $H_f(x, y)$ denotes the Hessian of the function f at (x, y). If $\varepsilon = 1$, then $(x_+(\alpha), y_+(\alpha))$ is a minimum and $(x_-(\alpha), y_-(\alpha))$ is a saddle point. So $H_{f_\alpha}(x_+(\alpha), y_+(\alpha)) > 0$ and $H_{f_\alpha}(x_-(\alpha), y_-(\alpha)) < 0$. If $\varepsilon = -1$, then $(x_+(\alpha), y_+(\alpha))$ is a saddle point and $(x_-(\alpha), y_-(\alpha))$ is a maximum. So $H_{f_\alpha}(x_+(\alpha), y_+(\alpha)) < 0$ and $H_{f_\alpha}(x_-(\alpha), y_-(\alpha)) > 0$. In both cases, we have

$$\pm\varepsilon H_{f_\alpha}(x_\pm(\alpha), y_\pm(\alpha)) = |H_{f_\alpha}(x_\pm(\alpha), y_\pm(\alpha))|. \tag{21.47}$$

Finally, we shall derive the asymptotic expansion of the integral

$$J(\lambda; \alpha) = \iint_{\mathbf{R}^2} \varphi_0(u, v; \alpha) e^{i\lambda[\varepsilon u^2 + v^3/3 + \xi v]} du dv, \tag{21.48}$$

in (21.41). Since $\varphi_0(u,v;\alpha)$ is a smooth function, the quotient

$$\psi_{01}(u,v;\alpha) = \frac{\varphi_0(u,v;\alpha) - \varphi_0(0,v;\alpha)}{2u\varepsilon} \tag{21.49}$$

is also a smooth function even at $u = 0$. For v and α small, we may write

$$\varphi_0(0,v;\alpha) = (v^2 + \xi(\alpha))\psi_{02}(v;\alpha) + a_0(\alpha) + b_0(\alpha)v, \tag{21.50}$$

where a_0, b_0 and ψ_{02} are all smooth functions. Putting $v = \pm\sqrt{-\xi(\alpha)}$ in (21.50), we obtain

$$\varphi_0(0, \pm\sqrt{-\xi(\alpha)};\alpha) = a_0(\alpha) \pm \sqrt{-\xi(\alpha)}b_0(\alpha),$$

which in turn gives

$$a_0(\alpha) = \frac{\varphi_0(0,\sqrt{-\xi(\alpha)};\alpha) + \varphi_0(0,-\sqrt{-\xi(\alpha)};\alpha)}{2}, \tag{21.51}$$

$$b_0(\alpha) = \frac{\varphi_0(0,\sqrt{-\xi(\alpha)};\alpha) - \varphi_0(0,-\sqrt{-\xi(\alpha)};\alpha)}{2\sqrt{-\xi(\alpha)}}. \tag{21.52}$$

Clearly, $a_0(0) = \varphi_0(0,0;0)$ and $b_0(0) = (\partial\varphi_0/\partial v)(0,0;0)$. With $a_0(\alpha)$ and $b_0(\alpha)$ chosen, we can rewrite (21.50) as

$$\psi_{02}(v;\alpha) = \frac{\varphi_0(0,v;\alpha) - a_0(\alpha) - b_0(\alpha)v}{v^2 + \xi(\alpha)}.$$

This shows that $\psi_{02}(v;\alpha)$ can be smoothly extended to all of $\mathbf{R}$. Coupling (21.49) and (21.50), we obtain

$$\varphi_0(u,v;\alpha) = 2u\varepsilon\psi_{01}(u,v;\alpha) + (v^2 + \xi(\alpha))\psi_{02}(v;\alpha) + a_0(\alpha) + b_0(\alpha)v. \tag{21.53}$$

If we use the notation

$$\Psi_0(u,v;\alpha) := (\psi_{01}(u,v;\alpha), \psi_{02}(v;\alpha))$$

and use the gradient operator $\nabla = (\partial/\partial u, \partial/\partial v)$, then (21.53) can be written as

$$\varphi_0 = \Psi_0 \cdot \nabla(F - \eta) + a_0 + b_0 v; \tag{21.54}$$

see (21.40). Here and in what follows, we shall often suppress the independent variable in order to simplify notations. Let us put $h(\lambda;u,v,\alpha) = e^{i\lambda(\varepsilon u^2 + v^3/3 + \xi v)}$. Clearly, $\nabla h = i\lambda h \nabla(F - \eta)$. From (21.54), it follows that

$$\varphi_0 h = \frac{1}{i\lambda}\Psi_0 \cdot \nabla h + (a_0 + b_0 v)h. \tag{21.55}$$

Note that

$$\nabla \cdot (\Psi_0 h) = \Psi_0 \cdot \nabla h + (\nabla \cdot \Psi_0)h.$$

Coupling this with (21.55) gives

$$\varphi_0 h = \frac{1}{i\lambda}\nabla\cdot(\Psi_0 h) - \frac{1}{i\lambda}(\nabla\cdot\Psi_0)h + (a_0 + b_0 v)h. \tag{21.56}$$

Substituting (21.56) into (21.48) with $h = e^{i\lambda(F-\eta)}$, we obtain

$$\begin{aligned} J(\lambda;\alpha) = &\iint_{\mathbf{R}^2}(a_0 + b_0 v)h\,du\,dv + \frac{1}{i\lambda}\iint_{\mathbf{R}^2}\nabla\cdot(\Psi_0 h)\,du\,dv \\ &- \frac{1}{i\lambda}\iint_{\mathbf{R}^2}(\nabla\cdot\Psi_0)h\,du\,dv =: J_1 + J_2 + J_3. \end{aligned} \tag{21.57}$$

The first integral on the right-hand side of (21.57) can be evaluated explicitly in terms of the Airy function $\mathrm{Ai}(x)$, and the result is (see Olver (1974a, p. 53))

$$\begin{aligned} J_1 &= \iint_{\mathbf{R}^2}(a_0 + b_0 v)e^{i\lambda(\varepsilon u^2 + v^3/3 + \xi v)}du\,dv \\ &= \sqrt{\frac{\pi}{\lambda}}e^{i\varepsilon\pi/4}\left[a_0\int_{\mathbf{R}} e^{i\lambda(v^3/3+\xi v)}dv + b_0\int_{\mathbf{R}} v e^{i\lambda(v^3/3+\xi v)}dv\right] \\ &= \frac{2\pi^{3/2}e^{i\varepsilon\pi/4}}{\lambda^{1/2}}\left[\frac{a_0}{\lambda^{1/3}}\mathrm{Ai}(\lambda^{2/3}\xi) + \frac{b_0}{i\lambda^{2/3}}\mathrm{Ai}'(\lambda^{2/3}\xi)\right]. \end{aligned} \tag{21.58}$$

By Green's theorem, we have

$$\begin{aligned} J_2 &= \frac{1}{i\lambda}\lim_{R\to\infty}\int_{\partial D_R}\psi_{01}h\,dv - \psi_{02}h\,du \\ &= \frac{1}{i\lambda}\lim_{R\to\infty}\int_{-R}^{R}[\psi_{01}(R,v;\alpha) - \psi_{01}(-R,v;\alpha)]e^{i\lambda\varepsilon R^2 + i\lambda(v^3/3+\xi v)}dv \\ &\quad + \frac{1}{i\lambda}\lim_{R\to\infty}\int_{-R}^{R}[\psi_{02}(-R;\alpha) - \psi_{02}(R;\alpha)]e^{i\lambda\varepsilon u^2 + i\lambda(R^3/3+\xi R)}du, \end{aligned} \tag{21.59}$$

where D_R is the rectangle $\{(u,v)\in\mathbf{R}^2 : -R \le u, v \le R\}$ and ∂D_R denotes its boundary. To estimate the integral J_2, we need to know the behavior of ψ_{01} and ψ_{02} as $u, v \to \pm\infty$. Since $\varphi_0(u,v;\alpha)$ has compact support in $\mathbf{R}^2$, $\varphi_0(u,v;\alpha) = 0$ when $|u| \ge R_0$ or $|v| \ge R_0$ for some fixed R_0. Hence when $|v| \ge R_0$,

$$\psi_{01}(u,v;\alpha) = \frac{\varphi_0(u,v;\alpha) - \varphi_0(0,v;\alpha)}{2u\varepsilon} = 0, \tag{21.60}$$

$$\psi_{02}(v;\alpha) = \frac{\varphi_0(0,v;\alpha) - a_0(\alpha) - b_0(\alpha)v}{v^2 + \xi(\alpha)} = -\frac{a_0 + b_0 v}{v^2 + \xi}. \tag{21.61}$$

Therefore,

$$|J_2| \le \frac{1}{\lambda} \lim_{R\to\infty} \left[\frac{1}{2R} \int_{-R_0}^{R_0} |\varphi_0(R, v; \alpha) - \varphi_0(-R, v; \alpha)| dv \right.$$
$$\left. + \frac{2(|a_0| + |b_0| R)}{R^2 + |\xi|} \int_{-R}^{R} \left| e^{i\lambda\varepsilon u^2} \right| du \right] = 0.$$

Let $\varphi_1(u, v; \alpha) = -\nabla \cdot \Psi_0$ and note that

$$\varphi_1(u, v; \alpha) = -\frac{\partial \psi_{01}}{\partial u} - \frac{\partial \psi_{02}}{\partial v}. \tag{21.62}$$

When $|v| \ge R_0$, we have from (21.60) and (21.61)

$$\frac{\partial \psi_{01}}{\partial u} = 0, \quad \frac{\partial \psi_{02}}{\partial v} = O(v^{-2}).$$

Both these quantities are independent of the variable u. Hence when $|v| \ge R_0$, $\varphi_1(u, v; \alpha)$ is independent of u, and is $O(v^{-2})$. These certainly imply that the integral

$$J_3 = \frac{1}{i\lambda} \iint_{\mathbf{R}^2} \varphi_1(u, v; \alpha) e^{i\lambda(\varepsilon u^2 + v^3/3 + \xi v)} du dv \tag{21.63}$$

is convergent. Summarizing the above results, we have the first approximation

$$J(\lambda; \alpha) = \frac{2\pi^{3/2} e^{i\varepsilon\pi/4}}{\lambda^{5/6}} \left[a_0 \operatorname{Ai}(\lambda^{2/3}\xi) + \frac{b_0}{i\lambda^{1/3}} \operatorname{Ai}'(\lambda^{2/3}\xi) \right]$$
$$+ \frac{1}{i\lambda} \iint_{\mathbf{R}^2} \varphi_1(u, v; \alpha) e^{i\lambda(\varepsilon u^2 + v^3/3 + \xi v)} du dv. \tag{21.64}$$

Note that the last integral is exactly of the same form as that in (21.48). Hence, the above procedure can be repeated. Define inductively

$$a_k(\alpha) = \frac{\varphi_k(0, \sqrt{-\xi(\alpha)}; \alpha) + \varphi_k(0, -\sqrt{-\xi(\alpha)}; \alpha)}{2}, \tag{21.65}$$

$$b_k(\alpha) = \frac{\varphi_k(0, \sqrt{-\xi(\alpha)}; \alpha) - \varphi_k(0, -\sqrt{-\xi(\alpha)}; \alpha)}{2\sqrt{-\xi(\alpha)}}, \tag{21.66}$$

$$\psi_{k1}(u, v; \alpha) = \frac{\varphi_k(u, v; \alpha) - \varphi_k(0, v; \alpha)}{2u\varepsilon}, \tag{21.67}$$

$$\psi_{k2}(v; \alpha) = \frac{\varphi_k(0, v; \alpha) - a_k(\alpha) - b_k(\alpha) v}{v^2 + \xi(\alpha)}, \tag{21.68}$$

$$\varphi_{k+1}(u, v; \alpha) = -\frac{\partial \psi_{k1}}{\partial u}(u, v; \alpha) - \frac{\partial \psi_{k2}}{\partial v}(v; \alpha). \tag{21.69}$$

We obtain

$$J(\lambda;\alpha) = \frac{2\pi^{3/2}e^{i\varepsilon\pi/4}}{\lambda^{5/6}} \sum_{k=0}^{n-1} \frac{1}{(i\lambda)^k}\left[a_k(\alpha)\operatorname{Ai}(\lambda^{2/3}\xi) + \frac{b_k(\alpha)}{i\lambda^{1/3}}\operatorname{Ai}'(\lambda^{2/3}\xi)\right] + J_n(\lambda;\alpha), \tag{21.70}$$

where

$$J_n(\lambda;\alpha) = \frac{1}{(i\lambda)^n}\iint_{\mathbf{R}^2} \varphi_n(u,v;\alpha)e^{i\lambda(\varepsilon u^2+v^3/3+\xi v)}dudv. \tag{21.71}$$

Here use has been made of the fact that $\varphi_k(u,v;\alpha)$ is independent of u when $|v| \geq R_0$, and hence also that $\psi_{k1}(u,v;\alpha) = 0$ when $|v| \geq R_0$ and $\partial\psi_{k2}(v;\alpha)/\partial v = O(v^{-2})$.

For (21.70) to be a compound uniform asymptotic expansion, we must prove that there exist positive constants C_n and D_n, independent of λ and α such that the integral (21.71) satisfies

$$|J_n(\lambda;\alpha)| \leq \frac{C_n}{\lambda^{n+5/6}}|\operatorname{Ai}(\lambda^{2/3}\xi)| + \frac{D_n}{\lambda^{n+7/6}}|\operatorname{Ai}'(\lambda^{2/3}\xi)|. \tag{21.72}$$

By using the recursive formula (21.65)-(21.69) one more time, we have

$$J_n(\lambda;\alpha) = \frac{2\pi^{3/2}e^{i\varepsilon\pi/4}}{(i\lambda)^n\lambda^{5/6}}\left[a_n\operatorname{Ai}(\lambda^{2/3}\xi) + \frac{b_n}{i\lambda^{1/3}}\operatorname{Ai}'(\lambda^{2/3}\xi)\right] + \frac{1}{(i\lambda)^{n+1}}\iint_{\mathbf{R}^2}\varphi_{n+1}(u,v;\alpha)e^{i\lambda(\varepsilon u^2+v^3/3+\xi v)}dudv. \tag{21.73}$$

It is sufficient to show that the integral

$$J^* = \frac{1}{i\lambda}\iint_{\mathbf{R}^2}\varphi_{n+1}(u,v;\alpha)e^{i\lambda(\varepsilon u^2+v^3/3+\xi v)}dudv \tag{21.74}$$

satisfies

$$|J^*| \leq \frac{M_n}{\lambda^{5/6}}|\operatorname{Ai}(\lambda^{2/3}\xi)| + \frac{N_n}{\lambda^{7/6}}|\operatorname{Ai}'(\lambda^{2/3}\xi)| \tag{21.75}$$

for some constants M_n and N_n independent of λ and α.

Without loss of generality, we first consider the case $\lambda \to \infty$ and $\varepsilon = 1$. If $\lambda^{2/3}\xi$ is bounded, say $|\lambda^{2/3}\xi| \leq \rho$, then the change of the variables $s = \lambda^{1/2}u$ and $t = \lambda^{1/3}v$ give

$$J^* = \frac{1}{i\lambda^{11/6}}\iint_{\mathbf{R}^2}\varphi_{n+1}(\lambda^{-1/2}s,\lambda^{-1/3}t;\alpha)e^{i(s^2+t^3/3+\lambda^{2/3}\xi t)}dsdt. \tag{21.76}$$

Since φ_{n+1} is smooth and bounded in $\mathbf{R}^2$, it follows that the integral on the right-hand side of (21.76) has a bound K_{n+1} independent of λ and α such that

$$|J^*| \leq \frac{K_{n+1}}{\lambda^{11/6}}. \tag{21.77}$$

Furthermore, since the zeros of $\mathrm{Ai}(z)$ and $\mathrm{Ai}'(z)$ do not coincide, (21.75) follows immediately from (21.77).

If $\lambda^{2/3}\xi$ is unbounded, say $|\lambda^{2/3}\xi| \geq \rho > 0$, then $\lambda^{2/3}(-\xi) \geq \rho$ since $\xi < 0$. Making the change of variables $s = (-\xi)^{-3/4}u$ and $t = (-\xi)^{-1/2}v$ gives

$$J^* = \frac{(-\xi)^{5/4}}{i\lambda} \iint_{\mathbf{R}^2} \varphi_{n+1}((-\xi)^{3/4}s, (-\xi)^{1/2}t; \alpha) e^{i\lambda(-\xi)^{3/2}(s^2+t^3/3-t)} dsdt. \tag{21.78}$$

The phase function in the last integral has two fixed stationary points, one at $(0,1)$ and the other at $(0,-1)$. By the principle of stationary phase for double integrals, we have (see Wong (1989, p. 428-435))

$$J^* \sim \frac{\pi}{i^{1/2}\lambda^2(-\xi)^{1/4}} \Big[\varphi_{n+1}(0, \sqrt{-\xi}; \alpha) e^{-i(2\lambda(-\xi)^{3/2}/3-\pi/4)} + \varphi_{n+1}(0, -\sqrt{-\xi}; \alpha) e^{i(2\lambda(-\xi)^{3/2}/3-\pi/4)}\Big]. \tag{21.79}$$

In view of the well-known results (see Poston and Stewart (1978, p. 116, p. 392))

$$\mathrm{Ai}(\lambda^{2/3}\xi) \sim \frac{\cos[2\lambda(-\xi)^{3/2}/3 - \pi/4]}{\pi^{1/2}\lambda^{1/6}(-\xi)^{1/4}}, \tag{21.80}$$

$$\mathrm{Ai}'(\lambda^{2/3}\xi) \sim \frac{\sin[2\lambda(-\xi)^{3/2}/3 - \pi/4]}{\pi^{1/2}\lambda^{-1/6}(-\xi)^{-1/4}}, \tag{21.81}$$

we again have (21.75).

Theorem 21.3. *If the phase function in* (21.38) *satisfies all conditions in Theorem 21.1, we then have the uniform asymptotic expansion*

$$I(\lambda; \alpha) \sim \frac{2\pi^{3/2}\lambda^{-5/6}}{e^{-i[\lambda\eta(\alpha)+\varepsilon\pi/4]}} \sum_{k=0}^{\infty} \frac{1}{(i\lambda)^k} \left[a_k(\alpha)\,\mathrm{Ai}(\lambda^{2/3}\xi) + \frac{b_k(\alpha)}{i\lambda^{1/3}}\,\mathrm{Ai}'(\lambda^{2/3}\xi)\right], \tag{21.82}$$

as $\lambda \to \infty$, *where* $\xi = \xi(\alpha)$ *and* $\eta = \eta(\alpha)$ *are given explicitly in* (21.44) *and* (21.45), *and where* $a_k(\alpha)$ *and* $b_k(\alpha)$ *are defined inductively by* (21.65)-(21.69).

The two leading coefficients $a_0(\alpha)$ and $b_0(\alpha)$ can be obtained as follows. From (21.46) and (21.47), we have (exercise)

$$\left|\frac{\partial(x,y)}{\partial(u,v)}\right|_{(0,\pm\sqrt{-\xi})} = \frac{2(-\xi)^{1/4}}{|H_{f_\alpha}(x_\pm, y_\pm)|^{1/2}}. \tag{21.83}$$

Inserting (21.83) into (21.42) gives

$$\varphi_0(0, \pm\sqrt{-\xi}; \alpha) = 2(-\xi)^{1/4} \frac{g(x_\pm, y_\pm; \alpha)}{|H_{f_\alpha}(x_\pm, y_\pm)|^{1/2}}. \tag{21.84}$$

A combination of (21.51), (21.52) and (21.84) yields

$$\begin{aligned} a_0(\alpha) =& (-\xi)^{1/4} \left[\frac{g(x_+, y_+; \alpha)}{|H_{f_\alpha}(x_+, y_+)|^{1/2}} + \frac{g(x_-, y_-; \alpha)}{|H_{f_\alpha}(x_-, y_-)|^{1/2}} \right], \\ b_0(\alpha) =& (-\xi)^{-1/4} \left[\frac{g(x_+, y_+; \alpha)}{|H_{f_\alpha}(x_+, y_+)|^{1/2}} - \frac{g(x_-, y_-; \alpha)}{|H_{f_\alpha}(x_-, y_-)|^{1/2}} \right]. \end{aligned}$$

Thus, a two-term asymptotic expansion for the integral in (21.38) is given by

$$\begin{aligned} I(\lambda; \alpha) \sim & \frac{2\pi^{3/2}\lambda^{-5/6}(-\xi)^{1/4}}{e^{-i[\lambda\eta(\alpha)+\varepsilon\pi/4]}} \\ & \times \left\{ \left[\frac{g(x_+, y_+; \alpha)}{|H_{f_\alpha}(x_+, y_+)|^{1/2}} + \frac{g(x_-, y_-; \alpha)}{|H_{f_\alpha}(x_-, y_-)|^{1/2}} \right] \mathrm{Ai}(\lambda^{2/3}\xi) \right. \\ & \left. + \frac{(-\xi)^{-1/2}}{i\lambda^{1/3}} \left[\frac{g(x_+, y_+; \alpha)}{|H_{f_\alpha}(x_+, y_+)|^{1/2}} - \frac{g(x_-, y_-; \alpha)}{|H_{f_\alpha}(x_-, y_-)|^{1/2}} \right] \mathrm{Ai}'(\lambda^{2/3}\xi) \right\}. \end{aligned} \tag{21.85}$$

21.4 Derivation of asymptotic expansions: coalescence at a boundary point

In this section we consider the case in which the phase function $f(x, y; \alpha)$ in (21.4) has two stationary points $(x_\pm(\alpha), y_\pm(\alpha))$ which coalesce at a boundary point $(x_0, y_0) \in \Gamma$ as α approaches α_0. By Theorem 21.2, with the conditions in Theorem 21.1, there exists a change of variables which transforms $f_\alpha(x, y) = f(x, y; \alpha)$ to the canonical form

$$F_\alpha(u, v) = F(u, v; \alpha) = \varepsilon u^2 + \frac{v^3}{3} + \tilde{\xi}(\alpha)v + \tilde{\eta}(\alpha) \tag{21.86}$$

and maps the boundary curve $\Gamma = \{(x, y) : h(x, y) = 0\}$ to the straight line

$$v = \zeta(\alpha). \tag{21.87}$$

As shown in Sections 21.2 and 21.3, we have $\tilde{\xi}(\alpha) = \xi(\alpha)$ and $\tilde{\eta}(\alpha) = \eta(\alpha)$ given by (21.44) and (21.45), respectively. In what follows we shall for convenience assume that $\varepsilon = 1$.

First, we need to determine the value $\zeta(\alpha)$ in (21.87). From Lemma 21.4, we know that the restriction of $f_\alpha(x, y)$ to Γ has a simple stationary point $(\bar{x}, \bar{y}) = (\bar{x}(\alpha), \bar{y}(\alpha))$ on Γ. Since it must be mapped into the boundary stationary point $(0, \zeta(\alpha))$ of $F_\alpha(u,v)$ on the line $v = \zeta(\alpha)$, the relation $f_\alpha(x,y) = F_\alpha(u,v)$ yields

$$f_\alpha(\bar{x},\bar{y}) = \frac{\zeta^3}{3} + \xi(\alpha)\zeta + \eta(\alpha). \tag{21.88}$$

If $(\bar{x}(\alpha), \bar{y}(\alpha))$ coincides with one of the stationary points $(x_\pm(\alpha), y_\pm(\alpha))$, then $\zeta(\alpha)$ is equal to either $\sqrt{-\xi(\alpha)}$ or $-\sqrt{-\xi(\alpha)}$ according to what type of stationary point it is. Otherwise, equation (21.88) has one or three real roots. If it has only one real root $\zeta = \zeta(\alpha)$, then $\zeta(\alpha)$ is uniquely determined. If (21.88) has three roots, say $\zeta_1 \le \zeta_2 \le \zeta_3$, arranged in increasing order, then we must have the following relationship between the ζ_j and the stationary points $\pm\sqrt{-\xi}$:

$$\zeta_1 \le -\sqrt{-\xi} \le \zeta_2 \le \sqrt{-\xi} \le \zeta_3.$$

Since Γ is smooth and the restriction of $f_\alpha(x,y)$ to Γ has an extreme point at $(\bar{x},\bar{y})$, Γ is tangent to the level curve of $f_\alpha(x,y)$ at $(\bar{x},\bar{y})$ (see Ursell (1980, p. 426)). The directional derivative of $f_\alpha(x,y)$ at $(\bar{x},\bar{y})$ along the outer normal vector is either positive or negative. When $f_\alpha(x,y)$ is mapped into $F_\alpha(u,v)$ and $(\bar{x},\bar{y})$ is mapped into $(0,\zeta(\alpha))$, the same property must also hold for $F_\alpha(u,v)$ at $(0,\zeta(\alpha))$. For this reason, to determine $\zeta(\alpha)$ we only need to consider the following three cases.

Case 1. One of the stationary points of $f_\alpha(x,y)$ lies inside D, and the other lies outside D. In this case, we have $\zeta(\alpha) = \zeta_2$.

Case 2. Both stationary points of $f_\alpha(x,y)$ are inside D. If the directional derivative of $f_\alpha(x,y)$ along the outer normal vector of D at $(\bar{x},\bar{y})$ is positive, then $\zeta(\alpha) = \zeta_3$; otherwise, $\zeta(\alpha) = \zeta_1$.

Case 3. None of the stationary points of $f_\alpha(x,y)$ lies inside D. If the directional derivative of $f_\alpha(x,y)$ along the outer normal vector of D at $(\bar{x},\bar{y})$ is positive, then $\zeta(\alpha) = \zeta_1$; otherwise, $\zeta(\alpha) = \zeta_3$.

Let us denote by $\tilde{\Delta}$ the image of Δ_δ. We also need to determine the image of $\Delta_\delta \cap D$ under the mapping $(x,y) \to (u,v)$. In case 1, if the extreme point, say (x_+, y_+), of $f_\alpha(x,y)$ lies inside D, then the image of $\Delta_\delta \cap D$ is $\tilde{\Delta} \cap \{v : v \ge \zeta(\alpha)\}$; otherwise, $\Delta_\delta \cap D$ is mapped into $\tilde{\Delta} \cap \{v : v \le \zeta(\alpha)\}$. In case 2, if $\zeta(\alpha) = \zeta_3$, then the image of $\Delta_\delta \cap D$ is $\tilde{\Delta} \cap \{v : v \le \zeta(\alpha)\}$; if $\zeta(\alpha) = \zeta_1$, then the image of $\Delta_\delta \cap D$ is $\tilde{\Delta} \cap \{v : v \ge \zeta(\alpha)\}$. In case 3, if $\zeta(\alpha) = \zeta_3$, then $\Delta_\delta \cap D$ is mapped into $\tilde{\Delta} \cap \{v : v \ge \zeta(\alpha)\}$, if $\zeta(\alpha) = \zeta_1$, then $\Delta_\delta \cap D$ is mapped into $\tilde{\Delta} \cap \{v : v \le \zeta(\alpha)\}$.

In the following, we first consider the case in which the image of $\Delta_\delta \cap D$ is $\tilde{\Delta} \cap \{v : v \ge \zeta(\alpha)\}$. For convenience, we let $R_\zeta = \{v : v \ge \zeta(\alpha)\}$.

After making the change of variables $(x, y) \to (u, v)$, the integral in (21.6) becomes

$$I(\lambda; \alpha) = e^{i\lambda\eta(\alpha)} \iint_{\tilde{\Delta} \cap R_\zeta} \varphi_0(u, v; \alpha) e^{i\lambda(u^2 + v^3/3 + \xi v)} du dv, \tag{21.89}$$

where $\varphi_0(u, v; \alpha)$ is as defined in (21.42) and the Jacobian there satisfies (21.46). As before, we can extend the definition of φ_0 to the whole plane by setting $\varphi_0 = 0$ outside $\tilde{\Delta}$. Then

$$I(\lambda; \alpha) = e^{i\lambda\eta(\alpha)} \iint_{R_\zeta} \varphi_0(u, v; \alpha) e^{i\lambda(u^2 + v^3/3 + \xi v)} du dv.$$

We shall now derive an asymptotic expansion for the integral

$$J(\lambda; \alpha) = \iint_{R_\zeta} \varphi_0(u, v; \alpha) e^{i\lambda(u^2 + v^3/3 + \xi v)} du dv, \tag{21.90}$$

in terms of the incomplete Airy function (see Levey and Felsen (1969)):

$$\mathcal{J}(x, y) = \frac{1}{2\pi} \int_y^\infty e^{i(u^3/3 + xu)} du.$$

It is readily seen that $\mathcal{J}(x, y)$ satisfies the following differential equation:

$$\frac{\partial^2 \mathcal{J}}{\partial x^2} - x\mathcal{J} = \frac{1}{2\pi i} e^{i(y^3/3 + xy)}.$$

Define $\psi_{01}(u, v; \alpha)$, $\psi_{02}(v; \alpha)$, $a_0(\alpha)$ and $b_0(\alpha)$ as in (21.49)-(21.52). The same discussion given in (21.53)-(21.57) yields

$$\begin{aligned} J(\lambda; \alpha) = &\iint_{R_\zeta} (a_0 + b_0 v) h du dv + \frac{1}{i\lambda} \iint_{R_\zeta} \nabla \cdot (\Psi_0 h) du dv \\ &- \frac{1}{i\lambda} \iint_{R_\zeta} (\nabla \cdot \Psi_0) h du dv =: J_1 + J_2 + J_3, \end{aligned} \tag{21.91}$$

where $\Psi_0 = (\psi_{01}, \psi_{02})$ and $h = e^{i\lambda(u^2 + v^3/3 + \xi v)}$. Straightforward calculation gives

$$J_1 = 2\pi \sqrt{\frac{\pi}{\lambda}} e^{i\pi/4} \left[\frac{a_0}{\lambda^{1/3}} \mathcal{J}(\lambda^{2/3}\xi, \lambda^{1/3}\zeta) + \frac{b_0}{i\lambda^{2/3}} \mathcal{J}'(\lambda^{2/3}\xi, \lambda^{1/3}\zeta) \right], \tag{21.92}$$

where $\mathcal{J}'(x, y)$ denotes the derivative of $\mathcal{J}$ with respect to the first variable x. By Green's theorem, we have, as in (21.59),

$$\begin{aligned} J_2 = &\frac{1}{i\lambda} \lim_{R \to \infty} \int_\zeta^R [\psi_{01}(R, v; \alpha) - \psi_{01}(-R, v; \alpha)] e^{i\lambda(R^2 + v^3/3 + \xi v)} dv \\ &+ \frac{1}{i\lambda} \lim_{R \to \infty} \int_{-R}^R \psi_{02}(R; \alpha) e^{i\lambda(u^2 + R^3/3 + \xi R)} du \\ &- \frac{1}{i\lambda} \lim_{R \to \infty} \int_{-R}^R \psi_{02}(\zeta; \alpha) e^{i\lambda(u^2 + \zeta^3/3 + \xi\zeta)} du. \end{aligned} \tag{21.93}$$

Since the first two terms on the right-hand side of (21.93) are equal to zero, (21.93) reduces to

$$J_2 = \frac{1}{i\lambda}\sqrt{\frac{\pi}{\lambda}}e^{i\pi/4}\psi_{02}(\zeta;\alpha)e^{i\lambda(\zeta^3/3+\xi\zeta)}. \tag{21.94}$$

Let $\varphi_1(u,v;\alpha)$ be given as in (21.62). Then (cf. (21.63))

$$J_3 = \frac{1}{i\lambda}\iint_{R_\zeta}\varphi_1(u,v;\alpha)e^{i\lambda(u^2+v^3/3+\xi v)}dudv. \tag{21.95}$$

A combination of (21.91), (21.92), (21.94) and (21.95) gives the first approximation

$$\begin{aligned} J(\lambda;\alpha) =&\sqrt{\frac{\pi}{\lambda}}e^{i\pi/4}\left[\frac{2\pi a_0(\alpha)}{\lambda^{1/3}}\mathcal{J}(\lambda^{2/3}\xi,\lambda^{1/3}\zeta)+\frac{2\pi b_0(\alpha)}{i\lambda^{2/3}}\mathcal{J}'(\lambda^{2/3}\xi,\lambda^{1/3}\zeta)\right.\\ &\left.+\frac{c_0(\alpha)}{i\lambda}e^{i\lambda(\zeta^3/3+\xi\zeta)}\right]+\frac{1}{i\lambda}\iint_{R_\zeta}\varphi_1(u,v;\alpha)e^{i\lambda(u^2+v^3/3+\xi v)}dudv, \end{aligned} \tag{21.96}$$

where $c_0(\alpha) = \psi_{02}(\zeta(\alpha);\alpha)$. With the definitions given in (21.65)-(21.69) and

$$c_k(\alpha) = \psi_{k2}(\zeta(\alpha);\alpha), \tag{21.97}$$

we obtain

$$\begin{aligned} J(\lambda;\alpha) =&\sqrt{\frac{\pi}{\lambda}}e^{i\pi/4}\sum_{k=0}^{n-1}\frac{1}{(i\lambda)^k}\left[\frac{2\pi a_k(\alpha)}{\lambda^{1/3}}\mathcal{J}(\lambda^{2/3}\xi,\lambda^{1/3}\zeta)\right.\\ &\left.+\frac{2\pi b_k(\alpha)}{i\lambda^{2/3}}\mathcal{J}'(\lambda^{2/3}\xi,\lambda^{1/3}\zeta)+\frac{c_k(\alpha)}{i\lambda}e^{i\lambda(\zeta^3/3+\xi\zeta)}\right]+J_n(\lambda;\alpha), \end{aligned} \tag{21.98}$$

where

$$J_n(\lambda;\alpha) = \frac{1}{(i\lambda)^n}\iint_{R_\zeta}\varphi_n(u,v;\alpha)e^{i\lambda(u^2+v^3/3+\xi v)}dudv.$$

To show that (21.98) provides a uniform asymptotic expansion, we need to prove that

$$|J_n(\lambda;\alpha)| \le \frac{1}{\lambda^{n+1/2}}\left[\frac{D_n|\mathcal{J}(\lambda^{2/3}\xi,\lambda^{1/3}\zeta)|}{\lambda^{1/3}}+\frac{E_n|\mathcal{J}'(\lambda^{2/3}\xi,\lambda^{1/3}\zeta)|}{\lambda^{2/3}}+\frac{F_n}{\lambda}\right], \tag{21.99}$$

for some constants D_n, E_n and F_n independent of λ and α. As in (21.73), by using the recursive formula (21.65)-(21.69) and (21.97) one more time, it is sufficient to prove that the integral

$$J^* = \frac{1}{i\lambda}\iint_{R_\zeta}\varphi_{n+1}(u,v;\alpha)e^{i\lambda(u^2+v^3/3+\xi v)}dudv$$

satisfies

$$|J^*| \le \frac{1}{\lambda^{1/2}}\left[\frac{L_n}{\lambda^{1/3}}|\mathcal{J}(\lambda^{2/3}\xi, \lambda^{1/3}\zeta)| + \frac{M_n}{\lambda^{2/3}}|\mathcal{J}'(\lambda^{2/3}\xi, \lambda^{1/3}\zeta)| + \frac{N_n}{\lambda}\right]$$

for constants L_n, M_n and N_n independent of λ and α. In fact, we only need to show that

$$|J^*| \le \frac{N_n}{\lambda^{3/2}}. \tag{21.100}$$

By making the change of variables $s = \lambda^{1/2}u$ and $t = \lambda^{1/3}v$, we have

$$J^* = \frac{1}{i\lambda^{11/6}}\int_{-\infty}^{\infty}\int_{\lambda^{1/3}\zeta}^{\infty} \varphi_{n+1}(\lambda^{-1/2}s, \lambda^{-1/3}t; \alpha)e^{i(s^2+t^3/3+\lambda^{2/3}\xi t)}dsdt.$$

Comparing it with (21.76), we see that if $\lambda^{2/3}\xi$ is bounded, say $|\lambda^{2/3}\xi| \le \rho$, then there exists a constant K_{n+1} independent of λ and α such that

$$|J^*| \le K_{n+1}/\lambda^{11/6}$$

whether $\lambda^{1/3}\zeta$ is bounded or not. The desired result (21.100) now follows immediately.

If $\lambda^{2/3}\xi$ is unbounded, say $|\lambda^{2/3}\xi| \ge \rho$, then making the change of variables $u = (-\xi)^{3/4}s$ and $v = (-\xi)^{1/2}t$ gives

$$J^* = \frac{(-\xi)^{5/4}}{i\lambda}\int_{-\infty}^{\infty}\int_{(-\xi)^{-1/2}\zeta}^{\infty} \varphi_{n+1}((-\xi)^{3/4}s, (-\xi)^{1/2}t; \alpha) \times e^{i\lambda(-\xi)^{3/2}(s^2+t^3/3-t)}dsdt. \tag{21.101}$$

Now we consider the integral

$$\Phi(k, y; \varphi) = \int_y^{\infty} \varphi(x; \alpha)e^{ik(x^3/3-x)}dx, \tag{21.102}$$

where φ is a smooth function and $k \to \infty$. Taking $\varphi(x; \alpha) = 1$ yields

$$\mathcal{J}(\lambda^{2/3}\xi, \lambda^{1/3}\zeta) = \frac{\lambda^{1/3}(-\xi)^{1/2}}{2\pi}\Phi(\lambda(-\xi)^{3/2}, (-\xi)^{-1/2}\zeta; 1), \tag{21.103}$$

and taking $\varphi(x; \alpha) = x$ gives

$$\mathcal{J}'(\lambda^{2/3}\xi, \lambda^{1/3}\zeta) = \frac{\lambda^{2/3}\xi}{2\pi i}\Phi(\lambda(-\xi)^{3/2}, (-\xi)^{-1/2}\zeta; x). \tag{21.104}$$

The double integral J^* in (21.101) can be viewed as the Fourier transform of the function $\Phi(k, y; \varphi)$ in (21.102) with

$$k = \lambda(-\xi)^{3/2}, \quad \eta = (-\xi)^{1/2}\zeta, \quad \varphi = \varphi_{n+1}((-\xi)^{3/4}s, (-\xi)^{1/2}t; \alpha).$$

Thus, when $\lambda(-\xi)^{3/2}$ is large,

$$J^* \sim e^{i\pi/4}\frac{(-\xi)^{1/2}}{i\lambda^{3/2}}\sqrt{\pi}\Phi(\lambda(-\xi)^{3/2},(-\xi)^{-1/2}\zeta;\tilde{\varphi}), \tag{21.105}$$

where $\tilde{\varphi}(x;\alpha)=\varphi_{n+1}(0,(-\xi)^{1/2}x;\alpha)$.

To obtain the asymptotic behavior of $\Phi(k,y;\varphi)$ as $k\to\infty$, we consider the value of y in five separate cases: (1) $y\geq 1+\tilde{\varepsilon}>1$; (2) y close to 1; (3) $-1+\tilde{\varepsilon}\leq y<1-\tilde{\varepsilon}$, $\tilde{\varepsilon}>0$; (4) y close to -1; and (5) $y\leq -1-\tilde{\varepsilon}<-1$.

Case 1. $y\geq 1+\tilde{\varepsilon}>1$. The phase function is monotonically increasing if $x\geq y$. So contribution comes from an endpoint, and we have (see Wong (1989, p. 130))

$$\Phi(k,y;\varphi)\sim e^{i\pi/2+ik(y^3/3-y)}\frac{\varphi(y;\alpha)}{(y^2-1)k}.$$

Hence, when $(-\xi)^{1/2}\zeta\geq 1+\varepsilon>1$, we have from (21.105)

$$J^*\sim e^{i\pi/4+i\lambda(\zeta^3/3+\xi\zeta)}\frac{\sqrt{\pi}}{\zeta^2+\xi}\varphi_{n+1}(0,\zeta;\alpha)\frac{1}{\lambda^{5/2}}, \tag{21.106}$$

which of course implies

$$|J^*|\leq\frac{K_n}{|\zeta^2+\xi|\lambda^{5/2}},$$

where K_n is a constant independent of λ and α. For convenience, we shall from here on use K_n as a generic symbol to denote a constant independent of λ and α in the estimate of $|J^*|$. The value of K_n may change from time to time.

Note that since $(-\xi)^{1/2}\zeta\geq 1+\tilde{\varepsilon}$ and $|\lambda^{2/3}\xi|\geq\rho$, it follows that

$$|J^*|\leq\frac{K_n}{2\varepsilon\rho\lambda^{11/6}}\leq\frac{N_n}{\lambda^{3/2}}; \tag{21.107}$$

i.e., (21.100) holds.

Case 2. y near 1. When the stationary point 1 is close to the endpoint y, we can use the method presented in Wong (1989, Chapter VII, Section 3). The first two terms in the expansion of the integral (21.102) are given by

$$\Phi(k,y;\varphi)\sim e^{ik(y^3/3-y)}\left[\frac{p_1}{k^{1/2}}U(k^{1/2}a)+\frac{iq_1}{k}U'(k^{1/2}a)\right], \tag{21.108}$$

where

$$U(x)=\int_0^\infty e^{i(t^2/2-xt)}dt, \tag{21.109}$$

$$a=(1-y)(2y/3+4/3)^{1/2}, \tag{21.110}$$

$$p_1=\varphi(y;\alpha)\frac{a}{1-y^2}, \tag{21.111}$$

$$q_1=\frac{\varphi(1;\alpha)/\sqrt{2}-p_1}{a}. \tag{21.112}$$

Note that as $y \to 1$, we have

$$p_1 \to \frac{\varphi(1;\alpha)}{\sqrt{2}}, \quad q_1 \to \frac{\varphi'(1;\alpha)}{2} - \frac{\varphi(1;\alpha)}{6}.$$

The function $U(x)$ has the asymptotic formulas (exercise)

$$U(x) \sim \sqrt{2\pi} e^{-i(x^2/2-\pi/4)}, \ U(-x) \sim \frac{i}{x}, \tag{21.113}$$

$$U'(x) \sim \sqrt{2\pi} x e^{-i(x^2/2+\pi/4)}, \ U(-x) \sim \frac{i}{x^2}, \tag{21.114}$$

as $x \to \infty$. Hence, when $(-\xi)^{-1/2}\zeta$ is close to 1, we have, from (21.105) and (21.108) with $y = (-\xi)^{-1/2}\zeta$, $k = \lambda(-\xi)^{3/2}$ and $\varphi = \varphi_{n+1}(0, (-\xi)^{1/2}x; \alpha)$,

$$\begin{aligned} J^* \sim & \frac{\sqrt{\pi}}{\lambda^2(-\xi)^{1/4}} e^{i\lambda(\zeta^3/3+\xi\zeta)-i\pi/4} \tilde{p}_1 U(\lambda^{1/2}(-\xi)^{3/4}\tilde{a}) \\ & - \frac{\sqrt{\pi}}{\lambda^{5/2}\xi} e^{i\lambda(\zeta^3/3+\xi\zeta)+i\pi/4} \tilde{q}_1 U'(\lambda^{1/2}(-\xi)^{3/4}\tilde{a}), \end{aligned}$$

where

$$\begin{aligned} \tilde{a} =& (1 - (-\xi)^{-1/2}\zeta)(2(-\xi)^{-1/2}\zeta/3 + 4/3)^{1/2}, \\ \tilde{p}_1 =& \varphi_{n+1}(0, \zeta; \alpha) \frac{\tilde{a}}{1 - [(-\xi)^{1/2}\zeta]^2}, \\ \tilde{q}_1 =& \frac{\varphi_{n+1}(0, (-\xi)^{1/2}; \alpha)/\sqrt{2} - \tilde{p}_1}{\tilde{a}_1}. \end{aligned}$$

Using the asymptotic formula in (21.113) and (21.114), we obtain

$$|J^*| \le \frac{K_n}{\lambda^2(-\xi)^{1/4}} \le \frac{K_n}{\rho^{1/4}\lambda^{11/6}} \le \frac{N_n}{\lambda^{3/2}}, \tag{21.115}$$

thus proving (21.100).

Case 3. $-1 + \tilde{\varepsilon} \le y \le 1 - \tilde{\varepsilon}$. Here, the phase function has an interior stationary point at $x = 1$ and the principle of stationary phase gives (exercise)

$$\Phi(k, y; \alpha) \sim \frac{\sqrt{\pi}}{k^{1/2}} e^{-2ik/3+i\pi/4} \varphi(1; \alpha). \tag{21.116}$$

Therefore, when $-1 + \tilde{\varepsilon} \le (-\xi)^{-1/2}\zeta \le 1 - \tilde{\varepsilon}$, we have

$$J^* \sim \frac{\pi}{\lambda^2(-\xi)^{1/4}} e^{-2i\lambda(-\xi)^{3/2}/3} \varphi_{n+1}(0, (-\xi)^{1/2}; \alpha). \tag{21.117}$$

This also yields

$$|J^*| \le \frac{K_n}{\lambda^2(-\xi)^{1/4}} \le \frac{K_n}{\rho^{1/4}\lambda^{11/6}} \le \frac{N_n}{\lambda^{3/2}}. \tag{21.118}$$

Case 4. y near -1. In this case, the two stationary points are separated, but one of them is close to the endpoint. Employing a combination of the arguments used in cases 2 and 3, we obtain

$$\Phi(k,y;\varphi) \sim \frac{\sqrt{\pi}}{k^{1/2}} e^{-2ik/3+i\pi/4}\varphi(1;\alpha) + e^{ik(y^3/3-y)}\left[\frac{p_2}{k^{1/2}}V(k^{1/2}b) + \frac{iq_2}{k}V'(k^{1/2}b)\right],$$

where

$$V(x) = \int_0^\infty e^{-i(t^2/2-xt)}dt, \; b = -(1+y)(-2y/3+4/3)^{1/2},$$

$$p_2 = -\varphi(y;\alpha)\frac{b}{1-y^2}, \; q_2 = \frac{\varphi(-1;\alpha)/\sqrt{2}-p_2}{b};$$

cf. (21.108) and (21.116). Hence, by the same argument as used in cases 2 and 3, when $(-\xi)^{-1/2}\zeta$ is close to -1, we also have

$$|J^*| \le \frac{K_n}{\lambda^2(-\xi)^{1/4}} \le \frac{N_n}{\lambda^{3/2}}.$$

Case 5. $y \le -1-\tilde{\varepsilon} \le -1$. Contribution to the asymptotic approximation of the integral (21.102) comes from the two stationary points 1 and -1, and the principle of stationary phase gives

$$\Phi(k,y;\varphi) \sim \frac{\sqrt{\pi}}{k^{1/2}} e^{-2ik/3+i\pi/4}\varphi(1;\alpha) + \frac{\sqrt{\pi}}{k^{1/2}} e^{2ik/3-i\pi/4}\varphi(-1;\alpha).$$

From this and (21.105), it follows that

$$J^* \sim \frac{\pi}{\lambda^2(-\xi)^{1/4}}\Big[e^{-2i\lambda(-\xi)^{3/2}/3}\varphi_{n+1}(0,(-\xi)^{1/2};\alpha) - ie^{2i\lambda(-\xi)^{3/2}/3}\varphi_{n+1}(0,-(-\xi)^{1/2};\alpha)\Big], \tag{21.119}$$

and (21.100) is obtained immediately. Coupling (21.98) and (21.100) gives

$$J(\lambda;\alpha) \sim \sqrt{\frac{\pi}{\lambda}} e^{i\pi/4}\sum_{k=0}^\infty \frac{1}{(i\lambda)^k}\left[\frac{2\pi a_k(\alpha)}{\lambda^{1/3}}\mathcal{J}(\lambda^{2/3}\xi,\lambda^{1/3}\zeta) + \frac{2\pi b_k(\alpha)}{i\lambda^{2/3}}\mathcal{J}'(\lambda^{2/3}\xi,\lambda^{1/3}\zeta) + \frac{c_k(\alpha)}{i\lambda}e^{i\lambda(\zeta^3/3+\xi\zeta)}\right]. \tag{21.120}$$

If the domain $\Delta_\delta \cap D$ is mapped into $\tilde{\Delta} \cap \{v: \; v \le \zeta(\alpha)\}$ then, instead of $\mathcal{J}(x,y)$, we should use the other incomplete Airy function

$$\mathcal{I}(x,y) = \frac{1}{2\pi}\int_{-\infty}^y e^{i(u^3/3+xu)}du.$$

These two functions are related via the identity

$$\mathcal{I}(x,y)+\mathcal{J}(x,y)=\mathrm{Ai}(x).$$

Moreover, in this case, (21.94) becomes

$$J_2=-\frac{1}{i\lambda}\sqrt{\frac{\pi}{\lambda}}e^{i\pi/4}\psi_{02}(\zeta;\alpha)e^{i\lambda(\zeta^3/3+\xi\zeta)}.$$

Hence, the asymptotic expansion of $J(\lambda;\alpha)$ is given by

$$\begin{aligned}J(\lambda;\alpha)\sim&\sqrt{\frac{\pi}{\lambda}}e^{i\pi/4}\sum_{k=0}^{\infty}\frac{1}{(i\lambda)^k}\left[\frac{2\pi a_k(\alpha)}{\lambda^{1/3}}\mathcal{I}(\lambda^{2/3}\xi,\lambda^{1/3}\zeta)\right.\\&\left.+\frac{2\pi b_k(\alpha)}{i\lambda^{2/3}}\mathcal{I}'(\lambda^{2/3}\xi,\lambda^{1/3}\zeta)-\frac{c_k(\alpha)}{i\lambda}e^{i\lambda(\zeta^3/3+\xi\zeta)}\right].\end{aligned}\tag{21.121}$$

When $\varepsilon=-1$, we can also divide the discussion into two cases. Note that changing to the case $\varepsilon=-1$, we will make use of the result

$$\int_{\mathbf{R}}e^{-i\lambda u^2}du=\sqrt{\frac{\pi}{\lambda}}e^{-i\pi/4},$$

instead of

$$\int_{\mathbf{R}}e^{i\lambda u^2}du=\sqrt{\frac{\pi}{\lambda}}e^{i\pi/4}.$$

This is the only change in the derivation of the asymptotic expansion of $J(\lambda;\alpha)$. Thus we have

$$\begin{aligned}J(\lambda;\alpha)\sim&\sqrt{\frac{\pi}{\lambda}}e^{-i\pi/4}\sum_{k=0}^{\infty}\frac{1}{(i\lambda)^k}\left[\frac{2\pi a_k(\alpha)}{\lambda^{1/3}}\mathcal{J}(\lambda^{2/3}\xi,\lambda^{1/3}\zeta)\right.\\&\left.+\frac{2\pi b_k(\alpha)}{i\lambda^{2/3}}\mathcal{J}'(\lambda^{2/3}\xi,\lambda^{1/3}\zeta)+\frac{c_k(\alpha)}{i\lambda}e^{i\lambda(\zeta^3/3+\xi\zeta)}\right]\end{aligned}\tag{21.122}$$

if $\Delta_\delta\cap D$ is mapped into $\tilde{\Delta}\cap\{v:\ v\geq\zeta(\alpha)\}$, and

$$\begin{aligned}J(\lambda;\alpha)\sim&\sqrt{\frac{\pi}{\lambda}}e^{-i\pi/4}\sum_{k=0}^{\infty}\frac{1}{(i\lambda)^k}\left[\frac{2\pi a_k(\alpha)}{\lambda^{1/3}}\mathcal{I}(\lambda^{2/3}\xi,\lambda^{1/3}\zeta)\right.\\&\left.+\frac{2\pi b_k(\alpha)}{i\lambda^{2/3}}\mathcal{I}'(\lambda^{2/3}\xi,\lambda^{1/3}\zeta)-\frac{c_k(\alpha)}{i\lambda}e^{i\lambda(\zeta^3/3+\xi\zeta)}\right].\end{aligned}\tag{21.123}$$

if $\Delta_\delta\cap D$ is mapped into $\tilde{\Delta}\cap\{v:\ v\leq\zeta(\alpha)\}$.

Summarizing the above results, we have the following theorem.

Theorem 21.4. *Under the conditions of Theorem 21.2, the integral in (21.6) has either the uniform asymptotic expansion*

$$I(\lambda;\alpha) \sim \sqrt{\frac{\pi}{\lambda}} e^{i[\lambda\eta(\alpha)+\varepsilon\pi/4]} \sum_{k=0}^{\infty} \frac{1}{(i\lambda)^k} \left[\frac{2\pi a_k(\alpha)}{\lambda^{1/3}} \mathcal{J}(\lambda^{2/3}\xi, \lambda^{1/3}\zeta) + \frac{2\pi b_k(\alpha)}{i\lambda^{2/3}} \mathcal{J}'(\lambda^{2/3}\xi, \lambda^{1/3}\zeta) + \frac{c_k(\alpha)}{i\lambda} e^{i\lambda(\zeta^3/3+\xi\zeta)} \right]$$

or

$$I(\lambda;\alpha) \sim \sqrt{\frac{\pi}{\lambda}} e^{i[\lambda\eta(\alpha)+\varepsilon\pi/4]} \sum_{k=0}^{\infty} \frac{1}{(i\lambda)^k} \left[\frac{2\pi a_k(\alpha)}{\lambda^{1/3}} \mathcal{I}(\lambda^{2/3}\xi, \lambda^{1/3}\zeta) + \frac{2\pi b_k(\alpha)}{i\lambda^{2/3}} \mathcal{I}'(\lambda^{2/3}\xi, \lambda^{1/3}\zeta) - \frac{c_k(\alpha)}{i\lambda} e^{i\lambda(\zeta^3/3+\xi\zeta)} \right]$$

as $\lambda \to \infty$, *where* $\xi = \xi(\alpha)$ *and* $\eta = \eta(\alpha)$ *are defined by* (21.44) *and* (21.45), $\zeta = \zeta(\alpha)$ *is defined by* (21.88), *and the coefficients are defined by* (21.65)-(21.69) *and* (21.97).

21.5 Exercises

1. Let f be a smooth two-variable function near the origin $(0,0)$. For any (x,y) close to $(0,0)$, we have $f(x,y) = P_n(x,y) + R_n(x,y)$, where
$$P_n(x,y) = \sum_{k=0}^{n} \sum_{i+j=k} \frac{x^i y^j}{i!j!} f^{(i,j)}(0,0),$$
$$R_n(x,y) = \sum_{i+j=n+1} \frac{x^i y^j}{i!j!} f^{(i,j)}(\xi x, \xi y),$$
and $\xi \in (0,1)$ is some constant depending on x and y (Hint: fix x and y, and define $g(t) = f(tx, ty)$).
2. Verify (21.44) and (21.45).
3. Verify (21.46).
4. Verify (21.83).
5. Verify (21.113) and (21.114).
6. Verify (21.116).
7. Find the leading term in the asymptotic expansion of the integral
$$I(\lambda) = \iint_D e^{i\lambda \cos x \cos y} dxdy$$
as $\lambda \to \infty$, where D is the unit disk
$$\{(x,y): \ x^2 + y^2 \le 1\}.$$

8. Derive the asymptotic expansion of the integral

$$I(\lambda) = \iint_D e^{i\lambda(\cos x + \cos y)} dxdy$$

as $\lambda \to \infty$, up to the order $O(\lambda^{-2})$, where D is the square

$$\{(x, y) : \ |x| < \pi/2, \ |y| < \pi/2\}.$$

9. Consider integrals of the form

$$I(\lambda) = \iint_D g(x, y) e^{i\lambda f(x,y)} dxdy,$$

where D is a bounded domain, $f(x, y)$ is a real-valued function, and λ is a large positive parameter. For simplicity, we assume that both f and g are infinitely differentiable in the closure of D. As in the one-dimensional case, the main contributions to the asymptotic expansion of $I(\lambda)$ come only from certain *critical points.* These include (i) *stationary points* of $f(x, y)$, i.e., points in D or on its boundary Γ which satisfy $f_x = f_y = 0$, (ii) points on the boundary at which a level curve of f is tangent to Γ, and (iii) points where Γ has a discontinuously turning tangent. Stationary points are also called *critical points of the first kind.* Critical points of other kinds are on the boundary of D. Representing Γ parametrically, they are just one-dimensional integrals, which we have considered in Section 17.4. From this, we know that contributions to the asymptotic expansion of these integrals come only from (i) points on the boundary at which $\nabla f \neq (0, 0)$, but at which f, considered along Γ as a function of a single variable, has a stationary point, and (ii) points on the boundary at which some parametrization of Γ has a discontinuous derivative of some order. The former are known as *critical points of the second kind* and the latter as *critical points of the third kind.* Critical points of the second kind are also points on the boundary where a level curve $f(x, y) = c$ is tangent to Γ for some c. To prove this, we suppose that Γ is given by $h(x, y) = 0$, where h is a smooth function with $(x(t), y(t))$ as a solution to the equation. We also suppose that $\nabla h \neq (0, 0)$, so that $h(x, y) = 0$ guarantees locally a smooth solution; i.e., Γ is locally a smooth curve. Stationary points of $f|_\Gamma$, the restriction of f to Γ, are exactly those points where

$$f_x \dot{x}(t) + f_y \dot{y}(t) = 0.$$

Since

$$h_x \dot{x} + h_y \dot{y} = \frac{d}{dt} h(x, y) = 0,$$

a stationary point (x_0, y_0) of $f|_\Gamma$ is one for which

$$\det\begin{pmatrix} f_x & f_y \\ h_x & h_y \end{pmatrix} = f_x h_y - f_y h_x = 0.$$

Thus it follows that $\nabla f(x_0, y_0)$ and $\nabla h(x_0, y_0)$ are linearly dependent. Geometrically, this means that these two vectors are parallel, which in turn implies that the level curve $f(x, y) = f(x_0, y_0)$ is tangent to the boundary $h(x, y) = 0$ at (x_0, y_0). This completes our discussion on the classification of critical points.

(a) Consider the double integral

$$I(\lambda) = \int_{-1}^{1}\int_{-1/2}^{1/2} e^{i\lambda(x^2+xy+y^2)(1-x/2)}\,dxdy.$$

Show that there are three critical points of the second kind, and that they are located at $(-1/3, 1)$, $(-1/2, -1/4)$ and $(-1/2, 1/4)$. Write down the leading-term contribution from each of these points.

(b) Show that $(0, 0)$ is the only critical point of the first kind for the integral in (a), and that its contribution is

$$\frac{2\pi}{\sqrt{3}}\left[\frac{i}{\lambda} - \frac{261}{512\lambda^2} + \cdots\right].$$

(c) Show that $(0, 0)$ is a boundary stationary point for the integral

$$\iint_D e^{i\lambda(x^2+xy+y^2)(1-x/2)}\,dxdy, \quad \lambda > 0,$$

where D is the region bounded by the circle $(x-1)^2 + (y-3)^2 = 10$. Also show that the first two terms in the asymptotic expansion from this critical point are given by

$$\frac{\pi i}{\sqrt{3}\lambda} - \frac{17i\sqrt{2\pi}e^{i\pi/4}}{2(14\lambda)^{3/2}}.$$

Chapter 22

Asymptotics of Pollaczek Polynomials

22.1 Introduction

The three classical orthogonal polynomials (namely, Jacobi, Laguerre and Hermite) all satisfy a second-order differential equation. Thus, their asymptotic expansions can be obtained by using methods in differential equation theory. But, there are many semi-classical orthogonal polynomials such as Charlier, Meixner, Krawtchouk, Hahn, etc. These polynomials do not satisfy any ordinary differential equations; so we have to use asymptotic methods for integrals to derive their behaviors.

Unlike the two examples (Bessel function and Hermite polynomials) in Chapter 19, the example we are going to present in this chapter is a relatively new special function; namely, the Pollaczek polynomials $P_n(x; a, b)$. In 1949, Pollaczek introduced this remarkable generalization of the Legendre polynomials, which shows in many respects a singular behavior; see Szegö (1975, p. 393-396). They are most easily defined by the generating function

$$\begin{aligned} f(x,w) =& f(\cos\theta, w) = \sum_{n=0}^{\infty} P_n(x; a, b) w^n \\ =& (1 - we^{i\theta})^{-1/2+ih(\theta)}(1 - we^{-i\theta})^{-1/2-ih(\theta)}, \end{aligned} \tag{22.1}$$

where

$$h(\theta) = \frac{a\cos\theta + b}{2\sin\theta}, \quad a > \pm b. \tag{22.2}$$

They reduce to the Legendre polynomials when $a = b = 0$.

In 1954, Novikoff (1954) wrote a well-known thesis on the asymptotics of these polynomials and their zeros. More specifically, he investigated the asymptotic behavior of $P_n(x; a, b)$ as $n \to \infty$, where $x = \cos(t/\sqrt{n})$ and

$t > 0$ is fixed. He divided the positive t-axis into two regions $0 < t < (a+b)^{1/2}$ and $t > (a+b)^{1/2}$, and derived an asymptotic formula in each of the two regions. Furthermore, if the zeros of these polynomials are denoted by $\cos\theta$, where $0 < \theta_{n,1} < \cdots < \theta_{n,n} < \pi$, then for any fixed ν

$$\lim_{n\to\infty} n^{1/2}\theta_{n,\nu} = (a+b)^{1/2}. \tag{22.3}$$

More recently, R. A. Askey conjectured that the next term in the asymptotic expansion of $\theta_{n,\nu}$ will involve zeros of a certain transcendental function. In an attempt to prove this conjecture, Ismail (1994) derived a two-term expansion for $\theta_{n,\nu}$, whose second term involves a zero of the entire function

$$F(\xi) = \int_0^1 (1-v^2)^{-1/2} e^{(a+b)v^2} \cos(\xi v) dv.$$

A drawback of Ismail's result is that $F(\xi)$ is not one of the familiar special functions of mathematical physics. Consequently, not much is known about the zeros of this function.

In this chapter, we will present an infinite asymptotic expansion for $P_n(\cos(t/\sqrt{n}); a, b)$, involving the Airy function and its derivative, which holds uniformly for $0 < \varepsilon \le t \le M < \infty$. Moreover, we show that $\theta_{n,\nu}$ has the two-term asymptotic expansion

$$\theta_{n,\nu} = \sqrt{\frac{a+b}{n}} + \frac{(a+b)^{1/6}(-a_\nu)}{2n^{5/6}} + O(n^{-7/6}), \tag{22.4}$$

where a_ν is the ν-th negative zero of the Airy function $\mathrm{Ai}(\cdot)$. This result contradicts that of Ismail (1994).

Our approach here is in spirit similar to that used in the derivation of uniform asymptotic expansions of the Bessel function and the Hermite polynomials. However, the details of analysis in this chapter, and especially in the discussion of the cubic transformation (22.29) in Section 22.3, are quite difficult, and in fact are more difficult than those in the two previous examples. The material presented in this chapter is taken from Bo and Wong (1996).

22.2 Novikoff's results

From (22.1), we have by the Cauchy integral formula

$$P_n(x; a, b) = \frac{1}{2\pi i}\int_C \frac{f(\cos\theta, z)}{z^{n+1}} dz, \tag{22.5}$$

where C is a positively oriented simple closed curve surrounding $z = 0$ not containing the branch points $z = e^{\pm i\theta}$. In view of the reflection formula [Novikoff (1954), p. 7]

$$P_n(x; a, b) = (-1)^n P_n(-x; a, -b), \tag{22.6}$$

we may assume without loss of generality that $0 < \theta < \pi/2$. Novikoff (1954) took C to be contour C_1 shown in Figure 22.1. This contour consists of a large circle of radius $R >> 1$, two small circles of radius $\delta << 1$, and two doubly traversed straight line segments, one vertical and one horizontal. The large circle is oriented in the counterclockwise direction, and the two small circles are oriented in the clockwise direction, so that $z = 0$ lies inside the region bounded by C_1, and $z = e^{\pm i\theta}$ lie outside this region. It can be shown that the integral along the large circle $|z| = R$ and the two small circles $|z - e^{\pm i\theta}| = \delta$ vanishes as $R \to \infty$ and $\delta \to 0$; see Novikoff (1954). The contribution along the horizontal line segment $z = \cos\theta + \tau$, $0 \le \tau \le R$ is also zero, since $f(x, w)$ has the same value on this portion of the contour in either direction. Thus, the only contribution to the integral in (22.5), which does not vanish when $R \to \infty$ and $\delta \to 0$, comes from that portion of the integration path lying along the two sides of the cut $z = \cos\theta + i\alpha \sin\theta$, $-1 < \alpha < 1$, joining the two branch points $e^{i\theta}$ and $e^{-i\theta}$. Novikoff showed that along the left edge of the cut,

$$f(x, z) = \frac{e^{(\pi - 2\theta)h(\theta)}}{\sin\theta\sqrt{1 - \alpha^2}} \left(\frac{1 + \alpha}{1 - \alpha}\right)^{ih(\theta)}, \tag{22.7}$$

where $\arg[(1+\alpha)/(1-\alpha)] = 0$ if $-1 < \alpha < 1$, and that this value is changed by a factor $-e^{-2\pi h(\theta)}$ along the right edge of cut. As a result, he obtained the integral representation

$$P_n(\cos\theta; a, b) = \frac{\cosh(\pi h(\theta))}{\pi e^{2\theta h(\theta)}} \int_{-1}^{1} \frac{[(1 + \alpha)/(1 - \alpha)]^{ih(\theta)} d\alpha}{(\cos\theta + i\alpha \sin\theta)^{n+1}\sqrt{1 - \alpha^2}}, \tag{22.8}$$

where $\ln[(1 + \alpha)/(1 - \alpha)]$ has to be taken as real for $-1 < \alpha < 1$. This integral is an analogue of the Laplace integral for the Legendre polynomial, which it includes as a special case if $a = b = 0$.

To approximate $P_n(\cos\theta; a, b)$ when $\theta = t/\sqrt{n}$ and n is large, Novikoff noted that

$$h(\theta) = h(t/\sqrt{n}) = \frac{a + b}{2t}\sqrt{n} + O(1/\sqrt{n}), \tag{22.9}$$

and

$$-(n + 1)\ln\left(\cos\frac{t}{\sqrt{n}} + i\alpha \sin\frac{t}{\sqrt{n}}\right) = -i\alpha t\sqrt{n} + \frac{1 - \alpha^2}{2}t^2 + O(1/\sqrt{n}). \tag{22.10}$$

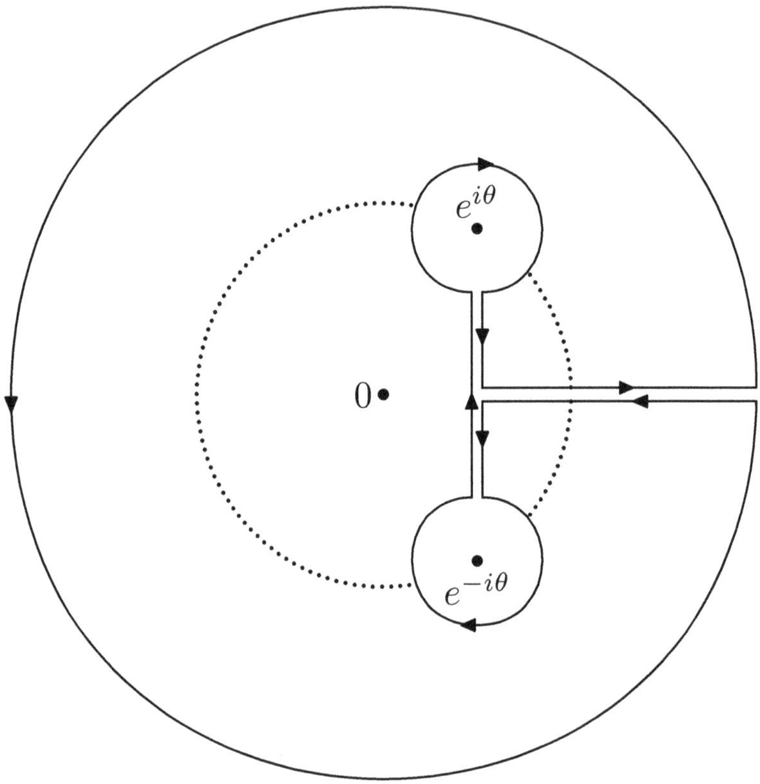

Fig. 22.1: Contour C_1.

Hence by putting

$$h(\alpha, t) := \alpha t - \frac{a+b}{2t} \ln \frac{1+\alpha}{1-\alpha}, \tag{22.11}$$

the last integral can be approximated by the simplified integral

$$J = \int_{-1}^{1} \exp\{-i\sqrt{n}h(\alpha, t) + (1-\alpha^2)t^2/2\}(1-\alpha^2)^{-1/2} d\alpha. \tag{22.12}$$

To the integral J, Novikoff applied the method of steepest descent (see Wong (1989, p. 84)), which consists of finding the saddle points of $h(\alpha, t)$, i.e., the zeros of

$$\frac{dh}{d\alpha} = t - \frac{a+b}{(1-\alpha^2)t},$$

and deforming the linear path of integration into an appropriate one passing through the saddle points. The saddle points of h are given by

$$\alpha_{\pm} = \pm(1 - (a+b)/t^2)^{1/2}. \tag{22.13}$$

They are distinct as long as $t^2 \neq a+b$. If $t^2 > a+b$ then α_+ and α_- are real. If $t^2 < a+b$ then α_+ and α_- are conjugate imaginary. In both cases, α_+ and α_- are symmetrically located with respect to the origin $\alpha = 0$. By considering these two cases separately, Novikoff obtained the following results:

$$P_n(\cos(t/\sqrt{n}); a, b) = \frac{e^{\sqrt{n}[\pi(a+b)/(2t)+\lambda(t)]-(a+b)/2}}{2\sqrt{\pi}[n(a+b-t^2)]^{1/4}}\left[1+O(n^{-1/4})\right], \tag{22.14}$$

with

$$\lambda(t) = a_1 t - \frac{a+b}{t}\arctan a_1, \tag{22.15}$$

$$a_1 = \frac{\sqrt{a+b-t^2}}{t}, \tag{22.16}$$

for $0 < t < \sqrt{a+b}$; and

$$\begin{aligned} P_n(\cos(t/\sqrt{n}); a, b) =& \frac{e^{\sqrt{n}\pi(a+b)/(2t)-(a+b)/2}\cos[\pi/4-\sqrt{n}\mu(t)]}{\sqrt{\pi}[n(t^2-a-b)]^{1/4}} \\ &+ O(n^{-1/2}), \end{aligned} \tag{22.17}$$

with

$$\mu(t) = \alpha_1 t - \frac{a+b}{2t}\ln\frac{1+\alpha_1}{1-\alpha_1}, \tag{22.18}$$

$$\alpha_1 = \frac{\sqrt{t^2-a-b}}{t}, \tag{22.19}$$

for $t > \sqrt{a+b}$.

22.3 Reduction to a canonical integral

For our purpose, we take C in (22.5) to be the contour C_2 shown in Figure 22.2, which consists of a large circle oriented in the counterclockwise direction, two small circles oriented in the clockwise direction, and two doubly traversed vertical infinite half-lines emitting from the branch points $z = e^{i\theta}$ and $z = e^{-i\theta}$.

As in Novikoff (1954), it can be shown that the integral along the large circle vanishes as $R \to \infty$. For the integral along the remaining portion of the contour, we make the change of variable $z = \cos\theta + i\alpha\sin\theta$, which is simply a linear transformation consisting of a translation, a rotation, and a rescaling. The two vertical cuts in the z-plane are now mapped into the

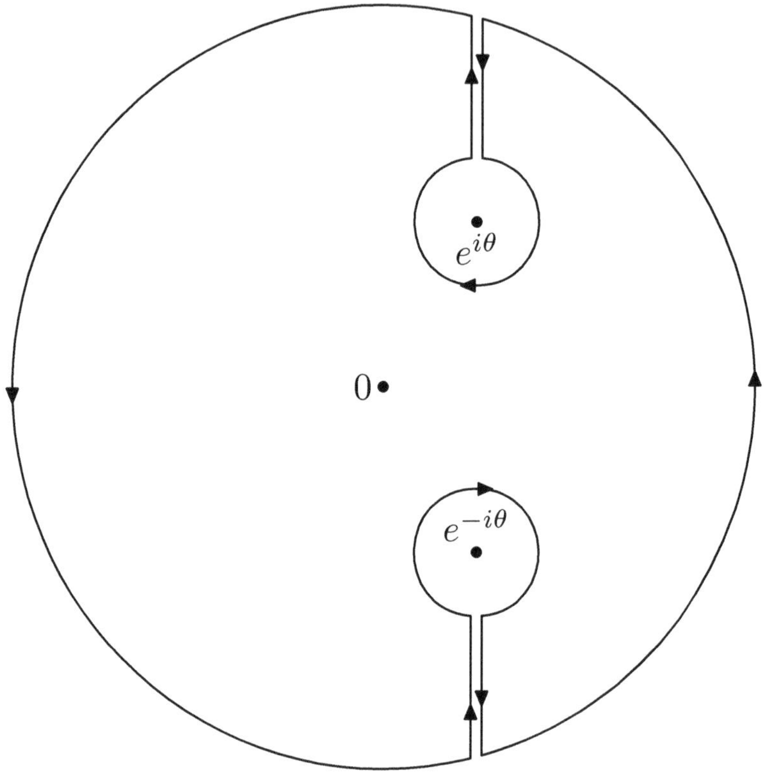

Fig. 22.2: Contour C_2.

two horizontal half-lines along the real axis in the α-plane, one extending from $\alpha = 1$ to $\alpha = \infty$ and the other from $\alpha = -1$ to $\alpha = -\infty$. In the cut α-plane, it can be shown that

$$f(\cos\theta, z) = \frac{e^{(\pi-2\theta)h(\theta)}}{\sin\theta\sqrt{1-\alpha^2}}\left(\frac{1+\alpha}{1-\alpha}\right)^{ih(\theta)}, \tag{22.20}$$

where $\arg[(1+\alpha)/(1-\alpha)] = 0$ if $-1 < \alpha < 1$; compare with (22.7). Consequently, the integral (22.5) can be written as

$$P_n(\cos\theta; a, b) = \frac{-e^{(\pi-2\theta)h(\theta)}}{2\pi} \times \left[\int_{\infty}^{(1^+)} + \int_{-\infty}^{(-1^+)}\right] \frac{[(1+\alpha)/(1-\alpha)]^{ih(\theta)} d\alpha}{(\cos\theta + i\alpha\sin\theta)^{n+1}(1-\alpha^2)^{1/2}}, \tag{22.21}$$

Fig. 22.3: α-plane.

where the loop paths of integration are as indicated in Figure 22.3.

Put $\theta = t/\sqrt{n}$, and rewrite (22.21) in the form

$$P_n(\cos\theta; a, b) = -ie^{(\pi-2\theta)h(\theta)} I, \tag{22.22}$$

where

$$I = \frac{1}{2\pi i}\left[\int_{\infty}^{(1^+)} + \int_{-\infty}^{(-1^+)}\right] e^{\sqrt{n}\tilde{g}(\alpha,t)} \frac{d\alpha}{(1-\alpha^2)^{1/2}}, \tag{22.23}$$

and

$$\begin{aligned}\tilde{g}(\alpha, t) =& \frac{i[a\cos(t/\sqrt{n}) + b]}{2\sqrt{n}\sin(t/\sqrt{n})} \ln\frac{1+\alpha}{1-\alpha} \\ &- \left(\sqrt{n} + \frac{1}{\sqrt{n}}\right) \ln\left(\cos\frac{t}{\sqrt{n}} + i\alpha\sin\frac{t}{\sqrt{n}}\right).\end{aligned} \tag{22.24}$$

Motivated by (22.11) and (22.12), we set

$$g(\alpha, t) = i\left(\frac{a+b}{2t}\ln\frac{1+\alpha}{1-\alpha} - \alpha t\right) \tag{22.25}$$

and write (22.23) as

$$I = \frac{1}{2\pi i}\left[\int_{\infty}^{(1^+)} + \int_{-\infty}^{(-1^+)}\right] e^{\sqrt{n}g(\alpha,t)} \bar{K}(\alpha, t, 1/\sqrt{n}) d\alpha, \tag{22.26}$$

where

$$\bar{K}(\alpha, t, 1/\sqrt{n}) = \frac{\exp\{\sqrt{n}[\tilde{g}(\alpha, t) - g(\alpha, t)]\}}{(1-\alpha^2)^{1/2}}. \tag{22.27}$$

Note that the saddle points of $g(\alpha, t)$ are exactly those of $h(\alpha, t)$ given in (22.13), i.e.,

$$\alpha_\pm = \begin{cases} \pm\sqrt{1 - \frac{a+b}{t^2}}, & t > \sqrt{a+b}, \\ \pm i\sqrt{\frac{a+b}{t^2} - 1}, & t < \sqrt{a+b}. \end{cases} \tag{22.28}$$

As $t \to \sqrt{a+b}$, the two points α_+ and α_- coalesce at $\alpha = 0$. This suggests the use of the cubic transformation introduced by Chester *et al.* (1957); see (19.6) and (19.50) in Chapter 19. Since $g(\alpha, t)$ is an odd function, we set

$$g(\alpha, t) = i\left(\frac{a+b}{2t}\ln\frac{1+\alpha}{1-\alpha} - \alpha t\right) = \frac{u^3}{3} - B^2(t)u. \tag{22.29}$$

For this transformation to be analytic in the region of interest, we must have $d\alpha/du \neq 0$ or ∞. Differentiating both sides of (22.29) gives

$$g_\alpha(\alpha, t)\frac{d\alpha}{du} = u^2 - B^2(t).$$

Since the right-hand side vanishes at $u = \pm B(t)$, we must make $\alpha = \alpha_\pm$ correspond to $u = \pm B(t)$. This gives (exercise)

$$B(t) = \begin{cases} i[3\beta(t)/2]^{1/3}, & t \geq \sqrt{a+b}, \\ [3\gamma(t)/2]^{1/3}, & t < \sqrt{a+b}, \end{cases} \tag{22.30}$$

where

$$\beta(t) = \frac{a+b}{t}\operatorname{arccosh}\frac{t}{\sqrt{a+b}} - \sqrt{t^2-a-b}, \tag{22.31}$$

$$\gamma(t) = \frac{a+b}{t}\arccos\frac{t}{\sqrt{a+b}} - \sqrt{a+b-t^2}. \tag{22.32}$$

Note that $\beta(t) \leq 0$ and $\gamma(t) \geq 0$ and that $B(t)$ is a C^∞ function in $0 < t < \infty$.

In Section 22.4, it will be shown that with the choice of $B(t)$ given in (22.30), the transformation (22.29) is one-to-one and analytic along an infinite path of integration for the contour integral (22.26). Here we first investigate the shapes and directions of the steepest descent curves through the saddle points α_+ and α_-.

Case (i): $t > \sqrt{a+b}$. The steepest paths through $\alpha = \alpha_+$ are given by

$$\operatorname{Im} g(\alpha, t) = \operatorname{Im} g(\alpha_+, t). \tag{22.33}$$

We write $\alpha = x + iy$. Then (22.33) gives (exercise)

$$y = \pm\sqrt{\frac{(1+x)^2 - (1-x)^2 e^{F(x,t)}}{e^{F(x,t)} - 1}}, \tag{22.34}$$

where

$$F(x,t) = \frac{4t^2 x}{a+b} + 2\ln\frac{t+\sqrt{t^2-a-b}}{t-\sqrt{t^2-a-b}} - \frac{4t\sqrt{t^2-a-b}}{a+b}. \tag{22.35}$$

In the case of + sign in (22.34), it can be shown that

$$\lim_{\alpha\to\alpha_+^+} y'(\alpha) = 1, \quad \lim_{\alpha\to\alpha_+^-} y'(\alpha) = -1. \tag{22.36}$$

In the case of − sign in (22.34), we have

$$\lim_{\alpha\to\alpha_+^+} y'(\alpha) = -1, \quad \lim_{\alpha\to\alpha_+^-} y'(\alpha) = 1. \tag{22.37}$$

The steepest curves are depicted in Figure 22.4, where the dotted lines represent the ascent curves and the solid lines represent the descent curves. From (22.36) and (22.37), it is readily seen that the slope of the steepest descent curve at α_+ is 1. Also, it can be proved that there is a vertical asymptote at $x = x_0$ and that the tangent to the curves at $x = x_1$ is vertical. The steepest curves at $\alpha = \alpha_-$ can be analyzed in an entirely similar manner. For instance, it can be proved that the slope of the steepest descent curve at α_- is -1. However, it should be noted that although the two steepest curves at α_+ and α_- are similar, they are not symmetric with respect to the y-axis.

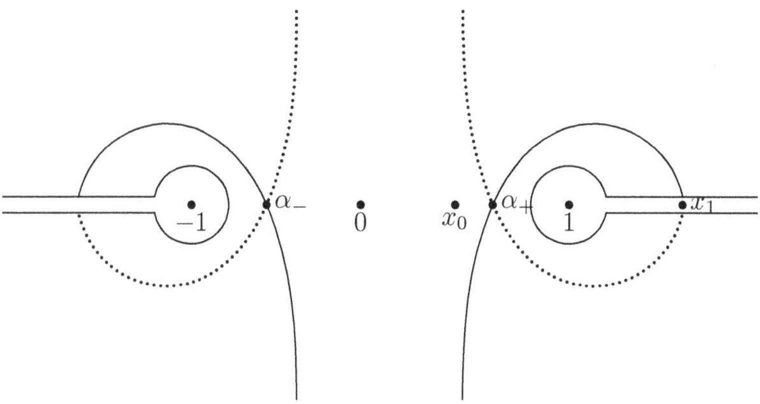

Fig. 22.4: Steepest curves in the α-plane ($t > \sqrt{a+b}$).

Next we deform the path of integration in (22.26). Referring to Figure 22.5, we see that this path consists of two infinite loops $HGCH'$ and $\bar{H}'\bar{C}\bar{G}\bar{H}$. By Cauchy's theorem, they can be replaced by the curves $HGBJ$ and $\bar{J}\bar{B}\bar{G}\bar{H}$, since it can be shown that the integral along the circular arcs $H'J$ and $\bar{J}\bar{H}'$ vanishes as the radius of the circle tends to infinity. Furthermore, by the same argument, the portions of the contour BJ and $\bar{J}\bar{B}$ can be replaced by the straight line segment $BA\bar{B}$. We have thus deformed the two loop paths of integration into the contour $HGBA\bar{B}\bar{G}\bar{H}$. To further deform the contour, we note that the line segment CH' lies on the lower edge of the cut $\alpha = 1 + \rho$, $\rho > 0$. Hence we may envisage that the deformation of the integration path from G to C to H' takes place on a sheet of the Riemann surface for $\sqrt{1-\alpha^2}$. Now the line segment GH lies on the upper edge of this cut. Hence by a similar reasoning, we can deform this portion of the contour to the vertical line segment GE on the other sheet of the Riemann surface for $\sqrt{1-\alpha^2}$; see Figure 22.6. We can repeat this argument on the left half-plane $\operatorname{Re}\alpha < 0$, i.e., replace the horizontal line segment $\bar{G}\bar{H}$ by the vertical line segment $\bar{G}\bar{E}$. The final path of integration is the curve $EGBA\bar{B}\bar{G}\bar{E}$, which we denote by Γ_1; see Figure 22.6.

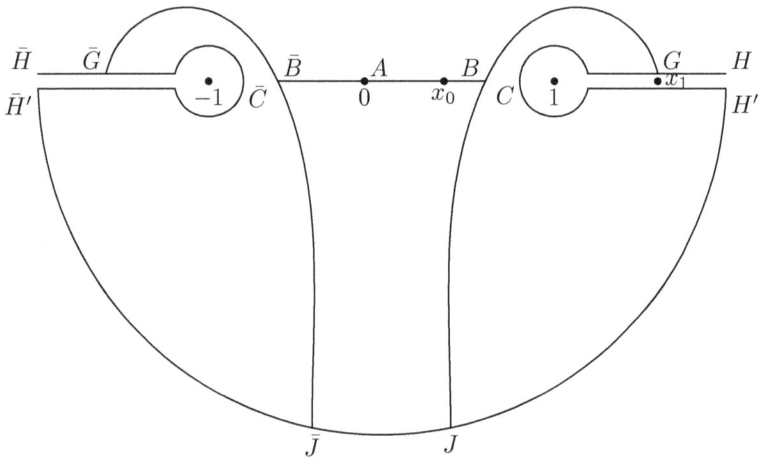

Fig. 22.5: Deformation of contour in the α-plane ($t > \sqrt{a+b}$).

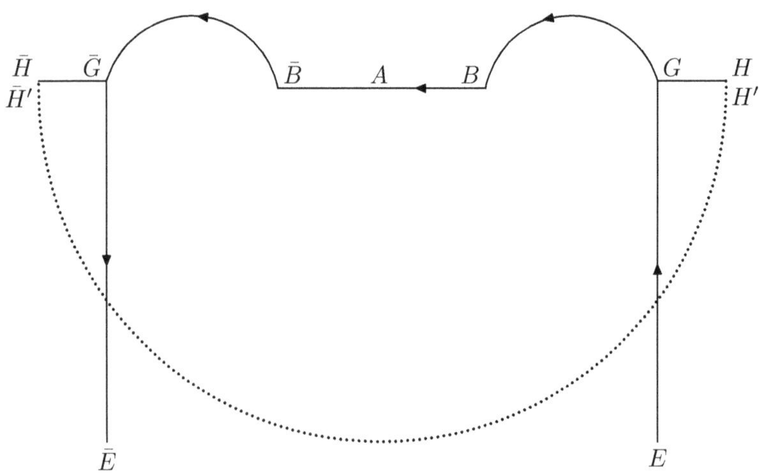

Fig. 22.6: Contour Γ_1.

The integral in (22.26) now becomes

$$I = \frac{1}{2\pi i}\int_{\Gamma_1} e^{\sqrt{n}g(\alpha,t)}\bar{K}(\alpha,t,1/\sqrt{n})d\alpha, \quad t > \sqrt{a+b}. \tag{22.38}$$

Case (ii): $t < \sqrt{a+b}$. The steepest descent paths through $\alpha = \alpha_+$ are given by equation (22.33). With $\alpha = x + iy$ and α_+ given by (22.28), this equation reduces to (exercise)

$$\frac{a+b}{4t}\ln\frac{(1+x)^2+y^2}{(1-x)^2+y^2} - xt = 0, \tag{22.39}$$

which has two solutions $x = 0$ and

$$y = \sqrt{\frac{(1+x)^2-(1-x)^2e^{4xt^2/(a+b)}}{e^{4xt^2/(a+b)}-1}}. \tag{22.40}$$

Equation (22.40) gives the steepest descent curve, whereas $x = 0$ is the steepest ascent curve. Similarly, it can be shown that the steepest descent path through $\alpha = \alpha_-$ is $x = 0$ and that the steepest ascent curve is given by

$$y = -\sqrt{\frac{(1+x)^2-(1-x)^2e^{4xt^2/(a+b)}}{e^{4xt^2/(a+b)}-1}}; \tag{22.41}$$

see Figure 22.7. It should again be noted that the curves in Figure 22.7 are not symmetric with respect to the y-axis.

Let Ω denote the closed curve $GCH'\bar{H}'\bar{C}\bar{G}BG$. By Cauchy's theorem, the integral in (22.26) along Ω is zero; i.e.,

$$\int_\Omega e^{\sqrt{n}g(\alpha,t)}\bar{K}(\alpha,t,1/\sqrt{n})d\alpha = 0,$$

where $g(\alpha,t)$ and $\bar{K}(\alpha,t,1/\sqrt{n})$ are as given in (22.25) and (22.27). Since the integral on the circular arc $\bar{H}'H'$ vanishes as $R \to \infty$, we may replace the portion of the loop paths of integration in (22.26) along the curves GCH' and $\bar{H}'\bar{C}\bar{G}$ by the curve $GB\bar{G}$. Again we envisage that this deformation takes place on one sheet of the Riemann surface for $\sqrt{1-\alpha^2}$. On the other sheet, we deform the horizontal line segments GH and $\bar{G}\bar{H}$ to vertical line segments EG and $\bar{G}\bar{E}$ as in case (i) (see Figure 22.8). The final path of integration is the curve $EGB\bar{G}\bar{E}$. We shall denote this curve by Γ_2, and the integral (22.26) becomes

$$I = \frac{1}{2\pi i}\int_{\Gamma_2} e^{\sqrt{n}g(\alpha,t)}\bar{K}(\alpha,t,1/\sqrt{n})d\alpha, \quad t < \sqrt{a+b}. \tag{22.42}$$

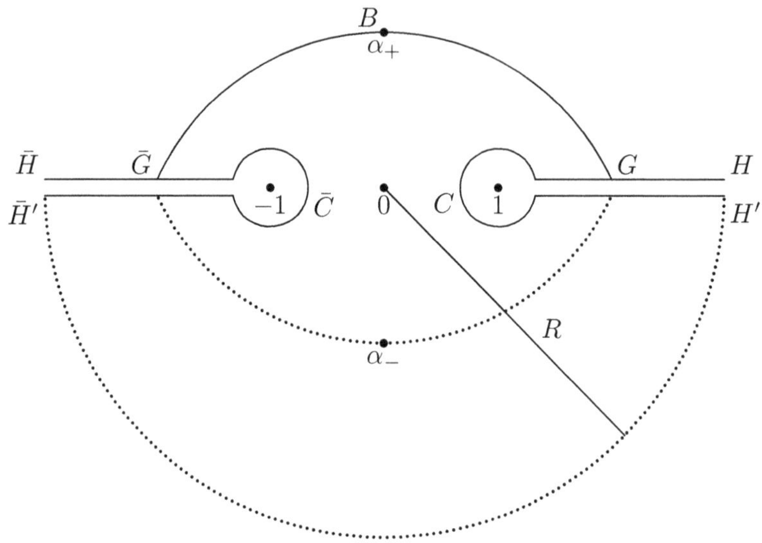

Fig. 22.7: Steepest curves in the α-plane ($t < \sqrt{a+b}$).

In the integrals in (22.38) and (22.42), we make the cubic change of variable given in (22.29). This leads to

$$I = \frac{1}{2\pi i} \int_{L_1} K(u) e^{\sqrt{n}[u^3/3 - B^2(t)u]} du, \tag{22.43}$$

for $t > \sqrt{a+b}$, and

$$I = \frac{1}{2\pi i} \int_{L_2} K(u) e^{\sqrt{n}[u^3/3 - B^2(t)u]} du, \tag{22.44}$$

for $t < \sqrt{a+b}$, where

$$K(u) = \bar{K}(\alpha(u), t, 1/\sqrt{n}) \frac{d\alpha}{du} \tag{22.45}$$

and L_1 and L_2 are the images of Γ_1 and Γ_2, respectively. In the following section, it is shown that both L_1 and L_2 begin at $\infty e^{-i\pi/3}$ and end at $\infty e^{i\pi/3}$, L_1 passes through $u = \pm B(t)$, and L_2 passes through $u = B(t)$. It is also shown that the transformation $\alpha \leftrightarrow u$ given in (22.29) is one-to-one and analytic along the entire path of integration Γ_1 and Γ_2.

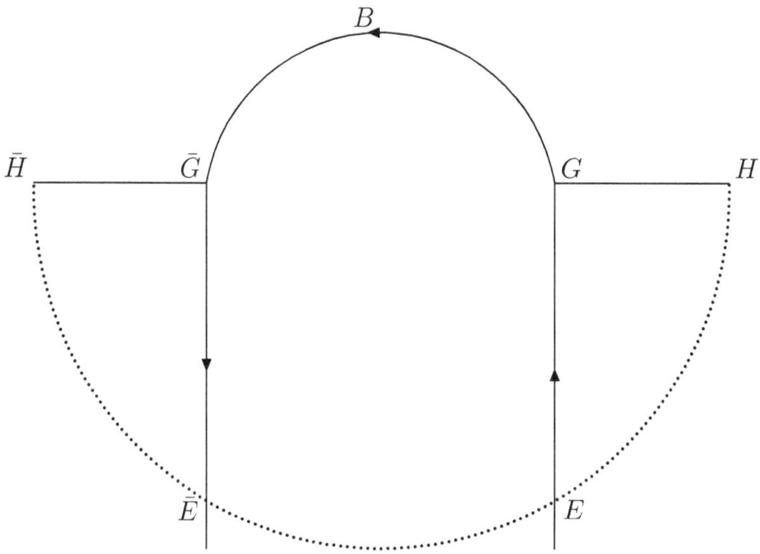

Fig. 22.8: Contour Γ_2.

22.4 The cubic transformation

The properties of the mapping between α and u are best seen by introducing an intermediate variable Z defined by

$$i\left(\frac{a+b}{2t}\ln\frac{1+\alpha}{1-\alpha}-\alpha t\right)=Z=\frac{u^3}{3}-B^2(t)u. \tag{22.46}$$

We first restrict ourselves to the case $t>\sqrt{a+b}$. In Figure 22.9, the region bounded by the curve $ABCC'DEFA$ contains half of the contour Γ_1. As α traverses once along the indented boundary of this region, Z also traverses exactly once along the corresponding curve in Figure 22.10. Here we treat the straight line AB and BC in Figure 22.10 as distinct parts of the boundary; see Ursell (1970, p. 375, lines 22-25). Hence, by Theorem 4.5 in Markushevich (1977, Vol. 2, p. 118), the function $g(\alpha,t)$ given in (22.29), i.e., the function on the left-hand side of (22.46), is one-to-one in the interior of the region bounded by this curve. The steepest descent curve BG in the Z-plane may be quite complicated. However, by the above argument, we know that it is not self-intersecting. To analyze the image of the vertical line GE in the Z-plane, we write $\alpha=x_1+iy$, $y\le 0$, and

$Z = X + iY$ in (22.46). This gives

$$X = yt - \frac{a+b}{2t}\left(\arctan\frac{y}{1+x_1} + \arctan\frac{y}{1-x_1}\right),$$
$$Y = \frac{a+b}{4t}\ln\frac{(1+x_1)^2+y^2}{(1-x_1)^2+y^2} - x_1 t.$$

As $y \to -\infty$, we have $X \to -\infty$ and $Y \to -x_1 t$. This demonstrates that the image of GE in the Z-plane has a horizontal asymptote $Y = -x_1 t$.

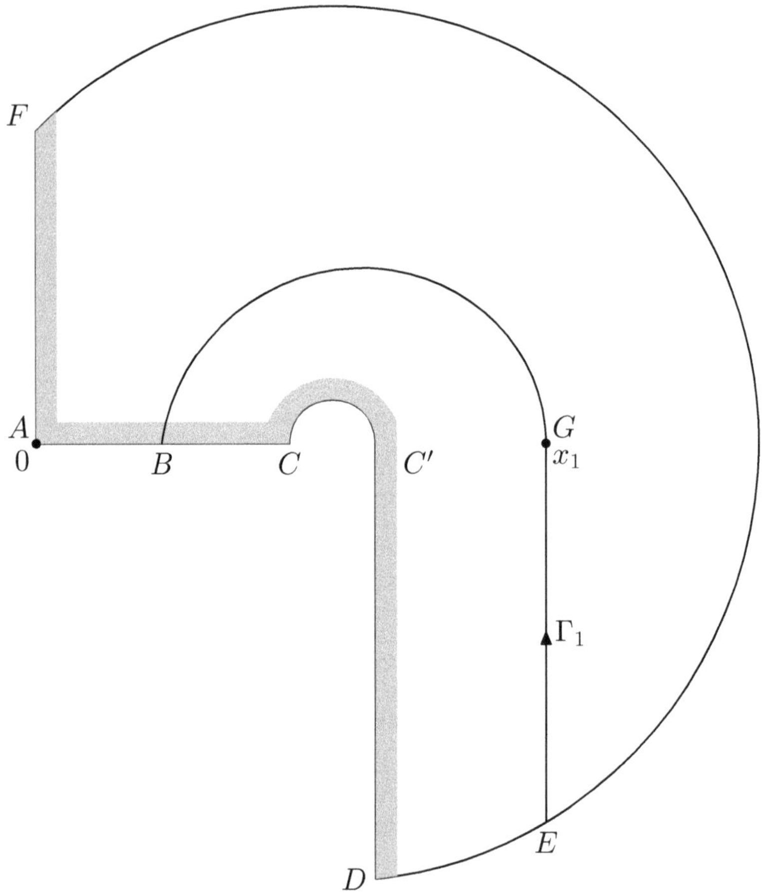

Fig. 22.9: α-plane ($t > \sqrt{a+b}$).

We now consider the mapping $u \leftrightarrow Z$ defined by $Z = u^3/3 - B^2(t)u$, which is a cubic equation and has three solutions. Since we require $\alpha = \alpha_\pm$

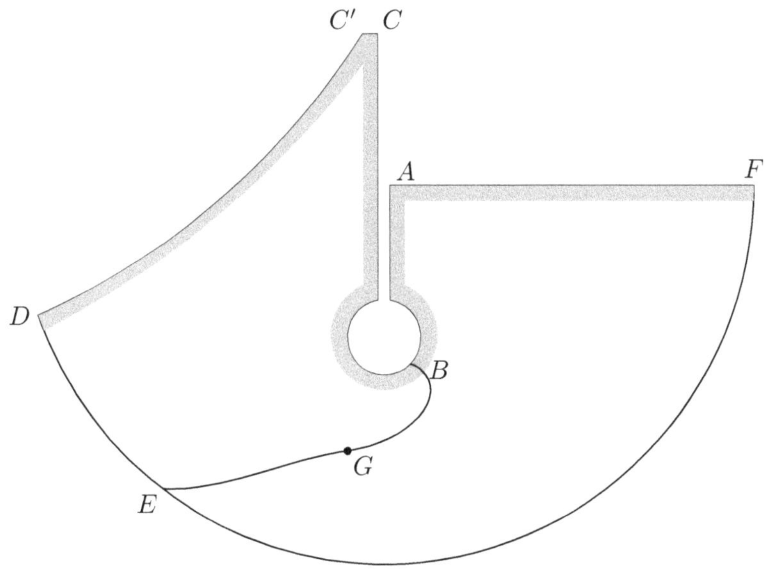

Fig. 22.10: Z-plane ($t > \sqrt{a+b}$).

to correspond to $u = \pm B(t)$, it can be shown that the required solution is given by

$$u(x,t) = 2B(t)\sin\frac{\phi}{3}, \quad \frac{2}{3}B^3(t)\sin\phi = -Z; \tag{22.47}$$

see Copson (1965, p. 109). From this exact solution, it is readily seen that the mapping $u \leftrightarrow Z$ is one-to-one in the domain bounded by the curve $ABCC'DEFA$. The equation for the curve BGE in Figure 22.11 is very complicated, but there is no need to have the precise graph of this curve. What should be noted is that this curve actually ends at $\infty e^{-i\pi/3}$; see Figure 22.11.

The transformation $\alpha \leftrightarrow u$ is obtained by composing the transformation $\alpha \leftrightarrow Z$ and $Z \leftrightarrow u$. Since the latter two are one-to-one in the region $ABCC'DEFA$, so is $\alpha \leftrightarrow u$. The mapping of a corresponding region containing the left half of the contour Γ_1 can be discussed in a similar manner. Although the contour L_1 is not symmetric with respect to the horizontal axis, it begins at $\infty e^{-i\pi/3}$, as mentioned above, and ends at $\infty e^{i\pi/3}$.

We next consider the case $t < \sqrt{a+b}$. In Figure 22.12, the region bounded by the curve $ACC'DEFBA$ contains half of the contour Γ_2. Its images in the Z-plane and u-plane are shown in Figures 22.13 and 22.14,

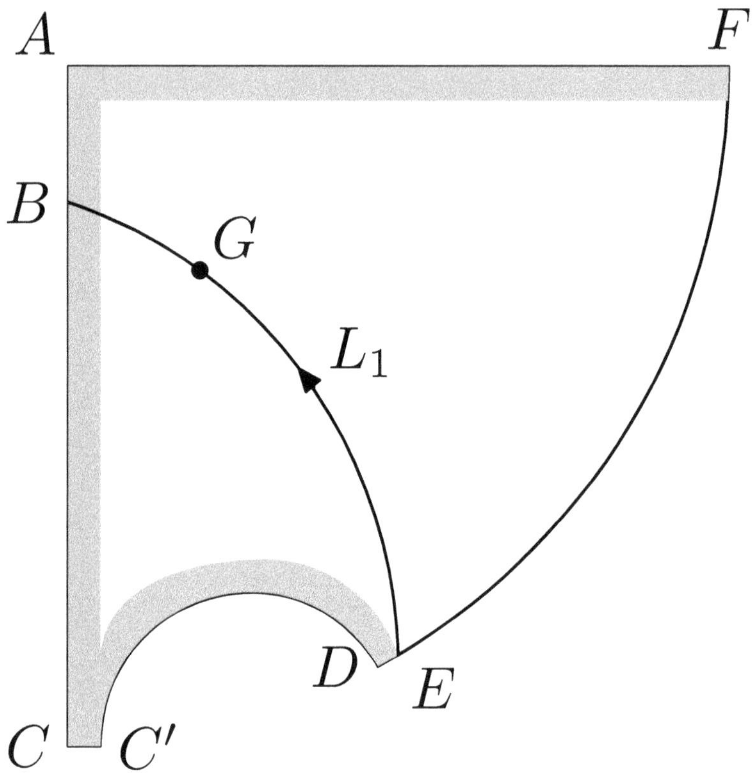

Fig. 22.11: u-plane ($t > \sqrt{a+b}$).

respectively. It can be shown, as in the case $t > \sqrt{a+b}$, that the mapping $\alpha \leftrightarrow u$ is one-to-one in the region $ACC'DEFBA$ and that the contour Γ_2 in α-plane is mapped into L_2 in the u-plane. Furthermore, the contour L_2 begins at $u = \infty e^{-i\pi/3}$ and ends at $u = \infty e^{i\pi/3}$.

Let L denote either L_1 or L_2, depending on whether $t > \sqrt{a+b}$ or $t < \sqrt{a+b}$. Then (22.43) and (22.44) can be combined into a single equation

$$I = \frac{1}{2\pi i} \int_L K(u) e^{\sqrt{n}[u^3/3 - B^2(t)u]} du. \tag{22.48}$$

Since $\alpha \leftrightarrow u$ is one-to-one and analytic along the entire path of integration Γ_1 when $t > \sqrt{a+b}$ and the entire path Γ_2 when $t < \sqrt{a+b}$, equation (22.48) is established for both cases $t > \sqrt{a+b}$ and $t < \sqrt{a+b}$. The fact that it holds also at $t = \sqrt{a+b}$ follows from continuity. To show that $u(\alpha, t)$ is infinitely continuously differentiable in $0 < t < \infty$, we recall the exact solution given in (22.47). Since $B(t)$ given in (22.30) and

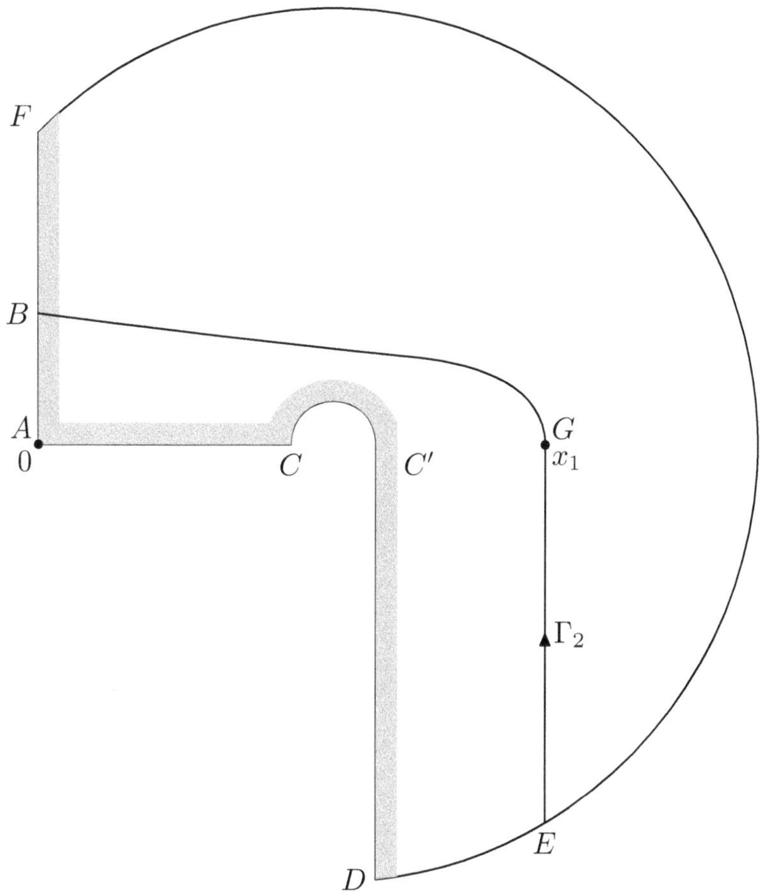

Fig. 22.12: α-plane ($t < \sqrt{a+b}$).

$Z = Z(\alpha, t)$ as given by the first equation in (22.46) are continuous in t, so is $u(\alpha, t)$. As $t \to \sqrt{a+b}$, we have $B(t) \to 0$. By expressing $\sin\phi$ in terms of $\sin(\phi/3)$, it can be shown from the second equation in (22.47) that $\sin(\phi/3) = O(1/B(t))$.

22.5 Derivation of the expansion

To derive the asymptotic expansion of the integral in (22.48), we put $K_0(u) = K(u)$ and write

$$K_0(u) = \alpha_0 + \beta_0 u + [u^2 - B^2(t)]G_0(u), \tag{22.49}$$

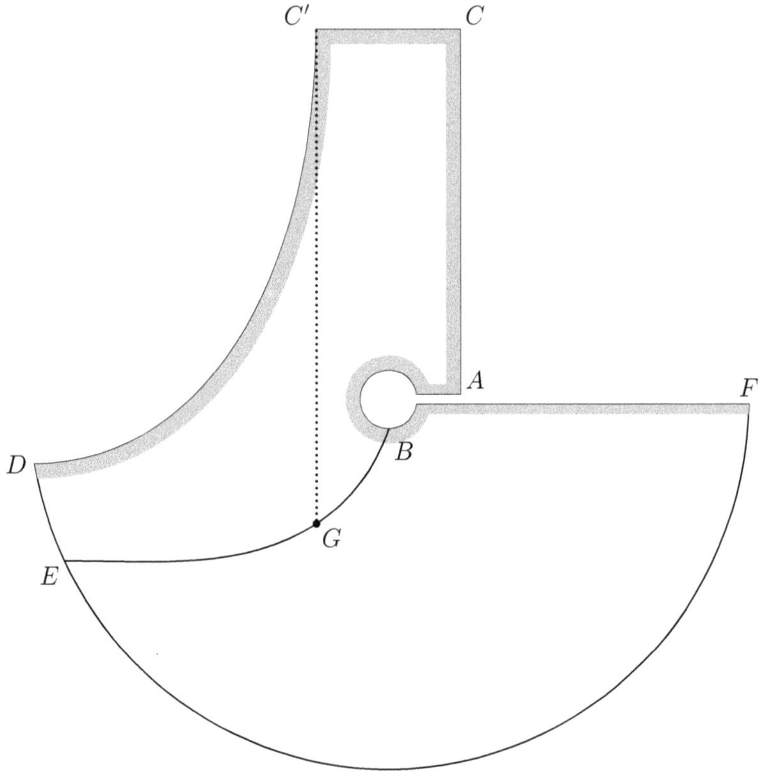

Fig. 22.13: Z-plane ($t < \sqrt{a+b}$).

where

$$\alpha_0 = \frac{K_0(B) + K_0(-B)}{2} = \frac{K(B) + K(-B)}{2}, \tag{22.50}$$

$$\beta_0 = \frac{K_0(B) - K_0(-B)}{2B} = \frac{K(B) - K(-B)}{2B}, \tag{22.51}$$

$$G_0(u) = \frac{K_0(u) - \alpha_0 - \beta_0 u}{u^2 - B^2}. \tag{22.52}$$

By rewriting $G_0(u)$ as

$$G_0(u) = \frac{K_0(u) - K_0(\pm B)}{u^2 - B^2} + \frac{K_0(\pm B) - K_0(\mp B)}{2(u^2 - B^2)}(1 \mp u/B), \tag{22.53}$$

it is clear that $G_0(u)$ has removable singularities at $u = \pm B(t)$ and consequently that it is analytic in the domain of $K_0(u)$. We now substitute (22.49) in (22.48) and express the first and second integrals in terms of the

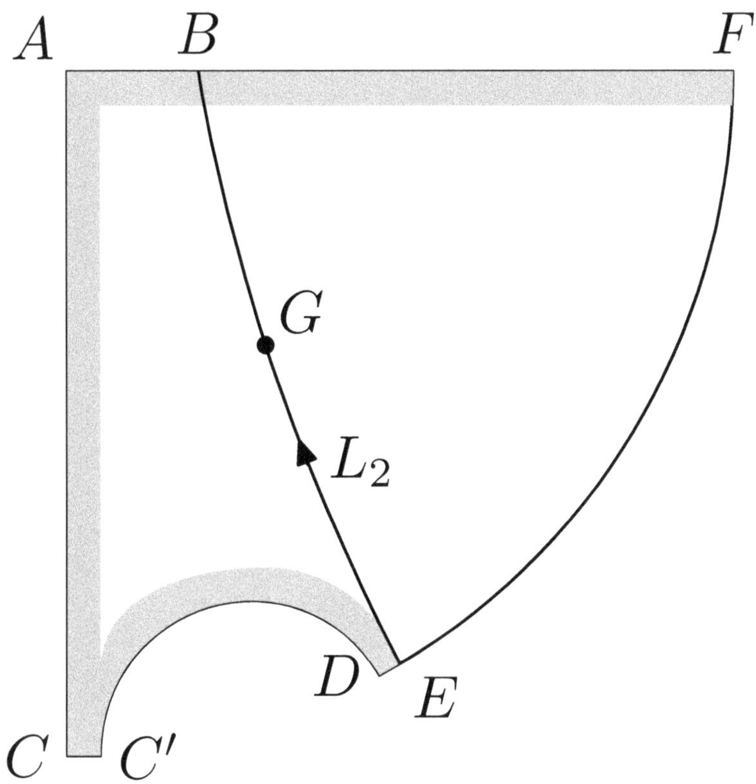

Fig. 22.14: u-plane ($t < \sqrt{a+b}$).

Airy function and its derivative, respectively. The third integral can be integrated by parts; the integrated term vanishes since $G_0(u)$ has at most algebraic growth at infinity and $\exp\{\sqrt{n}[u^3/3 - B^2(t)u]\}$ is exponentially decreasing. The final result is

$$I = \frac{\alpha_0 \operatorname{Ai}(n^{1/3}B^2)}{n^{1/6}} - \frac{\beta_0 \operatorname{Ai}'(n^{1/3}B^2)}{n^{1/3}} - \frac{1}{2\pi i\sqrt{n}} \int_L G_0'(u) e^{\sqrt{n}[u^3/3 - B^2(t)u]} du. \tag{22.54}$$

Since $G_0'(u)$ is analytic in u, the last integral is exactly of the same form as the original integral I. Hence the above procedure can be repeated. As in Chapter 19, we define inductively, for $m = 0, 1, 2, \cdots$,

$$K_m(u) = \alpha_m + \beta_m u + [u^2 - B^2(t)]G_m(u), \quad K_{m+1}(u) = G_m'(u). \tag{22.55}$$

Then we obtain

$$I = \frac{\mathrm{Ai}(n^{1/3}B^2(t))}{n^{1/6}} \sum_{k=0}^{p-1} \frac{(-1)^k \alpha_k}{n^{k/2}} - \frac{\mathrm{Ai}'(n^{1/3}B^2(t))}{n^{1/3}} \sum_{k=0}^{p-1} \frac{(-1)^k \beta_k}{n^{k/2}} + \varepsilon_p, \tag{22.56}$$

where

$$\varepsilon_p = \frac{(-1)^p}{2\pi i n^{p/2}} \int_L K_p(u) e^{\sqrt{n}[u^3/3 - B^2(t)u]} du. \tag{22.57}$$

From (22.55) it is easily seen that

$$\alpha_m(t) = \frac{K_m(B(t)) + K_m(-B(t))}{2}, \tag{22.58}$$

$$\beta_m(t) = \frac{K_m(B(t)) - K_m(-B(t))}{2B(t)}. \tag{22.59}$$

By induction, it can be proved that $G_m(u)$ and $K_m(u)$ are analytic in the domain of $K(u)$. Since the range of $B(t)$ is contained in the domain of $K(u)$, for each m, both $\alpha_m(t)$ and $\beta_m(t)$ are continuous in $0 < t < \infty$.

To estimate ε_p, we introduce the auxiliary functions

$$\widetilde{\mathrm{Ai}}(z) = \begin{cases} \mathrm{Ai}(z), & z \geq 0, \\ \sqrt{[\mathrm{Ai}(z)]^2 + [\mathrm{Bi}(z)]^2}, & z < 0, \end{cases} \tag{22.60}$$

$$\widetilde{\mathrm{Ai}}'(z) = \begin{cases} \mathrm{Ai}'(z), & z \geq 0, \\ \sqrt{[\mathrm{Ai}'(z)]^2 + [\mathrm{Bi}'(z)]^2}, & z < 0; \end{cases} \tag{22.61}$$

see Wong (1989, p. 381). The behavior of $\widetilde{\mathrm{Ai}}(z)$ and $\widetilde{\mathrm{Ai}}'(z)$ mimics the behavior of $\mathrm{Ai}(z)$ and $\mathrm{Ai}'(z)$, respectively. We now show that for $0 < \varepsilon \leq t \leq M < \infty$,

$$|\varepsilon_p| \leq \frac{M_p}{n^{p/2+1/6}} |\widetilde{\mathrm{Ai}}(n^{1/3}B^2(t))| + \frac{N_p}{n^{p/2+1/3}} |\widetilde{\mathrm{Ai}}'(n^{1/3}B^2(t))|, \tag{22.62}$$

where M_p and N_p are constants not depending on t and n.

Since $K_p(u)$ is analytic along the contour L, and since L passes through the saddle points, it is natural to use the saddle point method as described in an earlier chapter. It should be pointed out that the function $K_p(u)$ depends on the parameters t and n; see (22.45). However, since it is analytic in u for all relevant values of t and n, the dependence on t and n does not affect the final result; cf. Olver (1974a, Chapter 8, Section 2). We may therefore continue to suppress the dependence in the following discussion.

When $t \le \sqrt{a+b}$, L passes through only the saddle point $u = B(t)$; see Figure 22.14. Hence

$$\varepsilon_p \sim \frac{(-1)^p K_p(B) e^{-2\sqrt{n}B^3/3}}{2\sqrt{\pi B} n^{p/2+1/4}}. \tag{22.63}$$

When $t > \sqrt{a+b}$, L passes through both saddle points $u = \pm B(t)$; see Figure 22.11. Thus we have

$$\varepsilon_p \sim \frac{(-1)^p}{2\sqrt{\pi} n^{p/2+1/4}} \left\{ \frac{K_p(B) e^{-2\sqrt{n}B^3/3}}{B^{1/2}} + \frac{K_p(-B) e^{2\sqrt{n}B^3/3}}{(-B)^{1/2}} \right\}. \tag{22.64}$$

Since $K_p(B)$ and $K_p(-B)$ are continuous in the closed interval $[\varepsilon, M]$, in view of the large $|z|$ behavior of $\widetilde{\mathrm{Ai}}(z)$ and $\widetilde{\mathrm{Ai}}'(z)$, z being real, the result in (22.62) now follows immediately from (22.63) and (22.64).

22.6 Leading terms and local behavior

The leading coefficients α_0 and β_0 in (22.56) can be calculated as follows. From (22.50), (22.51) and (22.45) we have

$$\alpha_0 = \frac{1}{2} \left[\bar{K}(\alpha_+) \frac{d\alpha}{du}\bigg|_{u=B(t)} + \bar{K}(\alpha_-) \frac{d\alpha}{du}\bigg|_{u=-B(t)} \right], \tag{22.65}$$

$$\beta_0 = \frac{1}{2B(t)} \left[\bar{K}(\alpha_+) \frac{d\alpha}{du}\bigg|_{u=B(t)} - \bar{K}(\alpha_-) \frac{d\alpha}{du}\bigg|_{u=-B(t)} \right]. \tag{22.66}$$

Inserting (22.28) in (22.27) gives

$$\begin{aligned} \bar{K}(\alpha_\pm) = & \frac{t}{\sqrt{a+b}} \exp\left\{ \sqrt{n} \left[i \frac{a\cos(t/\sqrt{n}) + b}{2\sqrt{n}\sin(t/\sqrt{n})} \ln \frac{1 \pm \sqrt{1-(a+b)/t^2}}{1 \mp \sqrt{1-(a+b)/t^2}} \right. \right. \\ & - \left(\sqrt{n} + \frac{1}{\sqrt{n}} \right) \ln \left(\cos\frac{t}{\sqrt{n}} \pm i\sqrt{1 - \frac{a+b}{t^2}} \sin\frac{t}{\sqrt{n}} \right) \\ & \left. \left. - i \left(\frac{a+b}{2t} \ln \frac{1 \pm \sqrt{1-(a+b)/t^2}}{1 \mp \sqrt{1-(a+b)/t^2}} \mp t\sqrt{1 - \frac{a+b}{t^2}} \right) \right] \right\} \end{aligned}$$

if $t > \sqrt{a+b}$, and

$$\bar{K}(\alpha_\pm) = \exp\left\{ -(n+1) \ln\cos\sqrt{\frac{a+b}{n}} \right\}$$

if $t = \sqrt{a+b}$, and

$$\begin{aligned}\bar{K}(\alpha_\pm) =& \frac{t}{\sqrt{a+b}} \exp\left\{ \sqrt{n} \left[i \frac{a\cos(t/\sqrt{n}) + b}{2\sqrt{n}\sin(t/\sqrt{n})} \ln \frac{1 \pm i\sqrt{(a+b)/t^2 - 1}}{1 \mp i\sqrt{(a+b)/t^2 - 1}} \right.\right. \\ & - \left(\sqrt{n} + \frac{1}{\sqrt{n}}\right) \ln\left(\cos\frac{t}{\sqrt{n}} \mp \sqrt{\frac{a+b}{t^2} - 1} \sin\frac{t}{\sqrt{n}} \right) \\ & \left.\left. - i\left(\frac{a+b}{2t} \ln \frac{1 \pm i\sqrt{(a+b)/t^2 - 1}}{1 \mp i\sqrt{(a+b)/t^2 - 1}} \mp it\sqrt{\frac{a+b}{t^2} - 1} \right) \right] \right\}\end{aligned}$$

if $0 < t < \sqrt{a+b}$. By expanding $\sin(t/\sqrt{n})$, $\cos(t/\sqrt{n})$ and $\cos\sqrt{(a+b)/n}$ into Taylor series with n large, we can simplify the last three equations to (exercise)

$$\bar{K}(\alpha_\pm) = \begin{cases} \frac{te^{(a+b)/2}}{\sqrt{a+b}} \left[1 \pm \frac{i\psi_1(t)}{\sqrt{n}} + O(n^{-1}) \right], & t > \sqrt{a+b}, \\ e^{(a+b)/2} \left[1 + O(n^{-1}) \right], & t = \sqrt{a+b}, \\ \frac{te^{(a+b)/2}}{\sqrt{a+b}} \left[1 \pm \frac{\psi_2(t)}{\sqrt{n}} + O(n^{-1}) \right], & t < \sqrt{a+b}, \end{cases} \tag{22.67}$$

where

$$\begin{aligned}\psi_1(t) =& \frac{(b-2a)t}{12} \ln \frac{1 + \sqrt{1 - (a+b)/t^2}}{1 - \sqrt{1 - (a+b)/t^2}} - \frac{3+a+b}{3}\sqrt{t^2 - a - b}, \\ \psi_2(t) =& \frac{(2a-b)t}{6} \arctan\sqrt{\frac{a+b}{t^2} - 1} + \frac{3+a+b}{3}\sqrt{a+b-t^2}.\end{aligned}$$

Since ψ_1 and ψ_2 vanish at $t = \sqrt{a+b}$, (22.67) reassures us that $\bar{K}(\alpha_\pm(t))$ is continuous there. Differentiating (22.29) twice, we obtain

$$\left(\frac{d\alpha}{du}\right)^2 \bigg|_{u=\pm B(t)} = \frac{\pm B(t)}{i\left[\frac{a+b}{t} \cdot \frac{\alpha_\pm}{(1-\alpha_\pm^2)^2}\right]}.$$

Hence,

$$\left(\frac{d\alpha}{du}\right)^2 \bigg|_{u=\pm B(t)} = \frac{\left[\frac{3}{2}\left(\frac{a+b}{t} \operatorname{arccosh}\frac{t}{\sqrt{a+b}} - \sqrt{t^2-a-b}\right)\right]^{1/3}}{[t^2/(a+b)]\sqrt{t^2-a-b}}$$

if $t > \sqrt{a+b}$, and

$$\left(\frac{d\alpha}{du}\right)^2 \bigg|_{u=\pm B(t)} = -(a+b)^{-1/3}$$

if $t = \sqrt{a+b}$, and

$$\left(\frac{d\alpha}{du}\right)^2 \bigg|_{u=\pm B(t)} = \frac{\left[\frac{3}{2}\left(\frac{a+b}{t} \arccos\frac{t}{\sqrt{a+b}} - \sqrt{a+b-t^2}\right)\right]^{1/3}}{[-t^2/(a+b)]\sqrt{a+b-t^2}}$$

if $t < \sqrt{a+b}$. The second equation above is obtained by taking limits of the first or the third equation. We now take square root in these equations and make proper choice of the $\pm$ signs. This yields

$$\left.\frac{d\alpha}{du}\right|_{u=\pm B(t)} = \begin{cases} \dfrac{i\left[\frac{3}{2}\left(\sqrt{t^2-a-b}-\frac{a+b}{t}\operatorname{arccosh}\frac{t}{\sqrt{a+b}}\right)\right]^{1/6}}{(t/\sqrt{a+b})(t^2-a-b)^{1/4}}, & t > \sqrt{a+b}, \\ i(a+b)^{-1/6}, & t = \sqrt{a+b}, \\ \dfrac{i\left[\frac{3}{2}\left(\frac{a+b}{t}\arccos\frac{t}{\sqrt{a+b}}-\sqrt{a+b-t^2}\right)\right]^{1/6}}{(t/\sqrt{a+b})(a+b-t^2)^{1/4}}, & t < \sqrt{a+b}. \end{cases} \tag{22.68}$$

Substituting (22.67) and (22.68) in (22.65) and (22.66), we have (exercise)

$$\alpha_0(t) = \begin{cases} \dfrac{i\left[\frac{3}{2}\left(\sqrt{t^2-a-b}-\frac{a+b}{t}\operatorname{arccosh}\frac{t}{\sqrt{a+b}}\right)\right]^{1/6}}{e^{-(a+b)/2}(t^2-a-b)^{1/4}}[1+O(n^{-1})], & t > \sqrt{a+b}, \\ i(a+b)^{-1/6}e^{(a+b)/2}[1+O(n^{-1})], & t = \sqrt{a+b}, \\ \dfrac{i\left[\frac{3}{2}\left(\frac{a+b}{t}\arccos\frac{t}{\sqrt{a+b}}-\sqrt{a+b-t^2}\right)\right]^{1/6}}{e^{-(a+b)/2}(a+b-t^2)^{1/4}}[1+O(n^{-1})], & t < \sqrt{a+b}, \end{cases} \tag{22.69}$$

and

$$\beta_0(t) = \begin{cases} \dfrac{-\psi_1(t)\left[\frac{3}{2}\left(\sqrt{t^2-a-b}-\frac{a+b}{t}\operatorname{arccosh}\frac{t}{\sqrt{a+b}}\right)\right]^{-1/6}}{\sqrt{n}e^{-(a+b)/2}(t^2-a-b)^{1/4}} + O\left(\frac{1}{n}\right), & t > \sqrt{a+b}, \\ O(n^{-1/2}), & t = \sqrt{a+b}, \\ \dfrac{i\psi_2(t)\left[\frac{3}{2}\left(\frac{a+b}{t}\arccos\frac{t}{\sqrt{a+b}}-\sqrt{a+b-t^2}\right)\right]^{-1/6}}{\sqrt{n}e^{-(a+b)/2}(a+b-t^2)^{1/4}} + O\left(\frac{1}{n}\right), & t < \sqrt{a+b}. \end{cases} \tag{22.70}$$

We now take the leading term in (22.56) and insert it in (22.22). This leads to

$$P_n(\cos(t/\sqrt{n}); a, b) \sim -i\alpha_0(t)e^{(\pi-2t/\sqrt{n})h(t/\sqrt{n})}\operatorname{Ai}(n^{1/3}B^2(t))n^{-1/6}. \tag{22.71}$$

In view of (22.9) and (22.69), we obtain

$$P_n\left(\cos\frac{t}{\sqrt{n}}; a, b\right) \sim \exp\left\{\left(\pi - \frac{2t}{\sqrt{n}}\right)\left[\frac{a+b}{2t}\sqrt{n} + O\left(\frac{1}{\sqrt{n}}\right)\right] + \frac{a+b}{2}\right\} \\ \times \frac{\left[\frac{3}{2}\left(\sqrt{t^2-a-b} - \frac{a+b}{t}\operatorname{arccosh}\frac{t}{\sqrt{a+b}}\right)\right]^{1/6}\operatorname{Ai}(n^{1/3}B^2(t))}{n^{1/6}(t^2-a-b)^{1/4}} \tag{22.72}$$

if $\sqrt{a+b} < t \le M < \infty$, and

$$P_n\left(\cos\frac{t}{\sqrt{n}}; a, b\right) \sim \exp\left\{\left(\pi - \frac{2t}{\sqrt{n}}\right)\left[\frac{a+b}{2t}\sqrt{n} + O\left(\frac{1}{\sqrt{n}}\right)\right] + \frac{a+b}{2}\right\} \\ \times \frac{\left[\frac{3}{2}\left(\frac{a+b}{t}\arccos\frac{t}{\sqrt{a+b}} - \sqrt{a+b-t^2}\right)\right]^{1/6}\operatorname{Ai}(n^{1/3}B^2(t))}{n^{1/6}(a+b-t^2)^{1/4}} \tag{22.73}$$

if $0 < \varepsilon \le t < \sqrt{a+b}$.

If $t \geq \sqrt{a+b}(1+\delta)$, then $n^{1/3}B^2(t) \to -\infty$. Hence by Olver *et al.* (2010, p. 198)

$$\operatorname{Ai}(n^{1/3}B^2(t)) \sim \frac{n^{-1/12}\cos\left[\frac{\pi}{4} - \sqrt{n}\left(\alpha_+ t - \frac{a+b}{2t}\ln\frac{1+\alpha_+}{1-\alpha_-}\right)\right]}{\sqrt{\pi}\left[\frac{3}{2}\left(\sqrt{t^2-a-b} - \frac{a+b}{t}\operatorname{arccosh}\frac{t}{\sqrt{a+b}}\right)\right]^{1/6}}. \tag{22.74}$$

If $t \leq \sqrt{a+b}(1-\delta)$, then $n^{1/3}B^2(t) \to +\infty$. Consequently, again from Olver *et al.* (2010, p. 198)

$$\operatorname{Ai}(n^{1/3}B^2(t)) \sim \frac{n^{-1/12}\exp\left[\sqrt{n}\left(|\alpha_+|t - \frac{a+b}{t}\arctan|\alpha_+|\right)\right]}{2\sqrt{\pi}\left[\frac{3}{2}\left(\frac{a+b}{t}\arccos\frac{t}{\sqrt{a+b}} - \sqrt{a+b-t^2}\right)\right]^{1/6}}. \tag{22.75}$$

Inserting (22.74) and (22.75) in (22.72) and (22.73), respectively, we obtain (exercise)

$$P_n\left(\cos\frac{t}{\sqrt{n}}; a, b\right) \sim \frac{\cos\left[\frac{\pi}{4} - \sqrt{n}\left(\alpha_+ t - \frac{a+b}{2t}\ln\frac{1+\alpha_+}{1-\alpha_-}\right)\right] e^{(a+b)\left(\frac{\sqrt{n}\pi}{2t} - \frac{1}{2}\right)}}{\sqrt{\pi}(t^2-a-b)^{1/4}n^{1/4}} \tag{22.76}$$

if $\sqrt{a+b} < t \leq M < \infty$, and

$$P_n\left(\cos\frac{t}{\sqrt{n}}; a, b\right) \sim \frac{e^{\sqrt{n}\left(|\alpha_+|t - \frac{a+b}{t}\arctan|\alpha_+|\right) + (a+b)\left(\frac{\sqrt{n}\pi}{2t} - \frac{1}{2}\right)}}{2\sqrt{\pi}(a+b-t^2)^{1/4}n^{1/4}} \tag{22.77}$$

if $0 < \varepsilon \leq t < \sqrt{a+b}$. These results agree with those obtained by Novikoff.

22.7 Zeros

Let $t_\nu = t_\nu(n)$, $\nu = 1, 2, \cdots$, denote the zeros of $P_n(\cos(t/\sqrt{n}); a, b)$ and arrange them as $0 < t_1 < t_2 < \cdots$. Novikoff has shown that $t_\nu(n) \geq \sqrt{a+b}$ and

$$\lim_{n\to\infty} t_\nu(n) = \sqrt{a+b}; \tag{22.78}$$

see Novikoff (1954, p. 38). Hence for sufficiently large n, $t_\nu(n)$ is bounded away from 0 and $+\infty$.

From (22.22) and (22.56), we have

$$P_n\left(\cos\frac{t}{\sqrt{n}}; a, b\right) = -ie^{(\pi - 2t/\sqrt{n})h(t/\sqrt{n})} \times \left[\frac{\alpha_0(t)\operatorname{Ai}(n^{1/3}B^2(t))}{n^{1/6}} - \frac{\beta_0(t)\operatorname{Ai}'(n^{1/3}B^2(t))}{n^{1/3}} + \varepsilon_1\right],$$

where $\varepsilon_1 = O(n^{-2/3})$. Since $\beta_0(t) = O(n^{-1/2})$ by (22.70), we can rewrite the last equation as

$$P_n\left(\cos\frac{t}{\sqrt{n}}; a, b\right) = -\, ie^{(\pi-2t/\sqrt{n})h(t/\sqrt{n})}\left[\frac{\alpha_0(t)\,\mathrm{Ai}(n^{1/3}B^2(t))}{n^{1/6}} + \bar{\varepsilon}_1\right], \tag{22.79}$$

where $\varepsilon_1 = O(n^{-2/3})$. From (22.69) it is clear that for $0 < \varepsilon \le t < \sqrt{a+b}$ and n sufficiently large, $\alpha_0(t) \neq 0$. Hence it follows from (22.79) that $t_\nu(n)$ is a root of the equation

$$\mathrm{Ai}(n^{1/3}B^2(t)) + \varepsilon(t;n) = 0, \tag{22.80}$$

where

$$\varepsilon(t;n) = O(n^{-1/2}).$$

Hethcote (1968) showed that if a_ν is the ν-th negative zero of the Airy function, ρ is small enough so that $m = \min|\,\mathrm{Ai}'(x)| > 0$, minimum being taken over $[a_\nu - \rho, a_\nu + \rho]$, and n is sufficiently large so that

$$|\varepsilon(t;n)| \le \frac{K}{n^{1/2}} < \min\{|\,\mathrm{Ai}(a_\nu - \rho)|, |\,\mathrm{Ai}(a_\nu + \rho)|\},$$

then

$$|n^{1/3}B^2(t) - a_\nu| \le \frac{K}{mn^{1/2}}; \tag{22.81}$$

see also Hethcote (1970). Motivated by (22.78), we put

$$t_\nu(n) = \sqrt{a+b} + \delta_\nu(n), \tag{22.82}$$

where $\delta_\nu(n) > 0$ and

$$\lim_{n\to\infty} \delta_\nu(n) = 0. \tag{22.83}$$

Substituting (22.82) in (22.81), we have from (22.30)

$$-\left[\frac{3}{2}\beta(\sqrt{a+b} + \delta_\nu)\right]^{2/3} = \frac{a_\nu}{n^{1/3}} + O(n^{-5/6}). \tag{22.84}$$

In (22.84), we take square root on both sides. Since $\beta(t)$ is negative, we obtain

$$\beta(\sqrt{a+b} + \delta_\nu) = -\frac{2}{3}\left[-\frac{a_\nu}{n^{1/3}} + O(n^{-5/6})\right]^{3/2}. \tag{22.85}$$

Equation (22.31) gives

$$\beta(\sqrt{a+b} + \delta_\nu) = \frac{a+b}{\sqrt{a+b} + \delta_\nu}\,\mathrm{arccosh}\,\frac{\sqrt{a+b} + \delta_\nu}{\sqrt{a+b}} - \sqrt{2\delta_\nu\sqrt{a+b} + \delta_\nu^2}.$$

Clearly,

$$\sqrt{2\delta_\nu\sqrt{a+b}+\delta_\nu^2}=\sqrt{2\sqrt{a+b}\delta_\nu}+\frac{\sqrt{2}\delta_\nu^{3/2}}{4(a+b)^{1/4}}-\frac{\sqrt{2}\delta_\nu^{5/2}}{32(a+b)^{3/4}}+\cdots.$$

Using the identity $\operatorname{arccosh} x=\ln(x+\sqrt{x^2+1})$, $x\geq 1$, we also have

$$\operatorname{arccosh}\frac{\sqrt{a+b}+\delta_\nu}{\sqrt{a+b}}=\frac{\sqrt{2\delta_\nu}}{(a+b)^{1/4}}-\frac{\sqrt{2}\delta_\nu^{3/2}}{12(a+b)^{3/4}}+\frac{3\sqrt{2}\delta_\nu^{5/2}}{160(a+b)^{5/4}}+\cdots.$$

Thus

$$\beta(\sqrt{a+b}+\delta_\nu)=-\frac{4\sqrt{2}\delta_\nu^{3/2}}{3(a+b)^{1/4}}+\frac{17\sqrt{2}\delta_\nu^{5/2}}{15(a+b)^{3/4}}+\cdots. \tag{22.86}$$

Substituting (22.86) in (22.85) yields

$$-\frac{4\sqrt{2}\delta_\nu^{3/2}}{3(a+b)^{1/4}}+\frac{17\sqrt{2}\delta_\nu^{5/2}}{15(a+b)^{3/4}}+\cdots=-\frac{2(-a_\nu)^{3/2}}{3n^{1/2}}+O(n^{-1}). \tag{22.87}$$

On balancing the leading order terms on both sides, we obtain

$$\delta_\nu=\frac{-a_\nu(a+b)^{1/6}}{2n^{1/3}}+\sigma(n), \tag{22.88}$$

where $\sigma(n)=o(n^{-1/3})$. Inserting (22.88) in (22.87) shows that $\sigma(n)=O(n^{-2/3})$. Therefore it follows from (22.82) and (22.88) that

$$t_\nu(n)=\sqrt{a+b}+\frac{-a_\nu(a+b)^{1/6}}{2n^{1/3}}+O(n^{-2/3}). \tag{22.89}$$

Since $\theta_{n,\nu}=t_\nu(n)/\sqrt{n}$, (22.89) is equivalent to (22.4):

$$\theta_{n,\nu}=\sqrt{\frac{a+b}{n}}+\frac{(a+b)^{1/6}(-a_\nu)}{2n^{5/6}}+O(n^{-7/6}). \tag{22.90}$$

22.8 Exercises

1. Verify (22.30).
2. Verify (22.34).
3. Verify (22.39).
4. Verify (22.67).
5. Verify (22.69) and (22.70).
6. Verify (22.76) and (22.77).

7. (a) The hypergeometric function is defined by

$$F(a,b,c;x) = \sum_{n=0}^{\infty} \frac{(a)_n (b)_n}{(c)_n} x^n, \quad |x| < 1.$$

Verify the integral formula, $\operatorname{Re} a > 0$ and arbitrary complex b, c:

$$F(a,b,c;x) = \frac{\Gamma(c)\Gamma(1+a-c)}{2\pi i \Gamma(a)} \int_C s^{a-1}(s-1)^{c-a-1}(1-xs)^{-b} ds,$$

where C is a counter-clockwise loop that passes through the origin and encloses the point $s = 1$. (Hint: assume first that $\operatorname{Re} c > \operatorname{Re} a > 0$ and change the contour to run along the interval $[0,1]$ and back.)
(b) The Meixner polynomials are defined by

$$\begin{aligned} m_n(x;b,c) &= (-c)^{-n} \Psi_n(x) \\ &= \sum_{k=0}^{n} \binom{n}{k} (x+b)_{n-k}(x-k+1)_k(-c)^{-k}. \end{aligned}$$

Prove that for $\operatorname{Re}(b+x) > 0$, we have

$$m_n(x;b,c) = \frac{(b)_n \Gamma(x+1)\Gamma(b)}{2\pi i c^n \Gamma(b+x)} \int_C \frac{t^{b+x-1}[1-(1-c)t]^n}{(t-1)^{x+1}} dt,$$

where C is a counter-clockwise loop through $t = 0$ that encloses $t = 1$. Thus,

$$m_n(x;b,c) = \frac{(b)_n \Gamma(x+1)\Gamma(b)}{2\pi i c^n \Gamma(b+x)} \int_C \frac{t^{b-1}}{t-1} e^{-nf(t)} dt,$$

where $x = n\alpha$, and

$$f(t) = -\alpha \ln t + \alpha \ln(t-1) - \ln[1-(1-c)t].$$

(c) Show that for $0 < c < 1$, fixed x (i.e., $\alpha = O(n^{-1})$), and large n, the phase function $f(t)$ has two saddle points

$$t_+ \sim 1 - \frac{c\alpha}{1-c}, \quad t_- \sim \frac{\alpha}{1-c}$$

in the interval $(0,1)$. Note that for large n (i.e., small α), the two saddle points are well separated. Deform the contour in (b) to run from $t = 0$ to $t = t_+$ through t_- and with $\arg(t-1) = -\pi$, and from t_+ in the lower half-plane $\operatorname{Im} t < 0$ to $t_c = 1/(1-c)$, where the integrand vanishes; from t_c the contour runs in the half-plane $\operatorname{Im} t > 0$ to t_+, and returns to the origin with $\arg(t-1) = \pi$; see the figure below.

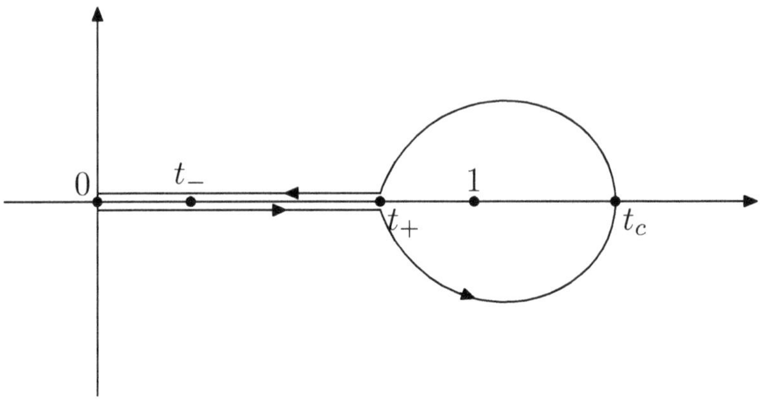

Fig. 22.15: The deformed contour.

Denote the sum of the integrals along $[0, t_+]$ by I_1 and the remaining portion of the contour by I_2. Show that

$$m_n(n\alpha; b, c) = \frac{(b)_n \Gamma(x+1)\Gamma(b)}{c^n \Gamma(b+x)}(I_1 + I_2),$$

where

$$I_1 = -\frac{\sin(\pi x)}{\pi} \int_0^{t_+} \frac{t^{b+x-1}}{(1-t)^{x+1}} [1-(1-c)t]^n dt,$$

$$I_2 = \frac{1}{2\pi i} \int_{t_+}^{(1^+)} \frac{t^{b+x-1}}{(1-t)^{x+1}} [1-(1-c)t]^n dt.$$

(d) With $x = n\alpha$, show that

$$I_1 = -\frac{\sin(\pi x)}{\pi} \int_0^{t_+} \frac{t^{b-1}}{1-t} e^{-nf(t)} dt,$$

where

$$f(t) = -\alpha \ln t + \alpha \ln(1-t) - \ln[1-(1-c)t].$$

This function has the same saddle points as the phase function given in (a). Note that on the interval $(0, t_+)$, this function has only one saddle point which occurs at $t = t_-$. Furthermore, as $\alpha \to 0$, this point coalesces with the end point $t = 0$. In fact, this point disappears when $\alpha = 0$, since the function $\ln[1-(1-c)t]$ has no saddle point in the interval $(0, t_+)$. Thus, the steepest descent method does not work. Show that the integral

$$\int_0^\infty s^{n\alpha} e^{-ns} ds = \int_0^\infty e^{-n(s-\alpha \ln s)} ds$$

has the same asymptotic phenomenon as I_1, namely, the phase function $\phi(s) = s - \alpha \ln s$ also has a saddle point (i.e., $s = \alpha$) which coincides with the endpoint $s = 0$ as $\alpha \to 0$. This suggests the change of variable $t \to s$ defined implicitly by

$$f(t) = -\alpha \ln s + s + A,$$

where A is a constant independent of t or s. Show that to make the mapping $t \to s$ one-to-one and analytic in the interval $0 < t < t_+$, one should choose A so that $t = t_-$ corresponds to $s = a$, i.e.,

$$A = f(t_-) + \alpha \ln \alpha - \alpha.$$

This gives

$$I_1 = -\frac{\sin(\pi x)}{\pi} e^{-nA} \int_0^{s_+} h(s) s^{b+x-1} e^{-ns} ds,$$

where

$$h(s) = \left(\frac{t}{s}\right)^{b-1} \frac{1}{1-t} \frac{dt}{ds},$$

and s_+ is implicitly defined via the equation $g(t) = -\alpha \ln s + s + A$ by substituting $t = t_+$ and $s = s_+$.

(e) Show that the function $t = t(s)$ defined in (d) satisfies

$$\left.\frac{dt}{ds}\right|_{s=\alpha} = \frac{1}{\sqrt{\alpha f''(t_-)}}.$$

Furthermore, the integral I_1 in (c) has the leading order approximation

$$I_1 \sim -\frac{\sin(\pi x)}{\pi} e^{-nA} h(\alpha) \frac{\Gamma(b+\alpha)}{n^{b+x}}$$

as $n \to \infty$, uniformly with respect to $\alpha \in [0, a_0]$, $a_0 > 0$. Deduce from this the asymptotic formula

$$m_n(x; b, c) \sim -\frac{\Gamma(b+n)\Gamma(x+1)}{c^n (1-c)^{x+b} n^{b+c}} \frac{\sin(\pi x)}{\pi}.$$

The derivation of this formula, as presented in this exercise, is due to N. M. Temme (private communication, 2009).

8. The Charlier polynomials are defined by

$$C_n(x; a) = \sum_{k=0}^{n} \frac{(-n)_k (-x)_k}{k!} \left(-\frac{1}{a}\right)^k.$$

By following the same arguments as in the previous exercise, prove the asymptotic formula

$$C_n(x; a) \sim -\frac{n! e^a}{a^n} \frac{\sin(\pi x)}{\pi} \frac{\Gamma(1+x)}{n^{x+1}} = \frac{n! e^a}{a^n \Gamma(-x) n^{x+1}}$$

as $n \to \infty$; see Beals and Wong (2016, p. 152).

Chapter 23

Error Bounds for the Steepest Descent Method

23.1 Airy function

In a previous chapter, we have already encountered the Airy function defined by the integral

$$\mathrm{Ai}(z) = \frac{1}{2\pi i}\int_L \exp\left(\frac{t^3}{3} - zt\right) dt, \tag{23.1}$$

where L is any contour which begins at infinity in the sector $-\pi/2 < \arg t < -\pi/6$ and ends at infinity in the sector $\pi/6 < \arg t < \pi/2$. From (23.1), it can be easily verified that $Ai(z)$ has the Maclaurin expansion (exercise)

$$\mathrm{Ai}(z) = 3^{-2/3}\sum_{n=0}^{\infty}\frac{z^{3n}}{3^{2n}n!\Gamma(n+3/2)} - 3^{-4/3}\sum_{n=0}^{\infty}\frac{z^{3n+1}}{3^{2n}n!\Gamma(n+4/3)}. \tag{23.2}$$

Thus, $\mathrm{Ai}(z)$ is an entire function. By using the steepest descent method, we have shown in Chapter 17 that

$$\mathrm{Ai}(z) \sim \frac{e^{-2z^{3/2}/3}}{2\pi z^{1/4}}\sum_{m=0}^{\infty}\frac{(-1)^m\Gamma(3m+1/2)z^{-3m/2}}{(2m)!9^m}, \tag{23.3}$$

as $z \to \infty$, uniformly in $\arg z$ for $|\arg z| \le \pi/3 - \delta$, $\delta > 0$.

In this chapter, we are concerned with the estimation of the error term upon truncation of the series in (23.3). To do this, we retrace some of the steps involved in the derivation of (23.3). First, we make the change of variable $t = z^{1/2}u$. Assuming temporarily that z is real and positive, the integral in (23.1) becomes

$$\mathrm{Ai}(z) = \frac{z^{1/2}}{2\pi i}\int_L \exp\left[z^{3/2}(u^3/3 - u)\right] du.$$

Once this identity has been established, the restriction on z can be removed by analytic continuation. Put

$$\xi = \frac{2}{3}z^{3/2}, \quad f(u) = \frac{u^3 - 3u}{2} \tag{23.4}$$

so that

$$\mathrm{Ai}(z) = \frac{(3\xi/2)^{1/3}}{2\pi i}\int_L e^{\xi f(u)}du. \tag{23.5}$$

The saddle points of $f(u)$ are located at $u = \pm 1$. Clearly, $f(\pm 1) = \mp 1$. Let

$$\theta := \arg\xi, \tag{23.6}$$

and consider the steepest descent curves

$$\Gamma_{\pm 1}(\theta): \ \arg\{\xi[f(\pm 1) - f(u)]\} = \arg\{e^{i\theta}[\mp 1 - f(u)]\} = 0. \tag{23.7}$$

It is readily seen that only the saddle point at $u = 1$ is relevant; see Wong (1989, p. 92). Deform L into $\Gamma_1(\theta)$, and write

$$\mathrm{Ai}(z) = \frac{e^{-\xi}(3\xi/2)^{1/3}}{2\pi i}\int_{\Gamma_1(\theta)} e^{\xi[f(u)+1]}du. \tag{23.8}$$

Furthermore, we introduce the notations

$$I^{(\pm 1)}(\xi) = \xi^{1/2}\int_{\Gamma_1(\theta)} e^{\xi[f(u)\pm 1]}du. \tag{23.9}$$

In terms of $I^{(1)}(\xi)$, (23.8) can be written as

$$\mathrm{Ai}(z) = \frac{e^{-\xi}(3/2)^{1/3}I^{(1)}(\xi)}{2\pi i\xi^{1/6}}. \tag{23.10}$$

In (23.9), we make the change of variable

$$\tau = \xi[-1 - f(u)]. \tag{23.11}$$

For $u \in \Gamma_1(\theta)$, τ is real and positive; see (23.7). Expanding $f(u)$ into a Taylor series at $u = 1$, we have from (23.11)

$$-\tau = \frac{3}{2}\xi[(u-1)^2 + (u-1)^3/3],$$

from which it follows that

$$\pm i\sqrt{\frac{2\tau}{3\xi}} = (u-1)[1 + (u-1)/3]^{1/2}$$

where proper branches of square roots have been chosen and $[\cdots]^{1/2}$ is that branch which reduces to 1 at $u = 1$. By Lagrange's formula for the reversion of series

$$u^{\pm} = 1 + \sum_{n=0}^{\infty}\alpha_n\left(\pm i\sqrt{\frac{2\tau}{3\xi}}\right)^n. \tag{23.12}$$

Breaking the integration path $\Gamma_1(\theta)$ in (23.9) at $u = 1$, we can rewrite the integral $I^{(1)}(\xi)$ as

$$I^{(1)}(\xi) = \xi^{1/2} \int_0^\infty \left(\frac{du^+}{d\tau} - \frac{du^-}{d\tau} \right) e^{-\tau} d\tau. \tag{23.13}$$

Coupling (23.12) and (23.13) gives the asymptotic expansion

$$I^{(1)}(\xi) \sim \sum_{n=0}^{\infty} \frac{d_n}{\xi^n}, \tag{23.14}$$

where the coefficients d_n can be given explicitly. For details, we refer to Wong (1989, p. 93).

23.2 A modification of the steepest descent method

Instead of (23.13), Berry and Howls (1990) used the equivalent representation

$$I^{(1)}(\xi) = \xi^{-1/2} \int_0^\infty \left[\frac{1}{f'(u^-(\tau))} - \frac{1}{f'(u^+(\tau))} \right] e^{-\tau} d\tau. \tag{23.15}$$

They further observed that the integrand in (23.15) can be expressed as a residue. Indeed, they gave

$$\frac{1}{f'(u^-(\tau))} - \frac{1}{f'(u^+(\tau))} = \frac{\xi^{3/2}}{2\pi i \tau^{1/2}} \int_{C_1(\theta)} \frac{[-1 - f(u)]^{1/2}}{\xi[-1 - f(u)] - \tau} du, \tag{23.16}$$

where $C_1(\theta)$ is a positively oriented curve surrounding the steepest descent path $\Gamma_1(\theta)$. (Since $\Gamma_1(\theta)$ is an infinite contour, $C_1(\theta)$ actually consists of two infinite curves embracing $\Gamma_1(\theta)$; see Figure 23.1.) To establish (23.16), we recall the formula

$$\operatorname{Res}\left\{ \frac{Q(u)}{P(u)}; u_0 \right\} = \frac{Q(u_0)}{P'(u_0)},$$

where $P(u)$ and $Q(u)$ are analytic functions with $P(u_0) = 0$, $P'(u_0) \neq 0$ and $Q(u_0) \neq 0$. Take $P(u) = \xi[-1 - f(u)] - \tau$, $Q(u) = [-1 - f(u)]^{1/2}$ and $u_0 = u^\pm(\tau)$. Then

$$\frac{Q(u^\pm(\tau))}{P'(u^\pm(\tau))} = \operatorname{Res}\left\{ \frac{Q(u)}{P(u)}; u^\pm(\tau) \right\} = \frac{1}{2\pi i} \int_{C^\pm} \frac{Q(u)}{P(u)} du, \tag{23.17}$$

where the two contours C^+ and C^- are shown in Figure 23.1.

The desired result (23.16) now follows from (23.17) and (23.11). Inserting (23.16) into (23.15) gives a double integral, and, upon interchanging the order of integration, one obtains

$$I^{(1)}(\xi) = \frac{1}{2\pi i} \int_{C_1(\theta)} [-1 - f(u)]^{-1/2} \int_0^\infty \frac{e^{-\tau} \tau^{-1/2}}{1 - \frac{\tau}{\xi[-1-f(u)]}} d\tau du. \tag{23.18}$$

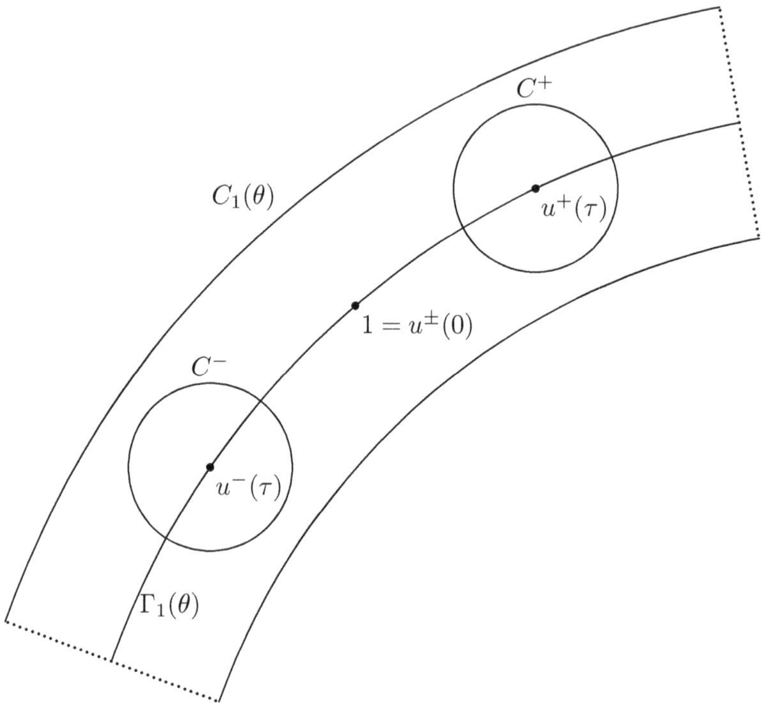

Fig. 23.1: Contour $C_1(\theta)$.

The geometric series

$$\frac{1}{1-x} = \sum_{s=0}^{N-1} x^s + \frac{x^N}{1-x}$$

then gives the asymptotic expansion

$$I^{(1)}(\xi) = \sum_{s=0}^{N-1} c_s \xi^{-s} + R_N(\xi), \tag{23.19}$$

where

$$c_s = \frac{\Gamma(s+1/2)}{2\pi i} \int_{C_1(\theta)} [-1-f(u)]^{-s-1/2} du, \tag{23.20}$$

and

$$R_N(\xi) = \frac{\xi^{-N}}{2\pi i} \int_0^\infty e^{-\tau} \tau^{N-1/2} \int_{C_1(\theta)} \frac{[-1-f(u)]^{-N-1/2}}{1 - \frac{\tau}{\xi[-1-f(u)]}} du d\tau. \tag{23.21}$$

The coefficients c_s can be evaluated exactly, and we have (exercise)

$$c_s = \frac{(-1)^s i\Gamma(3s+1/2)}{(2s)!9^s(3/2)^{s+1/2}}. \tag{23.22}$$

Now consider all steepest descent paths $\Gamma_1(\theta)$ passing through $u = 1$ for different values of θ; see Figure 23.2. Since $f(1) - f(-1) = -2$, the path

$$\Gamma_1(\pi): \quad \arg\{e^{i\pi}[f(1) - f(u)]\} = 0 \tag{23.23}$$

runs into the saddle point $u = -1$. Berry and Howl called $u = -1$ an *adjacent saddle* of $u = 1$, and the steepest descent path

$$\Gamma_{-1}(\pi): \quad \arg\{e^{i\pi}[f(-1) - f(u)]\} = 0 \tag{23.24}$$

an *adjacent contour*; see also Boyd (1993).

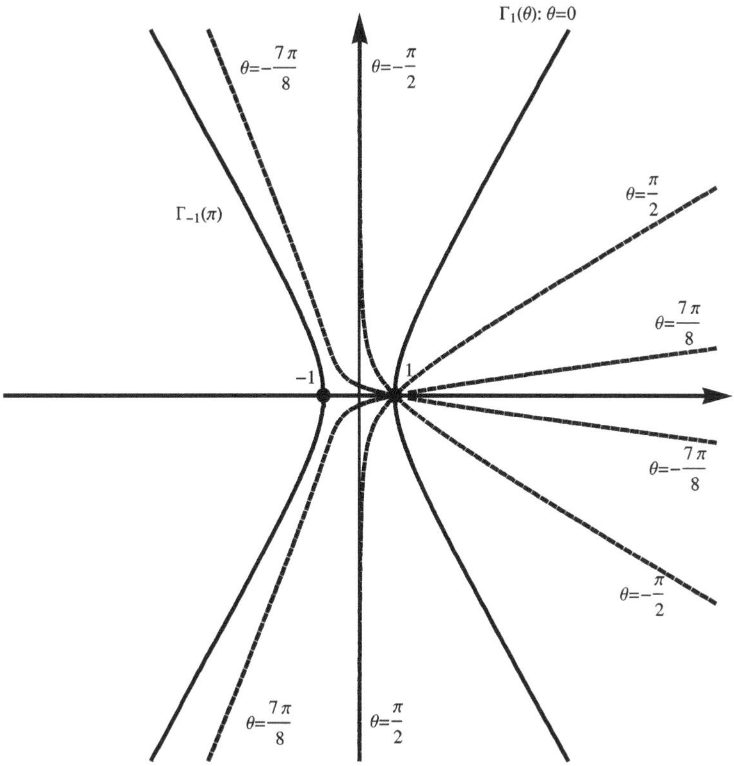

Fig. 23.2: Contours $\Gamma_1(\theta)$ with $-\pi < \theta < \pi$.

The next step is to deform $C_1(\theta)$ into $\Gamma_{-1}(\pi)$ so that (23.21) becomes

$$R_N(\xi) = \frac{\xi^{-N}}{2\pi i} \int_0^\infty e^{-\tau} \tau^{N-1/2} \int_{\Gamma_{-1}(\theta)} \frac{[-1-f(u)]^{-N-1/2}}{1 - \frac{\tau}{\xi[-1-f(u)]}} du d\tau. \quad (23.25)$$

In the last equation, we make the change of variable

$$\tau = \frac{t[f(1)-f(u)]}{f(1)-f(-1)}; \quad (23.26)$$

recall that $f(1) = -1$. Since

$$\frac{f(1)-f(u)}{f(1)-f(-1)} = 1 + \frac{f(-1)-f(u)}{f(1)-f(-1)}, \quad (23.27)$$

and since the quotient on the right-hand side is real and positive when $u \in \Gamma_{-1}(\pi)$, the quotient on the left-hand side is also real and positive for $u \in \Gamma_{-1}(\pi)$. Dingle (1973) called the quantity $f(1)-f(-1)$ a *singulant*; see also Berry and Howls (1991). Substituting (23.26) in (23.25), and making use of (23.27), we obtain (exercise)

$$\begin{aligned} R_N(\xi) =& \frac{\xi^{-N}(-2)^{-N-1/2}}{2\pi i} \int_0^\infty \frac{e^{-\tau}\tau^{N-1/2}}{1+t/(2\xi)} \\ &\times \int_{\Gamma_{-1}(\pi)} \exp\left\{-\frac{t[f(-1)-f(u)]}{f(1)-f(-1)}\right\} du dt. \end{aligned} \quad (23.28)$$

In the inner integral, we write $u = -w$. Since $f(u)$ is an odd function and $f(1) - f(-1) = -2$, it follows that

$$\int_{\Gamma_{-1}(\pi)} \exp\left\{-\frac{t[f(-1)-f(u)]}{f(1)-f(-1)}\right\} du = -\int_{\Gamma_1(0)} \exp\left\{\frac{t}{2}[f(w)-f(1)]\right\} dw.$$

The last integral can be expressed in terms of the integral $I^{(1)}(\xi)$ given in (23.9). Indeed, we have

$$\int_{\Gamma_{-1}(\pi)} \exp\left\{-\frac{t[f(-1)-f(u)]}{f(1)-f(-1)}\right\} du = -(t/2)^{-1/2} I^{(1)}(t/2). \quad (23.29)$$

Inserting (23.29) into (23.28) gives

$$R_N(\xi) = \frac{(-2\xi)^{-N}}{2\pi} \int_0^\infty \frac{e^{-t}t^{N-1}}{1+t/(2\xi)} I^{(1)}(t/2) dt. \quad (23.30)$$

Equations (23.19) and (23.30) coupled together is known as a *resurgence formula*, since the integral $I^{(1)}(\xi)$ on the left-hand side of (23.19) appears again in the remainder term $R_N(\xi)$ given in (23.30).

23.3 Error estimates

To estimate the remainder $R_N(\xi)$ in (23.25), we first observe that if $\zeta = |\zeta|e^{i\varphi}$ then

$$|1-\zeta| \geq \begin{cases} |\sin\varphi|, & |\varphi| < \pi/2, \\ 1, & \pi/2 \leq |\varphi| \leq \pi. \end{cases} \tag{23.31}$$

For convenience, let us introduce the function

$$\csc_1(\varphi) = \begin{cases} |\csc\varphi|, & |\varphi| < \pi/2, \\ 1, & \pi/2 \leq |\varphi| \leq \pi. \end{cases} \tag{23.32}$$

In terms of this function, (23.31) can be written as

$$1/|1-\zeta| \leq \csc_1(\varphi). \tag{23.33}$$

We shall extend $\csc_1(\varphi)$ into a 2π-periodic function by defining

$$\csc_1(\varphi + 2\pi) = \csc_1(\varphi). \tag{23.34}$$

Note that this is an even function, and that it is unbounded at $0, \ \pm 2\pi, \ \pm 4\pi, \cdots$. For $u \in \Gamma_{-1}(\pi)$, we know that the quotient on the left-hand side of (23.27) is real and positive. Since $f(\pm 1) = \mp 1$, it follows that $\arg\{e^{-i\pi}[-1 - f(u)]\} = 0$ or, equivalently, $\arg\{-1 - f(u)\} = \pi$. If we take $\zeta = \tau/\{\xi[-1 - f(u)]\}$, then $\arg\zeta = -\pi - \theta$, and we have from (23.33)

$$\left|1 - \frac{\tau}{\xi[-1 - f(u)]}\right|^{-1} \leq \csc_1(\pi + \varphi) \tag{23.35}$$

for $u \in \Gamma_{-1}(\pi)$. A direct application of (23.35) to (23.25) gives

$$|R_N(\xi)| \leq \frac{A_N}{|\xi|^N} \csc_1(\theta + \pi), \tag{23.36}$$

where A_N is a constant given by

$$A_N = \frac{\Gamma(N + 1/2)}{2\pi} \int_{\Gamma_{-1}(\pi)} |[-1 - f(u)]^{-N-1/2} du|. \tag{23.37}$$

There is another error estimate which is probably easier to compute. To obtain that estimate, we first note that $I^{(1)}(\xi) = R_0(\xi)$; see (23.19). Hence, by (23.36)

$$|I^{(1)}(t/2)| \leq A_0 \csc_1(\pi). \tag{23.38}$$

Next, we return to (23.30). With $\zeta = -t/(2\xi)$, it follows from (23.33)

$$|R_N(\xi)| \leq \frac{B_N}{|\xi|^N} \csc_1(\pi - \theta), \tag{23.39}$$

where

$$B_N = \frac{A_0\Gamma(N)}{2^{N+1}\pi}. \tag{23.40}$$

Since the function $\csc_1(\varphi)$ becomes unbounded at $\varphi = 0, \pm 2\pi, \pm 4\pi, \cdots$, both bounds in (23.36) and (23.38) break down when $\theta = \pm\pi, \pm 3\pi, \cdots$.

If $\operatorname{Re}\xi > 0$, then a simpler and sharper error bound can be given. Indeed, since $|1+t/(2\xi)| \geq 1 + (t/2)\operatorname{Re}(1/\xi) \geq 1$ for all $t \geq 0$, from (23.30) we have

$$\begin{aligned} |-i(-\xi)^N R_N(\xi)| &= \left| \frac{2^{-N}}{2\pi} \int_0^\infty e^{-t} t^{N-1} \frac{-iI^{(1)}(t/2)}{1+t/(2\xi)} dt \right| \\ &\leq \frac{2^{-N}}{2\pi} \int_0^\infty e^{-t} t^{N-1} [-iI^{(1)}(t/2)] dt \\ &= \lim_{\xi\to\infty} [-i(-\xi)^N R_N(\xi)] = -i(-1)^N c_N; \end{aligned} \tag{23.41}$$

in reaching the inequality above we have made use of the fact that $-iI^{(1)}(\xi) > 0$ for all $\xi > 0$ (cf. (23.10)). The last equality in (23.41) follows from (23.20) and (23.21). Thus we have shown that the remainder term in (23.19) is bounded by the first neglected term in this asymptotic expansion.

Before we conclude this section, we wish to make two observations. First, we can derive the asymptotic expansion in (23.19) with the error estimate in (23.41) by simply using the formula

$$\begin{aligned} \operatorname{Ai}(z) &= \frac{3z^{5/4} e^{-2z^{3/2}/3}}{4\pi} \int_0^\infty \frac{t^{-3/4} e^{-2t^{3/2}/3} \operatorname{Ai}(t)}{z^{3/2}+t^{3/2}} dt \\ &= \frac{3z^{-1/4} e^{-2z^{3/2}/3}}{4\pi} \int_0^\infty \frac{t^{-3/4} e^{-2t^{3/2}/3} \operatorname{Ai}(t)}{1+(t/z)^{3/2}} dt \end{aligned} \tag{23.42}$$

for $|\arg z| < 2\pi/3$; this formula is given in Olver *et al.* (2024, (9.10.18)). Let $N \geq 0$ be a nonnegative integer. We have

$$\frac{1}{1+(t/z)^{3/2}} = \sum_{k=0}^{N-1} \frac{(-1)^k t^{3k/2}}{z^{3k/2}} + \frac{(-1)^N t^{3N/2}}{z^{3N/2}[1+(t/z)^{3/2}]}. \tag{23.43}$$

Substituting (23.43) into (23.42) yields

$$\frac{4\pi \operatorname{Ai}(z) z^{1/4}}{3e^{-2z^{3/2}/3}} = \sum_{k=0}^{N-1} \frac{(-1)^k a_k}{z^{3k/2}} + \frac{(-1)^N E_N(z)}{z^{3N/2}}, \tag{23.44}$$

where

$$a_k := \int_0^\infty t^{3k/2-3/4} e^{-2t^{3/2}/3} \operatorname{Ai}(t) dt > 0, \tag{23.45}$$

for $k = 0, 1, \cdots$, and

$$E_N(z) := \int_0^\infty \frac{t^{3N/2-3/4} e^{-2t^{3/2}/3} \operatorname{Ai}(t)}{1 + (t/z)^{3/2}} dt, \tag{23.46}$$

for $N = 0, 1, \cdots$. If z is real and positive, say $z = x$, then it is readily seen that $E_N(x) < a_N$; namely, the error term is bounded by the first neglected term in the asymptotic expansion (23.44). If z is complex but $\operatorname{Re} z > 0$, then the argument for (23.41) can again be used to show that the remainder $E_N(z)$ in (23.46) is bounded by the first neglected term in the asymptotic expansion (23.44).

Our second observation is that one can obtain the same result as above by using the integral representation

$$\operatorname{Ai}(z) = \frac{e^{-2z^{3/2}/3}}{\pi} \int_0^\infty e^{-z^{1/2}t^2} \cos(t^3/3) dt \tag{23.47}$$

for $|\arg z| < \pi$. Expanding the cosine function into a Maclaurin series and integrating term-by-term, one obtains the result (exercise)

$$\operatorname{Ai}(z) \sim \frac{e^{-2z^{3/2}/3}}{2\pi z^{1/4}} \sum_{k=0}^{\infty} \frac{(-1)^k \Gamma(3k + 1/2)}{3^{2k}(2k)! z^{3k/2}} \tag{23.48}$$

as $|z| \to \infty$ in $|\arg z| < \pi$; see Wong (1989, p. 47).

23.4 Exponential asymptotics

One of the most important properties of the Airy function $\operatorname{Ai}(z)$ defined by the integral in (23.1) is that it satisfies the differential equation

$$\frac{d^2 w}{dz^2} - zw = 0; \tag{23.49}$$

see Wong (1989, p. 91). Clearly, $\operatorname{Ai}(\omega z)$ and $\operatorname{Ai}(\omega^2 z)$ are also solutions of this differential equation, where $\omega = e^{2\pi i/3}$. With the aid of Cauchy's theorem, we can show from (23.1) that these solutions are connected by the relation (exercise)

$$\operatorname{Ai}(z) + \omega \operatorname{Ai}(\omega z) + \omega^2 \operatorname{Ai}(\omega^2 z) = 0; \tag{23.50}$$

see Wong (1989, p. 91). To obtain an asymptotic expansion for $\operatorname{Ai}(z)$ in a sector that includes $\arg z = \pi$, one can use (23.50). Indeed, if $\pi/3 <$

$\arg z < 5\pi/3$, then $-\pi/3 < \arg(\omega^{-1}z) < \pi$ and $-\pi < \arg(\omega^{-2}z) < \pi/3$. From (23.50) and (23.48) it follows that

$$\begin{aligned}\mathrm{Ai}(z) &= -\,\omega\,\mathrm{Ai}(\omega^{-2}z) - \omega^2\,\mathrm{Ai}(\omega^{-1}z)\\ &\sim \frac{e^{-2z^{3/2}/3}}{2\pi z^{1/4}}\sum_{k=0}^{\infty}\frac{(-1)^k\Gamma(3k+1/2)}{3^{2k}(2k)!z^{3k/2}} + \frac{ie^{2z^{3/2}/3}}{2\pi z^{1/4}}\sum_{k=0}^{\infty}\frac{\Gamma(3k+1/2)}{3^{2k}(2k)!z^{3k/2}},\end{aligned} \tag{23.51}$$

an expansion that is uniformly valid in $\pi/3 + \delta \le \arg z \le 5\pi/3 - \delta$. If we denote by $u_-(z)$ and $u_+(z)$ the first and second formal series in (23.51), then the asymptotic behavior of $\mathrm{Ai}(z)$ is given by

$$\mathrm{Ai}(z) \sim u_-(z) \tag{23.52}$$

as $z \to \infty$ in $-\pi < \arg z < \pi$; see (23.48). (Clearly, this result cannot be valid in a wider section, since $\mathrm{Ai}(z)$ is a single-valued function, and the factor multiplying the infinite series in $u_-(z)$ is not.) On the other hand, as $z \to \infty$ in $\pi/3 < \arg z < 5\pi/3$, we have the compound expansion

$$\mathrm{Ai}(z) \sim u_-(z) + u_+(z). \tag{23.53}$$

In the common sector of validity, given by $\pi/3 < \arg z < \pi$, either (23.52) or (23.53) can be used since the contribution of $u_+(z)$ is exponentially small compared with that of $u_-(z)$; hence it is mandatory to drop $u_+(z)$ from the asymptotic expansion of $\mathrm{Ai}(z)$. By introducing a constant (coefficient) C, which is 0 for $\arg z \in (-\pi/3, \pi/3)$ and 1 for $\arg z \in (\pi/3, 5\pi/3)$, to the asymptotic series $u_+(z)$, the two results (23.52) and (23.53) can be combined into one

$$\mathrm{Ai}(z) \sim u_-(z) + Cu_+(z), \tag{23.54}$$

as $z \to \infty$ for $\arg z \in (-\pi/3, 5\pi/3)$. The coefficient C is called a *Stokes multiplier*, and is domain dependent. The discontinuous change of the coefficient C, when the argument of z changes in a continuous manner, is known as *Stokes phenomenon*.

Returning to the series $u_+(z)$ and $u_-(z)$ in (23.51), we let

$$S_\pm(z) = \pm\frac{2}{3}z^{3/2}.$$

Note that the behaviors of $u_\pm(z)$ are most unequal on the curves given by

$$\mathrm{Im}\{S_+(z) - S_-(z)\} = 0, \tag{23.55}$$

and they are nearly equal on the curves given by

$$\mathrm{Re}\{S_+(z) - S_-(z)\} = 0. \tag{23.56}$$

The curves given in (23.55) and (23.56) are known, respectively, as the Stokes and anti-Stokes lines. In the case of Airy function, it is easily seen that the rays $\arg z = 0, \ \pm 2\pi/3$ are the Stokes lines and the rays $\arg z = \pm\pi/3, \ \pm\pi$ are the anti-Stokes lines.

Since we are dealing with a continuous (in fact, analytic) function, it is rather unsatisfactory to have a discontinuous coefficient in the asymptotic expansion (23.54). Berry (1989) wrote an innovative paper, in which he adopted a different interpretation of the Stokes phenomenon. In his view, the coefficient of the series $u_+(z)$ should be a continuous function of $\arg z$, instead of a discontinuous constant. He gave a beautiful, although not mathematically rigorous, demonstration of this theory by truncating the series $u_-(z)$ at its optimal number of terms and the series $u_+(z)$ at its first term. Berry's theory has since become known variously as "exponential asymptotics" or "superasymptotics", and has been successfully applied to several known asymptotic expansions in a mathematically rigorous manner (see, for example, Boyd (1990); Olver (1991a,b); Olde Daalhuis and Olver (1994); Wong and Zhao (1999a,b)).

To present Berry's idea, we again use the Airy function as an illustration. To obtain an error estimate near the Stokes lines $\theta = \pm\pi$ or, equivalently, $\arg z = \pm 2\pi/3$, we follow a device suggested by Boyd (1990); see (23.4), (23.6) and the statement following (23.56). Rotating the path of integration in (23.30) by an angle $\eta \in (-\pi/2, 0)$ gives

$$R_N(\xi) = \frac{(-2\xi)^{-N}}{2\pi} \int_0^{\infty e^{i\eta}} \frac{e^{-t} t^{N-1} I^{(1)}(t/2)}{1 + t/(2\xi)} dt. \tag{23.57}$$

On the path of integration in (23.57), we have $\arg t = \eta$. Hence, it follows from (23.36)

$$|I^{(1)}(t/2)| \le A_0 \csc_1(\eta + \pi) = A_0; \tag{23.58}$$

see (23.38). The equality in (23.58) follows from the fact the $\pi/2 < \eta + \pi < \pi$. On the other hand, we also have from (23.33)

$$|1 + t/(2\xi)|^{-1} \le \csc_1(\pi + \eta - \theta). \tag{23.59}$$

A combination of (23.57), (23.58) and (23.59) yields

$$|R_N(\xi)| \le \frac{C\Gamma(N) \csc_1(\pi + \eta - \theta)}{(2|\xi| \cos\eta)^N}, \tag{23.60}$$

where $C = A_0/(2\pi)$.

If $0 < \theta < \pi$ and θ is close to π, then for any sufficiently small $\eta \in (-\pi/2, 0)$ we have $-\pi/2 < \pi + \eta - \theta < 0$. Similarly, if $-\pi < \theta < 0$ and

θ is close to $-\pi$, then for any sufficiently small $\eta \in (-\pi/2, 0)$ we have $3\pi/2 < \pi + \eta - \theta < 2\pi$. Since the function $\csc_1(\varphi)$ defined in (23.32) and (23.34) is 2π-periodic, from (23.60) we obtain

$$|R_N(\xi)| \le \frac{C\Gamma(N)}{(2|\xi|\cos\eta)^N |\sin(\pm\pi + \eta - \theta)|}, \tag{23.61}$$

where + sign is taken when $\theta \to \pi^+$ and $-$ sign is taken when $\theta \to -\pi^-$. As $\theta \to \pm\pi$, we have $\sin(\pm\pi + \eta - \theta) \sim \sin\eta$. Since η is arbitrary, we may choose it so that

$$\cos\eta \sim 1 - \frac{1}{N}, \quad \sin\eta \sim \sqrt{\frac{2}{N}}.$$

Thus, for θ near $\pm\pi$, we can find a constant C, such that

$$|R_N(\xi)| \le \frac{C_1\Gamma(N)\sqrt{N/2}}{[2|\xi|(1 - 1/N)]^N}.$$

(An explicit value for C_1 can be given if desired.) Since $(1 - 1/N)^{-N} \le e^{N/(N-1)}$ and

$$\Gamma(N) < \sqrt{2\pi} N^{N-1/2} e^{-N+1/(12N)}; \tag{23.62}$$

see Artin (1964), we obtain

$$|R_N(\xi)| \le C_2 e^{N(-1+\ln N - \ln|2\xi|)}, \tag{23.63}$$

where $C_2 = \sqrt{\pi}e^2 C_1$. The minimum value of the exponential function on the right-hand side of (23.63) is attained when

$$\frac{d}{dN}(-N + N\ln N - N\ln|2\xi|) = \ln N - \ln|2\xi| = 0.$$

Therefore, optimal truncation occurs near

$$N = N^* = 2|\xi|.$$

With N given by this value, we obtain from (23.62)

$$|R_N(\xi)| \le C_2 e^{-|2\xi|}, \tag{23.64}$$

as $|\xi| \to \infty$, when $\arg\xi = \theta$ is near the Stokes lines $\theta = \pm\pi$. Applying (23.62) to the error estimate in (23.39), one readily sees that inequality (23.64) also holds for ξ away from the Stokes lines. Olver (1991b) called the expansion (23.19) with error given by (23.64) a *uniform, exponentially improved*, asymptotic expansion in the sector $|\theta| \le \pi$. Optimally truncated asymptotic expansions have now been also called *superasymptotic expansions* by Berry and Howls (1990).

Returning to (23.30), we now replace the function $I^{(1)}(t/2)$ by its asymptotic expansion (23.19). Termwise integration gives a series of integrals which can be expressed in terms of Dingle's *terminant function*

$$\int_0^\infty \frac{t^{k-1}e^{-t}}{1+t/\zeta}dt = 2\pi i(-\zeta)^k e^\zeta T_k(\zeta); \tag{23.65}$$

see Olver (1991b). More precisely, we have

$$R_N(\xi) = ie^{2\xi}\sum_{r=0}^{N'-1}(-1)^r c_r \xi^{-r} T_{N-r}(2\xi) + R_{N,N'}(\xi), \tag{23.66}$$

where

$$R_{N,N'}(\xi) = \frac{(-2|\xi|)^{-N}}{2\pi}\int_0^\infty \frac{e^{-t}t^{N-1}R_{N'}(t/2)}{1+t/(2\xi)}dt. \tag{23.67}$$

The idea of re-expanding the remainder terms in the optimally truncated asymptotic series was introduced by Berry and Howls (1990). They called this theory *hyperasymptotics*; see also Olver (1993); Olde Daalhuis and Olver (1994).

With $\arg\zeta = \phi$, $u = t/|\zeta|$ and $k = |\zeta| + \alpha$, α being a positive and bounded quantity, the integral in (23.65) can be written as

$$\int_0^\infty \frac{t^{k-1}e^{-t}}{1+t/\zeta}dt = |\zeta|^k \int_0^\infty \frac{u^{\alpha-1}e^{-|\zeta|(u-\ln u)}}{1+e^{-i\phi}u}du.$$

The integrand on the right-hand side has a pole at $u = -e^{i\phi}$, which coalesces with the saddle point at $u = 1$ when $\phi = \pm\pi$. An existing theory on uniform asymptotic expansions (see Wong (1989, p. 356-358)) can now be used to show that as $|\zeta| \to \infty$,

$$T_k(\zeta) \sim \frac{\operatorname{erfc}(Z)}{2} - \frac{ie^{-Z^2}}{\sqrt{2\pi|\zeta|}}\sum_{s=0}^\infty (1/2)_s g_{2s}(\phi,\alpha)(2/|\zeta|)^s \tag{23.68}$$

uniformly with respect to $\phi \in [-\pi+\varepsilon, 3\pi-\delta]$, where erfc is the complementary error function, $Z := c(\phi)\sqrt{|\zeta|/2}$, and

$$\frac{[c(\phi)]^2}{2} := -e^{i(\phi-\pi)} + i(\phi-\pi) + 1.$$

Near $\phi = \pi$, the Taylor series expansion of $c(\phi)$ begins

$$c(\phi) = -(\phi-\pi) - \frac{i}{6}(\phi-\pi)^2 + \frac{1}{36}(\phi-\pi)^3 + \cdots.$$

The coefficient $g_{2s}(\phi,\alpha)$ can be given explicitly and the first one is

$$g_0(\phi,\alpha) = \frac{e^{i\alpha(\pi-\phi)}}{1+e^{-i\phi}} - \frac{i}{c(\phi)}.$$

This result is taken from Olver (1991b). He has also shown that Z lies in the sector $-\pi/4 \le \arg Z \le 0$ when $-\pi \le \phi \le \pi$, and in the sector $0 \le \arg(-Z) \le \pi/4$ when $\pi \le \phi \le 3\pi$. As ϕ increases from values below π to values above π, Z moves from the first sector to the second sector through the origin. Since $\operatorname{erfc}(Z) = O(e^{-Z^2})$ uniformly throughout the first sector and $\operatorname{erfc}(Z) = 2 + O(e^{-Z^2})$ uniformly throughout the second sector (see Wong (1989, p. 42)), if $|\zeta|$ is large and fixed and ϕ increases continuously from π^- to π^+, then (23.68) shows that $T_k(\zeta)$ changes rapidly, but smoothly, from being exponentially small to being exponentially close to one.

Coupling (23.19) and (23.66) gives

$$I^{(1)}(\xi) = \sum_{s=0}^{N-1} c_s \xi^{-s} + ie^{2\xi} \sum_{r=0}^{N'-1} (-1)^r c_r \xi^{-r} T_{N-r}(2\xi) + R_{N,N'}(\xi). \quad (23.69)$$

The remainder $R_{N,N'}(\xi)$ is given in (23.67), and can be estimated as before. Of course, it is expected to be of lower order of magnitude, and can be neglected. Inserting (23.68) into (23.69), we obtain from (23.10)

$$\operatorname{Ai}(z) \sim \frac{(3/2)^{1/2}}{2\pi i z^{1/2}} \left[e^{-\xi} \sum_{s=0}^{2|\xi|-1} \frac{c_s}{\xi^s} + \frac{ie^{\xi} \operatorname{erfc}[c(\theta)|\xi|^{1/2}]}{2} \sum_{r=0}^{N'-1} (-1)^r \frac{c_r}{\xi^r} \right] \quad (23.70)$$

where $\xi = 2z^{3/2}/3$. Note that in (23.70), we have truncated the first series at an optimal place. When θ is near π, $\operatorname{erfc}[c(\theta)|\xi|^{1/2}]$ will have an abrupt but smooth change. In Berry's terminology, this function is called a *Stokes multiplier*. A similar result holds for θ near $-\pi$.

23.5 Exercises

1. Verify (23.2).
2. Prove (23.22).
3. Verify (23.28).
4. Verify (23.44).
5. Prove (23.48).
6. Prove (23.50) by applying Cauchy's theorem to (23.1).
7. Consider

$$P(x) := \int_{\mathbf{R}} e^{-xf(t)} dt, \quad f(t) := \frac{t^4}{4} + \frac{t^2}{2}. \quad (23.71)$$

Since $f'(t) = t^3 + t$, the point $t_0 = 0$ is a saddle point, and the real axis $\mathbf{R}$ is its steepest descent path. There are two adjacent saddle points $t_\pm = \pm i$, each associated with an adjacent curve:

$$\Gamma_\pm := \{z = re^{i\theta} : \quad r = \sqrt{-1/\cos(2\theta)}, \quad \pm\theta \in (\pi/4, 3\pi/4)\}. \quad (23.72)$$

(a) Draw the graphs $\Gamma_\pm$. For $z \in \Gamma_\pm$, prove that $\operatorname{Im} f(z) = 0$ and $f(z) \leq -1/4$. Let $\Gamma = \Gamma_+ \cup \Gamma_-$, oriented counterclockwise. Verify that the open region enclosed by Γ is

$$\begin{aligned}\Omega :=&\{z = re^{i\theta} : \quad r \leq \sqrt{-1/\cos(2\theta)}, \quad \pm\theta \in (\pi/4, 3\pi/4)\}\\ &\cup \{z = re^{i\theta} : \quad |\theta| \notin (\pi/4, 3\pi/4)\}.\end{aligned}$$

For each $u \geq 0$, there exists a unique $t = g_+(u) \geq 0$ such that $u = f(g_+(u))$. For convenience, define $g_-(u) = -g_+(u) \leq 0$. Splitting the integral defining $P(x)$ and making the changes of variables $t = g_\pm(u)$, we obtain

$$\begin{aligned}P(x) &= \int_0^\infty e^{-xf(t)}dt + \int_{-\infty}^0 e^{-xf(t)}dt\\ &= \int_0^\infty e^{-xu}\left[\frac{1}{f'(g_+(u))} - \frac{1}{f'(g_-(u))}\right]du.\end{aligned}$$

(b) Use the Cauchy integral formula to show that the difference in the brackets can be written as

$$\frac{1}{f'(g_+(u))} - \frac{1}{f'(g_-(u))} = \frac{1}{2\pi i}\int_\Gamma \frac{[f(z)/u]^{1/2}}{f(z) - u}dz; \qquad (23.73)$$

See Section 23.2. The function $h(z) := [f(z)]^{1/2}$ is analytic in a small neighborhood of $z = 0$ and satisfies $h(z) \sim z/\sqrt{2}$ as $z \to 0$. By analytic continuation, $h(z)$ is analytic throughout Ω. In particular, for $t \geq 0$, we have $h(t) = \pm\sqrt{f(t)}$. Combining the above representations yields the double integral formula

$$P(x) = \frac{1}{2\pi i}\int_0^\infty u^{-1/2}e^{-xu}\int_\Gamma \frac{h(z)}{f(z) - u}dzdu. \qquad (23.74)$$

(c) Use $f(z) = [h(z)]^2$ and the expansion

$$\frac{1}{f(z) - u} = \sum_{k=0}^{n-1}\frac{u^k}{[f(z)]^{k+1}} + \frac{u^n}{[f(z) - u][f(z)]^n}$$

to prove

$$\begin{aligned}P(x) &= \sum_{k=0}^{n-1} \frac{1}{2\pi i}\int_0^\infty u^{k-1/2}e^{-xu}\int_\Gamma \frac{1}{[h(z)]^{2k+1}}dzdu \\ &\quad + \frac{1}{2\pi i}\int_0^\infty u^{n-1/2}e^{-xu}\int_\Gamma \frac{h(z)}{[f(z)-u][f(z)]^n}dzdu \\ &= \sum_{k=0}^{n-1}\frac{a_k}{x^{k+1/2}} + \frac{R_n(x)}{x^{n+1/2}},\end{aligned} \tag{23.75}$$

where

$$a_k = \frac{\Gamma(k+1/2)}{2\pi i}\int_\Gamma \frac{1}{[h(z)]^{2k+1}}dz, \tag{23.76}$$

and

$$R_n(x) = \frac{x^{n+1/2}}{2\pi i}\int_0^\infty u^{n-1/2}e^{-xu}\int_\Gamma \frac{h(z)}{[f(z)-u][f(z)]^n}dzdu. \tag{23.77}$$

(d) Show that $1 - u/f(z) \geq 1$ for all $u \geq 0$ and $z \in \Gamma$. Prove the error estimate

$$\begin{aligned}|R_n(x)| &\leq \frac{x^{n+1/2}}{2\pi}\int_0^\infty u^{n-1/2}e^{-xu}\int_\Gamma \frac{|h(z)|}{|f(z)|^{n+1}}|dz|du \\ &= \frac{\Gamma(n+1/2)}{2\pi}\int_\Gamma \frac{1}{|h(z)|^{2n+1}}|dz|.\end{aligned}$$

Chapter 24

Error Bounds for the Asymptotic Expansions of the Hermite Polynomials

24.1 Introduction

The Hermite polynomials $H_n(x)$ are defined by the generating function

$$e^{2xt-t^2} = \sum_{n=0}^{\infty} \frac{H_n(x)}{n!} t^n, \quad |t| < \infty.$$

These polynomials are definitely one of the most frequently mentioned orthogonal polynomials. By the Cauchy integral formula, these polynomials have the integral representation

$$H_n(x) = \frac{n!}{2\pi i} \int_C \frac{e^{2xw-w^2}}{w^{n+1}} dw,$$

where C is a closed contour encircling the origin in the counterclockwise direction. By adding two infinite line segments along the negative real axis, the path of integration C can be deformed into an infinite loop Γ, which starts from $-\infty$ in the lower half-plane, turning around at the origin, and goes back to $-\infty$ in the upper half-plane. To be more specific, we may choose Γ to be the contour

$$\Gamma = \{\lambda - i\delta : \lambda < 0\} \cup \{\delta e^{i\theta} : -\pi/2 \le \theta \le \pi/2\} \cup \{\lambda + i\delta : \lambda < 0\},$$

where $\delta > 0$ is any fixed positive number; see Figure 24.1.

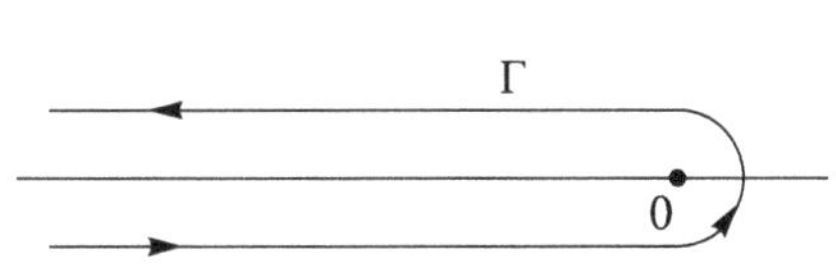

Fig. 24.1: The contour Γ with counter-clockwise orientation.

In this chapter, we consider the behavior of the Hermite polynomials, as $n \to \infty$, with scaled variable $\sqrt{2n+1}x$. With $N = 2n+1$, the above integral can be written in the form

$$H_n(\sqrt{N}x) = \frac{n!}{2\pi i N^{n/2}} \int_\Gamma e^{-N[t^2-2xt+(\ln t)/2]} t^{-1/2} dt. \tag{24.1}$$

Since $H_n(-x) = (-1)^n H_n(x)$, we may assume that $x \geq 0$; see Beals and Wong (2016, p. 103). Furthermore, we shall denote the phase function in the above integral by

$$f(t;x) = t^2 - 2xt + \frac{1}{2}\ln t. \tag{24.2}$$

The saddle points, i.e., the zeros of $f'(t;x)$, are

$$t_\pm = \frac{x \pm \sqrt{x^2-1}}{2} = \begin{cases} e^{\pm\beta}/2, & x = \cosh\beta > 1,\ \beta > 0; \\ e^{\pm i\alpha}/2, & x = \cos\alpha \in [0,1),\ \alpha \in (0,\pi/2]; \\ 1/2, & x = 1. \end{cases}$$

Note that the two saddle points coincide when $x = 1$; i.e., $t_+ = t_-$. For convenience, we shall drop the dependence of f on x; i.e., write $f(t) = f(t;x)$. We have three cases to consider; namely,

Case I: $x = \cosh\beta > 1$ with $\beta > 0$. This is called the non-oscillatory region.

Case II: $x = \cos\alpha \in [0,1)$ with $\alpha \in (0,\pi/2]$. This is known as the oscillatory region.

Case III: $x = 1$. This is called a turning point.

In the following, we will derive the asymptotic expansions with error bounds for each case in a separate section. The main results are stated at the end of these sections. The material in this chapter is taken from Shi *et al.* (2023).

24.2 Case I: $x = \cosh\beta > 1$ with $\beta > 0$

The saddle points are $t_\pm = e^{\pm\beta}/2$ with $0 < t_- < t_+$. Moreover, $f''(t_-) = 2(1-e^{2\beta}) < 0$ and $f''(t_+) = 2(1-e^{-2\beta}) > 0$. We deform the contour of integration Γ to the path of steepest descent passing through the saddle point t_-. To describe this steepest descent contour, we introduce the analytic function

$$w(z) = [f(z) - f(t_-)]^{1/2}, \tag{24.3}$$

where the branch of the square root function is chosen so that

$$\begin{aligned} w(z) &= -\,i(z-t_-)\sqrt{-f''(t_-)/2} + O((z-t_-)^2) \\ &= -\,i(z-t_-)\sqrt{e^{2\beta}-1} + O((z-t_-)^2), \end{aligned}$$

for z in a small complex neighborhood of t_-. Let $z = re^{i\theta}$ with $r > 0$ and $\theta \in (-\pi, \pi)$. We investigate the equation $\operatorname{Im}[f(z) - f(t_-)] = \operatorname{Im} f(z) = 0$; namely,

$$r^2 \sin(2\theta) - 2r \cosh\beta \sin\theta + \frac{\theta}{2} = 0. \tag{24.4}$$

Clearly, $\theta = 0$ satisfies the above equation. By symmetry, we can restrict $\theta \in (0, \pi)$ and find two solutions

$$\begin{aligned} r_-(\theta) &= \frac{\theta/(2\sin\theta)}{\cosh\beta + \sqrt{\cosh^2\beta - \theta/\tan\theta}}, \quad \theta \in (0,\pi), \\ r_+(\theta) &= \frac{\cosh\beta + \sqrt{\cosh^2\beta - \theta/\tan\theta}}{2\cos\theta}, \quad \theta \in (0, \pi/2). \end{aligned} \tag{24.5}$$

It is readily seen that $r_\pm(\theta) \to t_\pm = e^{\pm\beta}/2$ as $\theta \to 0$. For consistency, we define $r_\pm(0) = t_\pm$. Consequently, we can regard $r_-(\theta)$ as a symmetric function on $(-\pi, \pi)$ and $r_+(\theta)$ a symmetric function on $(-\pi/2, \pi/2)$. Define two contours

$$\Gamma_- = \{r_-(\theta)e^{i\theta} : \theta \in (-\pi, \pi)\}, \quad \Gamma_+ = \{r_+(\theta)e^{i\theta} : \theta \in (-\pi/2, \pi/2)\}.$$

It follows that the solution of $\operatorname{Im} f(z) = 0$ is the union $\Gamma_- \cup \Gamma_+ \cup \mathbf{R}^+$, where $\mathbf{R}^+$ is the positive real line. The contours $\Gamma_\pm$ are depicted in Figure 24.2.

Since $f'(z)$ has no zeros other than $t_\pm$, it follows from the asymptotic behavior $f(z) \sim z^2$ as $z \to \infty$ that Γ_- is a steepest descent contour passing through t_-, while Γ_+ is a steepest ascent contour passing through t_+. Moreover, the solutions of $f(z) - f(t_-) = 0$ are real and positive. Consider $f(t) = t^2 - 2t\cosh\beta + (\ln t)/2$ for $t > 0$. Since $f(t) \to -\infty$ as $t \to 0^+$ and $f(t) \to \infty$ as $t \to \infty$, we obtain from $f'(t_\pm) = 0$ that the equation $f(t) - f(t_-) = 0$ has exactly two positive roots: one is t_-; and the other, denoted by t_c, is larger than t_+. Hence, we can choose $[t_c, \infty)$ as a branch cut and extend the definition of $w(z)$ on $\mathbf{C} \setminus (-\infty, 0] \cup [t_c, \infty)$ by analytic continuation.

Now, we deform the contour of integration from Γ to the steepest descent contour Γ_- with counter-clockwise orientation and then make the change

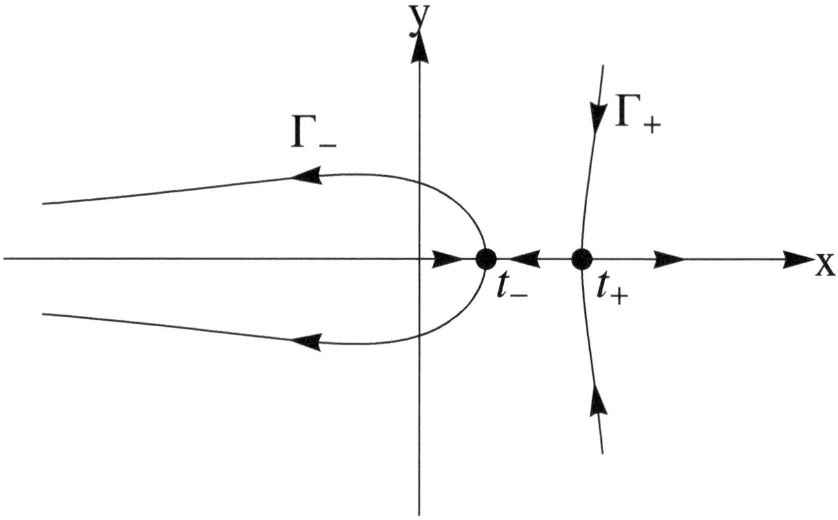

Fig. 24.2: The contours $\Gamma_\pm$. The arrows indicate the increasing direction of the phase function f.

of variable $u = w(t)$:

$$\begin{aligned} H_n(\sqrt{N}\cosh\beta) &= \frac{n! e^{-Nf(t_-)}}{2\pi i N^{n/2}} \int_{\Gamma_-} e^{-N[f(t)-f(t_-)]} t^{-1/2} dt \\ &= \frac{n! e^{-Nf(t_-)}}{2\pi i N^{n/2}} \int_{\mathbf{R}} e^{-Nu^2} \frac{[w^{-1}(u)]^{-1/2}}{w'(w^{-1}(u))} du. \end{aligned} \tag{24.6}$$

Recall that Γ_+ is the steepest ascent contour passing through t_+. We also denote by Γ_0 the contour in Figure 24.1 with opposite orientation; namely,

$$\Gamma_0 = \{\lambda + i\delta : \lambda < 0\} \cup \{\delta e^{i\theta} : -\pi/2 \le \theta \le \pi/2\} \cup \{\lambda - i\delta : \lambda < 0\},$$

where $\delta > 0$ is a small positive number such that $\delta < t_c$. Since $|w(z)|/|z| \to 1$ as $z \to \infty$, we have from the Cauchy integral formula

$$\frac{[w^{-1}(u)]^{-1/2}}{w'(w^{-1}(u))} = \frac{1}{2\pi i} \int_{\Gamma_0 \cup \Gamma_+} \frac{z^{-1/2}}{w(z) - u} dz$$

for any $u \in \mathbf{R}$. Substituting this expression into (24.6) gives a double integral representation

$$H_n(\sqrt{N}\cosh\beta) = \frac{n! e^{-Nf(t_-)}}{(2\pi i)^2 N^{n/2}} \int_{\mathbf{R}} e^{-Nu^2} \int_{\Gamma_0 \cup \Gamma_+} \frac{z^{-1/2}}{w(z) - u} dz du.$$

For any $p \geq 0$, it follows from

$$\frac{1}{w(z)-u} = \sum_{k=1}^{2p} \frac{u^{k-1}}{w(z)^k} + \frac{u^{2p}}{w(z)^{2p}[w(z)-u]}$$

and

$$\int_{\mathbf{R}} e^{-Nu^2} u^{k-1} du = \begin{cases} 0, & k = 2j, \\ \Gamma(j+1/2)/N^{j+1/2}, & k = 2j+1 \end{cases} \tag{24.7}$$

that

$$\begin{aligned} H_n(\sqrt{N}\cosh\beta) &= \frac{n! e^{-Nf(t_-)}}{(2\pi i)^2 N^{n/2}} \left\{ \sum_{k=1}^{2p} \int_{\mathbf{R}} e^{-Nu^2} u^{k-1} du \int_{\Gamma_0 \cup \Gamma_+} \frac{z^{-1/2}}{w(z)^k} dz \right. \\ &\quad \left. + \int_{\mathbf{R}} e^{-Nu^2} u^{2p} \int_{\Gamma_0 \cup \Gamma_+} \frac{z^{-1/2}}{w(z)^{2p}[w(z)-u]} dz du \right\} \\ &= \frac{n! e^{-Nf(t_-)}}{2\sqrt{\pi \sinh\beta} N^{(n+1)/2}} \left\{ \sum_{j=0}^{p-1} \frac{A_j(\coth\beta)}{N^j} + \varepsilon_p(N,\beta) \right\}, \end{aligned} \tag{24.8}$$

where

$$A_j(\coth\beta) = -\frac{\sqrt{\sinh\beta}\,\Gamma(j+1/2)}{2\pi^{3/2}} \int_{\Gamma_0 \cup \Gamma_+} \frac{z^{-1/2}}{w(z)^{2j+1}} dz, \tag{24.9}$$

$$\varepsilon_p(N,\beta) = -\frac{\sqrt{N \sinh\beta}}{2\pi^{3/2}} \int_{\mathbf{R}} e^{-Nu^2} u^{2p} \int_{\Gamma_0 \cup \Gamma_+} \frac{z^{-1/2}}{w(z)^{2p}[w(z)-u]} dz du. \tag{24.10}$$

To find an upper bound for the remainder $\varepsilon_p(N,\beta)$, we need to estimate the integral

$$c_p(u) = \int_{\Gamma_0 \cup \Gamma_+} \frac{z^{-1/2}}{w(z)^{2p}[w(z)-u]} dz, \quad u \in \mathbf{R}. \tag{24.11}$$

If there exists $C_p > 0$ such that $|c_p(u)| \leq C_p$ for all $u \in \mathbf{R}$, then we obtain from (24.7) with $k = 2p+1$ and (24.10) that

$$|\varepsilon_p(N,\beta)| \leq \frac{\sqrt{\sinh\beta}}{2\pi^{3/2}} \frac{C_p \Gamma(p+1/2)}{N^p}. \tag{24.12}$$

First, we estimate the integral

$$I_0 = \int_{\Gamma_0} \frac{z^{-1/2}}{w(z)^{2p}[w(z)-u]} dz.$$

Recall that the contour Γ_0 encircles the negative real line having a distance $\delta > 0$ from it. By analyticity of the integrand, the value I_0 remains the same if we let $\delta \to 0^+$. Hence,

$$\begin{aligned}|I_0| \le &\int_0^{+\infty} \frac{s^{-1/2}}{|w(se^{i\pi})|^{2p}|w(se^{i\pi}) - u|} ds \\ &+ \int_0^{+\infty} \frac{s^{-1/2}}{|w(se^{-i\pi})|^{2p}|w(se^{-i\pi}) - u|} ds, \end{aligned} \tag{24.13}$$

where $w(se^{\pm i\pi})$ is defined as the limit of $w(se^{i\theta})$ as $\theta \to \pm\pi$. Note from (24.2) and (24.3) that

$$w(se^{\pm i\pi})^2 = s^2 + 2s\cosh\beta + (\ln s)/2 - f(t_-) \pm i\pi/2.$$

Thus, we obtain

$$|w(se^{\pm i\pi})| = \{[s^2 + 2s\cosh\beta + (\ln s)/2 - f(t_-)]^2 + (\pi/2)^2\}^{1/4} =: R(s), \tag{24.14}$$

where the last equality defines $R(s)$, the modulus of $w(se^{\pm i\pi})$. Let $\theta_\pm(s)$ be the phase of $w(se^{\pm i\pi})$. We then have

$$R(s)^2\cos[2\theta_\pm(s)] = \mathrm{Re}[w(se^{\pm i\pi})^2] = s^2 + 2s\cosh\beta + (\ln s)/2 - f(t_-),$$

which implies that

$$\begin{aligned}|w(se^{\pm i\pi}) - u| \ge &|\,\mathrm{Im}[w(se^{\pm i\pi})]| = R(s)|\sin[\theta_\pm(s)]| \\ =&R(s)\sqrt{\frac{1 - \cos[2\theta_\pm(s)]}{2}} \\ =&\sqrt{\frac{R(s)^2 - [s^2 + 2s\cosh\beta + (\ln s)/2 - f(t_-)]}{2}} \\ =&\sqrt{\frac{(\pi/2)^2/2}{R(s)^2 + [s^2 + 2s\cosh\beta + (\ln s)/2 - f(t_-)]}} \\ \ge&\sqrt{\frac{(\pi/2)^2}{4R(s)^2}} = \frac{\pi}{4R(s)}. \end{aligned}$$

Assume that $p \ge 1$. Substituting this inequality into (24.13) yields

$$|I_0| \le \frac{8}{\pi}\int_0^{+\infty} \frac{s^{-1/2}}{R(s)^{2p-1}} ds.$$

To estimate the integral on the right-hand side of the above inequality, we note from (24.14) that $R(s) \ge \sqrt{\pi/2}$ for $s \in [0, 1]$ and $R(s) \ge \sqrt{2s\cosh\beta}$ for $s \ge 1 > t_-$. Thus,

$$\begin{aligned}|I_0| \le &\frac{8}{\pi}\int_0^1 \frac{s^{-1/2}}{(\pi/2)^{p-1/2}} ds + \frac{8}{\pi}\int_1^{+\infty} \frac{s^{-1/2}}{(2s\cosh\beta)^{p-1/2}} ds \\ =&\frac{2^{p+7/2}}{\pi^{p+1/2}} + \frac{1}{\pi(p-1)2^{p-7/2}(\cosh\beta)^{p-1/2}}, \end{aligned} \tag{24.15}$$

where we have assumed $p \ge 2$ to ensure convergence of the second integral.

It remains to estimate the integral

$$I_c = \int_{\Gamma_+} \frac{z^{-1/2}}{w(z)^{2p}[w(z)-u]} dz.$$

For $z \in \Gamma_+$, we have $\operatorname{Im} f(z) = 0$ and $\operatorname{Re}[f(z) - f(t_-)] < f(t_+) - f(t_-) < 0$. It then follows from (24.3) that $w(z)$ is purely imaginary and $|w(z) - u| \geq |w(z)|$ for $z \in \Gamma_+$ and $u \in \mathbf{R}$. Now, we parameterize Γ_+ as $z = z_+(\theta) = r_+(\theta)e^{i\theta}$ with $\theta \in (-\pi/2, \pi/2)$, where $r_+(\theta)$ is given in (24.5). Since $dz/d\theta = [r'_+(\theta) + ir_+(\theta)]e^{i\theta}$ and $|w(z_+(\theta))|^2 = |f(z_+(\theta)) - f(t_-)| = f(t_-) - f(z_+(\theta))$, we deduce

$$\begin{aligned} |I_c| &\leq \int_{-\pi/2}^{\pi/2} \frac{r_+(\theta)^{-1/2}\sqrt{[r'_+(\theta)]^2 + [r_+(\theta)]^2}}{|w(z_+(\theta))|^{2p+1}} d\theta \\ &= \int_0^{\pi/2} \frac{2r_+(\theta)^{-1/2}\sqrt{[r'_+(\theta)]^2 + [r_+(\theta)]^2}}{[f(t_-) - f(z_+(\theta))]^{p+1/2}} d\theta. \end{aligned}$$

As both $\theta/\tan\theta$ and $\cos\theta$ are decreasing functions for $\theta \in (0, \pi/2)$, it follows from (24.5) that $r_+(\theta)$ is an increasing function for $\theta \in (0, \pi/2)$; namely, $r'_+(\theta) \geq 0$. To estimate $r'_+(\theta)$, we apply an implicit differentiation on (24.4) and obtain

$$r'_+(\theta) = \frac{-2r_+(\theta)^2\cos(2\theta) + 2r_+(\theta)\cosh\beta\cos\theta - 1/2}{2r_+(\theta)\sin(2\theta) - 2\cosh\beta\sin\theta}.$$

A simple calculation together with (24.5) gives

$$\begin{aligned} 2r_+(\theta)\sin(2\theta) - 2\cosh\beta\sin\theta =& 2\sin\theta[2r_+(\theta)\cos\theta - \cosh\beta] \\ =& 2\sin\theta\sqrt{\cosh^2\beta - \theta/\tan\theta} \end{aligned}$$

and

$$\begin{aligned} & 2r_+(\theta)\cosh\beta\cos\theta - 2r_+(\theta)^2\cos^2\theta \\ =& \cosh\beta(\cosh\beta + \sqrt{\cosh^2\beta - \theta/\tan\theta}) - \frac{(\cosh\beta + \sqrt{\cosh^2\beta - \theta/\tan\theta})^2}{2} \\ =& \frac{\theta}{2\tan\theta}. \end{aligned} \tag{24.16}$$

Consequently,

$$\begin{aligned} 0 \leq & r'_+(\theta) = \frac{2r_+(\theta)^2\sin^2\theta + \theta/(2\tan\theta) - 1/2}{2\sin\theta\sqrt{\cosh^2\beta - \theta/\tan\theta}} \\ \leq & \frac{r_+(\theta)^2\sin\theta}{\sqrt{\cosh^2\beta - \theta/\tan\theta}} \leq \frac{r_+(\theta)\tan\theta}{\tanh\beta}, \end{aligned}$$

and

$$|I_c| \le \frac{2}{\tanh\beta}\int_0^{\pi/2} \frac{r_+(\theta)^{1/2}/\cos\theta}{[f(t_-) - f(z_+(\theta))]^{p+1/2}} d\theta. \tag{24.17}$$

Let $\theta_0 = \arccos(1/4)$. If $\theta \in (0, \theta_0)$, then $\cos\theta > 1/4$, $f(t_-) - f(z_+(\theta)) > f(t_-) - f(t_+) = \sinh(2\beta)/2 - \beta$ and $r_+(\theta) < 4\cosh\beta$. Hence,

$$\int_0^{\theta_0} \frac{r_+(\theta)^{1/2}/\cos\theta}{[f(t_-) - f(z_+(\theta))]^{p+1/2}} d\theta \le \frac{2\pi(4\cosh\beta)^{1/2}}{[\sinh(2\beta)/2 - \beta]^{p+1/2}}. \tag{24.18}$$

If $\theta \in (\theta_0, \pi/2)$, then $r_+(\theta) > 2e^\beta = 4t_+ > 2\cosh\beta > 2$. This together with (24.16) implies that

$$\begin{aligned} -f(z_+(\theta)) &= -r_+(\theta)^2\cos(2\theta) + 2r_+(\theta)\cosh\beta\cos\theta - \frac{1}{2}\ln r_+(\theta) \\ &= r_+(\theta)^2 + \frac{\theta}{2\tan\theta} - \frac{1}{2}\ln r_+(\theta) \\ &\ge \frac{4}{5}r_+(\theta)^2 + \frac{1}{5}r_+(\theta)^2 - \frac{1}{2}\ln r_+(\theta) \ge \frac{4}{5}r_+(\theta)^2, \end{aligned}$$

and

$$-f(t_-) = \frac{e^{-2\beta}}{4} + \frac{\beta}{2} + \frac{1+\ln 2}{2} \le \frac{e^{2\beta}}{4} + \frac{19}{20} \le t_+^2 + \frac{19}{80}r_+(\theta)^2 \le \frac{3}{10}r_+(\theta)^2.$$

Coupling the above two inequalities gives

$$f(t_-) - f(z_+(\theta)) \ge \frac{1}{2}r_+(\theta)^2.$$

Therefore,

$$\begin{aligned} &\int_{\theta_0}^{\pi/2} \frac{r_+(\theta)^{1/2}/\cos\theta}{[f(t_-) - f(z_+(\theta))]^{p+1/2}} d\theta \le \int_{\theta_0}^{\pi/2} \frac{r_+(\theta)^{1/2}\cdot r_+(\theta)/t_+}{r_+(\theta)^{2p+1}/2^{p+1/2}} d\theta \\ &= \frac{2^{p+1/2}}{t_+}\int_{\theta_0}^{\pi/2} r_+(\theta)^{-2p+1/2} d\theta \le \frac{2^{p+1/2}}{t_+}\int_{\pi/3}^{\pi/2} (4t_+)^{-2p+1/2} d\theta \\ &= \frac{2^{-3p+1/2}\pi}{3t_+^{2p+1/2}}, \end{aligned} \tag{24.19}$$

where we have assumed $p \ge 1$. Substituting (24.18) and (24.19) into (24.17) yields

$$|I_c| \le \frac{2}{\tanh\beta}\left\{\frac{4\pi\sqrt{\cosh\beta}}{[\sinh(2\beta)/2 - \beta]^{p+1/2}} + \frac{2^{-p+1}\pi}{3e^{(2p+1/2)\beta}}\right\}, \quad p \ge 1. \tag{24.20}$$

A combination of (24.11), (24.15) and (24.20) gives

$$\begin{aligned} |c_p(u)| \le |I_0| + |I_c| \le \frac{2^{p+7/2}}{\pi^{p+1/2}} &+ \frac{1}{\pi(p-1)2^{p-7/2}(\cosh\beta)^{p-1/2}} \\ &+ \frac{2}{\tanh\beta}\left\{\frac{4\pi\sqrt{\cosh\beta}}{[\sinh(2\beta)/2 - \beta]^{p+1/2}} + \frac{2^{-p+1}\pi}{3e^{(2p+1/2)\beta}}\right\}, \end{aligned}$$

for $p \geq 2$. Let C_p be the number on the right-hand side of the above inequality. The error bound (24.12) holds for $p \geq 2$. Note from (24.8) that

$$\varepsilon_p(N,\beta) = \frac{A_p(\coth\beta)}{N^p} + \varepsilon_{p+1}(N,\beta).$$

Hence, we may improve the error bound (24.12) to

$$|\varepsilon_p(N,\beta)| \leq \frac{|A_p(\coth\beta)|}{N^p} + \frac{\sqrt{\sinh\beta}}{2\pi^{3/2}}\frac{C_{p+1}\Gamma(p+3/2)}{N^{p+1}} = \frac{\tilde{C}_p}{N^p},$$

for $p \geq 1$, where

$$\tilde{C}_p = |A_p(\coth\beta)| + \frac{\sqrt{\sinh\beta}}{2\pi^{3/2}}\frac{C_{p+1}\Gamma(p+3/2)}{N}.$$

Using the explicit value

$$-f(t_-) = \frac{e^{-2\beta}}{4} + \frac{\beta}{2} + \frac{1+\ln 2}{2}$$

in (24.8), we can summarize our results as follows.

Theorem 24.1. *Let $\beta > 0$ and $N = 2n+1 \geq 1$. Then for any $p \geq 1$, we have*

$$H_n(\sqrt{N}\cosh\beta) = \frac{2^n n! e^{N(e^{-2\beta}+2\beta+2)/4}}{\sqrt{2\pi\sinh\beta}N^{(n+1)/2}}\left\{\sum_{j=0}^{p-1}\frac{A_j(\coth\beta)}{N^j} + \varepsilon_p(N,\beta)\right\}, \tag{24.21}$$

where

$$|\varepsilon_p(N,\beta)| \leq \frac{\tilde{C}_p}{N^p}, \tag{24.22}$$

with

$$\tilde{C}_p = |A_p(\coth\beta)| + \frac{\sqrt{\sinh\beta}}{2\pi^{3/2}}\frac{C_{p+1}\Gamma(p+3/2)}{N},$$

and

$$\begin{aligned} C_{p+1} =& \frac{2^{p+9/2}}{\pi^{p+3/2}} + \frac{1}{\pi p 2^{p-5/2}(\cosh\beta)^{p+1/2}} \\ &+ \frac{2}{\tanh\beta}\left\{\frac{4\pi\sqrt{\cosh\beta}}{[\sinh(2\beta)/2-\beta]^{p+3/2}} + \frac{2^{-p}\pi}{3e^{(2p+5/2)\beta}}\right\}. \end{aligned}$$

In particular, $\tilde{C}_p \to |A_p(\coth\beta)|$ as $N \to +\infty$; namely, the error bound is close to the absolute value of the first neglected term when N is large.

24.3 Case II: $x = \cos\alpha \in [0, 1)$ with $\alpha \in (0, \pi/2]$

The saddle points are $t_\pm = e^{\pm i\alpha}/2$, and we have

$$f''(t_\pm) = 2 - 2e^{\mp 2i\alpha} = 4\sin\alpha e^{\pm i(\pi/2-\alpha)}.$$

Now, we define an analytic function $w(z)$:

$$w(z) = [f(z) - f(t_+)]^{1/2}, \tag{24.23}$$

where the branch of the square root function is chosen so that

$$w(z) = \sqrt{2\sin\alpha}e^{i(\pi/4-\alpha/2)}(z - t_+) + O((z - t_+)^2)$$

for z in a small complex neighborhood of t_+. The steepest descent contour passing through t_+ is given by the equation $\mathrm{Im}[f(z) - f(t_+)] = 0$. By using polar coordinates $z = re^{i\theta}$, this equation becomes

$$r^2\sin(2\theta) - 2r\cos\alpha\sin\theta + \frac{\theta - \theta_0}{2} = 0,$$

with

$$\theta_0 = \alpha - \sin(2\alpha)/2 \in (0, \alpha).$$

We are only interested in solutions in the upper half-plane; namely, $\theta \in (0, \pi)$. By the quadratic formula, there are exactly two solutions:

$$r_{d+}(\theta) = \begin{cases} \dfrac{(\theta-\theta_0)/(2\sin\theta)}{\cos\alpha + \sqrt{\cos^2\alpha - (\theta-\theta_0)/\tan\theta}}, & \theta \in [\alpha, \pi), \\[2ex] \dfrac{\cos\alpha + \sqrt{\cos^2\alpha - (\theta-\theta_0)/\tan\theta}}{2\cos\theta}, & \theta \in (0, \alpha], \end{cases}$$

and

$$r_{a+}(\theta) = \begin{cases} \dfrac{\cos\alpha + \sqrt{\cos^2\alpha - (\theta-\theta_0)/\tan\theta}}{2\cos\theta}, & \theta \in [\alpha, \pi/2), \\[2ex] \dfrac{\cos\alpha - \sqrt{\cos^2\alpha - (\theta-\theta_0)/\tan\theta}}{2\cos\theta}, & \theta \in (\theta_0, \alpha], \end{cases}$$

where $r_{d+}(\theta)$ and $r_{a+}(\theta)$ are defined as piecewise functions so that they are differentiable at $\theta = \alpha$. This is because the function $\sqrt{\cos^2\alpha - (\theta-\theta_0)/\tan\theta}$ is not differentiable at $\theta = \alpha$. For any θ in a small real neighborhood of α, we have, by Taylor expansion, $\cos^2\alpha -$

$(\theta-\theta_0)/\tan\theta = (\theta-\alpha)^2 + O((\theta-\alpha)^3)$ and $\sqrt{\cos^2\alpha-(\theta-\theta_0)/\tan\theta} = |\theta-\alpha| + O((\theta-\alpha)^2)$. Hence,

$$\begin{aligned} r_{d+}(\theta) &= \frac{\cos\alpha-(\theta-\alpha)}{2[\cos\alpha-(\theta-\alpha)\sin\alpha]} + O((\theta-\alpha)^2) \\ &= \frac{1}{2} - \frac{1-\sin\alpha}{2\cos\alpha}(\theta-\alpha) + O((\theta-\alpha)^2), \\ r_{a+}(\theta) &= \frac{\cos\alpha+(\theta-\alpha)}{2[\cos\alpha-(\theta-\alpha)\sin\alpha]} + O((\theta-\alpha)^2) \\ &= \frac{1}{2} + \frac{1+\sin\alpha}{2\cos\alpha}(\theta-\alpha) + O((\theta-\alpha)^2). \end{aligned}$$

In particular, we have $r_{d+}(\alpha) = r_{a+}(\alpha) = 1/2$, and

$$r'_{d+}(\alpha) = -\frac{1}{2}\tan\left(\frac{\pi}{4}-\frac{\alpha}{2}\right), \quad r'_{a+}(\alpha) = \frac{1}{2}\cot\left(\frac{\pi}{4}-\frac{\alpha}{2}\right).$$

Now, we define two contours

$$\Gamma_{d+} = \{r_{d+}(\theta)e^{i\theta} : \theta \in (0,\pi)\}, \quad \Gamma_{a+} = \{r_{a+}(\theta)e^{i\theta} : \theta \in (\theta_0, \pi/2)\}.$$

Recall that $-Nf(t)$ is the phase function in the integral representation of $H_n(\sqrt{N}\cos\alpha)$. It is readily seen from the asymptotic behaviors $f(z) \sim z^2$ as $z \to \infty$ and $2f(z) \sim \ln z$ as $z \to 0$ that Γ_{d+} is a steepest descent contour and Γ_{a+} is a steepest ascent contour passing through the saddle point $t_+ = e^{i\alpha}/2$. Moreover, the only zero of $f(z) - f(t_+)$ in the upper half-plane is t_+. Hence, $w(z)$ defined in (24.23) is analytic for all $\operatorname{Im} z > 0$ and positive $z > 0$. For $z = se^{i\pi}$ with $s > 0$, we define

$$w(se^{i\pi}) = \lim_{\theta\to\pi^-} w(se^{i\theta}).$$

We can define in a similar manner the contours Γ_{d-} and Γ_{a-} which, together with Γ_{d+} and Γ_{a+}, are illustrated in Figure 24.3.

Now, we return to the integral representation of $H_n(\sqrt{N}\cos\alpha)$ in (24.1) and deform the contour of integration to the steepest descent contours $\Gamma_{d+} \cup \Gamma_{d-}$ oriented with increasing parameter θ (i.e., counter-clockwise direction):

$$\begin{aligned} H_n(\sqrt{N}\cos\alpha) =& \frac{n!}{2\pi i N^{n/2}} \int_{\Gamma_{d+}} e^{-Nf(t)} t^{-1/2} dt \\ &+ \frac{n!}{2\pi i N^{n/2}} \int_{\Gamma_{d-}} e^{-Nf(t)} t^{-1/2} dt =: I_+ + I_-, \end{aligned}$$

where the last equality defines the two integrals I_+ and I_-, respectively. Since the contours Γ_{d+} and Γ_{d-} are symmetric with respect to the real

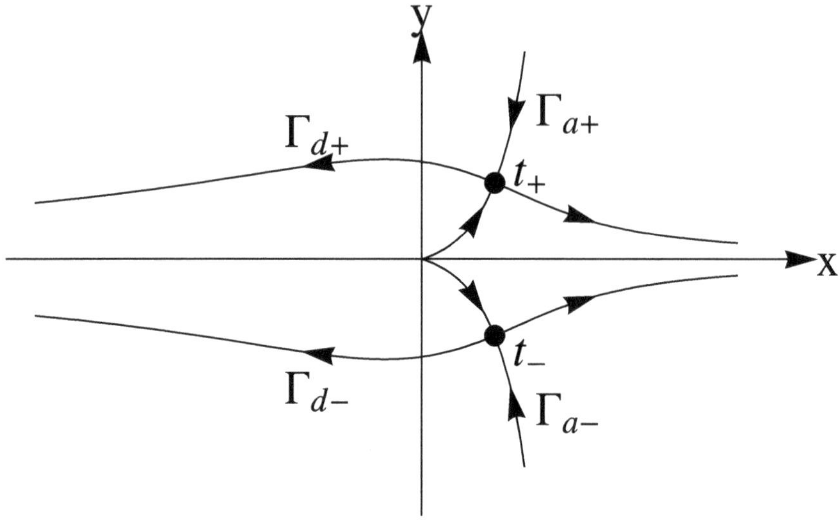

Fig. 24.3: The contours $\Gamma_{d\pm}$ and $\Gamma_{a\pm}$. The arrows indicate the increasing direction of the phase function f.

axis, have opposite orientations, and $f(t) = \overline{f(\bar{t})}$, the integrals I_+ and I_- are complex conjugates of each other. Therefore, it suffices to study the integral I_+. First, we introduce the change of variable $u = w(t)$:

$$I_+ = -\frac{n!e^{-Nf(t_+)}}{2\pi i N^{n/2}} \int_{\mathbf{R}} e^{-Nu^2} \frac{[w^{-1}(u)]^{-1/2}}{w'(w^{-1}(u))} du.$$

Since $w(z)$ is analytic in the upper half-plane and $|w(z)|/|z| \to 1$ as $z \to \infty$, we can infer from the Cauchy integral formula that

$$\frac{[w^{-1}(u)]^{-1/2}}{w'(w^{-1}(u))} = \frac{1}{2\pi i} \int_{\Gamma_0^+ \cup \mathbf{R}^+} \frac{z^{-1/2}}{w(z) - u} dz,$$

where $\mathbf{R}^+$ is the positive real line and Γ_0^+ is the negative real line with argument π:

$$\Gamma_0^+ = \lim_{\delta \to 0^+} \{\lambda + i\delta : \lambda < 0\} = \{se^{i\pi} : s > 0\}.$$

Both $\mathbf{R}^+$ and Γ_0^+ are oriented from left to right. Now, we have the following double integral representation:

$$I_+ = -\frac{n!e^{-Nf(t_+)}}{(2\pi i)^2 N^{n/2}} \int_{\mathbf{R}} e^{-Nu^2} \int_{\Gamma_0^+ \cup \mathbf{R}^+} \frac{z^{-1/2}}{w(z) - u} dz du.$$

For any $p \geq 0$, we use the identities (24.7) and

$$\frac{1}{w(z)-u} = \sum_{k=1}^{2p} \frac{u^{k-1}}{w(z)^k} + \frac{u^{2p}}{w(z)^{2p}[w(z)-u]}$$

to derive

$$I_+ = \frac{n!e^{-Nf(t_+)}}{2\sqrt{\pi \sin \alpha} N^{(n+1)/2}} \left\{ \sum_{j=0}^{p-1} \frac{A_j(i\cot\alpha)}{N^j} e^{\pi i/4} + \widetilde{\varepsilon}_p(N,\alpha) \right\}, \tag{24.24}$$

where

$$A_j(i\cot\alpha) = e^{-\pi i/4} \frac{\sqrt{\sin\alpha}\,\Gamma(j+1/2)}{2\pi^{3/2}} \int_{\Gamma_0^+ \cup \mathbf{R}^+} \frac{z^{-1/2}}{w(z)^{2j+1}} dz, \tag{24.25}$$

$$\widetilde{\varepsilon}_p(N,\alpha) = \frac{\sqrt{N\sin\alpha}}{2\pi^{3/2}} \int_{\mathbf{R}} e^{-Nu^2} u^{2p} \int_{\Gamma_0^+ \cup \mathbf{R}^+} \frac{z^{-1/2}}{w(z)^{2p}[w(z)-u]} dz du. \tag{24.26}$$

To find an upper bound for the remainder $\widetilde{\varepsilon}_p(N,\alpha)$, we need to estimate the integral

$$c_p(u) = \int_{\Gamma_0^+ \cup \mathbf{R}^+} \frac{z^{-1/2}}{w(z)^{2p}[w(z)-u]} dz, \quad u \in \mathbf{R}. \tag{24.27}$$

By expressing $w(z)$ in polar coordinates, one can easily show that

$$|w(z)-u| \geq |\operatorname{Im} w(z)| \geq \frac{|\operatorname{Im} w(z)^2|}{2|w(z)|}.$$

Note from (24.23) that

$$|\operatorname{Im} w(z)^2| = |\operatorname{Im} f(z) - \theta_0/2| = \begin{cases} \theta_0/2, & z \in \mathbf{R}^+, \\ (\pi-\theta_0)/2, & z \in \Gamma_0^+. \end{cases}$$

Assume that $p \geq 1$. Substituting the above two formulas into (24.27) yields

$$|c_p(u)| \leq \frac{4}{\theta_0} \int_0^{+\infty} \frac{s^{-1/2}}{|w(s)|^{2p-1}} ds + \frac{4}{\pi-\theta_0} \int_0^{+\infty} \frac{s^{-1/2}}{|w(se^{i\pi})|^{2p-1}} ds.$$

For $s \in (0,2)$, we have

$$|w(s)| \geq \sqrt{|\operatorname{Im} w(s)^2|} = \sqrt{\theta_0/2},$$
$$|w(se^{i\pi})| \geq \sqrt{|\operatorname{Im} w(se^{i\pi})^2|} = \sqrt{(\pi-\theta_0)/2},$$

and for $s > 2$, we have

$$\begin{aligned}|w(s)|^2 &\geq |\operatorname{Re} w(s)^2| = s^2 - 2s\cos\alpha + \frac{\ln s}{2} + \frac{\cos^2\alpha + \ln 2 + 1/2}{2}\\ &\geq (s-1)^2,\\ |w(se^{i\pi})|^2 &\geq |\operatorname{Re} w(se^{i\pi})^2| = s^2 + 2s\cos\alpha + \frac{\ln s}{2} + \frac{\cos^2\alpha + \ln 2 + 1/2}{2}\\ &\geq s^2.\end{aligned}$$

Accordingly,

$$\begin{aligned}|c_p(u)| &\leq \frac{4}{\theta_0}\left\{\int_0^2 \frac{s^{-1/2}}{(\theta_0/2)^{p-1/2}}ds + \int_2^{+\infty} \frac{s^{-1/2}}{(s-1)^{2p-1}}ds\right\}\\ &\quad + \frac{4}{\pi-\theta_0}\left\{\int_0^2 \frac{s^{-1/2}}{[(\pi-\theta_0)/2]^{p-1/2}}ds + \int_2^{+\infty}\frac{s^{-1/2}}{s^{2p-1}}ds\right\}\\ &\leq \frac{4}{\theta_0}\left\{\frac{2^{3/2}}{(\theta_0/2)^{p-1/2}} + \frac{1}{2p-3/2}\right\}\\ &\quad + \frac{4}{\pi-\theta_0}\left\{\frac{2^{3/2}}{[(\pi-\theta_0)/2]^{p-1/2}} + \frac{2^{3/2-2p}}{2p-3/2}\right\}.\end{aligned}$$

Denote the right-hand side of the above inequality by C_p. It then follows from (24.7), (24.26) and (24.27) that

$$|\widetilde{\varepsilon}_p(N,\alpha)| \leq \frac{\sqrt{\sin\alpha}}{2\pi^{3/2}}\frac{C_p\Gamma(p+1/2)}{N^p},$$

provided $p \geq 1$. We define

$$\varepsilon_p(N,\alpha) := \operatorname{Re}(\widetilde{\varepsilon}_p(N,\alpha)e^{-iN\theta_0/2}),$$

and observe that

$$\begin{aligned}|\varepsilon_p(N,\alpha)| &\leq \left|\operatorname{Re}\left(A_p(i\cot\alpha)e^{-i(\theta_0 N-\pi/2)/2}\right)\right|\frac{1}{N^p} + |\operatorname{Re}\widetilde{\varepsilon}_{p+1}(N,\alpha)|\\ &\leq \left|\operatorname{Re}\left(A_p(i\cot\alpha)e^{-i(\theta_0 N-\pi/2)/2}\right)\right|\frac{1}{N^p} + |\widetilde{\varepsilon}_{p+1}(N,\alpha)|\end{aligned}$$

for any $p \geq 0$. Using the explicit value

$$-f(t_+) = \frac{\cos(2\alpha)}{4} + \frac{1+\ln 2}{2} - i\frac{\theta_0}{2}$$

in (24.24), and the fact that $H_n(\sqrt{N}\cos\alpha) = I_+ + I_- = I_+ + \overline{I_+} = 2\operatorname{Re} I_+$, we can summarize our results as follows.

Theorem 24.2. *Let* $\alpha \in (0, \pi/2]$, $N = 2n+1$ *and* $\theta_0 = \alpha - \sin(2\alpha)/2$. *Then for any* $p \geq 0$, *we have*

$$\begin{aligned}
&H_n(\sqrt{N}\cos\alpha) \\
&= \frac{2^{n+1} n! e^{N(\cos(2\alpha)+2)/4}}{\sqrt{2\pi \sin\alpha} N^{(n+1)/2}} \left\{ \sum_{j=0}^{p-1} \frac{\mathrm{Re}\left(A_j(i\cot\alpha) e^{-i(\theta_0 N - \pi/2)/2}\right)}{N^j} + \varepsilon_p(N,\alpha) \right\} \\
&= \frac{2^{n+1} n! e^{N(\cos(2\alpha)+2)/4}}{\sqrt{2\pi \sin\alpha} N^{(n+1)/2}} \left\{ \sum_{j=0}^{p-1} \frac{\mathrm{Re}(A_j(i\cot\alpha))}{N^j} \cos\left(\frac{\theta_0 N}{2} - \frac{\pi}{4}\right) \right. \\
&\quad \left. + \sum_{j=1}^{p-1} \frac{\mathrm{Im}(A_j(i\cot\alpha))}{N^j} \sin\left(\frac{\theta_0 N}{2} - \frac{\pi}{4}\right) + \varepsilon_p(N,\alpha) \right\},
\end{aligned}$$

where

$$|\varepsilon_p(N,\alpha)| \leq \frac{\tilde{C}_p}{N^p}, \tag{24.28}$$

with

$$\tilde{C}_p = \left| \mathrm{Re}\left(A_p(i\cot\alpha) e^{-i(\theta_0 N - \pi/2)/2}\right) \right| + \frac{\sqrt{\sin\alpha}}{2\pi^{3/2}} \frac{C_{p+1}\Gamma(p+3/2)}{N},$$

and

$$\begin{aligned}
C_{p+1} =& \frac{4}{\theta_0} \left\{ \frac{2^{3/2}}{(\theta_0/2)^{p+1/2}} + \frac{1}{2p+1/2} \right\} \\
&+ \frac{4}{\pi - \theta_0} \left\{ \frac{2^{3/2}}{[(\pi-\theta_0)/2]^{p+1/2}} + \frac{2^{-1/2-2p}}{2p+1/2} \right\}.
\end{aligned}$$

In particular, $\tilde{C}_p \to |\mathrm{Re}(A_p(i\cot\alpha) e^{-i(\theta_0 N - \pi/2)/2})|$ *as* $N \to +\infty$; *namely, the error bound is close to the absolute value of the first omitted term for large* N.

24.4 Case III: $x = 1$

The saddle points coincide $t_s = t_\pm = 1/2$, and we have $f(t_s) = -3/4 - (\ln 2)/2$, $f'(t_s) = f''(t_s) = 0$ and $f'''(t_s) = 8$. We introduce the analytic function

$$w(z) = [f(z) - f(t_s)]^{1/3}, \tag{24.29}$$

where the branch of the cube root is chosen so that

$$w(z) = (4/3)^{1/3}(z - t_s)e^{-2i\pi/3} + O((z - t_s)^2) \tag{24.30}$$

for z in a small complex neighborhood of t_s. We extend the definition of $w(z)$ by analytic continuation to $\mathbf{C} \setminus (-\infty, 0]$. First, we investigate the equation $\operatorname{Im}[f(z) - f(t_s)] = \operatorname{Im} f(z) = 0$, which in polar coordinates $z = re^{i\theta}$ is

$$r^2 \sin(2\theta) - 2r \sin\theta + \theta/2 = 0. \tag{24.31}$$

This equation is clearly satisfied by all $z > 0$ (i.e., $\theta = 0$). There are two further solutions:

$$r_+(\theta) = \begin{cases} \dfrac{\theta/(2\sin\theta)}{1 + \sqrt{1 - \theta\cot\theta}}, & \theta \in (0, \pi), \\[2ex] \dfrac{1}{2}, & \theta = 0, \\[2ex] \dfrac{1 + \sqrt{1 - \theta\cot\theta}}{2\cos\theta}, & \theta \in (-\pi/2, 0), \end{cases}$$

and

$$r_-(\theta) = \begin{cases} \dfrac{1 + \sqrt{1 - \theta\cot\theta}}{2\cos\theta}, & \theta \in (0, \pi/2), \\[2ex] \dfrac{1}{2}, & \theta = 0, \\[2ex] \dfrac{\theta/(2\sin\theta)}{1 + \sqrt{1 - \theta\cot\theta}}, & \theta \in (-\pi, 0), \end{cases}$$

where we have defined $r_\pm(\theta)$ as piecewise functions so that they are differentiable at $\theta = 0$. Note that since $1 - \theta\cot\theta = \theta^2/3 + O(\theta^3)$ for small $|\theta| > 0$, we have

$$\sqrt{1 - \theta\cot\theta} = \frac{|\theta|}{\sqrt{3}} + O(\theta^2)$$

and

$$r_\pm(\theta) = \frac{1}{2} \mp \frac{\theta}{2\sqrt{3}} + O(\theta^2)$$

for small values of θ. Especially,

$$r_\pm(0) = \frac{1}{2}, \quad r'_\pm(0) = \mp\frac{1}{2\sqrt{3}}. \tag{24.32}$$

Now, we introduce two contours

$$\Gamma_1 = \{r_+(\theta)e^{i\theta} : \theta \in (-\pi/2, \pi)\}, \quad \Gamma_2 = \{r_-(\theta)e^{i\theta} : \theta \in (-\pi, \pi/2)\},$$

which we illustrate in Figure 24.4.

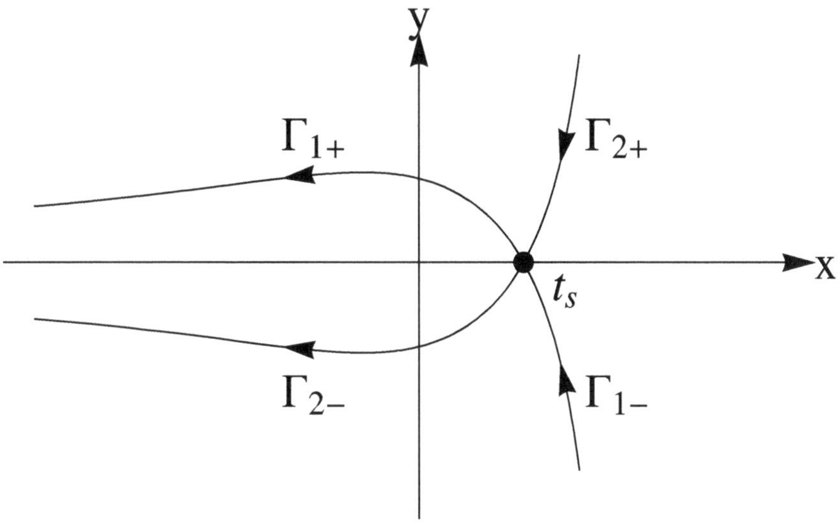

Fig. 24.4: The contours Γ_1 and Γ_2. The arrows indicate the increasing direction of the phase function f.

We have the following lemma.

Lemma 24.1. *Let $z_+(\theta) = r_+(\theta)e^{i\theta}$ for $\theta \in (-\pi/2, \pi)$; the function $f(z_+(\theta)) - f(t_s)$ increases from $-\infty$ to $+\infty$ as θ increases from $-\pi/2$ to π. Let $z_-(\theta) = r_-(\theta)e^{i\theta}$ for $\theta \in (-\pi, \pi/2)$; the function $f(z_-(\theta)) - f(t_s)$ decreases from $+\infty$ to $-\infty$ as θ increases from $-\pi$ to $\pi/2$.*

Proof. For small θ, we have from $f'(t_s) = f''(t_s) = 0$, $f'''(t_s) = 8$ and (24.32) that

$$z_+(\theta) - t_s = [r'_+(0) + ir_+(0)]\theta + O(\theta^2) = \frac{e^{2i\pi/3}}{\sqrt{3}}\theta + O(\theta^2),$$

and

$$f(z_+(\theta)) - f(t_s) = \frac{4}{3}(z_+(\theta) - t_s)^3 + O((z_+(\theta) - t_s)^4) = \frac{4}{9\sqrt{3}}\theta^3 + O(\theta^4).$$

Hence, the function $f(z_+(\theta)) - f(t_s)$ is increasing for small θ. We claim that $f(z_+(\theta)) - f(t_s)$ is increasing for all $\theta \in (-\pi/2, \pi)$. Assume to the contrary that

$$0 = \frac{d}{d\theta}f(z_+(\theta)) = f'(z_+(\theta))[r'_+(\theta) + ir_+(\theta)]e^{i\theta}$$

for some $\theta \in (-\pi/2, 0) \cup (0, \pi)$. Since $f'(z) \neq 0$ for any $z \neq t_s$, we have $r'_+(\theta) = r_+(\theta) = 0$. This contradicts the equation in (24.31). Hence, the function $f(z_+(\theta)) - f(t_s)$ is increasing for all $\theta \in (-\pi/2, \pi)$. Note that $r_+(\theta) \to \infty$ as θ tends to $-\pi/2$ from the right or π from the left. It is readily seen that $f(z_+(\theta)) - f(t_s) = r_+(\theta)^2 \cos(2\theta) - 2r_+(\theta)\cos\theta + (1/2)\ln r_+(\theta) - f(t_s)$ increases from $-\infty$ to $+\infty$ as θ increases from $-\pi/2$ to π.

The second statement of the lemma follows in a similar manner. □

Since the solutions of $\operatorname{Im} f(z) = 0$ are all the points in the union of $\Gamma_1 \cup \Gamma_2 \cup \mathbf{R}^+$, the equation $f(z) - f(t_s) = 0$ has a unique solution $z = t_s$ in $\mathbf{C} \setminus (-\infty, 0]$. This implies that $w(z)$ defined in (24.29) can be analytically continued to $\mathbf{C} \setminus (-\infty, 0]$. Moreover, it follows from (24.30) that $w(z_+(\theta))$ increases from $-\infty$ to ∞ as θ increases from $-\pi/2$ to π.

Next, we let Γ_{1+} be the portion of Γ_1 in the upper half-plane and Γ_{2-} be the portion of Γ_2 in the lower half-plane; namely,

$$\Gamma_{1+} = \{r_+(\theta)e^{i\theta} : \theta \in [0, \pi)\}, \quad \Gamma_{2-} = \{r_-(\theta)e^{i\theta} : \theta \in (-\pi, 0]\}.$$

Both Γ_{1+} and Γ_{2-} are oriented with increasing θ (counterclockwise direction). We now deform the contour of integration for $H_n(\sqrt{N})$ in (24.1) from Γ to $\Gamma_{1+} \cup \Gamma_{2-}$:

$$\begin{aligned} H_n(\sqrt{N}) =& \frac{n!}{2\pi i N^{n/2}} \int_{\Gamma_{1+}} e^{-Nf(t)} t^{-1/2} dt + \frac{n!}{2\pi i N^{n/2}} \int_{\Gamma_{2-}} e^{-Nf(t)} t^{-1/2} dt \\ =:& I_+ + I_-, \end{aligned}$$

where the last equality defines the two integrals I_+ and I_-, respectively. Since the contours Γ_{1+} and Γ_{2-} are symmetric with respect to the real axis, have opposite orientations, and $f(t) = \overline{f(\bar{t})}$, the integrals I_+ and I_- are complex conjugates of one another. Thus, it is sufficient to study the integral I_+. First, we introduce the change of variable $u = w(t)$:

$$I_+ = \frac{n! e^{-Nf(t_s)}}{2\pi i N^{n/2}} \int_0^{+\infty} e^{-Nu^3} \frac{[w^{-1}(u)]^{-1/2}}{w'(w^{-1}(u))} du. \tag{24.33}$$

We denote by $\Gamma_0^\pm$ the negative real lines with arguments $\pm\pi$. We also denote by Γ_* the union of Γ_0^+ oriented to the right and Γ_0^- oriented to the left. Since $w(z)$ is analytic in $\mathbf{C} \setminus (-\infty, 0]$ and $|w(z)|/|z|^{2/3} \to 1$ as $z \to \infty$, we can assert from the Cauchy integral formula that

$$\frac{[w^{-1}(u)]^{-1/2}}{w'(w^{-1}(u))} = \frac{1}{2\pi i} \int_{\Gamma_*} \frac{z^{-1/2}}{w(z) - u} dz.$$

Substituting this into (24.33) gives

$$I_+ = \frac{n!e^{-Nf(t_s)}}{(2\pi i)^2 N^{n/2}} \int_0^{+\infty} e^{-Nu^3} \int_{\Gamma_*} \frac{z^{-1/2}}{w(z)-u} dz du.$$

For any $p \geq 1$, we use the identities

$$\frac{1}{w(z)-u} = \sum_{j=1}^{p-1} \frac{u^{j-1}}{w(z)^j} + \frac{u^{p-1}}{w(z)^{p-1}[w(z)-u]}$$

and

$$\int_0^{+\infty} e^{-Nu^3} u^{j-1} du = \frac{\Gamma(j/3)}{3N^{j/3}} \tag{24.34}$$

to find

$$I_+ = \frac{2^{1/2} n! e^{-Nf(t_s)}}{6\pi N^{n/2}} \left\{ \sum_{j=1}^{p-1} \left(\frac{3}{4}\right)^{j/3} D_j e^{2\pi i j/3 - \pi i/2} \frac{\Gamma(j/3)}{N^{j/3}} + \widetilde{\varepsilon}_p(N) \right\}, \tag{24.35}$$

where

$$D_j = \left(\frac{4}{3}\right)^{j/3} \frac{e^{-2\pi i j/3}}{2^{3/2}\pi i} \int_{\Gamma_*} \frac{z^{-1/2}}{w(z)^j} dz, \tag{24.36}$$

$$\widetilde{\varepsilon}_p(N) = -\frac{3}{2^{3/2}\pi} \int_{\mathbf{R}} e^{-Nu^3} u^{p-1} \int_{\Gamma_*} \frac{z^{-1/2}}{w(z)^{p-1}[w(z)-u]} dz du. \tag{24.37}$$

To find an upper bound for the remainder $\widetilde{\varepsilon}_p(N)$, we need to estimate the integral

$$c_p(u) = \int_{\Gamma_*} \frac{z^{-1/2}}{w(z)^{p-1}[w(z)-u]} dz, \quad u \in \mathbf{R}. \tag{24.38}$$

By expressing $w(z)$ in polar coordinates, one can easily show that

$$|w(z)-u| \geq |\operatorname{Im} w(z)| \geq \frac{|\operatorname{Im} w(z)^3|}{3|w(z)|^2}.$$

We also note from (24.29) that

$$|\operatorname{Im} w(z)^3| = |\operatorname{Im} f(z)| = \pi/2,$$

for $z = se^{\pm i\pi} \in \Gamma_*$. Assume that $p \geq 4$. Substituting the above two formulas into (24.38) yields

$$|c_p(u)| \leq \frac{6}{\pi} \int_0^{+\infty} \frac{s^{-1/2}}{|w(se^{i\pi})|^{p-3}} ds + \frac{6}{\pi} \int_0^{+\infty} \frac{s^{-1/2}}{|w(se^{-i\pi})|^{p-3}} ds.$$

For $z = se^{\pm i\pi} \in \Gamma_*$ with $s = |z| \in (0, 1)$, we have

$$|w(z)| \geq |\operatorname{Im} w(z)^3|^{1/3} = (\pi/2)^{1/3}.$$

For $z = se^{\pm i\pi} \in \Gamma_*$ with $s = |z| > 1$, we have

$$|w(z)|^3 \geq |\operatorname{Re} w(z)^3| = s^2 + 2s + \frac{\ln s}{2} + \frac{\ln 2 + 3/2}{2} \geq s^2.$$

Consequently,

$$\begin{aligned} |c_p(u)| &\leq \frac{12}{\pi}\left\{\int_0^1 \frac{s^{-1/2}}{(\pi/2)^{p/3-1}}ds + \int_1^{+\infty} \frac{s^{-1/2}}{s^{2p/3-2}}ds\right\} \\ &\leq \frac{12}{\pi}\left\{\frac{2}{(\pi/2)^{p/3-1}} + \frac{6}{4p-15}\right\}. \end{aligned}$$

Denote the right-hand side of the above inequality by C_p. It then follows from (24.34), (24.37) and (24.38) that

$$|\widetilde{\varepsilon}_p(N)| \leq \frac{1}{2^{3/2}\pi}\frac{C_p\Gamma(p/3)}{N^{p/3}},$$

provided $p \geq 4$. We define

$$\varepsilon_p(N) := \operatorname{Re}\widetilde{\varepsilon}_p(N),$$

and observe that

$$\begin{aligned} |\varepsilon_p(N)| &\leq \left|\left(\frac{3}{4}\right)^{p/3} D_p \sin\left(\frac{2\pi p}{3}\right)\right| \frac{\Gamma(p/3)}{N^{p/3}} + |\operatorname{Re}\widetilde{\varepsilon}_{p+1}(N)| \\ &\leq \left|\left(\frac{3}{4}\right)^{p/3} D_p \sin\left(\frac{2\pi p}{3}\right)\right| \frac{\Gamma(p/3)}{N^{p/3}} + |\widetilde{\varepsilon}_{p+1}(N)| \end{aligned}$$

for any $p \geq 3$. Using the explicit value $-f(t_s) = 3/4 + (\ln 2)/2$ in (24.35), and the fact that $H_n(\sqrt{N}) = I_+ + I_- = I_+ + \overline{I_+} = 2\operatorname{Re} I_+$, we can summarize our results as follows.

Theorem 24.3. *Denote $N = 2n+1$. Then for any $p \geq 3$, we have*

$$H_n(\sqrt{N}) = \frac{2^{n+1}n!e^{3N/4}}{3\pi N^{n/2}}\left\{\sum_{j=1}^{p-1}\left(\frac{3}{4}\right)^{j/3} D_j \sin\left(\frac{2\pi j}{3}\right)\frac{\Gamma(j/3)}{N^{j/3}} + \varepsilon_p(N)\right\},$$

where

$$|\varepsilon_p(N)| \leq \frac{\tilde{C}_p}{N^{p/3}}, \tag{24.39}$$

with

$$\tilde{C}_p = \left|\left(\frac{3}{4}\right)^{p/3} D_p \sin\left(\frac{2\pi p}{3}\right)\right| \Gamma(p/3) + \frac{1}{2^{3/2}\pi}\frac{C_{p+1}\Gamma((p+1)/3)}{N^{1/3}},$$

and

$$C_{p+1} = \frac{12}{\pi}\left\{\frac{2}{(\pi/2)^{(p-2)/3}} + \frac{6}{4p-11}\right\}.$$

In particular, if p *is not divisible by* 3, *we have*

$$\tilde{C}_p \to \left|(3/4)^{p/3} D_p \sin(2\pi p/3)\right| \Gamma(p/3)$$

as $N \to +\infty$*; namely, the error bound is close to the absolute value of the first neglected term when* N *is large.*

24.5 Exercises

1. Let

$$h_{j,-}(z) := z^{-1/2}\left[\frac{-i(z-t_-)}{w(z)}\right]^{2j+1},$$

where $w(z)$ is given as in (24.3) and $t_- = e^{-\beta/2}$.

(a) By using the formula after (24.3), show that this function is analytic near t_-.

(b) Use the Cauchy integral formula to show that

$$\frac{1}{2\pi}\int_{\Gamma_0\cup\Gamma_+} \frac{z^{-1/2}}{[w(z)]^{2j+1}} dz = \frac{(-1)^{j+1} h_{j,-}^{(2j)}(t_-)}{(2j)!}.$$

(c) Rewrite (24.9) as

$$A_j(\coth\beta) = (-1)^j \sqrt{\frac{\sinh\beta}{\pi}} \frac{\Gamma(j+1/2) h_{j,-}^{(2j)}(t_-)}{(2j)!}.$$

2. Let

$$h_{j,+}(z) := z^{-1/2}\left[\frac{e^{i\pi/4}(z-t_+)}{w(z)}\right]^{2j+1},$$

where $w(z)$ is given as in (24.23) and $t_+ = e^{i\alpha/2}$.

(a) By using the formula after (24.23), show that this function is analytic near t_+.

(b) Use the Cauchy integral formula to show that

$$\frac{1}{2\pi}\int_{\Gamma_0^+\cup\mathbf{R}^+} \frac{z^{-1/2}}{[w(z)]^{2j+1}} dz = \frac{(-1)^{j+1} e^{i\pi/4} h_{j,+}^{(2j)}(t_+)}{(2j)!}.$$

(c) Rewrite (24.25) as

$$A_j(i\cot\alpha) = (-1)^j \sqrt{\frac{\sin\alpha}{\pi}} \frac{\Gamma(j+1/2)h_{j,+}^{(2j)}(t_-)}{(2j)!}.$$

3. Let

$$h_{j,s}(z) := z^{-1/2}\left[\frac{e^{-2i\pi/3}(z-t_s)}{w(z)}\right]^j,$$

where $w(z)$ is given as in (24.29) and $t_s = 1/2$.

(a) By using the formula in (24.30), show that this function is analytic near t_+.

(b) Use the Cauchy integral formula to show that

$$\frac{1}{2\pi i}\int_{\Gamma_*} \frac{z^{-1/2}}{[w(z)]^j} dz = \frac{e^{2ij\pi/3}h_{j,s}^{(j-1)}(t_s)}{(j-1)!}.$$

(c) Rewrite (24.36) as

$$D_j = \frac{(4/3)^{j/3}h_{j,s}^{(j-1)}(t_s)}{\sqrt{2}(j-1)!}.$$

Chapter 25

Generalized Bessel Functions

25.1 Introduction

In finding the asymptotic expansions of $p_k(n)$ (the number of partitions of n into perfect kth powers), Wright was led to study the asymptotic behavior of the entire function

$$\phi(z) = \phi(\rho, \beta; z) = \sum_{l=0}^{\infty} \frac{z^l}{\Gamma(l+1)\Gamma(\rho l + \beta)}, \tag{25.1}$$

where ρ is real and positive and β is any number, real or complex, see Wright (1934a) and Wright (1934b). When $\rho = 1$, this function can be expressed in terms of the Bessel function $J_\nu(t)$. More explicitly, we have

$$(t/2)^{1-\beta} J_{\beta-1}(t) = \phi(1, \beta; -t^2/4) = \phi(-t^2/2).$$

For this reason, $\phi(z)$ has been known as the generalized Bessel function.

To state the results in Wright (1934a), we let $z = |z|e^{i\theta}$ and

$$Z = (\rho|z|)^{1/(\rho+1)} e^{i\theta/(\rho+1)}. \tag{25.2}$$

Wright proved that for $|\theta| \le \pi$,

$$\phi(z) = H(Z), \tag{25.3}$$

where

$$H(Z) = Z^{1/2-\beta} e^{(1+1/\rho)Z} \left\{ \sum_{m=0}^{M} \frac{(-1)^m a_m}{Z^m} + O\left(\frac{1}{|Z|^{M+1}}\right) \right\}, \tag{25.4}$$

a_m being some explicitly determined coefficients. Let $\xi = \arg(-z)$ and

$$Z_1 = (\rho|z|)^{1/(\rho+1)} e^{i(\xi+\pi)/(\rho+1)},$$
$$Z_2 = (\rho|z|)^{1/(\rho+1)} e^{i(\xi-\pi)/(\rho+1)}. \tag{25.5}$$

Wright also proved that for $|\xi| \leq \pi$,

$$\phi(z) = H(Z_1) + H(Z_2), \tag{25.6}$$

where $H(Z)$ is again given in (25.4).

If $\theta \in [0, \pi]$ and $\xi = \theta - \pi \in [-\pi, 0]$, then (25.6) can be written as

$$\phi(z) = H(Z) + H(Ze^{-2\pi i/(\rho+1)}). \tag{25.7}$$

Similarly, if $\theta \in [-\pi, 0]$ and $\xi = \theta + \pi \in [0, \pi]$, then (25.6) becomes

$$\phi(z) = H(Z) + H(Ze^{2\pi i/(\rho+1)}). \tag{25.8}$$

Expansions (25.3) and (25.7) have a common sector of validity, namely, $0 \leq \theta \leq \pi$. Note from (25.2) that $|Z| = (\rho|z|)^{1/(\rho+1)} \to \infty$ as $|z| \to \infty$. In view of (25.4), we have

$$\frac{H(Ze^{-2\pi i/(\rho+1)})}{H(Z)} \sim e^{-2\pi i(\frac{1}{2}-\beta)/(\rho+1)+(1+1/\rho)[Ze^{-2\pi i/(\rho+1)}-Z]}, \tag{25.9}$$

which is exponentially small. Hence, in the sense of Poincaré, the term $H(Ze^{-2\pi i/(\rho+1)})$ can be neglected, and there is no inconsistency. However, the subdominant series $H(Ze^{-2\pi i/(\rho+1)})$ becomes of equal magnitude as the dominant series $H(Z)$, when the argument of z is near π; the line $\theta = \pi$ is known as the *anti-Stokes line*. (For definitions of Stokes and anti-Stokes lines, we refer to Olver (1974a, pp. 503-504) and Bender and Orszag (1978, pp. 115-117).) If we denote by $C(\theta)$ the constant multiplying the subdominant series $H(Ze^{-2\pi i/(\rho+1)})$ in (25.3) and (25.7), then we immediately notice that $C(\theta)$ changes abruptly from 0 to 1 when θ reaches π from below. The discontinuous change in $C(\theta)$, called a *Stokes multiplier*, is known as the Stokes phenomenon; see Copson (1965, pp. 103-104) and Olver (1974a, pp. 238-240). Similar remarks also apply to the expansion (25.8), and $\theta = -\pi$ is an anti-Stokes line for this expansion.

The view on Stokes phenomenon has changed dramatically in recent years when Berry (1989) came up with the new interpretation that for fixed $|z|$, Stokes multipliers should be regarded not as a discontinuous, but as a continuous, function of θ. Although Berry's arguments were highly innovative, they were quite formal. However, this theory has been successfully applied to several known asymptotic expansions in a mathematically rigorous manner (see, e.g., Boyd (1990); Olver (1991a,b, 1993); Olde Daalhuis and Olver (1994).) For an excellent exposition of Berry's theory, see Paris and Wood (1995).

The purpose of this chapter is to provide a further application of Berry's theory to the generalized Bessel function $\phi(z)$; that is, to study the smoothing of Stokes discontinuity for the function $\phi(z)$, or, equivalently, the continuous transition across the Stokes line of the expansions (25.3), (25.7) and (25.8). Our approach is based on a modified version of the steepest descent method introduced in Berry and Howls (1991). Although their method seems to have been developed only for analytic integrands, in actual fact, it can sometimes be applied to cases with singularities in the integrand, as we shall see in the following section. Smoothing of Stokes discontinuity can also be done from the differential equation point of view (see, for example, Olver (1993), Olde Daalhuis and Olver (1994)). The reason that we have taken the integral approach is because we are not aware of any second-order differential equation which is satisfied by the generalized Bessel function. The material in this chapter is taken from Wong and Zhao (1999a).

From the Hankel loop integral

$$\frac{1}{\Gamma(\mu)} = \frac{1}{2\pi i}\int_{-\infty}^{(0^+)} u^{-\mu} e^{u} du, \tag{25.10}$$

one readily sees that the generalized Bessel function $\phi(z)$ in (25.1) has the contour integral representation

$$\phi(z) = \frac{1}{2\pi i}\int_{-\infty}^{(0^+)} u^{-\beta} e^{u+z/u^{\rho}} du. \tag{25.11}$$

Here the path of integration can be any curve which begins at $u = -\infty$, encircles $u = 0$ once in the positive sense, and returns to its starting point. Let us assume for a moment that z is real and positive. Then, by making the change of variable $u = Zt$ with Z being given by (25.2), we obtain

$$\phi(z) = \frac{Z^{1-\beta}}{2\pi i}\int_{-\infty}^{(0^+)} t^{-\beta} e^{Z(t+t^{-\rho}/\rho)} dt. \tag{25.12}$$

The validity of this identity can now be extended to the sector $|\arg z| < (1+\rho)\pi/2$.

25.2 The case when $0 < \rho < 1$

In this section, we shall restrict ourselves to the case $0 < \rho < 1$. The steepest descent path of the function $t + t^{-\rho}/\rho$ through the saddle point $t = 1$ is just the loop contour in (25.12). Making a further change of variable $t = e^{\tau}$, (25.12) becomes

$$\phi(z) = \frac{Z^{1-\beta}}{2\pi i}\int_{C} e^{(1-\beta)\tau + Z(e^{\tau}+e^{-\rho\tau}/\rho)} d\tau, \tag{25.13}$$

where C is the image of the loop contour under the transformation that connects $\infty - i\pi/2$ to $\infty + i\pi/2$ in the τ-plane. Let

$$g(\tau) := e^{(1-\beta)\tau}, \quad f(\tau) := e^{\tau} + e^{-\rho\tau}/\rho. \tag{25.14}$$

The mapping $f(\tau)$ is illustrated in Figure 25.1. For convenience, we denote

$$\bar{\theta} := \frac{(1-\rho)\pi}{2(1+\rho)}, \quad \theta_Z := \arg Z = \frac{\theta}{1+\rho}. \tag{25.15}$$

Upon deforming the contour C into the steepest descent curve

$$\Gamma_0(\theta_Z): \quad \arg\{Z[f(0) - f(\tau)]\} = 0, \tag{25.16}$$

we can rewrite (25.13) as

$$\phi(z) = \frac{1}{2\pi i} Z^{1/2-\beta} e^{(1+1/\rho)Z} T(Z), \tag{25.17}$$

where

$$T(Z) = Z^{1/2} \int_{\Gamma_0(\theta_Z)} g(\tau) e^{Z[f(\tau)-f(0)]} d\tau. \tag{25.18}$$

There are infinitely many saddle points of the phase function $f(\tau)$ in (25.14): $\tau_k = 2k\pi i/(1+\rho)$ with $k = 0, \pm 1, \pm 2, \cdots$. The two saddle points nearby $\tau_0 = 0$ are $\tau_\pm := \pm 2\pi i/(1+\rho)$. Let $\Gamma_\pm$ be the paths passing through $\tau_\pm$ and satisfying

$$\arg\left[\frac{f(\tau_\pm) - f(\tau)}{F_\pm}\right] = 0, \tag{25.19}$$

where

$$F_\pm = f(0) - f(\tau_\pm) = \left[2\left(1 + \frac{1}{\rho}\right)\sin\frac{\pi}{1+\rho}\right] e^{\pm i\bar{\theta}}. \tag{25.20}$$

Note that for τ on $\Gamma_\pm$, we also have

$$\arg\left[\frac{f(0) - f(\tau)}{F_\pm}\right] = 0. \tag{25.21}$$

The two saddle points τ_+ and τ_- are called *adjacent saddles* of $\tau_0 = 0$ by Berry and Howls (1990), and the paths Γ_+ and Γ_- are known as *adjacent contours* (see Boyd (1993)). The steepest descent curve $\Gamma_0(\theta_Z)$ (with $\theta_Z = 0$) and the adjacent contours $\Gamma_\pm$ are illustrated in Figure 25.2.

The method of Berry and Howls (1991, (11)-(12)) then gives

$$\phi(z) = \frac{1}{2\pi i} Z^{1/2-\beta} e^{(1+1/\rho)Z} \left[\sum_{s=0}^{N-1} \frac{c_s}{Z^s} + R_N(Z)\right], \tag{25.22}$$

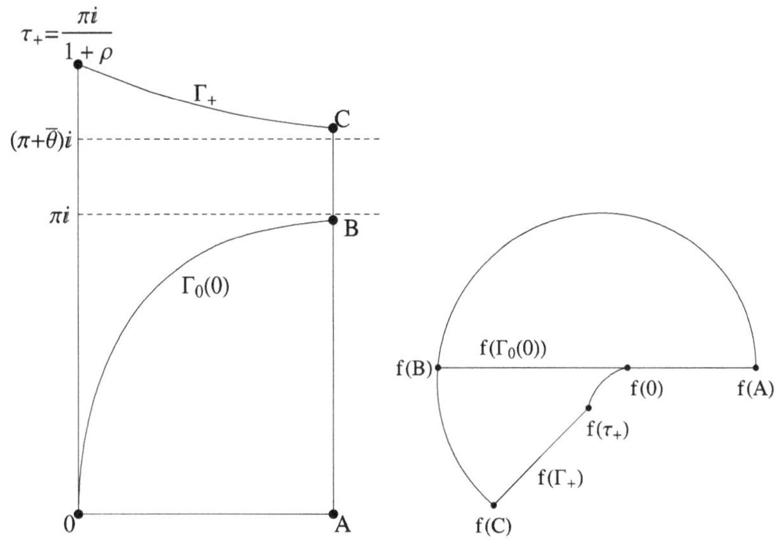

Fig. 25.1: The mapping $f(\tau)$ from the first quadrant of τ-plane to the f-plane $(0 < \rho < 1)$.

with

$$R_N(Z) = \frac{1}{2\pi i Z^N} \int_0^\infty u^{N-1/2} e^{-u} \int_{C_0(\theta_Z)} \frac{g(\tau)[f(0)-f(\tau)]^{-N-1/2}}{1-uZ^{-1}[f(0)-f(\tau)]^{-1}} d\tau du, \tag{25.23}$$

where $C_0(\theta_Z)$ is a positively oriented curve surrounding the steepest descent path $\Gamma_0(\theta_Z)$. Since $\Gamma_0(\theta_Z)$ is an infinite curve, $C_0(\theta_Z)$ actually consists of two pieces (see Boyd (1993)). The coefficients in (25.22), of course, agree with those in (25.4), after making the appropriate notational changes.

We now deform the contour $C_0(\theta_Z)$ in (25.23) into the union of Γ_+ and Γ_-. This can be done provided that the integrand in the inner integral decays rapidly near infinity, which can be achieved by choosing N so that $(N+1/2)\rho > |1 - \operatorname{Re}\beta|$. Thus we obtain

$$R_N(Z) = R_N^{(+)}(Z) + R_N^{(-)}(Z), \tag{25.24}$$

where

$$R_N^{(\pm)}(Z) = \frac{1}{2\pi i Z^N} \int_0^\infty u^{N-1/2} e^{-u} \times \int_{\Gamma_\pm} \frac{g(\tau)}{[f(0)-f(\tau)]^{N+1/2}} \left\{1 - \frac{u}{Z[f(0)-f(\tau)]}\right\}^{-1} d\tau du. \tag{25.25}$$

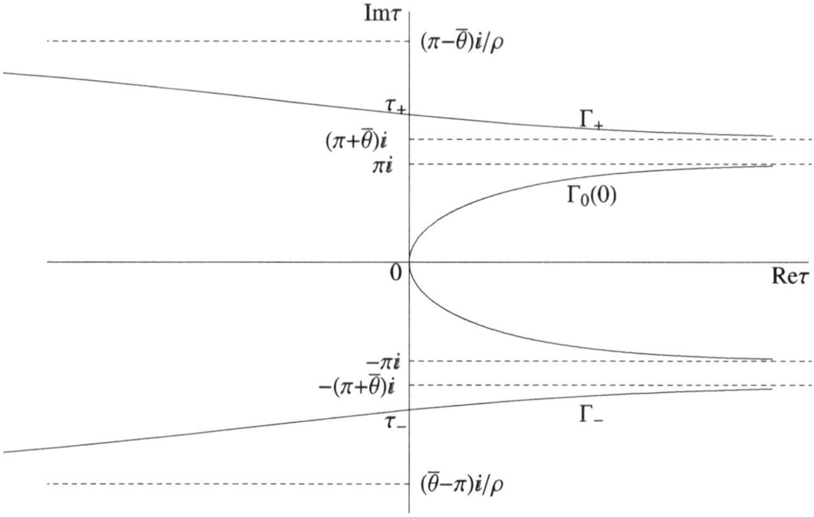

Fig. 25.2: The steepest descent curve $\Gamma_0(\theta_Z)$ (with $\theta_Z = 0$) and the adjacent contours $\Gamma_\pm$ in the τ-plane ($0 < \rho < 1$).

The square root in (25.23) and (25.25) is defined as follows. Let D be the domain bounded by the curves Γ_+ and Γ_-. Since Γ_+ and Γ_- are two adjacent contours of the saddle point $\tau = 0$, the domain D is the envelope of curves $\Gamma_0(\theta_Z)$ with varying θ_Z (see Boyd (1993)). Thus, each point in D can be reached from the origin by a steepest descent path $\Gamma_0(\theta_Z)$ for some value of θ_Z, and consequently $f(0) - f(\tau)$ has no zero in D, except for a double zero at $\tau = 0$. Therefore, we can define the square root so that $[f(0) - f(\tau)]^{1/2}$ is single valued, and specify the sign of $[f(0) - f(\tau)]^{1/2}$ when $[f(0) - f(\tau)] > 0$. We choose the sign so that $[f(0) - f(\tau)]^{1/2}$ is positive on the half of $\Gamma_0(\theta_Z)$ emanating from the origin when $\theta_Z = 0$.

By making changes of variables, the double integrals in (25.25) can be expressed as single integrals passing through the adjacent saddles τ_+ and τ_- (see Berry and Howls (1991, (16)-(17))). Indeed, we have from (25.25)

$$\begin{aligned} R_N^{(\pm)}(Z) = & \frac{e^{\pm(1-2\beta)\pi i/(1+\rho)}}{2\pi i Z^N F_\pm^N} \int_0^\infty e^{-v} v^{N-1} \left(1 - \frac{v}{ZF_\pm}\right)^{-1} \\ & \times T\left(\frac{v}{F_\pm} e^{\pm 2\pi i/(\rho+1)}\right) dv. \end{aligned} \tag{25.26}$$

A combination of (25.17), (25.22) and (25.24) gives

$$T(Z) = \sum_{s=0}^{N-1} \frac{c_s}{Z^s} + R_N^{(+)}(Z) + R_N^{(-)}(Z). \tag{25.27}$$

In view of the reappearance of the original integral $T(Z)$ in the remainder, (25.26)-(25.27) is called a *resurgence formula.* From (25.17), it of course also follows that

$$\phi(z) = \frac{1}{2\pi i} Z^{1/2-\beta} e^{(1+1/\rho)Z} \left\{ \sum_{s=0}^{N-1} \frac{c_s}{Z^s} + R_N^{(+)}(Z) + R_N^{(-)}(Z) \right\}, \tag{25.28}$$

with $R_N^{(+)}(Z)$ and $R_N^{(-)}(Z)$ being given as in (25.26).

To show that the resurgence formula (25.27) incorporates the concept of smoothing the Stokes discontinuity, we let

$$F = |F_+| = |F_-| = 2\left(1 + \frac{1}{\rho}\right) \sin(\frac{\pi}{1+\rho})$$

and $V = v/F$. By Laplace's method (17.76),

$$T\left(\frac{v}{F_+} e^{2\pi i/(\rho+1)}\right) = T\left(V e^{i(\pi+\bar{\theta})}\right) \sim i\sqrt{\frac{2\pi}{1+\rho}}, \tag{25.29}$$

as $V \to \infty$. To obtain the small-V behavior, one can first express $T(Ve^{i(\pi+\bar{\theta})})$ in the form of a Laplace transform by using a simple change of variable, and then apply a result on a Laplace transform near the origin (see Wong (1989, p. 322); see also Chapter 17). The result is that for small and finite values of V, we have

$$|T(Ve^{i(\pi+\bar{\theta})})| \leq CV^{-\kappa}, \tag{25.30}$$

where κ and C are positive constants. A similar result holds for

$$T\left(\frac{v}{F_-} e^{-2\pi i/(\rho+1)}\right) = T\left(\frac{v}{F} e^{-i(\pi+\bar{\theta})}\right).$$

Let

$$h_\pm(v) \equiv T\left(\frac{v}{F} e^{\pm i(\pi+\bar{\theta})}\right)$$

and $\zeta_\pm \equiv ZF_\pm$. Then

$$|R_N^{(\pm)}(Z)| \leq \frac{C^{(\pm)}}{|ZF|^N} \int_0^\infty \frac{e^{-v} v^{N-1}}{|1 - v/\zeta_\pm|} |h_\pm(v)| dv,$$

where $C^{(+)}$ and $C^{(-)}$ are positive constants. If $\phi_\pm$ denotes the argument of $v/\zeta_\pm$, then a simple estimation yields

$$|1 - v/\zeta_\pm| \geq |\sin(\pi - \phi_\pm)| = |\sin\phi_\pm|.$$

Hence, by (25.29) and (25.30), we obtain

$$|R_N(Z)| \leq C(\rho,\beta)\frac{\Gamma(N)}{|ZF|^N}\left\{ \left|\csc\left(\theta_Z+\bar{\theta}\right)\right| + \left|\csc\left(\theta_Z-\bar{\theta}\right)\right| \right\}, \tag{25.31}$$

where $C(\rho,\beta)$ is a constant depending only on ρ and β, and where $|\theta_Z| < \bar{\theta}$. When θ_Z gets close to $\pm\bar{\theta}$, the right-hand side of (25.31) becomes infinite. So we need a modification of this inequality. Since $\phi_+ = -\arg\zeta_+ = -\theta_Z - \bar{\theta}$, $-\phi_+$ decreases to zero when θ_Z decreases to $-\bar{\theta}$. Following Boyd (1990), we now rotate the path of integration in $R_N^{(+)}(Z)$ through an angle $\eta \in (-\pi/2, 0)$. Clearly, we have

$$\left|1 - \frac{v}{|\zeta_+|}e^{i(\eta+\phi_+)}\right| \geq |\sin(\eta+\phi_+)|.$$

Thus a straightforward estimation gives

$$|R_N^{(+)}(Z)| \leq \frac{C^{(+)}}{|ZF|^N}\frac{1}{\sin(-\phi_+-\eta)}\int_0^{\infty e^{i\eta}} |e^{-v}v^{N-1}h(v)dv|.$$

Again using (25.29) and (25.30), we obtain

$$|R_N^{(+)}(Z)| \leq \frac{C_1^{(+)}}{|ZF|^N}\frac{1}{\sin(-\phi_+-\eta)}\frac{\Gamma(N)}{(\cos\eta)^N},$$

where $C_1^{(+)}$ is still a constant. Since $\eta \in (-\pi/2, 0)$ is rather arbitrary, we may choose it so that $\cos\eta = 1 - 1/N$. Note that for $-\phi_+ \in [0, \varepsilon_0]$ and $\eta \in [-\pi/2 + \varepsilon_0, 0)$,

$$\sin(-\phi_+ - \eta) \geq \sin(-\eta) \sim \sqrt{\frac{2}{N}}.$$

Hence

$$|R_N^{(+)}(Z)| \leq \frac{C_2^{(+)}}{|ZF|^N}N^{1/2}(1-\frac{1}{N})^{-N}\Gamma(N). \tag{25.32}$$

A similar argument leads to

$$|R_N^{(-)}(Z)| \leq \frac{C_2^{(-)}}{|ZF|^N}N^{1/2}(1-\frac{1}{N})^{-N}\Gamma(N). \tag{25.33}$$

From Stirling's formula, it follows that a combination of (25.31), (25.32) and (25.33) yields

$$|R_N(Z)| \leq \widetilde{C}(\rho,\beta)e^{N(-1+\log N - \log|ZF|)} \tag{25.34}$$

for large value of N and θ_Z in the sector $|\theta_Z| \leq \bar{\theta}$, where $\widetilde{C}(\rho,\beta)$ is a constant independent of N and Z. This estimate suggests that an optimal truncation occurs at

$$N - |ZF| = 2(1+\frac{1}{\rho})|Z|\sin\frac{\pi}{1+\rho}. \tag{25.35}$$

When the expansion (25.22) is optimally truncated, the error is of the order

$$R_N(Z) = O(e^{-N}) = O(e^{-|ZF|}); \tag{25.36}$$

we thus obtain what Olver (1991b) called a *uniform, exponentially improved*, asymptotic expansion in the sector $|\theta_Z| \leq \bar{\theta}$. Optimally truncated asymptotic series have now been also called *superasymptotic expansions* by Berry and Howls (1990); see Chapter 23.

We next consider how to smooth Stokes discontinuity. To this end, we return to (25.28) and discuss the behaviour of the remainder terms $R_N^{(+)}(Z)$ and $R_N^{(-)}(Z)$ given in (25.26) near

$$\theta_Z = \pm\bar{\theta}.$$

By replacing the function $T((v/F)e^{\pm 2\pi i/(\rho+1)})$ in the integrand of $R_N^{(\pm)}(Z)$ by its asymptotic expansion (25.27), we obtain a series of integrals which can be expressed in terms of Dingle's *terminant functions*

$$\int_0^\infty \frac{t^{m-1}e^{-t}}{1+t/\zeta}dt = 2\pi i(-\zeta)^m e^{\zeta} T_m(\zeta); \tag{25.37}$$

see Olver (1991b). More precisely, we have

$$\begin{aligned} R_N(z) =& e^{(1-2\beta)\pi i/(\rho+1)-ZF_+} \sum_{r=0}^{N'-1} c_r e^{-2\pi ri/(\rho+1)} Z^{-r} T_{N-r}(-ZF_+) \\ &+ R_{N,N'}^{(+)}(Z) + e^{-(1-2\beta)\pi i/(\rho+1)-ZF_-} \\ &\times \sum_{r=0}^{N''-1} c_r e^{2\pi ri/(\rho+1)} Z^{-r} T_{N-r}(-ZF_-) + R_{N,N''}^{(-)}(Z), \end{aligned} \tag{25.38}$$

where

$$R_{N,N'}^{(+)}(Z) = \frac{e^{(1/2-\beta)\frac{2\pi i}{\rho+1}}}{2\pi i(ZF_+)^N} \int_0^\infty \frac{e^{-v}v^{N-1}}{1-v/(ZF_+)} R_{N'}\left(\frac{v}{F_+}e^{\frac{2\pi i}{\rho+1}}\right) dv \tag{25.39}$$

and

$$R_{N,N''}^{(-)}(Z) = \frac{e^{-(1/2-\beta)\frac{2\pi i}{\rho+1}}}{2\pi i(ZF_-)^N} \int_0^\infty \frac{e^{-v}v^{N-1}}{1-v/(ZF_-)} R_{N''}\left(\frac{v}{F_-}e^{\frac{-2\pi i}{\rho+1}}\right) dv. \tag{25.40}$$

The idea of re-expanding the remainder terms in the optimally truncated asymptotic series was introduced by Berry and Howls (1990). They call this theory *hyper-asymptotics* (see also Olver (1993), Olde Daalhuis and Olver (1994). For more related work on this subject, we mention Howls (1997) and Olde Daalhuis (1998).

In Olver (1991b, Theorem 1), it was proved that with $\zeta = |\zeta|e^{i\widetilde{\theta}}$ and $N = |\zeta| + r + \alpha$, α being bounded, we have

$$T_{N-r}(\zeta) \sim \frac{1}{2}\operatorname{erfc}\left\{c(\widetilde{\theta})\sqrt{\frac{|\zeta|}{2}}\right\} - i\frac{\exp\{-\frac{1}{2}|\zeta|[c(\widetilde{\theta})]^2\}}{(2\pi|\zeta|)^{1/2}}\sum_{s=0}^{\infty}\frac{(\frac{1}{2})_s 2^s g_{2s}(\widetilde{\theta},\alpha)}{|\zeta|^s} \tag{25.41}$$

uniformly with respect to $\widetilde{\theta} \in [-\pi+\delta, 3\pi-\delta]$. The function $c(\widetilde{\theta})$ in (25.41) is defined by

$$\frac{1}{2}\{c(\widetilde{\theta})\}^2 = -e^{i(\widetilde{\theta}-\pi)} + i(\widetilde{\theta}-\pi) + 1. \tag{25.42}$$

Near $\widetilde{\theta} = \pi$, the Taylor series expansion of $c(\widetilde{\theta})$ begins

$$c(\widetilde{\theta}) = -(\widetilde{\theta}-\pi) - \frac{1}{6}i(\widetilde{\theta}-\pi)^2 + \frac{1}{36}(\widetilde{\theta}-\pi)^3 + \cdots. \tag{25.43}$$

The coefficient $g_{2s}(\widetilde{\theta},\alpha)$ can be given explicitly and the first one is

$$g_0(\widetilde{\theta},\alpha) = \frac{e^{i\alpha(\pi-\widetilde{\theta})}}{1+e^{-i\widetilde{\theta}}} - \frac{i}{c(\widetilde{\theta})}. \tag{25.44}$$

Olver (1991b, p. 1471) has also shown that $\xi = c(\widetilde{\theta})\sqrt{\frac{1}{2}|\zeta|}$ lies in the sector $-\frac{1}{4}\pi \le \arg\xi \le 0$ when $-\pi \le \widetilde{\theta} \le \pi$, and in the sector $0 \le \arg(-\xi) \le \frac{1}{4}\pi$ when $\pi \le \widetilde{\theta} \le 3\pi$. Since $\operatorname{erfc}(\xi) = O(e^{-\xi^2})$ uniformly throughout the first sector and $\operatorname{erfc}(\xi) = 2 + O(e^{-\xi^2})$ uniformly throughout the second sector (see Wong (1989, p. 42)), if $|\zeta|$ is large and fixed and $\widetilde{\theta}$ increases continuously from values just below π to values just above π, then (25.41) shows that $T_{N-r}(\zeta)$ changes rapidly, but smoothly, from being exponentially small to being exponentially close to one.

By using this result in (25.38), we shall show in the following that if $|z|$ is large and the asymptotic expansion (25.28) is truncated at its optimal number of terms given in (25.35), then there is a rapid, but smooth, transition from the form (25.3) to the form (25.7) in the neighborhood of the Stokes line $\arg z = (1-\rho)\pi/2$, i.e., $\theta_Z := \arg Z = \bar{\theta}$. To this end, we restrict θ_Z to the interval $(\bar{\theta}-\varepsilon, \bar{\theta}+\varepsilon)$, with $0 < \varepsilon < 1$. (To be more precise, one can choose ε to be less than the minimum of $\bar{\theta}$ and $2\pi\rho/(1+\rho)$.) Let $\zeta'_+ \equiv -ZF_+, \zeta'_- \equiv -ZF_-, \theta'_+ \equiv \arg\zeta'_+$ and $\theta'_- \equiv \arg\zeta'_-$. Clearly, $\theta'_\pm = \mp\pi + \theta_Z \pm \bar{\theta}$ with $\theta'_- \in (\pi-\varepsilon, \pi+\varepsilon)$ and $\theta'_+ \in (-\pi+\delta, 0)$ for some small $\delta > 0$. With θ_Z restricted to this interval, we have, from (25.22),

(25.38) with $N' = 0$ and (25.41),

$$\begin{aligned}\phi(z) =& \frac{1}{2\pi i} Z^{1/2-\beta} e^{(1+1/\rho)Z} \sum_{s=0}^{N-1} \frac{c_s}{Z^s} \\ &+ \frac{\operatorname{erfc}\{c(\theta'_-)\sqrt{\frac{1}{2}|\zeta'_-|}\}}{4\pi i} \left[Z_-^{1/2-\beta} e^{(1+1/\rho)Z_-} \sum_{r=0}^{N''-1} \frac{c_r}{Z_-^r} \right] + R_{N,N''}(Z),\end{aligned} \tag{25.45}$$

where $Z_- = Ze^{-2\pi i/(\rho+1)}$, $N \sim |ZF|$, and

$$\begin{aligned}R_{N,N''}(Z) =& \frac{1}{2\pi i} \Big\{ Z^{1/2-\beta} e^{(1+1/\rho)Z} R^{(+)}_{N,0}(Z) \\ &+ Z^{1/2-\beta} e^{(1+1/\rho)Z} R^{(-)}_{N,N''}(Z) \\ &+ Z_-^{1/2-\beta} e^{(1+1/\rho)Z_- - 1/2|\zeta'|\{c(\theta'_-)\}^2} O(|Z|^{-1/2}) \Big\}.\end{aligned} \tag{25.46}$$

We now estimate each of the three terms in (25.46). Noting that $R_0(Z) = T(Z)$ by (25.27), it follows from (25.39) that

$$\begin{aligned}R^{(+)}_{N,0}(Z) =& \frac{e^{(1-2\beta)\pi i/(\rho+1)}}{2\pi i (ZF_+)^N} \\ &\times \int_0^\infty \frac{e^{-v} v^{N-1}}{1 - v/ZF_+} T\left(\frac{v}{F_+} e^{2\pi i/(\rho+1)} \right) dv.\end{aligned} \tag{25.47}$$

Since $|1 + v/\zeta'_+| \geq 1$ for $\theta'_+ \in [-\pi/2, 0)$ and $|1 + v/\zeta'_+| \geq \sin\delta$ for $\theta'_+ \in (-\pi + \delta, -\pi/2)$, we have

$$|R^{(+)}_{N,0}(Z)| \leq \frac{C_1(\rho,\beta,\delta)}{|ZF|^N} \int_0^\infty e^{-v} v^{N-1} \left| T\left(\frac{v}{F} e^{i(\pi+\bar{\theta})} \right) \right| dv,$$

where $F = |F_+| = |F_-|$ and $C_1(\rho,\beta,\delta)$ is a constant. By (25.29) and (25.30), we further obtain

$$|R^{(+)}_{N,0}(Z)| \leq C_2(\rho,\beta,\delta) \frac{\Gamma(N)}{(|Z|F)^N},$$

$C_2(\rho,\beta,\delta)$ being again a constant. In view of Stirling's formula and the fact that $N \sim |ZF|$, we conclude

$$|R^{(+)}_{N,0}(Z)| \leq C(\rho,\beta,\delta) e^{-|Z|F} |Z|^{-1/2}. \tag{25.48}$$

The constant $C(\rho,\beta,\delta)$ is independent of N and Z.

We next estimate the third term in (25.46). From (25.42), we have

$$-\frac{1}{2}|\zeta'_-|\{c(\theta'_-)\}^2 = -\zeta'_- - |\zeta'_-| - i|\zeta'_-|(\theta'_- - \pi).$$

Since $\zeta'_- = -ZF_-$, it follows that

$$|\exp(-\frac{1}{2}|\zeta'_-|\{c(\theta'_-)\}^2)| = \exp(-|Z|F + \operatorname{Re} ZF_-).$$

Furthermore, by (25.20)

$$(1+\frac{1}{\rho})Z_- + ZF_- = Zf(0) + (1+\frac{1}{\rho})Z.$$

Hence,

$$\begin{aligned} Z^{1/2-\beta} \exp\left[(1+\frac{1}{\rho})Z_- - \frac{1}{2}|\zeta'_-|\{c(\theta'_-)\}^2\right] O(|Z|^{-1/2}) \\ = O\left\{|Z|^{-\operatorname{Re}\beta} \exp\left[(1+\frac{1}{\rho})\operatorname{Re} Z - |Z|F\right]\right\}. \end{aligned} \tag{25.49}$$

Finally, we come to the more complicated case, namely, the estimation of the second term in (25.46). Returning to (25.40), we first note that

$$R_{N''}\left(\frac{v}{F_-}e^{-2\pi i/(\rho+1)}\right) = \begin{cases} O(v^{-N''}), & 1 \le v < \infty, \\ O(v^{-\max(K,N''-1)}), & 0 < v < 1, \end{cases} \tag{25.50}$$

by virtue of (25.27) and (25.30). We shall choose $N'' \ge K+1$ and N considerably bigger than N'', and apply the argument used to obtain (25.32) and (25.33). Recall that for

$$\theta_Z \in (\bar{\theta} - \varepsilon, \bar{\theta}],$$

we have $\theta'_- \in (\pi - \varepsilon, \pi]$. By rotating the path of integration in (25.40) through an angle $\eta \in (0, \pi/2 - \varepsilon)$, it can be shown, as in (25.32) and (25.33), that

$$|R^{(-)}_{N,N''}(Z)| \le \widetilde{C}_1(\rho,\beta)\frac{1}{|ZF_-|^N}\frac{1}{|\sin(\eta - \theta'_-)|}\frac{\Gamma(N-N''+1)}{(\cos\eta)^{N-N''+1}},$$

$\widetilde{C}_1(\rho,\beta)$ being a constant. Here use has been made of (25.50). Choosing η so that $\cos\eta = 1 - 1/N$ and $N'' \ge 2 + \frac{1}{2}$, we have, by Stirling's formula,

$$|R^{(-)}_{N,N''}(Z)| \le \widetilde{C}_2(\rho,\beta)|Z|^{-1/2}e^{-|Z|F}, \tag{25.51}$$

for

$$\theta_Z \in (\bar{\theta} - \varepsilon, \bar{\theta}].$$

We now study how the behavior of $R^{(-)}_{N,N''}(Z)$ changes when θ_Z crosses the Stokes line $\theta_Z = \bar{\theta}$ or, equivalently, when θ'_- goes beyond π. First we let $v = \zeta'_- t$ in (25.40) and recall that $\zeta'_- = -ZF_-$. This gives

$$\begin{aligned} R^{(-)}_{N,N''}(Z) = & \frac{(-1)^N e^{-(1-2\beta)\pi i/(\rho+1)}}{2\pi i} \\ & \times \int_0^{\infty e^{-i\theta'_-}} \frac{e^{-\zeta'_- t} t^{N-1}}{1+t} R_{N''}\left(-Zte^{-2\pi i/(\rho+1)}\right) dt. \end{aligned}$$

Note that $\theta'_- \in (\pi-\varepsilon,\pi)$. Using Cauchy's theorem, we can deform the radial line $\arg t = -\theta'_-$ to the contour C_- which goes from $t=0$ to $t=-\infty$ along the negative real-axis with an indentation at the pole $t=-1$. Let C_+ be the reflection of C_- with respect to the real-axis; see Figure 25.3.

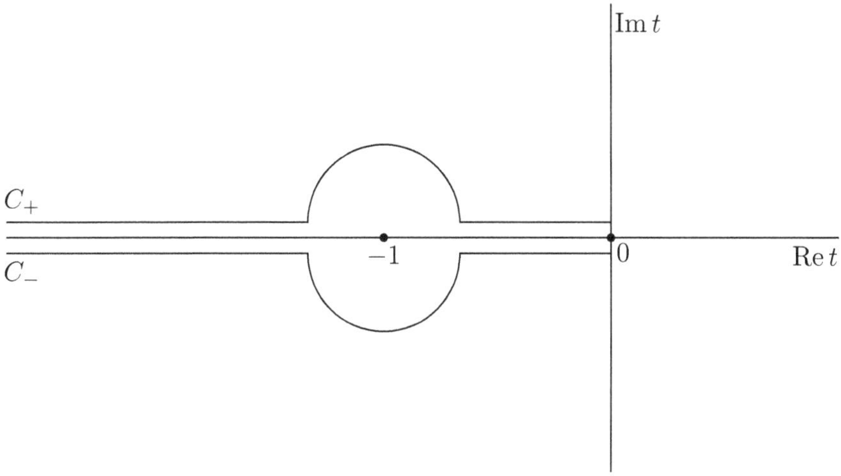

Fig. 25.3: The contours $C_\pm$.

Replacing C_- with C_+, we pick up a contribution from the residue at $t=-1$, and obtain

$$\begin{aligned} R^{(-)}_{N,N''}(Z) =& e^{-(1-2\beta)\pi i/(\rho+1)+\zeta'_-} R_{N''}(Ze^{-2\pi i/(\rho+1)}) \\ &+ \frac{(-1)^N}{2\pi i} e^{-(1-2\beta)\pi i/(\rho+1)} \\ &\times \int_{C_+} \frac{e^{-\zeta'_- t} t^{N-1}}{1+t} R_{N''}(-Zte^{-2\pi i/(\rho+1)})dt \end{aligned} \tag{25.52}$$

for $\theta_Z \in \left(\bar\theta-\varepsilon,\bar\theta\right]$. The right-hand side of this equation can be used to analytically continue $R^{(-)}_{N,N''}(Z)$ beyond $\theta_Z = \bar\theta$ or, equivalently, beyond $\theta'_- = \pi$. Using Cauchy's theorem again, the contour C_+ in (25.52) can be deformed into the radial line $\arg t = -\theta'_-$. Note that for $\theta'_- \in [\pi,\pi+\varepsilon]$, this radial line lies in the second, and not in the third quadrant of the complex t-plane. Since $|e^{\zeta'_-}| \le 1$ for $\theta'_- \in [\pi,\pi+\varepsilon]$, the first term on the right-hand side of (25.52) is $O(|Z|^{-N''})$. The integral in (25.52), with C_+ replaced by the ray $\arg t = -\theta'_-$, can be estimated in a manner similar to that leading to (25.51). In fact, the simple change of variable $v = \zeta'_- t$ puts this integral back into the form (25.40). Hence we obtain

$$R^{(-)}_{N,N''}(Z) = O(|Z|^{-N''}) + O(e^{-|Z|F}|Z|^{-1/2}), \tag{25.53}$$

for $\theta_Z \in (\bar{\theta}, \bar{\theta}+\varepsilon)$.

To summarize, we have from (25.46), (25.48), (25.49) and (25.51)

$$R_{N,N''}(Z) = O(|Z|^{-\operatorname{Re}\beta} e^{(1+1/\rho)\operatorname{Re} Z - |Z|F}),$$

for $\theta_Z \in (\bar{\theta}-\varepsilon, \bar{\theta})$, and from (25.46), (25.48), (25.49) and (25.53)

$$R_{N,N''}(Z) = O(|Z|^{1/2-\operatorname{Re}\beta - N''} e^{(1+1/\rho)\operatorname{Re} Z}),$$

for $\theta_Z \in (\bar{\theta}, \bar{\theta}+\varepsilon)$, In view of the statements following (25.44), it is readily seen that the factor $\frac{1}{2}\operatorname{erfc}\{c(\theta'_-)\sqrt{\frac{1}{2}|\zeta'|}\}$ in the second term of (25.45) is exponentially small when $\theta_Z < \bar{\theta}$ (i.e., $\theta'_- < \pi$), takes the value 1, with an exponentially small error, when $\theta_Z > \bar{\theta}$, and possesses a smooth transition between these two types of behavior. Thus we have achieved the smoothing of Stokes discontinuity when θ_Z passes through the Stokes line $\theta_Z = \bar{\theta}$. The case of the other Stokes line $\theta_Z = -\bar{\theta}$ can be handled in a similar manner, and the same smoothing is obtained.

25.3 The case when $\rho > 1$

We return to the integral representation (25.13),

$$\phi(z) = \frac{Z^{1-\beta}}{2\pi i}\int_C e^{(1-\beta)\tau}\exp\{Z(e^\tau + \frac{1}{\rho}e^{-\rho\tau})\}d\tau, \tag{25.54}$$

where C is the image of the loop path in (25.12) under the transformation $t = e^\tau$. The saddle points of the phase function $f(\tau) = e^\tau + (1/\rho)e^{-\rho\tau}$ are located at

$$\tau_s = \frac{2s\pi i}{1+\rho}, \quad s = 0, \pm 1, \pm 2, \cdots .$$

Let $\Gamma_s(\theta_Z)$ denote the steepest descent path through τ_s, i.e.,

$$\Gamma_s(\theta_Z): \quad \arg\{Z[f(\tau_s) - f(\tau)]\} = 0, \tag{25.55}$$

and let m be the positive integer such that

$$m - \frac{1}{2} \le \frac{1+\rho}{4} < m + \frac{1}{2}, \quad \rho > 1. \tag{25.56}$$

In this section, we shall prove that when $\rho > 1$,

$$\phi(z) = \sum_{s=-m}^{m} \frac{Z^{1-\beta}}{2\pi i}\int_{\Gamma_s(\theta_Z)} g(\tau)e^{Zf(\tau)}d\tau, \tag{25.57}$$

where $g(\tau)$ is as given in (25.14). Clearly, it suffices to show that this equation holds for $\theta_Z = 0$, since analytic continuation will extend its validity to other values of Z. For convenience, we will simply write Γ_s for $\Gamma_s(\theta_Z)$ when $\theta_Z = 0$. The contours will be examined in detail, and their graphs are shown in Figure 25.4. Since each integral in (25.57) can be treated in exactly the same manner as that in (25.18), there is no need to repeat the analysis of smoothing the Stokes discontinuity in the present case.

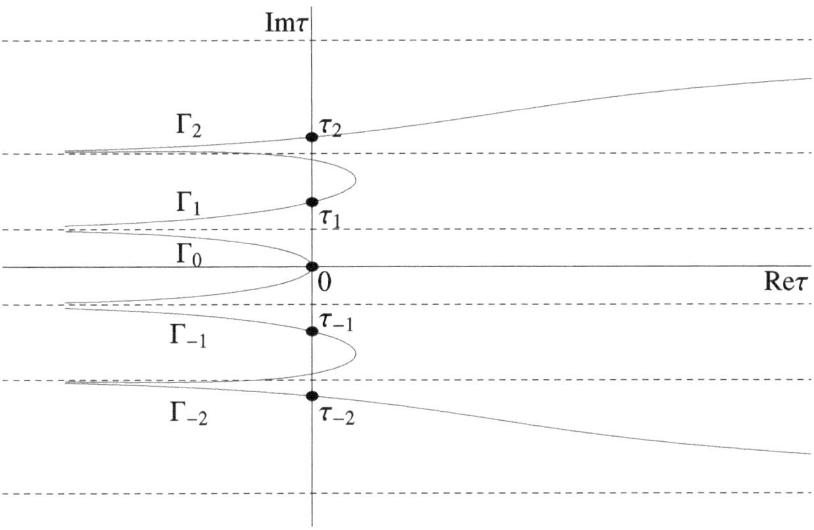

Fig. 25.4: The steepest descent curves Γ_s ($\rho = 6$ and $m = 2$).

We introduce two functions

$$R(x, y) := \operatorname{Re} f(x + iy) = e^x \cos y + \frac{1}{\rho} e^{-\rho x} \cos(\rho y), \tag{25.58}$$

$$I(x, y) := \operatorname{Im} f(x + iy) = e^x \sin y - \frac{1}{\rho} e^{-\rho x} \sin(\rho y), \tag{25.59}$$

for $x \in \mathbf{R}$ and $y \in (-\pi, \pi)$. For each $s = 0, \pm 1, \cdots, \pm m$, we have

$$f(\tau_s) = (1 + 1/\rho) e^{i\lambda_s},$$

where

$$\lambda_s := \frac{2s\pi}{1 + \rho}. \tag{25.60}$$

We also use r_s and i_s, respectively, to denote the real and imaginary parts of $f(\tau_s)$; namely,

$$r_s := \operatorname{Re} f(\tau_s) = (1 + 1/\rho)\cos\lambda_s, \tag{25.61}$$

$$i_s := \operatorname{Im} f(\tau_s) = (1 + 1/\rho)\sin\lambda_s. \tag{25.62}$$

Hence, Γ_s is the curve on the (x, y) plane which goes through $(0, \lambda_s)$ and satisfies $I(x, y) = i_s$ and $R(x, y) \le r_s$. For simplicity, we shall also denote

$$\mu_t := \frac{t\pi}{\rho}, \quad t \in \mathbf{R}. \tag{25.63}$$

The following lemma shows that within each strip $\{z \in \mathbf{Z} : \operatorname{Im} z \in (\lambda_{s-1}, \lambda_{s+1})\}$ the equation $f(z) = f(\tau_s)$ has exactly one root $\tau_s = i\lambda_s$.

Lemma 25.1. *Given $s = 0, \pm1, \cdots, \pm m$. Let $x \in \mathbf{R}$ and $\lambda_{s-1} < y < \lambda_{s+1}$. We have $R(x, y) = r_s$ and $I(x, y) = i_s$ if and only if $x = 0$ and $y = \lambda_s$.*

Proof. Solving the linear system

$$\begin{pmatrix} \cos y & \cos(\rho y)/\rho \\ \sin y & -\sin(\rho y)/\rho \end{pmatrix} \begin{pmatrix} e^x \\ e^{-\rho x} \end{pmatrix} = \begin{pmatrix} (1 + 1/\rho)\cos\lambda_s \\ (1 + 1/\rho)\sin\lambda_s \end{pmatrix}$$

gives

$$\begin{aligned} \begin{pmatrix} e^x \\ e^{-\rho x} \end{pmatrix} &= \frac{\rho}{\sin[(1+\rho)y]} \begin{pmatrix} \sin(\rho y)/\rho & \cos(\rho y)/\rho \\ \sin y & -\cos y \end{pmatrix} \begin{pmatrix} (1 + 1/\rho)\cos\lambda_s \\ (1 + 1/\rho)\sin\lambda_s \end{pmatrix} \\ &= \frac{1+\rho}{\sin[(1+\rho)y]} \begin{pmatrix} \sin(\rho y + \lambda_s)/\rho \\ \sin(y - \lambda_s) \end{pmatrix}. \end{aligned}$$

Since $(1 + \rho)\lambda_s = 2s\pi$, we have

$$\sin(\rho y + \lambda_s) = \sin[\rho(y - \lambda_s)],$$

and

$$\sin[(1 + \rho)y] = \sin[(1 + \rho)(y - \lambda_s)].$$

Denote

$$\varphi(y, p) := \sin[p(y - \lambda_s)]/p$$

for $p \ge 1$. We then have

$$e^x = \frac{\varphi(y, \rho)}{\varphi(y, 1 + \rho)}, \quad e^{-\rho x} = \frac{\varphi(y, 1)}{\varphi(y, 1 + \rho)}.$$

Note that the partial derivative

$$\varphi_y(y, p) = \cos[p(y - \lambda_s)] = \cos[p|y - \lambda_s|]$$

is strictly decreasing in p for $p|y-\lambda_s| \in (0,\pi)$. In particular, we have

$$\varphi_y(y,p) - \varphi_y(y,1+\rho) > 0$$

for any $p \in [1, 1+\rho)$ and $(1+\rho)(y-\lambda_s) \in (-\pi,0) \cup (0,\pi)$. Hence, the difference function $\varphi(y,p) - \varphi(y,1+\rho)$ is strictly increasing in y. This together with $\varphi(\lambda_s,p) = \varphi(\lambda_s,1+\rho) = 0$ implies that

$$\varphi(y,p) > \varphi(y,1+\rho) > 0$$

for $(1+\rho)(y-\lambda_s) \in (0,\pi)$, and

$$\varphi(y,p) < \varphi(y,1+\rho) < 0$$

for $(1+\rho)(y-\lambda_s) \in (-\pi,0)$. If $(1+\rho)(y-\lambda_s) \in (-\pi,0) \cup (0,\pi)$, we obtain

$$\frac{\varphi(y,p)}{\varphi(y,1+\rho)} > 1.$$

By choosing $p = \rho$ and $p = 1$ respectively, we obtain $e^x > 1$ and $e^{-\rho x} > 1$, which leads to a contradiction:

$$1 = (e^x)^\rho e^{-\rho x} > 1.$$

If $(1+\rho)(y-\lambda_s) \in (-2\pi,\pi] \cup [\pi,2\pi)$ then $e^{-\rho x} = \varphi(y,1)/\varphi(y,1+\rho)$ is either undefined (when $(1+\rho)(y-\lambda_s) = \pm\pi$) or negative, again a contradiction. Hence, the only solution in the strip $y \in (\lambda_{s-1}, \lambda_{s+1})$ is $y = \lambda_s$, which in turns gives $x = 0$. This completes the proof. □

Since $\tau_0 = 0$, the following result is obvious.

Proposition 25.1. *The steepest descent curve Γ_0 is the curve connecting $-\infty + i\mu_{-1}$ to $-\infty + i\mu_1$ and satisfying the equation*

$$x = \frac{1}{1+\rho} \ln \frac{\sin(\rho y)}{\rho \sin y}, \quad y \in (\mu_{-1}, \mu_1), \tag{25.64}$$

where the right-hand side takes the limiting value 0 at $y = 0$.

Now, for each $s = \pm 1, \cdots, \pm(m-1)$, we describe Γ_s as four connected pieces.

Proposition 25.2. *For each $s = \pm 1, \cdots, \pm(m-1)$, the steepest descent curve Γ_s through the saddle point τ_s connects $-\infty + i\mu_{2s-1}$ to $-\infty + i\mu_{2s+1}$.*

Proof. By symmetry, we may assume without loss of generality that $s > 0$. Since $s \le m - 1$, we have (exercise)

$$\mu_{2s-1} < \mu_{2s-1/2} < \lambda_s < \mu_{2s} < \frac{\pi}{2}. \tag{25.65}$$

Moreover, $\mu_{2s+1} < \mu_{2s+3/2} < \pi$. For any $x < 0$, the function $I(x, y)$ is strictly increasing for y in the interval $[\mu_{2s-1}, \mu_{2s-1/2}]$ and concave down for $y \in (\mu_{2s-1/2}, \lambda_s)$ with

$$I(x, \mu_{2s-1}) = e^x \sin \mu_{2s-1} < i_s,$$
$$I(x, \lambda_s) = (e^x + e^{-\rho x}/\rho) \sin \lambda_s > i_s.$$

Thus, there exists a unique $y_s^-(x) \in [\mu_{2s-1}, \lambda_s]$ such that $I(x, y_s^-(x)) = i_s$. Moreover, $y_s^-(0) = \lambda_s$ and $y_s^-(-\infty) = \mu_{2s-1}$. The curve $y = y_s^-(x)$ defines a steepest descent curve that connects $-\infty + i\mu_{2s-1}$ to τ_s.

Next, we define

$$x_{s1} := \ln \frac{i_s}{\sin \mu_{2s}}, \tag{25.66}$$

which is positive (exercise). For any $y \in [\lambda_s, \mu_{2s}]$, we have (exercise)

$$I(0, y) = \sin y - \sin(\rho y)/\rho \le i_s, \tag{25.67}$$

and $I(\infty, y) = \infty$. Moreover, the partial derivative

$$I_x(x, y) = e^x \sin y + e^{-\rho x} \sin(\rho y) \tag{25.68}$$

is strictly positive for $x > 0$ because $y \ge 2s\pi - \rho y > 0$. Hence, there exists a unique $x_s^-(y) \ge 0$ such that $I(x_s^-(y), y) = i_s$ for $y \in [\lambda_s, \mu_{2s}]$. It is easily seen that $x_s^-(\lambda_s) = 0$ and $x_s^-(\mu_{2s}) = x_{s1}$. This implies that the steepest descent curve defined by $x = x_s^-(y)$ connects τ_s to $x_{s1} + i\mu_{2s}$.

For $\mu_{2s} \le y \le \mu_{2s+1}$, we have $\rho y \in [2s\pi, (2s+1)\pi]$ and the partial derivative $I_x(x, y)$ is strictly positive for all $x \in \mathbf{R}$. Since $I(\infty, y) = \infty$ and $I(-\infty, y) = -\infty$, there exists a unique $x_s^+(y) \in \mathbf{R}$ such that $I(x_s^+(y), y) = i_s$. In particular, $x_s^+(\mu_{2s}) = x_{s1}$. Let

$$x_{s2} := \ln \frac{i_s}{\sin \mu_{2s+1}}.$$

We have $x_s^+(\mu_{2s+1}) = x_{s2}$. Hence, the steepest descent curve defined by $x = x_s^+(y)$ connects $x_{s1} + i\mu_{2s}$ to $x_{s2} + i\mu_{2s+1}$.

Finally, for each $x < x_{s2}$, we have

$$\begin{aligned} I(x, \mu_{2s+3/2}) &= e^x \sin \mu_{2s+3/2} + e^{-\rho x}/\rho > (e^x + e^{-\rho x}/\rho) \sin \lambda_s > i_s, \\ I(x, \mu_{2s+1}) &= e^x \sin \mu_{2s+1} < i_s, \end{aligned}$$

and the second-order partial derivative

$$I_{yy}(x,y) = -e^x \sin y + \rho e^{-\rho x} \sin(\rho y) \tag{25.69}$$

is strictly negative for $y \in (\mu_{2s+1}, \mu_{2s+3/2})$. Hence, there exists a unique $y_s^+(x) \in (\mu_{2s+1}, \mu_{2s+3/2})$ such that $I(x, y_s^+(x)) = i_s$. Moreover, $y_s^+(x_{s2}) = y_s^+(-\infty) = \mu_{2s+1}$, which implies that the steepest descent curve defined by $y = y_s^+(x)$ connects $x_{s2} + i\mu_{2s+1}$ to $-\infty + i\mu_{2s+1}$.

Connecting the above four curves defines the steepest descent curve Γ_s through the saddle point τ_s. Moreover, Γ_s connects $-\infty + i\mu_{2s-1}$ to $-\infty + i\mu_{2s+1}$. □

Proposition 25.3. *For $s = \pm m$, the steepest descent curve $\Gamma_{\pm m}$ through the saddle point $\tau_{\pm m}$ connects $-\infty + i\mu_{\pm(2m-1)}$ to $\infty \pm i\pi$.*

Proof. By symmetry, we only need to prove the case when $s = m$. Recall that $4m - 3 \le \rho < 4m + 1$, we have (exercise)

$$\rho \sin \frac{(\rho + 1 - 2m)\pi}{\rho} < (1 + \rho) \sin \frac{(\rho + 1 - 2m)\pi}{1 + \rho}, \tag{25.70}$$

and hence, $\sin \mu_{2m-1} < i_m$. Similar as in the proof of Proposition 25.2, for any $x < 0$, there exists a unique $y_m^-(x) \in [\mu_{2m-1}, \lambda_m]$ satisfying $I(x, y_m^-(x)) = i_m$ with $y_m^-(0) = \lambda_m$ and $y_m^-(-\infty) = \mu_{2m-1}$. The curve $y = y_m^-(x)$ defines a steepest descent curve that connects $-\infty + i\mu_{2m-1}$ to τ_m.

For each integer $k \in [m, (\rho - 1)/2)$, we have $\lambda_m < \mu_{2k} < \mu_{2k+1} < \pi$. Denote

$$x_{k1} := \ln \frac{i_m}{\sin \mu_{2k}}, \quad x_{k2} := \ln \frac{i_m}{\sin \mu_{2k+1}}. \tag{25.71}$$

For any $y \in [\mu_{2k}, \mu_{2k+1}]$, we have $\rho y \in [2k\pi, (2k+1)\pi]$ and $I_x(x,y) > 0$. Moreover, $I(\ln i_m, y) \le i_m \sin y < i_m$ and $I(\infty, y) = \infty$. Hence, there exists a unique $x_k^+(y) > \ln i_m$ such that $I(x_k^+(y), y) = i_m$. In particular, $x_k^+(\mu_{2k}) = x_{k1}$ and $x_k^+(\mu_{2k+1}) = x_{k2}$. Therefore, the curve defined by $x = x_k^+(y)$ connects $x_{k1} + i\mu_{2k}$ to $x_{k2} + i\mu_{2k+1}$.

For each integer $k \in [m+1, \rho/2)$, we have $\pi/2 < \mu_{2k-1} < \mu_{2k} < \pi$. For any $y \in (\mu_{2k-1}, \mu_{2k})$, the equation $I_x(x,y) = 0$ has a unique root

$$x_k^c(y) := \frac{1}{1+\rho} \ln \frac{\sin(2k\pi - \rho y)}{\sin y} \tag{25.72}$$

It can be verified that

$$I(x_k^c(y), y) = (1 + 1/\rho)[\sin(2k\pi - \rho y)(\sin y)^\rho]^{1/(1+\rho)} < i_m. \tag{25.73}$$

Since $I(-\infty, y) = I(\infty, y) = \infty$, there exists a unique $x_k^-(y) > x_k^c(y)$ such that $I(x_k^-(y), y) = i_m$. Moreover, we have $x_k^-(\mu_{2k-1}) = x_{k-1,2}$ and $x_k^-(\mu_{2k}) = x_{k1}$, where x_{k1} and $x_{k-1,2}$ are defined in (25.71). The curve defined by $x = x_k^-(y)$ connects $x_{k-1,2} + i\mu_{2k-1}$ to $x_{k1} + i\mu_{2k}$. For $y \in (\lambda_m, \mu_{2m})$, we can use a similar argument to define $x_m^c(y)$ as in (25.72) (with $k = m$) and find a unique $x_m^-(y) > x_m^c(y)$ such that $I(x_m^-(y), y) = i_m$. Moreover, we have $x_m^-(\lambda_m) = 0$ and $x_m^-(\mu_{2m}) = x_{m1}$. Thus, the curve defined by $x = x_m^-(y)$ connects τ_m to $x_{m1} + i\mu_{2m}$.

Finally, we will find a steepest descent curve connecting $\infty + i\pi$. There are two cases to consider, depending on whether the smallest integer no less than ρ is even or odd. If $2K < \rho \leq 2K + 1$ for some positive integer K, then we have $\mu_{2K} < \pi \leq \mu_{2K+1}$. For any $y \in [\mu_{2K}, \pi)$, we have $\rho y \in [2K\pi, \rho pi)$ and hence $I_x(x, y) > 0$ for all $x > 0$. Moreover, $I(\infty, y) = \infty$ and $I(x_{K1}) \leq e^{x_{K1}} \sin y \leq i_m$. Hence, there exists a unique $x_K^+(y) \geq x_{K1}$ such that $I(x_K^+(y), y) = i_m$. In particular, $x_K^+(\mu_{2K}) = x_{K1}$ and $x_K^+(\pi) = \infty$. Therefore, the curve defined by $x = x_K^+(y)$ connects $x_{K1} + i\mu_{2K}$ to $\infty + i\pi$. Now, we consider the second case when $2K - 1 < \rho \leq 2K$ for some positive integer K. For any $y \in [\mu_{2K-1}, \pi)$, we define $x_K^c(y)$ as the unique root of $I_x(x, y)$; see (25.72). Similar as in (25.73), we have $I(x_K^c(y), y) < i_m$. On account of $I(\infty, y) = \infty$, we can find a unique $x_K^+(y) > x_K^c(y)$ such that $I(x_K^+(y), y) = i_m$. It is easily seen that $x_K^+(\mu_{2K-1}) = x_{K-1,2}$. Moreover, in view of (exercise)

$$I(x, \pi) = -\frac{1}{\rho} e^{-\rho x} \sin(\rho\pi) < i_m, \tag{25.74}$$

we obtain $x_K^+(\pi) = \infty$. This implies that the curve defined by $x = x_K^+(y)$ connects $x_{K-1,2} + i\mu_{2K-1}$ to $\infty + i\pi$. □

We now show that the contours Γ_s and Γ_{s-1} do not intersect, except in the case when $s = m$ and $m - \frac{1}{2} = \frac{1}{4}(1 + \rho)$. The exceptional case is illustrated in Figure 25.5. Hence we consider only the other cases. Suppose $\tau \in \Gamma_s \cap \Gamma_{s-1}$. Then

$$\arg[f(\tau_s) - f(\tau)] = 0 \quad \text{and} \quad \arg[f(\tau_{s-1}) - f(\tau)] = 0,$$

which imply that $f(\tau_s) - f(\tau_{s-1})$ is real or, equivalently,

$$\sin \frac{2s\pi}{1+\rho} - \sin \frac{2(s-1)\pi}{1+\rho} = 0.$$

This is impossible, since the left-hand side of this equation is equal to

$$2 \sin \frac{\pi}{1+\rho} \cos \frac{2s-1}{1+\rho}\pi,$$

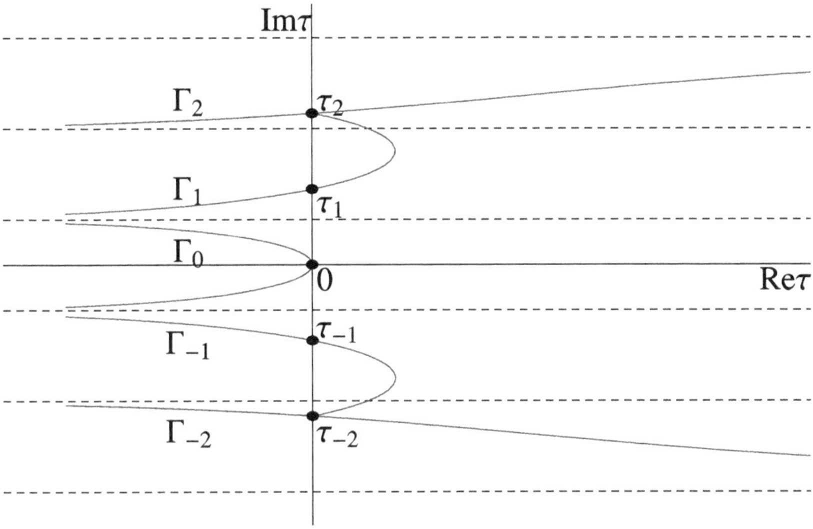

Fig. 25.5: The steepest descent curves Γ_s ($\rho = 5$ and $m = 2$).

and both $\pi/(1+\rho)$ and $\pi(2s-1)/(1+\rho)$ lie in $(0, \pi/2)$.

There are, in total, $2m+1$ integrals in (25.57), where Γ_0 is given in Proposition 25.1, $\Gamma_{\pm s}$ (with $s = 1, \cdots, m-1$) is given in Proposition 25.2, and $\Gamma_{\pm m}$ is given in Proposition 25.3. Let us write

$$\begin{aligned}\phi_s(Z) &= \frac{Z^{1-\beta}}{2\pi i}\int_{\Gamma_s(\theta_Z)} g(\tau)e^{Zf(\tau)}d\tau\\ &= \frac{Z^{1-\beta}}{2\pi i}e^{Zf(\tau_s)}\int_{\Gamma_s(\theta_Z)} g(\tau)e^{Z[f(\tau)-f(\tau_s)]}d\tau,\end{aligned}$$

$s = 0, \pm 1, \cdots, \pm m$. By shifting the contour vertically downward, the last integral can be put nearly in the same form as $T(Z)$ given in (25.18), which is asymptotically equal to a constant for large values of $|Z|$. Hence the magnitude of $\phi_s(Z)$ is characterized by $|e^{Zf(\tau_s)}|$ or, equivalently,

$$\cos(\theta_Z + \frac{2s\pi}{1+\rho}),$$

since $f(\tau_s) = e^{\tau_s} f(0)$. From this, it is readily seen that

(i) $\phi_s(Z)$ is dominant in the sector

$$-\frac{2s+1}{1+\rho}\pi < \theta_Z < -\frac{2s-1}{1+\rho}\pi \quad \text{for} \quad -m < s < m,$$

(*ii*) $\phi_s(Z)$ is dominant in the sector

$$-\pi < \theta_Z < -\frac{2m-1}{1+\rho}\pi,$$

(*iii*) $\phi_s(Z)$ is dominant in the sector

$$\frac{2m-1}{1+\rho}\pi < \theta_Z < \pi.$$

On the ray

$$\theta_Z = -\frac{2s+1}{1+\rho}\pi, \quad s = 0, \pm 1, \cdots, \pm(m-1),$$

$\phi_s(Z)$ and $\phi_{s+1}(Z)$ become of equal magnitude.

Although each integral $\phi_s(Z)$ can be studied in exactly the same manner as $T(Z)$, in view of the representation (25.57), the problem of smoothing the discontinuity in Stokes phenomenon becomes quite complicated due to competing singularity and error functions. However, each $\phi_s(Z)$ is associated with only two adjacent saddles. Thus, as far as transition is concerned, we need to find only two Stokes lines for each integral. Among all these lines, the ones which are closest to $\theta_Z = 0$ give the earliest transition. We summarize our findings below.

(*i*) When $m - \frac{1}{2} \le \frac{1}{4}(1+\rho) < m$, the earliest transitions occur at

$$\theta_Z = \pm\left(\frac{1}{2}\pi - \frac{2m-1}{1+\rho}\pi\right)$$

and the associated integrals are $\phi_{\pm(m-1)}$.

(*ii*) When $\frac{1}{4}(1+\rho) = m$, the earliest transitions occur at

$$\theta_Z = \pm\left(\frac{1}{2}\pi - \frac{2m-1}{1+\rho}\pi\right)$$

and the associated integrals are $\phi_{\pm(m-1)}$ and $\phi_{\pm m}$.

(*iii*) When $m < \frac{1}{4}(1+\rho) < m + \frac{1}{2}$, the earliest transitions occur at

$$\theta_Z = \pm\left(-\frac{1}{2}\pi + \frac{2m+1}{1+\rho}\pi\right)$$

and the associated integrals are $\phi_{\pm m}$.

These results were obtained by observing the fact that for $-m \le s \le m$; the Stokes lines of $\phi_s(Z)$ close to $\theta_Z = 0$ are

$$\theta_Z = -\frac{2s}{1+\rho}\pi \pm \left(\frac{1}{2}\pi - \frac{1}{1+\rho}\pi\right).$$

This fact can be verified by using the singulants

$$f(\tau_s) - f(\tau_{s+1}) =2(1+\frac{1}{\rho})\sin\frac{\pi}{1+\rho} \times \exp\left\{i\left[\frac{2s\pi}{1+\rho} - \left(\frac{1}{2}\pi - \frac{\pi}{1+\rho}\right)\right]\right\}$$

and

$$f(\tau_s) - f(\tau_{s-1}) =2(1+\frac{1}{\rho})\sin\frac{\pi}{1+\rho} \times \exp\left\{i\left[\frac{2s\pi}{1+\rho} + \left(\frac{1}{2}\pi - \frac{\pi}{1+\rho}\right)\right]\right\}.$$

Note that if the argument of a singulant is α, then $\theta_Z = -\alpha$ is a Stokes line.

We also note from (25.56) that

$$-\frac{1}{2}\pi + \frac{2m+1}{1+\rho}\pi > 0 \tag{25.75}$$

and

$$\frac{1}{2}\pi - \frac{2s+1}{1+\rho}\pi \geq 0 \quad \text{for} \quad 0 \leq s < m, \tag{25.76}$$

with equality holding only when $s = m-1$ and $m - \frac{1}{2} = \frac{1}{4}(1+\rho)(i.e.,\ \rho = 4m-3)$. The case of $\rho = 4m-3$ is considered exceptional in the sense that Γ_m and Γ_{m-1} intersect; see Wong and Zhao (1999a, p. 1395 and Fig. 5b).

The inequalities in (25.75) and (25.76) in turn give

$$\frac{1}{2}\pi - \frac{2m+1}{1+\rho}\pi < 0$$

and

$$-\frac{1}{2}\pi - \frac{2s-1}{1+\rho}\pi < 0 \quad \text{for} \quad -m < s < 0.$$

The results in (*i*)-(*iii*) now follow in a straight forward manner.

To conclude our discussion, we mention two recent paper by Berry (1991) and Boyd (1994), in which they investigated the asymptotic behavior of the Gamma function. Berry found that there are infinitely many Stokes smoothing in the function $\ln\Gamma(z)$, whereas Boyd showed that although there are infinitely many adjacent saddles in the contour integral representation of $\Gamma(z)$, by a careful choice of adjacent contours, it suffices to consider only the two closest ones to the origin. Our problem here is different in that we have neither infinitely many Stokes transitions nor an infinite number of adjacent saddles. What we have is a finite sum of contour integrals, each of which has two adjacent saddles. To replace the original contour integral in (25.13) by a sum of $2m+1$ contour integrals in (25.57), we had to make a very detailed study of the topology of the phase function $f(\tau)$ in (25.14).

25.4 Exercises

1. Verify (25.9) and prove that the ratio $H(Ze^{-2\pi i/(\rho+1)})/H(Z)$ is exponentially small.
2. Verify (25.11).
3. Let $\Gamma_0(\theta_Z)$ be defined as in (25.16). Verify that $\Gamma_0(0)$ is the same as the curve

$$x = \frac{1}{1+\rho}\ln\frac{\sin(\rho y)}{\rho\sin y}, \quad y \in (-\pi, \pi),$$

   where $\sin(\rho y)/(\rho \sin y)$ is defined as 1 at $y = 0$.
4. Prove that x_{s1} defined in (25.66) is positive.
5. Verify (25.65).
6. Verify (25.67).
7. Verify (25.70).
8. Let $k \in [m+1, \rho/2)$ and $x_k^c(y)$ be defined as in (25.72). Prove that (25.73) holds for all $y \in (\mu_{2k-1}, \mu_{2k})$.
9. Verify (25.74).

Part 5

Singular Perturbations

Chapter 26

A Singular Perturbation Problem

26.1 Introduction

In applied mathematics, one frequently encounters differential equations containing a small parameter ε which tends to 0. The subject that deals with such problems is known as singular perturbation. While it is clear that this topic belongs to asymptotic analysis, strangely enough very few experts in asymptotic analysis have actually tried to make the formal mathematics in this area rigorous, except for a few (e.g., Wasow (1960) and Erdélyi (1962a,b, 1963, 1976)) in their early days. The most familiar name in this subject is O'Malley Jr. (1974, 1976, 1991, 2014), but most of his works are formal.

In this chapter and subsequent chapters, we shall study some particular problems with great details and show why rigorous mathematics is needed even in applied area. The material in this chapter is taken from Ng and Wong (1994). The material in the Exercise section is taken from Bush (1992, Chapter 5) and Bender and Orszag (1978, Chapter 9).

26.2 A linear oscillator with damping

In Simmonds and Mann Jr. (1986), the following open problem is given in an exercise, and the authors indicate that they do not know the complete answer; construct a uniformly valid first approximation to the linear oscillator with damping proportional to time

$$\ddot{v} + \varepsilon t\dot{v} + v = 0, \quad v(0) = 0, \quad \dot{v}(0) = 1. \tag{26.1}$$

If one naively proceeds with the perturbation series

$$v(t, \varepsilon) = v_0(t) + \varepsilon v_1(t) + \varepsilon^2 v_2(t) + \cdots,$$

as suggested by them, one readily finds that $v_0(t) = \sin t$ and

$$v_1(t) = \frac{1}{4}\sin t - \frac{1}{4}t\cos t - \frac{1}{4}t^2\sin t.$$

But, by the usual method of multiplying the differential equation in (26.1) by $\dot{v}$ and integrating from 0 to t, one can easily show that the solution v is bounded for all positive t. Since $v_1(t)$ contains secular terms $t\cos t$ and $t^2\sin t$, the error due to truncation after two terms cannot be uniformly small in t. If one tries the multiple scale method, and assumes a perturbation series of the form

$$v(t,\varepsilon) = v_0(t,\tau) + \varepsilon v_1(t,\tau) + \varepsilon^2 v_2(t,\tau) + \cdots$$

with $\tau = \varepsilon t$, one obtains

$$\frac{\partial^2 v_0}{\partial t^2} + v_0 = 0,$$

$$\frac{\partial^2 v_1}{\partial t^2} + v_1 = -\frac{2\partial^2 v_0}{\partial\tau\partial t} - t\frac{\partial v_0}{\partial t}.$$

The solution to the first equation is $v_0 = A(\tau)\sin t + B(\tau)\cos t$. Substituting this expression in the right-hand side of the second equation gives

$$\frac{\partial^2 v_1}{\partial t^2} + v_1 = -[2A'(\tau) + tA(\tau)]\cos t + [2B'(\tau) + tB(\tau)]\sin t.$$

To eliminate secular terms, the usual procedure is now to set the quantities inside the square brackets to zero. Since these quantities involve the variable t, the solution $A(\tau)$ and $B(\tau)$ to the resulting equations depend on t, contradicting the fact that $A(\tau)$ and $B(\tau)$ are functions of τ only. This indicates that the choice of the scale $\tau = \varepsilon t$ is inappropriate. A similar consequence results, if one chooses the scale $\tau = \sqrt{\varepsilon}t$.

In this chapter we will show that the exact solution to equation (26.1) can be obtained in terms of parabolic cylinder functions, and that a uniformly valid first approximation to the solution can be given in terms of Airy functions. As a consequence, we will prove that as $\varepsilon \to 0$,

$$v(t,\varepsilon) \sim e^{-\varepsilon t^2/4}\sin t \quad \text{if} \quad \varepsilon t \to 0, \tag{26.2}$$

and

$$v(t,\varepsilon) \sim (\sin\frac{\pi}{2\varepsilon})e^{-1/2\varepsilon}\frac{\sqrt{2}}{(\varepsilon t)^{1/\varepsilon}} \quad \text{if} \quad \varepsilon t \to \infty. \tag{26.3}$$

Furthermore, we will show that there is a turning point at $t = 2/\varepsilon$, and that in a neighborhood of the turning point we have

$$\begin{aligned} v(t,\varepsilon) \sim &\sqrt{\pi}e^{-\varepsilon t^2/4}\varepsilon^{-\frac{1}{6}}\Big\{\sin\frac{\pi}{2\varepsilon}\cdot \mathrm{Bi}[\frac{1}{2}\varepsilon^{1/3}(t-\frac{2}{\varepsilon})] \\ &- \cos\frac{\pi}{2\varepsilon}\cdot \mathrm{Ai}[\frac{1}{2}\varepsilon^{1/3}(t-\frac{2}{\varepsilon})]\Big\}, \end{aligned} \tag{26.4}$$

as $\varepsilon \to 0$. More precisely, equation (26.4) holds when $|t - \frac{2}{\varepsilon}| = o(\varepsilon^{-\frac{2}{3}})$.

26.3 Exact solution

Let v be the solution of (26.1), and put

$$v(t,\varepsilon) = e^{-\varepsilon t^2/4} W(t,\varepsilon). \tag{26.5}$$

It is easily verified that W satisfies the equation

$$\ddot{W} + \left[(1 - \frac{\varepsilon}{2}) - \frac{\varepsilon^2}{4} t^2\right] W = 0. \tag{26.6}$$

Set $x = \sqrt{\varepsilon} t$ and $u(x,\varepsilon) = W(x/\sqrt{\varepsilon}, \varepsilon)$. Then u satisfies Weber's equation

$$\frac{d^2 u}{dx^2} - \left(\frac{1}{4}x^2 + a\right) u = 0 \tag{26.7}$$

and the initial conditions $u(0,\varepsilon) = 0$ and $(du/dx)(0,\varepsilon) = 1/\sqrt{\varepsilon}$. In (26.7), the parameter a is given by

$$a = a_\varepsilon = \frac{1}{2} - \frac{1}{\varepsilon}. \tag{26.8}$$

It is well known that two linearly independent solutions to (26.7) are $U(a,x)$ and $V(a,x)$, characterized by the asymptotic behavior

$$U(a,x) \sim x^{-a-\frac{1}{2}} e^{-x^2/4}, \quad \text{as } x \to \infty,$$

$$V(a,x) \sim \sqrt{\frac{2}{\pi}} x^{a-\frac{1}{2}} e^{x^2/4}, \quad \text{as } x \to \infty.$$

In terms of Whittaker's notation (see Olver (1974a, p. 207)) for parabolic cylinder functions $D_n(x)$, we have $U(a,x) = D_{-a-\frac{1}{2}}(x)$ and

$$V(a,x) = \frac{1}{\pi}\Gamma(\frac{1}{2} + a)\left\{\sin \pi a \cdot D_{-a-\frac{1}{2}}(x) + D_{-a-\frac{1}{2}}(-x)\right\}.$$

For the definition and properties of parabolic cylinder functions, we refer to Abramowitz and Stegun (1964, Chapter 19) or Miller (1955, p. 134). Since in our case the parameter a in equation (26.8) is negative, we will take the suggestion of Olver (1959, p. 134), and use in place of $V(a,x)$ its multiple

$$\begin{aligned} \bar{U}(a,x) &= \Gamma(\frac{1}{2} - a) V(a,x) \\ &= \tan \pi a \cdot U(a,x) + \sec \pi a \cdot U(a,-x). \end{aligned} \tag{26.9}$$

Since $u(x,\varepsilon)$ is a solution of (26.7), it is expressible as a linear combination of $U(a_\varepsilon, x)$ and $\bar{U}(a_\varepsilon, x)$. In view of the initial conditions, it is found that

$$u(x,\varepsilon) = -\frac{1}{2\sqrt{\varepsilon}} \frac{(1 + \sin \pi a_\varepsilon) U(a_\varepsilon, x) - \cos \pi a_\varepsilon \cdot \bar{U}(a_\varepsilon, x)}{U(a_\varepsilon - 1, 0)}, \tag{26.10}$$

where use has been made of the result $(dU/dt)(a,0) = -U(a-1,0)$. Since $u(x,\varepsilon) = W(t,\varepsilon)$, the exact solution to (26.1) is given by

$$v(t,\varepsilon) = e^{-\varepsilon t^2/4} u(x,\varepsilon), \quad x = \sqrt{\varepsilon} t, \tag{26.11}$$

on account of (26.5).

Before leaving this section, we recall two asymptotic formulas for $U(a,x)$ and $\bar{U}(a,x)$. For $x \geq 0$ we write

$$x = 2\sqrt{|a|}\xi, \quad \eta = (4|a|)^{2/3}\tau, \tag{26.12}$$

and

$$\tau = \begin{cases} -\left[\frac{3}{2}\left(\frac{1}{4}\cos^{-1}\xi - \frac{1}{4}\xi\sqrt{1-\xi^2}\right)\right]^{2/3}, & 0 \leq \xi \leq 1, \\ \left[\frac{3}{2}\left(\frac{1}{4}\xi\sqrt{\xi^2-1} - \frac{1}{4}\cosh^{-1}\xi\right)\right]^{2/3}, & \xi \geq 1. \end{cases} \tag{26.13}$$

Then for $x \geq 0$ and as $a \to -\infty$,

$$U(a,x) \sim 2^{-\frac{1}{4}-\frac{1}{2}a}\Gamma\left(\frac{1}{4} - \frac{1}{2}a\right)\left(\frac{\eta}{\xi^2-1}\right)^{\frac{1}{4}} \mathrm{Ai}(\eta), \tag{26.14}$$

$$\bar{U}(a,x) \sim 2^{-\frac{1}{4}-\frac{1}{2}a}\Gamma\left(\frac{1}{4} - \frac{1}{2}a\right)\left(\frac{\eta}{\xi^2-1}\right)^{\frac{1}{4}} \mathrm{Bi}(\eta); \tag{26.15}$$

see Abramowitz and Stegun (1964, p. 689) or Olver (1959, (8.11) and (11.22)).

26.4 Asymptotic behavior

To obtain a uniformly valid asymptotic approximation to the solution $v(t,\varepsilon)$, we first note that when $x = 0$ and a is replaced by $a_\varepsilon - 1$, we have, from (26.8) and (26.12), $\xi = 0$ and

$$\eta = -\left[\frac{3\pi}{4}\left(\frac{1}{\varepsilon} + \frac{1}{2}\right)\right]^{2/3}.$$

Hence, $\eta \to -\infty$ and

$$\mathrm{Ai}(\eta) \sim \frac{1}{\pi^{1/2}(-\eta)^{1/4}} \cos(\frac{\pi}{2\varepsilon}), \quad \text{as} \quad \varepsilon \to 0^+; \tag{26.16}$$

see Olver (1974a, p. 103). From (26.14), it follows that

$$U(a_\varepsilon - 1, 0) \sim 2^{\frac{1}{4}-\frac{1}{2}a_\varepsilon}\Gamma\left(\frac{1}{4} - \frac{1}{2}a_\varepsilon + \frac{1}{2}\right)\frac{1}{\sqrt{\pi}} \cos(\frac{\pi}{2\varepsilon}). \tag{26.17}$$

Since $1 + \sin \pi a_\varepsilon = 2\cos^2(\frac{\pi}{2\varepsilon})$ and $\cos \pi a_\varepsilon = 2\sin(\frac{\pi}{2\varepsilon})\cos(\frac{\pi}{2\varepsilon})$, inserting (26.14), (26.15) and (26.17) in (26.10), we obtain from (26.11)

$$v(t,\varepsilon) \sim \sqrt{\pi} e^{-\varepsilon t^2/4} \left(\frac{\eta}{\xi^2 - 1}\right)^{\frac{1}{4}} \left[\sin \frac{\pi}{2\varepsilon} \cdot \mathrm{Bi}(\eta) - \cos \frac{\pi}{2\varepsilon} \cdot \mathrm{Ai}(\eta)\right], \quad (26.18)$$

where η and ξ are given in (26.12) and (26.13). In (26.18), we have also made use of the result

$$\frac{\Gamma(x+a)}{\Gamma(x+b)} \sim x^{a-b}, \quad \text{as } x \to \infty; \quad (26.19)$$

see Olver (1974a, p. 119). The asymptotic formulas (26.2), (26.3) and (26.4) will now be deduced from (26.18).

First, we prove (26.2). If $\varepsilon t \to 0$, then $\sqrt{\varepsilon} x \to 0$ and $\xi \to 0$. Consequently, it follows from (26.8) and (26.12) that

$$\eta \sim -\left(\frac{3}{2}\right)^{2/3} \left(\frac{\pi - 4\xi}{2\varepsilon} - \frac{\pi}{4}\right)^{2/3}, \quad \text{as } \varepsilon \to 0.$$

In particular, we have $\eta \to -\infty$,

$$\mathrm{Ai}(\eta) \sim \frac{1}{\pi^{1/2}(-\eta)^{1/4}} \sin\left(\frac{\pi - 4\xi}{2\varepsilon}\right) \quad (26.20)$$

and

$$\mathrm{Bi}(\eta) \sim \frac{1}{\pi^{1/2}(-\eta)^{1/4}} \cos\left(\frac{\pi - 4\xi}{2\varepsilon}\right); \quad (26.21)$$

see Olver (1974a, pp. 393-394). Since $\xi \to 0$, inserting (26.20) and (26.21) in (26.18) gives

$$v(t,\varepsilon) \sim e^{-\varepsilon t^2/4} \left[\sin \frac{\pi}{2\varepsilon} \cdot \cos\left(\frac{\pi}{2\varepsilon} - \frac{4\xi}{2\varepsilon}\right) - \cos \frac{\pi}{2\varepsilon} \cdot \sin\left(\frac{\pi}{2\varepsilon} - \frac{4\xi}{2\varepsilon}\right)\right].$$

The desired result in (26.2) now follows from the addition formula for the sine function and the fact that $2\xi/\varepsilon \sim t$.

Next, we prove (26.3). If $\varepsilon t \to \infty$, then $\sqrt{\varepsilon} x \to \infty$ and $\xi \to \infty$. Since

$$\cosh^{-1} \xi \sim \ln 2\xi - \frac{1}{4}\xi^{-2} - \frac{3}{32}\xi^{-4} - \cdots$$

as $\xi \to \infty$ (see Abramowitz and Stegun (1964, p. 88)), we have

$$\eta \sim \left[\frac{3}{2}|a|(\xi^2 - \ln 2\xi - \frac{1}{2})\right]^{2/3}. \quad (26.22)$$

From the asymptotic formulas in Olver (1974a, pp. 393-394)

$$\mathrm{Ai}(\eta) \sim \frac{1}{2\pi^{1/2}\eta^{1/4}} \exp(-\frac{2}{3}\eta^{3/2}), \quad \text{as } \eta \to \infty,$$

and

$$\mathrm{Bi}(\eta) \sim \frac{1}{\pi^{1/2}\eta^{1/4}} \exp(\frac{2}{3}\eta^{3/2}), \quad \text{as} \quad \eta \to \infty,$$

it follows from (26.18) that

$$v(t, \varepsilon) \sim e^{-\varepsilon t^2/4} \xi^{-1/2} \left[\sin \frac{\pi}{2\varepsilon} \exp(\frac{2}{3}\eta^{3/2}) - \frac{1}{2} \cos \frac{\pi}{2\varepsilon} \exp(-\frac{2}{3}\eta^{3/2}) \right]. \tag{26.23}$$

Since $x = \sqrt{\varepsilon} t$, equation (26.22) gives

$$\frac{2}{3}\eta^{3/2} \sim \frac{1}{4}\varepsilon t^2 + a \log \sqrt{\varepsilon} t - \frac{a}{2} \log(-a) + \frac{a}{2}.$$

Disregarding the exponentially small term in (26.23), the required result (26.3) is now obtained by noting that $\xi = \sqrt{\varepsilon} t / 2\sqrt{|a|}$.

Finally, we consider the case when t is in a neighborhood of the turning point $t = 2/\varepsilon$. Let

$$f(\xi) = \xi \sqrt{\xi^2 - 1} - \cosh^{-1} \xi, \quad \xi \geq 1.$$

A simple calculation gives $f'(\xi) = 2\sqrt{\xi^2 - 1}$. From (26.13), we have

$$\tau = \left[\frac{3}{8} f(\xi) \right]^{2/3}$$

and

$$\frac{d\tau}{d\xi} = \frac{1}{3^{1/3}} \frac{\sqrt{\xi^2 - 1}}{[f(\xi)]^{1/3}}.$$

By l'Hospital's rule, it can be shown that

$$\left(\frac{d\tau}{d\xi} \right)^3 \Bigg|_{\xi = 1^+} = \frac{1}{2}.$$

In the same manner, it can be shown that

$$\left(\frac{d\tau}{d\xi} \right)^3 \Bigg|_{\xi = 1^-} = \frac{1}{2}.$$

Thus, expanding τ into a Taylor series near $\xi = 1$, we have from (26.12)

$$\eta = (4|a|)^{2/3} \left\{ \tau(1) + \frac{1}{2^{1/3}}(\xi - 1) + \cdots \right\}.$$

Since $\tau(1) = 0$, if $|a|^{2/3}(\xi - 1)^2$ is small, or equivalently if $t = \frac{2}{\varepsilon}[1 + o(\varepsilon^{1/3})]$, then it follows from (26.18) that

$$v(t, \varepsilon) \sim \sqrt{\pi} e^{-\varepsilon t^2/4} |a|^{1/6} \left\{ \sin \frac{\pi}{2\varepsilon} \cdot \mathrm{Bi} \left[2|a|^{2/3}(\xi - 1) \right] - \cos \frac{\pi}{2\varepsilon} \cdot \mathrm{Ai} \left[2|a|^{2/3}(\xi - 1) \right] \right\}. \tag{26.24}$$

The required asymptotic approximation (26.4) is obtained by observing that

$$|a|^{2/3}(\xi - 1) = \frac{1}{2}\varepsilon^{1/3}(t - \frac{2}{\varepsilon}) + O(\varepsilon^{1/3}), \quad \text{as} \quad \varepsilon \to 0.$$

26.5 Remarks

(1) The asymptotic formula (26.2) can be obtained by using the WKB analysis. Indeed, if we make the change of variable $\tau = \varepsilon t/\sqrt{4-2\varepsilon}$ in equation (26.6) and put $W(t) = y(\tau)$, then it is easily verified that y satisfies

$$\frac{d^2 y}{d\tau^2} + \lambda_\varepsilon^2 (1-\tau^2) y = 0,$$

where $\lambda_\varepsilon = \frac{2}{\varepsilon} - 1$. The WKB approximation (see Simmonds and Mann Jr. (1986, p. 79, (6.19))) gives

$$\begin{aligned} y(\tau) \sim &(1-\tau^2)^{-1/4}\Bigg[A\cos(\lambda_\varepsilon \int_0^\tau \sqrt{1-s^2}ds) \\ &+ B\sin(\lambda_\varepsilon \int_0^\tau \sqrt{1-s^2}ds)\Bigg], \end{aligned} \tag{26.25}$$

as $\lambda_\varepsilon \to +\infty, |\tau| < 1$. The initial conditions $y(0) = 0$ and $(dy/d\tau)(0) = 2\sqrt{\frac{1}{\varepsilon} - \frac{1}{2}}/\sqrt{\varepsilon}$ imply $A = 0$ and $B = (1 - \frac{\varepsilon}{2})^{-1/2}$. Since $B \sim 1$ and $\lambda_\varepsilon \tau \sim t$ as $\varepsilon \to 0$, and since

$$\sin\left(\lambda_\varepsilon \int_0^\tau \sqrt{1-s^2}ds\right) \sim \sin\lambda_\varepsilon\tau \quad \text{as} \quad \varepsilon t \to 0,$$

formula (26.2) follows immediately from (26.25).

(2) The above method can also be applied to the initial value problem

$$\ddot{v} - \varepsilon t\dot{v} + v = 0, \quad v(0) = 0, \quad \dot{v}(0) = 1. \tag{26.26}$$

The exact solution is

$$v(t,\varepsilon) = -e^{\varepsilon t^2/4}\frac{(1+\sin\pi a)U(a,\sqrt{\varepsilon}t) - \cos\pi a\cdot \bar{U}(a,\sqrt{\varepsilon}t)}{2\sqrt{\varepsilon}U(a-1,0)}, \tag{26.27}$$

where

$$a = a_\varepsilon = -\frac{1}{\varepsilon} - \frac{1}{2}. \tag{26.28}$$

The uniformly valid first approximation is given by

$$v(t,\varepsilon) \sim \sqrt{\pi}e^{\varepsilon t^2/4}\left(\frac{\eta}{\xi^2-1}\right)^{1/4}\left[\sin\frac{\pi}{2\varepsilon}\cdot \mathrm{Ai}(\eta) + \cos\frac{\pi}{2\varepsilon}\cdot \mathrm{Bi}(\eta)\right], \tag{26.29}$$

where ξ and η are defined as in (26.12) with a being given by (26.28). The above material is taken from Ng and Wong (1994).

26.6 Exercises

1. Consider the differential equation

$$\varepsilon y'' + a(x)y' + b(x)y = 0, \quad x_1 \le x \le x_2, \tag{26.30}$$

with boundary conditions

$$y(x_1) = A, \quad y(x_2) = B. \tag{26.31}$$

(a) Show that for $a(x) > 0$, the solution is given by

$$\begin{aligned} y(x) =& B\exp\left[\int_x^{x_2} \frac{b(t)}{a(t)}dt\right][1+O(\varepsilon)] \\ &+ \frac{a(x_1)}{a(x)}\left\{A - B\exp\left[\int_{x_1}^{x_2}\frac{b(t)}{a(t)}dt\right]\right\}\exp\left[\int_{x_1}^{x}\frac{b(t)}{a(t)}dt\right] \\ &\times \exp\left[-\frac{1}{\varepsilon}\int_{x_1}^{x} a(t)dt\right][1+O(\varepsilon)] \end{aligned} \tag{26.32}$$

if $B \neq 0$, and

$$\begin{aligned} y(x) =& -A\frac{a(x_1)}{a(x_2)}\exp\left[\int_{x_1}^{x_2}\frac{b(t)}{a(t)}dt\right]\exp\left[-\frac{1}{\varepsilon}\int_{x_1}^{x_2}a(t)dt\right] \\ &\times \exp\left[\int_x^{x_2}\frac{b(t)}{a(t)}dt\right][1+O(\varepsilon)] \\ &+ A\frac{a(x_1)}{a(x)}\exp\left[\int_{x_1}^{x}\frac{b(t)}{a(t)}dt\right]\exp\left[-\frac{1}{\varepsilon}\int_{x_1}^{x}a(t)dt\right][1+O(\varepsilon)] \end{aligned} \tag{26.33}$$

if $B = 0$.

(b) If $a(x) < 0$, let $s = x_2 - x$ and $Y(s) = y(x_2 - s) = y(x)$. Show that the problem is then changed to

$$\begin{aligned} &\varepsilon\frac{d^2Y}{ds^2} + \tilde{a}(s)\frac{dY}{ds} + \tilde{b}(s)Y = 0, \quad s_1 \le s \le s_2, \\ &Y(s_1) = B, \quad Y(s_2) = A, \end{aligned} \tag{26.34}$$

in which $\tilde{a}(s) = -a(x_2 - s)$, $\tilde{b}(s) = b(x_2 - s)$, $s_1 = 0$ and $s_2 = x_2 - x_1$. Using (26.32) and (26.33), show that if $a(x) < 0$, the solution is given by

$$\begin{aligned} y(x) =& A\exp\left[-\int_{x_1}^{x}\frac{b(t)}{a(t)}dt\right][1+O(\varepsilon)] \\ &+ \frac{a(x_2)}{a(x)}\left\{B - A\exp\left[-\int_{x_1}^{x_2}\frac{b(t)}{a(t)}dt\right]\right\}\exp\left[-\int_x^{x_2}\frac{b(t)}{a(t)}dt\right] \\ &\times \exp\left[\frac{1}{\varepsilon}\int_x^{x_2}a(t)dt\right][1+O(\varepsilon)] \end{aligned} \tag{26.35}$$

if $A \neq 0$, and

$$y(x) = -B\frac{a(x_2)}{a(x_1)}\exp\left[-\int_{x_1}^{x_2}\frac{b(t)}{a(t)}dt\right]\exp\left[\frac{1}{\varepsilon}\int_{x_1}^{x_2}a(t)dt\right] \times \exp\left[-\int_{x_1}^{x}\frac{b(t)}{a(t)}dt\right][1+O(\varepsilon)] + B\frac{a(x_2)}{a(x)}\exp\left[-\int_{x}^{x_2}\frac{b(t)}{a(t)}dt\right]\exp\left[\frac{1}{\varepsilon}\int_{x}^{x_2}a(t)dt\right][1+O(\varepsilon)] \tag{26.36}$$

if $A = 0$.

2. (a) Use (26.35) to show that the solution of the problem

$$\begin{aligned} &\varepsilon y'' + (x-2)y' + y = 0, \quad x \in [0,1], \\ &y(0) = 3, \quad y(1) = 2, \end{aligned} \tag{26.37}$$

has the asymptotic approximation

$$y(x) = \frac{6}{2-x}[1+O(\varepsilon)] - 4e^{(x-1)(3-x)/(2\varepsilon)}[1+O(\varepsilon)]. \tag{26.38}$$

(b) Let $X = (x-2)/\sqrt{\varepsilon}$ and $y(x) = Y(X)e^{-X^2/4}$. Show that in terms of these new variables, the equation in (26.37) becomes

$$\frac{d^2Y}{dX^2} = \left(\frac{X^2}{4} - \frac{1}{2}\right)Y, \tag{26.39}$$

which is the standard form of the parabolic cylinder equation. Its two linearly independent solutions are given by the parabolic cylinder functions $Y_1(X) = U(-1/2, X)$ and $Y_2(X) = V(-1/2, X)$. Therefore, two linearly independent solutions of our original equation in (a) are

$$y_1(x) = U\left(-\frac{1}{2}, \frac{x-2}{\sqrt{\varepsilon}}\right)e^{-(x-2)^2/(4\varepsilon)},$$

$$y_2(x) = V\left(-\frac{1}{2}, \frac{x-2}{\sqrt{\varepsilon}}\right)e^{-(x-2)^2/(4\varepsilon)}.$$

Suppose that the solution of the problem in (a) is

$$y(x) = c_1y_1(x) + c_2y_2(x).$$

Using the boundary conditions in (26.37) and the asymptotic behavior of the parabolic cylinder functions (see Olver *et al.* (2010, p. 309)), show that

$$c_1 = -4e^{1/(2\varepsilon)}[1+O(\varepsilon)], \quad c_2 = -\frac{3\sqrt{2\pi}}{\sqrt{\varepsilon}}[1+O(\varepsilon)]. \tag{26.40}$$

Thus, the solution is

$$\begin{aligned}y(x) =& -4e^{1/(2\varepsilon)}U\left(-\frac{1}{2}, \frac{x-2}{\sqrt{\varepsilon}}\right)e^{-(x-2)^2/(4\varepsilon)}[1+O(\varepsilon)] \\ &-\frac{3\sqrt{2\pi}}{\sqrt{\varepsilon}}V\left(-\frac{1}{2}, \frac{x-2}{\sqrt{\varepsilon}}\right)e^{-(x-2)^2/(4\varepsilon)}[1+O(\varepsilon)]. \qquad (26.41)\end{aligned}$$

By again using the asymptotic formulas of the parabolic cylinder functions in the above formula, show that

$$y(x) = \frac{6}{2-x}[1+O(\varepsilon)] - 4e^{(x-1)(3-x)/(2\varepsilon)}[1+O(\varepsilon)]$$

3. Consider the boundary value problem

$$\begin{aligned}&\varepsilon y'' - xy' + y = 0, \quad x \in [2,4], \\ &y(2) = 0, \quad y(4) = 1.\end{aligned}$$

(a) Using the result in (26.36), show that the solution has the asymptotic approximation

$$y(x) = -2xe^{-6/\varepsilon}[1+O(\varepsilon)] + 16x^{-2}e^{(x^2-16)/(2\varepsilon)}[1+O(\varepsilon)].$$

Derive the asymptotic approximation of $y(3)$.
(b) Let $X = x/\sqrt{\varepsilon}$ and $y(x) = Y(X)e^{X^2/4}$. Show that the above equation is reduced to

$$\frac{d^2Y}{dX^2} = \left(\frac{X^2}{4} - \frac{3}{2}\right)Y.$$

Two linearly independent solutions to the original equation are

$$\begin{aligned}y_1(x) &= U\left(-\frac{3}{2}, \frac{x}{\sqrt{\varepsilon}}\right)e^{x^2/(4\varepsilon)}, \\ y_2(x) &= V\left(-\frac{3}{2}, \frac{x}{\sqrt{\varepsilon}}\right)e^{x^2/(4\varepsilon)}.\end{aligned}$$

Suppose that our solution is

$$y(x) = c_1y_1(x) + c_2y_2(x).$$

Show that

$$c_1 = -2\sqrt{\varepsilon}e^{-6/\varepsilon}[1+O(\varepsilon)], \quad c_2 = \frac{8\sqrt{2\pi}}{\varepsilon}e^{-8/\varepsilon}[1+O(\varepsilon)],$$

and that

$$y(x) = -2xe^{-6/\varepsilon}[1+O(\varepsilon)] + 16x^{-2}e^{(x^2-16)/(2\varepsilon)}[1+O(\varepsilon)].$$

4. Consider the problem

$$\varepsilon y'' + (1+x)y' + y = 0, \quad x \in [0,1],$$
$$y(0) = 1, \quad y(1) = 1.$$

(a) By using (26.32), show that the solution has the asymptotic approximation

$$y(x) = \frac{2}{1+x}[1+O(\varepsilon)] - e^{-(x^2+2x)/(2\varepsilon)}[1+O(\varepsilon)].$$

Derive the asymptotic approximation of $y(1/2)$.
(b) To find the exact solution, let $X = (1+x)/\sqrt{\varepsilon}$ and $y(x) = Y(X)e^{-X^2/4}$. Show that the original equation is reduced to

$$\frac{d^2Y}{dX^2} = \left(\frac{X^2}{4} - \frac{1}{2}\right)Y.$$

Therefore, the two independent solutions of the original equation are

$$y_1(x) = U\left(-\frac{1}{2}, \frac{1+x}{\sqrt{\varepsilon}}\right)e^{-(1+x)^2/(4\varepsilon)},$$
$$y_2(x) = V\left(-\frac{1}{2}, \frac{1+x}{\sqrt{\varepsilon}}\right)e^{-(1+x)^2/(4\varepsilon)}.$$

Suppose the solution of the problem is

$$y(x) = c_1y_1(x) + c_2y_2(x).$$

Show that

$$c_1 = -e^{-1/(2\varepsilon)}[1+O(\varepsilon)], \quad c_2 = \frac{\sqrt{2\pi}}{\sqrt{\varepsilon}}[1+O(\varepsilon)],$$

and that

$$y(x) = \frac{2}{1+x}[1+O(\varepsilon)] - e^{-(x^2+2x)/(2\varepsilon)}[1+O(\varepsilon)].$$

Chapter 27

An Internal Boundary Layer Problem

27.1 Introduction

A typical problem in singular perturbation is to consider the differential equation

$$\varepsilon y'' + a(x)y' + b(x)y = 0, \quad x_- \leq x \leq x_+, \ x_- < 0 < x_+, \tag{27.1}$$

with boundary conditions

$$y(x_-) = A \ \text{ and } \ y(x_+) = B, \tag{27.2}$$

where $a(x)$ and $b(x)$ are some sufficiently smooth functions. It is now known that an asymptotic approximation to the solution is given by

$$y(x) = y_{unif}(x) + O(\varepsilon), \tag{27.3}$$

where

$$y_{unif}(x) = B \exp\left[\int_x^{x_+} \frac{b(t)}{a(t)} dt\right] + \left\{A - B\left[\int_{x_-}^{x_+} \frac{b(t)}{a(t)} dt\right]\right\} e^{a(x_-)(x_- - x)/\varepsilon}, \tag{27.4}$$

if $a(x) > 0$ in $[x_-, x_+]$, and

$$\begin{aligned} y_{unif}(x) =& A \exp\left[-\int_{x_-}^{x} \frac{b(t)}{a(t)} dt\right] \\ &+ \left\{B - A\left[-\int_{x_-}^{x_+} \frac{b(t)}{a(t)} dt\right]\right\} e^{a(x_+)(x_+ - x)/\varepsilon}, \end{aligned} \tag{27.5}$$

if $a(x) < 0$ in $[x_-, x_+]$. The O-term in (27.3) is uniform with respect to x in $[x_-, x_+]$. Heuristic derivations of this solution can be found in Bender and Orszag (1978, p. 425) and Olver (1974a, p. 94). For a rigorous proof of (27.3), we refer to O'Malley Jr. (1991, Chapter 3) and Smith (1985,

Section 8.3). Note that the asymptotic solution (27.4) is rapidly varying in a small region near $x = x_-$, and the asymptotic solution (27.5) is rapidly varying in a region near $x = x_+$. These small intervals are known as the *boundary layers*.

The problem of finding a uniformly valid asymptotic approximation for the exact solution becomes considerably more complicated when the coefficient function $a(x)$ has a zero in the interval $[x_-, x_+]$. For convenience, we assume that the zero occurs at $x = 0$, and that near $x = 0$ we have

$$a(x) \sim \alpha x \quad \text{and} \quad b(x) \sim \beta, \tag{27.6}$$

where α is a nonzero constant. By using matching techniques, Bender and Orszag (1978, p. 458) have constructed a relatively simple asymptotic solution in the typical case when $x_- = -1$ and $x_+ = 1$. Indeed, they showed that if $\alpha > 0$ and $\beta/\alpha \neq 1, 2, 3, \cdots$, then

$$\begin{aligned} y_{unif}(x) =& \frac{\Gamma(1-\beta/\alpha)}{\sqrt{2\pi}} (\sqrt{\alpha/\varepsilon})^{\beta/\alpha} e^{-\alpha x^2/4\varepsilon} \\ &\cdot \left(A \exp\left\{ \int_x^{-1} \left[\frac{b(t)}{a(t)} - \frac{\beta}{\alpha t} \right] dt \right\} D_{\beta/\alpha - 1}(x\sqrt{\alpha/\varepsilon}) \right. \\ &+ B \exp\left\{ \int_x^{1} \left[\frac{b(t)}{a(t)} - \frac{\beta}{\alpha t} \right] dt \right\} D_{\beta/\alpha - 1}(-x\sqrt{\alpha/\varepsilon}), \end{aligned} \tag{27.7}$$

where $D_v(z)$ is the parabolic cylinder function in the notation of Whittaker and Watson (1927). Since the argument used by Bender and Orszag (1978) is only formal, it naturally raises the question whether one can justify their result by providing a mathematically rigorous proof. This is exactly the aim of the present chapter.

Searching through the literature, we found that this problem has been addressed by numerous people, including Pearson (1968), Ackerberg and O'Malley Jr. (1970), Watts (1971), and Olver (1978). While the arguments used by most of these authors are again formal, the analysis given by Olver is completely rigorous. Olver's approach is based on the theory of uniform asymptotic approximation which he has developed for second-order linear differential equations with two coalescing turning points; see Olver (1975). However, it is not easily seen that the relatively simple asymptotic solution given in (27.7) can be deduced from the approximate solution constructed by Olver (1978, (5.28) and (5.16)). In a later section of this chapter we shall present a more detailed discussion of Olver's result. Our approach to this singularly perturbed boundary value problem differs entirely from that of Olver. In a sense, our analysis is completely elementary; that is, we do not make use of any results from the theory of uniform asymptotic expansions for second-order differential equations. Also, we require less stringent conditions on the smoothness of coefficient functions $a(x)$ and $b(x)$.

There is a corresponding result in Bender and Orszag (1978, p. 460, (9.6.19)) for the case $\alpha < 0$, and $\beta/\alpha \neq 0, -1, -2, \cdots$. However, this result does not seem to be correct. For instance, it gives the leading-order uniform asymptotic approximation

$$y_{unif}(x) = 2e^{-2(x+1)/\varepsilon} + e^{-2(1-x)/\varepsilon} \tag{27.8}$$

for the boundary value problem

$$\varepsilon y'' - 2xy' + (1+x^2)y = 0, \quad y(-1) = 2, \ y(1) = 1. \tag{27.9}$$

Clearly, this approximate solution is positive in the whole interval $[-1, 1]$. But, the exact solution to (27.9) is actually negative near $x = 0$; see the graph shown in Bender and Orszag (1978, p. 460). The material in this chapter is taken from Wong and Yang (2002b). The material in the Exercise section is taken from Bender and Orszag (1978, Chapter 9).

A detailed study of the case $\alpha < 0$ and $\beta/\alpha \neq 0, 1, 2, \cdots$ is given in Wong and Yang (2002a). Furthermore, in Wong and Yang (2003) asymptotic solutions are presented for the exceptional cases: (i) $\alpha > 0$ and $\beta/\alpha = 1, 2, 3, \cdots$; (ii) $\alpha < 0$ and $\beta/\alpha = 0, -1, -2, \cdots$.

27.2 Preliminary transformations

Throughout this section, it is assumed that $a(x)$ and $b(x)$ are real-valued functions which are, respectively, thrice and twice continuously differentiable in $[x_-, x_+]$. Furthermore, we assume that $a(x)$ does not vanish in $[x_-, x_+]$ except at $x = 0$, and that near $x = 0$ we have

$$a(x) \sim \alpha x \quad \text{and} \quad b(x) \sim \beta. \tag{27.10}$$

By rescaling the small parameter ε, we may also assume that $\alpha = \pm 1$.

Motivated by the approximate solution in (27.7), we will try to derive an equation which is a perturbation of Weber's equation for parabolic cylinder functions. To this end, we make the transformation

$$y(x) = \exp(-\int_0^x h(\tau)d\tau)u(x), \tag{27.11}$$

where $h(\tau)$ is to be determined later. By straightforward calculation, we have

$$y'(x) = [u'(x) - h(x)u(x)] \exp(-\int_0^x h(\tau)d\tau)$$

and

$$y''(x) = \{u''(x) - 2h(x)u'(x) + [h^2(x) - h'(x)]u(x)\}\exp(-\int_0^x h(\tau)d\tau).$$

Substituting these equations into (27.1) gives

$$\varepsilon u''(x) + a(x)u'(x) + [b(x) - h(x)a(x)]u(x) = \varepsilon\{2h(x)u'(x) + [h'(x) - h^2(x)]u(x)\}. \tag{27.12}$$

We shall introduce new independent and dependent variables

$$t = t(x) \quad \text{and} \quad U(t) = u(x) \tag{27.13}$$

so that (27.12) becomes

$$\varepsilon\ddot{U} + \alpha t\dot{U} + \beta U = \varepsilon\left[f(t)\dot{U} + g(t)U\right], \tag{27.14}$$

where "$\cdot$" denotes the derivative with respect to t. From (27.13), we have

$$\frac{du}{dx} = \frac{dt}{dx}\dot{U} \quad \text{and} \quad \frac{d^2u}{dx^2} = \frac{d^2t}{dx^2}\dot{U} + \left(\frac{dt}{dx}\right)^2\ddot{U}.$$

Inserting these into (27.12) yields

$$\varepsilon\left(\frac{dt}{dx}\right)^2\ddot{U} + a(x)\frac{dt}{dx}\dot{U} + [b(x) - h(x)a(x)]\,U = \varepsilon\left\{\left[2h(x)\frac{dt}{dx} - \frac{d^2t}{dx^2}\right]\dot{U} + [h'(x) - h^2(x)]U\right\}.$$

A comparison of this equation with equation (27.14) gives

$$\frac{a(x)}{dt/dx} = \alpha t, \tag{27.15}$$

$$\frac{b(x) - h(x)a(x)}{(dt/dx)^2} = \beta, \tag{27.16}$$

$$f(t) = \frac{2h(x)(dt/dx) - (d^2t/dx^2)}{(dt/dx)^2}, \tag{27.17}$$

and

$$g(t) = \frac{h'(x) - h^2(x)}{(dt/dx)^2}. \tag{27.18}$$

Solving (27.15), we obtain

$$t^2 = 2\alpha\int_0^x a(\tau)d\tau. \tag{27.19}$$

Note that $\alpha a(x) \sim x$ as $x \to 0$; see the statement following (27.10). Hence, $\alpha a(x) < 0$ if $x < 0$ and $\alpha a(x) > 0$ if $x > 0$. From this, it follows that $\alpha \int_0^x a(\tau)d\tau \geq 0$ for all x. The variable t is thus defined explicitly by

$$t(x) = \begin{cases} -\left[2\alpha \int_0^x a(\tau)d\tau\right]^{1/2}, & x < 0, \\ \left[2\alpha \int_0^x a(\tau)d\tau\right]^{1/2}, & x > 0. \end{cases} \tag{27.20}$$

Using l'Hospital's rule, we find

$$\lim_{x\to 0} \frac{t^2}{x^2} = \lim_{x\to 0} \frac{2\alpha a(x)}{2x} = 1.$$

Therefore

$$t(x) \sim x \quad \text{as} \quad x \to 0 \tag{27.21}$$

and

$$\lim_{x\to 0} \frac{dt}{dx} = \lim_{x\to 0} \frac{\alpha a(x)}{t} = 1. \tag{27.22}$$

From (27.15), we also have

$$\frac{dt}{dx} = \frac{\alpha a(x)}{t} > 0 \quad \text{for all} \quad x \neq 0; \tag{27.23}$$

see the statement following (27.10). As a consequence, $t = t(x)$ is a well-defined and one-to-one mapping from $[x_-, x_+]$ to $[t_-, t_+]$, where

$$t_- = t(x_-) \quad \text{and} \quad t_+ = t(x_+). \tag{27.24}$$

Solving (27.16), we obtain from (27.15)

$$h(x) = \frac{b(x)}{a(x)} - \frac{\beta(dt/dx)^2}{a(x)} = \frac{b(x)}{a(x)} - \frac{\beta a(x)}{t^2}. \tag{27.25}$$

The boundary value problem (27.1)-(27.2) is now transformed into the boundary value problem

$$\varepsilon \ddot{U} + \alpha t \dot{U} + \beta U = \varepsilon[f(t)\dot{U} + g(t)U], \tag{27.26}$$

$$U(t_-) = \lambda \quad \text{and} \quad U(t_+) = \mu, \tag{27.27}$$

where

$$\lambda = A \exp\left(\int_0^{x_-} h(\tau)d\tau\right), \quad \mu = B \exp\left(\int_0^{x_+} h(\tau)d\tau\right). \tag{27.28}$$

Lemma 27.1. *Under our conditions on $a(x)$ and $b(x)$ stated at the beginning of this section, the functions $f(t)$ and $g(t)$ defined in (27.17) and (27.18) are continuous in $[t_-, t_+]$.*

Proof. From (27.22) and (27.23), it is evident that dt/dx is continuous and positive in $[x_-, x_+]$. Thus, in view of (27.17) and (27.18), it suffices to prove that $h(x)$ is continuously differentiable in $[x_-, x_+]$ and $t(x)$ is twice continuously differentiable there. By our conditions on $a(x)$ and $b(x)$, we have

$$a(x) = \alpha x[1 + a_1 x + a_2 x^2 + o(x^2)], \tag{27.29}$$

$$b(x) = \beta + b_1 x + b_2 x^2 + o(x^2), \tag{27.30}$$

and these expansions can be termwise differentiated to give approximations to the derivatives of $a(x)$ and $b(x)$. From (27.19), it also follows that

$$t^2 = x^2 + \frac{2}{3}a_1 x^3 + \frac{1}{2}a_2 x^4 + o(x^4). \tag{27.31}$$

First, we show that $t''(x)$ is continuous in $[x_-, x_+]$. When $x \neq 0$, differentiation of (27.15) gives

$$t''(x) = \alpha \frac{ta'(x) - a(x)t'}{t^2} = \alpha \frac{t^2 a'(x) - \alpha a^2(x)}{t^3}. \tag{27.32}$$

On account of (27.21), we obtain by using l'Hospital's rule

$$\begin{aligned}
\lim_{x\to 0} t''(x) &= \alpha \lim_{x\to 0} \frac{t^2 a'(x) - \alpha a^2(x)}{x^3} \\
&= \alpha \lim_{x\to 0} \frac{2\alpha a(x)a'(x) + t^2 a''(x) - 2\alpha a(x)a'(x)}{3x^2} \\
&= \alpha \lim_{x\to 0} \frac{t^2 a''(x)}{3x^2} = \frac{\alpha}{3}a''(0).
\end{aligned} \tag{27.33}$$

When $x = 0$, we have from (27.22) and (27.23)

$$t''(0) = \lim_{x\to 0} \frac{t'(x) - t'(0)}{x} = \lim_{x\to 0} \frac{\alpha a(x) - t}{tx}.$$

Again by using (27.21), we can write

$$t''(0) = \lim_{x\to 0} \frac{a^2(x) - t^2}{x^2[\alpha a(x) + t]} = \lim_{x\to 0} \frac{a^2(x) - t^2}{2x^3}.$$

An application of (27.29) and (27.31) gives

$$t''(0) = \frac{2}{3}a_1. \tag{27.34}$$

Since $a''(0) = 2\alpha a_1$ by (27.29), coupling (27.33) and (27.34) shows that $t''(x)$ is continuous at $x = 0$, and hence in $[x_-, x_+]$.

Next, we demonstrate that $h(x)$ and $h'(x)$ are continuous in $[x_-, x_+]$. By (27.25),

$$h(x) = \frac{b(x)t^2 - \beta a^2(x)}{a(x)t^2}.$$

From (27.29)-(27.31), we obtain

$$h(x) = \frac{(b_1 - (4/3)\beta a_1)x^3 + o(x^3)}{a(x)t^2} \tag{27.35}$$

and

$$\lim_{x\to 0} h(x) = \frac{1}{\alpha}(b_1 - \frac{4}{3}\beta a_1). \tag{27.36}$$

Thus, $h(x)$ can be made continuous in $[x_-, x_+]$ by defining $h(0)$ to be the value given in (27.36). In a similar manner, we get

$$h'(x) = \frac{a(x)b'(x) - a'(x)b(x)}{a^2(x)} - \beta\frac{t^2a'(x) - 2tt'(x)a(x)}{t^4}$$

and

$$h'(x) = \frac{((11/9)\alpha\beta a_1^2 - (3/2)\alpha\beta a_2 + \alpha b_2 - \alpha a_1 b_1)x^6 + o(x^6)}{t^4 a^2(x)}.$$

By defining

$$h'(0) = \frac{11}{9}\alpha\beta a_1^2 - \frac{3}{2}\alpha\beta a_2 + \alpha b_2 - \alpha a_1 b_1,$$

we also have $h'(x)$ continuous in $[x_-, x_+]$. This completes the proof of the lemma. □

Remark 27.1. Since $f(t)$ and $g(t)$ are continuous in $[t_-, t_+]$, we can assign a positive number γ_0 such that

$$|f(t)| \le \gamma_0 \quad \text{and} \quad |g(t)| \le \gamma_0 \tag{27.37}$$

for all $t \in [t_-, t_+]$.

27.3 Conversion to an integral equation

Equation (27.26) can be considered as a perturbation of the homogeneous equation

$$\varepsilon W'' + \alpha t W' + \beta W = 0. \tag{27.38}$$

If we let

$$\xi = \frac{t}{\sqrt{\varepsilon}} \quad \text{and} \quad w(\xi) = e^{\alpha\xi^2/4}W(t),$$

then (27.38) becomes

$$\frac{d^2w}{d\xi^2} = \left(\frac{1}{4}\xi^2 - \beta + \frac{\alpha}{2}\right)w,$$

which is the standard form of Weber's equation; see Olver (1974a, p. 147). Thus two linearly independent solutions of (27.38) are given by

$$U(a, \frac{t}{\sqrt{\varepsilon}})e^{-\alpha t^2/4\varepsilon} \quad \text{and} \quad U(a, -\frac{t}{\sqrt{\varepsilon}})e^{-\alpha t^2/4\varepsilon} \tag{27.39}$$

with $a = -\beta + \alpha/2$ and $a \neq -n - \frac{1}{2}, n = 0, 1, 2, \cdots$, where $U(a, \xi)$ is a parabolic cylinder function; see Abramowitz and Stegun (1989, p. 687). The following properties are well known:

$$U(a, \xi) \sim \xi^{-a-\frac{1}{2}} e^{-\xi^2/4}, \quad \xi \to \infty, \tag{27.40}$$

$$U'(a, \xi) \sim -\frac{1}{2}\xi^{-a+\frac{1}{2}} e^{-\xi^2/4}, \quad \xi \to \infty; \tag{27.41}$$

for $a \neq -n - \frac{1}{2}, n = 0, 1, 2, \cdots$.

$$U(a, \xi) \sim \frac{\sqrt{2\pi}}{\Gamma(a + \frac{1}{2})} |\xi|^{a-\frac{1}{2}} e^{\xi^2/4}, \quad \xi \to -\infty, \tag{27.42}$$

$$U'(a, \xi) \sim -\frac{1}{2}\frac{\sqrt{2\pi}}{\Gamma(a + \frac{1}{2})} |\xi|^{a+\frac{1}{2}} e^{\xi^2/4}, \quad \xi \to -\infty, \tag{27.43}$$

We also have the recurrence relations

$$U'(a, \xi) - \frac{1}{2}\xi U(a, \xi) = -U(a - 1, \xi) \tag{27.44}$$

and

$$U'(a, \xi) + \frac{1}{2}\xi U(a, \xi) = -(a + \frac{1}{2})U(a + 1, \xi). \tag{27.45}$$

The Wronskian of $U(a, \xi)$ and $U(a, -\xi)$ is given by

$$W\{U(a, \xi), U(a, -\xi)\} = \frac{\sqrt{2\pi}}{\Gamma(a + 1/2)}. \tag{27.46}$$

We now construct two linearly independent solutions to (27.38), when $\alpha = 1$. Denote them by $W_1(t)$ and $W_2(t)$, such that

$$W_1(t_-) = 1, \quad W_1(t_+) = 0, \tag{27.47}$$

$$W_2(t_-) = 0, \quad W_2(t_+) = 1. \tag{27.48}$$

Since they must be linear combinations of the solutions given in (27.39), we put

$$W_i(t) = C_i U(a, \frac{t}{\sqrt{\varepsilon}})e^{-t^2/4\varepsilon} + D_i U(a, -\frac{t}{\sqrt{\varepsilon}})e^{-t^2/4\varepsilon}, \quad i = 1, 2. \tag{27.49}$$

By (27.47), we have

$$C_1U(a, \frac{t_-}{\sqrt{\varepsilon}})e^{-t_-^2/4\varepsilon} + D_1U(a, -\frac{t_-}{\sqrt{\varepsilon}})e^{-t_-^2/4\varepsilon} = 1,$$
$$C_1U(a, \frac{t_+}{\sqrt{\varepsilon}})e^{-t_+^2/4\varepsilon} + D_1U(a, -\frac{t_+}{\sqrt{\varepsilon}})e^{-t_+^2/4\varepsilon} = 0.$$

Upon solving these two equations, we get

$$C_1 = \frac{1}{\Delta(\varepsilon)}U(a, -\frac{t_+}{\sqrt{\varepsilon}})e^{t_-^2/4\varepsilon}, \quad D_1 = -\frac{1}{\Delta(\varepsilon)}U(a, \frac{t_+}{\sqrt{\varepsilon}})e^{t_-^2/4\varepsilon}.$$

where

$$\Delta(\varepsilon) = U(a, \frac{t_-}{\sqrt{\varepsilon}})U(a, -\frac{t_+}{\sqrt{\varepsilon}}) - U(a, -\frac{t_-}{\sqrt{\varepsilon}})U(a, \frac{t_+}{\sqrt{\varepsilon}}). \tag{27.50}$$

In a similar manner, we obtain

$$C_2 = -\frac{1}{\Delta(\varepsilon)}U(a, -\frac{t_-}{\sqrt{\varepsilon}})e^{t_+^2/4\varepsilon}, \quad D_2 = \frac{1}{\Delta(\varepsilon)}U(a, \frac{t_-}{\sqrt{\varepsilon}})e^{t_+^2/4\varepsilon}.$$

Therefore, we may write

$$W_1(t) = \frac{V_1(t)}{\Delta(\varepsilon)}e^{(t_-^2-t^2)/4\varepsilon}, \quad W_2(t) = \frac{V_2(t)}{\Delta(\varepsilon)}e^{(t_+^2-t^2)/4\varepsilon} \tag{27.51}$$

with

$$V_1(t) = U(a, -\frac{t_+}{\sqrt{\varepsilon}})U(a, \frac{t}{\sqrt{\varepsilon}}) - U(a, \frac{t_+}{\sqrt{\varepsilon}})U(a, -\frac{t}{\sqrt{\varepsilon}}),$$
$$V_2(t) = U(a, \frac{t_-}{\sqrt{\varepsilon}})U(a, -\frac{t}{\sqrt{\varepsilon}}) - U(a, -\frac{t_-}{\sqrt{\varepsilon}})U(a, \frac{t}{\sqrt{\varepsilon}}). \tag{27.52}$$

From (27.51), it also follows that

$$W_1'(t) = \frac{\bar{V}_1(t)}{\Delta(\varepsilon)}e^{(t_-^2-t^2)/4\varepsilon}, \quad W_2'(t) = \frac{\bar{V}_2(t)}{\Delta(\varepsilon)}e^{(t_+^2-t^2)/4\varepsilon}, \tag{27.53}$$

where

$$\bar{V}_i(t) = V_i'(t) - \frac{1}{2\varepsilon}tV_i(t), \quad i = 1, 2. \tag{27.54}$$

A straightforward calculation gives

$$\mathcal{W}\{W_1(t), W_2(t)\} = \frac{1}{\Delta^2(\varepsilon)}e^{(t_-^2+t_+^2-2t^2)/4\varepsilon}\begin{vmatrix} V_1(t), & V_2(t) \\ \bar{V}_1(t), & \bar{V}_2(t) \end{vmatrix}. \tag{27.55}$$

It can be shown from (27.54) that

$$\begin{vmatrix} V_1(t), & V_2(t) \\ \bar{V}_1(t), & \bar{V}_2(t) \end{vmatrix} = \begin{vmatrix} V_1(t), & V_2(t) \\ V_1'(t), & V_2'(t) \end{vmatrix}. \tag{27.56}$$

Substituting (27.52) in (27.56), we obtain from (27.46)

$$\begin{vmatrix} V_1(t), V_2(t) \\ \bar{V}_1(t), \bar{V}_2(t) \end{vmatrix} = \frac{\Delta(\varepsilon)}{\sqrt{\varepsilon}} \mathcal{W}\{U(a,\xi), U(a,-\xi)\}$$

$$= \frac{\sqrt{2\pi}}{\Gamma(a+1/2)} \frac{\Delta(\varepsilon)}{\sqrt{\varepsilon}}. \tag{27.57}$$

Coupling (27.55) and (27.57) yields

$$\mathcal{W}\{W_1(t), W_2(t)\} = \frac{\sqrt{2\pi}}{\Gamma(a+1/2)} \frac{1}{\sqrt{\varepsilon}\Delta(\varepsilon)} e^{(t_-^2+t_+^2-2t^2)/4\varepsilon}. \tag{27.58}$$

By considering the right-hand side of (27.26) as a nonhomogeneous term of the homogeneous equation (27.38), we can convert the boundary value problem (27.26)-(27.27) into the integral equation

$$U(t) = \lambda W_1(t) + \mu W_2(t) - \int_{t_-}^{t_+} G(t,s)[f(s)U'(s) + g(s)U(s)]ds, \tag{27.59}$$

where $G(t,s)$ is the Green's function defined by

$$G(t,s) = \begin{cases} \frac{W_1(t)W_2(s)}{\mathcal{W}\{W_1(s),W_2(s)\}}, & t_- \le s \le t \le t_+, \\ \frac{W_1(s)W_2(t)}{\mathcal{W}\{W_1(s),W_2(s)\}}, & t_- \le t \le s \le t_+. \end{cases} \tag{27.60}$$

In view of (27.51) and (27.58), $G(t,s)$ can be written as

$$G(t,s) = \begin{cases} \frac{\Gamma(a+1/2)}{\sqrt{2\pi}} \varepsilon^{1/2} W_1(t)V_2(s) e^{\frac{s^2-t_-^2}{4\varepsilon}}, & t_- \le s \le t \le t_+, \\ \frac{\Gamma(a+1/2)}{\sqrt{2\pi}} \varepsilon^{1/2} W_2(t)V_1(s) e^{\frac{s^2-t_+^2}{4\varepsilon}}, & t_- \le t \le s \le t_+, \end{cases} \tag{27.61}$$

from which it also follows

$$\frac{\partial G}{\partial t}(t,s) = \begin{cases} \frac{\Gamma(a+1/2)}{\sqrt{2\pi}} \varepsilon^{1/2} W_1'(t)V_2(s) e^{\frac{s^2-t_-^2}{4\varepsilon}}, & t_- \le s \le t \le t_+, \\ \frac{\Gamma(a+1/2)}{\sqrt{2\pi}} \varepsilon^{1/2} W_2'(t)V_1(s) e^{\frac{s^2-t_+^2}{4\varepsilon}}, & t_- \le t \le s \le t_+. \end{cases} \tag{27.62}$$

27.4 Behavior of $W_i(t)$ and $W_i'(t)(i = 1, 2)$ as $\varepsilon \to 0$

We recall that here, we consider only the case $\alpha = 1$ and $\beta \neq 1, 2, 3, \cdots$. Thus, $a = -\beta + \alpha/2 \neq -n - \frac{1}{2}, n = 0, 1, 2, \cdots$. Using (27.40) and (27.42), we obtain

$$\Delta(\varepsilon) = U(a, \frac{t_-}{\sqrt{\varepsilon}})U(a, -\frac{t_+}{\sqrt{\varepsilon}}) \left[1 + O(\varepsilon^{2a} e^{-(t_-^2+t_+^2)/2\varepsilon})\right], \ \varepsilon \to 0, \tag{27.63}$$

$$V_1(t) = U(a, -\frac{t_+}{\sqrt{\varepsilon}})U(a, \frac{t}{\sqrt{\varepsilon}}) \left[1 + O((\frac{\varepsilon}{|t|})^{2a} e^{-(t^2+t_+^2)/2\varepsilon})\right] \tag{27.64}$$

if $t/\sqrt{\varepsilon} \to -\infty$, and

$$V_1(t) = U(a, -\frac{t_+}{\sqrt{\varepsilon}})U(a, \frac{t}{\sqrt{\varepsilon}}) \left[1 + O(t^{2a}e^{(t^2-t_+^2)/2\varepsilon})\right] \tag{27.65}$$

if $t/\sqrt{\varepsilon} \to \infty$. In (27.65), the O-term is exponentially small if $t_+ - t \gg \varepsilon$, and is bounded if $t_+ - t = O(\varepsilon)$. Thus, in any case, we have from (27.64) and (27.65)

$$V_1(t) = O\left\{U(a, -\frac{t_+}{\sqrt{\varepsilon}})U(a, \frac{t}{\sqrt{\varepsilon}})\right\} \tag{27.66}$$

as $t/\sqrt{\varepsilon} \to \pm\infty$. By virtue of the behavior of $U(a, \xi)$ near $\xi = 0$ (see Abramowitz and Stegun (1989, p. 687, (19.3.1))), (27.66) also holds when $t = O(\sqrt{\varepsilon})$ and hence uniformly for all $t \in [t_-, t_+]$.

In exactly the same manner, one can show that

$$V_2(t) = U(a, \frac{t_-}{\sqrt{\varepsilon}})U(a, -\frac{t}{\sqrt{\varepsilon}}) \left[1 + O(|t|^{2a}e^{(t^2-t_-^2)/2\varepsilon})\right] \tag{27.67}$$

as $t/\sqrt{\varepsilon} \to -\infty$, and

$$V_2(t) = U(a, \frac{t_-}{\sqrt{\varepsilon}})U(a, -\frac{t}{\sqrt{\varepsilon}}) \left[1 + O((\frac{\varepsilon}{t})^{2a}e^{-(t^2+t_-^2)/2\varepsilon})\right] \tag{27.68}$$

as $t/\sqrt{\varepsilon} \to \infty$. Furthermore, uniformly for all $t \in [t_-, t_+]$,

$$V_2(t) = O\left\{U(a, \frac{t_-}{\sqrt{\varepsilon}})U(a, -\frac{t}{\sqrt{\varepsilon}})\right\}. \tag{27.69}$$

A combination of (27.51), (27.63), (27.66) and (27.69) gives the following result.

Lemma 27.2. *For $a \neq -n - \frac{1}{2}$, we have as $\varepsilon \to 0$*

$$W_1(t) = O\left\{U(a, \frac{t}{\sqrt{\varepsilon}})e^{(t_-^2-t^2)/4\varepsilon} \Big/ U(a, \frac{t_-}{\sqrt{\varepsilon}})\right\} \tag{27.70}$$

and

$$W_2(t) = O\left\{U(a, -\frac{t}{\sqrt{\varepsilon}})e^{(t_+^2-t^2)/4\varepsilon} \Big/ U(a, -\frac{t_+}{\sqrt{\varepsilon}})\right\}. \tag{27.71}$$

To obtain estimates for $W_1'(t)$ and $W_2'(t)$, we note from (27.52) that

$$V_1'(t) = \frac{1}{\sqrt{\varepsilon}}\left[U(a, -\frac{t_+}{\sqrt{\varepsilon}})U'(a, \frac{t}{\sqrt{\varepsilon}}) + U(a, \frac{t_+}{\sqrt{\varepsilon}})U'(a, -\frac{t}{\sqrt{\varepsilon}})\right]$$

and

$$V_2'(t) = -\frac{1}{\sqrt{\varepsilon}}\left[U(a, \frac{t_-}{\sqrt{\varepsilon}})U'(a, -\frac{t}{\sqrt{\varepsilon}}) + U(a, -\frac{t_-}{\sqrt{\varepsilon}})U'(a, \frac{t}{\sqrt{\varepsilon}})\right].$$

From (27.54), it follows that

$$\bar{V}_1(t) = -\frac{1}{\sqrt{\varepsilon}}\left[U(a, -\frac{t_+}{\sqrt{\varepsilon}})U(a-1, \frac{t}{\sqrt{\varepsilon}}) + U(a, \frac{t_+}{\sqrt{\varepsilon}})U(a-1, -\frac{t}{\sqrt{\varepsilon}})\right]$$

and

$$\bar{V}_2(t) = \frac{1}{\sqrt{\varepsilon}}\left[U(a, \frac{t_-}{\sqrt{\varepsilon}})U(a-1, -\frac{t}{\sqrt{\varepsilon}}) + U(a, -\frac{t_-}{\sqrt{\varepsilon}})U(a-1, \frac{t}{\sqrt{\varepsilon}})\right].$$

where use has been made of (27.44). Again, by (27.40) and (27.42), we obtain

$$\bar{V}_1(t) = -\frac{1}{\sqrt{\varepsilon}}U(a, -\frac{t_+}{\sqrt{\varepsilon}})U(a-1, \frac{t}{\sqrt{\varepsilon}})\left[1 + O(\varepsilon^{2a-1}|t|^{2(1-a)}e^{-(t_+^2+t^2)/2\varepsilon})\right] \tag{27.72}$$

if $t/\sqrt{\varepsilon} \to -\infty$, and

$$\bar{V}_1(t) = -\frac{1}{\sqrt{\varepsilon}}U(a, -\frac{t_+}{\sqrt{\varepsilon}})U(a-1, \frac{t}{\sqrt{\varepsilon}})\left[1 + O(\varepsilon t^{2(a-1)}e^{(t^2-t_+^2)/2\varepsilon})\right] \tag{27.73}$$

if $t/\sqrt{\varepsilon} \to \infty$. Note that the O-term in (27.72) is exponentially small and the O-term in (27.73) is $O(\varepsilon)$. Coupling (27.72) and (27.73), we conclude that as $t/\sqrt{\varepsilon} \to \pm\infty$,

$$\bar{V}_1(t) = O\left\{\frac{1}{\sqrt{\varepsilon}}U(a, -\frac{t_+}{\sqrt{\varepsilon}})U(a-1, \frac{t}{\sqrt{\varepsilon}})\right\}. \tag{27.74}$$

The boundedness of $U(a-1, \xi)$ near $\xi = 0$ ensures that (27.74) also holds when $t = O(\sqrt{\varepsilon})$ and hence uniformly for all $t \in [t_-, t_+]$.

The same argument gives that

$$\bar{V}_2(t) = \frac{1}{\sqrt{\varepsilon}}U(a, \frac{t_-}{\sqrt{\varepsilon}})U(a-1, -\frac{t}{\sqrt{\varepsilon}})\left[1 + O(\varepsilon|t|^{2(a-1)}e^{(t^2-t_-^2)/2\varepsilon})\right] \tag{27.75}$$

if $t/\sqrt{\varepsilon} \to -\infty$, and

$$\bar{V}_2(t) = \frac{1}{\sqrt{\varepsilon}}U(a, \frac{t_-}{\sqrt{\varepsilon}})U(a-1, -\frac{t}{\sqrt{\varepsilon}})\left[1 + O(\varepsilon^{2a-1}t^{2(1-a)}e^{-(t^2+t_-^2)/2\varepsilon})\right] \tag{27.76}$$

if $t/\sqrt{\varepsilon} \to \infty$. Moreover, uniformly for all $t \in [t_-, t_+]$,

$$\bar{V}_2(t) = O\left\{\frac{1}{\sqrt{\varepsilon}}U(a, \frac{t_-}{\sqrt{\varepsilon}})U(a-1, -\frac{t}{\sqrt{\varepsilon}})\right\}. \tag{27.77}$$

On account of (27.63), (27.74) and (27.77), we have from (27.53)

Lemma 27.3. *For $a \neq -n - \frac{1}{2}$ and as $\varepsilon \to 0$,*

$$W_1'(t) = O\left\{\frac{1}{\sqrt{\varepsilon}}e^{(t_-^2-t^2)/4\varepsilon}U(a-1, \frac{t}{\sqrt{\varepsilon}})\Big/U(a, \frac{t_-}{\sqrt{\varepsilon}})\right\} \tag{27.78}$$

and

$$W_2'(t) = O\left\{\frac{1}{\sqrt{\varepsilon}}e^{(t_+^2-t^2)/4\varepsilon}U(a-1, -\frac{t}{\sqrt{\varepsilon}})\Big/U(a, -\frac{t_+}{\sqrt{\varepsilon}})\right\} \tag{27.79}$$

uniformly for $t \in [t_-, t_+]$.

By making use of the behavior of $V_i(t)$ and $\bar{V}_i(t)$ as $t/\sqrt{\varepsilon} \to \pm\infty, i = 1, 2,$ we also obtain

$$\frac{V_2(t)}{V_1(t)} = O\left\{\left(\frac{|t|}{\sqrt{\varepsilon}}\right)^{-2a} e^{(t_-^2 - t_+^2 - 2t^2)/4\varepsilon}\right\}$$

and

$$\frac{\bar{V}_2(t)}{\bar{V}_1(t)} = O\left\{\left(\frac{|t|}{\sqrt{\varepsilon}}\right)^{2(1-a)} e^{(t_-^2 - t_+^2 - 2t^2)/4\varepsilon}\right\}$$

as $t/\sqrt{\varepsilon} \to -\infty$,

$$\frac{V_1(t)}{V_2(t)} = O\left\{\left(\frac{t}{\sqrt{\varepsilon}}\right)^{-2a} e^{(t_+^2 - t_-^2 - 2t^2)/4\varepsilon}\right\}$$

and

$$\frac{\bar{V}_1(t)}{\bar{V}_2(t)} = O\left\{\left(\frac{t}{\sqrt{\varepsilon}}\right)^{2(1-a)} e^{(t_+^2 - t_-^2 - 2t^2)/4\varepsilon}\right\}$$

as $t/\sqrt{\varepsilon} \to \infty$,

A combination of the last four estimates with (27.51) and (27.53) gives

Lemma 27.4. *As $\varepsilon \to 0$, there exists a positive constant ρ such that*

$$W_2(t) = W_1(t) \cdot O\left\{\left(\frac{|t|}{\sqrt{\varepsilon}}\right)^{-2a} e^{-t^2/2\varepsilon}\right\}, \quad t \in [t_-, -\rho\sqrt{\varepsilon}], \tag{27.80}$$

$$W_1(t) = W_2(t) \cdot O\left\{\left(\frac{t}{\sqrt{\varepsilon}}\right)^{-2a} e^{-t^2/2\varepsilon}\right\}, \quad t \in [\rho\sqrt{\varepsilon}, t_+], \tag{27.81}$$

$$W_2'(t) = W_1'(t) \cdot O\left\{\left(\frac{|t|}{\sqrt{\varepsilon}}\right)^{2(1-a)} e^{-t^2/2\varepsilon}\right\}, \quad t \in [t_-, -\rho\sqrt{\varepsilon}] \tag{27.82}$$

and

$$W_1'(t) = W_2'(t) \cdot O\left\{\left(\frac{t}{\sqrt{\varepsilon}}\right)^{2(1-a)} e^{-t^2/2\varepsilon}\right\}, \quad t \in [\rho\sqrt{\varepsilon}, t_+] \tag{27.83}$$

hold.

27.5 Construction of the solution

To show that the integral equation (27.59) has a solution, we define a sequence $\{U_n(t)\}$, $n = 0, 1, 2, \cdots$, such that

$$U_0(t) = \lambda W_1(t) + \mu W_2(t) \tag{27.84}$$

and

$$U_n(t) = U_0(t) - \int_{t_-}^{t_+} G(t,s)[f(s)U'_{n-1}(s) + g(s)U_{n-1}(s)]ds \tag{27.85}$$

for $n \geq 1$. Differentiation yields that

$$U'_0(t) = \lambda W'_1(t) + \mu W'_2(t) \tag{27.86}$$

and

$$U'_n(t) = U'_0(t) - \int_{t_-}^{t_+} \frac{\partial G}{\partial t}(t,s)[f(s)U'_{n-1}(s) + g(s)U_{n-1}(s)]ds \tag{27.87}$$

for $n \geq 1$.

To prove that the series $\sum_{n=1}^{\infty}[U_n(t) - U_{n-1}(t)]$ converges uniformly in $[t_-, t_+]$, we need the following lemma.

Lemma 27.5. *As $\varepsilon \to 0$, we have*

$$\int_{t_-}^{t_+} |G(t,s)| \left(|W'_1(s)| + |W'_2(s)| + |W_1(s)| + |W_2(s)|\right) ds$$
$$= \left(|W_1(t)| + |W_2(t)|\right) O(\sqrt{\varepsilon}) \tag{27.88}$$

and

$$\int_{t_-}^{t_+} \left|\frac{\partial G}{\partial t}(t,s)\right| \left(|W'_1(s)| + |W'_2(s)| + |W_1(s)| + |W_2(s)|\right) ds$$
$$= \left(|W'_1(t)| + |W'_2(t)|\right) O(\sqrt{\varepsilon}) \tag{27.89}$$

uniformly for $t \in [t_-, t_+]$.

Proof. Let

$$I(\varepsilon, t) = \int_{t_-}^{t_+} |G(t,s)| \left(|W'_1(s)| + |W'_2(s)| + |W_1(s)| + |W_2(s)|\right) ds \tag{27.90}$$

and

$$J(\varepsilon, t) = \int_{t_-}^{t_+} \left|\frac{\partial G}{\partial t}(t,s)\right| \left(|W'_1(s)| + |W'_2(s)| + |W_1(s)| + |W_2(s)|\right) ds. \tag{27.91}$$

By making use of (27.61) and (27.62), one can write $I(\varepsilon,t)$ and $J(\varepsilon,t)$ as

$$I(\varepsilon,t)=\frac{\Gamma(a+1/2)}{\sqrt{2\pi}}\sqrt{\varepsilon}\Big[(I_1+I_2)e^{-t_-^2/4\varepsilon}|W_1(t)| \\ +(I_3+I_4)e^{-t_+^2/4\varepsilon}|W_2(t)|\Big] \tag{27.92}$$

and

$$|J(\varepsilon,t)|=\frac{\Gamma(a+1/2)}{\sqrt{2\pi}}\sqrt{\varepsilon}\Big[(I_1+I_2)e^{-t_-^2/4\varepsilon}|W_1'(t)| \\ +(I_3+I_4)e^{-t_+^2/4\varepsilon}|W_2'(t)|\Big] \tag{27.93}$$

where

$$I_1=\int_{t_-}^{t_+}|V_2(s)|(|W_1'(s)|+|W_1(s)|)e^{s^2/4\varepsilon}ds, \tag{27.94}$$

$$I_2=\int_{t_-}^{t_+}|V_2(s)|(|W_2'(s)|+|W_2(s)|)e^{s^2/4\varepsilon}ds, \tag{27.95}$$

$$I_3=\int_{t_-}^{t_+}|V_1(s)|(|W_1'(s)|+|W_1(s)|)e^{s^2/4\varepsilon}ds, \tag{27.96}$$

and

$$I_4=\int_{t_-}^{t_+}|V_1(s)|(|W_2'(s)|+|W_2(s)|)e^{s^2/4\varepsilon}ds. \tag{27.97}$$

Thus, all we need to do is to estimate the quantities $I_i, i=1,2,3,4$.

First, we consider I_1. By (27.69) and the results in Lemmas 27.2 and 27.3, we have

$$I_1=O(\varepsilon^{-1/2}e^{t_-^2/4\varepsilon})\int_{t_-}^{t}|U(a,-\frac{s}{\sqrt{\varepsilon}})| \\ \cdot\left(|U(a-1,\frac{s}{\sqrt{\varepsilon}})|+\sqrt{\varepsilon}|U(a,\frac{s}{\sqrt{\varepsilon}})|\right)ds. \tag{27.98}$$

For $t\in[t_-,-\rho\sqrt{\varepsilon}]$, the behavior of the parabolic cylinder function gives

$$\begin{aligned}I_1&=O(\varepsilon^{-1/2}e^{t_-^2/4\varepsilon})\int_{t_-}^{t}\left(\frac{|s|}{\sqrt{\varepsilon}}\right)^{-a-1/2}\left(\left(\frac{|s|}{\sqrt{\varepsilon}}\right)^{a-3/2}+\sqrt{\varepsilon}\left(\frac{|s|}{\sqrt{\varepsilon}}\right)^{a-1/2}\right)ds\\&=O(\varepsilon^{-1/2}e^{t_-^2/4\varepsilon})\int_{t_-}^{t}\left(\left(\frac{|s|}{\sqrt{\varepsilon}}\right)^{-2}+\sqrt{\varepsilon}\left(\frac{|s|}{\sqrt{\varepsilon}}\right)^{-1}\right)ds\\&=O(e^{t_-^2/4t}).\end{aligned} \tag{27.99}$$

For $t \in [-\rho\sqrt{\varepsilon}, \rho\sqrt{\varepsilon}]$, we write I_1 as follows

$$\begin{aligned}
I_1 =& O(\varepsilon^{-1/2} e^{t_-^2/4\varepsilon}) \int_{t_-}^{-\rho\sqrt{\varepsilon}} |U(a, -\frac{s}{\sqrt{\varepsilon}})| \\
&\cdot \left(|U(a-1, \frac{s}{\sqrt{\varepsilon}})| + \sqrt{\varepsilon}|U(a, \frac{s}{\sqrt{\varepsilon}})| \right) ds \\
&+ O(\varepsilon^{-1/2} e^{t_-^2/4\varepsilon}) \int_{-\rho\sqrt{\varepsilon}}^{t} |U(a, -\frac{s}{\sqrt{\varepsilon}})| \\
&\cdot \left(|U(a-1, \frac{s}{\sqrt{\varepsilon}})| + \sqrt{\varepsilon}|U(a, \frac{s}{\sqrt{\varepsilon}})| \right) ds. \qquad (27.100)
\end{aligned}$$

The first term is $O(e^{t_-^2/4\varepsilon})$ by (27.99). In view of the behavior of the parabolic cylinder function $U(a, \xi)$ near $\xi = 0$, the integrand in the second term is bounded. Hence, for $t \in [-\rho\sqrt{\varepsilon}, \rho\sqrt{\varepsilon}]$

$$I_1 = O(e^{t_-^2/4\varepsilon}) + O(\varepsilon^{-1/2} e^{t_-^2/4\varepsilon}) \int_{-\rho\sqrt{\varepsilon}}^{t} ds = O(e^{t_-^2/4\varepsilon}). \qquad (27.101)$$

For $t \in [\rho\sqrt{\varepsilon}, t_+]$, I_1 can be expressed as

$$\begin{aligned}
I_1 =& O(\varepsilon^{-1/2} e^{t_-^2/4\varepsilon}) \int_{t_-}^{\rho\sqrt{\varepsilon}} |U(a, -\frac{s}{\sqrt{\varepsilon}})| \\
&\cdot \left(|U(a-1, \frac{s}{\sqrt{\varepsilon}})| + \sqrt{\varepsilon}|U(a, \frac{s}{\sqrt{\varepsilon}})| \right) ds \\
&+ O(\varepsilon^{-1/2} e^{t_-^2/4\varepsilon}) \int_{\rho\sqrt{\varepsilon}}^{t} |U(a, -\frac{s}{\sqrt{\varepsilon}})| \\
&\cdot \left(|U(a-1, \frac{s}{\sqrt{\varepsilon}})| + \sqrt{\varepsilon}|U(a, \frac{s}{\sqrt{\varepsilon}})| \right) ds. \qquad (27.102)
\end{aligned}$$

By (27.101), the first term on the right-hand side of (27.102) is $O(e^{t_-^2/4\varepsilon})$. Using the behavior of the parabolic cylinder function, it can be shown that the second term on the right-hand side is $O(\frac{1}{\sqrt{\varepsilon}} e^{t_-^2/4\varepsilon})$. Coupling these two results gives that

$$I_1 = O(e^{t_-^2/4\varepsilon})(1 + \frac{t}{\sqrt{\varepsilon}}), \quad t \in [\rho\sqrt{\varepsilon}, t_+]. \qquad (27.103)$$

Therefore, it follows that as $\varepsilon \to 0$

$$I_1 = \begin{cases} O(e^{t_-^2/4\varepsilon}), & t \in [t_-, \rho\sqrt{\varepsilon}], \\ O(e^{t_-^2/4\varepsilon})(1 + \frac{t}{\sqrt{\varepsilon}}), & t \in [\rho\sqrt{\varepsilon}, t_+], \end{cases} \qquad (27.104)$$

Now, we give the estimation of I_2. Again by (27.69) and the results in Lemmas 27.2 and 27.3, we have

$$\begin{aligned} I_2 =& O(\varepsilon^{-1/2} e^{t_-^2/4\varepsilon}) |U(a, \frac{t_-}{\sqrt{\varepsilon}})| \Big/ |U(a, -\frac{t_+}{\sqrt{\varepsilon}})| \\ &\times \int_{t_-}^{t} |U(a, -\frac{s}{\sqrt{\varepsilon}})| \left(|U(a-1, -\frac{s}{\sqrt{\varepsilon}})| + \sqrt{\varepsilon} |U(a, -\frac{s}{\sqrt{\varepsilon}})| \right) ds. \end{aligned} \tag{27.105}$$

Using (27.42), (27.105) can be simplified to

$$\begin{aligned} I_2 =& O(\varepsilon^{-1/2} e^{t_-^2/4\varepsilon}) \int_{t_-}^{t} |U(a, -\frac{s}{\sqrt{\varepsilon}})| \\ &\times \left(|U(a-1, -\frac{s}{\sqrt{\varepsilon}})| + \sqrt{\varepsilon} |U(a, -\frac{s}{\sqrt{\varepsilon}})| \right) ds. \end{aligned} \tag{27.106}$$

Similar to the consideration of I_1, the behavior of the parabolic cylinder function will be used to estimate I_2 in three separate cases: (i) $t \in [t_-, -\rho\sqrt{\varepsilon}]$, (ii) $t \in [-\rho\sqrt{\varepsilon}, \rho\sqrt{\varepsilon}]$, and (iii) $t \in [\rho\sqrt{\varepsilon}, t_+]$. For $t \in [t_-, -\rho\sqrt{\varepsilon}]$, (27.40) gives

$$I_2 = O(\varepsilon^{-1/2} e^{t_-^2/4\varepsilon}) \int_{t_-}^{t} \left[\left(\frac{|s|}{\sqrt{\varepsilon}} \right)^{-2a} + \sqrt{\varepsilon} \left(\frac{|s|}{\sqrt{\varepsilon}} \right)^{-2a-1} \right] e^{-s^2/2\varepsilon} ds. \tag{27.107}$$

By an integration by parts, one can show that as $\varepsilon \to 0$

$$\int_{t_-}^{t} \left(\frac{|s|}{\sqrt{\varepsilon}} \right)^{l} e^{-s^2/2\varepsilon} ds \sim \sqrt{\varepsilon} \left(\frac{|s|}{\sqrt{\varepsilon}} \right)^{l-1} e^{-t^2/2\varepsilon} \tag{27.108}$$

for any constant l and all $t \in [t_-, -\rho\sqrt{\varepsilon}]$. Substituting (27.108) into (27.107), we obtain

$$\begin{aligned} I_2 &= O(e^{(t_-^2 - 2t^2)/4\varepsilon}) \left[\left(\frac{|t|}{\sqrt{\varepsilon}} \right)^{-2a-1} + \sqrt{\varepsilon} \left(\frac{|t|}{\sqrt{\varepsilon}} \right)^{-2a-2} \right] \\ &= O\left(\left(\frac{|t|}{\sqrt{\varepsilon}} \right)^{-2a-1} e^{(t_-^2 - 2t^2)/4\varepsilon} \right) = O\left(e^{t_-^2/4\varepsilon} \right) \end{aligned} \tag{27.109}$$

for $t \in [t_-, -\rho\sqrt{\varepsilon}]$. In the interval $-\rho\sqrt{\varepsilon} \le t \le \rho\sqrt{\varepsilon}$, the same argument for (27.101) can be used to show that

$$I_2 = O\left(e^{t_-^2/4\varepsilon} \right). \tag{27.110}$$

For $t \in [\rho\sqrt{\varepsilon}, t_+]$, I_2 can again be written as

$$\begin{aligned} I_2 =& O(\varepsilon^{-1/2} e^{t_-^2/4\varepsilon}) \int_{t_-}^{\rho\sqrt{\varepsilon}} |U(a, -\frac{s}{\sqrt{\varepsilon}})| \\ & \times \left(|U(a-1, -\frac{s}{\sqrt{\varepsilon}})| + \sqrt{\varepsilon}|U(a, -\frac{s}{\sqrt{\varepsilon}})| \right) ds \\ & + O(\varepsilon^{-1/2} e^{t_-^2/4\varepsilon}) \int_{\rho\sqrt{\varepsilon}}^{t} |U(a, -\frac{s}{\sqrt{\varepsilon}})| \\ & \times \left(|U(a-1, -\frac{s}{\sqrt{\varepsilon}})| + \sqrt{\varepsilon}|U(a, -\frac{s}{\sqrt{\varepsilon}})| \right) ds. \end{aligned} \tag{27.111}$$

By using (27.110), it can be shown that the first term on the right-hand side is $O(e^{t_-^2/4\varepsilon})$. Applying (27.42) to the integrand of the second term in (27.111), we have

$$\begin{aligned} & O(\varepsilon^{-\frac{1}{2}} e^{t_-^2/4\varepsilon}) \int_{\rho\sqrt{\varepsilon}}^{t} |U(a, -\frac{s}{\sqrt{\varepsilon}})| \\ & \qquad \times \left(|U(a-1, -\frac{s}{\sqrt{\varepsilon}})| + \sqrt{\varepsilon}|U(a, -\frac{s}{\sqrt{\varepsilon}})| \right) ds \\ & = O(\varepsilon^{-\frac{1}{2}} e^{t_-^2/4\varepsilon}) \int_{\rho\sqrt{\varepsilon}}^{t} \left[\left(\frac{s}{\sqrt{\varepsilon}}\right)^{a-3/2} + \sqrt{\varepsilon} \left(\frac{s}{\sqrt{\varepsilon}}\right)^{a-1/2} \right] e^{s^2/2\varepsilon} ds \\ & = O(\varepsilon^{-\frac{1}{2}} e^{t_-^2/4\varepsilon}) \int_{\rho\sqrt{\varepsilon}}^{t} \left[\left(\frac{s}{\sqrt{\varepsilon}}\right)^{2(a-1)} + \sqrt{\varepsilon} \left(\frac{s}{\sqrt{\varepsilon}}\right)^{2a-1} \right] e^{s^2/2\varepsilon} ds \end{aligned} \tag{27.112}$$

An integration by parts gives

$$\int_{\rho\sqrt{\varepsilon}}^{t} \left(\frac{s}{\sqrt{\varepsilon}}\right)^{l} e^{s^2/2\varepsilon} ds = O\left(\sqrt{\varepsilon} \left(\frac{t}{\sqrt{\varepsilon}}\right)^{l-1} e^{t^2/2\varepsilon} \right) \quad \text{as } \varepsilon \to 0, \tag{27.113}$$

for any constant l and all $t \in [\rho\sqrt{\varepsilon}, t_+]$. Substituting (27.113) into (27.112) yields

$$\begin{aligned} & O(\varepsilon^{-\frac{1}{2}} e^{t_-^2/4\varepsilon}) \int_{\rho\sqrt{\varepsilon}}^{t} |U(a, -\frac{s}{\sqrt{\varepsilon}})| \\ & \qquad \times \left(|U(a-1, -\frac{s}{\sqrt{\varepsilon}})| + \sqrt{\varepsilon}|U(a, -\frac{s}{\sqrt{\varepsilon}})| \right) ds \\ & = O(e^{(t_-^2+2t^2)/4\varepsilon}) \left[(\frac{t}{\sqrt{\varepsilon}})^{2a-3} + \sqrt{\varepsilon}(\frac{t}{\sqrt{\varepsilon}})^{2(a-1)} \right] \\ & = O\left((\frac{t}{\sqrt{\varepsilon}})^{2(a-1)} e^{(t_-^2+2t^2)/4\varepsilon} \right). \end{aligned} \tag{27.114}$$

Thus, for $t \in [\rho\sqrt{\varepsilon}, t_+]$, we obtain

$$I_2 = O(e^{t_-^2/4\varepsilon}) \left[1 + \left(\frac{t}{\sqrt{\varepsilon}}\right)^{2(a-1)} e^{t^2/2\varepsilon}\right]. \qquad (27.115)$$

Therefore, for $\varepsilon \to 0$, the estimation of I_2 is given by

$$I_2 = \begin{cases} O(e^{t_-^2/4\varepsilon}), & t \in [t_-, \rho\sqrt{\varepsilon}], \\ O(e^{t_-^2/4\varepsilon}) \left[1 + \left(\frac{t}{\sqrt{\varepsilon}}\right)^{2(a-1)} e^{t^2/2\varepsilon}\right], & t \in [\rho\sqrt{\varepsilon}, t_+], \end{cases} \qquad (27.116)$$

In a similar manner, one can show that the estimations of I_3 and I_4 are given by

$$I_3 = \begin{cases} O(e^{t_+^2/4\varepsilon}) \left[1 + \left(\frac{|t|}{\sqrt{\varepsilon}}\right)^{2(a-1)} e^{t^2/2\varepsilon}\right], & t \in [t_-, -\rho\sqrt{\varepsilon}], \\ O(e^{t_+^2/4\varepsilon}) & t \in [-\rho\sqrt{\varepsilon}, t_+], \end{cases} \qquad (27.117)$$

and

$$I_4 = \begin{cases} O(e^{t_+^2/4\varepsilon}) \left(1 + \frac{|t|}{\sqrt{\varepsilon}}\right), & t \in [t_-, -\rho\sqrt{\varepsilon}], \\ O(e^{t_+^2/4\varepsilon}) & t \in [-\rho\sqrt{\varepsilon}, t_+], \end{cases} \qquad (27.118)$$

Substituting (27.104), (27.116), (27.117) and (27.118) into (27.92), we obtain

$$I(\varepsilon, t) = (|W_1(t)| + |W_2(t)|)O(\sqrt{\varepsilon}) + |W_2(t)| \left(\frac{|t|}{\sqrt{\varepsilon}}\right)^{2(a-1)} e^{t^2/2\varepsilon} O(\sqrt{\varepsilon})$$

for $t \in [t_-, -\rho\sqrt{\varepsilon}]$,

$$I(\varepsilon, t) = (|W_1(t)| + |W_2(t)|)O(\sqrt{\varepsilon})$$

for $t \in [-\rho\sqrt{\varepsilon}, \rho\sqrt{\varepsilon}]$, and

$$I(\varepsilon, t) = (|W_1(t)| + |W_2(t)|)O(\sqrt{\varepsilon}) + |W_1(t)| \left(\frac{t}{\sqrt{\varepsilon}}\right)^{2(a-1)} e^{t^2/2\varepsilon} O(\sqrt{\varepsilon})$$

for $t \in [\rho\sqrt{\varepsilon}, t_+]$. Similarly, it follows from (27.93)

$$J(\varepsilon, t) = (|W_1'(t)| + |W_2'(t)|)O(\sqrt{\varepsilon}) + |W_2'(t)| \left(\frac{t}{\sqrt{\varepsilon}}\right)^{2(a-1)} e^{t^2/2\varepsilon} O(\sqrt{\varepsilon})$$

for $t \in [t_-, -\rho\sqrt{\varepsilon}]$,

$$J(\varepsilon, t) = (|W_1'(t)| + |W_2'(t)|)O(\sqrt{\varepsilon})$$

for $t \in [-\rho\sqrt{\varepsilon}, \rho\sqrt{\varepsilon}]$, and

$$J(\varepsilon, t) = (|W_1'(t)| + |W_2'(t)|)O(\sqrt{\varepsilon}) + |W_1'(t)| \left(\frac{t}{\sqrt{\varepsilon}}\right)^{2(a-1)} e^{t^2/2\varepsilon} O(\sqrt{\varepsilon})$$

for $t \in [\rho\sqrt{\varepsilon}, t_+]$. By Lemma 27.4

$$|W_2(t)| \left(\frac{|t|}{\sqrt{\varepsilon}}\right)^{2(a-1)} e^{t^2/2\varepsilon} = |W_1(t)| O(\frac{\varepsilon}{t^2}) = |W_1(t)| O(1)$$

and

$$|W_2'(t)| \left(\frac{|t|}{\sqrt{\varepsilon}}\right)^{2(a-1)} e^{t^2/2\varepsilon} = |W_1'(t)| O(1)$$

for $t \in [t_-, -\rho\sqrt{\varepsilon}]$, and

$$|W_1(t)| \left(\frac{t}{\sqrt{\varepsilon}}\right)^{2(a-1)} e^{t^2/2\varepsilon} = |W_2(t)| O(\frac{\varepsilon}{t^2}) = |W_2(t)| O(1)$$

and

$$|W_1'(t)| \left(\frac{|t|}{\sqrt{\varepsilon}}\right)^{2(a-1)} e^{t^2/2\varepsilon} = |W_2'(t)| O(1)$$

for $t \in [\rho\sqrt{\varepsilon}, t_+]$.

Hence, as $\varepsilon \to 0$,

$$I(\varepsilon, t) = (|W_1(t)| + |W_2(t)|) O(\sqrt{\varepsilon}) \tag{27.119}$$

and

$$J(\varepsilon, t) = (|W_1'(t)| + |W_2'(t)|) O(\sqrt{\varepsilon}) \tag{27.120}$$

uniformly for all $t \in [t_-, t_+]$, thus proving the lemma. □

Lemma 27.5 implies that there exist constants ε_0 and γ_1 such that for $0 < \varepsilon \le \varepsilon_0$,

$$\begin{aligned} \int_{t_-}^{t_+} &|G(t,s)| \left(|W_1'(s)| + |W_2'(s)| + |W_1(s)| + |W_2(s)|\right) ds \\ &\le \left(|W_1(t)| + |W_2(t)|\right) \gamma_1 \sqrt{\varepsilon} \end{aligned} \tag{27.121}$$

and

$$\begin{aligned} \int_{t_-}^{t_+} &|\frac{\partial G}{\partial t}(t,s)| \left(|W_1'(s)| + |W_2'(s)| + |W_1(s)| + |W_2(s)|\right) ds \\ &\le \left(|W_1'(t)| + |W_2'(t)|\right) \gamma_1 \sqrt{\varepsilon}. \end{aligned} \tag{27.122}$$

Now, we shall prove that the sequence of functions $\{U_n(t)\}$ and $\{U_n'(t)\}$ defined in (27.85) and (27.87) are convergent, and this will be done by showing inductively that for $0 < \varepsilon \le \varepsilon_0, t \in [t_-, t_+]$ and $n = 0, 1, 2 \cdots$,

$$|U_n(t) - U_{n-1}(t)| \le k_0(|W_1(t)| + |W_2(t)|)(k_0\sqrt{\varepsilon})^n, \tag{27.123}$$

$$|U_n'(t) - U_{n-1}'(t)| \le k_0(|W_1'(t)| + |W_2'(t)|)(k_0\sqrt{\varepsilon})^n, \tag{27.124}$$

where $k_0 = \max\{\gamma_0\gamma_1, |\lambda|, |\mu|\}$, $U_{-1}(t) = 0$ and $U_{-1}'(t) = 0$.

For $n = 0$, it is evident from (27.84) and (27.86) that

$$|U_0(t)| \le k_0(|W_1(t)| + |W_2(t)|) \tag{27.125}$$

and

$$|U_0'(t)| \le k_0(|W_1'(t)| + |W_2'(t)|). \tag{27.126}$$

Assume that (27.123) and (27.124) hold for $n = k$; we need to prove that they also hold for $n = k + 1$. From (27.85) and (27.87), we have

$$|U_{k+1}(t) - U_k(t)| \le \int_{t_-}^{t_+} |G(t,s)| \Big[|f(s)||U_k'(s) - U_{k-1}'(s)| + |g(s)||U_k(s) - U_{k-1}(s)| \Big] ds$$

and

$$|U_{k+1}'(t) - U_k'(t)| \le \int_{t_-}^{t_+} \left| \frac{\partial G}{\partial t}(t,s) \right| \Big[|f(s)||U_k'(s) - U_{k-1}'(s)| + |g(s)||U_k(s) - U_{k-1}(s)| \Big] ds.$$

By (27.37), the last two inequalities are simplified to

$$|U_{k+1}(t) - U_k(t)| \le \gamma_0 \int_{t_-}^{t_+} |G(t,s)|[|U_k'(s) - U_{k-1}'(s)| + |U_k(s) - U_{k-1}(s)|]ds$$

and

$$|U_{k+1}'(t) - U_k'(t)| \le \gamma_0 \int_{t_-}^{t_+} \left| \frac{\partial G}{\partial t}(t,s) \right| [|U_k'(s) - U_{k-1}'(s)| + |U_k(s) - U_{k-1}(s)|]ds$$

From the induction hypothesis, it follows that

$$|U_{k+1}(t) - U_k(t)| \le k_0\gamma_0(k_0\sqrt{\varepsilon})^k \int_{t_-}^{t_+} |G(t,s)|(|W_1'(s)| + |W_2'(s)| + |W_1(s)| + |W_2(s)|)ds$$

and

$$|U_{k+1}'(t) - U_k'(t)| \le k_0\gamma_0(k_0\sqrt{\varepsilon})^k \int_{t_-}^{t_+} \left| \frac{\partial G}{\partial t}(t,s) \right| (|W_1'(s)| + |W_2'(s)| + |W_1(s)| + |W_2(s)|)ds$$

Substituting (27.121) and (27.122) in the above inequalities gives

$$|U_{k+1}(t) - U_k(t)| \leq k_0(|W_1(t)| + |W_2(t)|)(k_0\sqrt{\varepsilon})^{k+1}$$

and

$$|U'_{k+1}(t) - U'_k(t)| \leq k_0(|W'_1(t)| + |W'_2(t)|)(k_0\sqrt{\varepsilon})^{k+1}.$$

Therefore, (27.123) and (27.124) hold for $n = k + 1$, and hence for all $n = 0, 1, 2, \cdots$.

The estimates in (27.123) and (27.124) in fact show that for $0 < \varepsilon \leq \varepsilon_0$, the series

$$U(t) = \lim_{n\to\infty} U_n(t) = U_0(t) + \sum_{n=1}^{\infty}[U_n(t) - U_{n-1}(t)] \tag{27.127}$$

defines a continuously differentiable function $U(t)$ in $[t_-, t_+]$. Furthermore,

$$U'(t) = \lim_{n\to\infty} U'_n(t).$$

Since

$$\sum_{n=1}^{\infty} |U_n(t) - U_{n-1}(t)| \leq \frac{k_0^2\sqrt{\varepsilon}}{1 - k_0\sqrt{\varepsilon}}(|W_1(t)| + |W_2(t)|),$$

we have from (27.127)

$$U(t) = \lambda W_1(t) + \mu W_2(t) + (|W_1(t)| + |W_2(t)|)O(\sqrt{\varepsilon}). \tag{27.128}$$

Taking limit in (27.85), we find that $U(t)$ satisfies the integral equation (27.59). Thus, $U(t)$ is the unique solution of the boundary value problem (27.26) and (27.27).

Using (27.11), (27.13) and (27.28), $U(t)$ in (27.128) can be expressed in terms of the original variables x and y. Indeed, an asymptotic solution to the boundary value problem (27.1)-(27.2) when $\alpha = 1$ (cf. (27.6)) is given by

$$y(x) = A\exp\left[\int_x^{x_-} h(\tau)d\tau\right] W_1(t(x)) + B\exp\left[\int_x^{x_+} h(\tau)d\tau\right] W_2(t(x)) + (|W_1(t(x))| + |W_2(t(x))|)O(\sqrt{\varepsilon}), \tag{27.129}$$

where $W_1(t)$ and $W_2(t)$ are the two linearly independent solutions of (27.38) satisfying boundary conditions (27.47) and (27.48),

$$t(x) = \begin{cases} -\left[2\alpha\int_0^x a(\tau)d\tau\right]^{1/2}, & x < 0 \\ \left[2\alpha\int_0^x a(\tau)d\tau\right]^{1/2}, & x > 0 \end{cases}$$

and

$$h(x) = \frac{b(x)}{a(x)} - \frac{\beta a(x)}{t^2},$$

cf. (27.20) and (27.25). In summary, we have the following theorem.

Theorem 27.1. *Suppose that $a(x)$ and $b(x)$ satisfy the smoothness conditions imposed at the beginning of Section 2, and that $\alpha = 1$ and $\beta \neq 1, 2, \cdots$ in* (27.10)*. Then there exists a constant $\varepsilon_0 > 0$ such that for $0 < \varepsilon \leq \varepsilon_0$, the boundary value problem* (27.1) *and* (27.2) *has a unique solution $y(x)$, whose asymptotic behavior is given in* (27.129).

If we make use of the behaviors of $W_1(t(x))$ and $W_2(t(x))$ given in Section 4, then (27.129) can further be written as

$$\begin{aligned} y(x) \sim & \frac{\Gamma(1-\beta)}{\sqrt{2\pi}} \varepsilon^{-\beta/2} \exp\left[-\frac{1}{2\varepsilon}\int_0^x a(\tau)d\tau\right] \\ & \times \left\{ A|t_-|^\beta \exp\left[\int_x^{x_-} h(\tau)d\tau\right] U(-\beta + \frac{1}{2}, \frac{t(x)}{\sqrt{\varepsilon}}) \right. \\ & \left. + Bt_+^\beta \exp\left[\int_x^{x_+} h(\tau)d\tau\right] U(-\beta + \frac{1}{2}, -\frac{t(x)}{\sqrt{\varepsilon}}) \right\}. \end{aligned} \tag{27.130}$$

One can now readily see the resemblance of this asymptotic solution with that of Bender and Orszag given in (27.7). Recall that $U(a,t) = D_{-a-\frac{1}{2}}(t)$; see Abramowitz and Stegun (1989, p. 687).

From (27.130), we can deduce some nonuniform formulas for the solution. First, by (27.23) and (27.25), it is easily found that

$$\exp\left[\int_x^{x_-} h(\tau)d\tau\right] = \left(\frac{|t(x)|}{|t_-|}\right)^\beta \exp\left[-\int_{x_-}^x \frac{b(\tau)}{a(\tau)}d\tau\right], \quad x < 0,$$

and

$$\exp\left[\int_x^{x_+} h(\tau)d\tau\right] = \left(\frac{|t(x)|}{|t_+|}\right)^\beta \exp\left[\int_x^{x_+} \frac{b(\tau)}{a(\tau)}d\tau\right], \quad x > 0.$$

Next, the one-to-one relationship between t and x shows that $t/\sqrt{\varepsilon} \to \pm\infty$ if and only if $x/\sqrt{\varepsilon} \to \pm\infty$. Thus, by making use of the behavior of the parabolic cylinder functions, we obtain that

$$y(x) \sim A \exp\left[-\int_{x_-}^x \frac{b(\tau)}{a(\tau)}d\tau\right], \quad x/\sqrt{\varepsilon} \to -\infty, \tag{27.131}$$

$$y(x) \sim B \exp\left[\int_x^{x_+} \frac{b(\tau)}{a(\tau)}d\tau\right], \quad x/\sqrt{\varepsilon} \to \infty, \tag{27.132}$$

and

$$y(x) = O(\varepsilon^{-\beta/2}), \ x = O(\sqrt{\varepsilon}). \tag{27.133}$$

Formulas (27.131)-(27.133) reveal that an internal boundary layer occurs when $x = O(\sqrt{\varepsilon})$.

By using (27.22), it can be shown that (27.130) implies (27.7) when $x = O(\sqrt{\varepsilon})$. When $x/\sqrt{\varepsilon} \to \pm\infty$, (27.7) agrees with (27.131) and (27.132). Thus, Bender and Orszag's result in (27.7) is correct.

27.6 Olver's method

Olver's approach in [6] is to first make the change of variable

$$w(x) = \exp(\frac{1}{2\varepsilon}\int_0^x a(t)dt)y(x) \tag{27.134}$$

so that the term involving y' in (27.1) is eliminated. The function $w(x)$ satisfies an equation of the form

$$w'' = \{\varepsilon^{-2}f(\varepsilon, x) + g(\varepsilon, x)\}w, \tag{27.135}$$

where both $f(\varepsilon, x)$ and $g(\varepsilon, x)$ are smooth functions of ε and x, and $f(\varepsilon, x)$ has only two zeros $x = z_1(\varepsilon)$ and $x = z_2(\varepsilon)$, with $z_1(\varepsilon)$ being nonpositive and increasing in ε, and $z_2(\varepsilon)$ being nonnegative and decreasing in ε; further

$$z_1(0) = z_2(0) = 0. \tag{27.136}$$

Then he used the Liouville-Green transformation

$$\dot{x}^2 f(\varepsilon, x) = \zeta^2 - \alpha^2(\varepsilon), \ \ W(\zeta) = \dot{x}^{-1/2}w(x) \tag{27.137}$$

to transform (27.135) into the equation

$$\frac{d^2W}{d\zeta^2} = \{\varepsilon^{-2}(\zeta^2 - \alpha^2(\varepsilon)) + \psi(\alpha(\varepsilon), \zeta)\}W, \tag{27.138}$$

where $\psi(\alpha(\varepsilon), \zeta)$ is a continuous function of $\alpha(\varepsilon)$ and ζ. In (27.137), the dot denotes differentiation with respect to ζ, and $\alpha(\varepsilon)$ is a nonnegative real number chosen to make $\zeta = -\alpha(\varepsilon)$ correspond to $x = z_1(\varepsilon)$ and $\zeta = \alpha(\varepsilon)$ correspond to $x = z_2(\varepsilon)$; thus $\alpha(\varepsilon)$ is given by

$$\int_{z_1}^{z_2} \{-f(\varepsilon, x)\}^{1/2}dx = \int_{-\alpha(\varepsilon)}^{\alpha(\varepsilon)} [\alpha^2(\varepsilon) - \zeta^2]^{1/2}d\zeta = \frac{1}{2}\pi\alpha^2(\varepsilon). \tag{27.139}$$

Calculation shows

$$\alpha^2(\varepsilon) = k\varepsilon + O(\varepsilon^2) \ \text{ as } \ \varepsilon \to 0, \tag{27.140}$$

where $k = 2\beta - 1 > 0$ and $k \neq 2n - 1, n = 1, 2, \cdots$. Finally, the applied his earlier result on the coalescence of two turning points (see Olver (1975)) to write down the asymptotic formula

$$y(x) = \exp\left[-\frac{1}{2\varepsilon}\int_0^x a(t)dt\right]\dot{x}^{1/2}(\zeta)W(\zeta), \tag{27.141}$$

where

$$W(\zeta)\sim\begin{cases}[A_1 + A_2\cot\beta\pi]U(\zeta\sqrt{2/\varepsilon}) - A_2\csc\beta\pi U(-\zeta\sqrt{2/\varepsilon}),\ 0 \le x \le x_+,\\ [A_3 + A_4\cot\beta\pi]U(-\zeta\sqrt{2/\varepsilon}) - A_4\csc\beta\pi U(\zeta\sqrt{2/\varepsilon}),\ x_- \le x \le 0,\end{cases} \tag{27.142}$$

and

$$U(\zeta\sqrt{2/\varepsilon}) = U(-\beta + \frac{1}{2}, \zeta\sqrt{2/\varepsilon}),\ U(-\zeta\sqrt{2/\varepsilon}) = U(-\beta + \frac{1}{2}, -\zeta\sqrt{2/\varepsilon}).$$

The coefficients A_i, $i = 1, 2, 3, 4$, in (27.142) all depend on ε, and are determined by the boundary conditions corresponding to (27.2) and the matching conditions of $W(\zeta)$ and $W'(\zeta)$ at $\zeta = 0$; see Olver (1978, (5.23)-(5.27)). The expressions for the coefficients are very complicated. Furthermore, the Liouviue-Green variable ζ defined in (27.137) cannot be expressed explicitly in terms of the original independent variable x. Thus, it is nearly impossible to write down a simple asymptotic formula such as (27.7) or (27.130) from Olver's result (27.141).

27.7 Exercises

1. Consider the differential equation

$$\varepsilon y'' + (1 + \varepsilon)y' + y = 0,\quad y(0) = 0,\ y(1) = 1. \tag{27.143}$$

Show that the exact solution is

$$y(x) = \frac{e^{-x} - e^{-x/\varepsilon}}{e^{-1} - e^{-1/\varepsilon}}. \tag{27.144}$$

Plot the curves of this function with $\varepsilon = 0.025$ and $\varepsilon = 0.1$. Note that for very small ε, the solution $y(x)$ is slowly varying for $\varepsilon \ll x \le 1$. On the small interval $0 \le x \le O(\varepsilon)$ $(\varepsilon \to 0^+)$, the solution undergoes an abrupt and rapid change. The small interval of rapid change is called a *boundary layer.* (The notation $0 \le x \le O(\varepsilon)$ means that the thickness of the boundary layer is proportional to ε as $\varepsilon \to 0^+$.) The region of slow variation of $y(x)$ is called the *outer region* and the boundary layer region is called the *inner region.*

2. (a) The outer limit of the solution in (27.143) is

$$y_{out}(x) = \lim_{\varepsilon \to 0^+} y(x) = e^{1-x} \tag{27.145}$$

for $\delta \leq x \leq 1$. Show that $|y(x) - y_{out}(x)|$ is exponentially small.
(b) For fixed x, by letting $\varepsilon \to 0^+$ the equation in (27.143) simplifies to

$$y'_{out}(x) + y_{out}(x) = 0, \tag{27.146}$$

which has (27.145) as its solution. Since the outer limit of (27.143) is a first-order differential equation, its solution cannot be required to satisfy both boundary conditions $y(0) = 0$ and $y(1) = 1$. Hence we can only require the outer limit of (27.143) to satisfy $y(1) = 1$. That is, the small-ε limit of the solution is not everywhere close to the solution of the unperturbed differential equation (27.146); i.e., equation (27.143) with $\varepsilon = 0$. Therefore, the boundary layer problem (27.143) is a singular perturbation problem. The singular behavior occurs because the highest-order derivative in (27.143) disappeared when $\varepsilon = 0$.

3. (a) We now consider the *inner limit* of the solution in (27.144); that is, let "$\varepsilon \to 0^+$ with $x = O(\varepsilon)$". In this case, x lies inside the boundary layer at $x = 0$. For convenience, we let $x = \varepsilon X$ with X finite and fixed. Show from (27.144)

$$y_{in}(x) = Y_{in}(X) = \lim_{\varepsilon \to 0^+} y(\varepsilon X) = e - e^{1-X}. \tag{27.147}$$

(b) Let $Y(X) \equiv y(x)$. Use

$$\varepsilon \frac{dy}{dx} = \frac{dY}{dX}, \quad \varepsilon^2 \frac{d^2y}{dx^2} = \frac{d^2Y}{dX^2} \tag{27.148}$$

to show that the differential equation in (27.143) can be written as

$$\frac{d^2Y}{dX^2} + (1+\varepsilon)\frac{dY}{dX} + \varepsilon Y = 0. \tag{27.149}$$

Taking the inner limit $\varepsilon \to 0^+$, we obtain

$$\frac{d^2Y_{in}}{dX^2} + \frac{dY_{in}}{dX} = 0. \tag{27.150}$$

Note that the inner limit function (27.147) does satisfy equation (27.150) and the boundary condition $Y_{in}(0) = 0$.

4. (a) How did we know that the inner and outer limits of the differential equation (27.143) have a common region of validity? To see this, we now assume a complete expansion for the solution of equation (27.143)

$$y_{out}(x) \sim \sum_{n=0}^{\infty} y_n(x)\varepsilon^n, \quad \varepsilon \to 0^+, \tag{27.151}$$

and restate the boundary condition $y(1) = 1$. Show that

$$y_0(1) = 1, \quad y_1(1) = y_2(1) = y_3(1) = \cdots = 0. \qquad (27.152)$$

Note that $y_{out}(x)$ in (27.151) is not the same as $y_{out}(x)$ in (27.145); $y_{out}(x)$ in (27.145) is the first term $y_0(x)$ in (27.151).
Substituting (27.151) into (27.143) and collecting powers of ε gives a sequence of differential equations

$$y_0' + y_0 = 0, \quad y_0(1) = 1,$$
$$y_n' + y_n = -y_{n-1}'' - y_{n-1}', \quad y_n(1) = 0, \ n \geq 1.$$

Show that the solution to these equations is

$$y_0 = e^{1-x}, \quad y_n = 0, \ n \geq 1. \qquad (27.153)$$

Note that the leading-order outer solution, $y_{out}(x) = e^{1-x}$, is correct to *all* orders in perturbation theory. This is the reason why in the outer region, $x \gg \varepsilon$, the difference between $y(x)$ and $y_{out}(x)$ is at most exponentially small (subdominant): $|y - y_{out}| = O(\varepsilon^n)$ for *all* n as $\varepsilon \to 0^+$.
(b) Next, we assume a similar expansion for the inner solution

$$Y_{in}(X) \sim \sum_{n=0}^{\infty} Y_n(X)\varepsilon^n, \quad \varepsilon \to 0^+, \qquad (27.154)$$

with the boundary condition $Y_{in}(0) = y(0) = 0$; i.e.,

$$Y_n(0) = 0, \quad n \geq 0. \qquad (27.155)$$

Show that

$$Y_0'' + Y_0' = 0, \quad Y_0(0) = 0,$$
$$Y_n'' + Y_n' = -Y_{n-1}' - Y_{n-1}, \quad Y_n(0) = 0, \ n \geq 1.$$

Furthermore, show that

$$Y_0(X) = A_0(1 - e^{-X}), \quad Y_n(X) = \int_0^X [A_n e^{-t} - Y_{n-1}(t)]dt, \quad n \geq 1, \qquad (27.156)$$

where A_n are arbitrary integration constants.
(c) To see whether this inner solution matches asymptotically, order by order in powers of ε, to $y_{out}(x)$, we substitute $x = \varepsilon X$ into y_{out} in (27.153) and expand in powers of ε:

$$y_{out}(x) = e^{1-x} = e\left(1 - \varepsilon X + \frac{\varepsilon^2 X^2}{2!} - \frac{\varepsilon^3 X^3}{3!} + \cdots\right). \qquad (27.157)$$

Returning to (27.156), we take X large ($X \to \infty$) and obtain $Y_0(X) \sim A_0$ ($X \to \infty$). Comparing with the first term in (27.157), we have $A_0 = e$. With Y_0 now known, show from (27.156) that

$$Y_1(X) = (A_1 + A_0)(1 - e^{-X}) - eX.$$

By asymptotic matching with y_{out} [comparing $Y_1(X)$, as $X \to \infty$, with the second term in (27.157)], show that $A_1 = -e$ and $Y_1(X) = -eX$. Similarly, show that $Y_n(X) = e[(-1)^n/n!]X^n$. Thus, the full inner expansion is

$$Y_{in}(X) = e\sum_{n=0}^{\infty} \frac{(-1)^n \varepsilon^n X^n}{n!} - e^{1-X} = e^{1-\varepsilon X} - e^{1-X}. \tag{27.158}$$

It is now clear that the inner expansion is a valid asymptotic expansion not only for values of X inside the boundary layer [$X = O(1)$] but also for large values of X [$X = O(\varepsilon^{-\alpha})$, $0 < \alpha < 1$]. At the same time the expansion for $y_{out}(x)$ is valid for $\varepsilon \ll x \le 1$ ($\varepsilon \to 0^+$). [$y_{out}(x)$ is not valid for $x = O(\varepsilon)$ because it does not satisfy the boundary condition $y(0) = 0$; nor does it have the boundary layer term e^{1-X} which is present in $Y_{in}(X)$.] We conclude that *to all orders in powers of* ε it is possible to match asymptotically the inner and outer expansions because they have a common region of validity: $\varepsilon \ll x \ll 1$ (as $\varepsilon \to 0^+$).

5. Consider the boundary layer problem

$$\varepsilon y'' + (1 + x)y' + y = 0, \quad y(0) = 0, \ y(1) = 1. \tag{27.159}$$

We wish to find an asymptotic approximation for its solution.

(a) We first use the outer solution in the form of a series in powers of ε:

$$y_{out}(x) \sim y_0(x) + y_1(x)\varepsilon + y_2(x)\varepsilon^2 + \cdots, \quad \varepsilon \to 0^+. \tag{27.160}$$

Since the coefficient function $1 + x$ is positive for $0 \le x \le 1$, we expect a boundary layer only at $x = 0$. Thus, the outer solution must satisfy the boundary condition $y_{out}(1) = 1$ and we must require that

$$y_0(1) = 1, \quad y_n(1) = 0, \ n \ge 1. \tag{27.161}$$

Substituting (27.160) in (27.159) and equating coefficients of like powers of ε, we obtain

$$(1 + x)y_0' + y_0 = 0, \quad (1 + x)y_n' + y_n = -y_{n-1}'', \ n \ge 1.$$

(b) Show that the solution to these equations which satisfy the boundary conditions (27.161) are

$$\begin{aligned} y_0(x) =& 2(1+x)^{-1}, \\ y_1(x) =& 2(1+x)^{-3} - \frac{1}{2}(1+x)^{-1}, \\ y_2(x) =& 6(1+x)^{-5} - \frac{1}{2}(1+x)^{-3} - \frac{1}{4}(1+x)^{-1}. \end{aligned}$$

Thus,

$$\begin{aligned} y_{out}(x) \sim & \frac{2}{1+x} + \left[\frac{2}{(1+x)^3} - \frac{1}{2(1+x)}\right]\varepsilon \\ & + \left[\frac{6}{(1+x)^5} - \frac{1}{2(1+x)^3} - \frac{1}{4(1+x)}\right]\varepsilon^2 + \cdots . \end{aligned} \tag{27.162}$$

This completes the determination of the outer solution to the second order in powers of ε.

(c) As expected, the outer solution (27.162) does not satisfy the boundary condition $y(0) = 1$, so a boundary layer at $x = 0$ is necessary, and we expect the thickness δ of the boundary layer to be $O(\varepsilon)$. Thus, we introduce the inner variable $X = x/\varepsilon$ and $Y_{in}(X) \equiv y_{in}(x)$. In terms of these variables, (27.159) becomes

$$\frac{d^2Y_{in}}{dX^2} + (1+\varepsilon X)\frac{dY_{in}}{dX} + \varepsilon Y_{in} = 0. \tag{27.163}$$

If we represent $Y_{in}(X)$ as a perturbation series in power of ε,

$$Y_{in}(X) \sim Y_0(X) + Y_1(X)\varepsilon + Y_2(X)\varepsilon^2 + \cdots, \quad \varepsilon \to 0^+, \tag{27.164}$$

show that the boundary condition $Y_{in}(0) = 1$ is translated into the series of boundary conditions

$$Y_0(0) = 1, \quad Y_1(0) = Y_2(0) = \cdots = 0. \tag{27.165}$$

Also, show that substituting (27.164) into (27.163) and equating coefficients of like powers of ε converts (27.163) into a sequence of second-order constant-coefficient equations:

$$\frac{d^2Y_0}{dX^2} + \frac{dY_0}{dX} = 0, \quad \frac{d^2Y_n}{dX^2} + \frac{dY_n}{dX} = -X\frac{dY_{n-1}}{dX} - Y_{n-1}, \; n \geq 1. \tag{27.166}$$

Show that the following functions are the solutions of (27.166) which satisfy the boundary condition (27.165):

$$\begin{aligned}Y_0(X) =&1 + A_0(e^{-X} - 1),\\ Y_1(X) =&- X + A_0\left(-\frac{1}{2}X^2e^{-X} + X\right) + A_1(e^{-X} - 1),\\ Y_2(X) =&X^2 - 2X + A_0\left(\frac{1}{8}X^4e^{-X} - X^2 + 2X\right)\\ &+ A_1\left(-\frac{1}{2}X^2e^{-X} + X\right) + A_2(e^{-X} - 1).\end{aligned}$$

This completes the determination of the inner solution to second order in powers of ε.

(d) To determine the constants A_0, A_1, $A_2, \cdots$, we match the inner and outer solutions asymptotically. The matching requires that the intermediate limits $[\varepsilon \to 0^+,\ x \to 0^+,\ X = x/\varepsilon \to \infty]$ of the inner and outer solutions agree. First, we consider the leading-order (zero-order in ε) match. As $\varepsilon \to 0^+$ in the outer solution (27.162), show that $y_{out}(x) = 2 + O(\varepsilon, x)$ as $\varepsilon \to 0^+$ and $x \to 0^+$, where the symbol $O(a, b, c, \cdots)$ means $O(a) + O(b) + O(c) + \cdots$. The error term $O(\varepsilon, x)$ indicates that we have neglected powers of ε higher than zero and that we have expanded the solution in a Taylor series in powers of x and have neglected all but the first term. On the other hand, show that in the limit $X \to \infty$ and $\varepsilon X = x \to 0^+$ the inner solution becomes

$$Y_{in}(X) = 1 - A_0 + O(\varepsilon X),\quad \varepsilon X \to 0^+,\ X \to \infty, \tag{27.167}$$

where the correction of order εX comes from the term $\varepsilon Y_1(X)$ in the expansion of $Y_{in}(X)$. The inner and outer solutions are required to match to lowest order [that is, $y_{out}(x) \sim Y_{in}(X)$ in the intermediate limit]. Thus, $2 = 1 - A_0$ or $A_0 = -1$. This completes the leading-order match. Observe that the match occurs for values of x for which $1 \ll X = x/\varepsilon$ as well as $x \ll 1$. Thus, the size of the overlap region is determined to leading order as $\varepsilon \ll x \ll 1$ $(\varepsilon \to 0^+)$.

(e) We match to first order in ε. By expanding the outer solution (27.160) and keeping terms of order ε and x but discarding terms of order ε^2, x^2 and εx, show that we have

$$y_{out}(x) = 2 - 2x + \frac{3}{2}\varepsilon + O(\varepsilon^2, \varepsilon x, x^2),\quad x \to 0^+,\ \varepsilon \to 0^+. \tag{27.168}$$

We also expand the inner solution (27.164) for $X \gg 1$ (x outside the boundary layer), but neglect terms of order $\varepsilon^2 X^2 = x^2$, $\varepsilon^2 X = \varepsilon x$ and

ε^2. Show that

$$Y_{in}(X) = 1 - A_0 - \varepsilon X + \varepsilon A_0 X - \varepsilon A_1 + O(x^2), \quad \varepsilon X \to 0^+, \ X \to \infty. \tag{27.169}$$

By matching (27.168) and (27.169), show that $A_0 = -1$, which we already know, and $A_1 = -3/2$.
Note that for a successful match to first order, it is necessary to neglect terms of order ε^2, x^2 and εx, compared with x and ε. If we had retained some of these terms, we would have found that there is no way to choose the constants of integration to make (27.168) agree with (27.169). Since matching now requires that $x^2 \ll \varepsilon$, the size of the overlap region is *smaller* than it was in the leading-order match. Its extent is $\varepsilon \ll x \ll \varepsilon^{1/2}$ $(\varepsilon \to 0^+)$.
To perform a second-order match, we expand the inner and outer solutions and neglect terms of order ε^3, $\varepsilon^2 x$, εx^2 and x^3. Show that

$$y_{out}(x) = 2 - 2x + 2x^2 + \frac{\varepsilon(3-11x)}{2} + \frac{21\varepsilon^2}{4} + O(\varepsilon^3, \varepsilon^2 x, \varepsilon x^2, x^3),$$

as $\varepsilon \to 0^+$ and $x \to 0^+$, and

$$\begin{aligned} Y_{in}(X) =& 1 - A_0 - \varepsilon X + \varepsilon A_0 X - \varepsilon A_1 \\ &- \varepsilon^2 X^2 - 2\varepsilon^2 X - \varepsilon^2 A_0 X^2 + 2\varepsilon^2 A_0 X + \varepsilon^2 A_1 X - \varepsilon^2 A_2 \\ &+ O(\varepsilon^3, \varepsilon^2 x, \varepsilon x^2, x^3), \end{aligned}$$

as $\varepsilon X \to 0^+$ and $X \to \infty$. For matching, show that

$$A_0 = -1, \quad A_1 = -\frac{3}{2}, \quad A_2 = -\frac{21}{4}.$$

Since one must neglect x^3 compared with ε^2 to obtain a match in second order, it follows that the size of the matching region, $\varepsilon \ll x \ll \varepsilon^{2/3}$ $(\varepsilon \to 0^+)$, is smaller than it was in the first-order calculation.
(f) The process of asymptotic matching may be carried out to all orders in powers of ε. As the order of perturbation theory increases, the size of the matching region continues to shrink. In nth order the common region of validity of the inner and outer expansions is $\varepsilon \ll x \ll \varepsilon^{n/(n+1)}$ $(\varepsilon \to 0^+)$. However, the extent of the matching region in terms of the inner variable is still infinite as $\varepsilon \to 0^+$: $1 \ll X \ll \varepsilon^{-1/(n+1)}$ $(\varepsilon \to 0^+)$.
Once the boundary layer solution is determined, one may construct a uniform approximation to the solution $y(x)$ by using the formula $y_{unif}(x) = y_{out}(x) + y_{in}(x) - y_{match}(x)$, where $y_{match}(x)$ is the expansion of either the inner or outer approximation in the matching region.

Show that to the third order in ε, we have

$$\begin{aligned} y_{unif,3}(x) = &\left(\frac{2}{1+x} - e^{-X}\right) + \varepsilon\left[\frac{4-(1+x)^2}{2(1+x)^3} + \frac{X^2-3}{2}e^{-X}\right] \\ &+ \varepsilon^2\left[\frac{24-2(1+x)^2-(1+x)^4}{4(1+x)^5} - \frac{X^4-6X^2+42}{8}e^{-X}\right] \\ &+ \varepsilon^3\left[\frac{480-24(1+x)^2-4(1+x)^4-5(1+x)^6}{16(1+x)^7}\right. \\ &\qquad \left. + \frac{3X^6-27X^4+378X^2-3898}{144}e^{-X}\right]. \end{aligned}$$

This expression is a uniform approximation to $y(x)$ over the entire region $0 \le x \le 1$:

$$|y_{unif,3}(x) - y(x)| = O(\varepsilon^4), \quad \varepsilon \to 0^+.$$

Also show that

$$|y'_{unif,3}(x) - y'(x)| = O(\varepsilon^3), \quad |y''_{unif,3}(x) - y''(x)| = O(\varepsilon^2),$$

as $\varepsilon \to 0^+$.

Chapter 28

A Boundary-Value Problem with Spurious Solutions

28.1 Introduction

Even more than 50 years ago, Carrier and Pearson (1968, pp. 202-205) already observed some strange phenomena in the solutions obtained by using the matched asymptotics. They used the simple example

$$\begin{cases} \varepsilon u''(x) + u^2(x) = 1, \quad -1 < x < 1, \\ u(-1) = u(1) = 0 \end{cases} \tag{28.1}$$

to show that the method of matched asymptotics can lead to spurious solutions. More precisely, they first show that an approximate solution to (28.1) is given by

$$\widetilde{u}_1(x, \varepsilon) = -1 + \frac{12e^{p_1}}{(1+e^{p_1})^2} + \frac{12e^{p_2}}{(1+e^{p_2})^2}, \tag{28.2}$$

where

$$p_1 = \sqrt{\frac{2}{\varepsilon}}(1+x) + 2\ln(\sqrt{3}+\sqrt{2}) \tag{28.3}$$

and

$$p_2 = \sqrt{\frac{2}{\varepsilon}}(1-x) + 2\ln(\sqrt{3}+\sqrt{2}). \tag{28.4}$$

Let $x_0 \in (-1, 1)$, $\zeta = (x - x_0)/\sqrt{\varepsilon}$, and

$$q(\zeta) = \frac{12e^{\sqrt{2}\zeta}}{(1+e^{\sqrt{2}\zeta})^2}. \tag{28.5}$$

They also noted that

$$\widetilde{u}(x, \varepsilon) = \widetilde{u}_1(x, \varepsilon) + q(\zeta) \tag{28.6}$$

appears to be a valid approximate solution if

$$1 - |x_0| \geq \sqrt{\varepsilon}. \tag{28.7}$$

However, by using the phase-plane method, they demonstrated that $\widetilde{u}(x, \varepsilon)$ in (28.6) could not be an approximate solution unless $x_0 = 0$. Thus, for most values of x_0, the solutions given in (28.6) cannot be valid, and they are referred to as the "spurious solutions". The function $q(\zeta)$ in (28.5) behaves like a spike near x_0 for sufficiently small ε.

Lange (1983) considered a situation in which there are n internal boundary layers, i.e., n spikes. For any n points satisfying

$$-1 < x_1 < x_2 < \cdots < x_n < 1, \tag{28.8}$$

the function

$$\widetilde{u}(x, \varepsilon) = \widetilde{u}_1(x, \varepsilon) + \sum_{j=1}^{n} q\left(\frac{x - x_j}{\sqrt{\varepsilon}}\right) \tag{28.9}$$

could be a valid approximate solution to (28.1), where $\widetilde{u}_1$ and q are given, respectively, by (28.2) and (28.5). Lange provided an explanation why the method of matched asymptotics fails to locate these internal boundary layers, and presented a modification of the method. Instead of Poincaré-type perturbation series, he used expansions in powers of exponentially small functions in ε, and allowed the high-order terms in these expansions to contain growing exponentials in the stretched variable, e.g., ζ in (28.5).

More recently, MacGillivray (1997) re-visited the problem (28.1) with a single spike. His idea is basically similar to that of Lange. By incorporating exponentially small terms in the matching, he was able to show directly that (28.6) could be a valid approximate solution only if $x_0 = 0$, without using the phase-plane method.

All results obtained in Carrier and Pearson (1968); Lange (1983); MacGillivray (1997) are based on heuristic arguments, and it is generally recognized that it is very difficult to establish them on mathematically rigorous grounds; see a comment in the first paragraph of Lange (1983). In this chapter, we are concerned with three mathematically more fundamental questions; namely, (i) in what sense does the function $\widetilde{u}_1$ in (28.2) approximate a solution of (28.1), (ii) for a given $\varepsilon > 0$, how many spikes can there be in the interval $(-1, 1)$ so that the function $\widetilde{u}$ in (28.9) is a valid approximate solution to (28.1), (iii) how well does $\widetilde{u}$ in (28.9) approximate a solution of (28.1)? An expected answer to question (i) is that there is a solution u_1 of (28.1) satisfying

$$u_1(x, \varepsilon) = \widetilde{u}_1(x; \varepsilon) + O(\varepsilon)$$

for all $x \in [-1,1]$; cf. Bender and Orszag (1978, p. 479, Problem 9.2). It turns out that we have the much better result

$$u_1(x,\varepsilon) = \widetilde{u}_1(x;\varepsilon) + O(e^{-\sqrt{2/\varepsilon}}) \tag{28.10}$$

for all $x \in [-1,1]$, the meaning of (28.10) being that there exist $\varepsilon_0 > 0$ and $K > 0$ such that for $0 < \varepsilon < \varepsilon_0$,

$$|u_1(x,\varepsilon) - \widetilde{u}_1(x,\varepsilon)| \le K e^{-\sqrt{2/\varepsilon}} \tag{28.11}$$

for all $x \in [-1,1]$. As to question (ii), i.e., the number N of (internal) spikes, we have the estimate

$$N \le \frac{0.41}{\sqrt{\varepsilon}} + 1. \tag{28.12}$$

Our answer to question (iii) is similar to that of question (i); namely, there exists a solution u of (28.1) satisfying

$$u(x,\varepsilon) = \widetilde{u}_1(x,\varepsilon) + \sum_{j=1}^{n} q\left(\frac{x-x_j}{\sqrt{\varepsilon}}\right) + O\left[\exp\left(-\frac{1}{n+1}\sqrt{\frac{2}{\varepsilon}}\right)\right] \tag{28.13}$$

for all $x \in [-1,1]$, where $\widetilde{u}_1$ and q are given in (28.2) and (28.5), respectively.

The material in this chapter is taken from Ou and Wong (2003). For a study of a nonlinear equation more general than that in (28.1), see Ou and Wong (2004). The material in the Exercises section is taken from Carrier and Pearson (1968, Chapter 18).

28.2 The shooting method

Our approach to the BVP (28.1) is through the shooting method. Thus, instead of (28.1), we consider the associated initial value problem

$$\begin{cases} \varepsilon u''(x) + u^2(x) = 1, \\ u(-1) = 0, \ u'(-1) = k, \end{cases} \tag{28.14}$$

where k is a real number. It is readily verified that this problem has a unique solution, which can be extended to a maximal interval $[-1, x_{max})$, where either $x_{max} = \infty$ or $|u(x)| + |u'(x)| \to \infty$ as $x \to x_{max}^+$.

As in Carrier and Pearson (1968, p. 203), we let

$$u(x) = -1 + v(x)$$

so that (28.1) is converted into the equivalent problem

$$\begin{cases} \varepsilon v'' - 2v + v^2 = 0, \\ v(-1) = v(1) = 1. \end{cases} \tag{28.15}$$

The initial value problem (28.14) now becomes

$$\begin{cases} \varepsilon v'' - 2v + v^2 = 0, \\ v(-1) = 1, \quad v'(-1) = k. \end{cases} \tag{28.16}$$

Multiplying both sides of the differential equation in (28.15) by v' and integrating from -1 to x, we obtain

$$\frac{\varepsilon}{2}(v')^2 - v^2 + \frac{v^3}{3} = \frac{\varepsilon}{2}k^2 - \frac{2}{3}. \tag{28.17}$$

For convenience, we define

$$F(v) := -v^2 + \frac{v^3}{3}, \tag{28.18}$$

which is negative for $v \in (0,3)$. In particular, we have $F(1) = -2/3$. The graph of $F(v)$ is depicted in Figure 28.1.

The results in the following lemmas will play an important role in the proof of our main theorems.

Lemma 28.1. *Let $u(x,k)$ denote the solution of the initial value problem (28.14). Depending on the values of k, we obtain the following results.*

(a) If $k < -\sqrt{4/3\varepsilon}$, then there exists a finite $x_{max} > -1$ such that the solution $u(x,k)$ exists and is decreasing for $x \in [-1, x_{max})$ with $\lim_{x \to x_{max}^+} u(x,k) = -\infty$.

(b) If $k > \sqrt{4/3\varepsilon}$, then there exists $x_{max} > \eta > -1$ such that the solution exists on the interval $[-1, x_{max})$ with $u(\eta,k) = 0$ and $\lim_{x \to x_{max}^+} u(x,k) = -\infty$. Moreover, $u(x,k)$ is increasing in $-1 < x < (\eta-1)/2$ and decreasing in $(\eta-1)/2 < x < x_{max}$.

(c) If $k = \pm\sqrt{4/3\varepsilon}$, then

$$u(x,k) = -1 + 3\,\mathrm{sech}^2\left[\mp\frac{x+1}{\sqrt{2\varepsilon}} + \ln(\sqrt{3}+\sqrt{2})\right]. \tag{28.19}$$

(d) If $|k| < \sqrt{4/3\varepsilon}$, then $u(x,k)$ is periodic and intersects the x-axis infinitely many times.

Proof. (a) If $k < -\sqrt{4/3\varepsilon}$, we claim that $v'(x) < 0$ for all $x \geq -1$. Assume the contrary. Then there exists a number $\alpha > -1$ such that $v'(x) < 0$ for all $x \in [-1, \alpha)$ and $v'(\alpha) = 0$. Hence, we have $v(\alpha) < v(-1) = 1$ and $F(v(\alpha)) \leq 0$. This implies that the left-hand side of (28.17) is non-positive at $x = \alpha$, a contradiction to the fact that the right-hand side of (28.17) is positive.

Let $[-1, x_{max})$ be the maximal interval of existence of solution. We shall prove that $x_{max} < \infty$. Otherwise, $v(x) \leq 1$ and $F(v(x)) \leq 0$ for all $x \geq -1$. By (28.17), v' is bounded above by a negative constant, which implies that $v(x)$ decreases to $-\infty$ as $x \to \infty$. In particular, there exists a large $\beta > -1$ such that $v(x) \leq -1$ for all $x \geq \beta$. It again follows from (28.17) that

$$v'(x) \leq -\sqrt{\frac{2}{3\varepsilon}}[-v(x)]^{3/2},$$

for all $x \geq \beta$. Dividing both sides of the inequality by $-[-v(x)]^{3/2}$ and then integrating over (β, x), we obtain

$$[-v(x)]^{-1/2} \geq [-v(\beta)]^{-1/2} + \frac{x - \beta}{\sqrt{6\varepsilon}}$$

for all $x \geq \beta$. As $x \to \infty$, we have $v(x) \to 0$, a contradiction to $v(x) \leq -1$ for all $x \geq \beta$. Hence, $x_{max} < \infty$ and the solution blows up as $x \to x_{max}^+$. On account of (28.17), we have $v(x) \to -\infty$ and $v'(x) \to -\infty$ as $x \to x_{max}^+$.

(b) When $k > \sqrt{4/3\varepsilon}$, we have $v'(-1) > 0$. Denote by $\bar{v}$ the unique real root of $F(v) = \varepsilon k^2/2 - 2/3$. It then follows from (28.17) that $v(x)$ is uniformly bounded above by $\bar{v}$. We claim that v' cannot be always positive. Otherwise, $v(x)$ is strictly increasing in x and remain in the interval $[1, \bar{v}]$. By (28.17), the function $[v'(x)]^2$ is also uniformly bounded above by a finite constant. Hence, the solution exists for all $x \geq -1$. As $x \to \infty$, the increasing and bounded function $v(x)$ converges. In view of (28.17), the positive function $v'(x)$ also converges. The limit $v'(\infty)$ should be zero. Consequently, the limit $v(\infty) = \bar{v}$. This together with the differential equation (28.16) implies that the function $v''(x)$ converges to a negative constant $\bar{v}(2 - \bar{v})/\varepsilon < 0$, which contradicts $v'(\infty) = 0$. Thus, there exists $\eta_0 > -1$ such that $v'(x) > 0$ for all $x \in [-1, \eta_0)$ and $v'(\eta_0) = 0$. By (28.17), we have $v(\eta_0) = \bar{v}$. Let $\eta = 2\eta_0 + 1$ and consider the initial value problem

$$v'(x) = w(x), \quad \varepsilon w'(x) = 2v(x) - [v(x)]^2, \quad x \in [\eta_0, \eta]$$
$$v(\eta_0) = \bar{v}, \quad w(\eta_0) = 0.$$

We obtain from the uniqueness of solution to the above initial value problem that (exercise)

$$v(x) = v(2\eta_0 - x), \; w(x) = -v'(2\eta_0 - x), \quad x \in [\eta_0, \eta]. \tag{28.20}$$

In particular, $v'(x) < 0$ for $x \in (\eta_0, \eta]$, $v(\eta) = v(-1) = 1$ and $v'(\eta) = -v'(-1) = -k < -\sqrt{4/3\varepsilon}$. A similar argument as in (a) shows that $v(x) \to -\infty$ as $x \to x_{max}^+$ for some finite $x_{max} > \eta$.

(c) When $k = \pm\sqrt{4/3\varepsilon}$, it is easy to verify that (exercise) (28.17) has the solutions

$$v = 3\,\mathrm{sech}^2\left[\mp\frac{x+1}{\sqrt{2\varepsilon}} + \ln(\sqrt{3}+\sqrt{2})\right]. \tag{28.21}$$

Consequently, (28.19) are the solutions to the initial value problem (28.14).

(d) When $|k| < \sqrt{4/3\varepsilon}$, the equation $F(v) = \varepsilon k^2/2 - 2/3$ has three distinct real roots $v_1 < 0 < v_2 < 1 < v_3 < 3$. The initial value problem (28.17) is equivalent to the planar system

$$\begin{aligned} &v'(x) = w(x), \quad \varepsilon w'(x) = 2v(x) - [v(x)]^2, \quad x \geq 0, \\ &v(-1) = 1, \quad w(-1) = k, \end{aligned}$$

with trajectory given by the equation (28.17); see Figure 28.2. This trajectory contains two parts that are symmetric about the horizontal line $v' = 0$. On the upper part, we have $v'(x) \geq 0$ and $v(x)$ increases from v_2 to v_3; while on the lower part, we have $v'(x) \leq 0$ and $v(x)$ decreases from v_3 to v_2. The solution is periodic in x with period (exercise)

$$\int_{v_2}^{v_3} \frac{\sqrt{6\varepsilon}dv}{\sqrt{(v-v_1)(v-v_2)(v_3-v)}}.$$

□

Lemma 28.2. *If* $|k| \geq \sqrt{4/3\varepsilon}$ *and* $\varepsilon < 1/3$, *then the boundary value problem* (28.1) *has no solution with* $u'(-1) = k$.

Proof. Assume to the contrary that u is a solution to the boundary value problem (28.1) with $u'(-1) = k$. Then $v = 1 + u$ is a solution to the boundary value problem (28.15) with $v'(-1) = k$. Moreover, v satisfies the initial value problem (28.16) for $x \in [-1, 1]$ and $v(1) = 1$. In the following, we shall use a contradiction argument to show that $v(1) = 1$ cannot hold.

If $k < -\sqrt{4/3\varepsilon}$, we have shown in Lemma 28.1 that $v(x)$ is strictly decreasing for $x > -1$, and hence $v(1) < 1$, a contradiction.

If $k = -\sqrt{4/3\varepsilon}$, then by (28.21),

$$v = 3\,\mathrm{sech}^2\left[\frac{x+1}{\sqrt{2\varepsilon}} + \ln(\sqrt{3}+\sqrt{2})\right]$$

is also strictly decreasing for $x > -1$, and hence $v(1) < 1$, a contradiction.

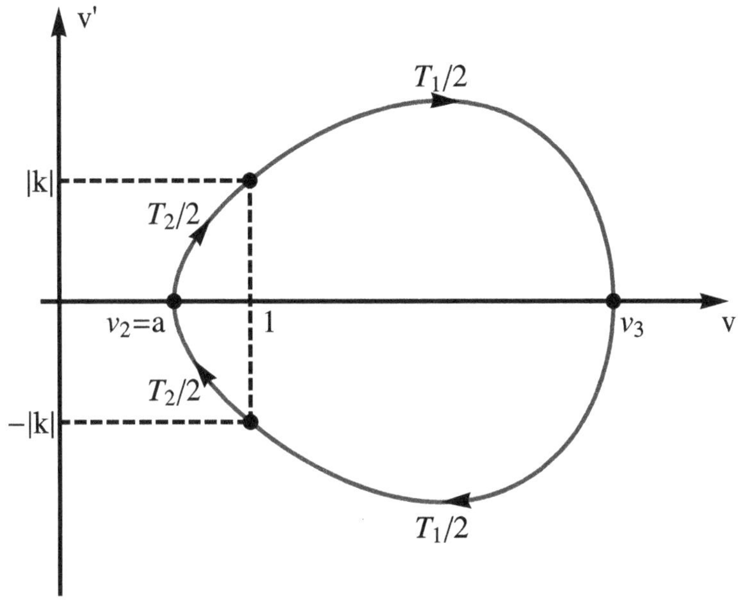

Fig. 28.1: The trajectory of periodic solution: $\varepsilon(v')^2/2 + F(v) = F(a) = \varepsilon k^2/2 - 2/3$ in the vv'-phase plane, where $a \in (0, 1]$ and $\varepsilon k^2 < 4/3$. The functions $T_1(a)/2$ and $T_2(a)/2$ denote the shortest travel times from $(1, |k|)$ to $(v_3, 0)$ and from $(v_2, 0)$ to $(1, |k|)$, respectively.

If $k > \sqrt{4/3\varepsilon}$, then the equation $F(v) = \varepsilon k^2/2 - 2/3$ has a unique real root $\bar{v}$. On account of (28.17), the solution $v(x)$ increases from 1 to $\bar{v}$ for $x \in [-1, -1 + \tau]$ and decreases from $\bar{v}$ to 1 for $x \in [-1 + \tau, -1 + 2\tau]$, where

$$\tau = \int_1^{\bar{v}} \frac{\sqrt{\varepsilon/2}dv}{\sqrt{F(\bar{v}) - F(v)}}.$$

Moreover, $v(-1 + 2\tau) = 1$ and $v'(-1 + 2\tau) = -k < -\sqrt{4/3\varepsilon}$. If we can show that $-1 + 2\tau < 1$, then as shown in Lemma 28.1, we have $v'(x) < 0$ for all $x \in [-1 + 2\tau, 1]$, and hence $v(1) < 1$, a contradiction. It remains to show that $\tau < 1$. Recall the definition of $F(v)$ in (28.18). It is easily seen that (exercise)

$$\frac{F(\bar{v}) - F(v)}{\bar{v} - v} \geq \frac{F(\bar{v}) - F(1)}{\bar{v} - 1} \geq \frac{\bar{v} - 1}{6}. \tag{28.22}$$

Hence,

$$\tau \leq \int_1^{\bar{v}} \frac{\sqrt{3\varepsilon}dv}{\sqrt{(\bar{v} - v)(\bar{v} - 1)}} = 2\sqrt{3\varepsilon} < 1.$$

The above argument holds by taking the limit $k \to \sqrt{4/3\varepsilon}$. In this case, $\bar{v} = 3$ and we still have $\tau < 1$ and $v(1) < 1$, a contradiction. □

From the above lemma, it is clear that to have $u(1) = 0$, we must choose k in the interval $(-\sqrt{4/3\varepsilon}, \sqrt{4/3\varepsilon})$, in which case by Lemma 28.1 the solution to the initial value problem (28.14) is periodic and intersects the x-axis infinitely many times.

We consider only the case $k \geq 0$, because the case $k < 0$ is entirely similar. Let x_1 be the first point to the right of -1 such that $u(x_1) = 0$ and $u(x) > 0$ for $-1 < x < x_1$, and x_2 be the first point to the right of x_1 such that $u(x_2) = 0$ and $u(x) < 0$ for $x_1 < x < x_2$. We denote by T_1 and T_2, respectively, the lengths of the interval $[-1, x_1]$ and $[x_1, x_2]$; see Figure 28.2.

For $0 < a \leq 1$, we define a one-to-one mapping from a to $k^2 \in [0, \frac{4}{3\varepsilon})$ such that

$$\frac{\varepsilon}{2}k^2 - \frac{2}{3} = \frac{a^3}{3} - a^2. \tag{28.23}$$

From (28.17), we have

$$\begin{aligned} \frac{\varepsilon}{2}(v')^2 &= \frac{1}{3}a^3 - a^2 - \frac{1}{3}v^3 + v^2 \\ &= \frac{1}{3}(v - v_1)(v - v_2)(v_3 - v), \end{aligned} \tag{28.24}$$

where $v_2 = a$,

$$v_1 = \frac{1}{2}(3 - a - \sqrt{9 + 6a - 3a^2}), \quad v_3 = \frac{1}{2}(3 - a + \sqrt{9 + 6a - 3a^2}). \tag{28.25}$$

It can be easily verified that for $0 < a \leq 1$, we also have

$$1 - \sqrt{3} \leq v_1 < 0 \quad \text{and} \quad 1 + \sqrt{3} \leq v_3 < 3. \tag{28.26}$$

Note that v_1, v_2 and v_3 all satisfy

$$\frac{a^3}{3} - a^2 = \frac{v_1^3}{3} - v_1^2 = \frac{v_2^3}{3} - v_2^2 = \frac{v_3^3}{3} - v_3^2. \tag{28.27}$$

From (28.24), it follows as in the proof of Lemma 28.2 that

$$T_1(a) = \sqrt{2\varepsilon} \int_1^{v_3} \frac{dv}{\sqrt{v^2 - \frac{1}{3}v^3 - a^2 + \frac{1}{3}a^3}} \tag{28.28}$$

and

$$T_2(a) = \sqrt{2\varepsilon} \int_a^1 \frac{dv}{\sqrt{v^2 - \frac{1}{3}v^3 - a^2 + \frac{1}{3}a^3}}, \tag{28.29}$$

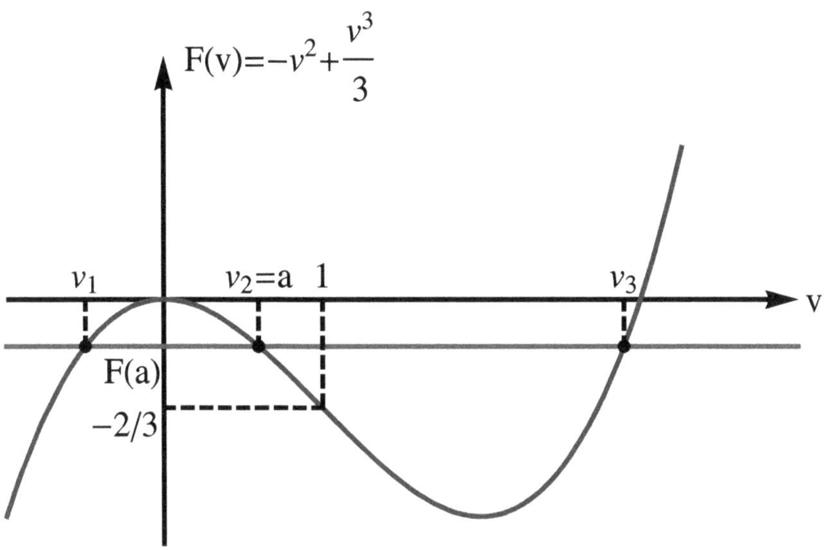

Fig. 28.2: The graph of $F(v) = v^3/3 - v^2$ and three roots of $F(v) = F(a) = \varepsilon k^2/2 - 2/3$ where $a \in (0, 1]$.

where T_1 and T_2 are the lengths of the intervals $(-1, x_1)$ and (x_1, x_2); see Figure 28.2. Obviously, $T_1(a)$ and $T_2(a)$ are continuous in a, and in fact differentiable with respect to a.

Lemma 28.3. *$T_1(a)$ is increasing and $T_2(a)$ is decreasing for $0 < a < 1$. In addition, we have*

$$\frac{d(T_1 + T_2)}{da} < 0, \quad \frac{d^2T_1}{da^2} > 0, \quad \textit{and} \quad \frac{d^2(T_1 + T_2)}{da^2} > 0 \tag{28.30}$$

for $0 < a < 1$.

Proof. In view of (28.27), we can replace a by v_3 in (28.28). Expanding $v^2 - \frac{1}{3}v^3$ into a Taylor series at $v = 1$ gives

$$v^2 - \frac{1}{3}v^3 = \frac{2}{3} + (v - 1) - \frac{1}{3}(v - 1)^3.$$

Similarly, we have

$$v_3^2 - \frac{1}{3}v_3^3 = \frac{2}{3} + (v_3 - 1) - \frac{1}{3}(v_3 - 1)^3.$$

Upon subtracting the second equation from the first, (28.28) becomes

$$T_1(a) = \sqrt{2\varepsilon} \int_1^{v_3} \left\{ (v - v_3) - \frac{1}{3}[(v - 1)^3 - (v_3 - 1)^3] \right\}^{-1/2} dv.$$

Now we make the change of variable $t = (v-1)/(v_3-1)$. This yields

$$T_1(a) = \sqrt{2\varepsilon}\int_0^1 \sqrt{\frac{1}{1-t}}\left\{-\frac{1}{v_3-1}+\frac{(v_3-1)}{3}(1+t+t^2)\right\}^{-1/2} dt.$$

Straightforward differentiation shows that

$$\frac{dT_1}{dv_3} < 0 \text{ and } \frac{d^2T_1}{dv_3^2} > 0.$$

Because

$$\frac{dv_3}{da} = \frac{1}{2}\left(-1+\frac{3-3a}{\sqrt{9+6a-3a^2}}\right) < 0 \tag{28.31}$$

and d^2v_3/da^2 is also negative for $0 < a < 1$, we have

$$\frac{dT_1}{da} = \frac{dT_1}{dv_3}\cdot\frac{dv_3}{da} > 0$$

and

$$\frac{d^2T_1}{da^2} = \frac{d^2T_1}{dv_3^2}\cdot\left(\frac{dv_3}{da}\right)^2 + \frac{dT_1}{dv_3}\cdot\frac{d^2v_3}{da^2} > 0$$

for $0 < a < 1$.

In a similar manner, we make the change of variable $t = (1-v)/(1-a)$ in (28.29), we obtain

$$T_2(a) = \sqrt{2\varepsilon}\int_0^1 \sqrt{\frac{1}{1-t}}\left\{\frac{1}{1-a}-\frac{1-a}{3}(1+t+t^2)\right\}^{-1/2} dt. \tag{28.32}$$

Again, straightforward differentiation shows that dT_2/da is negative for $0 < a < 1$.

From (28.24), (28.28) and (28.29), it follows that

$$T_1(a) + T_2(a) = \sqrt{6\varepsilon}\int_a^{v_3} \frac{dv}{\sqrt{(v-a)(v-v_1)(v_3-v)}}.$$

Set $v = a\cos^2\theta + v_3\sin^2\theta$. Then $v - a = (v_3 - a)\sin^2\theta, v_3 - v = (v_3 - a)\cos^2\theta$, and

$$\frac{dv}{\sqrt{(v-a)(v-v_1)(v_3-v)}} = \frac{2}{\sqrt{v-v_1}}d\theta.$$

Adding the two equations in (28.25) yields $v_1 + v_3 = 3 - a$. Hence

$$v - v_1 = a\cos^2\theta + v_3\sin^2\theta - (3 - a - v_3).$$

Changing the integration variable from v to θ gives

$$T_1(a) + T_2(a) = \sqrt{6\varepsilon}\int_0^{\pi/2} \frac{2}{\sqrt{-3+a(1+\cos^2\theta)+v_3(1+\sin^2\theta)}}d\theta.$$

For $\theta \in [0, \pi/2]$ and $a \in (0,1)$, it can be easily verified that

$$1 + \cos^2\theta + \frac{dv_3}{da}(1 + \sin^2\theta) > 0,$$

because $-1/2 \le dv_3/da < 0$; see (28.31). Now straightforward differentiation shows that

$$\frac{d(T_1 + T_2)}{da} < 0 \quad \text{and} \quad \frac{d^2(T_1 + T_2)}{da^2} > 0$$

for $a \in (0,1)$. In the last inequality, the result $d^2v_3/da^2 < 0$ has been used for $a \in (0,1)$. This completes the proof of the lemma. □

From (28.28) and (28.29), we also have

$$T_1(0) = \sqrt{6\varepsilon}\int_1^3 \frac{dv}{v\sqrt{3-v}} = 2\sqrt{2}\ln(\sqrt{3}+\sqrt{2})\sqrt{\varepsilon} \approx 3.242\sqrt{\varepsilon}, \tag{28.33}$$

$$T_1(1) = \sqrt{6\varepsilon}\int_1^{1+\sqrt{3}} \frac{dv}{\sqrt{(v-1+\sqrt{3})(v-1)(1+\sqrt{3}-v)}} \approx 4.88\sqrt{\varepsilon}, \tag{28.34}$$

$$T_2(1) = 0 \quad \text{and} \quad \lim_{a\to 0} T_2(a) = \infty. \tag{28.35}$$

To prove the second result in (28.35), we observe that the function $-a^2 + \frac{1}{3}a^3$ is decreasing in $0 < a < 1$; cf. Figure 28.1. Hence, from (28.29), we have

$$T_2(a) \ge \sqrt{6\varepsilon}\int_a^1 \frac{dv}{v\sqrt{3-v}}.$$

The integral on the right becomes infinite when a tends to 0.

Remark 28.1. For small values of ε, $T_1(a) \le T_1(1) \approx 4.88\sqrt{\varepsilon} < 2$. From this, it follows that the BVP (28.1) does not have a solution, which is positive in the entire interval $(-1,1)$. This explains why Carrier and Pearson chose the outer solution to be -1, instead of 1, in their matching process; see (28.2).

The discussion for the case $-\sqrt{4/3\varepsilon} < k < 0$ is similar, and $v(x)$ is again oscillatory about the straight line $v = 1$, except that it now starts at $x = -1$ with a negative slope. Hence, the graph of $v(x)$ is below the horizontal line $v = 1$ in the interval $(-1, x_1)$, where x_1 is the first point to the right of -1 at which $v(x)$ meets the line $v = 1$ again. Equivalently, we have $u(x) < 0$ in $(-1, x_1)$. Let x_2 denote the first point to the right of x_1 such that $u(x_2) = 0$ and $u(x) > 0$ for $x_1 < x < x_2$. Unlike before, we now reverse the order, and denote by T_2 and T_1 the lengths of the intervals $[-1, x_1]$ and $[x_1, x_2]$, respectively. The values of T_1 and T_2 again depend on the slope k. Note that the relationship between a and k given in (28.23) still holds, and formulas (28.28) and (28.29) remain valid. Hence, T_1 and T_2 are actually dependent on the values of k^2.

28.3 Number of spikes and solutions

We first recall that our objective is to use the shooting method to determine whether the BVP (28.1) has solutions. If it has, then we would like to give an estimate for the maximum number of spikes that solutions to this problem can have. This is equivalent to choosing appropriate values of k so that

$$v(1) = 1. \tag{28.36}$$

Because $v(x)$ is periodic and oscillatory about the line $v = 1$, Equation (28.36) can be expressed in the form

$$NT_1(a) + MT_2(a) = 2, \tag{28.37}$$

where M and N are nonnegative integers and represent, respectively, the number of intervals with lengths T_1 and T_2. Now, we look at the arches bounded by the graph of $v(x)$ and the line $v = 1$. Because every arch above $v = 1$ is followed by an arch below $v = 1$ and vice versa, the integers N and M are related in such a way that for every given N, there are only three possible values of M, namely $M = N - 1, N$ or $N + 1$.

From here on, our task is to choose appropriate values of k (or a), and determine possible integers N and M so that (28.37) holds.

For $\varepsilon > 0$, we define

$$N(\varepsilon) := \left[\frac{2}{T_1(1)}\right] \approx \left[\frac{0.41}{\sqrt{\varepsilon}}\right], \tag{28.38}$$

where $[x]$ denotes the largest integer less than or equal to x. The following results provide estimates for the number of spikes and solutions to the boundary value problem (28.1). Here, by "spikes" we mean those, which lie above the line $v = 1$.

Lemma 28.4. *If $N > N(\varepsilon) + 1$, then the boundary value problem* (28.1) *has no solution with N spikes. For any integer $0 \le N \le N(\varepsilon)$, there exists at least one solution for the problem* (28.1) *with N peaks (including the boundary and interval spikes). If $N = N(\varepsilon) + 1$, then either case can happen.*

Proof. When $N > N(\varepsilon)+1$, we have by the monotonicity of T_1+T_2

$$\begin{aligned} NT_1(a)+MT_2(a) &\geq [N(\varepsilon)+2]T_1(a)+[N(\varepsilon)+1]T_2(a) \\ &= [N(\varepsilon)+1][T_1(a)+T_2(a)]+T_1(a) \\ &> [N(\varepsilon)+1][T_1(1)+T_2(1)]. \end{aligned}$$

Because $T_2(1)=0$, by (28.38)

$$NT_1(a)+MT_2(a) > 2;$$

that is, (28.37) fails to hold and the BVP (28.1) has no solution.

For $0 \leq N \leq N(\varepsilon)$, we may choose $M = N+1$. Then

$$NT_1(1)+MT_2(1) = NT_1(1) \leq N(\varepsilon)T_1(1) \leq 2$$

by (28.38), and

$$\lim_{a\to 0}\{NT_1(a)+(N+1)T_2(a)\} = \infty$$

by (28.35). In view of the continuity of the function $NT_1(a)+(N+1)T_2(a)$, there exists at least one point $a_0 \in (0,1]$ such that

$$NT_1(a_0)+(N+1)T_2(a_0) = 2,$$

which means that there exists at least one solution to the BVP (28.1) with N arches with base length T_1 and $N+1$ arches with length T_2.

When $N = N(\varepsilon)+1$, both functions $NT_1(a)+NT_2(a)$ and $NT_1(a)+(N+1)T_2(a)$ are decreasing in a, and their values at $a=1$ satisfy

$$NT_1(1)+NT_2(1) = [N(\varepsilon)+1]T_1(1) > 2$$

and

$$NT_1(1)+(N+1)T_2(1) = [N(\varepsilon)+1]T_1(1) > 2.$$

Thus it is not possible to expect that

$$NT_1(a)+NT_2(a) = 2 \quad \text{or} \quad NT_1(a)+(N+1)T_2(a) = 2$$

for some $a \in (0,1)$. However, if we choose $M = N-1$, then the function

$$NT_1(a)+(N-1)T_2(a) = N(\varepsilon)[T_1(a)+T_2(a)]+T_1(a)$$

is convex with

$$NT_1(0)+MT_2(0) = \infty \quad \text{and} \quad NT_1(1)+MT_2(1) > 2.$$

Let $m := \min_{0<a\leq 1}[NT_1(a)+MT_2(a)]$. Clearly, m depends on ε. If $m > 2$, then $NT_1(a)+MT_2(a) > 2$ for all $a \in (0,1]$ and there is no solution to the BVP (28.1). When $m = 2$, there exists $a_0 \in (0,1]$ such that $NT_1(a_0)+MT_2(a_0) = 2$, i.e., there exists a solution to the BVP (28.1). When $m < 2$, then there exist two distinct points a_i, $i = 1,2$, such that $NT_1(a_i)+MT_2(a_i) = 2$. This completes the proof of the lemma. □

Theorem 28.1. *If $N(\varepsilon) \neq 2/T_1(1)$, then the number $n(\varepsilon)$ of solutions to the BVP (28.1) satisfies*

$$4N(\varepsilon) \leq n(\varepsilon) \leq 4N(\varepsilon) + 2. \tag{28.39}$$

If $N(\varepsilon) = 2/T_1(1)$, then it satisfies

$$4(N(\varepsilon) - 1) + 1 \leq n(\varepsilon) \leq 4(N(\varepsilon) - 1) + 2. \tag{28.40}$$

Proof. We first consider the case $N(\varepsilon) \neq 2/T_1(1)$. By Lemma 28.4, the BVP (28.1) has no solution with N spikes if $N > N(\varepsilon) + 1$. Let us now count the number of solutions with N spikes, where N ranges from 0 to $N(\varepsilon) + 1$.

(1) If $N = 0$, then we should look for solutions with arches whose base length is all $T_2(a)$. This can happen only if the initial slope k is negative. Because $T_2(0) = \infty$ and $T_2(1) = 0$, by the monotonicity and continuity of $T_2(a)$, there exists one and only one value $a \in (0, 1)$ such that (28.37) holds with $M = 1$. This, in turn, means that the BVP (28.1) has a unique solution. (This is the negative solution referred to in Carrier and Pearson (1968).)

(2) If $N = 1$, then we should look for solutions with only one arch whose base length is $T_1(a)$. These arches are exactly the ones which lie above the line $v = 1$. Possible values of M are 0,1, and 2. By Remark 1, the BVP (28.1) has no solution which is entirely above $v = 1$. Hence, $M \neq 0$ and the only possibilities are $M = 1$ and $M = 2$.

When $N = M = 1$, we have $NT_1(a) + MT_2(a) = T_1(a) + T_2(a)$. The function on the right-hand side of this equation is decreasing in $(0, 1)$ by Lemma 28.3. Because $T_1(0) + T_2(0) = \infty$ and $T_1(1) + T_2(1) = T_1(1) < 2$, there is one (and only one) values $a_0 \in (0, 1)$ such that $T_1(a_0) + T_2(a_0) = 2$. In view of (28.23), each value a_0 is associated with two values of k, namely, k_0 and $-k_0$ such that (28.36) holds; that is, when $N = M = 1$, there are two solutions to the BVP (28.1).

When $N = 1$ and $M = 2$, we look for solutions with one arch above the line $v = 1$ and two arches below $v = 1$. Because these arches alternate, the initial value k in (28.16) must be negative. Clearly, $T_1(0) + 2T_2(0) = \infty$ and $T_1(1) + 2T_2(1) = T_1(1) < 2$. By Lemma 28.3, there is one and only one a_0 such that $T_1(a_0) + 2T_2(a_0) = 2$. In view of the restriction that k must be negative, we conclude that there is only one solution to the BVP (28.1). In summary, there are three solutions to (28.1) when $N = 1$.

(3) For $2 \leq N \leq N(\varepsilon)$, possible choices for M are $M = N - 1, N$, and $N + 1$. If $M = N - 1$, then the solution must start at $x = -1$ with a positive

slope k. Again, we have $NT_1(0) + MT_2(0) = \infty$ and $NT_1(1) + MT_2(1) = NT_1(1) < 2$. Although $NT_1(a) + MT_2(a)$ is not necessarily decreasing in $(0,1)$, it is convex by Lemma 28.3. Hence, we can have only one value $a_0 \in (0,1)$ such that $NT_1(a_0)+MT_2(a_0) = 2$. If $M = N$, then the solution will have the same number of arches above and below the line $v = 1$, and it can start at $x = -1$ with either a positive or a negative slope. The values of $NT_1(a) + NT_2(a)$ at the end-points 0 and 1 are again ∞ and less than 2, respectively. Because this is a decreasing function, there exists only one value $a_0 \in (0,1)$ such that $NT_1(a_0)+NT_2(a_0) = 2$. In terms of k, there are two values k_0 and $-k_0$ such that $v(1,\varepsilon,k_0) = v(1,\varepsilon,-k_0) = 1$; i.e., we have two solutions to the BVP (28.1). If $M = N+1$, then the solution will have $N+1$ arches below $v = 1$ and N arches above $v = 1$. Thus, it must start at $x = -1$ with a negative slope k. As before, we have $NT_1(0)+(N+1)T_2(0) = \infty$ and $NT_1(1)+(N+1)T_2(1) = NT_1(1) < 2$. Because $NT_1(a)+(N+1)T_2(a)$ is decreasing in a by Lemma 28.3, there exists one and only one value $a_0 \in (0,1)$ such that $NT_1(a_0) + (N+1)T_2(a_0) = 2$. Furthermore, because k must be negative, there is only one value k_0 such that $v(1,\varepsilon,k_0) = 1$.

Therefore, we conclude that for each N in $2 \le N \le N(\varepsilon)$, there are four solutions to BVP (28.1). Adding them up, we have $4(N(\varepsilon) - 1)$ solutions when N various from 2 to $N(\varepsilon)$.

(4) For $N = N(\varepsilon)+1$, the only choice for M is $M = N-1$. To see this, we note that if $M = N$ or $N+1$, we have

$$NT_1(a) + MT_2(a) \ge N(T_1(a) + T_2(a)) \ge N(T_1(1) + T_2(1))$$

by Lemma 28.3. Because $T_2(1) = 0$ and $N(\varepsilon)+1 > 2/T_1(1)$ by the definition in (28.38), it follows that $NT_1(a)+MT_2(a) > 2$ for all $a \in [0,1]$. Hence, we only need to find out how many solutions there are for the BVP (28.1) when $N = N(\varepsilon)+1$ and $M = N(\varepsilon)$. Note that in the present case, the values of $NT_1(a) + MT_2(a)$ at the endpoints $a = 0$ and $a = 1$ are both greater then 2; i.e., $NT_1(0)+MT_2(0) > 2$ and $NT_1(1)+MT_2(1) > 2$. However, by Lemma 28.3, it is easily seen that this function is convex, i.e.,

$$\frac{d^2}{da^2}[NT_1(a) + MT_2(a)] \ge 0.$$

Therefore, it follows that there are at most two solutions for the BVP (28.1). Adding up all the solutions together, we obtain $1+3+4(N(\varepsilon)-1) \le n(\varepsilon) \le 1+3+4(N(\varepsilon)-1)+2$, thus proving (28.39).

Finally, we come to the proof of (28.40), when $N(\varepsilon) = 2/T_1(1)$ is an integer. As before, we consider the function $NT_1(a)+MT_2(a)$, and choose appropriate values of N, M, and a so that $NT_1(a)+MT_2(a)=2$. First, we claim that $N<N(\varepsilon)+1$. Indeed, if $N\geq N(\varepsilon)+1$ then $M\geq N(\varepsilon)$, and

$$\begin{aligned} NT_1(a)+MT_2(a) &\geq N(\varepsilon)(T_1(a)+T_2(a))+T_1(a) \\ &\geq N(\varepsilon)(T_1(1)+T_2(1))+T_1(a)>2, \end{aligned}$$

because $T_2(1)=0$ and $T_1(a)>0$. In a similar manner as in the earlier part of this proof, it can be shown that when N varies from 0 to $N(\varepsilon)-1$, there are exactly $4(N(\varepsilon)-1)$ solutions for the BVP (28.1).

For the last case $N=N(\varepsilon)=2/T_1(1)$, we also need to consider possible values of M and a such that $NT_1(a)+MT_2(a)=2$. Here, again, M may be equal to $N-1, N$, or $N+1$. When $N=N(\varepsilon)$ and $M=N-1=N(\varepsilon)-1$, the function $NT_1(a)+MT_2(a)$ is convex by Lemma 28.3 and its endpoint values are given by

$$NT_1(0)+MT_2(0)=\infty \quad \text{and} \quad NT_1(1)+MT_2(1)=2.$$

Hence, there are at most two values of $a\in[0,1]$ such that $NT_1(a)+MT_2(a)=2$. The right endpoint $a=1$ happens to correspond to a solution of the BVP (28.15), or equivalently (28.1). Let us denote this solution by v_1, and the other solution by v_2. Note that $a=1$ implies $k=0$. Hence, $v_1=v(x,\varepsilon,0)$.

When $N=M=N(\varepsilon)$, the function $NT_1(a)+MT_2(a)$ is decreasing in a with $NT_1(0)+MT_2(0)=\infty$ and $NT_1(1)+MT_2(1)=2$. Hence, there is only one value of a, namely, $a=1$, that yields a solution to the BVP (28.15). We denote this solution by v_3, and note that $v_3=v(x,\varepsilon,0)$.

When $N=N(\varepsilon)$ and $M=N+1=N(\varepsilon)+1$, the function $NT_1(a)+MT_2(a)$ is also decreasing in a with

$$NT_1(0)+MT_2(0)=\infty \quad \text{and} \quad NT_1(1)+MT_2(1)=2.$$

Hence, there is again only one solution, corresponding to the value $a=1$, which we denote by v_4. Note that $v_4=v(x,\varepsilon,0)$.

Because v_1, v_3 and v_4 are the same solution $v(x,\varepsilon,0)$, the number of solutions to (28.15) or (28.1), in the case when $N=N(\varepsilon)=2/T_1(1)$, is at most 2 and at least 1. Recall that there are $4(N(\varepsilon)-1)$ solutions for $0\leq N\leq N(\varepsilon)-1$. Adding the number $4(N(\varepsilon)-1)$ to the number of solutions when $N=N(\varepsilon)=2/T_1(1)$, we conclude that the number of solutions to (28.1) or, equivalently, (28.15) satisfies $4(N(\varepsilon)-1)+1\leq n(\varepsilon)\leq 4(N(\varepsilon)-1)+2$. This completes the proof of the theorem. □

Remark 28.2. From Theorem 28.1, it follows that $n(\varepsilon)$ has the asymptotic formula

$$n(\varepsilon)\sim\frac{1.64}{\sqrt{\varepsilon}} \quad \text{as} \quad \varepsilon\to 0^+. \tag{28.41}$$

28.4 Proof of formula (28.10)

First, we observe that the approximate solution $\widetilde{u}_1(x,\varepsilon)$ again in (28.2) can be expressed as

$$\widetilde{u}_1(x,\varepsilon) = -1 + 3\,\text{sech}^2\left(\frac{x+1}{\sqrt{2\varepsilon}} + \ln(\sqrt{3}+\sqrt{2})\right) + 3\,\text{sech}^2\left(\frac{1-x}{\sqrt{2\varepsilon}} + \ln(\sqrt{3}+\sqrt{2})\right). \tag{28.42}$$

It is known that there are three other approximate solutions of a similar type; they are

$$\widetilde{u}_2(x,\varepsilon) = -1 + 3\,\text{sech}^2\left(-\frac{x+1}{\sqrt{2\varepsilon}} + \ln(\sqrt{3}+\sqrt{2})\right) + 3\,\text{sech}^2\left(\frac{1-x}{\sqrt{2\varepsilon}} + \ln(\sqrt{3}+\sqrt{2})\right), \tag{28.43}$$

$$\widetilde{u}_3(x,\varepsilon) = -1 + 3\,\text{sech}^2\left(\frac{x+1}{\sqrt{2\varepsilon}} + \ln(\sqrt{3}+\sqrt{2})\right) + 3\,\text{sech}^2\left(-\frac{1-x}{\sqrt{2\varepsilon}} + \ln(\sqrt{3}+\sqrt{2})\right), \tag{28.44}$$

and

$$\widetilde{u}_4(x,\varepsilon) = -1 + 3\,\text{sech}^2\left(-\frac{x+1}{\sqrt{2\varepsilon}} + \ln(\sqrt{3}+\sqrt{2})\right) + 3\,\text{sech}^2\left(-\frac{1-x}{\sqrt{2\varepsilon}} + \ln(\sqrt{3}+\sqrt{2})\right), \tag{28.45}$$

Theorem 28.2. *For small values of ε, there exists exactly four solution $u_i(x,\varepsilon), i = 1,2,\cdots,4$, to the BVP* (28.1)*, which have boundary layer behavior near the endpoints of the interval and no internal spikes. These solutions are uniformly approximated by the functions $\widetilde{u}_i(x,\varepsilon), i = 1,2,\cdots,4$, given in* (28.42)-(28.45) *with an error estimate of the order $O(e^{-\sqrt{2/\varepsilon}})$, that is,*

$$u_i(x,\varepsilon) = \widetilde{u}_i(x,\varepsilon) + O(e^{-\sqrt{2/\varepsilon}}), \quad i = 1,2,\cdots,4, \tag{28.46}$$

as $\varepsilon \to 0^+$, uniformly for $x \in [-1,1]$.

Proof. (a) We first consider the case $i = 1$. Recall that $\lim_{a\to 0} T_2(a) = \infty, T_2(1) = 0$, and $T_2(a)$ is decreasing for $0 < a < 1$. Hence there is only one value $a_0 \in (0,1)$ such that

$$T_2(a_0) = 2, \tag{28.47}$$

which, in turn, means that if we choose

$$k_0 = -\sqrt{\frac{2}{\varepsilon}}\sqrt{\frac{2}{3} - \frac{a_0^3}{3} + a_0^2},$$

then the solution $u(x, \varepsilon, k_0)$ will satisfy $u(1, \varepsilon, k_0) = 0$ and is negative in the entire interval $-1 < x < 1$. First, we claim that $u(x, \varepsilon, k_0)$ or, equivalently, $v(x, \varepsilon, k_0)$ is symmetric with respect to the origin, i.e., $v(x, \varepsilon, k_0) = v(-x, \varepsilon, k_0)$ and $u'(0, \varepsilon, k_0) = v'(0, \varepsilon, k_0) = 0$. To see this, we note that if $v(x, \varepsilon, k_0)$ is a solution of BVP (28.1), then so is $v(-x, \varepsilon, k_0)$. But, by the monotonicity of $T_2(a)$, there is only one negative solution to the BVP (28.1). Hence, $v(x, \varepsilon, k_0) = v(-x, \varepsilon, k_0)$ and $v'(0, \varepsilon, k_0) = 0$. Furthermore, we have (exercise)

$$v(0, \varepsilon, k_0) = a_0. \tag{28.48}$$

Next, we give an estimate for a_0, which will play a key role in the proof of the theorem. From (28.29) and (28.47), we have

$$\int_{a_0}^{1} \left\{ v^2 - \frac{v^3}{3} - a_0^2 + \frac{a_0^3}{3} \right\}^{-1/2} dv = \sqrt{\frac{2}{\varepsilon}}, \tag{28.49}$$

which can be written as

$$\int_{a_0}^{1} \left\{ (v^2 - a_0^2) \left[1 - \frac{v^3 - a_0^3}{3(v^2 - a_0^2)} \right] \right\}^{-1/2} dv = \sqrt{\frac{2}{\varepsilon}}.$$

Because the quantity inside the square brackets is greater than $1/2$, the last equation gives

$$\int_{a_0}^{1} \frac{1}{\sqrt{v^2 - a_0^2}} dv > \sqrt{\frac{1}{\varepsilon}}$$

and hence

$$\cosh^{-1} \frac{1}{a_0} > \sqrt{\frac{1}{\varepsilon}},$$

which, in turn, yields

$$a_0 \le 2e^{-1/\sqrt{\varepsilon}}. \tag{28.50}$$

This is only a rough estimate of a_0. A more precise description of the behavior of a_0 can be obtained by using an asymptotic formula for the elliptic integral

$$R_F(x, y, z) = \frac{1}{2} \int_0^\infty \frac{dt}{\sqrt{(t + x)(t + y)(t + z)}}. \tag{28.51}$$

In Carlson and Gustafson (1985), it has been shown that for real x, y and z, we have

$$(1-\theta_1)R_F(x,y,z) = z^{-1/2}\log\frac{4z^{1/2}}{x^{1/2}+y^{1/2}}, \quad 0<\theta_1<\frac{x+y}{4z}. \tag{28.52}$$

Returning to (28.32), we now make the change of variable $1-\tau=t$ in the integral on the right-hand side. This gives

$$\int_{a_0}^{1}\left\{v^2-\frac{v^3}{3}-a_0^2+\frac{a_0^3}{3}\right\}^{-1/2}dv = \sqrt{\frac{3}{1-a_0}}\int_0^1\frac{1}{\sqrt{t}\sqrt{3t-t^2+b}}dt,$$

where

$$b=\frac{6a_0-3a_0^2}{(1-a_0^2)^2}.$$

Note that $b\sim 6a_0$ and hence $b\to 0$ as $a_0\to 0$. Now we make a further transformation $t=b/(b+y)$ in the last integral to obtain

$$\begin{aligned}&\int_{a_0}^{1}\left\{v^2-\frac{v^3}{3}-a_0^2+\frac{a_0^3}{3}\right\}^{-1/2}dv\\&\quad=\sqrt{\frac{3}{1-a_0}}\int_0^\infty\frac{dy}{\sqrt{(y+b)(y+c)(y+d)}},\end{aligned} \tag{28.53}$$

where

$$c=\frac{1}{2}(2b+3-\sqrt{4b+9})=\frac{2b}{3}+\frac{b^2}{27}+o(b^2), \quad \text{as } b\to 0,$$

and

$$d=\frac{1}{2}(2b+3+\sqrt{4b+9})=3+\frac{4b}{3}-\frac{b^2}{27}+o(b^2), \quad \text{as } b\to 0.$$

Applying (28.52) to (28.53) gives

$$\int_{a_0}^{1}\left\{v^2-\frac{v^3}{3}-a_0^2+\frac{a_0^3}{3}\right\}^{-\frac{1}{2}}dv \sim 2\sqrt{\frac{3}{1-a_0}}\frac{1}{\sqrt{d}}\log\frac{4\sqrt{d}}{\sqrt{b}+\sqrt{c}}$$

as $\varepsilon\to 0$. Inserting the last formula into (28.49) yields

$$a_0\sim\frac{24}{5+2\sqrt{6}}\exp\left(-\sqrt{\frac{2}{\varepsilon}}\right), \quad \varepsilon\to 0^+. \tag{28.54}$$

To prove (28.46) when $i=1$, we first restrict x to the interval $[-1,0]$ and put

$$v_L(x) := 3\,\mathrm{sech}^2\left(\frac{x+1}{\sqrt{2\varepsilon}}+\ln(\sqrt{3}+\sqrt{2})\right).$$

From the proof of case (c) in Lemma 28.1, it is readily seen that v_L satisfies

$$\varepsilon v_L'' - 2v_L + v_L^2 = 0$$

and

$$v_L(-1) = 1, \quad v_L'(-1) = -\sqrt{\frac{4}{3\varepsilon}}.$$

By definition, $v'(-1, \varepsilon, k_0) = k_0$. Hence, it follows from Lemma 28.1 (a) that

$$v'(-1, \varepsilon, k_0) = k_0 > -\sqrt{\frac{4}{3\varepsilon}} = v_L'(-1), \tag{28.55}$$

for otherwise one can deduce that $v(1, \varepsilon, k_0) \neq 1$. Now, let $R(x)$ denote the difference between $v_L(x)$ and the true solution $v(x, \varepsilon, k_0)$, i.e.,

$$R(x) := v(x, \varepsilon, k_0) - v_L(x). \tag{28.56}$$

In view of (28.54), we have

$$R(-1) = 0 \quad \text{and} \quad R'(-1) > 0. \tag{28.57}$$

We further claim that

$$R(x) > 0 \quad \text{and} \quad R''(x) > 0 \tag{28.58}$$

for all $x \in (-1, 0]$. Indeed, if $R(x) \leq 0$ for some $x \in (-1, 0]$, then there exists a point $\xi \in (-1, 0]$ such that $R(\xi) = 0$ and $R(x) > 0$ for $x \in (-1, \xi]$. Because $2v - v^2$ is an increasing function for $-\infty < v < 1$, we have

$$R''(x) = \frac{1}{\varepsilon}\left[2v(x, \varepsilon, k_0) - v^2(x, \varepsilon, k_0) - 2v_L(x) + v_L^2(x)\right] > 0 \tag{28.59}$$

for $x \in (-1, \xi)$. Furthermore, from (28.57), it follows that

$$R(\xi) = R(-1) + R'(-1)(\xi + 1) + \int_{-1}^{\xi}\int_{-1}^{s} R''(\tau)d\tau ds > 0,$$

which contradicts the assumption that $R(\xi) = 0$. Hence, we conclude that $R(x) > 0$ for all $x \in (-1, 0)$ and by (28.59), $R''(x) > 0$ for all $x \in (-1, 0]$, thus proving (28.58). A combination of the results in (28.57) and (28.58) shows that $R(x)$ is an increasing function in the interval $[-1, 0]$ and bounded by

$$R(0) = v(0, \varepsilon, k_0) - 3\operatorname{sech}^2\left(\frac{1}{\sqrt{2\varepsilon}} + \ln(\sqrt{3} + \sqrt{2})\right).$$

Because $v(0, \varepsilon, k_0) = a_0$ as shown in (28.48), we obtain

$$R(0) \sim \frac{12}{5 + 2\sqrt{6}} \exp\left(-\sqrt{\frac{2}{\varepsilon}}\right)$$

on account of (28.54). Thus, for $-1 \le x \le 0$,

$$0 \le R(x) \le R(0) \sim \frac{12}{5+2\sqrt{6}} \exp\left(-\sqrt{\frac{2}{\varepsilon}}\right).$$

Similarly, with

$$v_R(x) := 3\,\mathrm{sech}^2\left(\frac{1-x}{\sqrt{2\varepsilon}} + \ln(\sqrt{3}+\sqrt{2})\right),$$

one can show that $v(x,\varepsilon,k_0) - v_R(x)$ is a decreasing function on $[0,1]$ and bounded by $R(0)$. This also follows from the symmetry of the functions $v(x,\varepsilon,k_0)$ and $v_R(x)$, thus completing the proof of (28.46) when $i=1$.

(*b*) We next consider the case $i=2$. Let u_2 be a solution of the initial valued problem (28.14), which is positive in the interval $(-1,x_1)$ and negative in $(x_1,1)$, where $u_2(x_1)=0$. The existence of such a solution is established in Lemma 28.1 (d). As in the case of $i=1$, we need to estimate the value of $a_0 \in (0,1)$, which is uniquely determined by the equation

$$T_1(a_0) + T_2(a_0) = 2. \tag{28.60}$$

Denote $v = u+1$ and $k_0 = v'(-1) < 0$. We obtain from (28.23)

$$\frac{\varepsilon}{2}k_0^2 - \frac{2}{3} = \frac{a_0^3}{3} - a_0^2.$$

Moreover, we have

$$v(-1,\varepsilon,k_0) = v(x_1,\varepsilon,k_0) = v(1,\varepsilon,k_0) = 1,$$

and

$$v(x,\varepsilon,k_0) > 1 \quad \text{for } -1 < x < x_1,$$

and

$$v(x,\varepsilon,k_0) < 1, \quad \text{for } x_1 < x < 1;$$

see Figure 28.2 and Equation (28.60), and recall that $v = 1+u$.

By definition, $T_1(a_0)$ is the length of the interval $(-1,x_1)$, which is equal to x_1+1. By Lemma 28.3, $T_1(a_0) < T_1(1) = k_1\sqrt{\varepsilon}$, where $k_1 \approx 4.88$. Integrating the equation

$$\sqrt{\frac{\varepsilon}{2}}v' = \sqrt{v^2 - \frac{1}{3}v^3 - a_0^2 + \frac{1}{3}a_0^3}$$

from $v = \frac{1}{2}(1+x_1)$ to 1 gives

$$\int_{a_0}^{1} \frac{1}{\sqrt{v^2 - \frac{1}{3}v^3 - a_0^2 + \frac{1}{3}a_0^3}}dv = \sqrt{\frac{2}{\varepsilon}}\left(\frac{1-x_1}{2}\right) \ge \sqrt{\frac{2}{\varepsilon}}\left(1 - \frac{k_1}{2}\sqrt{\varepsilon}\right);$$

see the proof of Lemma 28.2. Using the same argument as in the previous case, one can show that

$$0 < a_0 \sim \frac{24}{5+2\sqrt{6}} e^{k_1/\sqrt{2}} \exp\left(-\sqrt{\frac{2}{\varepsilon}}\right)$$

for small values of ε, i.e.,

$$a_0 = O\left(\exp\left(-\sqrt{\frac{2}{\varepsilon}}\right)\right), \quad \varepsilon \to 0^+. \tag{28.61}$$

To estimate the value of x_1, we return to (28.28)

$$T_1(a_0) = \sqrt{2\varepsilon} \int_1^{v_3} \frac{dv}{\sqrt{v^2 - \frac{1}{3}v^3 - a_0^2 + \frac{1}{3}a_0^3}}$$

and obtain

$$T_1(a_0) = \sqrt{2\varepsilon} \int_0^1 \frac{1}{\sqrt{1-t}} \left\{-\frac{1}{v_3 - 1} + \frac{v_3 - 1}{3}(1 + t + t^2)\right\}^{-\frac{1}{2}} dt;$$

see the proof of Lemma 28.3. By (28.25), we have (exercise)

$$v_3 = 3 - \frac{1}{3}a_0^2 + o(a_0^2), \quad \varepsilon \to 0^+. \tag{28.62}$$

Coupling the last two results yields

$$T_1(a_0) = \sqrt{2\varepsilon} \int_0^1 \frac{1}{\sqrt{1-t}} \left\{-\frac{1}{2} + \frac{2}{3}(1 + t + t^2)\right\}^{-\frac{1}{2}} dt + O(\sqrt{\varepsilon}a_0^2).$$

The integral on the right can be evaluated explicitly, and its value is $2\ln(\sqrt{3} + \sqrt{2})$ (exercise). Hence

$$T_1(a_0) = 2\alpha + O(\sqrt{\varepsilon}a_0^2), \quad \varepsilon \to 0^+, \tag{28.63}$$

where

$$\alpha = \sqrt{2\varepsilon}\ln(\sqrt{3} + \sqrt{2}), \tag{28.64}$$

for which it follows that

$$x_1 = -1 + T_1(a_0) = -1 + 2\alpha + O(\sqrt{\varepsilon}a_0^2). \tag{28.65}$$

Now note that $u_2(x)$ is entirely negative in the interval $(x_1, 1)$. By repeating the argument in case (a), one can show that

$$u_2(x) = -1 + 3\,\mathrm{sech}^2\left(\frac{x - x_1}{\sqrt{2\varepsilon}} + \ln(\sqrt{3} + \sqrt{2})\right)$$
$$+ 3\,\mathrm{sech}^2\left(\frac{1 - x}{\sqrt{2\varepsilon}} + \ln(\sqrt{3} + \sqrt{2})\right) + O(e^{-\sqrt{2/\varepsilon}})$$

for $x \in [x_1, 1]$. Using (28.65), we further obtain

$$u_2(x) = -1 + 3\,\mathrm{sech}^2\left(-\frac{x+1}{\sqrt{2\varepsilon}} + \ln(\sqrt{3}+\sqrt{2})\right) + 3\,\mathrm{sech}^2\left(\frac{1-x}{\sqrt{2\varepsilon}} + \ln(\sqrt{3}+\sqrt{2})\right) + O(e^{-\sqrt{2/\varepsilon}}), \tag{28.66}$$

because sech is an even function.

In the interval $[-1, x_1]$, we make a change of variable

$$x = -1 + y\sqrt{\varepsilon}, \tag{28.67}$$

and denote

$$y_1 := \frac{x_1+1}{\sqrt{\varepsilon}} = \frac{T_1(a_0)}{\sqrt{\varepsilon}} = 2\sqrt{2}\ln(\sqrt{3}+\sqrt{2}) + O(a_0^2). \tag{28.68}$$

It follows that $v = 1$ at $y = 0$ and $y = y_1$. For $y \in (0, y_1)$, we have $v > 1$ and

$$\frac{d^2v}{dy^2} - 2v + v^2 = 0. \tag{28.69}$$

Moreover, v attains its maximum value v_3 at $y = y_1/2$. Now, we define

$$\bar{v}(y) := 3\,\mathrm{sech}^2\left(\frac{y - y_1/2}{\sqrt{2}}\right), \tag{28.70}$$

which also satisfies equation (28.69) for $y \in (0, y_1)$ (exercise). Moreover, $\bar{v}$ attains its maximum value 3 at $y = y_1/2$. Consider the difference $w := \bar{v} - v$, which satisfies the differential equation

$$\frac{d^2w}{dy^2} = (2 - v - \bar{v})w, \tag{28.71}$$

for $y \in (0, y_1)$, with $w'(y_1/2) = 0$ and

$$w(y_1/2) = 3 - v_3 = O(a_0^2). \tag{28.72}$$

Since $|v| \le 3$ and $|\bar{v}| < 3$ for $y \in [0, y_1]$, where $y_1 = O(1)$, it is easily seen that (exercise) $w(y) = O(a_0^2)$ uniformly for $y \in [0, y_1]$. Return to the original variable x, we obtain

$$v = 3\,\mathrm{sech}^2\left(\frac{x+1-\alpha}{\sqrt{2\varepsilon}}\right) + O(a_0^2) = 3\,\mathrm{sech}^2\left(-\frac{x+1}{\sqrt{2\varepsilon}} + \ln(\sqrt{3}+\sqrt{2})\right) + O(e^{-2\sqrt{2/\varepsilon}}). \tag{28.73}$$

Since $v = 1 + u$, we have

$$u(x, k_0) = -1 + 3\,\mathrm{sech}^2\left(-\frac{x+1}{\sqrt{2\varepsilon}} + \ln(\sqrt{3}+\sqrt{2})\right) + O(e^{-2\sqrt{2/\varepsilon}}), \tag{28.74}$$

as $\varepsilon \to 0$, for $x \in [-1, x_1]$. The desired result (28.46), when $i = 2$, follows from (28.66) and (28.74). Observe that there are three terms in the approximation in (28.66), whereas in (28.74) we have only two terms. The explanation for this difference is that the third term on the right-hand side of (28.66) is exponentially small in the interval $[-1, x_1]$.

(*c*) The proof of this case is essentially the same as that of case $i = 2$.

(*d*) Similar to the case $i = 1$, we consider the function $2T_1(a) + T_2(a)$. As $a \to 0$, this function approaches infinity; as $a \to 1$, it is $O(\sqrt{\varepsilon})$. By Lemma 28.3, we also know that this function is convex. Hence, there is a unique value $a_0 \in (0, 1)$ such that

$$2T_1(a_0) + T_2(a_0) = 2.$$

Furthermore, by Lemma 28.3 we have

$$T_2(a_0) \geq 2 - 2T_1(1) = 2 - 9.76\sqrt{\varepsilon}.$$

Using an argument similar to that in (*a*), we can also deduce that

$$a_0 = O(e^{-\sqrt{2/\varepsilon}});$$

see (28.54). Because $T_1(a_0) = x_1 + 1 = 1 - x_2$, we obtain

$$x_1 = -1 + 2\alpha + O(\sqrt{\varepsilon} a_0^2) \tag{28.75}$$

and

$$x_2 = 1 - 2\alpha + O(\sqrt{\varepsilon} a_0^2); \tag{28.76}$$

cf. (28.65). By the result in (*a*), one can write

$$\begin{aligned} u_4 = &-1 + 3\,\text{sech}^2\left(\frac{x - x_1}{\sqrt{2\varepsilon}} + \ln(\sqrt{3} + \sqrt{2})\right) \\ &+ 3\,\text{sech}^2\left(\frac{x_2 - x}{\sqrt{2\varepsilon}} + \ln(\sqrt{3} + \sqrt{2})\right) + O(e^{-\sqrt{2/\varepsilon}}), \end{aligned} \tag{28.77}$$

for $x \in [x_1, x_2]$. Inserting (28.75) and (28.76) into (28.77) gives

$$\begin{aligned} u_4 = &-1 + 3\,\text{sech}^2\left(-\frac{x + 1}{\sqrt{2\varepsilon}} + \ln(\sqrt{3} + \sqrt{2})\right) \\ &+ 3\,\text{sech}^2\left(-\frac{1 - x}{\sqrt{2\varepsilon}} + \ln(\sqrt{3} + \sqrt{2})\right) + O(e^{-\sqrt{2/\varepsilon}}). \end{aligned} \tag{28.78}$$

For $x \in [-1, x_1]$, we have by the result in case (*b*)

$$u_4 = -1 + 3\,\text{sech}^2\left(-\frac{x + 1}{\sqrt{2\varepsilon}} + \ln(\sqrt{3} + \sqrt{2})\right) + O(e^{-\sqrt{2/\varepsilon}}). \tag{28.79}$$

Similarly, for $x \in [x_2, 1]$, we have from case (c)

$$u_4 = -1 + 3\,\mathrm{sech}^2\left(-\frac{1-x}{\sqrt{2\varepsilon}} + \ln(\sqrt{3}+\sqrt{2})\right) + O(e^{-\sqrt{2/\varepsilon}}). \tag{28.80}$$

Note that the third and the second term in (28.78) are $O(e^{-2\sqrt{2/\varepsilon}})$ for $x \in [-1, x_1]$ and $[x_2, 1]$, respectively. Hence, upon comparing (28.78) with (28.79) and (28.80), we conclude (28.78) actually holds for all $x \in [-1, 1]$. This completes the proof of the theorem. □

Remark 28.3. In the above proof, we have on several occasions made use of a result corresponding to (28.46) with $i = 1$ when the interval $[-1, 1]$ is replaced by $[a, b]$, The precise approximation is given by

$$\begin{aligned} u_1(x,\varepsilon) = &-1 + 3\,\mathrm{sech}^2\left(\frac{x-a}{\sqrt{2\varepsilon}} + \ln(\sqrt{3}+\sqrt{2})\right) \\ &+ 3\,\mathrm{sech}^2\left(\frac{b-x}{\sqrt{2\varepsilon}} + \ln(\sqrt{3}+\sqrt{2})\right) + O\left\{\exp\left(-\frac{b-a}{\sqrt{2\varepsilon}}\right)\right\} \end{aligned} \tag{28.81}$$

uniformly for $x \in [a, b]$.

28.5 Solutions with spikes

We are now ready to establish the result stated in (28.13), i.e., the case with n internal spikes. In fact, we have the following formula corresponding to (28.46).

Theorem 28.3. *For each $i = 1, \cdots, 4$ and $n = 1, 2, \cdots$, there exists a unique solution $u_{i,n}(x,\varepsilon)$ satisfying*

$$u_{i,n}(x,\varepsilon) = \widetilde{U}_i(x,\varepsilon) + \sum_{j=1}^{n} q\left(\frac{x - x_{i,j}}{\sqrt{\varepsilon}}\right) + O\left[\exp\left(-\frac{1}{n+1}\sqrt{\frac{2}{\varepsilon}}\right)\right] \tag{28.82}$$

for all $x \in [-1, 1]$, where

$$x_{1,j} = -1 - \sqrt{2\varepsilon}\ln(\sqrt{3}+\sqrt{2}) + \frac{j}{n+1}[2 + 2\sqrt{2\varepsilon}\ln(\sqrt{3}+\sqrt{2})], \tag{28.83}$$

$$x_{2,j} = -1 + \frac{2j}{n+1} + \sqrt{2\varepsilon}\ln(\sqrt{3}+\sqrt{2}), \tag{28.84}$$

$$x_{3,j} = -1 + \frac{2j}{n+1} - \sqrt{2\varepsilon}\ln(\sqrt{3}+\sqrt{2}), \tag{28.85}$$

$$x_{4,j} = -1 + \sqrt{2\varepsilon}\ln(\sqrt{3}+\sqrt{2}) + \frac{j}{n+1}[2 - 2\sqrt{2\varepsilon}\ln(\sqrt{3}+\sqrt{2})], \tag{28.86}$$

for $j = 1, \cdots, n$.

Proof. We consider only the case $i = 1$; all other cases can be handled in a similar manner. As in the proof of Theorem 28.2, by using Lemma 28.3, it can be shown that there exists a unique value $a_0 \in (0,1)$ such that

$$nT_1(a_0) + (n+1)T_2(a_0) = 2. \tag{28.87}$$

Because $T_1(a_0) < T_1(1) = t_1\sqrt{\varepsilon}$ with

$$t_1 := \int_1^{1+\sqrt{3}} \frac{\sqrt{2}}{\sqrt{v^2 - v^3/2 - 2/3}} dv \approx 4.88,$$

we have

$$T_2(a_0) \geq \frac{2 - t_1 n\sqrt{\varepsilon}}{n+1}.$$

From (28.29), it follows that

$$\sqrt{2\varepsilon} \int_{a_0}^{1} \frac{dv}{\sqrt{v^2 - \frac{1}{3}v^3 + \frac{1}{3}a_0^3 - a_0^2}} \geq \frac{2 - t_1 n\sqrt{\varepsilon}}{n+1}.$$

The argument given in the proof of the case $i = 1$ in Theorem 28.2 then yields

$$a_0 = O\left\{\exp\left(-\frac{1}{n+1}\sqrt{\frac{2}{\varepsilon}}\right)\right\}. \tag{28.88}$$

On account of (28.63) and (28.87), we obtain respectively

$$T_1(a_0) = 2\alpha + O\left\{\sqrt{\varepsilon}\exp\left(-\frac{2}{n+1}\sqrt{\frac{2}{\varepsilon}}\right)\right\} \tag{28.89}$$

and

$$T_2(a_0) = \frac{2 - 2n\alpha}{n+1} + O\left\{\sqrt{\varepsilon}\exp\left(-\frac{2}{n+1}\sqrt{\frac{2}{\varepsilon}}\right)\right\}, \tag{28.90}$$

where $\alpha = \sqrt{2\varepsilon}\ln(\sqrt{3} + \sqrt{2})$.

Let $u_1(x, k_0)$ be the solution of the initial value problem (28.14) with k_0 related to a_0 via (28.23) and a_0 satisfying (28.87). Then $u_1(x, k_0)$ is also the unique solution to the BVP (28.1) with n internal spikes, i.e., $u_1(x, k_0) = u_{1,n}(x, \varepsilon)$. Let $\bar{x}_j \in (-1, 1), j = 1, \cdots, n$, denote the zeros of $u_1'(x, k_0)$. The position of $\bar{x}_j$ can be represented by

$$\bar{x}_j = -1 + T_2(a_0) + \frac{1}{2}T_1(a_0) + (j-1)[T_1(a_0) + T_2(a_0)],$$

$j = 1, \cdots, n$. From (28.89) and (28.90), we have

$$\bar{x}_j = -1 - \sqrt{2\varepsilon}\ln(\sqrt{3}+\sqrt{2}) + \frac{j}{n+1}[2 + 2\sqrt{2\varepsilon}\ln(\sqrt{3}+\sqrt{2})] + O\left\{\exp\left(-\frac{2}{n+1}\sqrt{\frac{2}{\varepsilon}}\right)\right\}.$$

With the notation $x_{1,j}$ defined in (28.83), the last equation can be written as

$$\bar{x}_j = x_{1,j} + O\left\{\exp\left(-\frac{2}{n+1}\sqrt{\frac{2}{\varepsilon}}\right)\right\}, \quad j = 1, \cdots, n. \tag{28.91}$$

Let β_j and $\gamma_j, j = 1, \cdots, n$, denote the zeros of $u_{1,n}(x, \varepsilon)$, which are closest to $\bar{x}_j$ so that $\bar{x}_j \in (\beta_j, \gamma_j)$ and $u_{1,n}(x, \varepsilon) > 0$ for $x \in (\beta_j, \gamma_j)$; see Figure 28.3. Note that the length of each of the intervals $(\beta_j, \bar{x}_j)$ and $(\bar{x}_j, \gamma_j)$ is $T_1(a_0)/2$. Hence, by virtue of (28.91),

$$\beta_j = x_{1,j} - \alpha + O\left\{\exp\left(-\frac{2}{n+1}\sqrt{\frac{2}{\varepsilon}}\right)\right\},$$

and

$$\gamma_j = x_{1,j} + \alpha + O\left\{\exp\left(-\frac{2}{n+1}\sqrt{\frac{2}{\varepsilon}}\right)\right\}.$$

Using the argument given in the proof of case (*b*) in Theorem 28.2, one can show that for $x \in (\beta_j, \gamma_j)$,

$$\begin{aligned} u_{1,n}(x,\varepsilon) &= -1 + 3\,\mathrm{sech}^2\left(\frac{x-\bar{x}_j}{\sqrt{2\varepsilon}}\right) + O\left\{\exp\left(-\frac{2}{n+1}\sqrt{\frac{2}{\varepsilon}}\right)\right\} \\ &= -1 + 3\,\mathrm{sech}^2\left(\frac{x-x_{1,j}}{\sqrt{2\varepsilon}}\right) + O\left\{\exp\left(-\frac{2}{n+1}\sqrt{\frac{2}{\varepsilon}}\right)\right\}; \end{aligned} \tag{28.92}$$

see (28.73), (28.74), (28.88), and (28.91). Similarly, by using the argument given in the proof of case (*a*) in Theorem 28.2, we obtain for $x \in (\gamma_j, \beta_{j+1})$

$$\begin{aligned} u_{1,n}(x,\varepsilon) = -1 &+ 3\,\mathrm{sech}^2\left(\frac{x-\gamma_j}{\sqrt{2\varepsilon}} + \ln(\sqrt{3}+\sqrt{2})\right) \\ &+ 3\,\mathrm{sech}^2\left(\frac{\beta_{j+1}-x}{\sqrt{2\varepsilon}} + \ln(\sqrt{3}+\sqrt{2})\right) + O\left\{\exp\left(-\frac{1}{n+1}\sqrt{\frac{2}{\varepsilon}}\right)\right\}, \end{aligned}$$

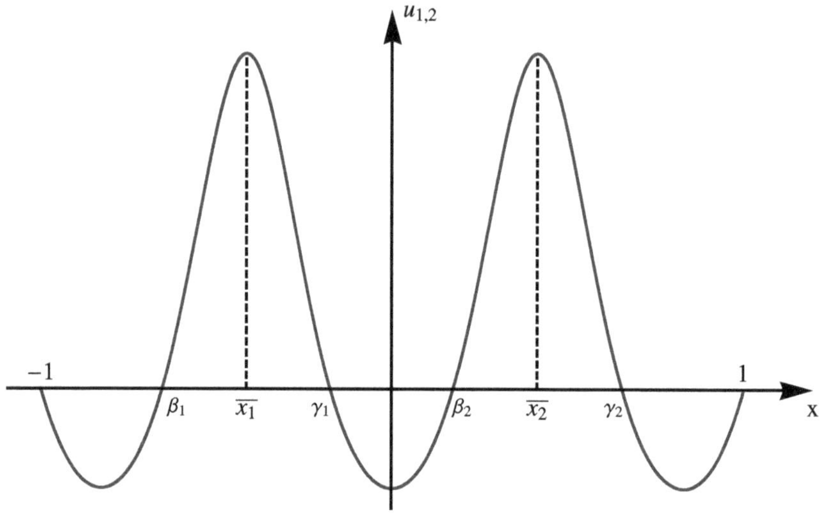

Fig. 28.3: The graph of $u_{1,n}$ with $n = 2$ internal spikes.

which in turn gives

$$u_{1,n}(x, \varepsilon) = -1 + 3\operatorname{sech}^2\left(\frac{x - x_{1,j}}{\sqrt{2\varepsilon}}\right) + 3\operatorname{sech}^2\left(\frac{x - x_{1,j+1}}{\sqrt{2\varepsilon}}\right) + O\left\{\exp\left(-\frac{1}{n+1}\sqrt{\frac{2}{\varepsilon}}\right)\right\}; \tag{28.93}$$

see the remark following the proof of Theorem 28.2. Furthermore, we have

$$u_{1,n}(x, \varepsilon) = -1 + 3\operatorname{sech}^2\left(\frac{x+1}{\sqrt{2\varepsilon}} + \ln(\sqrt{3} + \sqrt{2})\right) + 3\operatorname{sech}^2\left(\frac{x - x_{1,1}}{\sqrt{2\varepsilon}}\right) + O\left\{\exp\left(-\frac{1}{n+1}\sqrt{\frac{2}{\varepsilon}}\right)\right\} \tag{28.94}$$

for $x \in (-1, \beta_1)$, and

$$u_{1,n}(x, \varepsilon) = -1 + 3\operatorname{sech}^2\left(\frac{1-x}{\sqrt{2\varepsilon}} + \ln(\sqrt{3} + \sqrt{2})\right) + 3\operatorname{sech}^2\left(\frac{x - x_{1,n}}{\sqrt{2\varepsilon}}\right) + O\left\{\exp\left(-\frac{1}{n+1}\sqrt{\frac{2}{\varepsilon}}\right)\right\} \tag{28.95}$$

for $x \in (\gamma_n, 1)$. Let

$$\delta := \frac{1}{2}[T_1(a_0) + T_2(a_0)],$$

and note that

$$3\,\mathrm{sech}^2\left(\frac{x-x_j}{\sqrt{2\varepsilon}}\right) = O\left\{\exp\left(-\frac{1}{n+1}\sqrt{\frac{2}{\varepsilon}}\right)\right\}, \quad j=1,\cdots,n,$$

for $x \notin (x_j-\delta, x_j+\delta)$. Analogously, we have

$$3\,\mathrm{sech}^2\left(\frac{1+x}{\sqrt{2\varepsilon}} + \ln(\sqrt{3}+\sqrt{2})\right) = O\left\{\exp\left(-\frac{1}{n+1}\sqrt{\frac{2}{\varepsilon}}\right)\right\}$$

for $x \geq -1 + \frac{1}{2}T_1(a_0)$ and

$$3\,\mathrm{sech}^2\left(\frac{1-x}{\sqrt{2\varepsilon}} + \ln(\sqrt{3}+\sqrt{2})\right) = O\left\{\exp\left(-\frac{1}{n+1}\sqrt{\frac{2}{\varepsilon}}\right)\right\}$$

for $x \leq 1 - \frac{1}{2}T_1(a_0)$. The final result given in (28.82) with $i = 1$ now follows from a combination of (28.92)-(28.95). □

28.6 Exercises

1. Assume $b > 0$ and $c > 0$. Consider the initial value problem

$$v'(x) = -b[-v(x)]^{3/2},$$
$$v(0) = -c.$$

   Show that the solution blows up at a finite $x_{max} > 0$. Express x_{max} in terms of b and c.
2. Verify (28.20).
3. Verify that (28.21) are the solutions to the initial value problem (28.17) with $k = \pm\sqrt{4/3\varepsilon}$.
4. Assume $|k| < \sqrt{4/3\varepsilon}$. Let $F(v)$ be given as in (28.18). Show that the equation $F(v) = \varepsilon k^2/2 - 2/3$ has three distinct real roots $v_1 < 0 < v_2 < 1 < v_3 < 3$. Moreover, prove that the solution to the planar system

$$v'(x) = w(x), \quad \varepsilon w'(x) = 2v(x) - [v(x)]^2, \quad x \geq 0,$$
$$v(-1) = 1, \quad w(-1) = k,$$

   is periodic in x with period

$$\int_{v_2}^{v_3} \frac{\sqrt{6\varepsilon}dv}{\sqrt{(v-v_1)(v-v_2)(v_3-v)}}.$$

5. Assume $k < -\sqrt{4/3\varepsilon}$. Prove that the solution to the initial value problem (28.16) blows up at

$$t_{max} = -1 + \int_{-\infty}^{1} \frac{\sqrt{\varepsilon/2}dv}{\sqrt{\varepsilon k^2/2 - 2/3 + v^2 - v^3/3}}.$$

6. Assume $k > \sqrt{4/3\varepsilon}$. Let $F(v)$ be given as in (28.18). Show that the equation $F(v) = \varepsilon k^2/2 - 2/3$ has a unique real root $\bar{v}$. Prove that the solution to the initial value problem (28.16) blows up at

$$\begin{aligned} t_{max} = & -1 + \int_{-\infty}^{1} \frac{\sqrt{\varepsilon/2}dv}{\sqrt{\varepsilon k^2/2 - 2/3 + v^2 - v^3/3}} \\ & + \int_{1}^{\bar{v}} \frac{\sqrt{2\varepsilon}dv}{\sqrt{\varepsilon k^2/2 - 2/3 + v^2 - v^3/3}}. \end{aligned}$$

7. Verify (28.22).
8. Prove (28.48).
9. Verify (28.62).
10. Show that

$$\int_0^1 \frac{1}{\sqrt{1-t}} \left\{ -\frac{1}{2} + \frac{2}{3}(1+t+t^2) \right\}^{-\frac{1}{2}} dt = 2\ln(\sqrt{3}+\sqrt{2}).$$

11. As $\varepsilon \to 0$, prove that

$$\operatorname{sech}^2(x+\varepsilon) = \operatorname{sech}^2 x + O(\varepsilon),$$

uniformly for $x \in \mathbf{R}$.
12. Show that the function $\bar{v}(y)$ defined in (28.70) satisfies equation (28.69) for $y \in (0, y_1)$ and attains its maximum value 3 at $y = y_1/2$.
13. Let $a_0 > 0$ be a small number and $y_1 = y_1(a_0) \to \bar{y}_1 > 0$ as $a_0 \to 0^+$. Assume $w(y)$ satisfies the differential equation

$$w''(y) = f(y)w(y), \quad y \in (0, y_1)$$

with $w'(y_1/2) = 0$ and $w(y_1/2) = O(a_0^2)$ as $a_0 \to 0^+$, where f is continuous and uniformly bounded for all $y \in [0, y_1]$ and small $a_0 > 0$. Prove that $w(y) = O(a_0^2)$ uniformly in $y \in [0, y_1]$ as $a_0 \to 0^+$.
14. Consider the boundary value problem

$$\varepsilon u''(x) - (2 - x^2)u(x) = -1, \quad |x| < 1, \tag{28.96}$$

$$u(-1) = u(1) = 0, \tag{28.97}$$

where $0 < \varepsilon < 1$. If ε is sufficiently small, say $\varepsilon = 10^{-10}$, then it may look natural to suggest that the term $\varepsilon u''(x)$ could be neglected and a solution of the equation

$$(2 - x^2)u_1(x) = 1 \tag{28.98}$$

might be an approximation to the solution of the equation (28.96), i.e.,

$$u(x) \approx u_1(x) = \frac{1}{2 - x^2}. \tag{28.99}$$

But the function $u_1(x)$ cannot satisfy the boundary conditions in (28.97). This immediately tells us that the term $\varepsilon u''(x)$ cannot be omitted. The only possibility then is that $\varepsilon u''(x)$ is of the same order as the nonhomogeneous term $(2-x^2)u(x)$ in (28.96).

(a) If we replace the positive coefficient function $2-x^2$ in (28.96) by a positive constant a, where $1 < a < 2$ (this is the range of $2-x^2$ for $|x| < 1$), show that the solution of the substitute problem is

$$v(x) = \frac{1}{a}\left[1 - \frac{\cosh(x\sqrt{a/\varepsilon})}{\cosh\sqrt{a/\varepsilon}}\right].$$

Draw a graph of this function by using computer software.

(b) Show that $v''(x)$ is much greater than $v(x)$ near $x = 1$, $x = -1$, and nowhere else. Also, show that the boundary layers where v is steep and $\varepsilon v''$ is of order unity have a width of order $\varepsilon^{1/2}$.

15. It is reasonable to ask whether the solution of (28.96) with boundary conditions (28.97) can be approximated accurately by

$$u(x) = u_1(x) + p + q, \tag{28.100}$$

where p is a function which is "steep" near $x = -1$ and very small elsewhere, and q is "steep" near $x = 1$ and very small elsewhere. To display these characteristics of p and q, we let

$$p = p(y) = p\left(\frac{x+1}{\varepsilon^\alpha}\right), \quad q = q(z) = q\left(\frac{1-x}{\varepsilon^\alpha}\right), \tag{28.101}$$

and choose $\alpha > 0$ in such a way that $p(y)$, $p'(y)$, $p''(y), \cdots$ are all of comparable size. Such a choice has two advantages. It implies that significant changes of p occur in distances of order unity in the y variable and, more importantly, it implies that when additive combinations of $p, p', p'', \cdots$ appear in an equation, the relative importance of such terms is displayed explicitly by the size of their coefficients.

(a) By substituting equations (28.100) and (28.101) into (28.96) and (28.97), show that

$$\begin{aligned} &\varepsilon u_1''(x) - (2-x^2)u_1(x) + 1 + \varepsilon^{1-2\alpha}p''(y) - (2-x^2)p(y) \\ &+ \varepsilon^{1-2\alpha}q''(z) - (2-x^2)q(z) = 0. \end{aligned} \tag{28.102}$$

Since we expect $p(y)$, $q(z)$ and their derivatives to be very small except near $|x| = 1$, the contributions of p, p'', q, q'' cannot be significant except near $|x| = 1$. Accordingly (as we had already anticipated) $u_1(x)$ as given by equation (28.99) is potentially an excellent approximation to a solution of (28.102) away from $|x| = 1$.

(b) Near $x = -1$, where q and q'' are very small, show that $2 - x^2$ can be written as

$$2 - x^2 = 1 + 2\varepsilon^\alpha y - \varepsilon^{2\alpha} y^2,$$

$\varepsilon u_1''$ is of order ε compared to $p(y)$ (since $p(0) = 1$), and $\varepsilon^{2\alpha} y^2 \ll 2\varepsilon^\alpha y \ll 1$ in the range y is interest (since we know that $\alpha > 0$). Show that equation (28.102) is well approximated by

$$\varepsilon^{1-2\alpha} p''(y) - p(y) = 0. \tag{28.103}$$

Furthermore, show that if we choose $\alpha < 1/2$, then equation (28.103) degenerates to $p(y) = 0$ and we are back to where we started; alternatively, if we choose $\alpha > 1/2$, then the first term dominates equation (28.103) and we must use $p''(y) \approx 0$ instead of equation (28.103). Thus, with $\alpha > 1/2$, explain why there is no $p(y)$ which is small far from $y = -1$ except $p(y) \equiv 0$. Thus, *the only acceptable choice for* α *is* $\alpha = 1/2$. With this choice, we have

$$p(y) = Ae^{-y} + Be^{y}. \tag{28.104}$$

(c) Show that B must be zero in order that $p(y)$ be very small except in $0 < y < O(1)$. We always interpret the phrase "very small elsewhere" to mean $p(y) \to 0$ as $y \to \infty$. With $B = 0$, show that from (28.100) and (28.104) we have

$$u(-1) \approx u_1(-1) + p(0) = 1 + A = 0,$$

so that

$$A = -1. \tag{28.105}$$

Use symmetry to conclude that

$$u(x, \varepsilon) \approx \frac{1}{2 - x^2} - \exp\left(-\frac{x+1}{\sqrt{\varepsilon}}\right) - \exp\left(-\frac{1-x}{\sqrt{\varepsilon}}\right). \tag{28.106}$$

Draw the graph of this function for each of the two values $\varepsilon = 10^{-2}$ and $\varepsilon = 10^{-3}$.

(d) Verify carefully that, for every x in $|x| < 1$, each term in (28.102) that we have ignored is indeed small compared to those retained. Although the arguments given here are not rigorous, they do give readers some confidence that equation (28.106) provides an excellent approximation to $u(x, \varepsilon)$.

16. We have seen that in some problems where the differential equation $L(u, \varepsilon) = 0$ involves a small parameter ε, in such a way that, as $\varepsilon \to 0^+$, the order of the differential equation is lowered.
(i) There is hope that the solution $u(x, \varepsilon)$ is well approximated by $u_1(x)$, a solution of $L(u_1, 0) = 0$, except near points x_1 and x_2 at which the boundary conditions are prescribed for $L(u, \varepsilon) = 0$.
(ii) It may be convenient to write

$$u(x, \varepsilon) \approx u_1(x) + p\left(\frac{x - x_1}{\varepsilon^\alpha}\right) + q\left(\frac{x - x_2}{\varepsilon^\alpha}\right),$$

where, because of the change of coordinate, one can readily sort out the simplest differential equation for each of the functions p and q which will give a suitable approximation for $u(x, \varepsilon)$.
(a) Consider the boundary value problem

$$\varepsilon u''(x, \varepsilon) + (1 - x^2)u(x, \varepsilon) + u^2(x, \varepsilon) = 1, \quad |x| < 1, \tag{28.107}$$

with

$$u(-1, \varepsilon) = u(1, \varepsilon) = 0, \quad 0 < \varepsilon \ll 1. \tag{28.108}$$

We seek a one-signed solution of equation (28.107) with boundary conditions (28.108); that is, either $u(x) > 0$ for all x in $|x| < 1$ or $u(x) < 0$ for all x in $|x| < 1$. We specify a one-signed solution because, as we shall see later, the absence of this constraint allows the problem to have many solutions. Use the reason given in the previous exercises, show that $u(x, \varepsilon)$ may be well approximated by a solution of

$$u^2 + (1 - x^2)u = 1. \tag{28.109}$$

That is, $u(x, \varepsilon)$ may resemble

$$u_1(x) = \frac{1}{2}[x^2 - 1 + \sqrt{5 - 2x^2 + x^4}],$$

or

$$u_2(x) = \frac{1}{2}[x^2 - 1 - \sqrt{5 - 2x^2 + x^4}].$$

Thus, it is possible that equation (28.107) with boundary conditions (28.108) may have solutions, one resembles u_1, and one resembles u_2. Therefore, we try (using the symmetry from the outset)

$$u(x, \varepsilon) = u_1(x) + w(y) + w(z), \tag{28.110}$$

where

$$y = \frac{x + 1}{\varepsilon^\alpha}, \quad z = \frac{1 - x}{\varepsilon^\alpha}, \quad \alpha > 0.$$

(b) Show that near $y = 0$, we have (ignoring contribution from $w(z)$)

$$\varepsilon u_1''(x) + \varepsilon^{1-2\alpha} w''(y) + 2\varepsilon^{\alpha} y w(y) - \varepsilon^{2\alpha} y^2 w(y) + w^2(y) + 2u_1(-1 + \varepsilon^{a} y) w(y) = 0. \quad (28.111)$$

[The expression in the last term on the left-hand side of equation (28.111) means that u_1 is evaluated at $-1 + \varepsilon^{a} y$.] Show that choosing $\alpha = 1/2$ again and noting that

$$u_1(-1 + \varepsilon^{\alpha} y) = u_1(-1) + \varepsilon^{1/2} y u_1'(-1) + \cdots , \quad (28.112)$$

and an excellent approximation to equation (28.111) for $y = O(1)$ is

$$w'' + 2u_1(-1)w + w^2 = 0, \quad (28.113)$$

where

$$w(\infty) = 0. \quad (28.114)$$

Because of (28.108), we have

$$w(0) = -u_1(-1) = -1. \quad (28.115)$$

(c) Show that any other choice of α will lead to a modification of (28.113) which implies that $w''(y)/w(y)$ is *not* of order unity when $\varepsilon \ll 1$. Also, show that from (28.113), we have

$$(w')^2 + 2w^2 + 2w^3/3 = k \quad (28.116)$$

for some constant k. Since both $w'(y)$ and $w(y)$ must tend to zero as y gets large, $k = 0$ and

$$(w')^2 + 2w^2 + 2w^3/3 = 0. \quad (28.117)$$

Furthermore, since $w(0) = -1$, equation (28.117) implies

$$[w'(0)]^2 = -4/3,$$

which is impossible for $w'(0)$ to be a real number. Thus we conclude that there is no function $w(y)$ such that equation (28.110) is a good description of $u(x, \varepsilon)$.

(d) Turning to the second possibility, i.e., we seek $u(x, \varepsilon)$ in the form

$$u(x, \varepsilon) = u_2(x) + v(y) + v(z). \quad (28.118)$$

By following the manipulation leading formerly to equation (28.113), show that

$$v'' - 2v + v^2 = 0, \quad (28.119)$$

with

$$v(\infty) = 0, \quad v(0) = 1. \tag{28.120}$$

By continuing the process (as before), show that we have

$$(v')^2 - 2v^2 + \frac{2}{3}v^3 = 0,$$

which can be re-arranged to read

$$\frac{dv}{v\sqrt{3-v}} = -\sqrt{\frac{2}{3}}dy, \tag{28.121}$$

where we choose the minus sign, anticipating that v will decrease monotonically as y increases. (The other sign would also lead to a function $u(x, \varepsilon) = U$ which satisfies equation (28.107) with boundary conditions (28.108), but U would not be one-signed.) Show by elementary integration that

$$v = \frac{12e^p}{(1+e^p)^2}, \tag{28.122}$$

where

$$p = y\sqrt{2} + 2\ln(\sqrt{3} + \sqrt{2}). \tag{28.123}$$

Thus, equations (28.118) and (28.122) define an excellent approximation to a one-signed solution of the problem stated in (28.107) and (28.108).

17. By using the method presented above, find the function $u(x, \varepsilon)$ which satisfies

$$\varepsilon u''(x, \varepsilon) - u(x, \varepsilon) = \cos x, \quad 0 < x < \pi/2, \ 0 < \varepsilon \ll 1,$$
$$u(0, \varepsilon) = u(\pi/2, \varepsilon) = 1.$$

Compare the result obtained with the exact solution.

18. Compare the approximate solution and the exact solution of the problem

$$\varepsilon u''(x, \varepsilon) + u'(x, \varepsilon) + u(x, \varepsilon) = 0,$$
$$u(0, \varepsilon) = u'(0, \varepsilon) = 1.$$

19. Can the method in Exercises 14 and 15 be used on the problem $\varepsilon u'' + u = \cos x$ with $u(0, \varepsilon) = u(L, \varepsilon) = 0$? To what feature of the problem can this be attributed?

Chapter 29

A Nested Boundary Layer Problem

29.1 Introduction

In Chapter 27, we have considered a singular perturbation problem with internal boundary layers. In the current chapter, we will discuss another interesting problem; that is, singularly perturbed equations with *nested boundary layers*, which means that one boundary layer lies inside another one. A well-known example is the equation

$$\varepsilon^3 x y''(x) + x^2 y'(x) - (x^3 + \varepsilon) y(x) = 0 \tag{29.1}$$

with boundary conditions

$$y(0) = 1, \quad y(1) = \sqrt{e}; \tag{29.2}$$

see Bender and Orszag (1978, p. 453). By using the method of matched asymptotics, Bender and Orszag (1978) constructed the one term asymptotic solution

$$y_{unif} = \frac{2\sqrt{x}}{\varepsilon} K_1\left(\frac{2\sqrt{x}}{\varepsilon}\right) + e^{-\frac{\varepsilon}{x}} + e^{x^2/2} - 1, \tag{29.3}$$

where the first three terms represent the leading terms in the expansion of the *inner-inner solution, inner solution and outer solution*, respectively, and the fourth term results from the matching of the inner solution and outer solution.

The two boundary layers arise from the two corresponding distinguished limits ε and ε^2. Distinguished limits are those proper stretching which produce a nontrivial balance between two or more terms of the equation. To determine the distinguished limits, we let

$$y(x) = Y(X) \quad \text{and} \quad X = x/\delta.$$

With these new variables, (29.1) becomes

$$\frac{\varepsilon^3}{\delta}X\frac{d^2Y}{dX^2}+\delta X^2\frac{dY}{dX}-\delta^3X^3Y-\varepsilon Y=0. \tag{29.4}$$

For the scaling $\delta=\varepsilon$, the second and fourth terms of (29.4) are of comparable size while the first and third are smaller. Regarding the other choice $\delta=\varepsilon^2$, it is clear that the first and fourth terms are of the same order while the rest are smaller.

As we all know, for single boundary layer problems, the inner solution matches the outer solution and also satisfies one boundary condition. Unfortunately, the inner solution for the boundary value problem (29.1)-(29.2)

$$Y_{in}(X,\varepsilon)=e^{-1/X}\left[1+\varepsilon\left(1-\frac{2}{3X^3}+\frac{1}{4X^4}\right)\right]+\cdots, \tag{29.5}$$

where $X=x/\varepsilon$, does not satisfy the boundary condition $y(0)=1$ since $Y_{in}(X,\varepsilon)$ vanishes exponentially fast as $X\to 0^+$. Intuition suggests that an additional boundary layer very near $x=0$ might be required in order to satisfy the condition $y(0)=1$. Since $\delta=\varepsilon^2$ is the only other distinguished limit for (29.1), Bender and Orszag obtain the inner-inner solution

$$\bar{Y}(\bar{X};\varepsilon)=2\sqrt{\bar{X}}K_1(2\sqrt{\bar{X}})+\cdots, \tag{29.6}$$

where $\bar{X}=x/\varepsilon^2$. Combining (29.5) and (29.6) with the outer solution, one obtains the uniformly valid one-term asymptotic solution (29.3).

However, in lieu of matching the inner-inner solution $\bar{Y}(\bar{X};\varepsilon)$ with the inner solution $Y_{in}(X;\varepsilon)$, Bender and Orszag claimed that the two solutions match automatically because they both vanish exponentially. Indeed, they are exponentially small, but they may not be of the same order. Thus, the solution (29.3) is only valid up to $O(\varepsilon)$ and cannot give the leading term in the matching region where the exact solution may be exponentially small.

The treatment of the nested boundary layers problem is based on Prandtl's principle of matched asymptotics. However, the lack of rigor in Prandtl's boundary layer theory does raise concern from mathematicians who believe that arguments based on purely heuristic reasoning may lead to incorrect results. Examples of this nature can be found in Wong (2004) and Wong and Yang (2002a). With the same purpose, we are going to investigate the problem (29.1)-(29.2) and to derive a uniformly valid asymptotic solution by a mathematically rigorous argument. Moreover, we can find explicitly the exponentially small leading term in the interval where the exact solution exhibits such behavior.

Before proceeding, let us look at one more example: a triple-deck problem. For a nested boundary layer problem, the domain of validity of the outer solution, inner solution and inner-inner solution are called outer region, inner region and inner-inner region, accordingly. Furthermore, these layers are also known as decks, especially by physicists; in other words, they are called right deck, middle deck and left deck, respectively. Hence, a nested boundary layer problem is also referred to as a triple-deck problem. Several typical examples are given in Nayfeh (1981, 1985). In Nayfeh (1981, pp. 304-307), the author considered the boundary value problem

$$\begin{aligned} &\varepsilon^3 y''(x) + x^3 y'(x) + (x^3 - \varepsilon)y(x) = 0 \\ &y(0) = \alpha, \quad y(1) = \beta, \end{aligned} \tag{29.7}$$

and gave the composite expansion

$$y^c(x) = \beta e^{1-x} + \beta e^{1-\varepsilon/2x^2} + \alpha e^{-x/\varepsilon} - \beta e + \cdots . \tag{29.8}$$

To illustrate that $y^c(x)$ is close to the exact solution, Nayfeh presented a figure in Nayfeh (1981, p. 307) to compare $y^c(x)$ with the numerical solution. In Figure 29.1, we plot the numerical solution $y(x)$, Nayfeh's solution $y^c(x)$ and our rigorously established approximate solution $Y(x)$ given in (29.93).

The material in this chapter is taken from Liang and Wong (2009). The material in the Exercises section is taken from Nayfeh (1981, Chapter 12).

29.2 Preliminary transformations

For the general singularly perturbed two-point boundary value problem (B.V.P)

$$\varepsilon y''(x) + a(x)y' + b(x)y(x) = 0 \tag{29.9}$$

with the boundary conditions

$$y(0) = A \quad \text{and} \quad y(1) = B, \tag{29.10}$$

it is now well-known that if $a(x)$ is positive, then the asymptotic solution which holds uniformly in the interval $[0,1]$ is given by

$$y_{unif}(x) = B \exp\left(\int_x^1 \frac{b(t)}{a(t)} dt\right) + \left\{A - B \exp\left(\int_0^1 \frac{b(t)}{a(t)} dt\right)\right\} e^{-a(0)\frac{x}{\varepsilon}}. \tag{29.11}$$

This formula can be easily derived by using Prandtl's boundary layer theory; see, e.g., Bender and Orszag (1978, p. 425). Moreover, y_{unif} in (29.11) approximates the true solution of (29.9) in the sense that

$$y(x) = y_{unif}(x) + O(\varepsilon),$$

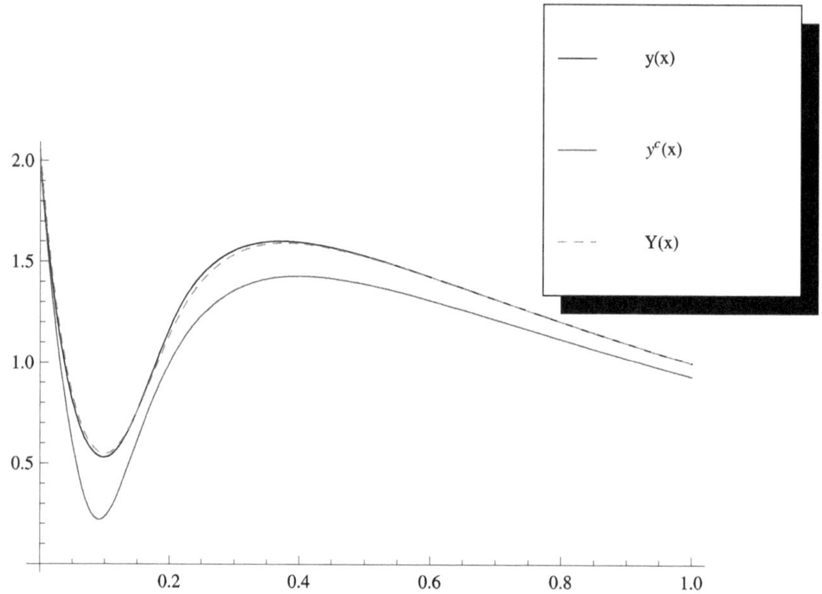

Fig. 29.1: Comparison of Nayfeh's solution $y^c(x)$ and our solution $Y(x)$ with the numerical solution $y(x)$ when $\alpha = 2.0, \beta = 1.0$ and $\varepsilon = 0.05$.

where the O-term is uniform with respect to $x \in [0,1]$; cf. Bender and Orszag (1978, p. 749). As we have mentioned before, heuristic reasoning may lead to incorrect results. For instance, if the boundary value B in (29.10) is zero, then (29.11) becomes

$$y_{unif}(x) = Ae^{-a(0)\frac{x}{\varepsilon}},$$

which is exponentially small for $x > 0$ and asymptotically equal to zero with respect to the error estimate $O(\varepsilon)$. A more accurate formula for the B.V.P (29.9)-(29.10) is

$$\begin{aligned} y(x) =& B\exp\left(\int_x^1 \frac{b(t)}{a(t)}dt\right)[1+O(\varepsilon)] \\ &+ \frac{a(0)}{a(x)}\left\{A - B\exp\left(\int_0^1 \frac{b(t)}{a(t)}dt\right)\right\}\exp\left(\int_0^x \frac{b(t)}{a(t)}dt\right) \\ &\times \exp\left(-\frac{1}{\varepsilon}\int_0^x a(t)dt\right)[1+O(\varepsilon)]. \end{aligned}$$

One can establish this result by using the Liouville-Green (WKB) approximation, which provides good asymptotic solutions to differential equations of the type

$$y''(x) + [\lambda^2 p(x) + q(x)]y(x) = 0, \tag{29.12}$$

where λ is a positive large parameter and $p(x)$ is positive. The basic idea in obtaining an asymptotic solution to (29.12) is to introduce the Liouville transformation

$$\xi = \int p^{1/2}(x)dx \quad \text{and} \quad w = p^{1/4}(x)y(x), \tag{29.13}$$

which convert (29.12) into (exercise)

$$\frac{d^2w}{d\xi^2} + [\lambda^2 + r(\xi)]w = 0, \quad r(\xi) = \frac{5[p'(x)]^2}{16[p(x)]^3} - \frac{p''(x)}{4[p(x)]^2} + \frac{q(x)}{p(x)}. \tag{29.14}$$

When the coefficient function $p(x)$ in (29.12) has a zero (turning point), say $x = x_0$, then there is an ambiguity in taking the square root of the function $p(x)$, and hence the Liouville transformation (29.13) is not well defined. However, a good approximation can be obtained by introducing the transformation of Langer (1949), which can also be used to obtain asymptotic solutions to internal boundary layer problem (e.g., when $a(x)$ in (29.9) has a zero in the interval $(0, 1)$; see Chapter 27).

A third situation arises when $p(x)$ in (29.12) has a simple pole at x_0. Again, the transformation for this case was first introduced by Langer, extended and popularized by Olver (1974a). Our treatment of (29.1) is based on this transformation. The key idea behind this transformation is to get a good approximation equation for (29.1), and to estimate the error term by using the method of successive approximation.

Motivated by the *inner-inner solution* in (29.3), we will try to derive an equation which is a perturbation of Weber's equation for modified Bessel functions. As we all know, the ordinary differential equation

$$\frac{d^2w}{dz^2} = \frac{1}{4z}w \tag{29.15}$$

has two linearly independent solutions

$$\sqrt{z}K_1(\sqrt{z}) \quad \text{and} \quad \sqrt{z}I_1(\sqrt{z}).$$

Hence, we will seek a transformation which converts (29.1) into a perturbed equation of (29.15). To this end, we introduce new independent and dependent variables

$$\zeta = \zeta(x) \quad \text{and} \quad y(x) = A(x)U(\zeta(x)). \tag{29.16}$$

Straightforward calculation gives

$$\frac{dy}{dx} = A'(x)U(\zeta(x)) + A(x)\frac{dU}{d\zeta}\frac{d\zeta}{dx}$$

and

$$\frac{d^2y}{dx^2} = A''(x)U(\zeta) + 2A'(x)\frac{dU}{d\zeta}\frac{d\zeta}{dx} + A(x)\left(\frac{dU}{d\zeta}\frac{d^2\zeta}{dx^2} + \frac{d^2U}{d\zeta^2}\left(\frac{d\zeta}{dx}\right)^2\right).$$

With the new variables, (29.1) becomes

$$\varepsilon^3 A(x)\left(\frac{d\zeta}{dx}\right)^2\frac{d^2U}{d\zeta^2} + \left(2\varepsilon^3 A'(x)\frac{d\zeta}{dx} + \varepsilon^3 A(x)\frac{d^2\zeta}{dx^2} + xA(x)\frac{d\zeta}{dx}\right)\frac{dU}{d\zeta} + \left(\varepsilon^3 A''(x) + xA'(x) - (x^2 + \frac{\varepsilon}{x})A(x)\right)U = 0 \tag{29.17}$$

To eliminate the first order derivative of $U(\zeta)$, we set

$$2\varepsilon^3 A'(x)\frac{d\zeta}{dx} + \varepsilon^3 A(x)\frac{d^2\zeta}{dx^2} + xA(x)\frac{d\zeta}{dx} = 0. \tag{29.18}$$

Upon solving (29.18), we have

$$A(x) = (\zeta'(x))^{-\frac{1}{2}}\exp\left\{-\frac{x^2}{4\varepsilon^3}\right\}. \tag{29.19}$$

Substituting (29.19) into (29.17) yields

$$\frac{d^2U}{d\zeta^2} - \frac{U}{4\zeta} = \left\{\frac{1}{\varepsilon^3}\frac{1}{(\zeta'(x))^2}\left(\frac{x^2}{4\varepsilon^3} + x^2 + \frac{1}{2} + \frac{\varepsilon}{x}\right) - \frac{1}{4\zeta}\right\}U + \left\{\frac{\zeta(x)}{(\zeta'(x))^3}\left(2\zeta'''(x) - \frac{3(\zeta''(x))^2}{\zeta'(x)}\right)\right\}\frac{U}{4\zeta}. \tag{29.20}$$

Again, set

$$\frac{1}{\varepsilon^3}\frac{1}{(\zeta'(x))^2}\left(\frac{x^2}{4\varepsilon^3} + x^2 + \frac{1}{2} + \frac{\varepsilon}{x}\right) - \frac{1}{4\zeta} = 0. \tag{29.21}$$

Solving (29.21), we obtain

$$\zeta^{\frac{1}{2}}(x) = \frac{1}{\varepsilon}\int_0^x \sqrt{\frac{1}{t} + \frac{t^2}{4\varepsilon^4} + \frac{1}{2\varepsilon}(1 + 2t^2)}dt. \tag{29.22}$$

As a consequence, $\zeta = \zeta(x)$ is a well-defined and one-to-one mapping from $[0, 1]$ into $[\zeta_-, \zeta_+]$, where

$$\zeta_- = \zeta(0) = 0 \quad \text{and} \quad \zeta_+ = \zeta(1).$$

Let us summarize the result in the following lemma.

Lemma 29.1. *Under the transformations* (29.16), (29.19) *and* (29.22), (29.1) *is converted into*

$$\frac{d^2U}{d\zeta^2} - \frac{U}{4\zeta} = \phi(\zeta)\frac{U}{4\zeta}, \tag{29.23}$$

where

$$\phi(\zeta) = \frac{\zeta(x)}{(\zeta'(x))^3}\left(2\zeta'''(x) - \frac{3(\zeta''(x))^2}{\zeta'(x)}\right). \tag{29.24}$$

29.3 Conversion to an integral equation

We construct two linearly independent solutions $W_1(\zeta)$ and $W_2(\zeta)$ to the associated homogeneous equation of (29.23), namely,

$$\frac{d^2U}{d\zeta^2} - \frac{U}{4\zeta} = 0, \tag{29.25}$$

such that

$$W_1(\zeta_-) = 1, \quad W_1(\zeta_+) = 0, \quad W_2(\zeta_-) = 0, \quad W_2(\zeta_+) = 1.$$

By straightforward calculation, we have

$$W_1(\zeta) = \sqrt{\zeta}K_1(\sqrt{\zeta}) - P(\sqrt{\zeta})I_1(\sqrt{\zeta}),$$
$$W_2(\zeta) = \frac{\sqrt{\zeta}I_1(\sqrt{\zeta})}{Q},$$

where

$$P = \frac{K_1(\sqrt{\zeta_+})}{I_1(\sqrt{\zeta_+})}, \quad Q = \sqrt{\zeta_+}I_1(\sqrt{\zeta_+}).$$

Moreover,

$$\mathcal{W}\{W_1(\zeta), W_2(\zeta)\} = \frac{1}{2Q}.$$

By considering the right-hand side of (29.23) as an inhomogeneous term of the homogeneous equation (29.25), we can convert (29.23) into the integral equation

$$U(\zeta) = \lambda W_1(\zeta) + \mu W_2(\zeta) + \int_{\zeta_-}^{\zeta_+} G(\zeta, s)\phi(s)\frac{U(s)}{4s}ds, \tag{29.26}$$

where λ, μ are arbitrary constants, and $G(\zeta, s)$ is the Green's function defined by

$$G(\zeta, s) = \begin{cases} \frac{W_1(\zeta)W_2(s)}{\mathcal{W}\{W_1(s), W_2(s)\}} = 2QW_1(\zeta)W_2(s), & \zeta_- \le s \le \zeta \le \zeta_+, \\ \frac{W_1(s)W_2(\zeta)}{\mathcal{W}\{W_1(s), W_2(s)\}} = 2QW_1(s)W_2(\zeta), & \zeta_- \le \zeta \le s \le \zeta_+. \end{cases} \tag{29.27}$$

29.4 Construction of the solution

Lemma 29.2. *Let ρ be a positive constant. We have the order estimate*

$$\phi(x) = O\left(\frac{x^2}{\varepsilon^2}\right) \quad \text{for} \;\; 0 \le x \le \rho\varepsilon^{4/3} \tag{29.28}$$

$$\phi(x) = O\left(\frac{\varepsilon^6}{x^4}\right) \quad \text{for} \;\; \rho\varepsilon^{4/3} \le x \le 1. \tag{29.29}$$

Proof. For convenience, we put

$$f(t) = \sqrt{\frac{1}{t} + \frac{t^2}{4\varepsilon^4} + \frac{1}{2\varepsilon}(1+2t^2)} \quad \text{for} \;\; 0 < t \le 1. \tag{29.30}$$

From (29.22), (29.24) and (29.30), it follows that

$$\phi(\zeta) = \phi_1(x) + \phi_2(x),$$

where

$$\phi_1(x) = -\frac{3}{4}\frac{1}{\zeta(x)} \tag{29.31}$$

and

$$\phi_2(x) = \varepsilon^2 f^{-6}(x)\left\{\frac{3}{16}\frac{1}{x^4} + \frac{1}{4\varepsilon}\frac{1}{x^3} + \frac{9}{4x}\left(\frac{1}{4\varepsilon^4} + \frac{1}{\varepsilon}\right) + \frac{1}{4\varepsilon}\left(\frac{1}{4\varepsilon^4} + \frac{1}{\varepsilon}\right) - \frac{3x^2}{4}\left(\frac{1}{4\varepsilon^4} + \frac{1}{\varepsilon}\right)^2\right\}. \tag{29.32}$$

Let us first consider the case $0 < t \le x \le \rho\varepsilon^{4/3}$. Assume temporarily that x is much smaller than $\varepsilon^{4/3}$ (i.e., $x \ll \varepsilon^{4/3}$), so that

$$f(t) = \frac{1}{\sqrt{t}}\left\{1 + \frac{1}{2}\left(\frac{t^3}{4\varepsilon^4} + \frac{t}{2\varepsilon}\right)\left[1 + O\left(\frac{t^3}{\varepsilon^4}, \frac{t}{\varepsilon}\right)\right]\right\}, \tag{29.33}$$

$$\zeta^{\frac{1}{2}}(x) = \frac{2\sqrt{x}}{\varepsilon}\left\{1 + \left(\frac{x^3}{56\varepsilon^4} + \frac{x}{12\varepsilon}\right)\left[1 + O\left(\frac{x^3}{\varepsilon^4}, \frac{x}{\varepsilon}\right)\right]\right\}, \tag{29.34}$$

and

$$\phi_1(x) = -\frac{3\varepsilon^2}{16x} + \left(\frac{3x^2}{448\varepsilon^2} + \frac{\varepsilon}{32}\right)\left[1 + O\left(\frac{x^3}{\varepsilon^4}, \frac{x}{\varepsilon}\right)\right], \tag{29.35}$$

where $O(h, k)$ is used to indicate that the error term is $O(h) + O(k)$. Coupling (29.32) and (29.33) yields

$$\phi_2(x) = \frac{3\varepsilon^2}{16x} + \left(\frac{27x^2}{64\varepsilon^2} - \frac{\varepsilon}{32}\right)\left[1 + O\left(\frac{x^3}{\varepsilon^4}, \frac{x}{\varepsilon}, \varepsilon^3\right)\right]. \tag{29.36}$$

Hence,

$$\phi(x) = \phi_1(x) + \phi_2(x) = \frac{3x^2}{7\varepsilon^2}\left[1 + O\left(\frac{x^3}{\varepsilon^4}, \frac{x}{2\varepsilon}, \varepsilon^3\right)\right] = O\left(\frac{x^2}{\varepsilon^2}\right). \quad (29.37)$$

Since we are only interested in the leading order estimate for $\phi(x)$, our earlier assumption "$x \ll \varepsilon^{4/3}$" can now be dropped because x^3/ε^4 is still bounded.

For the second case $\rho\varepsilon^{4/3} \leq x \leq 1$, (29.22) and (29.30) imply that

$$\zeta^{1/2}(x) = \frac{1}{\varepsilon}\int_0^x f(t)dt \geq \frac{1}{\varepsilon}\int_0^x \frac{t}{2\varepsilon^2}dt = \frac{x^2}{4\varepsilon^3}.$$

Thus, we have

$$|\phi_1(x)| = \left| -\frac{3}{4}\frac{1}{\zeta(x)} \right| \leq \frac{12\varepsilon^6}{x^4} = O\left(\frac{\varepsilon^6}{x^4}\right).$$

In a similar manner, it can be shown that $\phi_2(x) = O(\varepsilon^6/x^4)$. □

Lemma 29.2 implies that $\phi(x)$ is much smaller than 1, thus we can use the solution of the homogeneous equation (29.25) to approximate the solution of the inhomogeneous equation (29.23). In the following lemma, we are going to estimate the error term caused by discarding the inhomogeneous term in (29.23).

Lemma 29.3. *As $\varepsilon \to 0$, we have*

$$\int_{\zeta_-}^{\zeta_+} G(\zeta, s)\frac{|\phi(s)|}{4s}W_1(s)ds = W_1(\zeta)O(\varepsilon^{1/3}), \quad (29.38)$$

$$\int_{\zeta_-}^{\zeta_+} G(\zeta, s)\frac{|\phi(s)|}{4s}W_2(s)ds = W_2(\zeta)O(\varepsilon^{1/3}). \quad (29.39)$$

Proof. Define

$$M_1(\zeta, \varepsilon) := \int_{\zeta_-}^{\zeta} G(\zeta, s)\frac{|\phi(s)|}{4s}W_1(s)ds,$$

$$M_2(\zeta, \varepsilon) := \int_{\zeta}^{\zeta_+} G(\zeta, s)\frac{|\phi(s)|}{4s}W_1(s)ds,$$

$$N_1(\zeta, \varepsilon) := \int_{\zeta_-}^{\zeta} G(\zeta, s)\frac{|\phi(s)|}{4s}W_2(s)ds,$$

$$N_2(\zeta, \varepsilon) := \int_{\zeta}^{\zeta_+} G(\zeta, s)\frac{|\phi(s)|}{4s}W_2(s)ds.$$

First of all, we recall some asymptotic formulas of $K_1(z)$ and $I_1(z)$:

$$K_1(z) \sim \frac{1}{z} \quad \text{as} \quad z \to 0, \tag{29.40}$$

$$I_1(z) \sim \frac{z}{2} \quad \text{as} \quad z \to 0, \tag{29.41}$$

$$K_1(z) \sim \sqrt{\frac{\pi}{2z}}e^{-z} \quad \text{as} \quad z \to \infty, \tag{29.42}$$

$$I_1(z) \sim \frac{e^z}{\sqrt{2\pi z}} \quad \text{as} \quad z \to \infty. \tag{29.43}$$

By making the change of variable given in (29.22) with z and s replacing x and ζ, we obtain

$$\begin{aligned} M_1(\zeta, x) = O(1)W_1(\zeta) \int_0^x & I_1(\sqrt{s(z)})K_1(\sqrt{s(z)})|\phi(s(z))| \\ & \times 2\sqrt{s(z)}\frac{1}{\varepsilon}\sqrt{\frac{1}{z} + \frac{z^2}{4\varepsilon^4} + \frac{1}{2\varepsilon}(1+2z^2)}dz. \end{aligned} \tag{29.44}$$

To proceed further, we divide our discussion into three cases. Let k_1, k_2 be positive constants.

Case I: $0 < x \leq k_1\varepsilon^2$.

For $0 < z < x$, we have from (29.33), (29.34) and (29.37)

$$f(z) = \frac{1}{\sqrt{z}}\left[1 + O\left(\frac{z^3}{\varepsilon^4}, \frac{z}{\varepsilon}\right)\right], \tag{29.45}$$

$$s^{1/2}(z) = \frac{2\sqrt{z}}{\varepsilon}\left[1 + O\left(\frac{z^3}{\varepsilon^4}, \frac{z}{\varepsilon}\right)\right], \tag{29.46}$$

$$\phi(z) = \frac{3z^2}{7\varepsilon^2}\left[1 + O\left(\frac{z^3}{\varepsilon^4}, \frac{z}{\varepsilon}, \varepsilon^3\right)\right]. \tag{29.47}$$

A combination of (29.40), (29.41), (29.44), (29.45), (29.46) and (29.47) gives

$$\begin{aligned} M_1(\zeta, \varepsilon) &= O(1)W_1(\zeta)\int_0^x \frac{\sqrt{s}}{2}\frac{1}{\sqrt{s}}\frac{3}{7}\frac{z^2}{\varepsilon^2}\frac{4\sqrt{z}}{\varepsilon}\frac{1}{\varepsilon}\frac{1}{\sqrt{z}}dz \\ &= W_1(\zeta)O(\varepsilon^2). \end{aligned} \tag{29.48}$$

Case II: $k_1\varepsilon^2 \leq x \leq k_2\varepsilon^{4/3}$.

On account of (29.42), (29.43), (29.46), (29.47) and (29.48), we obtain

$$\begin{aligned} M_1(\zeta,\varepsilon) &= W_1(\zeta)O(\varepsilon^2) \\ &\quad + O(1)W_1(\zeta)\int_{k_1\varepsilon^2}^{x} \frac{e^{\sqrt{s}}}{\sqrt{2\pi}s^{1/4}}\sqrt{\frac{\pi}{2}}\frac{e^{-\sqrt{s}}}{s^{1/4}}\frac{3}{7}\frac{z^2}{\varepsilon^2}2\sqrt{s}\frac{1}{\varepsilon}\frac{1}{\sqrt{z}}dz \\ &= W_1(\zeta)O(\varepsilon^{1/3}). \end{aligned} \tag{29.49}$$

Case III: $k_2\varepsilon^{4/3} \le x \le 1$.

In a similar manner, we have

$$\begin{aligned} M_1(\zeta,\varepsilon) &= W_1(\zeta)O(\varepsilon^{1/3}) \\ &\quad + O(1)W_1(\zeta)\int_{k_2\varepsilon^{4/3}}^{x} \frac{e^{\sqrt{s}}}{\sqrt{2\pi}s^{1/4}}\sqrt{\frac{\pi}{2}}\frac{e^{-\sqrt{s}}}{s^{1/4}}\frac{\varepsilon^6}{z^4}2\sqrt{s}\frac{1}{\varepsilon}\frac{z}{2\varepsilon^2}dz \\ &= W_1(\zeta)O(\varepsilon^{1/3}). \end{aligned} \tag{29.50}$$

Therefore,

$$M_1(\zeta,\varepsilon) = W_1(\zeta)O(\varepsilon^{1/3}) \quad \text{for} \quad \zeta_- \le \zeta \le \zeta_+.$$

For real positive z, $K_1(z)$ is a monotonically decreasing function, while $I_1(z)$ is a strictly increasing function. Making use of these properties, we get

$$\begin{aligned} M_2(\zeta,\varepsilon) &= \frac{1}{2}W_1(\zeta)\int_{\zeta}^{\zeta_+} I_1(\sqrt{\zeta})|\phi(s)|\frac{|K_1(\sqrt{s}) - PI_1(\sqrt{s})|^2}{(K_1(\sqrt{\zeta}) - PI_1(\sqrt{\zeta}))}ds \\ &\le W_1(\zeta)\int_{\zeta}^{\zeta_+} I_1(\sqrt{s})|\phi(s)|(K_1(\sqrt{s}) - PI_1(\sqrt{s}))ds \\ &= W_1(\zeta)O(\varepsilon^{1/3}). \end{aligned}$$

Hence, as $\varepsilon \to 0$,

$$\int_{\zeta_-}^{\zeta_+} G(\zeta,s)\frac{|\phi(s)|}{4s}W_1(s)ds = M_1(\zeta,\varepsilon) + M_2(\zeta,\varepsilon) = W_1(\zeta)O(\varepsilon^{1/3}) \tag{29.51}$$

uniformly for all $\zeta \in [\zeta_-,\zeta_+]$.

Similar arguments show that

$$N_1(\zeta,\varepsilon) = W_1(\zeta)O(\varepsilon^{1/3}) \quad \text{and} \quad N_2(\zeta,\varepsilon) = W_2(\zeta)O(\varepsilon^{1/3}).$$

Thus, (29.39) follows. □

Remark 29.1. Lemma 29.3 implies that there exist positive constants ε_0 and γ_0 such that for $0 < \varepsilon \le \varepsilon_0$,

$$\int_{\zeta_-}^{\zeta_+} G(\zeta,s)\frac{|\phi(s)|}{4s}W_1(s)ds \le W_1(\zeta)\gamma_0\varepsilon^{1/3} \tag{29.52}$$

and

$$\int_{\zeta_-}^{\zeta_+} G(\zeta,s)\frac{|\phi(s)|}{4s}W_2(s)ds \le W_2(\zeta)\gamma_0\varepsilon^{1/3}. \tag{29.53}$$

Theorem 29.1. *There exists a constant $\varepsilon_0 > 0$ such that for $0 < \varepsilon \le \varepsilon_0$, equation (29.23) has two linearly independent solutions*

$$w_1(\zeta) = \sqrt{\zeta}K_1(\sqrt{\zeta})[1 + O(\varepsilon^{\frac{1}{3}})] \tag{29.54}$$

and

$$w_2(\zeta) = \sqrt{\zeta}I_1(\sqrt{\zeta})[1 + O(\varepsilon^{\frac{1}{3}})], \tag{29.55}$$

where the O-symbols hold uniformly with respect to $\zeta \in [\zeta_-, \zeta_+]$.

Proof. In (29.26), we let $\lambda = 1$ and $\mu = 0$. Then

$$U(\zeta) = W_1(\zeta) + \int_{\zeta_-}^{\zeta_+} G(\zeta, s)\frac{\phi(s)}{4s}U(s)ds. \tag{29.56}$$

We define a sequence $U_j(\zeta)$, $j = 0, 1, \cdots$, by $U_0(\zeta) = 0$ and

$$U_j(\zeta) = W_1(\zeta) + \int_{\zeta_-}^{\zeta_+} G(\zeta, s)\frac{\phi(s)}{4s}U_{j-1}(s)ds \quad (j \ge 1). \tag{29.57}$$

Clearly, $U_1(\zeta) = W_1(\zeta)$. Now suppose that for a particular value j, we have

$$|U_j(\zeta) - U_{j-1}(\zeta)| \le W_1(\zeta)(\gamma_0\varepsilon^{\frac{1}{3}})^{j-1}, \tag{29.58}$$

as indeed in the case when $j = 1$. From (29.57), we get

$$U_{j+1}(\zeta) - U_j(\zeta) = \int_{\zeta_-}^{\zeta_+} G(\zeta, s)\frac{\phi(s)}{4s}[U_j(s) - U_{j-1}(s)]ds, \quad j \ge 1. \tag{29.59}$$

Hence, (29.52) and (29.59) give

$$\begin{aligned} |U_{j+1}(\zeta) - U_j(\zeta)| &\le \int_{\zeta_-}^{\zeta_+} G(\zeta, s)\frac{|\phi(s)|}{4s}W_1(s)(\gamma_0\varepsilon^{\frac{1}{3}})^{j-1}ds \\ &= (\gamma_0\varepsilon^{\frac{1}{3}})^{j-1}\int_{\zeta_-}^{\zeta_+} G(\zeta, s)\frac{|\phi(s)|}{4s}W_1(s)ds \\ &\le W_1(\zeta)(\gamma_0\varepsilon^{\frac{1}{3}})^{j}. \end{aligned} \tag{29.60}$$

By induction, (29.58) holds for all $j \ge 1$, and the series

$$\widetilde{U}_1(\zeta) := \sum_{j=0}^{\infty}[U_{j+1}(\zeta) - U_j(\zeta)]$$

converges. From (29.60), it follows that

$$\begin{aligned} \widetilde{U}_1(\zeta) &\le U_1(\zeta) + \sum_{j=1}^{\infty}|U_{j+1}(\zeta) - U_j(\zeta)| \\ &\le W_1(\zeta) + \sum_{j=1}^{\infty}W_1(\zeta)(\gamma_0\varepsilon^{\frac{1}{3}})^{j} \\ &= W_1(\zeta)[1 + O(\varepsilon^{\frac{1}{3}})]. \end{aligned} \tag{29.61}$$

Since $\widetilde{U}_1(\zeta) = \lim_{n\to\infty} U_n(\zeta) = \lim_{n\to\infty}\sum_{j=0}^{n-1}[U_{j+1}(\zeta) - U_j(\zeta)]$, taking limits (as $j \to \infty$) on both sides of (29.57) shows that equation (29.23) has a solution $\widetilde{U}_1(\zeta)$ satisfying

$$\begin{aligned}\widetilde{U}_1(\zeta) &= W_1(\zeta)[1 + O(\varepsilon^{\frac{1}{3}})]\\ &= \left[\sqrt{\zeta}K_1(\sqrt{\zeta}) - P\sqrt{\zeta}I_1(\sqrt{\zeta})\right][1 + O(\varepsilon^{\frac{1}{3}})]. \qquad (29.62)\end{aligned}$$

Similar arguments show that equation (29.23) has another solution

$$\widetilde{U}_2(\zeta) = W_2(\zeta)[1 + O(\varepsilon^{\frac{1}{3}})] = \frac{\sqrt{\zeta}I_1(\sqrt{\zeta})}{Q}[1 + O(\varepsilon^{\frac{1}{3}})]. \qquad (29.63)$$

Since P and Q depend only on ε, and $w_1(\zeta)$ and $w_2(\zeta)$ are just linear combinations of $\widetilde{U}_1(\zeta)$ and $\widetilde{U}_2(\zeta)$, the result stated in the theorem is proved. □

Now, we are ready to construct the solution to boundary value problem (29.1)-(29.2). Coupling (29.19) and (29.22), we have

$$A(x) = \sqrt{\frac{\varepsilon}{2}}\left(\frac{1}{x} + \frac{x^2}{4\varepsilon^4} + \frac{1}{2\varepsilon}(1 + 2x^2)\right)^{-\frac{1}{4}} \exp\left\{-\frac{x^2}{4\varepsilon^3}\right\}\zeta^{-\frac{1}{4}}(x). \qquad (29.64)$$

By Theorem 29.1, it is straightforward to show that boundary value problem (29.1)-(29.2) has two linearly independent solutions

$$\begin{aligned}y_1(x) = &\sqrt{\frac{\varepsilon}{2}}\left(\frac{1}{x} + \frac{x^2}{4\varepsilon^4} + \frac{1}{2\varepsilon}(1 + 2x^2)\right)^{-\frac{1}{4}}\\ &\cdot \exp\left\{-\frac{x^2}{4\varepsilon^3}\right\}\zeta^{\frac{1}{4}}(x)K_1(\sqrt{\zeta})(1 + O(\varepsilon^{\frac{1}{3}})), \qquad (29.65)\end{aligned}$$

$$\begin{aligned}y_2(x) = &\sqrt{\frac{\varepsilon}{2}}\left(\frac{1}{x} + \frac{x^2}{4\varepsilon^4} + \frac{1}{2\varepsilon}(1 + 2x^2)\right)^{-\frac{1}{4}}\\ &\cdot \exp\left\{-\frac{x^2}{4\varepsilon^3}\right\}\zeta^{\frac{1}{4}}(x)I_1(\sqrt{\zeta})(1 + O(\varepsilon^{\frac{1}{3}})), \qquad (29.66)\end{aligned}$$

In view of (29.34), (29.40), (29.41) and (29.42), we obtain

$$\lim_{x\to 0} y_1(x) = \frac{\varepsilon}{2}(1 + O(\varepsilon^{\frac{1}{3}})), \quad \lim_{x\to 0} y_2(x) = 0, \qquad (29.67)$$

$$y_1(1) = \frac{\sqrt{\varepsilon\pi}}{2}\left(\frac{1}{4\varepsilon^4}\right)^{-\frac{1}{4}} \exp\left\{-\frac{1}{4\varepsilon^3}\right\}\exp\{-\sqrt{\zeta(1)}\}(1 + O(\varepsilon^{\frac{1}{3}})). \qquad (29.68)$$

The last estimate indicates that $y_1(1)$ is exponentially small as $\varepsilon \to 0$.

Let us define

$$Y_1(x) = \frac{2}{\varepsilon} y_1(x), \tag{29.69}$$

$$Y_2(x) = \frac{\sqrt{e}\, y_2(x)}{y_2(1)}. \tag{29.70}$$

The solution to the boundary value problem (29.1)-(29.2) is given by

$$Y(x) = Y_1(x) + \left(1 - \frac{Y_1(1)}{Y_2(1)}\right) Y_2(x). \tag{29.71}$$

Note that $Y_1(1)/Y_2(1) = 2y_1(1)/(\varepsilon\sqrt{e})$ is exponentially small as ε tends to 0.

In summary, we have the following theorem.

Theorem 29.2. *There exists a constant $\varepsilon_0 > 0$ such that for $0 < \varepsilon \leq \varepsilon_0$, the boundary value problem* (29.1)-(29.2) *has the unique solution*

$$Y(x) = Y_1(x) + Y_2(x), \tag{29.72}$$

where $Y_1(x)$ and $Y_2(x)$ are given by (29.69) *and* (29.70), *respectively.*

29.5 Simplification of our solution

To simplify our solution (29.72), we have to study the behaviour of $\zeta^{1/2}(x)$.

Theorem 29.3. *There is a positive number ρ_0 such that*

$$\zeta^{\frac{1}{2}}(x) = 2\frac{\sqrt{x}}{\varepsilon}\left(1 + O\left(\frac{x^3}{\varepsilon^4}, \varepsilon^{\frac{1}{3}}\right)\right), \tag{29.73}$$

for $0 \leq x \leq \rho_0 \varepsilon^{\frac{4}{3}}$, and

$$\begin{aligned}\zeta^{\frac{1}{2}}(x) =& \frac{c_0}{\varepsilon^{1/3}} + \frac{x^2}{4\varepsilon^3} + \frac{1}{2}\log x - \frac{2}{3}\log\varepsilon + \frac{x^2}{2} \\ & - \frac{\varepsilon}{x}\left(1 + O\left(\frac{\varepsilon^4}{x^3}, \varepsilon^{\frac{1}{3}}\right)\right) + O(\varepsilon^{\frac{1}{3}}\log\varepsilon),\end{aligned} \tag{29.74}$$

for $\rho_0 \varepsilon^{\frac{4}{3}} \leq x \leq 1$.

Proof. By the definition of $\zeta^{1/2}(x)$ in (29.22), we have

$$\zeta^{\frac{1}{2}}(x) = \frac{1}{\varepsilon}\int_0^x \sqrt{\left(\frac{t^2}{4\varepsilon^4} + \frac{1}{t}\right)\left(1 + \frac{\frac{1}{2\varepsilon}(1+2t^2)}{\frac{t^2}{4\varepsilon^4} + \frac{1}{t}}\right)}\, dt. \tag{29.75}$$

It is easy to verify that

$$\frac{\frac{1}{2\varepsilon}(1+2t^2)}{\frac{t^2}{4\varepsilon^4}+\frac{1}{t}} = O(\varepsilon^{\frac{1}{3}}) \quad \text{for} \quad 0 < t \leq 1. \tag{29.76}$$

By the binomial expansion, we have

$$\begin{aligned}
\zeta^{\frac{1}{2}}(x) =& \varepsilon^{-\frac{1}{3}} \int_0^{x/\varepsilon^{4/3}} \sqrt{\frac{t^2}{4}+\frac{1}{t}}dt \\
&+ \frac{1}{4}\int_0^{x/\varepsilon^{4/3}} \frac{1}{\sqrt{\frac{t^2}{4}+\frac{1}{t}}}dt \left[1+O(\varepsilon^{\frac{1}{3}})\right] \\
&+ \frac{1}{2\varepsilon^2}\int_0^{x} \frac{t^2}{\sqrt{\frac{t^2}{4\varepsilon^4}+\frac{1}{t}}}dt \left[1+O(\varepsilon^{\frac{1}{3}})\right].
\end{aligned} \tag{29.77}$$

The first two integrals on the right-hand side of (29.77) can be expressed in terms of hypergeometric function and inverse hyperbolic function, respectively. Hence, we have (exercise)

$$\begin{aligned}
\zeta^{\frac{1}{2}}(x) =& \frac{2\sqrt{x}}{\varepsilon}{}_2F_1(-\frac{1}{2},\frac{1}{6};\frac{7}{6};-\frac{x^3}{4\varepsilon^4}) \\
&+ \frac{1}{3}\operatorname{arcsinh}\left(\frac{x^{\frac{3}{2}}}{2\varepsilon^2}\right)\left[1+O(\varepsilon^{\frac{1}{3}})\right] \\
&+ \frac{1}{2\varepsilon^2}\int_0^{x} \frac{t^2}{\sqrt{\frac{t^2}{4\varepsilon^4}+\frac{1}{t}}}dt \left[1+O(\varepsilon^{\frac{1}{3}})\right].
\end{aligned} \tag{29.78}$$

This result can be also be verified by using integral representation of the above two special functions. Now the theorem follows from (29.78) and the Lemmas 29.4, 29.5 and 29.6. □

Lemma 29.4.

$$\frac{2\sqrt{x}}{\varepsilon}{}_2F_1(-\frac{1}{2},\frac{1}{6};\frac{7}{6};-\frac{x^3}{4\varepsilon^4}) = \frac{2\sqrt{x}}{\varepsilon}\left(1+O\left(\frac{x^3}{\varepsilon^4}\right)\right) \quad \textit{for} \;\; 0 \leq x \leq \rho_0\varepsilon^{4/3} \tag{29.79}$$

and

$$\frac{2\sqrt{x}}{\varepsilon}{}_2F_1(-\frac{1}{2},\frac{1}{6};\frac{7}{6};-\frac{x^3}{4\varepsilon^4}) = \frac{c_0}{\varepsilon^{1/3}}+\frac{x^2}{4\varepsilon^3}-\frac{\varepsilon}{x}\left(1+O\left(\frac{\varepsilon^4}{x^3}\right)\right) \quad \textit{for} \;\; \rho_0\varepsilon^{4/3} \leq x \leq 1, \tag{29.80}$$

where $c_0 = \frac{\Gamma(-\frac{2}{3})\Gamma(\frac{7}{6})2^{4/3}}{\Gamma(-\frac{1}{2})} \approx 2.64996$ *and* $\rho_0 = 2^{2/3}$.

Proof. We know that the hypergeometric series

$$ {}_2F_1(a,b;c;z) = \sum_{n=0}^{\infty} \frac{(a)_n(b)_n}{(c)_n}\frac{z^n}{n!} \tag{29.81} $$

is absolutely convergent for $|z| \leq 1$, provided that $\mathrm{Re}(c-a-b) > 0$. In particular, ${}_2F_1(-\frac{1}{2},\frac{1}{6};\frac{7}{6};-\frac{x^3}{4\varepsilon^4})$ is convergent for $|\frac{x^3}{4\varepsilon^4}| \leq 1$ and (29.79) follows.

Referring to Abramowitz and Stegun (1992, p. 559), we have the following connection formula

$$ \begin{aligned} {}_2F_1(a,b;c;z) =& \frac{\Gamma(c)\Gamma(b-a)}{\Gamma(b)\Gamma(c-a)}(-z)^{-a}{}_2F_1(a,1-c+a;1-b+a;\frac{1}{z}) \\ &+ \frac{\Gamma(c)\Gamma(a-b)}{\Gamma(a)\Gamma(c-b)}(-z)^{-b}{}_2F_1(b,1-c+b;1-a+b;\frac{1}{z}) \\ &\text{for } |\arg(-z)| < \pi. \end{aligned} \tag{29.82} $$

With $a=-\frac{1}{2}, b=\frac{1}{6}, c=\frac{7}{6}$ and $z=-\frac{x^3}{4\varepsilon^4}$, equation (29.82) becomes

$$ \begin{aligned} {}_2F_1\left(-\frac{1}{2},\frac{1}{6};\frac{7}{6};-\frac{x^3}{4\varepsilon^4}\right) =& \frac{\Gamma(\frac{7}{6})\Gamma(\frac{2}{3})}{\Gamma(\frac{1}{6})\Gamma(\frac{10}{6})}\left(\frac{x^3}{4\varepsilon^4}\right)^{\frac{1}{2}} {}_2F_1\left(-\frac{1}{2},-\frac{2}{3};\frac{1}{3};-\frac{4\varepsilon^4}{x^3}\right) \\ &+ \frac{\Gamma(\frac{7}{6})\Gamma(-\frac{2}{3})}{\Gamma(-\frac{1}{2})\Gamma(1)}\left(\frac{x^3}{4\varepsilon^4}\right)^{-\frac{1}{6}} {}_2F_1\left(\frac{1}{6},0;\frac{5}{3};-\frac{4\varepsilon^4}{x^3}\right). \end{aligned} \tag{29.83} $$

Furthermore, it can be simplified to

$$ {}_2F_1\left(-\frac{1}{2},\frac{1}{6};\frac{7}{6};-\frac{x^3}{4\varepsilon^4}\right) = \frac{1}{8}\frac{x^{3/2}}{\varepsilon^2}{}_2F_1\left(-\frac{1}{2},-\frac{2}{3};\frac{1}{3};-\frac{4\varepsilon^4}{x^3}\right) + c_0\frac{\varepsilon^{2/3}}{2x^{1/2}}, \tag{29.84} $$

where $c_0 = \frac{\Gamma(\frac{7}{6})\Gamma(-\frac{2}{3})2^{4/3}}{\Gamma(-\frac{1}{2})} \approx 2.64996$. Multiplying both sides of (29.84) by $\frac{2\sqrt{x}}{\varepsilon}$ and making use of (29.81), we obtain (29.80). □

Lemma 29.5.

$$ \frac{1}{3}\operatorname{arcsinh}\left(\frac{x^{3/2}}{2\varepsilon^2}\right) = \frac{x^{3/2}}{6\varepsilon^2}\left[1+O\left(\frac{x^3}{\varepsilon^4}\right)\right] \quad \textit{for } 0 \leq x \leq \rho_0\varepsilon^{\frac{4}{3}}, $$

and

$$ \frac{1}{3}\operatorname{arcsinh}\left(\frac{x^{3/2}}{2\varepsilon^2}\right) = \left(\frac{1}{2}\log x - \frac{2}{3}\log\varepsilon\right) + \frac{\varepsilon^4}{3x^3}\left(1+O\left(\frac{\varepsilon^4}{x^3}\right)\right) $$
$$ \textit{for } \rho_0\varepsilon^{4/3} \leq x \leq 1. $$

Lemma 29.6.

$$\frac{1}{2\varepsilon^2}\int_0^x \frac{t^2}{\sqrt{\frac{t^2}{4\varepsilon^4}+\frac{1}{t}}}dt = \frac{x^{7/2}}{7\varepsilon^2}\left(1+O\left(\frac{x^3}{\varepsilon^4}\right)\right) \quad \textit{for} \ \ 0 \le x \le \rho_0\varepsilon^{\frac{4}{3}},$$

and

$$\frac{1}{2\varepsilon^2}\int_0^x \frac{t^2}{\sqrt{\frac{t^2}{4\varepsilon^4}+\frac{1}{t}}}dt = \frac{x^2}{2} + O(\varepsilon^{\frac{8}{3}}) \quad \textit{for} \ \ \rho_0\varepsilon^{\frac{4}{3}} \le x \le 1.$$

Lemmas 29.5 and 29.6 can be proved by using some basic calculus and the asymptotic formulas of $\operatorname{arcsinh} z$; see Abramowitz and Stegun (1992, p. 88).

From Theorem 29.3 and the asymptotic formulas of $K_1(z)$ and $I_1(x)$, the solution $Y(x)$ in (29.72) can be simplified to

$$Y(x) = \begin{cases} \frac{2\sqrt{x}}{\varepsilon}K_1\left(\frac{2\sqrt{x}}{\varepsilon}(1+o(1))\right)\exp\left\{-\frac{x^2}{4\varepsilon^3}\right\}(1+o(1)) & \text{for } 0 < x \ll O(\varepsilon^{\frac{4}{3}}), \\ \exp\left[\frac{x^2}{2} - \frac{\varepsilon}{x}(1+o(1))\right](1+o(1)) & \text{for } O(\varepsilon^{\frac{4}{3}}) \ll x \ll 1. \end{cases} \tag{29.85}$$

We claim that, for $O(\varepsilon^2) \ll x \ll O(\varepsilon)$, the solution $Y(x)$ is exponentially small. To see this, we note from (29.85) that the terms $K_1(\frac{2\sqrt{x}}{\varepsilon})\exp\left(-\frac{x^2}{4\varepsilon^3}\right)$ for $O(\varepsilon^2) \ll x \ll O(\varepsilon^{4/3})$ and $\exp(\frac{x^2}{2} - \frac{\varepsilon}{x})$ for $O(\varepsilon^{4/3}) \ll x \ll O(\varepsilon)$ are both exponentially small.

To compare our solution with that of Bender and Orszag, we evaluate the solution at some particular points. For example, at $x = \varepsilon^{41/28}$, Bender and Orszag's solution (29.3) gives

$$\begin{aligned} y_{unif}(\varepsilon^{41/28}) =& \frac{2\sqrt{\varepsilon^{41/28}}}{\varepsilon}K_1\left(\frac{2\sqrt{\varepsilon^{41/28}}}{\varepsilon}\right) \\ &+ e^{-\frac{\varepsilon}{\varepsilon^{41/28}}} + e^{\frac{(\varepsilon^{41/28})^2}{2}} - 1 \sim \frac{\varepsilon^{41/14}}{2}, \end{aligned}$$

while our solution yields

$$\begin{aligned} Y(\varepsilon^{41/28}) &\sim \sqrt{\frac{\pi}{\varepsilon}}(\varepsilon^{41/28})^{\frac{1}{4}}\exp\left\{-\frac{(\varepsilon^{41/28})^2}{4\varepsilon^3} - \frac{2\sqrt{\varepsilon^{41/28}}}{\varepsilon}\right\} \\ &\sim \frac{\sqrt{\pi}}{e^{15/112}}\exp\left\{-\frac{2}{\varepsilon^{15/56}} - \frac{1}{4\varepsilon^{1/14}}\right\}. \end{aligned}$$

Moreover, at $x = \varepsilon^{7/6}$, Bender and Orszag's solution (29.3) gives

$$y_{unif}(\varepsilon^{7/6}) = \frac{2\sqrt{\varepsilon^{7/6}}}{\varepsilon} K_1\left(\frac{2\sqrt{\varepsilon^{7/6}}}{\varepsilon}\right) + e^{-\frac{\varepsilon}{\varepsilon^{7/6}}} + e^{\frac{(\varepsilon^{7/6})^2}{2}} - 1$$
$$\sim \frac{\varepsilon^{7/3}}{2},$$

while our solution gives

$$Y(\varepsilon^{7/6}) \sim \exp\left\{\frac{\varepsilon^{(7/6)^2}}{2} - \frac{\varepsilon}{\varepsilon^{7/6}}(1+o(1))\right\} \sim \exp\left\{-\frac{1}{\varepsilon^{1/6}}\right\}.$$

In conclusion, for $O(\varepsilon^2) \ll x \ll O(\varepsilon)$, Bender and Orszag's asymptotic solution (29.3) seems to be incorrect since it shows that the leading term in the asymptotic approximation of the exact solution is of algebraic order, whereas our solution (29.72) shows that the true solution decays exponentially, this phenomenon has not been mentioned elsewhere in the literature.

29.6 Triple-deck problem

Bases on the experience from studying problem (29.1)-(29.2), we believe that our method also works for problem (29.7). In fact, problem (29.7) is much simpler than problem (29.1)-(29.2) due to the lack of singularity at the origin. Since the techniques are the same, we present only a brief outline of the derivation of an asymptotic solution to the triple-deck problem (29.7).

In parallel to Lemma 29.1, we have

Lemma 29.7. *Under the transformations*

$$\zeta(x) = \frac{1}{2\varepsilon^3}\int_0^x \sqrt{t^6 + 4\varepsilon^4 + \varepsilon^3(6t^2 - 4t^3)}dt \tag{29.86}$$

and

$$y(x) = (\zeta'(x))^{-\frac{1}{2}} \exp\left\{-\frac{x^4}{8\varepsilon^3}\right\} U(\zeta(x)), \tag{29.87}$$

(29.7) *is converted to*

$$\frac{d^2U}{d\zeta^2} - U = \phi(\zeta)U, \tag{29.88}$$

where

$$\phi(\zeta(x)) = \frac{a}{2}\frac{\zeta'''(x)}{(\zeta'(x))^3} - \frac{3}{4}\frac{(\zeta''(x))^2}{(\zeta'(x))^4}. \tag{29.89}$$

The estimate $\phi(\zeta) = O(\varepsilon^{2/3})$ *implies that*

$$\frac{d^2U}{d\zeta^2} - U = 0 \tag{29.90}$$

can be used as a perturbed equation to (29.88), *and guarantees success of the successive approximation method used in Sect. 4. As before, one can show that equation* (29.7) *has two linearly independent solutions*

$$\begin{aligned} y_1(x) &= \sqrt{2\varepsilon^3}\left(x^6 + 4\varepsilon^4 + \varepsilon^3(6x^2 - 4x^3)\right)^{-\frac{1}{4}} \exp\left\{-\frac{x^4}{8\varepsilon^3}\right\} \\ &\quad \times \exp\left\{-\frac{1}{2\varepsilon^3}\int_0^x \sqrt{t^6 + 4\varepsilon^4 + \varepsilon^3(6t^2 - 4t^3)}dt\right\}(1 + O(\varepsilon^{\frac{1}{3}})) \end{aligned} \tag{29.91}$$

and

$$\begin{aligned} y_2(x) &= \sqrt{2\varepsilon^3}\left(x^6 + 4\varepsilon^4 + \varepsilon^3(6x^2 - 4x^3)\right)^{-\frac{1}{4}} \exp\left\{-\frac{x^4}{8\varepsilon^3}\right\} \\ &\quad \times \exp\left\{\frac{1}{2\varepsilon^3}\int_0^x \sqrt{t^6 + 4\varepsilon^4 + \varepsilon^3(6t^2 - 4t^3)}dt\right\}(1 + O(\varepsilon^{\frac{1}{3}})). \end{aligned} \tag{29.92}$$

The solution which satisfies the boundary conditions of (29.7) *is given by*

$$Y(x) = \frac{\alpha}{\sqrt{\varepsilon}}y_1(x) + \frac{\beta}{y_2(1)}y_2(x), \tag{29.93}$$

where $y_1(x)$ *and* $y_2(x)$ *are given in* (29.91) *and* (29.92), *respectively.*

29.7 Exercises

1. Verify (29.14).
2. Verify (29.78).
3. Consider the problem

$$\varepsilon^3 y'' + x^3 y' + (x^3 - \varepsilon)y = 0, \quad 0 < x < 1, \tag{29.94}$$

$$y(0) = \alpha, \quad y(1) = \beta. \tag{29.95}$$

Since the coefficient of y' is positive, the boundary layer is expected to be at the origin (see Section 27.1 or Section 28.2). Hence, the outer expansion must satisfy the boundary condition $y(1) = \beta$ and it is not expected to satisfy the boundary condition $y(0) = \alpha$.

(a) We seek an outer expansion in the form

$$y^o = y_0(x) + \varepsilon y_1(x) + \cdots .$$

By substituting y^o into (29.94) and $y(1) = \beta$ and equating the coefficient of ε^0 on both sides, show that

$$x^3 y_0' + x^3 y_0 = 0, \quad y_0(1) = \beta, \tag{29.96}$$

and that the solution for y_0 is

$$y_0 = \beta e^{1-x}.$$

In general $\alpha \neq \beta e$ and hence

$$y^o = \beta e^{1-x} + \cdots$$

does not satisfy the boundary condition at the origin.
(b) To investigate the neighborhood of the origin, we introduce the stretching transformation

$$x = \varepsilon^\nu \xi, \quad \nu > 0 \tag{29.97}$$

into (29.94). Show that

$$\varepsilon^{3-2\nu}\frac{d^2y}{d\xi^2} + \varepsilon^{2\nu}\xi^3\frac{dy}{d\xi} + (\varepsilon^{3\nu}\xi^3 - \varepsilon)y = 0. \tag{29.98}$$

As $\varepsilon \to 0^+$, the term $O(\varepsilon^{3\nu})$ is small compared with $\varepsilon^{2\nu}$ since $\nu > 0$. Hence the dominant part of (29.98) is

$$\varepsilon^{3-2\nu}\frac{d^2y}{d\xi^2} + \varepsilon^{2\nu}\xi^3\frac{dy}{d\xi} - \varepsilon y + \cdots = 0. \tag{29.99}$$

To balance the first and second terms, we require $3 - 2\nu = 2\nu$; i.e., $\nu = 3/4$. Then (29.98) becomes

$$\varepsilon^{3/2}\frac{d^2y}{d\xi^2} + \varepsilon^{3/2}\xi^3\frac{dy}{d\xi} + (\varepsilon^{9/4}\xi^3 - \varepsilon)y = 0;$$

the dominant term in this equation is a trivial case $y = 0$. Hence, discard this case.
(b) Balancing the first and third term in (29.99) gives $3 - 2\nu = 1$; i.e., $\nu = 1$. Show that equation (29.98) becomes

$$\varepsilon\frac{d^2y}{d\xi^2} + \varepsilon^2\xi^3\frac{dy}{d\xi} + (\varepsilon^3\xi^3 - \varepsilon)y = 0. \tag{29.100}$$

Since it is not clear which term is dominant, we must keep this case. Balancing the second and third terms in (29.99), we have $2\nu = 1$; i.e., $\nu = 1/2$. To distinguish this case from the proceeding one, we use the variable ζ instead of ξ so that $\xi = x/\varepsilon$ and $\zeta = x/\varepsilon^{1/2}$. Show that equation (29.98) then becomes

$$\varepsilon^2\frac{d^2y}{d\zeta^2} + \varepsilon\zeta^3\frac{dy}{d\zeta} + (\varepsilon^{3/2}\zeta^3 - \varepsilon)y = 0. \tag{29.101}$$

Since it is again not clear which part is dominant, this case must also be included. The stretching variable $\xi = x/\varepsilon$, corresponding to $\nu = 1$, describes a layer close to the origin. We shall call this layer the *left deck*. The stretching variable $\zeta = x/\varepsilon^{1/2}$, which corresponds to $\nu = 1/2$, describes a layer that is between the left deck and the outer expansion (*right deck*). As a consequence, we call this deck the *middle deck*. In the case of viscous-inviscid interactions, these decks are usually referred to as the lower, middle and upper decks, respectively.

(c) Show from (29.100) that the leading term Y_0 in the left-deck expansion y^l satisfies

$$\frac{d^2Y_0}{d\xi^2} - Y_0 = 0$$

and its general solution is

$$Y_0 = a_0 e^{-\xi} + b_0 e^{\xi},$$

where b_0 must be zero for otherwise Y_0 would grow exponentially with ξ, making it unmatchable with either the middle deck or the right deck. Since this deck is valid at the origin, it must satisfy the boundary condition $y(0) = \alpha$. But $x = 0$ corresponds to $\xi = 0$, hence $y^l(0) = \alpha$. Then, it follows that $a_0 = \alpha$, $Y_0 = \alpha e^{-\xi}$, and

$$y^l = \alpha e^{-\xi} + \cdots . \tag{29.102}$$

Show from (29.101) that the leading term $\tilde{Y}_0$ in the middle-deck expansion y^m is governed by

$$\zeta^3 \tilde{Y}_0' - \tilde{Y}_0 = 0,$$

which, by separation of variables, gives $\tilde{Y}_0 = d_0 e^{-1/(2\zeta^2)}$ and

$$y^m = d_0 e^{-1/(2\zeta^2)} + \cdots , \tag{29.103}$$

where d_0 is a constant. Since $x = 0$ corresponds to $\zeta = 0$, it follows from (29.103) that $y^m \to 0$ as $\zeta \to 0$, and hence, it cannot satisfy the boundary condition $y(0) = \alpha$. Therefore, the constant d_0 needs to be determined from matching y^m with either y^l (i.e., left-deck) or y^o (i.e., outer expansion). Show that $y^l(\infty) = 0$ and $y^o(0) = \beta e$. Hence, y^l cannot be matched directly with y^o, and y^m must be used to bridge the gap between y^l and y^o. Show that $y^m(\infty) = d_0$ and $y^o(0) = \beta e$. Hence, $d_0 = \beta e$ and

$$y^m = \beta e^{1-1/(2\zeta^2)} + \cdots . \tag{29.104}$$

Show that $y^m(0) = y^l(\infty)$; i.e., y^m and y^l are matchable.

Finally, form a composite expansion as follows:

$$y^c = y^o + y^m + y^l - y^m(0) - y^m(\infty).$$

Hence,

$$y^c = \beta e^{1-x} + \beta e^{1-1/(2\zeta^2)} + \alpha e^{-\xi} - \beta e + \cdots. \tag{29.105}$$

From (29.105), show that y^c is asymptotically equal to y^l, y^m, and y^o in the left deck, middle deck, and right deck, respectively. Based on the observations, it is reasonable to postulate that y^c is valid for all $x \in [0,1]$; see Figure 29.1 in Section 29.1. Despite what we have shown in Figure 29.1, the method of "matched asymptotics" is till very useful, especially when there is no rigorous method to construct an asymptotic solution to a singularly perturbed differential equation.

4. Consider the boundary value problem

$$y'' + \frac{2}{x}y' + \varepsilon y y' = 0, \quad x \geq 1,\ 0 < \varepsilon < 1, \tag{29.106}$$

$$y(1) = 0, \quad y(\infty) = 1. \tag{29.107}$$

Note that the small parameter ε does not appear in the term with the highest derivative, and that the domain is infinite.

We look for a two-term expansion of the form

$$y = y_0(x) + \varepsilon y_1(x) + \cdots. \tag{29.108}$$

By substituting (29.108) into (29.106) and (29.107), show that

$$y_0 = 1 - \frac{1}{x}, \tag{29.109}$$

and

$$y_1 = -\ln x - \frac{\ln x}{x} + b_1\left(1 - \frac{1}{x}\right), \tag{29.110}$$

where b_1 is a constant. Here, we have made use of the boundary condition $y_1(1) = 0$, which follows from (29.107). As $x \to \infty$, $y_1 \to -\ln x$, and hence, it cannot satisfy the second boundary condition $y_1(\infty) = 0$; see (29.107). Consequently, the straightforward expansion, which we will call the outer expansion and denote by the superscript 'o', is

$$y^o = 1 - \frac{1}{x} + \varepsilon\left[-\left(1 + \frac{1}{x}\right)\ln x + b_1\left(1 - \frac{1}{x}\right)\right] + \cdots, \tag{29.111}$$

and is not valid for large x.

To investigate the behavior of the solution for large x, we introduce the contraction transformation

$$\xi = x\varepsilon^{\nu}, \quad \nu > 0. \tag{29.112}$$

Show that (29.106) becomes

$$\frac{d^2y}{d\xi^2} + \frac{2}{\xi}\frac{dy}{d\xi} + \varepsilon^{1-\nu}y\frac{dy}{d\xi} = 0. \tag{29.113}$$

The distinguish limit corresponds to $\nu = 1$ and the resulting expansion is called the inner expansion. We use the superscript 'i' to indicate the inner expansion and rewrite the equation as

$$\frac{d^2y^i}{d\xi^2} + \frac{2}{\xi}\frac{dy^i}{d\xi} + y^i\frac{dy^i}{d\xi} = 0. \tag{29.114}$$

Instead of assuming the form of the expansion for y^i, finding and solving the equations for the different orders, and then matching with the outer expansion, one can use the outer expansion as a guide to determine the form of the inner expansion. For instance, in this case, the first term in (29.109) is valid for all x in $[1, \infty)$; hence, it is valid in the inner region. Thus, expressing it in terms of the inner variable, we have

$$y_0 = 1 - \frac{1}{x} = 1 - \frac{\varepsilon}{\xi}. \tag{29.115}$$

Then, letting $\varepsilon \to 0^+$ with ξ being fixed yields $y_0 = 1$, which is the first term in the inner expansion. Moreover, the form of y_0 in (29.115) suggests that y^o goes in power of ε. Hence, we seek an inner expansion in the form

$$y^i = 1 + \varepsilon Y_1(\xi) + \cdots. \tag{29.116}$$

Substituting (29.116) into (29.114) and equating the coefficient of ε to zero, we obtain

$$Y_1'' + \left(\frac{2}{\xi} + 1\right)Y_1' = 0. \tag{29.117}$$

Show that (29.117) gives

$$Y_1 = c_1 \int_{\infty}^{\xi} \frac{e^{-\tau}}{\tau^2} d\tau + d_1, \tag{29.118}$$

where c_1 and d_1 are arbitrary constants. The inner expansion must satisfy the boundary condition at infinity and this is the reason why we set the lower limit in (29.118) to ∞ because $x = \infty$ corresponds to

$\xi = \infty$. It follows from (29.107) and (29.116) that $Y_1(\infty) = 0$, and hence, it follows from (29.118) that $d_1 = 0$. Then, (29.118) can be rewritten as

$$Y_1 = c_1 \int_{\infty}^{\xi} \frac{e^{-\tau}}{\tau^2} d\tau$$

so that (29.116) becomes

$$y^i = 1 - \varepsilon c_1 \int_{\xi}^{\infty} \frac{e^{-\tau}}{\tau^2} d\tau + \cdots , \tag{29.119}$$

where the limits of integration have been interchanged so that the resulting integral is positive. The constant c_1 needs to be determined by matching the inner and outer expansion.

By using integration by parts, show that the two-term inner expansion (29.119) can be written as

$$y^i = 1 - \varepsilon \left(\frac{1}{\xi} + \ln \xi - \ln \varepsilon - b_1 \right) \tag{29.120}$$

and the two-term outer expansion (29.111) can be written as

$$y^o = 1 - \frac{c_1}{x} - \varepsilon c_1(-1 + \gamma + \ln x + \ln \varepsilon). \tag{29.121}$$

Equating (29.120) to equation (29.121), we have

$$1 - \frac{1}{x} - \varepsilon \ln x + \varepsilon b_1 = 1 - \frac{c_1}{x} - \varepsilon c_1(-1 + \gamma + \ln x + \ln \varepsilon). \tag{29.122}$$

Hence, $c_1 = 1$ and $b_1 = 1 - \gamma - \ln \varepsilon$ so that equation (29.111) becomes

$$y^o = 1 - \frac{1}{x} - \varepsilon \left[- \left(1 + \frac{1}{x} \right) \ln x + (1 - \gamma - \ln \varepsilon) \left(1 - \frac{1}{x} \right) \right] + \cdots , \tag{29.123}$$

and (29.119) becomes

$$y^i = 1 - \varepsilon \int_{\xi}^{\infty} \frac{e^{-\tau}}{\tau^2} d\tau + \cdots . \tag{29.124}$$

Finally, by adding the outer expansion (29.123) and the inner expansion (29.124) and subtracting the common part (29.121), show that the composite expansion is given by

$$y^c = 1 - \varepsilon \int_{\xi}^{\infty} \frac{e^{-\tau}}{\tau^2} d\tau - \varepsilon \left(\frac{\ln x}{x} + \frac{1 - \gamma - \ln \varepsilon}{x} \right) + \cdots . \tag{29.125}$$

5. Consider the problem

$$\varepsilon y'' + y' + y = 0,$$
$$y(0) = \alpha, \quad y(1) = \beta.$$

(a) Find the exact solution.
(b) Use the method of matched asymptotic expansions to determine a first-order uniform asymptotic solution. Compare the asymptotic solution with the exact solution.

6. Find the first-order (one term) uniform expansions for the solutions of

$$\varepsilon y'' \pm (3x + 1)y' = 1,$$
$$y(0) = \alpha, \quad y(1) = \beta.$$

7. Consider the problem

$$\varepsilon y''' - (2x + 1)y' = 1,$$
$$y(0) = \alpha, \quad y'(0) = \beta, \quad y(1) = \gamma.$$

Find a two-term uniform asymptotic expansion for the solution.

Chapter 30

Number of Solutions to Carrier's Problem

30.1 Introduction

In their classic textbook, Carrier and Pearson (1968) presented two singularly perturbed two-point boundary value problems. We have already encountered the first problem in Chapter 28, and studied it in great detail. In this chapter, we shall study the second one, which is the following:

$$\begin{cases} \varepsilon^2 y'' = 1 - y^2 - 2(1 - x^2)y, \\ y(-1) = y(1) = 0, \end{cases} \tag{30.1}$$

where ε is again a small positive parameter; see Carrier and Pearson (1968, p. 197). This problem is also mentioned in a survey article by Carrier (1970), and discussed in more detail in the well-known book by Bender and Orszag (1978, p. 464). (Bender and Orszag expressed that this problem is "extremely beautiful and intricate".) This problem has the same difficulty as the first one discussed in Chapter 28, that is, it also has spurious solutions involving spike-like functions. Bender and Orszag (1978, p. 467) used this example to show that "boundary-layer theory cannot predict the number or the location of these boundary layers. In fact, for a given positive value of ε there are exactly $4(N + 1)$ solutions which have from 0 to N internal boundary layers at definite locations, where N is a finite number depending on ε. Because the equation in (30.1) is nonautonomous, the difficulty with this problem is even greater. There are other studies on this problem in the literature; see, for instance, Kath (1985) and MacGillivray *et al.* (2000). In the latter reference, the authors also showed that the spikes will cluster in the neighborhood of $x = 0$. Here, we are concerned with the maximum number of spikes that a solution to problem (30.1) can have. A closely related study has been made by Ai (2003), who showed that problem (30.1) can have solutions whose number of spikes is of order $1/\varepsilon$. Here, we will give a more precise estimate for the maximum number of spikes. Let us denote it by N_ε. Then, we will show that

$$N_\varepsilon \sim \left[\frac{K}{\varepsilon}\right], \quad \text{as } \varepsilon \to 0, \tag{30.2}$$

where $K = 0.4725\cdots$ and $[\alpha]$ represents the greatest integer less than or equal to α. From (30.2), it can be deduced that the number of solutions to (30.1) lies between $4N_\varepsilon - 3$ and $4N_\varepsilon$. The material in this chapter is taken from Wong and Zhao (2008). The material in the Exercise section is taken from Bender and Orszag (1978, Chapter 9).

The equation in (30.1) is a particular case of the equation $\varepsilon^2 u'' = f(x,u)$, which describes the stationary solution of the diffusion equation

$$\varepsilon^2 \frac{\partial^2 u}{\partial x^2} - \frac{\partial u}{\partial t} = f(u,x), \tag{30.3}$$

where ε is a small diffusion coefficient. Equation (30.3) often appears as a mathematical model equation in many areas such as chemical and biological kinetics, etc. We hope that our approach will be helpful in the investigations of stationary solutions of (30.3) with boundary conditions.

30.2 The shooting method

To apply the argument used in Ou and Wong (2003), we consider, instead of (30.1), the associated initial-value problem

$$\begin{cases} \varepsilon^2 y'' = 1 - y^2 - 2(1-x^2)y, \\ y(-1) = 0, \quad y'(-1) = k, \end{cases} \tag{30.4}$$

where k is a real number. This initial-value problem has a unique solution; for a reference, see the corollary in Arnold (1992, p. 107).

In a neighborhood of $x = 1$, the coefficient function in (30.4) vanishes. Hence, it is expected that the solution to (30.4) can be approximated by the solution to the initial-value problem

$$\begin{cases} \varepsilon^2 u'' = 1 - u^2, \\ u(-1) = 0, \quad u'(-1) = k, \end{cases} \tag{30.5}$$

which has been studied in great detail by Ou and Wong (2003), where a proof of the following result is given; see also Chapter 28.

Lemma 30.1. *Let $u(x,k)$ denote the solution of the initial-value problem* (30.5).

(a) *If* $k \le -\frac{2}{\sqrt{3}\varepsilon}$, *then* $u(x,k)$ *is decreasing for* $x \ge -1$.
(b) *If* $k \ge \frac{2}{\sqrt{3}\varepsilon}$, *then there exists a number* $-1 < \eta < -1 + 4\sqrt{3}\varepsilon$ *such that* $u(\eta, k) = 0$, $u(x,k)$ *is increasing in* $-1 < x < (\eta - 1)/2$ *and decreasing for* $x > (\eta - 1)/2$.
(c) *If* $|k| < \frac{2}{\sqrt{3}\varepsilon}$, *then* $u(x,k)$ *is periodic and intersects the x-axis infinitely many times, and the period is of order* ε.

Motivated by this result, we now consider the problem

$$\begin{cases} \varepsilon^2 u'' = 1 - u^2 - lu, \\ u(0) = b, \quad u'(0) = 0, \end{cases} \tag{30.6}$$

where $b \le 0$ and l is a parameter which takes values in $0 \le l \le 2$. The interval $[0, 2]$ is exactly the range of the coefficient function $2(1 - x^2)$ in (30.4) for $x \in [-1, 1]$. Note that, without loss of generality, we have chosen the initial point $x = 0$, because (30.6) is an autonomous equation.

Lemma 30.2. *Let* $u(x,b)$ *denote the solution of the initial-value problem* (30.6).

(a) *If* $b < -\frac{1}{2}(l + \sqrt{4 + l^2})$, *then* $u(x,b)$ *is decreasing for* $x > 0$.
(b) *If* $b = -\frac{1}{2}(l + \sqrt{4 + l^2})$, *then* $u(x,b) = -\frac{1}{2}(l + \sqrt{4 + l^2})$.
(c) *If* $-\frac{1}{2}(l + \sqrt{4 + l^2}) < b < -\frac{1}{2}(l - \sqrt{4 + l^2})$, *then* $u(x,b)$ *is periodic and* b *is the minimum of* $u(x,b)$.
(d) *If* $b = -\frac{1}{2}(l - \sqrt{4 + l^2})$, *then* $u(x,b) = -\frac{1}{2}(l - \sqrt{4 + l^2})$.
(e) *If* $-\frac{1}{2}(l - \sqrt{4 + l^2}) < b < -\frac{1}{2}l + \sqrt{4 + l^2}$, *then* $u(x,b)$ *is periodic and* b *is the maximum of* $u(x,b)$.
(f) *If* $b \ge -\frac{1}{2}l + \sqrt{4 + l^2}$, *then* $u(x,b)$ *is decreasing for* $x > 0$.

Proof. (a) The energy equation of (30.6) is

$$\frac{\varepsilon^2}{2}(u')^2 = u - \frac{u^3}{3} - \frac{l}{2}u^2 - (b - \frac{b^3}{3} - \frac{l}{2}b^2), \tag{30.7}$$

which will be used throughout the proof. Consider the quadratic equation $u^2 + lu - 1 = 0$ obtained from (30.6) by letting the right-hand side of (30.6) equal to 0. The zeros of this equation are $-\frac{1}{2}(l \pm \sqrt{4 + l^2})$. At $x = 0$, we have $\varepsilon^2 u''(0,b) = 1 - b^2 - lb$. Hence, from Figure 30.1, it is readily seen that if $b < -\frac{1}{2}(l + \sqrt{4 + l^2})$, $u''(0,b) < 0$. Because $u'(0,b) = 0$, in the right neighborhood of $x = 0$ we have $u'(x,b) < 0$.

Define

$$H_l(u) = u - \frac{u^3}{3} - \frac{l}{2}u^2. \tag{30.8}$$

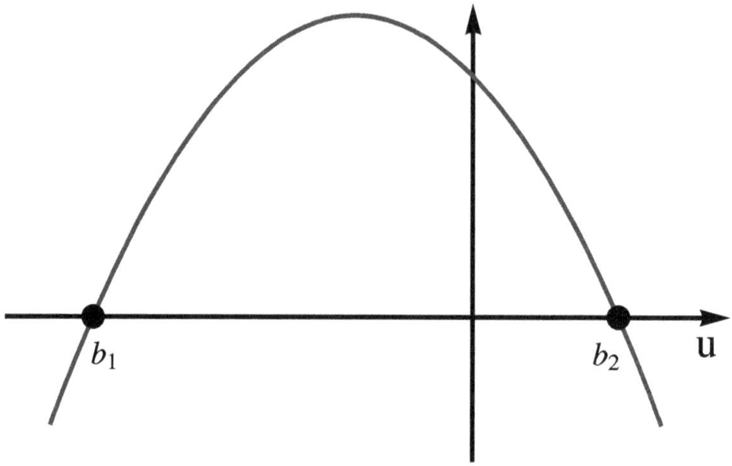

Fig. 30.1: The graph of $1-u^2-lu$ with $l \in [0,2]$, where $b_1 := -\frac{1}{2}(l+\sqrt{l^2+4})$ and $b_2 := -\frac{1}{2}(l-\sqrt{l^2+4})$ are two zeros.

Assume that there exists a point $\eta > 0$ such that $u'(\eta, b) = 0$ and $u'(x,b) < 0$ when $x \in (0, \eta)$. From (30.7), we have

$$H_l(u(\eta, b)) - H_l(b) = 0. \tag{30.9}$$

However, (30.9) cannot be true when $u(x,b) < b$; cf. Figure 30.2. Therefore, $u'(x,b) < 0$ for all $x > 0$ when $b < -\frac{1}{2}(l+\sqrt{4+l^2})$.

(b) When $b = -\frac{1}{2}(l+\sqrt{4+l^2})$, $1-b^2-lb = 0$. Thus, $u(x,b) = -\frac{1}{2}(l+\sqrt{4+l^2})$ is a solution of (30.6). By the uniqueness theorem, it is the only solution.

(c) When $-\frac{1}{2}(l+\sqrt{4+l^2}) < b < -\frac{1}{2}(l-\sqrt{4+l^2})$, Figure 30.1 shows that $u''(0,b) > 0$. As an equation in u, $H_l(u) - H_l(b) = 0$ has three solutions. Clearly, b is a solution. The other two solutions are

$$a_l = \frac{1}{4}\left(-2b - 3l + \sqrt{3}\sqrt{16 - 4b^2 - 4bl + 3l^2}\right) \tag{30.10}$$

and

$$c_l = \frac{1}{4}\left(-2b - 3l - \sqrt{3}\sqrt{16 - 4b^2 - 4bl + 3l^2}\right). \tag{30.11}$$

From Figure 30.3 (a), it is clear that

$$c_l < b < a_l. \tag{30.12}$$

From (30.7), it can be shown that there is a point $x = d_1 > 0$ such that

$$u(d_1, b) = a_l, \quad u'(d_1, b) = 0,$$

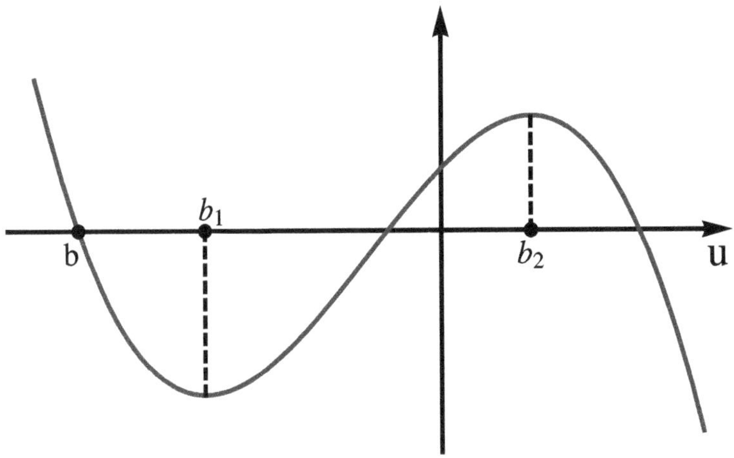

Fig. 30.2: The graph of $H_l(u) - H_l(b)$ when $b < b_1 := -\frac{1}{2}(l + \sqrt{l^2+4})$.

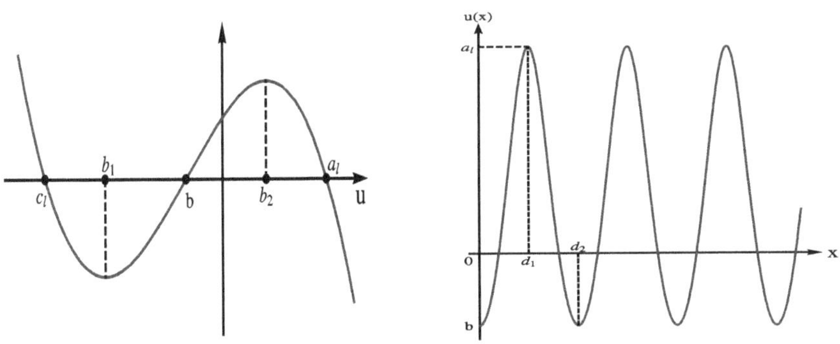

(a) The graph of $H_l(u) - H_l(b)$. (b) The periodic solution $u(x)$ of (30.6).

Fig. 30.3: (*a*) The graph of $H_l(u) - H_l(b)$ and (*b*) the periodic solution of (30.6), when $b \in (b_1, b_2)$ with $b_1 := -\frac{1}{2}(l + \sqrt{l^2+4})$ and $b_2 := -\frac{1}{2}(l - \sqrt{l^2+4})$.

and $u'(x,b) > 0$ when $x \in (0, d_1)$. Because $a_l > -\frac{1}{2}(l - \sqrt{4+l^2})$, it is evident from Figure 30.1 that $u''(d_1, b) < 0$ and hence $u(x,b)$ reaches its maximum at $x = d_1$. From (30.7), we have (exercise)

$$d_1 = \varepsilon \int_0^{a_l} \frac{du}{\sqrt{2}\sqrt{H_l(u) - H_l(b)}}. \tag{30.13}$$

Now consider the initial value problem (30.6) with 0 replaced by d_1. By the same argument, there exists a point $d_2 > d_1$ such that

$$u(d_2, b) = b, \quad u'(d_2, b) = 0,$$

and $u'(x, b) < 0$ when $x \in (d_1, d_2)$. Again from (30.7), we have (exercise)

$$d_2 = 2d_1 = 2\varepsilon \int_0^{a_l} \frac{du}{\sqrt{2}\sqrt{H_l(u) - H_l(b)}}. \tag{30.14}$$

At $x = d_2$, $u(x, b)$ has the same initial values as those at the starting point $x = 0$. Hence, $u(x, b)$ is periodic with period d_2; see Figure 30.3 (b).

(d) When $b = -\frac{1}{2}(l - \sqrt{4 + l^2})$, $1 - b^2 - lb = 0$. Thus, $u(x, b) = -\frac{1}{2}(l - \sqrt{4 + l^2})$ is a solution of (30.6). By the uniqueness theorem, it is the only solution.

(e) When $-\frac{1}{2}(l - \sqrt{4 + l^2}) < b < -\frac{1}{2}l + \sqrt{4 + l^2}$, Figure 30.1 shows that $u''(0, b) < 0$. Again, as an equation in u, $H_l(u) - H_l(b) = 0$ has three solutions b, a_l and c_l; cf. (30.10) and (30.11). It can be seen from Figure 30.4 (a) that

$$c_l < a_l < b. \tag{30.15}$$

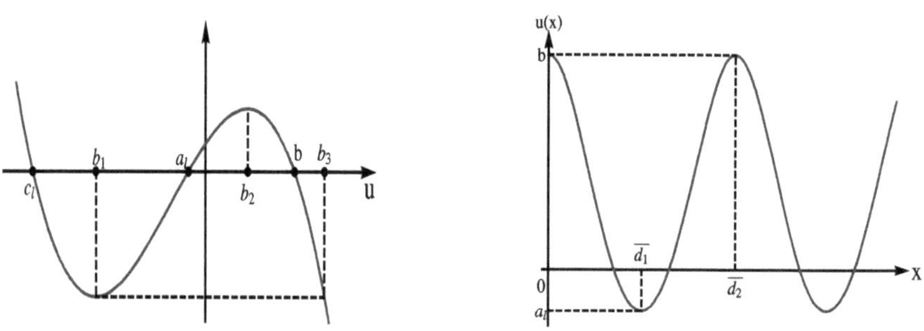

(a) The graph of $H_l(u) - H_l(b)$.

(b) The periodic solution $u(x)$ of (30.6).

Fig. 30.4: (*a*) The graph of $H_l(u) - H_l(b)$ and (*b*) the periodic solution of (30.6), when $b \in (b_2, b_3)$ with $b_2 := -\frac{1}{2}(l - \sqrt{l^2 + 4})$ and $b_3 := -\frac{1}{2}l + \sqrt{l^2 + 4}$.

By using a similar argument to that in (*c*), one can show that there are two points $0 < \bar{d}_1 < \bar{d}_2$ such that

$$u(\bar{d}_1, b) = a_l, \quad u'(\bar{d}_1, b) = 0,$$

and

$$u(\bar{d}_2, b) = b, \quad u'(\bar{d}_2, b) = 0.$$

The integral representations of $\bar{d}_1$ and $\bar{d}_2$ are (exercise)

$$\bar{d}_1 = \varepsilon \int_{a_l}^{b} \frac{du}{\sqrt{2}\sqrt{H_l(u) - H_l(b)}}, \tag{30.16}$$

$$\bar{d}_2 = 2\bar{d}_1 = 2\varepsilon \int_{a_l}^{b} \frac{du}{\sqrt{2}\sqrt{H_l(u) - H_l(b)}}. \tag{30.17}$$

Furthermore, $u'(x, b) < 0$ when $x \in (0, \bar{d}_1)$ and $u'(x, b) > 0$ when $x \in (\bar{d}_1, \bar{d}_2)$. Obviously, a_l is the minimum of $u(x, b)$ and b is the maximum of $u(x, b)$. At $x = \bar{d}_2$, $u(x, b)$ has the same initial values as those at the starting point $x = 0$. Hence, $u(x, b)$ is periodic with period $\bar{d}_2$; see Figure 30.4 (b).

(f) When $b \geq -\frac{1}{2}l + \sqrt{4 + l^2}$, it can be seen from Figure 30.1 that $u''(0, b) < 0$. As before, because $u'(0, b) = 0$, in the right neighborhood of $x = 0$ we have $u'(x, b) < 0$. Assume that there exists a point $\eta > 0$ such that $u'(\eta, b) = 0$ and $u'(x, b) < 0$ when $x \in (0, \eta)$. From (30.7), we have

$$H_l(u(\eta, b)) - H_l(b) = 0. \tag{30.18}$$

If $b = -\frac{1}{2}l + \sqrt{4 + l^2}$, then a_l and c_l in (30.10) and (30.11) coincide, and

$$H_l(u) - H_l(b) = \frac{1}{3}(b - u)(u - a_l)^2,$$

where $a_l = -\frac{1}{2}(l + \sqrt{4 + l^2}) < b$ and $u(\eta) = a_l$. From (30.7), we have

$$\eta = -\varepsilon \int_{b}^{a_l} \frac{du}{\sqrt{2}\sqrt{H_l(u) - H_l(b)}} = -\sqrt{\frac{3}{2}}\varepsilon \int_{b}^{a_l} \frac{du}{(u - a_l)\sqrt{b - u}} = \infty.$$

If $b > -\frac{1}{2}l + \sqrt{4 + l^2}$, then (30.18) cannot be true when $u(x, b) < b$; cf. Figure 30.5. Therefore, $u'(x, b) < 0$ for all $x > 0$ when $b \geq -\frac{1}{2}l + \sqrt{4 + l^2}$. □

Let us now focus on the case when b is the minimum of the solution to (30.6). From the above lemma, we know that when $-\frac{1}{2}(l + \sqrt{4 + l^2}) < b < -\frac{1}{2}(l - \sqrt{4 + l^2})$, the solution of (30.6) is periodic. Adding (30.13)-(30.14) gives the period $T_l(b)$, which can be written as

$$T_l(b) = \sqrt{2}\varepsilon \int_{b}^{a_l} \frac{du}{\sqrt{H_l(u) - H_l(b)}}. \tag{30.19}$$

Lemma 30.3. *When* $-\frac{1}{2}(l + \sqrt{4 + l^2}) < b < -\frac{1}{2}(l - \sqrt{4 + l^2})$,

$$\frac{\partial T_l}{\partial b} < 0. \tag{30.20}$$

For fixed b,

$$\frac{\partial T_l}{\partial l} < 0. \tag{30.21}$$

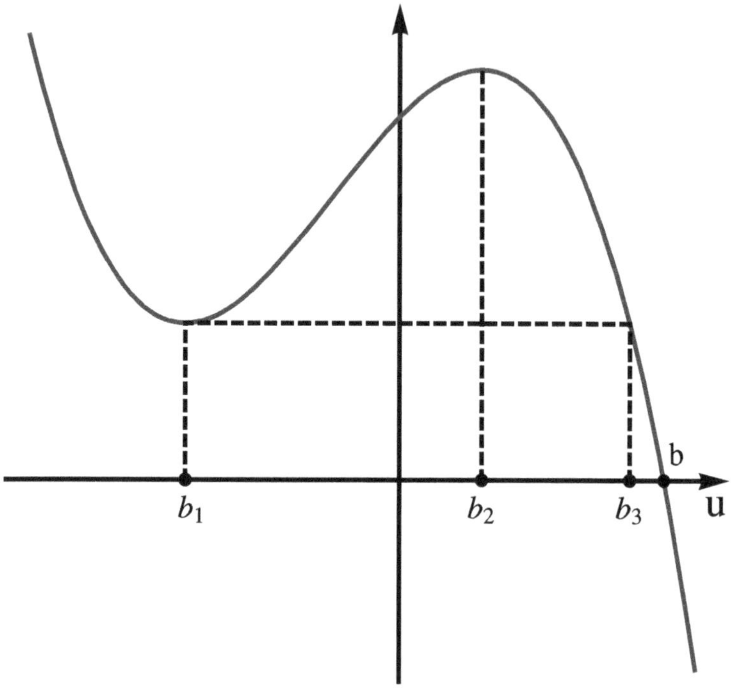

Fig. 30.5: The graph of $H_l(u) - H_l(b)$, when $b > b_3 := -\frac{1}{2}l + \sqrt{4 + l^2}$.

Proof. Because a_l, b and c_l are three solutions of $H_l(u) - H_l(b) = 0$, we can rewrite (30.19) as

$$T_l(b) = \sqrt{6}\varepsilon \int_b^{a_l} \frac{du}{\sqrt{(a_l - u)(u - b)(u - c_l)}};$$

cf. (30.8). Let $u = \theta a_l + (1 - \theta)b$. Because $a_l + b + c_l = -\frac{3}{2}l$, we have

$$T_l(b) = \sqrt{6}\varepsilon \int_0^1 \frac{d\theta}{\sqrt{(1 - \theta)\theta\left((1 + \theta)a_l + (2 - \theta)b + \frac{3}{2}l\right)}}. \tag{30.22}$$

Straightforward differentiation of (30.22) with respect to b gives

$$\begin{aligned}\frac{\partial T_l}{\partial b} =&\sqrt{6}\varepsilon\int_0^1 \frac{-\frac{1}{2}\left[(1-\theta)\theta((1+\theta)(\partial a_l/\partial b)+(2-\theta))\right]}{\left[(1-\theta)\theta((1+\theta)a_l+(2-\theta)b+\frac{3}{2}l\right]^{3/2}}d\theta\\ =&\sqrt{6}\varepsilon\int_0^{\frac{1}{2}} \frac{-\frac{1}{2}\left[(1-\theta)\theta((1+\theta)(\partial a_l/\partial b)+(2-\theta))\right]}{\left[(1-\theta)\theta((1+\theta)a_l+(2-\theta)b+\frac{3}{2}l\right]^{3/2}}d\theta\\ &+\sqrt{6}\varepsilon\int_0^{\frac{1}{2}} \frac{-\frac{1}{2}\left[(1-\theta)\theta((2-\theta)(\partial a_l/\partial b)+(1+\theta))\right]}{\left[(1-\theta)\theta((2-\theta)a_l+(1+\theta)b+\frac{3}{2}l\right]^{3/2}}d\theta. \end{aligned} \tag{30.23}$$

Because a_l is a solution of $H_l(u)-H_l(b)=0$, it is readily seen that

$$\begin{aligned} a_l-&\frac{a_l^3}{3}-\frac{l}{2}a_l^2-(b-\frac{b^3}{3}-\frac{l}{2}b^2)\\ &=-\frac{1}{3}(a_l-b)\left(a_l^2+(b+\frac{3}{2}l)a_l+b^2+\frac{3}{2}lb-3\right)=0. \end{aligned} \tag{30.24}$$

Taking the derivative with respect to b in (30.24) yields

$$2a_l\frac{\partial a_l}{\partial b}+a_l+(b+\frac{3}{2}l)\frac{\partial a_l}{\partial b}+\frac{3}{2}l=0.$$

Hence, we obtain

$$\frac{\partial a_l}{\partial b}=-\frac{2b+a_l+\frac{3}{2}l}{2a_l+b+\frac{3}{2}l}.$$

From (30.10), we know $2a_l+b+\frac{3}{2}l>0$ and $a_l>b$. Thus,

$$\frac{\partial a_l}{\partial b}>-1. \tag{30.25}$$

Note that in (30.23), $0\le\theta\le\frac{1}{2}$ and the numerator of the first integrand on the right-hand side satisfies

$$\begin{aligned} -\frac{1}{2}\left[(1-\theta)\theta(1+\theta)\frac{\partial a_l}{\partial b}+(2-\theta)\right] &< -\frac{1}{2}\left[(1-\theta)\theta(-(1+\theta)+(2-\theta))\right]\\ &=-\frac{1}{2}\left[(1-\theta)\theta(1-2\theta)\right]\le 0. \end{aligned}$$

Because $a_l>b$, the denominator of the corresponding integrand satisfies

$$\begin{aligned} &\left[(1-\theta)\theta\left((1+\theta)a_l+(2-\theta)b+\frac{3}{2}l\right)\right]^{\frac{1}{2}}\\ &\le\left[(1-\theta)\theta\left((2-\theta)a_l+(1+\theta)b+\frac{3}{2}l\right)\right]^{\frac{1}{2}}. \end{aligned}$$

Hence, we have from (30.23)

$$\begin{aligned}\frac{\partial T_l}{\partial b} \leq& \sqrt{6}\varepsilon \int_0^{\frac{1}{2}} \frac{-\frac{1}{2}\left[(1-\theta)\theta(1+\theta)(\partial a_l/\partial b)+(2-\theta)\right]}{\left[(1-\theta)\theta((2-\theta)a_l+(1+\theta)b+\frac{3}{2}l)\right]^{3/2}}d\theta \\ &+ \sqrt{6}\varepsilon \int_0^{\frac{1}{2}} \frac{-\frac{1}{2}\left[(1-\theta)\theta((2-\theta)(\partial a_l/\partial b)+(1+\theta))\right]}{\left[(1-\theta)\theta((2-\theta)a_l+(1+\theta)b+\frac{3}{2}l)\right]^{3/2}}d\theta \\ =& \sqrt{6}\varepsilon \int_0^{\frac{1}{2}} \frac{-\frac{1}{2}\left[(1-\theta)\theta(3\partial a_l/\partial b+3)\right]}{\left[(1-\theta)\theta((2-\theta)a_l+(1+\theta)b+\frac{3}{2}l)\right]^{3/2}}d\theta < 0,\end{aligned}$$

thus proving (30.20).

Returning to (30.22), we could obtain by straightforward differentiation with respect to l

$$\begin{aligned}\frac{\partial T_l}{\partial l} =& \sqrt{6}\varepsilon \int_0^{1} \frac{-\frac{1}{2}\left[(1-\theta)\theta((1+\theta)\frac{\partial a_l}{\partial l}+\frac{3}{2})\right]}{\left[(1-\theta)\theta((1+\theta)a_l+(2-\theta)b+\frac{3}{2}l)\right]^{1/2}}d\theta \\ =& \sqrt{6}\varepsilon \int_0^{\frac{1}{2}} \frac{-\frac{1}{2}\left[(1-\theta)\theta((1+\theta)\frac{\partial a_l}{\partial l}+\frac{3}{2})\right]}{\left[(1-\theta)\theta((1+\theta)a_l+(2-\theta)b+\frac{3}{2}l)\right]^{3/2}}d\theta \\ &+ \sqrt{6}\varepsilon \int_0^{\frac{1}{2}} \frac{-\frac{1}{2}\left[(1-\theta)\theta((2-\theta)\frac{\partial a_l}{\partial l}+\frac{3}{2})\right]}{\left[(1-\theta)\theta((2-\theta)a_l+(1+\theta)b+\frac{3}{2}l)\right]^{3/2}}d\theta. \qquad (30.26)\end{aligned}$$

Taking the derivative with respect to l in (30.24) gives

$$2a_l\frac{\partial a_l}{\partial l}+\frac{3}{2}a_l+(b+\frac{3}{2}l)\frac{\partial a_l}{\partial l}+\frac{3}{2}b=0,$$

and hence

$$\frac{\partial a_l}{\partial l}=\frac{\frac{3}{2}a_l+\frac{3}{2}b}{2a_l+b+\frac{3}{2}l}.$$

Because $\frac{a_l}{2}-\frac{b}{2}+\frac{3}{2}l>0$ and $a_l>b$, we have

$$\frac{\partial a_l}{\partial l}>-1.$$

The numerator of the first integrand on the right-hand side of (30.26) satisfies

$$-\frac{1}{2}\left[(1-\theta)\theta\left((1+\theta)\frac{\partial a_l}{\partial l}+\frac{3}{2}\right)\right]<-\frac{1}{2}\left[(1-\theta)\theta\left(-(1+\theta)+\frac{3}{2}\right)\right]\leq 0.$$

Because $a_l>b$, the denominator of the corresponding integrand satisfies

$$\begin{aligned}&\left[(1-\theta)\theta\left((1+\theta)a_l+(2-\theta)b+\frac{3}{2}l\right)\right]^{\frac{3}{2}} \\ &<\left[(1-\theta)\theta\left((2-\theta)a_l+(1+\theta)b+\frac{3}{2}l\right)\right]^{\frac{3}{2}}.\end{aligned}$$

From (30.26), it follows that

$$\begin{aligned}\frac{\partial T_l}{\partial l} \leq& \sqrt{6}\varepsilon\int_0^{\frac{1}{2}} \frac{-\frac{1}{2}\left[(1-\theta)\theta((1+\theta)(\partial a_l/\partial l)+\frac{3}{2})\right]}{\left[(1-\theta)\theta((2-\theta)a_l+(1+\theta)b+\frac{3}{2}l)\right]^{3/2}}d\theta\\ &+\sqrt{6}\varepsilon\int_0^{\frac{1}{2}} \frac{-\frac{1}{2}\left[(1-\theta)\theta((2-\theta)(\partial a_l/\partial l)+\frac{3}{2})\right]}{\left[(1-\theta)\theta((2-\theta)a_l+(1+\theta)b+\frac{3}{2}l)\right]^{3/2}}d\theta\\ =&\sqrt{6}\varepsilon\int_0^{\frac{1}{2}} \frac{-\frac{1}{2}\left[(1-\theta)\theta(3\partial a_l/\partial l+3)\right]}{\left[(1-\theta)\theta((2-\theta)a_l+(1+\theta)b+\frac{3}{2}l)\right]^{3/2}}d\theta < 0.\end{aligned}$$

This completes the proof. □

Until now, we have only investigated the properties of the solutions of (30.5) and (30.6). However, what we are interested in is the problem in (30.4). For that, we need the following two lemmas which connect (30.4) to (30.5) and (30.6).

Lemma 30.4. *Let $|k| < \frac{2}{\sqrt{3}\varepsilon}$ in (30.4) and (30.5), and let T_0 denote the period of the solution u in (30.5). Then the solution y to (30.4) satisfies*

$$|y-u|+\varepsilon|y'-u'| = O(\varepsilon) \tag{30.27}$$

for $x \in [-1, -1+\frac{3}{2}T_0]$.

If $y(x_{\varepsilon,1})$ is the first minimum of the solution y to (30.4) and $u(x^*_{\varepsilon,1})$ is the first minimum of solution u to (30.5), then

$$|x_{\varepsilon,1} - x^*_{\varepsilon,1}| = O(\varepsilon^2). \tag{30.28}$$

Proof. From Lemma 30.1, we know that u is periodic when $|k| < \frac{2}{\sqrt{3}\varepsilon}$ and its period T_0 is of the order $O(\varepsilon)$. Moreover, because the energy equation of (30.5) is

$$\varepsilon^2\frac{(u')^2}{2} = u - \frac{u^3}{3} + \frac{\varepsilon^2}{2}k^2, \tag{30.29}$$

we have $|u| + \varepsilon|u'| \leq \bar{M}$, where $\bar{M} \geq 1$ is a constant. Let $\widetilde{u} = y - u$. Then $\widetilde{u}$ satisfies

$$\begin{cases}\varepsilon^2\widetilde{u}'' = -\widetilde{u}(\widetilde{u}+2u) - 2(1-x^2)(\widetilde{u}+u),\\ \widetilde{u}(-1) = \widetilde{u}'(-1) = 0.\end{cases} \tag{30.30}$$

Let $\bar{T}_0 = \sup\{t \in (0, \frac{3T_0}{2}) : \ |\widetilde{u}| + \varepsilon|\widetilde{u}'| < 1 \text{ for } x \in [-1, -1+t]\}$, which is well defined. For $t \in (0, \bar{T}_0)$, we have from (30.30)

$$\begin{aligned}
\varepsilon^2|\widetilde{u}'| &\le \int_{-1}^{-1+t} |\widetilde{u}||\widetilde{u}+2u|dx + 2\int_{-1}^{-1+t}(1-x^2)|\widetilde{u}+u|dx \\
&\le \int_{-1}^{-1+t} 3\bar{M}|\widetilde{u}|dx + 4\bar{M}\int_{-1}^{-1+t}(1-x)(1+x)dx \\
&\le \int_{-1}^{-1+t} 3\bar{M}|\widetilde{u}|dx + 8\bar{M}\int_{0}^{\frac{3T_0}{2}} sds \\
&= \int_{-1}^{-1+t} 3\bar{M}|\widetilde{u}|dx + 9T_0^2\bar{M}. \qquad (30.31)
\end{aligned}$$

Furthermore, we have

$$|\widetilde{u}| \le \int_{-1}^{-1+t} |\widetilde{u}'|dx. \qquad (30.32)$$

Coupling (30.31) and (30.32) gives the integral inequality

$$|\widetilde{u}| + \varepsilon|\widetilde{u}'| \le \frac{1}{\varepsilon}\int_{-1}^{-1+t}\left[3\bar{M}|\widetilde{u}| + \varepsilon|\widetilde{u}'|\right]dx + \frac{9}{\varepsilon}T_0^2\bar{M}.$$

By Gronwall's inequality, we have

$$|\widetilde{u}| + \varepsilon|\widetilde{u}'| \le \frac{9\bar{M}T_0^2}{\varepsilon}\exp\left\{\frac{9\bar{M}T_0}{4\varepsilon}\right\}.$$

Recall that $T_0 = O(\varepsilon)$. Thus, we have

$$|\widetilde{u}| + \varepsilon|\widetilde{u}'| = O(\varepsilon) \ \text{ when } \ x \in [-1, -1 + \frac{3T_0}{2}],$$

which is equivalent to (30.27).

At the points $x_{\varepsilon,1}$ and $x^*_{\varepsilon,1}$, equation (30.29) can be written as

$$\varepsilon^2\frac{(u'(x_{\varepsilon,1}))^2}{2} = u(x_{\varepsilon,1}) - \frac{u^3(x_{\varepsilon,1})}{3} + \frac{\varepsilon^2}{2}k^2 \qquad (30.33)$$

and

$$0 = u(x^*_{\varepsilon,1}) - \frac{u^3(x^*_{\varepsilon,1})}{3} + \frac{\varepsilon^2}{2}k^2, \qquad (30.34)$$

respectively. Subtracting (30.33) from (30.34) yields

$$\begin{aligned}
&\varepsilon^2\frac{(\widetilde{u}'(x_{\varepsilon,1}))^2}{2} \\
=&(u(x_{\varepsilon,1}) - u(x^*_{\varepsilon,1}))\left(1 - \frac{1}{3}(u^2(x^*_{\varepsilon,1}) + u(x^*_{\varepsilon,1})u(x_{\varepsilon,1}) + u^2(x_{\varepsilon,1}))\right) \\
=&\left(u'(x_{\varepsilon,1})(x_{\varepsilon,1} - x^*_{\varepsilon,1}) + \frac{u''(x_{\varepsilon,1})}{2}(x_{\varepsilon,1} - x^*_{\varepsilon,1})^2 + \cdots\right) \\
&\times\left(1 - \frac{1}{3}(u^2(x^*_{\varepsilon,1}) + u(x^*_{\varepsilon,1})u(x_{\varepsilon,1}) + u^2(x_{\varepsilon,1}))\right).
\end{aligned}$$

Because $|u'(x_{\varepsilon,1}) - y'(x_{\varepsilon,1})| = O(1)$ from (30.27) and $y'(x_{\varepsilon,1}) = 0$, we have $|\widetilde{u}'(x_{\varepsilon,1})| = |u'(x_{\varepsilon,1})| = O(1)$, thus establishing (30.28). □

Now we claim that if $y(x_{\varepsilon,n}) = b$ is a minimum of $y(x)$, then $y''(x_{\varepsilon,n})$ should be positive. If $y''(x_{\varepsilon,n}) = 0$, then we have $y(x_{\varepsilon,n}) = -(1 - x^2_{\varepsilon,n}) \pm \sqrt{1 + (1 - x^2_{\varepsilon,n})^2}$ from (30.4). Because $y(x_{\varepsilon,n})$ is a minimum and $y''(x_{\varepsilon,n}) = 0$, it can be shown that $y'''(x_{\varepsilon,n}) = 0$ and $y^{(4)}(x_{\varepsilon,n}) \geq 0$. However, $y'''(x_{\varepsilon,n}) = 4x_{\varepsilon,n}y(x_{\varepsilon,n})/\varepsilon^2$. Therefore, we conclude that $x_{\varepsilon,n} = 0$. Furthermore, we have $y^{(4)}(x_{\varepsilon,n}) = 4y(x_{\varepsilon,n})/\varepsilon^2$; i.e., $y^{(4)}(0) = 4y(0)/\varepsilon^2$. From Ai (2003, Corollary 3.1), we know that $y(0) \leq 0$. Because $-1 + \sqrt{2} > 0$, it follows that $y(0) = -1 - \sqrt{2} < 0$. Therefore, we obtain $y^{(4)}(0) < 0$, which is a contradiction. Hence, $y''(x_{\varepsilon,n})$ must be positive if $y(x_{\varepsilon,n}) = b$ is a minimum of $y(x)$. From (30.4) we see that $1 - b^2 - 2(1 - x^2_{\varepsilon,n})b > 0$. By setting $l = 2(1 - x^2_{\varepsilon,n})$, we get from Figure 30.1

$$\begin{aligned} -(1 - x^2_{\varepsilon,n}) - \sqrt{1 + (1 - x^2_{\varepsilon,n})^2} &< b \\ &< -(1 - x^2_{\varepsilon,n}) + \sqrt{1 + (1 - x^2_{\varepsilon,n})^2}. \end{aligned} \tag{30.35}$$

Lemma 30.5. *Assume $y(x_{\varepsilon,n}) = b$ is a minimum of the solution $y(x)$ to (30.4), when $|k| < \frac{2}{\sqrt{3}\varepsilon}$, and let $u(x)$ be the solution of (30.6) with $l = 2(1 - x^2_{\varepsilon,n})$. Because (30.6) is autonomous, we may change the initial point $x = x_{\varepsilon,n}$ (see the sentence just before Lemma 30.2, i.e. set $u(x_{\varepsilon,n}) = b$ and $u'(x_{\varepsilon,n}) = 0$). If T_l is the period of $u(x)$, then we have*

$$|y - u| + \varepsilon|y' - u'| = O(\varepsilon) \tag{30.36}$$

for $x \in [x_{\varepsilon,n}, x_{\varepsilon,n} + \frac{3}{2}T_l]$.

If $y(x_{\varepsilon,n+1})$ is the next minimum of $y(x)$ following $y(x_{\varepsilon,n})$, and $u(x^*_{\varepsilon,n+1})$ is the next minimum of $u(x)$ following $u(x_{\varepsilon,n})$, then

$$|x_{\varepsilon,n+1} - x^*_{\varepsilon,n+1}| = O(\varepsilon^2) \tag{30.37}$$

Because the proof of Lemma 30.5 is similar to that of Lemma 30.4, we omit it altogether.

Remark 30.1. Because $x^*_{\varepsilon,n+1} - x_{\varepsilon,n}$ is the length of the period of the solution u of (30.6) with $l = 2(1 - x^2_{\varepsilon,n})$, from (30.19) and (30.37) it is readily seen that

$$\begin{aligned} x_{\varepsilon,n+1} - x_{\varepsilon,n} &= x^*_{\varepsilon,n+1} - x_{\varepsilon,n} + O(\varepsilon^2) = T_l(b) + O(\varepsilon^2) \\ &= \sqrt{2}\varepsilon \int_b^{a_l} \frac{du}{\sqrt{H_l(u) - H_l(b)}} + O(\varepsilon^2). \end{aligned} \tag{30.38}$$

When

$$-\frac{1}{2}(l+\sqrt{4+l^2})+\delta \le b \le -\frac{1}{2}(l-\sqrt{4+l^2})-\delta,$$

where δ is a small positive constant, we know from Lemma 30.3 that there exists a constant m such that

$$x_{\varepsilon,n+1}-x_{\varepsilon,n} \ge m\varepsilon + O(\varepsilon^2). \tag{30.39}$$

30.3 The envelope function

In this section, we construct the envelope function which describes the asymptotic behavior of the minima of the solutions to (30.4) when $|k| < \frac{2}{\sqrt{3}\varepsilon}$. First we define

$$\rho = \lim_{\varepsilon\to 0}\frac{\varepsilon^2}{2}k^2. \tag{30.40}$$

Obviously, $0 \le \rho < \frac{2}{3}$ if $|k| < \frac{2}{\sqrt{3}\varepsilon}$. Next, we introduce the piecewise linear function $b_{\varepsilon,\rho}(x)$ which joins the minima of the solution to (30.4). Clearly, it can be written as

$$b_{\varepsilon,\rho}(x) = y_{\varepsilon,n} + \frac{y_{\varepsilon,n+1}-y_{\varepsilon,n}}{x_{\varepsilon,n+1}-x_{\varepsilon,n}}(x-x_{\varepsilon,n}), \quad x\in[x_{\varepsilon,n},x_{\varepsilon,n+1}], \tag{30.41}$$

for $n = 1,2,3,\cdots,S_{\varepsilon,\rho}-1$, where $y_{\varepsilon,n} = y(x_{\varepsilon,n})$ denotes the n-th minimum of the solution to (30.4) and $S_{\varepsilon,\rho}$ is the number of minima that the solution to equation (30.4) could have for given ε and ρ. From (30.35), we know that

$$\begin{aligned}-(1-x_{\varepsilon,n}^2)-&\sqrt{1+(1-x_{\varepsilon,n}^2)^2}+\delta\\ &\le y_{\varepsilon,n} \le -(1-x_{\varepsilon,n}^2)+\sqrt{1+(1-x_{\varepsilon,n}^2)^2}-\delta,\end{aligned} \tag{30.42}$$

where δ is a sufficiently small positive number.

Lemma 30.6. *As $\varepsilon \to 0$, there exists a subsequence of $\{b_{\varepsilon,\rho}(x)\}$ which converges uniformly to function $b_\rho(x)$.*

Proof. By (30.42), $y_{\varepsilon,n}$ is bounded uniformly in ε and n because $x_{\varepsilon,n}\in[-1,1]$. Moreover, if $u_{\varepsilon,n} = u(x^*_{\varepsilon,n+1})$ and $y_{\varepsilon,n} = u_{\varepsilon,n}$, then we have

$$\begin{aligned}|y_{\varepsilon,n+1}-y_{\varepsilon,n}| &= |y_{\varepsilon,n+1}-u_{\varepsilon,n}| = |y_{\varepsilon,n+1}-u(x^*_{\varepsilon,n+1})|\\ &\le |y_{\varepsilon,n+1}-u(x_{\varepsilon,n+1})| + |u(x_{\varepsilon,n+1})-u(x^*_{\varepsilon,n+1})|\end{aligned} \tag{30.43}$$

From (30.36), we know that $|y_{\varepsilon,n+1} - u(x_{\varepsilon,n+1})| = O(\varepsilon)$. Because $\varepsilon u'$ is bounded by a constant (cf. (30.7)), it follows from the mean-value theorem and (30.37) that

$$|u(x_{\varepsilon,n+1}) - u(x^*_{\varepsilon,n+1})| = |u'(\xi)||x_{\varepsilon,n+1} - x^*_{\varepsilon,n+1}| = O(\varepsilon),$$

where ξ lies between $x_{\varepsilon,n+1}$ and $x^*_{\varepsilon,n+1}$. Thus, (30.43) can be written as

$$|y_{\varepsilon,n+1} - y_{\varepsilon,n}| = O(\varepsilon); \tag{30.44}$$

see Figure 30.6. Moreover, by (30.39), there exists a constant M independent of ε such that

$$\frac{|y_{\varepsilon,n+1} - y_{\varepsilon,n}|}{|x_{\varepsilon,n+1} - x_{\varepsilon,n}|} \leq M. \tag{30.45}$$

Hence, by the Arzelá-Ascoli theorem, there exists a subsequence of $\{b_{\varepsilon,\rho}(x)\}$ which converges uniformly to a function $b_\rho(x)$. □

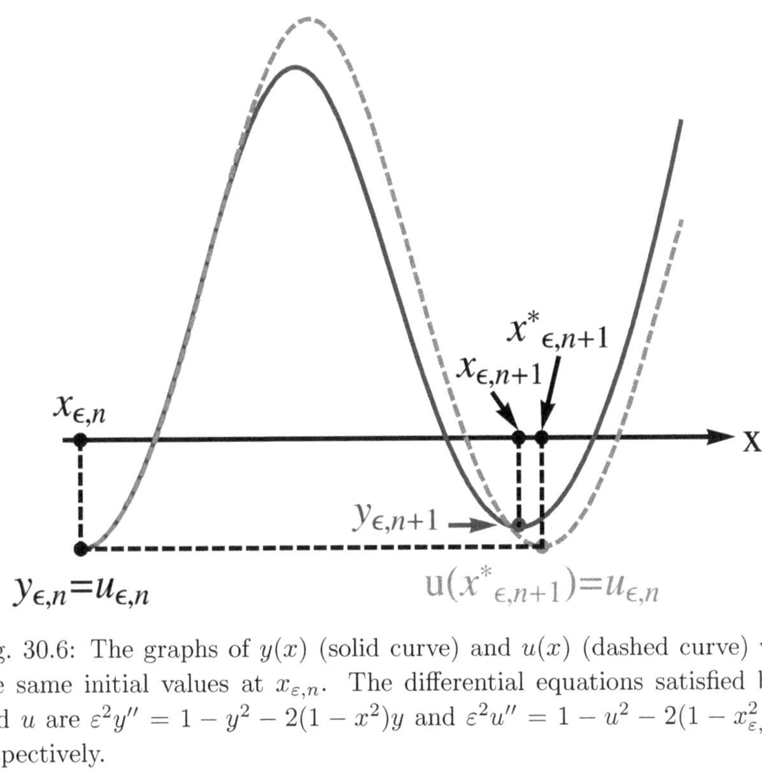

Fig. 30.6: The graphs of $y(x)$ (solid curve) and $u(x)$ (dashed curve) with the same initial values at $x_{\varepsilon,n}$. The differential equations satisfied by y and u are $\varepsilon^2 y'' = 1 - y^2 - 2(1 - x^2)y$ and $\varepsilon^2 u'' = 1 - u^2 - 2(1 - x^2_{\varepsilon,n})u$, respectively.

Remark 30.2. Because $y_{\varepsilon,n}$ depends on ε and n and $b_{\varepsilon,\rho}(x)$ depends on ε and ρ, the subsequence of $\{b_{\varepsilon,\rho}(x)\}$ converges to different functions for different values of ρ. Thus $b_\rho(x)$ is actually a one-parameter family of functions. From (30.42), we have

$$\begin{aligned}-(1-x^2)-\sqrt{1+(1-x^2)^2}+\delta \le b_\rho(x)\\ \le -(1-x^2)+\sqrt{1+(1-x^2)^2}-\delta,\end{aligned} \tag{30.46}$$

where δ is a sufficiently small positive constant. Furthermore, from (30.41) and (30.42), we also have

$$|b'_\rho(x)| \le M. \tag{30.47}$$

We now investigate the properties of the envelope function $b_\rho(x)$. Before stating our results, we need some definitions and notations. First, we define

$$G(u,x) = u - \frac{u^3}{3} - (1-x^2)u^2. \tag{30.48}$$

From (30.8), it is evident that $G(u,x) = H_l(u)$ with $l = 2(1-x^2)$. As an equation in u, $G(u,x) - G(b_\rho(x),x) = 0$ has three solutions when $b_\rho(x)$ satisfies (30.46). Clearly, $b_\rho(x)$ is a solution. The other two solutions are

$$\begin{aligned}a_\rho(x) =& \frac{1}{2}\Big(-3 - b_\rho(x) + 3x^2\\ &+ \sqrt{3}\sqrt{7-2b_\rho(x)-b_\rho^2(x)-6x^2+2b_\rho(x)x^2+3x^4}\Big)\end{aligned} \tag{30.49}$$

and

$$\begin{aligned}c_\rho(x) =& \frac{1}{2}\Big(-3 - b_\rho(x) + 3x^2\\ &- \sqrt{3}\sqrt{7-2b_\rho(x)-b_\rho^2(x)-6x^2+2b_\rho(x)+3x^4}\Big).\end{aligned} \tag{30.50}$$

From (30.10) and (30.11), it is readily seen that $a_\rho(x) = a_l$ with $l = 2(1-x^2)$ and $b = b_\rho(x), c_\rho(x) = c_l$ with $l = 2(1-x^2)$ and $b = b_\rho(x)$, respectively. Thus, just like (30.12), we have

$$a_\rho(x) > b_\rho(x) > c_\rho(x) \tag{30.51}$$

for $x \in [-1,1]$. Next, we introduce the initial-value problem

$$\begin{cases} v''(s) = 1 - v(s)^2 - 2(1-x^2)v(s), \\ v(0) = b_\rho(x), \quad v'(0) = 0, \end{cases} \tag{30.52}$$

where x and $b_\rho(x)$ are treated as parameters. Similar to Lemma 30.2 (c), one can show that the solution $v(s)$ of (30.52) is periodic with the period $T(b_\rho(x), x)$ if

$$-(1-x^2) - \sqrt{1+(1-x^2)^2} < b_\rho(x) < -(1-x^2) + \sqrt{1+(1-x^2)^2}.$$

In fact, with $l = 2(1-x^2)$ and $b = b_\rho(x)$ in (30.19), one also has

$$T(b_\rho(x), x) = \frac{T_l(b)}{\varepsilon} = \sqrt{2}\int_{b_\rho(x)}^{a_\rho(x)} \frac{du}{\sqrt{G(u,x) - G(b_\rho(x), x)}}. \tag{30.53}$$

Lemma 30.7. *For a given ρ, the envelope function $b_\rho(x)$ satisfies*

$$\int_{b_\rho(x)}^{a_\rho(x)} \sqrt{G(u,x) - G(b_\rho(x), x)}du = C_\rho, \tag{30.54}$$

where C_ρ is independent of x.

Proof. We multiply both sides of equation (30.4) by y' and then integrate it from -1 to $x_{\varepsilon,n}$. Upon an integration by parts, we obtain

$$-\frac{\varepsilon^2}{2}k^2 = G(y_{\varepsilon,n}, x_{\varepsilon,n}) - 2\int_{-1}^{x_{\varepsilon,n}} xy^2dx. \tag{30.55}$$

Because $y_{\varepsilon,n} = y(x_{\varepsilon,n})$, we have $y_{\varepsilon,n} \to b_\rho(x)$ and $x_{\varepsilon,n} \to x$ as $\varepsilon \to 0$; see (30.41). Hence, (30.55) gives

$$-\rho = G(b_\rho(x), x) - 2\lim_{\varepsilon\to 0}\int_{-1}^{x_{\varepsilon,n}} xy^2dx. \tag{30.56}$$

Let $v(s) = y(x) = y(x_{\varepsilon,j} + \varepsilon s)$ when $x \in [x_{\varepsilon,j}, x_{\varepsilon,j+1}]$. It can be readily verified that $v(s)$ satisfies (30.52). Moreover, $x_{\varepsilon,i+1} - x_{\varepsilon,i} = \varepsilon T(b_\rho(x_{\varepsilon,i}), x_{\varepsilon,i}) + O(\varepsilon^2)$ by (30.38) and (30.53). Note that $x_{\varepsilon,1}$ approaches -1 as $\varepsilon \to 0$ by Lemma 30.4. The integral on the right-hand side of (30.56) can be written as

$$\begin{aligned}
\lim_{\varepsilon\to 0}\int_{-1}^{x_{\varepsilon,n}} xy^2dx &= \lim_{\varepsilon\to 0}\sum_{i=1}^{n-1}\int_{x_{\varepsilon,i}}^{x_{\varepsilon,i+1}} xy^2dx \\
&= \lim_{\varepsilon\to 0}\sum_{i=1}^{n-1}\frac{\int_0^{T(b_\rho(x_{\varepsilon,i}),x_{\varepsilon,i})+O(\varepsilon)}(x_{\varepsilon,i}+\varepsilon s)v^2(s)ds}{T(b_\rho(x_{\varepsilon,i}), x_{\varepsilon,i}) + O(\varepsilon)}(x_{\varepsilon,i+1} - x_{\varepsilon,i}) \\
&= \int_{-1}^{x}\frac{t\int_0^{T(b_\rho(t),t)} v^2(s)ds}{T(b_\rho(t), t)}dt.
\end{aligned} \tag{30.57}$$

From (30.56), it follows that

$$-\rho = G(b_\rho(x), x) - 2\int_{-1}^{x}\frac{t\int_0^{T(b_\rho(t),t)} v^2(s)ds}{T(b_\rho(t), t)}dt. \tag{30.58}$$

Straightforward differentiation of (30.58) with respect to x gives

$$T(b_\rho(x), x)\frac{dG(b_\rho(x), x)}{dx} = \int_0^{T(b_\rho(x),x)} 2xv^2(s)ds. \tag{30.59}$$

Note that from (30.52), we have

$$\frac{(v')^2}{2} = G(v.x) - G(b_\rho(x), x).$$

Thus, by a similar analysis used in the proof of Lemma 30.2 (c), we can show that x varies from 0 to $\frac{1}{2}T(b_\rho(x), x)$ as v varies from b_ρ to a_ρ, and

$$\frac{dv}{ds} = \sqrt{2}\sqrt{G(v, x) - G(b_\rho(x), x)}. \tag{30.60}$$

Also, x varies from $\frac{1}{2}T(b_\rho(x), x)$ to $T(b_\rho(x), x)$ as v varies from a_ρ to b_ρ, and

$$\frac{dv}{ds} = -\sqrt{2}\sqrt{G(v, x) - G(b_\rho(x), x)}. \tag{30.61}$$

On account of (30.60) and (30.61), the right-hand side of (30.59) becomes

$$\int_0^{T(b_\rho(x),x)} 2xv^2(s)ds = 2\int_{b_\rho(x)}^{a_\rho(x)} \frac{\sqrt{2}xv^2}{\sqrt{G(v, x) - G(b_\rho(x), x)}}dv. \tag{30.62}$$

Coupling (30.59) and (30.62), and noting that $T(b_\rho(x), x)$ can be expressed as in (30.53), we obatin

$$\begin{aligned}\int_{b_\rho(x)}^{a_\rho(x)} &\frac{1}{\sqrt{G(u, x) - G(b_\rho(x), x)}}\frac{d}{dx}G(b_\rho(x), x)du \\ &= \int_{b_\rho(x)}^{a_\rho(x)} \frac{2xu^2}{\sqrt{G(u, x) - G(b_\rho(x), x)}}du. \end{aligned}\tag{30.63}$$

Taking derivative with respect to x in (30.54), the left-hand side of the equation becomes

$$\frac{1}{2}\int_{b_\rho(x)}^{a_\rho(x)} \frac{2xu^2 - \frac{dG(b_\rho(x),x)}{dx}}{\sqrt{G(u, x) - G(b_\rho(x), x)}}du \tag{30.64}$$

By (30.63), the last integral is equal to 0. Hence, C_ρ is independent of x. □

Lemma 30.8. *Given C_ρ in (30.54), there is a unique $b_\rho(x)$ satisfying this equation.*

Proof. Consider C_ρ as a function of $b_\rho(x)$, and differentiating (30.54) with respect to $b_\rho(x)$, we have

$$\frac{\partial C_\rho}{\partial b_\rho(x)} = \frac{1}{2}\int_{b_l(x)}^{a_l(x)} \frac{-\frac{\partial G(b_\rho(x),x)}{\partial b_\rho(x)}}{\sqrt{G(u,x)-G(b_\rho(x),x)}}du. \tag{30.65}$$

Direct calculation shows that

$$-\frac{\partial G(b_\rho(x),x)}{\partial b_\rho(x)} = -1 + b_\rho^2(x) + 2(1-x^2)b_\rho(x).$$

However we also have by straightforward calculation

$$-\frac{1}{3}(a_\rho(x)-b_\rho(x))(b_\rho(x)-c_\rho(x)) = -1 + b_\rho^2(x) + 2(1-x^2)b_\rho(x).$$

Thus,

$$-\frac{\partial G(b_\rho(x),x)}{\partial b_\rho(x)} = -\frac{1}{3}(a_\rho(x)-b_\rho(x))(b_\rho(x)-c_\rho(x)).$$

From (30.51) and (30.65), we obtain

$$\frac{\partial C_\rho}{\partial b_\rho(x)} = \frac{1}{6}\int_{b_\rho(x)}^{a_\rho(x)} \frac{-(a_\rho(x)-b_\rho(x))(b_\rho(x)-c_\rho(x))}{\sqrt{G(u,x)-G(b_\rho(x),x)}}du < 0. \tag{30.66}$$

Therefore, C_ρ is monotonic in $b_\rho(x)$, and the proof is complete. □

When $x = -1$, we have from (30.58)

$$-\rho = G(b_\rho(-1),-1). \tag{30.67}$$

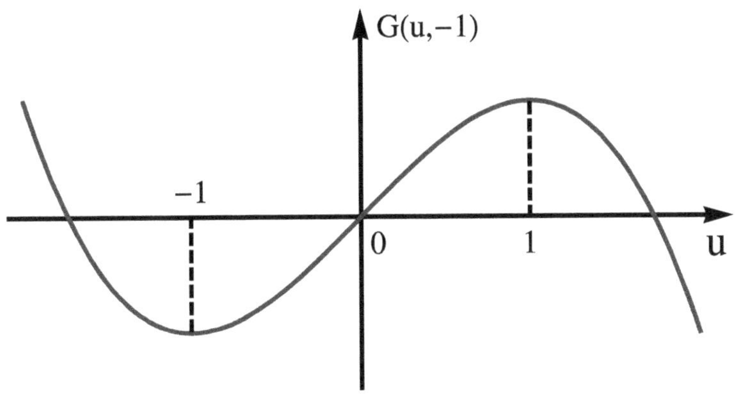

Fig. 30.7: The graph of $G(u,-1) = u - u^3/3$.

Assume that $\{b_{\varepsilon,\rho}(x)\}$ has another subsequence which converges uniformly to $b^*_\rho(x)$ as $\varepsilon \to 0$. It is readily seen that $b^*_\rho(x)$ also satisfies (30.67) when $x = -1$; that is, $-\rho = G(b^*_\rho(-1), -1)$. Thus, $G(b_\rho(-1), -1) = G(b^*_\rho(-1), -1)$. Moreover, because both $b_\rho(x)$ and $b^*_\rho(x)$ satisfy (30.46) when $x = -1$, both $G(b_\rho(-1), -1)$ and $G(b^*_\rho(-1), -1)$ are monotonic as $b_\rho(-1)$ and $b^*_\rho(-1)$ vary between -1 and 1 (see Figure 30.7), and we have $b_\rho(-1) = b^*_\rho(-1)$. From (30.54), we can calculate C_ρ and C^*_ρ, given $b_\rho(-1)$ and $b^*_\rho(-1)$, respectively. Obviously, $C_\rho = C^*_\rho$ because $b_\rho(-1) = b^*_\rho(-1)$. By Lemma 30.8, $b_\rho(x) = b^*_\rho(x)$. Hence, we have shown that all convergent subsequences of $\{b_{\varepsilon,\rho}(x)\}$ have the same limiting function $b_\rho(x)$. Furthermore, for any subsequence of $\{b_{\varepsilon,\rho}(x)\}$, by using the same argument as that in the proof of Lemma 30.6, one can show that it has a convergent subsequence. We have therefore proved the following lemma.

Lemma 30.9. *As $\varepsilon \to 0$, the sequence $\{b_{\varepsilon,\rho}(x)\}$ converges uniformly to the function $b_\rho(x)$.*

Lemma 30.10. *$b_\rho(x)$ is an entire function.*

Proof. We can write (30.49) as $a_\rho(x) = a_\rho(b_\rho(x), x)$, from which it can be readily seen that

$$a_\rho(-x) = a_\rho(b_\rho(-x), -x) = a_\rho(b_\rho(-x), x).$$

From (30.48), we know that $G(u, -x) = G(u, x)$. Thus, we have

$$\begin{aligned}
&\int_{b_\rho(-x)}^{a_\rho(b_\rho(-x),x)} \sqrt{G(u,x) - G(b_\rho(-x), x)}du \\
&\qquad = \int_{b_\rho(-x)}^{a_\rho(b_\rho(-x),-x)} \sqrt{G(u,-x) - G(b_\rho(-x), -x)}du \\
&\qquad = \int_{b_\rho(x)}^{a_\rho(b_\rho(x),x)} \sqrt{G(u,x) - G(b_\rho(x), x)}du.
\end{aligned}$$

The last equality follows from Lemma 30.7 because (30.54) is independent of x. By Lemma 30.8, $b_\rho(x) = b_\rho(-x)$. □

Remark 30.3. It has been shown in Ai (2003, Corollary 3.1) that all minima of the solution to (30.4) are negative when $x \in (-1, 0]$. Hence, $b_\rho(x) < 0$ when $x \in (-1, 0]$. Because $b_\rho(x)$ is an even function, we have $b_\rho(x) < 0$ when $x \in (-1, 1)$, which means that all minima of the solutions to (30.4) should be less or equal to zero when ε is sufficiently small. Thus, for any given ρ, we can choose suitable k in (30.4) so that the solution $y(x)$ satisfies $y(1) = 0$.

Theorem 30.1. *When $\rho = 0$, the solution of* (30.4) *has the maximum number of spikes in* $[-1, 1]$.

Proof. Differentiating (30.54) with respect to $G(b_\rho(x), x)$ gives

$$\frac{\partial C_\rho}{\partial G(b_\rho(x), x)} = -\frac{1}{2}\int_{b_\rho(x)}^{a_\rho(x)} \frac{du}{\sqrt{G(u, x) - G(b_\rho(x), x)}} < 0.$$

Thus, C_ρ is minimum when $G(b_\rho(x), x)$ is maximum. Because C_ρ is independent of x, we may take $x = -1$. Then, C_ρ is minimum when $G(b_\rho(-1), -1)$ is maximum. Note that from (30.67), we have $G(b_\rho(-1), -1) = -\rho$. Moreover, $\rho \in [0, \frac{2}{3})$. Thus, $G(b_\rho(-1), -1)$ is maximum when $\rho = 0$; i.e., when $\rho = 0$, C_ρ reaches its minimum.

Define $b_0(x)$ to be the envelope function when $\rho = 0$. From (30.66), we have $b_0(x) > b_\rho(x)$ for any given x in $[-1, 1]$, which means that $b_0(x)$ yields the highest envelope curve. Moreover, from (30.20) we know that for any fixed x, $\partial T_l(b)/\partial b < 0$. Thus, from (30.38), to get as many spikes as we can, we need $T_l(b)$ to be as small as possible, or equivalently, we require b, which represents the minimum of the solution to (30.4), to be as large as possible. Because $b_0(x)$ gives the highest envelope curve, the solution to (30.4) has the maximum number of spikes when $\rho = 0$. □

Remark 30.4. If we denote by N_ε the maximum number of spikes of the solution to (30.4) for $x \in [-1, 1]$, then the maximum number of spikes of the solution to (30.1) should also be N_ε.

30.4 Number of solutions

Based on Theorem 30.1, we now focus on the case $k = 0$ (and hence $\rho = 0$). The following result gives an estimate between $b_0(x_{\varepsilon,n})$ and $y_{\varepsilon,n}$.

Lemma 30.11. *When $k = 0$,*

$$|y_{\varepsilon,n} - b_0(x_{\varepsilon,n})| = O(\varepsilon). \tag{30.68}$$

Proof. First, we put

$$F(b_0(x), x) := \frac{2x\left[b_0^2(x) - \frac{1}{T(b_0(x),x)}\int_0^{T(b_0(x),x)} v^2(s)ds\right]}{b_0^2(x) - 1 + 2(1 - x^2)b_0(x)}. \tag{30.69}$$

Straightforward differentiation of (30.58) with respect to x when $k = 0$ gives

$$b_0'(x) = F(b_0(x), x). \tag{30.70}$$

Multiplying y' on both sides of (30.4), and integrating from $x_{\varepsilon,n}$ to $x_{\varepsilon,n+1}$, we obtain

$$\begin{aligned}
-\int_{x_{\varepsilon,n}}^{x_{\varepsilon,n+1}} 2xy^2dx &= \left(\frac{\varepsilon^2}{2}(y')^2+(1-x^2)y^2+\frac{y^3}{3}-y\right)\Bigg|_{x_{\varepsilon,n}}^{x_{\varepsilon,n+1}}\\
&=\frac{y_{\varepsilon,n+1}^3}{3}-\frac{y_{\varepsilon,n}^3}{3}-(y_{\varepsilon,n+1}-y_{\varepsilon,n})\\
&\quad+(1-x_{\varepsilon,n+1}^2)(y_{\varepsilon,n+1}^2-y_{\varepsilon,n}^2)-y_{\varepsilon,n}^2(x_{\varepsilon,n+1}^2-x_{\varepsilon,n}^2).
\end{aligned} \tag{30.71}$$

From (30.71), it follows that

$$\begin{aligned}
&\frac{y_{\varepsilon,n+1}-y_{\varepsilon,n}}{x_{\varepsilon,n+1}-x_{\varepsilon,n}}\\
&=\frac{y_{\varepsilon,n}^2(x_{\varepsilon,n+1}+x_{\varepsilon,n})-\frac{2}{x_{\varepsilon,n+1}-x_{\varepsilon,n}}\int_{x_{\varepsilon,n}}^{x_{\varepsilon,n+1}}xy^2dx}{\frac{1}{3}(y_{\varepsilon,n}^2+y_{\varepsilon,n}y_{\varepsilon,n+1}+y_{\varepsilon,n+1}^2)-1+(1-x_{\varepsilon,n+1}^2)(y_{\varepsilon,n+1}+y_{\varepsilon,n})}
\end{aligned} \tag{30.72}$$

Because

$$y_{\varepsilon,n+1}-y_{\varepsilon,n}=O(\varepsilon)$$

and

$$x_{\varepsilon,n+1}-x_{\varepsilon,n}=O(\varepsilon) \tag{30.73}$$

by (30.44) and (30.38), respectively, (30.72) can be written as

$$\begin{aligned}
\frac{y_{\varepsilon,n+1}-y_{\varepsilon,n}}{x_{\varepsilon,n+1}-x_{\varepsilon,n}} &= \frac{2y_{\varepsilon,n}^2x_{\varepsilon,n}+O(\varepsilon)}{y_{\varepsilon,n}^2-1+2(1-x_{\varepsilon,n}^2)y_{\varepsilon,n}+O(\varepsilon)}\\
&\quad-\frac{\frac{2}{T(y_{\varepsilon,n},x_{\varepsilon,n})+O(\varepsilon)}\int_0^{T(y_{\varepsilon,n},x_{\varepsilon,n})+O(\varepsilon)}(x_{\varepsilon,n}+\varepsilon s)v^2(s)ds}{y_{\varepsilon,n}^2-1+2(1-x_{\varepsilon,n}^2)y_{\varepsilon,n}+O(\varepsilon)}\\
&=F(y_{\varepsilon,n},x_{\varepsilon,n})+O(\varepsilon)
\end{aligned} \tag{30.74}$$

on account of (30.53), where $v(s)$ is the solution of the initial-value problem (30.52).

Next, we claim that $F(b_0(x),x)$ is Lipschitz continuous in both variables $b_0(x)$ and x. Taking the derivative with respect to $b_0(x)$ in (30.69) gives

$$\begin{aligned}
\frac{\partial F(b_0(x),x)}{\partial b_0(x)} &= \frac{2x}{b_0^2(x)-1+2(1-x^2)b_0(x)}\\
&\times\left(2b_0^2(x)-\frac{T(b_0(x),x)v^2(T(b_0(x),x))-\int_0^{T(b_0(x),x)}v^2(s)ds}{T^2(b_0(x),x)}\frac{\partial T(b_0(x),x)}{\partial b_0(x)}\right)\\
&-\frac{4x(b_0(x)+(1-x^2))}{b_0^2(x)-1+2(1-x^2)b_0(x)}\left(b_0^2(x)-\frac{1}{T(b_0(x),x)}\int_0^{T(b_0(x),x)}v^2(s)ds\right)
\end{aligned} \tag{30.75}$$

From (30.46), we know that the denominator of (30.75) is bounded away from zero. From (30.23) and (30.53), we also know that $\partial T(b_0(x),x)/\partial b_0(x)$ is bounded. Furthermore, $b_0(x), x$ and $T(b_0(x),x)$ are bounded. Hence, $\partial F(b_0(x),x)/\partial b_0(x)$ is bounded. Again, direct differentiation with respect to x in (30.69) gives

$$\begin{aligned}
&\frac{\partial F(b_0(x),x)}{\partial x}\\
&= \frac{2}{b_0^2(x)-1+2(1-x^2)b_0(x)}\left(b_0^2(x)-\frac{1}{T(b_0(x),x)}\int_0^{T(b_0(x),x)} v^2(s)ds\right)\\
&+\frac{2x}{b_0^2(x)-1+2(1-x^2)b_0(x)}\\
&\times\left(b_0^2(x)-\frac{T(b_0(x),x)v^2(T(b_0(x),x))-\int_0^{T(b_0(x),x)} v^2(s)ds}{T^2(b_0(x),x)}\frac{\partial T(b_0(x),x)}{\partial x}\right)\\
&+\frac{8x^2b_0(x)}{(b_0^2(x)-1+2(1-x^2)b_0(x))^2}\left(b_0^2(x)-\frac{1}{T(b_0(x),x)}\int_0^{T(b_0(x),x)} v^2(s)ds\right).
\end{aligned}$$

Note that here $b_0(x)$ has been treated as an independent variable. It is readily seen from (30.26) and (30.53) that $\partial T(b_0(x),x)/\partial x$ is bounded. Hence, $\partial F(b_0(x),x)/\partial x$ is bounded. This completes the proof of our claim. Let us denote by L and P, respectively, the Lipschitz constants with respect to $b_0(x)$ and x.

Now we estimate the error between $b_0(x_{\varepsilon,n})$ and $y_{\varepsilon,n}$. Define

$$r_n := \int_{x_{\varepsilon,n}}^{x_{\varepsilon,n+1}} \left[F(b_0(x),x)-F(b_0(x_{\varepsilon,n}),x_{\varepsilon,n})\right]dx + O(\varepsilon^2).$$

Taking absolute values on both sides of the above equation gives

$$\begin{aligned}
|r_n| =& \left|\int_{x_{\varepsilon,n}}^{x_{\varepsilon,n+1}} \left[F(b_0(x),x)-F(b_0(x_{\varepsilon,n}),x_{\varepsilon,n})\right]dx + O(\varepsilon^2)\right|\\
\leq& \int_{x_{\varepsilon,n}}^{x_{\varepsilon,n+1}} |F(b_0(x),x)-F(b_0(x_{\varepsilon,n}),x)|dx\\
&+\int_{x_{\varepsilon,n}}^{x_{\varepsilon,n+1}} |F(b_0(x_{\varepsilon,n}),x)-F(b_0(x_{\varepsilon,n}),x_{\varepsilon,n})|dx + O(\varepsilon^2)\\
\leq& L\int_{x_{\varepsilon,n}}^{x_{\varepsilon,n+1}} |b_0(x)-b_0(x_{\varepsilon,n})|dx + P\int_{x_{\varepsilon,n}}^{x_{\varepsilon,n+1}} (x-x_{\varepsilon,n})dx + O(\varepsilon^2)\\
\leq& L\int_{x_{\varepsilon,n}}^{x_{\varepsilon,n+1}} |b_0'(\lambda)|(x-x_{\varepsilon,n})dx + \frac{1}{2}Ph^2 + O(\varepsilon^2)\\
\leq& \frac{h^2}{2}(LM+P)+O(\varepsilon^2) := r, \quad x_{\varepsilon,n}<\lambda<x_{\varepsilon,n+1}.
\end{aligned} \tag{30.76}$$

where

$$h = \sup\{x_{\varepsilon,n+1} - x_{\varepsilon,n} : n = 1, 2, \cdots, S_{\varepsilon,0} - 1\},$$

$$M = \max_{-1\le x\le 1} |b_0'(x)|,$$

and $S_{\varepsilon,0}$ is the number of minima of the solution to equation (30.4) when $k = 0$; see (30.47) and the sentence following (30.41). By (30.73), $h = O(\varepsilon)$ and $r = O(\varepsilon^2)$. Define

$$R_n := b_0(x_{\varepsilon,n}) - y_{\varepsilon,n}.$$

From (30.70) we have

$$b_0(x_{\varepsilon,n+1}) = b_0(x_{\varepsilon,n}) + \int_{x_{\varepsilon,n}}^{x_{\varepsilon,n+1}} F(b_0(x), x)dx, \tag{30.77}$$

and from (30.73) and (30.74) we obtain

$$y_{\varepsilon,n+1} = y_{\varepsilon,n} + (x_{\varepsilon,n+1} - x_{\varepsilon,n})F(y_{\varepsilon,n}, x_{\varepsilon,n}) + O(\varepsilon^2). \tag{30.78}$$

Subtracting (30.78) from (30.77) gives

$$R_{n+1} = R_n + (x_{\varepsilon,n+1} - x_{\varepsilon,n})\left(F(b_0(x_{\varepsilon,n}), x_{\varepsilon,n}) - F(y_{\varepsilon,n}, x_{\varepsilon,n})\right) + r_n \tag{30.79}$$

By Lipschitz continuity, (30.79) yields

$$|R_{n+1}| \le (1 + Lh)|R_n| + |r_n| \le (1 + Lh)|R_n| + r,$$

where r is given in (30.76). Using the above inequality successively, we get

$$\begin{aligned} |R_{n+1}| &\le (1 + Lh)|R_n| + r \\ &\le (1 + Lh)^2|R_{n-1}| + r + (1 + hL)r \\ &\cdots\cdots \\ &\le (1 + Lh)^n|R_1| + r\sum_{i=0}^{n-1}(1 + hL)^i \\ &= (1 + Lh)^n|R_1| + \frac{r}{hL}\left((1 + hL)^n - 1\right). \end{aligned} \tag{30.80}$$

From (30.39), we know that the sequence $\{x_{\varepsilon,n+1} - x_{\varepsilon,n}\}$ has a lower bound $\widetilde{h}$, which is of the order ε. Because the length of the interval $[-1, 1]$ is 2, we have $n \le 2/\widetilde{h}$. Furthermore, because $h = O(\varepsilon)$, it can be shown that there exists a positive constant H such that $nh \le 2h/\widetilde{h} \le H$. Note that $(1 + hL)^n \le e^{nhL} \le e^{HL}$. Because $\rho = 0$, we have $b_0(-1) = 0$ from (30.67) and (30.48). Moreover, because $y_{\varepsilon,1} = 0$, it follows that $R_1 = 0$. Thus, replacing r in (30.80) by $\frac{h^2}{2}(LM + P) + O(\varepsilon^2)$ in (30.76), we obtain

$$|R_{n+1}| \le \frac{h}{2}(M + \frac{P}{L})(e^{HL} - 1) + O(\varepsilon) = O(\varepsilon).$$

This completes the proof. □

Once the envelope function $b_0(x)$ is known, we can begin to estimate the number of spikes of the solutions to (30.1).

Lemma 30.12. *Let N_ε denote the maximum number of spikes that a solution to* (30.1) *can have. Then, we have*

$$N_\varepsilon = \left[\frac{1}{\varepsilon}\int_{-1}^{1}\frac{dx}{T(b_0(x),x)}\right] + O(1), \tag{30.81}$$

where $T(b_0(x),x)$ is given in (30.53) *with $\rho = 0$ and $[\alpha]$ denotes the largest integer less than or equal to α.*

Proof. For convenience, we simply write $S = S_{\varepsilon,0}$; i.e., S denotes the maximum number of minima of the solutions to (30.4), and define $T_\varepsilon(x) = x_{\varepsilon,n+1} - x_{\varepsilon,n}$ when $x \in [x_{\varepsilon,n}, x_{\varepsilon,n+1}]$ and $n = 1, 2, \cdots, S$. Here, $x_{\varepsilon,1} = -1$ and we let $x_{\varepsilon,S+1} = 1$. Obviously,

$$S = \sum_{n=1}^{S}\int_{x_{\varepsilon,n}}^{x_{\varepsilon,n+1}}\frac{dx}{T_\varepsilon(x)} = \int_{-1}^{1}\frac{dx}{T_\varepsilon(x)}. \tag{30.82}$$

On the other hand, by (30.38) and (30.48), $T_\varepsilon(x)$ can be written as

$$\begin{aligned} T_\varepsilon(x) &= \sqrt{2}\varepsilon\int_{y_{\varepsilon,n}}^{y_{\varepsilon,n}^{+}}\frac{du}{\sqrt{G(u,x_{\varepsilon,n}) - G(y_{\varepsilon,n},x_{\varepsilon,n})}} + O(\varepsilon^2) \\ &= \sqrt{6}\varepsilon\int_{y_{\varepsilon,n}}^{y_{\varepsilon,n}^{+}}\frac{du}{\sqrt{(y_{\varepsilon,n}^{+} - u)(u - y_{\varepsilon,n})(u - y_{\varepsilon,n}^{-})}} + O(\varepsilon^2) \end{aligned} \tag{30.83}$$

when $x \in [x_{\varepsilon,n}, x_{\varepsilon,n+1}]$ and $n = 1, 2, \cdots, S-1$, where

$$y_{\varepsilon,n}^{+} = a_0(x_{\varepsilon,n}) \quad \text{and} \quad y_{\varepsilon,n}^{-} = c_0(x_{\varepsilon,n}) \tag{30.84}$$

with $a_0(x)$ and $c_0(x)$ being defined by (30.49) and (30.50), respectively. From (30.51), we note that

$$y_{\varepsilon,n}^{-} < y_{\varepsilon,n} < y_{\varepsilon,n}^{+};$$

these are three solutions to

$$G(u, x_{\varepsilon,n}) - G(y_{\varepsilon,n}, x_{\varepsilon,n}) = 0.$$

The last integral in (30.83) can be expressed in terms of the complete elliptic integral of the first kind (see Gradshteyn and Ryzhik (1994, p. 266, Section 3.131, eq. (5); p. 908, Section 8.112, eq. (1))); and we have

$$\int_{y_{\varepsilon,n}}^{y_{\varepsilon,n}^{+}}\frac{du}{\sqrt{(y_{\varepsilon,n}^{+} - u)(u - y_{\varepsilon,n})(u - y_{\varepsilon,n}^{-})}} = \frac{2}{\sqrt{y_{\varepsilon,n}^{+} - y_{\varepsilon,n}^{-}}}K\left(\sqrt{\frac{y_{\varepsilon,n}^{+} - y_{\varepsilon,n}}{y_{\varepsilon,n}^{+} - y_{\varepsilon,n}^{-}}}\right). \tag{30.85}$$

Similarly, we have

$$\begin{aligned}\sqrt{2}\varepsilon\int_{b_0(x)}^{a_0(x)}&\frac{du}{\sqrt{G(u,x)-G(b_0(x),x)}}\\&=\sqrt{6}\varepsilon\int_{b_0(x)}^{a_0(x)}\frac{du}{\sqrt{(a_0(x)-u)(u-b_0(x))(u-c_0(x))}}\\&=\frac{2\sqrt{6}\varepsilon}{\sqrt{a_0(x)-c_0(x)}}K\left(\sqrt{\frac{a_0(x)-b_0(x)}{a_0(x)-c_0(x)}}\right).\end{aligned}\tag{30.86}$$

Thus, with the help of (30.85) and (30.86), (30.83) can be written as

$$\begin{aligned}T_\varepsilon(x)=&\frac{2\sqrt{6}\varepsilon}{\sqrt{y_{\varepsilon,n}^+-y_{\varepsilon,n}^-}}K\left(\sqrt{\frac{y_{\varepsilon,n}^+-y_{\varepsilon,n}}{y_{\varepsilon,n}^+-y_{\varepsilon,n}^-}}\right)+O(\varepsilon^2)\\=&\sqrt{2}\varepsilon\int_{b_0(x)}^{a_0(x)}\frac{du}{\sqrt{G(u,x)-G(b_0(x),x)}}+O(\varepsilon^2)\\&+\frac{2\sqrt{6}\varepsilon}{\sqrt{y_{\varepsilon,n}^+-y_{\varepsilon,n}}}K\left(\sqrt{\frac{y_{\varepsilon,n}^+-y_{\varepsilon,n}}{y_{\varepsilon,n}^+-y_{\varepsilon,n}^-}}\right)\\&-\frac{2\sqrt{6}\varepsilon}{\sqrt{a_0(x)-c_0(x)}}K\left(\sqrt{\frac{a_0(x)-b_0(x)}{a_0(x)-c_0(x)}}\right).\end{aligned}\tag{30.87}$$

Lemma 30.11 tells us that

$$b_0(x_{\varepsilon,n})=y_{\varepsilon,n}+O(\varepsilon).\tag{30.88}$$

Because $b_0'(x)$ is bounded and

$$x-x_{\varepsilon,n}=O(\varepsilon)\tag{30.89}$$

when $x\in[x_{\varepsilon,n},x_{\varepsilon,n+1}]$ by (30.73), we have

$$b_0(x)=b_0(x_{\varepsilon,n})+O(\varepsilon).\tag{30.90}$$

From (30.88) and (30.90), it follows that

$$y_{\varepsilon,n}=b_0(x)+O(\varepsilon)\tag{30.91}$$

when $x\in[x_{\varepsilon,n},x_{\varepsilon,n+1}]$. Furthermore, we also have

$$y_{\varepsilon,n}^+=a_0(x)+O(\varepsilon)\quad\text{and}\quad y_{\varepsilon,n}^-=c_0(x)+O(\varepsilon)\tag{30.92}$$

on account of (30.84), (30.89) and (30.90). From (30.90) and (30.92), it can be shown that

$$\frac{1}{\sqrt{y_{\varepsilon,n}^+-y_{\varepsilon,n}^-}}=\frac{1}{\sqrt{a_0(x)-c_0(x)+O(\varepsilon)}}=\frac{1}{\sqrt{a_0(x)-c_0(x)}}+O(\varepsilon)\tag{30.93}$$

and

$$\sqrt{\frac{y_{\varepsilon,n}^+ - y_{\varepsilon,n}}{y_{\varepsilon,n}^+ - y_{\varepsilon,n}^-}} = \sqrt{\frac{a_0(x) - b_0(x) + O(\varepsilon)}{a_0(x) - c_0(x) + O(\varepsilon)}} = \sqrt{\frac{a_0(x) - b_0(x)}{a_0(x) - c_0(x)}} + O(\varepsilon) \tag{30.94}$$

with the help of (30.93) and (30.94), and noting that $K'(x)$ is bounded when $0 < x < 1$, we obtain

$$\begin{aligned}
&\frac{1}{\sqrt{y_{\varepsilon,n}^+ - y_{\varepsilon,n}^-}} K\left(\sqrt{\frac{y_{\varepsilon,n}^+ - y_{\varepsilon,n}}{y_{\varepsilon,n}^+ - y_{\varepsilon,n}^-}}\right) - \frac{1}{\sqrt{a_0(x) - c_0(x)}} K\left(\sqrt{\frac{a_0(x) - b_0(x)}{a_0(x) - c_0(x)}}\right) \\
&= \left(\frac{1}{\sqrt{y_{\varepsilon,n}^+ - y_{\varepsilon,n}^-}} - \frac{1}{\sqrt{a_0(x) - c_0(x)}}\right) K\left(\sqrt{\frac{y_{\varepsilon,n}^+ - y_{\varepsilon,n}}{y_{\varepsilon,n}^+ - y_{\varepsilon,n}^-}}\right) \\
&\quad + \frac{1}{\sqrt{a_0(x) - c_0(x)}} \left(K\left(\sqrt{\frac{y_{\varepsilon,n}^+ - y_{\varepsilon,n}}{y_{\varepsilon,n}^+ - y_{\varepsilon,n}^-}}\right) - K\left(\sqrt{\frac{a_0(x) - b_0(x)}{a_0(x) - c_0(x)}}\right)\right) \\
&= O(\varepsilon).
\end{aligned} \tag{30.95}$$

Coupling (30.95) and (30.87) yields

$$\begin{aligned}
T_\varepsilon(x) &= \sqrt{2}\varepsilon \int_{b_0(x)}^{a_0(x)} \frac{du}{\sqrt{G(u,x) - G(b_0(x),x)}} + O(\varepsilon^2) \\
&= \varepsilon T(b_0(x), x) + O(\varepsilon^2).
\end{aligned} \tag{30.96}$$

Because $T_\varepsilon(x)$ is of the order $O(\varepsilon)$ and $T(b_0(x), x)$ is of the order $O(1)$, we have from (30.96)

$$\frac{1}{T_\varepsilon(x)} = \frac{1}{\varepsilon T(b_0(x), x)} + O(1). \tag{30.97}$$

Moreover, because $1 - x_{\varepsilon,S} \le \varepsilon T(y_{\varepsilon,S}, x_{\varepsilon,S}) = O(\varepsilon)$ by (30.38) and (30.53), we have

$$\int_{x_{\varepsilon,S}}^1 \frac{dx}{\varepsilon T(b_0(x), x)} = O(1).$$

Integrating from -1 to 1 on both sides of (30.97) and noting that $S - 1 \le N_\varepsilon \le S$, we obtain (30.81); see (30.82). □

Theorem 30.2. *Let $n(\varepsilon)$ denote the number of solutions to the boundary value problem* (30.1). *Then*

$$4N_\varepsilon - 3 \le n(\varepsilon) \le 4N_\varepsilon. \tag{30.98}$$

Proof. We have already shown that the maximum number of spikes a solution to (30.1) could have is N_ε. If we denote by "i" the number of spikes of the solutions to (30.1), then according to Ai (2003, Theorem 1.1), i can range from 0 to N_ε, when $i \geq 2$, the solution to (30.1) with i spikes should be one of the following four types:

(a) $y'(-1) > 0$ and $y'(1) < 0$. There are $i - 2$ spikes in the neighborhood of $x = 0$, and 1 spike each near $x = -1$ and $x = 1$.
(b) $y'(-1) > 0$ and $y'(1) > 0$. There are $i - 1$ spikes in the neighborhood of $x = 0$ and 1 spike near $x = -1$.
(c) $y'(-1) < 0$ and $y'(1) > 0$. There are i spikes in the neighborhood of $x = 0$.
(d) $y'(-1) < 0$ and $y'(1) < 0$. There are $i - 1$ spikes in the neighborhood of $x = 0$ and 1 spike near $x = 1$.

In Figure 30.8, we give the four types of solutions to (30.1) when $\varepsilon = 0.1$ and the number of spikes is equal to 3. While $i = 1$, only case $(b), (c), (d)$ can happen; when $i = 0$, only case (c) can happen. However, there is a special situation when $i = N_\varepsilon$ and $y'(-1) = y'(1) = 0$, in which there is only one type of solution instead of four with N_ε spikes and all spikes are away from the end points. Thus, the number of solutions to (30.1) satisfies (30.98). □

30.5 Numerical results

In this section, we use (30.54) to get a numerical value for N_ε. When $\rho = 0, b_0(-1) = 0$ and $a_0(-1) = \sqrt{3}$. From (30.54), it follows that

$$c_0 = \frac{1}{\sqrt{3}} \int_0^{\sqrt{3}} \sqrt{(\sqrt{3} - u)(u + \sqrt{3})u}\,du = \frac{\pi^{\frac{1}{2}} \Gamma(\frac{7}{4})}{3^{\frac{1}{4}} \Gamma(\frac{9}{4})}. \tag{30.99}$$

Note that (30.54) can be expressed in terms of complete elliptic integral (see Gradshteyn and Ryzhik (1994, pp. 907-908)), and we have

$$\frac{4}{15}\sqrt{a_0(x) - c_0(x)}(a_0^2(x) + b_0^2(x) + c_0^2(x))E\left(\sqrt{\frac{a_0(x) - b_0(x)}{a_0(x) - c_0(x)}}\right)$$
$$- \frac{4}{15}\sqrt{a_0(x) - c_0(x)}(a_0(x)b_0(x) + a_0(x)c_0(x) + b_0(x)c_0(x))$$
$$\times E\left(\sqrt{\frac{a_0(x) - b_0(x)}{a_0(x) - c_0(x)}}\right)$$

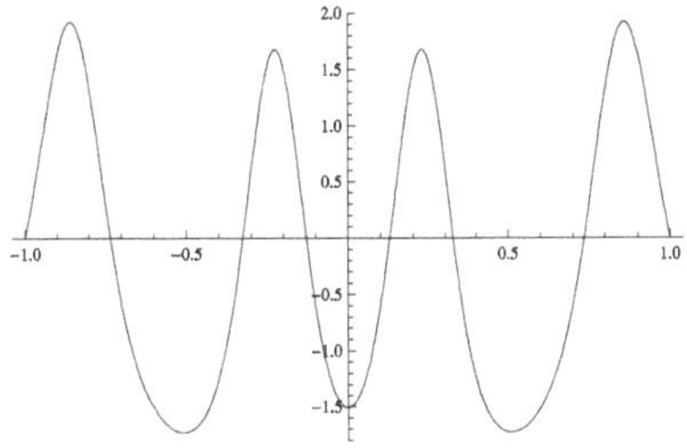

(a) $y_1(x)$.

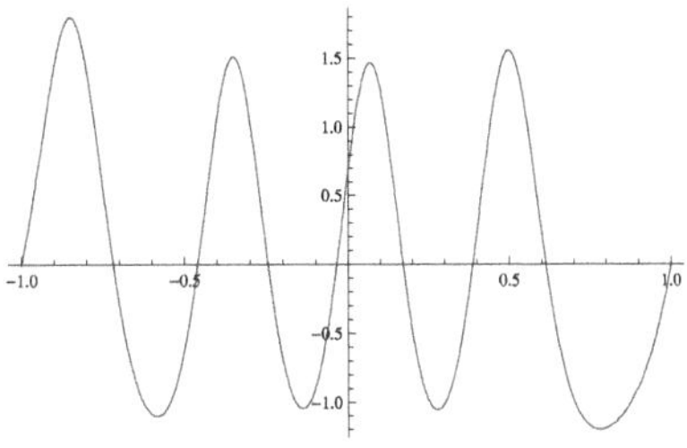

(b) $y_2(x)$.

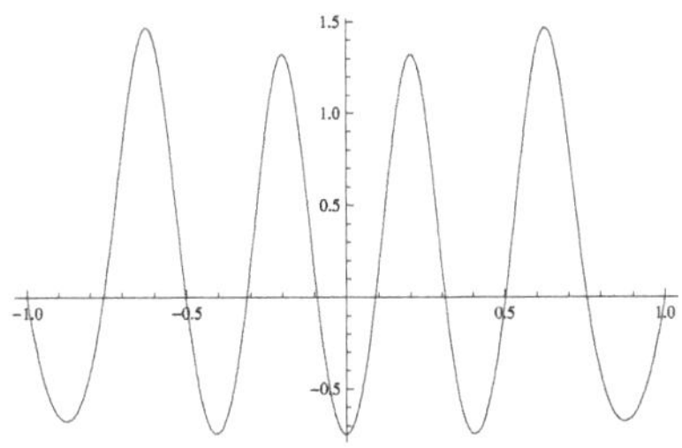

(c) $y_3(x)$.

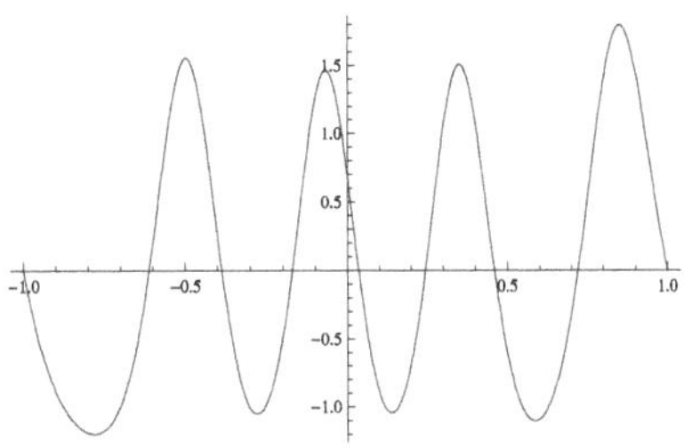

(d) $y_4(x)$.

Fig. 30.8: Four types of the solutions to (30.1) when $\varepsilon = 0.1$ and the number of spikes is 4.

$$
\begin{aligned}
&-\frac{2(a_0(x)+b_0(x)-2c_0(x))(a_0(x)-c_0(x))(b_0(x)-c_0(x))}{15\sqrt{a_0(x)-c_0(x)}} \\
&\times K\left(\sqrt{\frac{a_0(x)-b_0(x)}{a_0(x)-c_0(x)}}\right) \\
&= \frac{3^{\frac{1}{4}}\pi^{\frac{1}{2}}\Gamma(\frac{7}{4})}{\Gamma(\frac{9}{4})},
\end{aligned}
\tag{30.100}
$$

where $a_0(x)$ and $c_0(x)$ are defined in (30.49) and (30.50), respectively. In (30.100), $K(x)$ is the complete elliptic integral of the first kind and $E(x)$ is the complete elliptic integral of the second kind; for details, see Appendix.

We can solve for $b_0(x)$ in (30.100) numerically, and use the numerical values obtained to construct a sixth order polynomial which approximates $b_0(x)$ closely. We express the result as

$$
b_0(x) \doteq 0.2689x^6 + 0.0244x^4 + 0.1409x^2 - 0.4344. \tag{30.101}
$$

It can be seen from Figure 30.9 that the sixth order polynomial fits well with the minimum points of the solution to (30.4) when $k = 0$ with different values of ε. Replacing $b_0(x)$ in (30.81) by the polynomial in (30.101), and carrying out the integration in (30.81) numerically, we obtain

$$N_\varepsilon \sim \left[\frac{0.4725}{\varepsilon}\right].$$

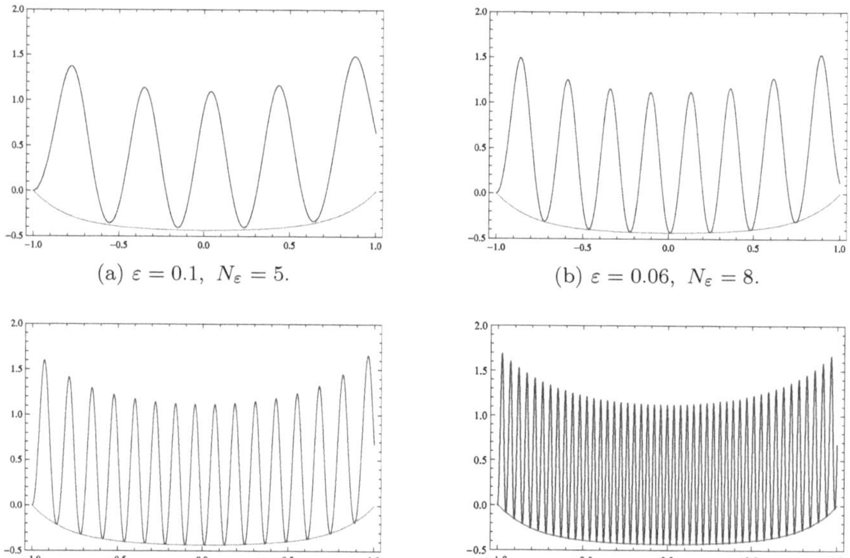

(a) $\varepsilon = 0.1,\ N_\varepsilon = 5$. (b) $\varepsilon = 0.06,\ N_\varepsilon = 8$. (c) $\varepsilon = 0.03,\ N_\varepsilon = 16$. (d) $\varepsilon = 0.01,\ N_\varepsilon = 47$.

Fig. 30.9: The envelope function in (30.101) and the solutions to (30.4) with $k = 0$ and different values of ε.

30.6 Appendix: The deduction of (30.100)

From Gradshteyn and Ryzhik (1994, p. 98, Section 2.260, eq. (1)), we have

$$\int x^{\frac{1}{2}}\sqrt{R}dx = \frac{x^{-\frac{1}{2}}\sqrt{R^3}}{\frac{5}{2}\gamma} - \frac{2\beta}{5\gamma}\int x^{-\frac{1}{2}}\sqrt{R}dx + \frac{\alpha}{5\gamma}\int x^{-\frac{3}{2}}\sqrt{R}dx, \tag{30.102}$$

where $R = \alpha + \beta x + \gamma x^2$. Let $x = u - c$ and $R = (a-u)(u-b)$ with $a > b > c$. Then, $\alpha = -c^2 + (a+b)c - ab, \beta = a + b - 2c$ and $\gamma = -1$. We can obtain from (30.102) that

$$
\begin{aligned}
&\int_b^a \sqrt{(a-u)(u-b)(u-c)}du \\
&=\frac{2(a+b-2c)}{5}\int_b^a \sqrt{\frac{(a-u)(u-b)}{u-c}}du \\
&\quad+\frac{c^2-(a+b)c+ab}{5}\int_b^a \sqrt{\frac{(a-u)(u-b)}{(u-c)^3}}du. \qquad (30.103)
\end{aligned}
$$

With the help of Gradshteyn and Ryzhik (1994, p. 283, Section 3.141, eq. (34)), the first integral on the right-hand side of (30.103) can be calculated as follows:

$$
\begin{aligned}
&\int_b^a \sqrt{\frac{(a-u)(u-b)}{u-c}}du \\
&=\frac{2}{3}\sqrt{a-c}\left[(a+b-2c)E(p)-2(b-c)K(p)\right] \\
&=\frac{2}{3}\sqrt{a-c}(a+b-2c)E(p)-\frac{4}{3}\sqrt{a-c}(b-c)K(p), \qquad (30.104)
\end{aligned}
$$

where $p=\sqrt{a-b}/\sqrt{a-c}$, $K(p)$ is the complete elliptic integral of the first kind and $E(p)$ is the complete elliptic integral of the second kind. The second integral on the right-hand of (30.103) can be written as

$$
\begin{aligned}
&\int_b^a \sqrt{\frac{(a-u)(u-b)}{(u-c)^3}}du \\
&=\int_b^a \sqrt{\frac{(a-u)(u-b)}{u-c}}du-(b-c)\int_b^a \sqrt{\frac{a-u}{(u-b)(u-c)^3}}du. \qquad (30.105)
\end{aligned}
$$

From Gradshteyn and Ryzhik (1994, p. 280, Section 3.141, eq. (4)), we have

$$\int_b^a \sqrt{\frac{(a-u)(u-b)}{u-c}}du=2\sqrt{a-c}\left[K(p)-E(p)\right].$$

Also, we have

$$\int_b^a \sqrt{\frac{a-u}{(u-b)(u-c)^3}}du=\frac{2\sqrt{a-c}}{b-c}E(p)-\frac{2}{\sqrt{a-c}}K(p)$$

on account of Gradshteyn and Ryzhik (1994, p. 284, Section 3.142, eq. (3)). Then, it can be obtained that

$$\int_b^a \sqrt{\frac{(a-u)(u-b)}{(u-c)^3}}du=\frac{2(a+b-2c)}{\sqrt{a-c}}K(p)-4\sqrt{a-c}E(p). \qquad (30.106)$$

Hence, combining (30.103), (30.104) and (30.106) gives (30.100).

30.7 Exercises

1. Verify (30.13) and (30.14).
2. Verify (30.16) and (30.17).
3. Consider the boundary value problem

$$\varepsilon y'' + 2y' + e^y = 0, \quad y(0) = y(1) = 0. \tag{30.107}$$

(a) Show that if e^y were a linear function of y, then there would be a boundary layer at $x = 0$ and no boundary layer at $x = 1$. Show also that this nonlinear problem has just one boundary layer at $x = 0$.

(b) Assume that the outer expansion has the form

$$y^o \sim y_0(x) + \varepsilon y_1(x) + \cdots, \quad \varepsilon \to 0^+. \tag{30.108}$$

Substituting (30.108) into (30.107) gives $2y_0' + e^{y_0} = 0$. Show that its solution is $y_0 = \ln[2/(x + c_0)]$. Assuming that there is no boundary layer at $x = 1$, we impose the boundary condition $y_0(1) = 0$. Show that

$$y_0(x) = \ln \frac{2}{1 + x}. \tag{30.109}$$

(c) Show that the boundary layer has thickness $\delta = \varepsilon$, and that setting $X = x/\varepsilon$ and $Y^i(X) = y(x)$ in (30.107) gives

$$\frac{d^2Y^i}{dX^2} + 2\frac{dY^i}{dX} = -\varepsilon e^{Y^i}.$$

Furthermore, show that assuming an inner expansion of the form $Y^i \sim Y_0 + \varepsilon Y_1 + \cdots$ gives in leading order

$$Y_0 = A_0 + B_0 e^{-2X}. \tag{30.110}$$

By the boundary condition $y(0) = 0$, we have $Y_0(0) = 0$, which implies $A_0 + B_0 = 0$. The asymptotic match of (30.110) with (30.109) in the intermediate limit $x \to 0^+$ and $X \to +\infty$ gives $A_0 = \ln 2$. Thus, $B_0 = -\ln 2$ and

$$Y_0(X) = (1 - e^{-2X}) \ln 2. \tag{30.111}$$

Combining (30.109) and (30.111), we obtain the single uniform asymptotic expansion

$$y^c(x) = \ln \frac{2}{1 + x} - e^{-2x/\varepsilon} \ln 2. \tag{30.112}$$

(d) By using computer software, give a graphical comparison between (30.112) and the exact solution to (30.107).

(e) Solve the equation in (30.107) correct to first order in ε.

4. The Rayleigh equation is

$$\varepsilon\frac{d^2y}{dt^2}-\left[\frac{dy}{dt}-\frac{1}{3}\left(\frac{dy}{dt}\right)^3\right]+y=0. \tag{30.113}$$

(a) Show that this autonomous equation can be rewritten as the system

$$\frac{dy}{dt}=z, \tag{30.114}$$

$$\varepsilon\frac{dz}{dt}=z-\frac{z^3}{3}-y. \tag{30.115}$$

We divide (30.115) by (30.114) to obtain

$$\varepsilon\frac{dz}{dy}=\frac{z-z^3/3-y}{z}, \tag{30.116}$$

and treat z as a function of the independent variable y. First, we look at the leading-order approximation to (30.116) as $\varepsilon\to 0$. The outer limit is obtained by simply setting $\varepsilon=0$. This gives the algebraic equation

$$z_{out}-\frac{z_{out}^3}{3}=y, \tag{30.117}$$

the graph of which is plotted as a dashed curve in Figure 30.10. Observe that for $|y|<2/3$, there are three possible values of z_{out}. As we shall see, it is consistent to have a boundary-layer solution which joins the point $A(y=2/3, z=1)$ to the point $B(y=2/3, z=-2)$ by the almost vertical line shown in Figure 30.10 and to have a second boundary-layer solution which joins the point $C(y=-2/3, z=-1)$ to the point $D(y=-2/3, z=2)$.

To obtain the boundary-layer approximation joining the outer solutions DA and BC from A to B, we introduce the inner variable $Y=(y-2/3)/\varepsilon$. Let $Z(Y)=z(y)$.

(b) Show that

$$\frac{dZ}{dY}=\frac{Z-Z^3/3-2/3-\varepsilon Y}{Z}, \tag{30.118}$$

and the leading-order approximation Z_0 to Z satisfies

$$\frac{dZ_0}{dY}=\frac{Z_0-Z_0^3/3-2/3}{Z_0}. \tag{30.119}$$

Show that the solution to this separable equation is

$$-\frac{2}{9}\ln|Z_0+2|+\frac{2}{9}\ln|Z_0-1|-\frac{1}{3(Z_0-1)}=-\frac{Y+c}{3}, \tag{30.120}$$

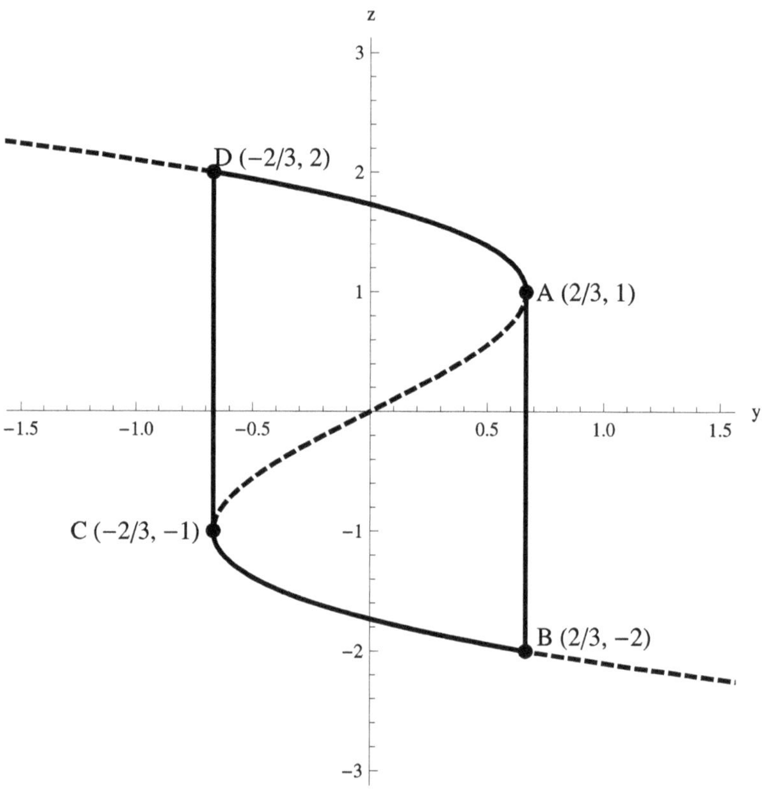

Fig. 30.10: A plot of the leading-order outer solution in (30.117) to the Rayleigh equation (30.113). The outer solution is given by the solid lines connecting the points D to A and B to C. The dash lines are additional segments of the graph of the algebraic function in (30.117). Boundary-layer solutions connect the points A to B and C to D along the curves $y = \pm 2/3$, respectively.

where c is a constant of integration. Furthermore, show that in the boundary-layer region this solution is a double-valued function of Y. Also verify that, as Z_0 decreases from 1, Y increases from $-\infty$; and that Y reaches a maximum at $Z_0 = 0$, $Y = 3c + \frac{2}{3}\ln 2 - 1$. Furthermore, show that as Z_0 decreases from 0 to -2, Y decreases from its maximum back to $-\infty$. Illustrate that asymptotic matching to lowest order is accomplished by the intermediate limits $(y, z_{out}) \to (\frac{2}{3}-, 1)$, $(Y, Z_0) \to (-\infty, 1)$ and $(y, z_{out}) \to (\frac{2}{3}-, -2)$, $(Y, Z_0) \to (-\infty, -2)$. These asymptotic matches

complete the lowest-order boundary-layer analysis. Note that the constant c remains undetermined in leading order.
(c) In the same way, it is possible to join the points C and D in Figure 30.10 by a boundary-layer approximation; this approximation is given by equation (30.120) with the signs of Z_0 and Y reversed. The solution obtained in this way is periodic in t because the trajectory is closed in phase space.
The leading-order boundary-layer solutions (30.117) and (30.120) can be used to compute a leading order approximation to the period of the limit cycle of the Rayleigh equation as $\varepsilon \to 0^+$. Since $z = dy/dt$, show that the period T of the limit cycle is given by

$$T = \oint \frac{dy}{dz}, \tag{30.121}$$

where the integral is taken over the full limit cycle. Show that the leading-order approximation to the period T is dominated by the contributions to the integral (30.121) from the outer solutions (30.117). The contribution to T from the boundary layer goes to 0 as $\varepsilon \to 0^+$, because the width of the boundary layers in y is $O(\varepsilon)$. Prove that

$$T \sim \int_{DA} \frac{dy}{z} + \int_{BC} \frac{dy}{z} = \int_2^1 \frac{1-z^2}{z} dz + \int_{-1}^{-2} \frac{1-z^2}{z} dz = 3 - 2\ln 2, \tag{30.122}$$

as $\varepsilon \to 0^+$.

5. Discuss the qualitative nature of the solutions to
(a) $\varepsilon y'' - y' + y^2 = 0$, $y(0) = y(1) = 1$, in the limit $\varepsilon \to 0^+$;
(b) $\varepsilon y'' - y' + 1/y = 0$, $y(0) = \sqrt{2}$, and $y(1) = 0$, in the limit $\varepsilon \to 0^+$.
6. Do the boundary-layer problems
(a) $\varepsilon y'' - (y')^2 + e^y = 0$, $y(0) = y(1) = 1$,
(b) $\varepsilon y'' - (y')^2 + y^2 = 0$, $y(0) = y(1) = 1$,
have solutions when $\varepsilon \to 0^+$? If they do, find a leading-order uniform approximation. (Hint: perform a boundary-layer analysis in the phase plane (y, y').)
7. Show that it is consistent for the solution to the boundary-layer problem

$$\varepsilon y'' + 2(1 - x^2)y + y^2 = 1, \quad y(-1) = y(1) = 0,$$

to exhibit an internal boundary layer at any value of x in $(-1, 1)$.

Bibliography

Abramowitz, M. and Stegun, I. A. (1964). *Handbook of Mathematical Functions* (NBS Appl. Math. Series 55, Washington, DC).

Abramowitz, M. and Stegun, I. A. (1970). *Handbook of Mathematical Functions, with Formulas, Graphs, and Mathematical Tables* (Dover Publications, Inc., New York).

Abramowitz, M. and Stegun, I. A. (1989). *Handbook of Mathematical Functions* (Dover, New York).

Abramowitz, M. and Stegun, I. A. (1992). *Handbook of Mathematical Functions with Formulas, Graphs, and Mathematical Tables* (Dover, New York).

Ackerberg, R. C. and O'Malley Jr., R. E. (1970). Boundary layer problems exhibiting resonance, *Stud. Appl. Math.* **49**, pp. 277–295.

Ai, S. B. (2003). Multi-bump solutions to Carrier's problem, *J. Math. Anal. Appl.* **277**, pp. 405–422.

Arnold, V. I. (1992). *Ordinary Differential Equations* (Springer-Verlag, Berlin Heidelberg).

Artin, E. (1964). *The Gamma Function* (Holt, Rinehart and Winston, New York-Toronto-London).

Askey, R. A. (1982). *The Collected Papers of Gabor Szegö. Vols. 1 and 2* (Birkhäuser, Cambridge, MA).

Bai, X.-X. and Zhao, Y.-Q. (2007). A uniform asymptotic expansion for Jacobi polynomials via uniform treatment of Darboux's method, *J. Approx. Theory* **148**, pp. 1–11.

Baik, J., Kriecherbauer, T., McLaughlin, K. T.-R., and Miller, P. D. (2003). Uniform asymptotics for polynomials orthogonal with respect to a general class of discrete weights and universality results for associated ensembles: announcement of results, *Int. Math. Res. Not.* **2003**, pp. 821–858.

Baik, J., Kriecherbauer, T., McLaughlin, K. T.-R., and Miller, P. D. (2007). *Discrete Orthogonal Polynomials. Asymptotics and Applications, Annals of Mathematics Studies*, Vol. 164 (Princeton University Press, Princeton, NJ).

Baik, J. and Suidan, T. M. (2007). Random matrix central limit theorems for nonintersecting random walks, *Ann. Probab.* **35**, pp. 1807–1834.

Baratella, P. and Gatteschi, L. (1988). The bounds for the error term of an asymptotic approximation of Jacobi polynomials, in M. Alfaro, J. S. Dehesa, F. J. Marcellán, J. L. R. de Francia, and J. Vinuesa (eds.), *Orthogonal Polynomials and Their Applications*, *Lecture Notes in Math.*, Vol. 1329 (Springer-Verlag, Berlin), pp. 203–221.

Beals, R. and Wong, R. (2016). *Special Functions and Orthogonal Polynomials*, Cambridge Studies in Advanced Mathematics (Cambridge University Press, Cambridge).

Beals, R. and Wong, R. (2020). *Explorations in Complex Functions*, *Graduate Texts in Mathematics*, Vol. 287 (Springer, Cham).

Bender, C. M. and Orszag, S. A. (1978). *Advanced Mathematical Methods for Scientists and Engineers*, International Series in Pure and Applied Mathematics (McGraw-Hill Book Co., New York).

Berry, M. V. (1989). Uniform asymptotic smoothing of Stokes's discontinuities, *Proc. R. Soc. Lond. Ser. A* **422**, pp. 7–21.

Berry, M. V. (1991). Infinitely many Stokes smoothings in the gamma function, *Proc. R. Soc. Lond. Ser. A* **434**, pp. 465–472.

Berry, M. V. and Howls, C. J. (1990). Hyperasymptotics, *Proc. R. Soc. Lond. Ser. A* **430**, pp. 653–668.

Berry, M. V. and Howls, C. J. (1991). Hyperasymptotics for integrals with saddles, *Proc. R. Soc. Lond. Ser. A* **434**, pp. 657–675.

Birkhoff, G. (ed.) (1973). *A Source Book in Classical Analysis* (Harvard University Press, Cambridge, MA).

Birkhoff, G. D. (1930). Formal theory of irregular linear difference equations, *Acta Math.* **54**, pp. 205–246.

Birkhoff, G. D. and Trjitzinsky, W. J. (1933). Analytic theory of singular difference equations, *Acta Math.* **60**, pp. 1–89.

Bleher, P. and Its, A. (1999). Semiclassical asymptotics of orthogonal polynomials, Riemann-Hilbert problem, and universality in the matrix model, *Ann. Math.* **150**, pp. 185–266.

Bleistein, N. (1966). Uniform asymptotic expansions of integrals with stationary point near algebraic singularity, *Comm. Pure Appl. Math.* **19**, pp. 353–370.

Bleistein, N. (1967). Uniform asymptotic expansions of integrals with many nearby stationary points and algebraic singularities, *J. Math. Mech.* **17**, pp. 533–559.

Bleistein, N. and Handelsman, R. A. (1967). Uniform asymptotic expansions of certain integrals, *SIAM J. Appl. Math.* **15**, pp. 1422–1433.

Bleistein, N. and Handelsman, R. A. (1969). Uniform asymptotic expansions of double integrals, *J. Math. Anal. Appl.* **27**, pp. 434–453.

Bo, R. and Wong, R. (1996). Asymptotic behavior of the Pollaczek polynomials and their zeros, *Stud. Appl. Math.* **96**, pp. 307–338.

Boyd, W. G. C. (1990). Stieltjes transforms and the Stokes phenomenon, *Proc. R. Soc. Lond. Ser. A* **429**, pp. 227–246.

Boyd, W. G. C. (1993). Error bounds for the method of steepest descents, *Proc. R. Soc. Lond. Ser. A* **440**, pp. 493–518.

Boyd, W. G. C. (1994). Gamma function asymptotics by an extension of the method of steepest descents, *Proc. R. Soc. Lond. Ser. A* **447**, pp. 609–630.

Bremmer, H. (1955). Diffraction problems of microwave optics, *IEEE Trans. Antennas Propag.* **3**, pp. 222–228.

Bush, A. W. (1992). *Perturbation Methods for Engineers and Scientists* (CRC Press, Boca Raton).

Cao, L.-H. and Li, Y.-T. (2014). Linear difference equations with a transition point at the origin, *Anal. Appl.* **12**, pp. 75–106.

Carlson, B. C. (1977). *Special Functions of Applied Mathematics* (Academic Press [Harcourt Brace Jovanovich, Publishers], New York-London).

Carlson, B. C. and Gustafson, J. L. (1985). Asymptotic expansion of the first elliptic integral, *SIAM J. Math. Anal.* **16**, pp. 1072–1092.

Carrier, G. F. (1970). Singular perturbation theory and geophysics, *SIAM Rev.* **12**, pp. 175–193.

Carrier, G. F. and Pearson, C. E. (1968). *Ordinary Differential Equations* (Blaisdell Publishing Co., Waltham, MA), reprinted by SIAM's Classics in Applied Mathematics series, vol. 6, SIAM, Philadelphia, PA, 1991.

Chako, N. (1965). Asymptotic expansions of double and multiple integrals occurring in diffraction theory, *J. Inst. Math. Appl.* **1**, pp. 372–422.

Chapman, C. J. (1992). The asymptotic theory of rapidly rotating sound fields, *Proc. R. Soc. Lond. Ser. A* **436**, pp. 511–526.

Chester, C., Friedman, B., and Ursell, F. (1957). An extension of the method of steepest descents, *Math. Proc. Cambridge Philos. Soc.* **53**, pp. 599–611.

Claeys, T., Kuijlaars, A. B. J., and Vanlessen, M. (2008). Multi-critical unitary random matrix ensembles and the general Painlevé II equation, *Ann. Math.* **168**, pp. 601–641.

Clarkson, P. A. and Jordaan, K. (2018). Properties of generalized Freud polynomials, *J. Approx. Theory* **225**, pp. 148–175.

Clarkson, P. A., Jordaan, K., and Kelil, A. (2016). A generalized Freud weight, *Stud. Appl. Math.* **136**, pp. 288–320.

Conrad, E. V. F. and Flajolet, P. (2006). The Fermat cubic, elliptic functions, continued fractions, and a combinatorial excursion, *Sem. Lothar. Combin.* **54**, p. B54g.

Copson, E. T. (1935). *Theory of Functions of a Complex Variable* (Oxford University Press, London).

Copson, E. T. (1965). *Asymptotic Expansions, Cambridge Tracts in Mathematics and Mathematical Physics*, Vol. 55 (Cambridge University Press, New York).

Dai, D., Ismail, M. E. H., and Wang, X.-S. (2014). Plancherel-Rotach asymptotic expansion for some polynomials from indeterminate moment problems, *Constr. Approx.* **40**, pp. 61–104.

Dai, D. and Wong, R. (2007). Global asymptotics of Krawtchouk polynomials - a Riemann-Hilbert approach, *Chin. Ann. Math. Ser. B* **28**, pp. 1–34.

de Kok, F. (1971). On the method of stationary phase for multiple integrals, *SIAM J. Math. Anal.* **2**, pp. 76–104.

Debye, P. (1909). Näherungsformeln für die Zylinderfunktionen für große Werte des Arguments und unbeschränkt veränderliche Werte des Index, *Math. Ann.* **67**, pp. 535–558.

Deift, P., Kriecherbauer, T., McLaughlin, K. T.-R., Venakides, S., and Zhou, X. (1999). Strong asymptotics of orthogonal polynomials with respect to exponential weights, *Comm. Pure Appl. Math.* **52**, pp. 1491–1552.

Deift, P. and Zhou, X. (1993). A steepest descent method for oscillatory Riemann-Hilbert problems. Asymptotics for the MKdV equation, *Ann. of Math.* **137**, pp. 295–368.

Deift, P. A. (1999). *Orthogonal Polynomials and Random Matrices: a Riemann-Hilbert Approach, Courant Lecture Notes in Mathematics*, Vol. 3 (American Mathematical Society, Providence, RI).

Dingle, R. B. (1973). *Asymptotic Expansions: Their Derivation and Interpretation* (Academic Press, London-New York).

Dingle, R. B. and Morgan, G. J. (1967a). WKB methods for difference equations I, *Appl. Sci. Res.* **18**, pp. 221–237.

Dingle, R. B. and Morgan, G. J. (1967b). WKB methods for difference equations II, *Appl. Sci. Res.* **18**, pp. 238–245.

Doetsch, G. (1974). *Introduction to the Theory and Application of the Laplace Transformation* (Springer-Verlag, New York-Heidelberg).

Dunster, T. M. (2025). Asymptotic expansions for Legendre functions via differential equations having coalescing turning points, *arXiv preprint arXiv:2504.19405* doi:10.48550/arXiv.2504.19405.

Erdélyi, A. (1955). Asymptotic representations of Fourier integrals and the method of stationary phase, *J. Soc. Indust. Appl. Math.* **3**, pp. 17–27.

Erdélyi, A. (1960). Asymptotic forms for Laguerre polynomials, *J. Indian Math. Soc.* **24**, pp. 235–250.

Erdélyi, A. (1962a). On a nonlinear boundary value problem involving a small parameter, *J. Aust. Math. Soc.* **2**, pp. 425–439.

Erdélyi, A. (1962b). Singular perturbations, *Bull. Amer. Math. Soc.* **68**, pp. 420–424.

Erdélyi, A. (1963). Singular perturbations of boundary value problems involving ordinary differential equations, *J. Soc. Ind. App. Math.* **11**, pp. 105–116.

Erdélyi, A. (1976). Singular perturbation, in G. Fishera (ed.), *Trends in Applications of Pure Mathematics to Mechanics, Monographs and Studies in Mathematics*, Vol. 2 (Pitman Publishing, London-San Francisco-Melbourne), pp. 53–62.

Erdélyi, A., Magnus, W., Oberhettinger, F., and Tricomi, F. G. (1953). *Higher transcendental functions. Vols. I and II* (McGraw-Hill Book Company, Inc., New York).

Erdélyi, A., Magnus, W., Oberhettinger, F., and Tricomi, F. G. (1954). *Tables of Integral Transforms*, Vol. I (McGraw-Hill Book Company, Inc., New York-Toronto-London).

Fedoryuk, M. V. (1964). The stationary phase method. Close saddle points in the multi-dimensional case, *Ž. Vyčisl. Mat i Mat. Fiz.* **4**, pp. 671–682, english translation in USSR Comput. Math. .Math. Phys. 4 (1964), 66–81.

Fields, J. L. (1967). A uniform treatment of Darboux's method, *Arch. Rational Mech. Anal.* **27**, pp. 289–305.

Fokas, A. S., Its, A. R., Kapaev, A. A., and Novokshenov, V. Y. (2006). *Painlevé transcendents, the Riemann-Hilbert Approach, Mathematical Surveys and Monographs*, Vol. 128 (American Mathematical Society, Providence, RI).

Fokas, A. S., Its, A. R., and Kitaev, A. V. (1991). Discrete Painlevé equations and their appearance in quantum gravity, *Comm. Math. Phys.* **142**, pp. 313–344.

Frenzen, C. L. and Wong, R. (1985a). A uniform asymptotic expansion of the Jacobi polynomials with error bounds, *Canad. J. Math.* **37**, pp. 979–1007.

Frenzen, C. L. and Wong, R. (1985b). The Lebesgue constants for Jacobi series, *Rend. Sem. Mat. Univ. Politec. Torino* , pp. 117–148.

Frenzen, C. L. and Wong, R. (1986). Asymptotic expansions of the Lebesgue constants for Jacobi series, *Pacific J. Math.* **122**, pp. 391–415.

Frenzen, C. L. and Wong, R. (1988). Uniform asymptotic expansions of Laguerre polynomials, *SIAM J. Math. Anal.* **19**, pp. 1232–1248.

Friedman, B. (1959). Stationary phase with neighboring critical points, *J. Soc. Indust. Appl. Math.* **7**, pp. 280–289.

Gasper, G. (1975). Formulas of the Dirichlet-Mehler type, in *Fractional calculus and its applications, Lecture Notes in Math.*, Vol. 457 (Springer-Verlag, Berlin-New York), pp. 207–215.

Gel'fand, I. M. and Shilov, G. E. (1964). *Generalized functions*, Vol. 1 (Academy Press, New York).

Gradshteyn, I. S. and Ryzhik, I. M. (1994). *Tables of Integrals, Series and Products*, Vol. 5th ed. (Academy Press, Boston).

Gronwall, T. H. (1913). Über die Laplacesche Reihe, *Math. Ann.* **74**, pp. 213–270.

Handelsman, R. A. and Lew, J. S. (1974). Asymptotic expansion of Laplace convolutions for large argument and tail densities for certain sums of random variables, *SIAM J. Math. Anal.* **5**, pp. 425–451.

Hethcote, H. W. (1968). *Asymptotic Approximations with Error Bounds for Zeros of Airy and Cylindrical Functions*, Ph.D. dissertation, University of Michigan, Ann Arbor.

Hethcote, H. W. (1970). Error bounds for asymptotic approximations of zeros of transcendental functions, *SIAM J. Math. Anal.* **1**, pp. 147–152.

Hörmander, L. (1990). *The Analysis of Linear Partial Differential Operators*, Vol. 1, 2nd edn. (Springer-Verlag, Berlin).

Howls, C. J. (1997). Hyperasymptotics for multidimensional integrals exact remainder terms and the global connection problem, *Proc. R. Soc. Lond. Ser. A* **453**, pp. 2271–2294.

Ismail, M. E. H. (1994). Asymptotics of Pollaczek polynomials and their zeros, *SIAM J. Math. Anal.* **25**, pp. 462–473.

Ismail, M. E. H. (2005). Asymptotics of q-orthogonal polynomials and a q-Airy function, *Int. Math. Res. Not.* , pp. 1063–1088.

Jin, X.-S. and Wong, R. (1998). Uniform asymptotic expansions for Meixner polynomials, *Constr. Approx.* **14**, pp. 113–150.

Jin, X.-S. and Wong, R. (1999). Asymptotic formulas for the zeros of the Meixner polynomials, *J. Approx. Theory* **96**, pp. 281–300.

Johansson, K. (2000). Shape fluctuations and random matrices, *Comm. Math. Phys.* **209**, pp. 437–476.

Jones, D. S. (1966). *Generalised Functions* (McGraw-Hill Book Co., New York-Toronto-London).

Jones, D. S. (1969). Generalized transforms and their asymptotic behaviour, *Philos. Trans. Roy. Soc. London Ser. A* **265**, pp. 1–43.

Jones, D. S. (1972). Asymptotic behavior of integrals, *SIAM Rev.* **14**, pp. 286–317.

Jones, D. S. (1976). The mathematical theory of noise shielding, *Prog. Aerospace Sci.* **17**, pp. 149–229.

Jones, D. S. (1979). Arthur erdélyi, 2 october 1908 - 12 december 1977, *Biogr. Mem. Fellows R. Soc.* , pp. 266 – 286.

Jones, D. S. (1980). Infinite integrals and convolution, *Proc. R. Soc. Lond. Ser. A* **371**, pp. 479–508.

Jones, D. S. (1982a). Generalised functions and their convolution, *Proc. Roy. Soc. Edinburgh Sect. A* **91**, pp. 213–233.

Jones, D. S. (1982b). *The Theory of Generalised Functions* (Cambridge University Press, Cambridge-New York).

Jones, D. S. and Kline, M. (1958). Asymptotic expansion of multiple integrals and the method of stationary phase, *J. Math. Phys.* **37**, pp. 1–28.

Kath, W. (1985). Slowly varying phase planes and boundary-layer theory, *Stud. Appl. Math.* **72**, pp. 221–239.

Koekoek, R., Lesky, P. A., and Swarttouw, R. F. (2010). *Hypergeometric Orthogonal Polynomials and Their q-Analogues*, Springer Monographs in Mathematics (Springer-Verlag, Berlin).

Kriecherbauer, T. and McLaughlin, K. T.-R. (1999). Strong asymptotics of polynomials orthogonal with respect to Freud weights, *Internat. Math. Res. Notices* **6**, pp. 299–333.

Lange, C. G. (1983). On spurious solutions of singular perturbation problems, *Stud. Appl. Math.* **68**, pp. 227–257.

Langer, R. E. (1949). The asymptotic solutions of ordinary linear differential equations of the second order, with special reference to a turning point, *Trans. Amer. Math. Soc.* **67**, pp. 461–490.

Laplace, P. S. de (1820). *Théorie analytique des probabilités*, 3rd ed. (Courcier, Paris).

Levey, L. and Felsen, L. B. (1969). On incomplete Airy functions and their application to diffraction problems, *Radio Sci.* **4**, pp. 959–969.

Li, K. and Wong, R. (2021). Asymptotic expansions for Wiener-Hopf equations, *Anal. Appl.* **19**, pp. 1059–1092.

Li, X. and Wong, R. (1994). Error bounds for asymptotic expansions of Laplace convolutions, *SIAM J. Math. Anal.* **25**, pp. 1537–1553.

Li, Y.-T., Wang, X.-S., and Wong, R. (2020). Asymptotics of the Wilson polynomials, *Anal. Appl.* **18**, pp. 237–270.

Li, Y.-T. and Wong, R. (2013). Global asymptotics of Stieltjes-Wigert polynomials, *Anal. Appl.* **11**, p. 1350028.

Liang, X. and Wong, R. (2009). On a nested boundary-layer problem, *Commun. Pure Appl. Anal.* **8**, pp. 419–433.

Lorch, L. (1959). The Lebesgue constants for Jacobi series. II, *Amer. J. Math.* **81**, pp. 875–888.

MacGillivray, A. D. (1997). A method for incorporating transcendentally small terms into the method of matched asymptotic expansions, *Stud. Appl. Math.* **99**, pp. 285–310.

MacGillivray, A. D., Braun, R. J., and Tanoğlu, G. (2000). Perturbation analysis of a problem of Carrier's, *Stud. Appl. Math.* **104**, pp. 293–311.

Maejima, M. and Van Assche, W. (1985). Probabilistic proofs of asymptotic formulas for some classical polynomial, *Math. Proc. Cambridge Philos. Soc.* **97**, pp. 499–510.

Markushevich, A. I. (1977). *Theory of Functions of a Complex Variable* (Chelsea Publishing Co., New York).

Miller, J. C. P. (1955). *Tables of Weber Parabolic Cylinder Functions* (H. M. Stationery Office, London).

Miller, R. K. (1971). *Nonlinear Volterra Integral Equations*, Mathematics Lecture Note Series (W. A. Benjamin, Inc., Menlo Park, CA).

Muki, R. and Sternberg, E. (1970). Note on an asymptotic property of solutions to a class of Fredholm integral equations, *Quart. Appl. Math.* **28**, pp. 277–281.

Nayfeh, A. H. (1981). *Introduction to Perturbation Techniques* (John Wiley & Sons, New York).

Nayfeh, A. H. (1985). *Problems in Perturbation* (John Wiley & Sons, New York).

Nemes, G. (2010a). New asymptotic expansion for the Gamma function, *Arch. Math. (Basel)* **95**, pp. 161–169.

Nemes, G. (2010b). On the coefficients of the asymptotic expansion of $n!$, *J. Integer Seq.* **13**, article 10.6.6.

Nemes, G. (2013). An explicit formula for the coefficients in Laplace's method, *Constructive Approximation* **38**, pp. 471–487.

Ng, K.-C. and Wong, R. (1994). On a singular perturbation problem, in *Approximation and Computation: A Festschrift in Honor of Walter Gautschi*, *International Series of Numerical Mathematics*, Vol. 119 (Birkhäuser Boston, Inc., Boston, MA), pp. 477–484.

Novikoff, A. (1954). *On a Special System of Orthogonal Polynomials*, Ph.D. dissertation, Stanford University.

Olde Daalhuis, A. B. (1998). Hyperasymptotic solutions of higher-order linear differential equations with a singularity of rank one, *Proc. R. Soc. Lond. Ser. A* **454**, pp. 1–29.

Olde Daalhuis, A. B. and Olver, F. W. J. (1994). Exponentially improved asymptotic solutions of ordinary differential equations. II. Irregular singularities of rank one, *Proc. R. Soc. Lond. Ser. A* **445**, pp. 39–56.

Olver, F. W. J. (1959). Uniform asymptotic expansions for Weber parabolic cylinder functions of large orders, *J. Res. Nat. Bur. Standards Sect. B* **63**, pp. 131–169.

Olver, F. W. J. (1974a). *Asymptotics and Special Functions* (Academic Press, New York), reprinted by A. K. Peters, Wellesley, MA, 1997.

Olver, F. W. J. (1974b). Error bounds for stationary phase approximations, *SIAM J. Math. Anal.* **5**, pp. 19–29.

Olver, F. W. J. (1975). Second-order linear differential equations with two turning points, *Philos. Trans. Roy. Soc. London Ser. A* **278**, pp. 137–174.

Olver, F. W. J. (1978). Sufficient conditions for Ackerberg-O'Malley resonance, *SIAM J. Math. Anal.* **9**, pp. 328–355.

Olver, F. W. J. (1991a). Uniform, exponentially improved, asymptotic expansions for the confluent hypergeometric function and other integral transforms, *SIAM J. Math. Anal.* **22**, pp. 1475–1489.

Olver, F. W. J. (1991b). Uniform, exponentially improved, asymptotic expansions for the generalized exponential integral, *SIAM J. Math. Anal.* **22**, pp. 1460–1474.

Olver, F. W. J. (1993). Exponentially-improved asymptotic solutions of ordinary differential equations. I. The confluent hypergeometric function, *SIAM J. Math. Anal.* **24**, pp. 756–767.

Olver, F. W. J. (2014). Mathematics that has intrigued me, *Anal. Appl.* **12**, pp. 341–354.

Olver, F. W. J., Lozier, D. W., Boisvert, R. F., and Clark, C. W. (2010). *NIST Handbook of Mathematical Functions* (Cambridge University Press, Cambridge).

Olver, F. W. J., Olde Daalhuis, A. B., Lozier, D. W., Schneider, B. I., Boisvert, R. F., Clark, C. W., Miller, B. R., Saunders, B. V., Cohl, H. S., , and McClain, M. A. (2024). NIST Digital Library of Mathematical Functions, `https://dlmf.nist.gov/`, Release 1.2.0 of 2024-03-15.

O'Malley Jr., R. E. (1974). *Introduction to Singular Perturbations* (Academic Press, New York).

O'Malley Jr., R. E. (1976). Phase-plane solutions to some singular perturbation problems, *J. Math. Anal. Appl.* **54**, pp. 449–466.

O'Malley Jr., R. E. (1991). *Singular Perturbation Methods for Oridinary Differential Equations* (Springer-Verlag, New York).

O'Malley Jr., R. E. (2014). *Historical Developments in Singular Perturbations* (Springer, New York-London).

Ou, C. and Wong, R. (2010). The Riemann-Hilbert approach to global asymptotics of discrete orthogonal polynomials with infinite nodes, *Anal. Appl.* **8**, pp. 247–286.

Ou, C. H. and Wong, R. (2003). On a two-point boundary-value problem with spurious solutions, *Stud. Appl. Math.* **111**, pp. 377–408.

Ou, C. H. and Wong, R. (2004). Shooting method for nonlinear singularly perturbed boundary-value problems, *Stud. Appl. Math.* **112**, pp. 161–200.

Paris, R. B. and Wood, A. D. (1995). Stokes' phenomenon demystified, *Bull. Inst. Math. Appl.* **31**, pp. 21–28.

Pearson, C. E. (1968). On a differential equation of boundary layer type, *J. Math. Phys.* **47**, pp. 134–154.

Pipkin, A. C. (1991). *A Course on Integral Equations, Texts in Applied Mathematics*, Vol. 9 (Springer-Verlag, New York).

Poston, T. and Stewart, I. (1978). *Catastrophe Theory and Its Applications, Surveys Reference Works Math.*, Vol. 2 (Pitman, Belmont, CA).

Prentice, P. R. (1994). Time-domain asymptotics. I. General theory for double integrals, *Proc. R. Soc. Lond. Ser. A* **446**, pp. 341–360.

Qiu, W.-Y. and Wong, R. (2000). Uniform asymptotic expansions of a double integral: coalescence of two stationary points, *Proc. R. Soc. Lond. Ser. A* **456**, pp. 407–431.

Qiu, W.-Y. and Wong, R. (2008). Asymptotic expansions for Riemann-Hilbert problems, *Anal. Appl.* **6**, pp. 269–298.

Qu, C. and Wong, R. (1988). Szegö's conjecture on Lebesgue constants for Legendre series, *Pacific J. Math.* **135**, pp. 157–188.

Rau, H. (1929). Über die Lebesgueschen Konstanten der Reihenentwicklungen nach Jacobischen Polynomen, *J. Reine Angew. Math.* **161**, pp. 237–254.

Riekstiņš, E. (1970). Asymptotic representation of certain types of convolution integral, in *Latvian Mathematical Yearbook*, Vol. 8 (Academy of Sciences of the Latvian SSR), pp. 223–239.

Rudin, W. (1973). *Functional Analysis* (McGraw-Hill, New York).

Rudin, W. (1974). *Real and Complex Analysis*, 2nd edn. (McGraw-Hill, New York).

Saff, E. B. and Totik, V. (1997). *Logarithmic potentials with external fields. Appendix B by Thomas Bloom, Grundlehren der Mathematischen Wissenschaften [Fundamental Principles of Mathematical Sciences]*, Vol. 316 (Springer-Verlag, Berlin).

Schwartz, L. (1966a). *Mathematics for the Physical Sciences* (Addison-Wesley Publishing Company, Reading, MA).

Schwartz, L. (1966b). *Théorie des Distributions* (Hermann, Paris).

Shi, W. (2008). A globally uniform asymptotic expansion of the Hermite polynomials, *Acta Math. Sci. Ser. B* **28**, pp. 834–842.

Shi, W., Nemes, G., Wang, X.-S., and Wong, R. (2023). Error bounds for the asymptotic expansions of the Hermite polynomials, *Proc. Roy. Soc. Edinburgh Sect. A* **153**, pp. 417–440.

Shivakumar, P. N. and Wong, R. (1979). Asymptotic expansion of multiple Fourier transforms, *SIAM J. Math. Anal.* **10**, pp. 1095–1104.

Simmonds, J. G. and Mann Jr., J. E. (1986). *A First Look at Perturbation Theory* (Robert E. Krieger Publishing Company, Malabar, FL).

Smith, D. R. (1985). *Singular Perturbation Theory* (Cambridge University Press, Cambridge).

Sneddon, I. N. (1972). *The Use of Integral Transforms* (McGraw-Hill, New York).

Szegö, G. (1933a). Asymptotische Entwicklungen der Jacobischen Polynome, *Schr. der König. Gelehr. Gesell.* **10**, pp. 33–112.

Szegö, G. (1933b). Über eine von Herrn S. Bernstein herrührende Abschätzung der Legendreschen Polynome, *Math. Ann.* **108**, pp. 360–369.

Szegö, G. (1934). Über Einige Asymptotische Entwicklungen der Legendreschen Funktionen, *Proc. Lond. Math. Soc.* **36**, pp. 427–450.

Szegö, G. (1975). *Orthogonal polynomials, American Mathematical Society, Colloquium Publications*, Vol. XXIII, 4th edn. (American Mathematical Society, Providence, RI).

Szegö, G. (1982). A Lebesgue-féle állandók elméletéhez, in R. A. Askey (ed.), *The Collected Papers of Gabor Szegö*, Vol. 1 (Birkhäuser, Cambricke, MA), pp. 777–783.

Temme, N. M. (2015). *Asymptotic Methods for Integrals, Series in Analysis*, Vol. 6 (World Scientific, Hackensack, NJ).

Thomson, W. (1887). On the waves produced by a single impulse in water of any depth, or in a dispersive medium, *Phil. Mag.* **23**, pp. 252–255.

Ursell, F. (1970). Integrals with a large parameter: Paths of descent and conformal mapping, *Math. Proc. Cambridge Philos. Soc.* **67**, pp. 371–381.

Ursell, F. (1980). Integrals with a large parameter: a double complex integral with four nearly coincident saddle-points, *Math. Proc. Cambridge Philos. Soc.* **87**, pp. 249–273.

van der Corput, J. G. (1935). Zur methode der stationären phase. erste mitteilung: Einfache integrale, *Compositio Math.* **1**, pp. 15–38.

van der Corput, J. G. (1936). Zur methode der stationären phase. zweite mitteilung: Wiederum einfache integrale, *Compositio Math.* **3**, pp. 328–372.

Wang, X.-S. and Wong, R. (2007). Discrete analogues of Laplace's approximation, *Asympt. Anal.* **54**, pp. 165–180.

Wang, X.-S. and Wong, R. (2011). Global asymptotics of the Meixner polynomials, *Asympt. Anal.* **75**, pp. 211–231.

Wang, X.-S. and Wong, R. (2012). Asymptotics of orthogonal polynomials via recurrence relations, *Anal. Appl.* **10**, pp. 215–235.

Wang, Z. and Wong, R. (2002). Uniform asymptotic expansion of $J_\nu(\nu a)$ via a difference equation, *Numer. Math.* **91**, pp. 147–193.

Wang, Z. and Wong, R. (2003). Asymptotic expansions for second-order linear difference equations with a turning point, *Numer. Math.* **94**, pp. 147–194.

Wang, Z. and Wong, R. (2005). Linear difference equations with transition points, *Math. Comp.* **74**, pp. 629–653.

Wang, Z. and Wong, R. (2006). Uniform asymptotics of the Stieltjes-Wigert polynomials via the Riemann-Hilbert approach, *J. Math. Pures Appl.* **85**, pp. 698–718.

Wasow, W. (1960). *Asymptotic Expansions for Differential Equations* (Wiley (Interscience), New York).

Watson, G. N. (1918). Harmonic functions associated with the parabolic cylinder, *Proc. Lond. Math. Soc.* **17**, pp. 116–148.

Watson, G. N. (1930). The constants of Landau and Lebesgue, *Quart. J. Math.* **os-1**, pp. 310–318.

Watson, G. N. (1944). *A Treatise on the Theory of Bessel Functions* (Cambridge University Press, London).

Watts, A. M. (1971). A singular perturbation problem with a turning point, *Bull. Austral. Math. Soc.* **5**, pp. 61–73.

Wen, Z.-T., Wong, R., and Xu, S.-X. (2018). Global asymptotics of orthogonal polynomials associated with a generalized Freud weight, *Chin. Ann. Math. Ser. B* **39**, pp. 553–596.

Whittaker, E. T. and Watson, G. N. (1927). *A Course of Modern Analysis*, 4th edn. (Cambridge University Press, Cambridge).

Widder, D. N. (1941). *The Laplace Transform* (Princeton University Press, Princeton, New Jersey).

Wigert, S. (1923). Sur les polynômes orthogonaux et l'approximation des fonctions continues, *Arkiv för matematik, astronomi och fysik* **17**, pp. 1–15.

Wimp, J. (1991). Current trends in asymptotics: some problems and some solutions, *J. Comput. Appl. Math.* **35**, pp. 53–79.

Wimp, J. and Zeilberger, D. (1985). Resurrecting the asymptotics of linear recurrences, *J. Math. Anal. Appl.* **111**, pp. 162–176.

Wolf, E. (1951). The diffraction theory of aberrations, *Rep. Prog. Phys.* **14**, pp. 95–120.

Wong, J. S. W. and Wong, R. (1976). On asymptotic solutions of the renewal equation, *J. Math. Anal. Appl.* **53**, pp. 243–250.

Wong, R. (1981). Asymptotic expansions of the Kontorovich-Lebedev transform, *Appl. Anal.* **12**, pp. 161–172.

Wong, R. (1989). *Asymptotic Approximations of Integrals* (Academic Press, Boston), reprinted by SIAM, Philadelphia, PA, 2001.

Wong, R. (2004). Five Lectures on Asymptotics, in C. Hua and R. Wong (eds.), *Differential Equations & Asymptotic Theory in Mathematical Physics* (World Scientific, Singapore), pp. 189–262.

Wong, R. and Li, H. (1992a). Asymptotic expansions for second-order linear difference equations, *J. Comput. Appl. Math.* **41**, pp. 65–94.

Wong, R. and Li, H. (1992b). Asymptotic expansions for second-order linear difference equations. II, *Stud. Appl. Math.* **87**, pp. 289–324.

Wong, R. and McClure, J. P. (1984). Generalized Mellin convolutions and their asymptotic expansions, *Canad. J. Math.* **36**, pp. 924–960.

Wong, R. and Yang, H. (2002a). On a boundary layer problem, *Stud. Appl. Math.* **108**, pp. 369–398.

Wong, R. and Yang, H. (2002b). On an interval boundary layer problem, *J. Comp. Appl. Math.* **144**, pp. 301–323.

Wong, R. and Yang, H. (2003). On the Ackerberg-O'Malley resonance, *Stud. Appl. Math.* **110**, pp. 157–179.

Wong, R. and Zhang, J.-M. (1996). A uniform asymptotic expansion for the Jacobi polynomials with explicit remainder, *Appl. Anal.* **61**, pp. 17–29.

Wong, R. and Zhang, L. (2007). Global asymptotics of Hermite polynomials via Riemann-Hilbert approach, *Discrete Contin. Dyn. Syst. Ser. B* **7**, pp. 661–682.

Wong, R. and Zhao, Y. (2008). On the number of solutions to Carrier's problem, *Stud. Appl. Math.* **120**, pp. 213–245.

Wong, R. and Zhao, Y.-Q. (1999a). Smoothing of Stokes's discontinuity for the generalized Bessel function, *Proc. R. Soc. Lond. Ser. A* **455**, pp. 1381–1400.

Wong, R. and Zhao, Y.-Q. (1999b). Smoothing of Stokes's discontinuity for the generalized Bessel function. II, *Proc. R. Soc. Lond. Ser. A* **455**, pp. 3065–3084.

Wong, R. and Zhao, Y.-Q. (2003). Estimates for the error term in a uniform asymptotic expansion of the Jacobi polynomials, *Anal. Appl.* **1**, pp. 213–241.

Wong, R. and Zhao, Y.-Q. (2005). On a uniform treatment of Darboux's method, *Constr. Approx.* **21**, pp. 225–255.

Wong, R. and Zhao, Y.-Q. (2016). Special functions, integral equations and a Riemann-Hilbert problem, *Proc. Amer. Math. Soc.* **144**, pp. 4367–4380.

Wright, E. M. (1934a). Asymptotic partition formulae. III. Partitions into kth powers, *Acta Math.* **63**, pp. 143–191.

Wright, E. M. (1934b). The asymptotic expansion of the generalized Bessel function, *Proc. Lond. Math. Soc.* **38**, pp. 257–270.

Wyman, M. (1964). The method of Laplace, *Trans. Roy. Soc. Canada* **2**, pp. 227–256.

Yosida, K. (1980). *Functional Analysis, Grundlehren der Mathematischen Wissenschaften [Fundamental Principles of Mathematical Sciences]*, Vol. 123, 6th edn. (Springer-Verlag, Berlin-New York).

Zhou, X. (1989). The Riemann-Hilbert problem and inverse scattering, *SIAM J. Math. Anal.* **20**, pp. 966–986.

Zygmund, A. (1935). *Trigonometric series, Monografie Matematyczne*, Vol. 5 (Seminarium Matematyczne Uniwersytetu Warszawskiego; Instytut Matematyczny PAN; Subwencji Funduszu Kultury Narodowej, Warszawa).

Index

www.ingramcontent.com/pod-product-compliance
Lightning Source LLC
LaVergne TN
LVHW070906300926
849162LV00010B/1838
* 9 7 8 9 8 1 9 8 2 4 6 8 7 *